Handbook of Optical Fibers

Gang-Ding Peng
Editor

Handbook of Optical Fibers

Volume 1

With 1544 Figures and 89 Tables

Editor
Gang-Ding Peng
Photonics and Optical Communications
School of Electrical Engineering and Telecommunications
University of New South Wales
Sydney, NSW, Australia

ISBN 978-981-10-7085-3   ISBN 978-981-10-7087-7 (eBook)
ISBN 978-981-10-7086-0 (print and electronic bundle)
https://doi.org/10.1007/978-981-10-7087-7

© Springer Nature Singapore Pte Ltd. 2019
This work is subject to copyright. All rights are reserved by the Publisher, whether the whole or part of the material is concerned, specifically the rights of translation, reprinting, reuse of illustrations, recitation, broadcasting, reproduction on microfilms or in any other physical way, and transmission or information storage and retrieval, electronic adaptation, computer software, or by similar or dissimilar methodology now known or hereafter developed.
The use of general descriptive names, registered names, trademarks, service marks, etc. in this publication does not imply, even in the absence of a specific statement, that such names are exempt from the relevant protective laws and regulations and therefore free for general use.
The publisher, the authors, and the editors are safe to assume that the advice and information in this book are believed to be true and accurate at the date of publication. Neither the publisher nor the authors or the editors give a warranty, express or implied, with respect to the material contained herein or for any errors or omissions that may have been made. The publisher remains neutral with regard to jurisdictional claims in published maps and institutional affiliations.

This Springer imprint is published by the registered company Springer Nature Singapore Pte Ltd.
The registered company address is: 152 Beach Road, #21-01/04 Gateway East, Singapore 189721, Singapore

Preface

This research- and application-oriented book covers main topical areas of optical fibers. The selection of the chapters is weighted on technological and application-specific topics, very much a reflection of where research is heading to and what researchers are looking for. Chapters are arranged in a user-friendly format essentially self-contained and with extensive cross-references. They are organized in the following sections:

Optical Fiber Communication
Solitons and Nonlinear Waves in Optical Fibers
Optical Fiber Fabrication
Active Optical Fibers
Special Optical Fibers
Optical Fiber Measurement
Optical Fiber Devices
Optical Fiber Device Measurement
Distributed Optical Fiber Sensing
Optical Fiber Sensors for Industrial Applications
Polymer Optical Fiber Sensing
Photonic Crystal Fiber Sensing
Optical Fiber Microfluidic Sensors

Several sections and chapters of the book show how diverse optical fiber technologies are becoming now. We envisage that many new optical fibers under development will find important future applications in telecommunications, sensing, and so on. Though we have been trying to cover most relevant and important topics in optical fibers, some topics may have not been presented.

All the authors are either pioneers or leading researchers in their respective areas. Their chapters have reflected well the excellent research work, technology deployment, and commercial application of their own and others. Hence this handbook, as a new entry to the Springer Nature's Major Reference Works (MRWs), will be useful for researchers, academics, engineers, and students to access expertly summarized specific topics on optical fibers for research, education, and learning purposes.

There could be technical and grammatical errors in this book. Please feel free to send your correction, advice, and feedback to us. One key feature of the Springer Nature's MRWs is to ensure continuing update and improvements.

I would take this opportunity to express by deepest gratitude to all my colleagues, either as section editors or authors, for their hard work and great contribution to this book. I also would like to thank the Springer Nature editors and staff, especially Dr. Stephen Siu Wai Yeung and Dr. Juby George, for their kind and professional support throughout this book project.

July 2019   Gang-Ding Peng

Contents

Volume 1

Part I   Optical Fiber for Communication 1

1   **Single-Mode Fibers for High Speed and Long-Haul Transmission**... 3
    John D. Downie, Ming-Jun Li, and Sergejs Makovejs

2   **Multimode Fibers for Data Centers** 41
    Xin Chen, Scott R. Bickham, John S. Abbott, J. Doug Coleman, and Ming-Jun Li

3   **Multi-core Fibers for Space Division Multiplexing** 99
    Tetsuya Hayashi

4   **Optical Coherent Detection and Digital Signal Processing of Channel Impairments** 147
    Ezra Ip

Part II   Solitons and Nonlinear Waves in Optical Fibers 219

5   **A Brief History of Fiber-Optic Soliton Transmission**............. 221
    Fedor Mitschke

6   **Perturbations of Solitons in Optical Fibers** 269
    Theodoros P. Horikis and Dimitrios J. Frantzeskakis

7   **Emission of Dispersive Waves from Solitons in Axially Varying Optical Fibers**... 301
    A. Kudlinski, A. Mussot, Matteo Conforti, and D. V. Skryabin

8   **Nonlinear Waves in Multimode Fibers**........................ 317
    I. S. Chekhovskoy, O. S. Sidelnikov, A. A. Reduyk,
    A. M. Rubenchik, O. V. Shtyrina, M. P. Fedoruk, S. K. Turitsyn,
    E. A. Zlobina, S. I. Kablukov, S. A. Babin, K. Krupa, V. Couderc,
    A. Tonello, A. Barthélémy, G. Millot, and S. Wabnitz

| 9 | Shock Waves | 373 |

Stefano Trillo and Matteo Conforti

| 10 | A Variety of Dynamical Settings in Dual-Core Nonlinear Fibers | 421 |

Boris A. Malomed

Part III Optical Fiber Fabrication 475

| 11 | Advanced Nano-engineered Glass-Based Optical Fibers for Photonics Applications | 477 |

M. C. Paul, S. Das, A. Dhar, D. Dutta, P. H. Reddy, M. Pal, and A. V. Kir'yanov

| 12 | Fabrication of Negative Curvature Hollow Core Fiber | 529 |

Muhammad Rosdi Abu Hassan

| 13 | Optimized Fabrication of Thulium Doped Silica Optical Fiber Using MCVD | 551 |

S. Z. Muhamad Yassin, Nasr Y. M. Omar, and Hairul Azhar Bin Abdul Rashid

| 14 | Microfiber: Physics and Fabrication | 587 |

Horng Sheng Lin and Zulfadzli Yusoff

| 15 | Flat Fibers: Fabrication and Modal Characterization | 623 |

Ghafour Amouzad Mahdiraji, Katrina D. Dambul, Soo Yong Poh, and Faisal Rafiq Mahamd Adikan

| 16 | 3D Silica Lithography for Future Optical Fiber Fabrication | 637 |

Gang-Ding Peng, Yanhua Luo, Jianzhong Zhang, Jianxiang Wen, Yushi Chu, Kevin Cook, and John Canning

Part IV Active Optical Fibers 655

| 17 | Rare-Earth-Doped Laser Fiber Fabrication Using Vapor Deposition Technique | 657 |

Sonja Unger, Florian Lindner, Claudia Aichele, and Kay Schuster

| 18 | Powder Process for Fabrication of Rare Earth-Doped Fibers for Lasers and Amplifiers | 679 |

Valerio Romano, Sönke Pilz, and Hossein Najafi

| 19 | Progress in Mid-infrared Fiber Source Development | 723 |

Darren D. Hudson, Alexander Fuerbach, and Stuart D. Jackson

| 20 | Crystalline Fibers for Fiber Lasers and Amplifiers | 757 |

Sheng-Lung Huang

| 21 | Cladding-Pumped Multicore Fiber Amplifier for Space Division Multiplexing | 821 |

Kazi S. Abedin

| 22 | Optical Amplifiers for Mode Division Multiplexing | 849 |

Yongmin Jung, Shaif-Ul Alam, and David J. Richardson

Volume 2

Part V Special Optical Fibers ... 875

| 23 | Optical Fibers for High-Power Lasers | 877 |

Xia Yu, Biao Sun, Jiaqi Luo, and Elizabeth Lee

| 24 | Multicore Fibers | 895 |

Ming Tang

| 25 | Polymer Optical Fibers | 967 |

Kishore Bhowmik and Gang-Ding Peng

| 26 | Optical Fibers in Terahertz Domain | 1019 |

Georges Humbert

| 27 | Optical Fibers for Biomedical Applications | 1069 |

Gerd Keiser

Part VI Optical Fiber Measurement ... 1097

| 28 | Basics of Optical Fiber Measurements | 1099 |

Mingjie Ding, Desheng Fan, Wenyu Wang, Yanhua Luo, and Gang-Ding Peng

| 29 | Measurement of Active Optical Fibers | 1139 |

Gui Xiao, Ghazal Fallah Tafti, Amirhassan Zareanborji, Anahita Ghaznavi, and Qiancheng Zhao

| 30 | Characterization of Specialty Fibers | 1177 |

Quan Chai, Yushi Chu, and Jianzhong Zhang

| 31 | Characterization of Distributed Birefringence in Optical Fibers | 1227 |

Yongkang Dong, Lei Teng, Hongying Zhang, Taofei Jiang, and Dengwang Zhou

| 32 | Characterization of Distributed Polarization-Mode Coupling for Fiber Coils | 1259 |

Jun Yang, Zhangjun Yu, and Libo Yuan

Part VII Optical Fiber Devices 1299

33 Materials Development for Advanced Optical Fiber Sensors and Lasers 1301
Peter Dragic and John Ballato

34 Optoelectronic Fibers 1335
Lei Wei

35 Fiber Grating Devices 1351
Christophe Caucheteur and Tuan Guo

36 CO_2-Laser-Inscribed Long Period Fiber Gratings: From Fabrication to Applications 1379
Yiping Wang and Jun He

37 Micro-/Nano-optical Fiber Devices 1425
Fei Xu

Part VIII Optical Fiber Device Measurement 1465

38 Measurement of Optical Fiber Grating 1467
Zhiqiang Song, Jian Guo, Haifeng Qi, and Weitao Wang

39 Measurement of Optical Fiber Amplifier 1485
Yanhua Luo, Binbin Yan, Jianxiang Wen, Jianzhong Zhang, and Gang-Ding Peng

40 Measurement of Optical Fiber Laser 1529
Haifeng Qi, Weitao Wang, Jian Guo, and Zhiqiang Song

Volume 3

Part IX Distributed Optical Fiber Sensing 1557

41 Distributed Rayleigh Sensing 1559
Xinyu Fan

42 Distributed Raman Sensing 1609
Marcelo A. Soto and Fabrizio Di Pasquale

43 Distributed Brillouin Sensing: Time-Domain Techniques 1663
Marcelo A. Soto

44 Distributed Brillouin Sensing: Frequency-Domain Techniques 1755
Aleksander Wosniok

45	**Distributed Brillouin Sensing: Correlation-Domain Techniques** .... Weiwen Zou, Xin Long, and Jianping Chen	1781

Part X Optical Fiber Sensors for Industrial Applications 1813

46	**Optical Fiber Sensors for Remote Condition Monitoring of Industrial Structures** .... Tong Sun OBE, M. Fabian, Y. Chen, M. Vidakovic, S. Javdani, K. T. V. Grattan, J. Carlton, C. Gerada, and L. Brun	1815
47	**Optical Fiber Sensor Network and Industrial Applications** .... Qizhen Sun, Zhijun Yan, Deming Liu, and Lin Zhang	1839
48	**Fiber Optic Sensors for Coal Mine Hazard Detection** .... Tongyu Liu	1885
49	**Optical Fiber Sensors in Ionizing Radiation Environments** .... Dan Sporea	1913

Part XI Polymer Optical Fiber Sensing 1955

50	**Polymer Optical Fiber Sensors and Devices** .... Ricardo Oliveira, Filipa Sequeira, Lúcia Bilro, and Rogério Nogueira	1957
51	**Solid Core Single-Mode Polymer Fiber Gratings and Sensors** ..... Kishore Bhowmik, Gang Ding Peng, Eliathamby Ambikairajah, and Ginu Rajan	1997
52	**Microstructured Polymer Optical Fiber Gratings and Sensors** .... Getinet Woyessa, Andrea Fasano, and Christos Markos	2037
53	**Polymer Fiber Sensors for Structural and Civil Engineering Applications** .... Sascha Liehr	2079

Part XII Photonic Crystal Fiber Sensing 2115

54	**Photonic Microcells for Sensing Applications** .... Chao Wang, Wei Jin, Hoi Lut Ho, and Fan Yang	2117
55	**Filling Technologies of Photonic Crystal Fibers and Their Applications** .... Chun-Liu Zhao, D. N. Wang, and Limin Xiao	2139

56 Photonic Crystal Fiber-Based Grating Sensors 2201
Changrui Liao, Feng Zhu, and Chupao Lin

57 Photonic Crystal Fiber-Based Interferometer Sensors 2231
Min Wang, Jiankun Peng, Weijia Wang, and Minghong Yang

Part XIII Optical Fiber Microfluidic Sensors 2281

58 Optical Fiber Microfluidic Sensors Based on Opto-physical Effects 2283
Chen-Lin Zhang, Chao-Yang Gong, Yuan Gong, Yun-Jiang Rao, and Gang-Ding Peng

59 Micro-/Nano-optical Fiber Microfluidic Sensors 2319
Lei Zhang

60 All Optical Fiber Optofluidic or Ferrofluidic Microsensors Fabricated by Femtosecond Laser Micromachining 2351
Hai Xiao, Lei Yuan, Baokai Cheng, and Yang Song

Index 2389

About the Editor

Gang-Ding Peng
Photonics and Optical Communications
School of Electrical Engineering
and Telecommunications
University of New South Wales
Sydney, NSW, Australia

Gang-Ding Peng received his B.Sc. degree in physics from Fudan University, Shanghai, China, in 1982, and the M.Sc. degree in applied physics and Ph.D. in electronic engineering from Shanghai Jiao Tong University, Shanghai, China, in 1984 and 1987, respectively. From 1987 through 1988 he was a lecturer at Jiao Tong University. He was a postdoctoral research fellow in the Optical Sciences Centre of the Australian National University, Canberra, from 1988 to 1991. He has been working at the University of NSW in Sydney, Australia, since 1991; was a Queen Elizabeth II Fellow from 1992 to 1996; and is currently a professor in the same university. He is a fellow and life member of both Optical Society of America (OSA) and The International Society for Optics and Photonics (SPIE). His research interests include silica and polymer optical fibers, optical fiber and waveguide devices, optical fiber sensors, and nonlinear optics.

He has worked in research and teaching in photonics and fiber optics for more than 30 years and maintained a high research profile internationally.

Section Editors

Part I: Optical Fiber Communication

Ming-Jun Li   Corning Incorporated, Corning, NY, USA

Chao LU   Department of Electronic and Information Engineering, The Hong Kong Polytechnic University, Hong Kong SAR, China

Part II: Solitons and Nonlinear Waves in Optical Fibers

Boris A. Malomed   Faculty of Engineering, Department of Physical Electronics, School of Electrical Engineering, Tel Aviv University, Tel Aviv, Israel

ITMO University, St. Petersburg, Russia

Part III: Optical Fiber Fabrication

Hairul Azhar Bin Abdul Rashid   Faculty of Engineering, Multimedia University, Cyberjaya, Malaysia

Part IV: Active Optical Fibers

Kyunghwan Oh   Department of Physics, Institute of Physics and Applied Physics, Yonsei University, Seoul, Republic of Korea

Part V: Special Optical Fibers

Perry Shum   Nanyang Technological University, Singapore, Singapore

Zhilin Xu Center for Gravitational Experiments, School of Physics, Huazhong University of Science and Technology, Wuhan, China

Part VI: Optical Fiber Measurement

Jianzhong Zhang Key Lab of In-fiber Integrated Optics, Ministry of Education, Harbin Engineering University, Harbin, China

Part VII: Optical Fiber Devices

John Canning interdisciplinary Photonics Laboratories (iPL), Global Big Data Technologies Centre (GBDTC), Tech Lab, School of Electrical and Data Engineering, University of Technology Sydney, Sydney, NSW, Australia

Tuan Guo Institute of Photonics Technology, Jinan University, Guangzhou, China

Part VIII: Optical Fiber Device Measurement

Yanhua Luo Photonics and Optical Communications, School of Electrical Engineering and Telecommunications, University of New South Wales, Sydney, NSW, Australia

Key Laboratory of Optoelectronic Devices and Systems of Ministry of Education and Guangdong Province, Shenzhen University, Shenzhen, China

Part IX: Distributed Optical Fiber Sensing

Yosuke Mizuno Institute of Innovative Research, Tokyo Institute of Technology, Yokohama, Japan

Part X: Optical Fiber Sensors for Industrial Applications

Tong Sun OBE   School of Mathematics, Computer Science and Engineering, City, University of London, London, UK

Part XI: Polymer Optical Fiber Sensing

Ginu Rajan   School of Electrical, Computer and Telecommunications Engineering, University of Wollongong, Wollongong, Australia

School of Electrical Engineering and Telecommunications, UNSW, Sydney, Australia

Part XII: Photonic Crystal Fiber Sensing

D. N. Wang College of Optical and Electrical Technology, China Jiliang University, Hangzhou, China

Part XIII: Optical Fiber Microfluidic Sensors

Yuan Gong Key Laboratory of Optical Fiber Sensing and Communications (Ministry of Education of China), University of Electronic Science and Technology of China, Chengdu, Sichuan, China

Contributors

John S. Abbott Corning Incorporated, Corning, NY, USA

Hairul Azhar Bin Abdul Rashid Faculty of Engineering, Multimedia University, Cyberjaya, Malaysia

Kazi S. Abedin OFS Laboratories, Somerset, NJ, USA

Muhammad Rosdi Abu Hassan Centre for Optical Fibre Technology (COFT), School of Electrical, Electronic Engineering, Nanyang Technological University, Singapore, Singapore, Singapore

Claudia Aichele Department of Fiber Optics, Leibniz Institute of Photonic Technology (Leibniz IPHT), Jena, Germany

Shaif-Ul Alam Optoelectronics Research Centre (ORC), University of Southampton, Southampton, UK

Eliathamby Ambikairajah School of Electrical Engineering and Telecommunications, UNSW, Sydney, Australia

Ghafour Amouzad Mahdiraji School of Engineering, Taylor's University, Subang Jaya, Selangor, Malaysia

Flexilicate Sdn. Bhd., University of Malaya, Kuala Lumpur, Malaysia

S. A. Babin Institute of Automation and Electrometry SB RAS, Novosibirsk, Russia

Novosibirsk State University, Novosibirsk, Russia

John Ballato Center for Optical Materials Science and Engineering Technologies (COMSET) and the Department of Materials Science and Engineering, Clemson University, Clemson, SC, USA

A. Barthélémy XLIM, UMR CNRS 7252, Université de Limoges, Limoges, France

Kishore Bhowmik HFC Assurance, Operate and Maintain Network, NBN, Melbourne, VIC, Australia

Scott R. Bickham Corning Incorporated, Corning, NY, USA

Lúcia Bilro Instituto de Telecomunicações, Campus Universitário de Santiago, Aveiro, Portugal

L. Brun Faiveley Brecknell Willis, Somerset, UK

John Canning interdisciplinary Photonics Laboratories (iPL), Global Big Data Technologies Centre (GBDTC), Tech Lab, School of Electrical and Data Engineering, University of Technology Sydney, Sydney, NSW, Australia

J. Carlton City, University of London, London, UK

Christophe Caucheteur Electromagnetism and Telecommunication Department, University of Mons, Mons, Belgium

Quan Chai Key Laboratory of In-Fiber Integrated Optics, Ministry Education of China, Harbin Engineering University, Harbin, China

I. S. Chekhovskoy Novosibirsk State University, Novosibirsk, Russia

Institute of Computational Technologies SB RAS, Novosibirsk, Russia

Jianping Chen State Key Laboratory of Advanced Optical Communication Systems and Networks, Department of Electronic Engineering, Shanghai Jiao Tong University, Shanghai, China

Xin Chen Corning Incorporated, Corning, NY, USA

Y. Chen City, University of London, London, UK

Baokai Cheng Department of Electrical and Computer Engineering, Center for Optical Materials Science and Engineering Technologies (COMSET), Clemson University, Clemson, SC, USA

Yushi Chu Key Laboratory of In-Fiber Integrated Optics, Ministry Education of China, Harbin Engineering University, Harbin, China

Photonics and Optical Communications, School of Electrical Engineering and Telecommunications, UNSW, Sydney, NSW, Australia

interdisciplinary Photonics Laboratories (iPL), Global Big Data Technologies Centre (GBDTC), Tech Lab, School of Electrical and Data Engineering, University of Technology Sydney, Sydney, NSW, Australia

J. Doug Coleman Corning Incorporated, Corning, NY, USA

Matteo Conforti CNRS, UMR 8523 – PhLAM – Physique des Lasers Atomes et Molécules, University of Lille, Lille, France

Kevin Cook interdisciplinary Photonics Laboratories (iPL), Global Big Data Technologies Centre (GBDTC), Tech Lab, School of Electrical and Data Engineering, University of Technology Sydney, Sydney, NSW, Australia

V. Couderc XLIM, UMR CNRS 7252, Université de Limoges, Limoges, France

Katrina D. Dambul Faculty of Engineering, Multimedia University, Cyberjaya, Selangor, Malaysia

S. Das Fiber Optics and Photonics Division, CSIR-Central Glass and Ceramic Research Institute, Kolkata, India

A. Dhar Fiber Optics and Photonics Division, CSIR-Central Glass and Ceramic Research Institute, Kolkata, India

Mingjie Ding Photonics and Optical Communications, School of Electrical Engineering and Telecommunications, University of New South Wales, Sydney, NSW, Australia

Fabrizio Di Pasquale Institute of Communication, Information and Perception Technologies (TECIP), Scuola Superiore Sant'Anna, Pisa, Italy

Yongkang Dong National Key Laboratory of Science and Technology on Tunable Laser, Harbin Institute of Technology, Harbin, China

John D. Downie Corning Incorporated, Corning, NY, USA

Peter Dragic Department of Electrical and Computer Engineering, University of Illinois at Urbana-Champaign, Urbana, IL, USA

D. Dutta Fiber Optics and Photonics Division, CSIR-Central Glass and Ceramic Research Institute, Kolkata, India

M. Fabian City, University of London, London, UK

Ghazal Fallah Tafti Photonics and Optical Communications, School of Electrical Engineering and Telecommunications, UNSW, Sydney, NSW, Australia

Desheng Fan Photonics and Optical Communications, School of Electrical Engineering and Telecommunications, University of New South Wales, Sydney, NSW, Australia

Xinyu Fan State Key Laboratory of Advanced Optical Communication Systems and Networks, Department of Electronic Engineering, Shanghai Jiao Tong University, Shanghai, China

Andrea Fasano DTU Mekanik, Department of Mechanical Engineering, Technical University of Denmark, Lyngby, Denmark

M. P. Fedoruk Novosibirsk State University, Novosibirsk, Russia

Institute of Computational Technologies SB RAS, Novosibirsk, Russia

Dimitrios J. Frantzeskakis Department of Physics, National and Kapodistrian University of Athens, Athens, Greece

Alexander Fuerbach MQ Photonics Research Centre, Department of Physics and Astronomy, Macquarie University, North Ryde, NSW, Australia

C. Gerada The University of Nottingham, Nottingham, UK

Anahita Ghaznavi Photonics and Optical Communications, School of Electrical Engineering and Telecommunications, UNSW, Sydney, NSW, Australia

Chao-Yang Gong Key Laboratory of Optical Fiber Sensing and Communications (Ministry of Education of China), University of Electronic Science and Technology of China, Chengdu, Sichuan, China

Yuan Gong Key Laboratory of Optical Fiber Sensing and Communications (Ministry of Education of China), University of Electronic Science and Technology of China, Chengdu, Sichuan, China

K. T. V. Grattan City, University of London, London, UK

Jian Guo Shandong Key Laboratory of Optical Fiber Sensing Technologies, Qilu Industry University (Laser Institute of Shandong Academy of Sciences), Jinan, China

Tuan Guo Institute of Photonics Technology, Jinan University, Guangzhou, China

Tetsuya Hayashi Optical Communications Laboratory, Sumitomo Electric Industries, Ltd., Yokohama, Kanagawa, Japan

Jun He Key Laboratory of Optoelectronic Devices and Systems of Ministry of Education and Guangdong Province, College of Physics and Optoelectronic Engineering, Shenzhen University, Shenzhen, China

Guangdong and Hong Kong Joint Research Centre for Optical Fibre Sensors, Shenzhen University, Shenzhen, China

Hoi Lut Ho Department of Electrical Engineering, The Hong Kong Polytechnic University, Hong Kong, China

Theodoros P. Horikis Department of Mathematics, University of Ioannina, Ioannina, Greece

Sheng-Lung Huang Graduate Institute of Photonics and Optoelectronics, and Department of Electrical Engineering, National Taiwan University, Taipei, Taiwan

Darren D. Hudson MQ Photonics Research Centre, Department of Physics and Astronomy, Macquarie University, North Ryde, NSW, Australia

Georges Humbert XLIM Research Institute, UMR 7252 CNRS, University of Limoges, Limoges, France

Ezra Ip NEC Laboratories America, Princeton, NJ, USA

Stuart D. Jackson Department of Engineering, MQ Photonics Research Centre, School of Engineering, Macquarie University, North Ryde, NSW, Australia

S. Javdani City, University of London, London, UK

Taofei Jiang National Key Laboratory of Science and Technology on Tunable Laser, Harbin Institute of Technology, Harbin, China

Wei Jin Department of Electrical Engineering, The Hong Kong Polytechnic University, Hong Kong, China

Yongmin Jung Optoelectronics Research Centre (ORC), University of Southampton, Southampton, UK

S. I. Kablukov Institute of Automation and Electrometry SB RAS, Novosibirsk, Russia

Gerd Keiser Boston University, Boston, MA, USA

A. V. Kir'yanov Centro de Investigaciones en Optica, Guanajuato, Mexico

K. Krupa Department of Information Engineering, University of Brescia, Brescia, Italy

A. Kudlinski CNRS, UMR 8523 – PhLAM – Physique des Lasers Atomes et Molécules, University of Lille, Lille, France

Elizabeth Lee Precision Measurements Group, Singapore Institute of Manufacturing Technology, Singapore, Singapore

Ming-Jun Li Corning Incorporated, Corning, NY, USA

Changrui Liao College of Optoelectronic Engineering, Shenzhen University, Shenzhen, China

Sascha Liehr Division 8.6 "Fibre Optic Sensors", Bundesanstalt für Materialforschung und –prüfung (BAM), Berlin, Germany

Chupao Lin College of Optoelectronic Engineering, Shenzhen University, Shenzhen, China

Horng Sheng Lin Universiti Tunku Abdul Rahman, Sungai Long Campus, Kajang, Malaysia

Florian Lindner Department of Fiber Optics, Leibniz Institute of Photonic Technology (Leibniz IPHT), Jena, Germany

Deming Liu School of Optical and Electronic Information, Next Generation Internet Access National Engineering Laboratory (NGIAS), Huazhong University of Science and Technology, Wuhan, Hubei, P. R. China

Tongyu Liu Laser Institute, Qilu University of Technology-Shandong Academy of Science, Jinan, Shandong, China

Xin Long State Key Laboratory of Advanced Optical Communication Systems and Networks, Department of Electronic Engineering, Shanghai Jiao Tong University, Shanghai, China

Jiaqi Luo Precision Measurements Group, Singapore Institute of Manufacturing Technology, Singapore, Singapore

Yanhua Luo Photonics and Optical Communications, School of Electrical Engineering and Telecommunications, University of New South Wales, Sydney, NSW, Australia

Key Laboratory of Optoelectronic Devices and Systems of Ministry of Education and Guangdong Province, Shenzhen University, Shenzhen, China

Faisal Rafiq Mahamd Adikan Flexilicate Sdn. Bhd., University of Malaya, Kuala Lumpur, Malaysia

Integrated Lightwave Research Group, Department of Electrical Engineering, Faculty of Engineering, University of Malaya, Kuala Lumpur, Malaysia

Sergejs Makovejs Corning Incorporated, Ewloe, UK

Boris A. Malomed Faculty of Engineering, Department of Physical Electronics, School of Electrical Engineering, Tel Aviv University, Tel Aviv, Israel

ITMO University, St. Petersburg, Russia

Christos Markos DTU Fotonik, Department of Photonics Engineering, Technical University of Denmark, Lyngby, Denmark

G. Millot ICB, UMR CNRS 6303, Université de Bourgogne, Dijon, France

Fedor Mitschke Institut für Physik, Universität Rostock, Rostock, Germany

S. Z. Muhamad Yassin Photonics Laboratory, Telekom Research and Development, Cyberjaya, Malaysia

A. Mussot CNRS, UMR 8523 – PhLAM – Physique des Lasers Atomes et Molécules, University of Lille, Lille, France

Hossein Najafi Institute for Applied Laser, Photonics and Surface Technologies (ALPS), Bern University of Applied Sciences, Burgdorf, Switzerland

Rogério Nogueira Instituto de Telecomunicações, Campus Universitário de Santiago, Aveiro, Portugal

Ricardo Oliveira Instituto de Telecomunicações, Campus Universitário de Santiago, Aveiro, Portugal

Nasr Y. M. Omar Faculty of Engineering, Multimedia University, Cyberjaya, Malaysia

M. Pal Fiber Optics and Photonics Division, CSIR-Central Glass and Ceramic Research Institute, Kolkata, India

M. C. Paul Fiber Optics and Photonics Division, CSIR-Central Glass and Ceramic Research Institute, Kolkata, India

Gang-Ding Peng Photonics and Optical Communications, School of Electrical Engineering and Telecommunications, University of New South Wales, Sydney, NSW, Australia

Jiankun Peng National Engineering Laboratory for Fiber Optic Sensing Technology (NEL-FOST), Wuhan University of Technology, Wuhan, China

Sönke Pilz Institute for Applied Laser, Photonics and Surface Technologies (ALPS), Bern University of Applied Sciences, Burgdorf, Switzerland

Soo Yong Poh Integrated Lightwave Research Group, Department of Electrical Engineering, Faculty of Engineering, University of Malaya, Kuala Lumpur, Malaysia

Haifeng Qi Shandong Key Laboratory of Optical Fiber Sensing Technologies, Qilu Industry University (Laser Institute of Shandong Academy of Sciences), Jinan, China

Ginu Rajan School of Electrical, Computer and Telecommunications Engineering, University of Wollongong, Wollongong, Australia

School of Electrical Engineering and Telecommunications, UNSW, Sydney, Australia

Yun-Jiang Rao Key Laboratory of Optical Fiber Sensing and Communications (Ministry of Education of China), University of Electronic Science and Technology of China, Chengdu, Sichuan, China

P. H. Reddy Academy of Scientific and Innovative Research (AcSIR), IR-CGCRI Campus, Kolkata, India

A. A. Reduyk Novosibirsk State University, Novosibirsk, Russia

David J. Richardson Optoelectronics Research Centre (ORC), University of Southampton, Southampton, UK

Valerio Romano Institute for Applied Laser, Photonics and Surface Technologies (ALPS), Bern University of Applied Sciences, Burgdorf, Switzerland

Institute of Applied Physics (IAP), University of Bern, Bern, Switzerland

A. M. Rubenchik Lawrence Livermore National Laboratory, Livermore, CA, USA

Kay Schuster Department of Fiber Optics, Leibniz Institute of Photonic Technology (Leibniz IPHT), Jena, Germany

Filipa Sequeira Instituto de Telecomunicações, Campus Universitário de Santiago, Aveiro, Portugal

O. V. Shtyrina Novosibirsk State University, Novosibirsk, Russia

Institute of Computational Technologies SB RAS, Novosibirsk, Russia

O. S. Sidelnikov Novosibirsk State University, Novosibirsk, Russia

D. V. Skryabin Department of Nanophotonics and Metamaterials, ITMO University, St Petersburg, Russia

Department of Physics, University of Bath, Bath, UK

Yang Song Department of Electrical and Computer Engineering, Center for Optical Materials Science and Engineering Technologies (COMSET), Clemson University, Clemson, SC, USA

Zhiqiang Song Shandong Key Laboratory of Optical Fiber Sensing Technologies, Qilu Industry University (Laser Institute of Shandong Academy of Sciences), Jinan, China

Marcelo A. Soto Institute of Electrical Engineering, EPFL Swiss Federal Institute of Technology, Lausanne, Switzerland

Dan Sporea National Institute for Laser, Plasma and Radiation Physics, Center for Advanced Laser Technologies, Măgurele, Romania

Biao Sun Precision Measurements Group, Singapore Institute of Manufacturing Technology, Singapore, Singapore

Qizhen Sun School of Optical and Electronic Information, Next Generation Internet Access National Engineering Laboratory (NGIAS), Huazhong University of Science and Technology, Wuhan, Hubei, P. R. China

Tong Sun OBE School of Mathematics, Computer Science and Engineering, City, University of London, London, UK

Ming Tang Wuhan National Lab for Optoelectronics (WNLO) and National Engineering Laboratory for Next Generation Internet Access System (NGIA), School of Optical and Electronic Information, Huazhong University of Science and Technology (HUST), Wuhan, China

Lei Teng National Key Laboratory of Science and Technology on Tunable Laser, Harbin Institute of Technology, Harbin, China

A. Tonello XLIM, UMR CNRS 7252, Université de Limoges, Limoges, France

Stefano Trillo Department of Engineering, University of Ferrara, Ferrara, Italy

S. K. Turitsyn Novosibirsk State University, Novosibirsk, Russia

Aston Institute of Photonic Technologies, Aston University, Birmingham, UK

Sonja Unger Department of Fiber Optics, Leibniz Institute of Photonic Technology (Leibniz IPHT), Jena, Germany

M. Vidakovic City, University of London, London, UK

S. Wabnitz Novosibirsk State University, Novosibirsk, Russia

Department of Information Engineering, University of Brescia, Brescia, Italy

National Institute of Optics INO-CNR, Brescia, Italy

Chao Wang School of Electrical Engineering, Wuhan University, Wuhan, Hubei, China

D. N. Wang College of Optical and Electrical Technology, China Jiliang University, Hangzhou, China

Min Wang National Engineering Laboratory for Fiber Optic Sensing Technology (NEL-FOST), Wuhan University of Technology, Wuhan, China

School of Electronic and Electrical Engineering, Wuhan Textile University, Wuhan, China

Weijia Wang National Engineering Laboratory for Fiber Optic Sensing Technology (NEL-FOST), Wuhan University of Technology, Wuhan, China

Weitao Wang Shandong Key Laboratory of Optical Fiber Sensing Technologies, Qilu Industry University (Laser Institute of Shandong Academy of Sciences), Jinan, China

Wenyu Wang Photonics and Optical Communications, School of Electrical Engineering and Telecommunications, University of New South Wales, Sydney, NSW, Australia

Yiping Wang Key Laboratory of Optoelectronic Devices and Systems of Ministry of Education and Guangdong Province, College of Physics and Optoelectronic Engineering, Shenzhen University, Shenzhen, China

Guangdong and Hong Kong Joint Research Centre for Optical Fibre Sensors, Shenzhen University, Shenzhen, China

Lei Wei School of Electrical and Electronic Engineering, Nanyang Technological University, Singapore, Singapore

Jianxiang Wen Key Laboratory of Specialty Fiber Optics and Optical Access Networks, Shanghai University, Shanghai, China

Aleksander Wosniok 8.6 Fibre Optic Sensors, Federal Institute for Materials Research and Testing (BAM), Berlin, Germany

Getinet Woyessa DTU Fotonik, Department of Photonics Engineering, Technical University of Denmark, Lyngby, Denmark

Gui Xiao Photonics and Optical Communications, School of Electrical Engineering and Telecommunications, UNSW, Sydney, NSW, Australia

Hai Xiao Department of Electrical and Computer Engineering, Center for Optical Materials Science and Engineering Technologies (COMSET), Clemson University, Clemson, SC, USA

Limin Xiao Advanced Fiber Devices and Systems Group, Key Laboratory of Micro and Nano Photonic Structures (MoE), Department of Optical Science and Engineering Fudan University, Shanghai, China

Key Laboratory for Information Science of Electromagnetic Waves (MoE), Fudan University, Shanghai, China

Shanghai Engineering Research Center of Ultra-Precision Optical Manufacturing, Fudan University, Shanghai, China

Fei Xu National Laboratory of Solid State Microstructures and College of Engineering and Applied Sciences, Nanjing University, Nanjing, Jinagsu, P. R. China

Binbin Yan State Key Laboratory of Information Photonics and Optical Communications, Beijing University of Posts and Telecommunications, Beijing, China

Zhijun Yan School of Optical and Electronic Information, Next Generation Internet Access National Engineering Laboratory (NGIAS), Huazhong University of Science and Technology, Wuhan, Hubei, P. R. China

Fan Yang Department of Electrical Engineering, The Hong Kong Polytechnic University, Hong Kong, China

Jun Yang Key Lab of In-Fiber Integrated Optics, Ministry Education of China, Harbin Engineering University, Harbin, China

College of Science, Harbin Engineering University, Harbin, China

Minghong Yang National Engineering Laboratory for Fiber Optic Sensing Technology (NEL-FOST), Wuhan University of Technology, Wuhan, China

Xia Yu Precision Measurements Group, Singapore Institute of Manufacturing Technology, Singapore, Singapore

Zhangjun Yu Key Lab of In-Fiber Integrated Optics, Ministry Education of China, Harbin Engineering University, Harbin, China

College of Science, Harbin Engineering University, Harbin, China

Lei Yuan Department of Electrical and Computer Engineering, Center for Optical Materials Science and Engineering Technologies (COMSET), Clemson University, Clemson, SC, USA

Libo Yuan Key Lab of In-Fiber Integrated Optics, Ministry Education of China, Harbin Engineering University, Harbin, China

College of Science, Harbin Engineering University, Harbin, China

Zulfadzli Yusoff Multimedia University, Persiaran Multimedia, Cyberjaya, Malaysia

Amirhassan Zareanborji Photonics and Optical Communications, School of Electrical Engineering and Telecommunications, UNSW, Sydney, NSW, Australia

Chen-Lin Zhang Key Laboratory of Optical Fiber Sensing and Communications (Ministry of Education of China), University of Electronic Science and Technology of China, Chengdu, Sichuan, China

Hongying Zhang Institute of Photonics and Optical Fiber Technology, Harbin University of Science and Technology, Harbin, China

Jianzhong Zhang Key Lab of In-fiber Integrated Optics, Ministry of Education, Harbin Engineering University, Harbin, China

Lei Zhang College of Optical Science and Engineering, Zhejiang University, Hangzhou, China

Lin Zhang Aston Institute of Photonic Technologies, Aston University, Birmingham, UK

Chun-Liu Zhao College of Optical and Electrical Technology, China Jiliang University, Hangzhou, China

Qiancheng Zhao Photonics and Optical Communications, School of Electrical Engineering and Telecommunications, UNSW, Sydney, NSW, Australia

Dengwang Zhou National Key Laboratory of Science and Technology on Tunable Laser, Harbin Institute of Technology, Harbin, China

Feng Zhu College of Optoelectronic Engineering, Shenzhen University, Shenzhen, China

E. A. Zlobina Institute of Automation and Electrometry SB RAS, Novosibirsk, Russia

Weiwen Zou State Key Laboratory of Advanced Optical Communication Systems and Networks, Department of Electronic Engineering, Shanghai Jiao Tong University, Shanghai, China

Part I
Optical Fiber for Communication

Single-Mode Fibers for High Speed and Long-Haul Transmission

John D. Downie, Ming-Jun Li, and Sergejs Makovejs

Contents

Introduction	4
Background and History of Optical Fiber	5
History of Fiber Evolution (1966–1987)	5
History of Fiber Evolution (1987–2007)	6
History of Fiber Evolution (2007 Onwards)	8
Optical Fiber Designs for Long-Haul Transmission	11
Quantification of System Level Performance	16
Long-Haul and Ultra-Long-Haul Transmission Systems	17
Raman Gain Considerations	20
Unrepeatered Span Transmission Systems	23
Transmission System Modeling and Experiments	27
Other Factors and Considerations	30
Splice Loss	30
Practical Benefits of Ultra-Low Attenuation and Large Effective Area Fibers	31
Potential Future Directions	35
Conclusions	36
References	38

Abstract

The design and manufacture of optical fibers have evolved over time as optical system technologies and data rates have changed. Fiber characteristics and parameters that were important for previous system generations may be different

---

J. D. Downie (✉) · M.-J. Li
Corning Incorporated, Corning, NY, USA
e-mail: downiejd@corning.com; lim@corning.com

S. Makovejs
Corning Incorporated, Ewloe, UK
e-mail: makovejss@corning.com

now in the era of coherent transmission systems with data rates of 100 Gb/s and higher for long-haul (LH) and ultra-long-haul (ULH) transmission links. New systems are designed with no in-line optical dispersion compensation so that all dispersion compensation is performed digitally in the transmitter and receiver. In this system approach, attenuation and nonlinear tolerance are the fiber characteristics that have the largest impact on overall system performance. In this chapter, we examine the history of single-mode fiber designs and quantify differences in performance of various fibers. This is done mainly in the context of conventional repeatered LH and ULH systems, with a brief consideration of the special case of long single-span unrepeatered systems. Practical aspects of different fibers are also considered, including bend performance and splice loss.

Introduction

Fiber-optic communication systems are comprised of several separate, but interdependent, parts. The overall performance of a system depends on the quality and characteristics of each element. The transmitter and receiver equipment are active components of the system that determine the bit rate, modulation format, spectral efficiency, capacity, and other system aspects. A second important part of communication systems is optical amplification. The type of optical amplifiers, and the noise figure and spectral bandwidth of those amplifiers also impact the performance and total capacity that the system can carry. These devices are active parts of the system in the sense that they are electrically powered and may have active control of their operation and performance. A third major element of fiber-optic systems is the transmission medium, optical fiber, a purely passive part of the system. There are also other smaller passive components such as couplers, taps, optical multiplexers, and de-multiplexers, but these play a much smaller role in determining the performance, capacity, and reach of optical communication systems. In this chapter, we examine the properties of single-mode optical fibers that promote the best performance in modern coherent transmission systems. With respect to fiber, the highest or best system performance generally translates into longest reach before regeneration is required and largest total capacity that can be carried by a fiber over a given distance. We study the role of fiber characteristics here largely in this context of system performance, which can be evaluated in quantities such as Q-factor (inversely related to bit error rate or BER) and fiber figure of merit. We also touch on practical performance issues associated with optical fibers such as macro-bend loss, micro-bend loss, splice loss, and Raman gain. We will show that for today's coherent systems, the fiber characteristics that have the greatest influence on overall system performance are attenuation and nonlinear tolerance, mainly governed by the fiber effective area.

The rest of this chapter is organized in the following manner; in the first section, we first take a look at the evolution of optical fiber in a historical context and describe the four generations of optical transmission systems and the fibers used in them. In section "Optical Fiber Designs for Long-haul Transmission," we discuss design aspects of optical fibers and the constraints and interdependencies of various fiber parameters such as dispersion, effective area, and cutoff wavelength. In the third section, we look at means to quantify system level performance with regard to fiber characteristics. This is done mainly with regard to long-haul systems with optical amplifiers at the end of each fiber span, but we briefly explore which fiber characteristics have most effect in unrepeatered span systems as well. Modeling and experimental data are shown, comparing the reach lengths of different single-mode optical fiber types under comparable system conditions and configurations. In the fourth section, splice loss considerations and issues are discussed, along with some other practical benefits that accrue from the use of high-performing fibers with low attenuation and large effective area. Finally, we mention one possible future direction of fibers that stretches the definition from single-mode to quasi-single-mode.

Background and History of Optical Fiber

History of Fiber Evolution (1966–1987)

The proposal of low loss silica-based optical fiber by Charles Kao in 1966 (Kao and Hockham 1966) marked the debut of optical communications. A major milestone in developing optical fibers was the demonstration of optical fiber with attenuation less than 20 dB/km in 1970 (Kapron et al. 1970), which first enabled the use of optical fibers for practical transmission applications. Since then, optical fiber, components, and transmission system technologies have advanced rapidly to increase the transmission capacity of fibers and cables during the past five decades. In fact, the transmission capacity of a single fiber has increased by a factor of approximately ten every 4 years.

The long-haul optical fiber transmission system has gone through four generations. The first fiber transmission system generation utilized multimode optical fibers and light emitting diode (LED) sources operating in the 850 nm wavelength region (Sanferrare 1987). Multimode fibers have a large core size and high numerical aperture, which facilitated coupling of the light from the LED source into the fiber with low cost. However, the transmission rate and distance afforded by multimode fiber-based systems are limited by the fundamental bandwidth limitation of multimode fibers due to intermodal dispersion.

The solution to eliminating intermodal dispersion is to use a single-mode optical fiber, i.e., a fiber that supports only one transverse mode. With the development of semiconductor lasers (Joyce et al. 1977) and the opening of long wavelength transmission windows (Miya et al. 1979) as well as advances in single-mode fiber coupling technology (Tynes and Derosier 1977) in the later 1970s, single-mode fiber transmission systems became possible. The second generation of optical fiber communication systems employed standard single-mode fiber and single-mode lasers operating at 1310 nm. Standard single-mode fiber has lower attenuation than multimode fiber and exhibits nearly zero chromatic dispersion in the 1310 nm wavelength region, enabling longer transmission distance with higher data rates.

Although the attenuation of a single-mode optical fiber is lowest in the 1550 nm wavelength window, the chromatic dispersion in this wavelength window is rather large (about $+17$ ps/km/nm) due to silica glass material dispersion. If uncompensated, dispersion of this order can be a limitation for high data rate systems. In order to overcome the dispersion limitation, dispersion-shifted optical fiber (DSF) was proposed (Cohen et al. 1979). In a dispersion shifted fiber, the material dispersion of silica glass is compensated by waveguide dispersion through refractive index profile designs, resulting in zero total dispersion at 1550 nm. This shift in fiber design allowed the use of conventional lasers exhibiting relatively large spectral width of several nm, which enabled the third generation optical fiber transmission system operating at 1550 nm. The dispersion shifted fiber was designed for single wavelength transmission at 1550 nm before multi-wavelength transmission systems were considered.

History of Fiber Evolution (1987–2007)

In the late 1980s, the fourth generation high capacity optical fiber transmission system was driven by the development of the erbium doped fiber amplifier (EDFA) (Desurvire et al. 1987) and wavelength-division multiplexing (WDM) (Taga et al. 1988). WDM technology allows the simultaneous transmission of multiple wavelengths in a single fiber. For WDM transmission, dispersion shifted fibers proved unsuitable because the crosstalk between two neighboring channels (due to the nonlinear effect of four-wave mixing) is strongest when fiber dispersion is zero. The four-wave mixing effect can be suppressed effectively with a certain amount of chromatic dispersion. On the other hand, the dispersion should be small enough to minimize pulse broadening. Considering nonlinear effects, dispersion, and their interplay, a new type of fiber known as non-zero dispersion-shifted fiber (NZDSF) was proposed (Tkach et al. 1995). Non-zero dispersion-shifted fibers have typical dispersion values of 3–8 ps/nm/km at 1550 nm with effective areas of about 50 μm^2. Because nonlinear effects and impairments are proportional to power density in the core (power divided by the effective area), fibers with larger effective areas are beneficial in terms of reducing the nonlinear effects. To increase the effective area, index profile designs with large effective area were proposed and NZDSF fiber with an effective area of about 72 μm^2 was developed for WDM transmission (Liu 1997). NZDSFs have since been widely deployed worldwide for high capacity WDM networks, in large part for systems with 10 Gb/s channel data rates.

Wavelength division multiplexing technology added a new dimension to increase the transmission capacity of optical fiber and has been an ingrained feature of fiber communication systems since the 1990s. In conjunction with WDM development, channel rates have increased to meet rising traffic demands. This was largely made possible through bandwidth improvements of electrical and opto-electronic

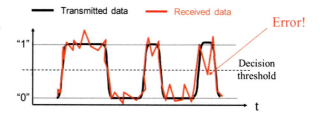

Fig. 1 Impact of noise on signal quality and an error in bit detection

devices such as modulators, photodiodes, and RF components. The channel rate of long-haul systems increased from 2.5 to 10 Gb/s using intensity modulation and direct detection. These systems were based on amplitude modulation, in which the information is encoded as either "1" or "0," and the receiver uses a direct detection technique to determine whether "1" or "0" was transmitted. Under normal circumstances with high quality optical signals, the receiver can easily differentiate "1" from "0" on the other side of the transmission line. However, when the accumulated noise on a signal is high (e.g., very long transmission links with a series of optical amplifiers), a receiver is more likely to make a detection error. For example, an error can result from the receiver's interpretation that the transmitted bit was "0," when in reality the transmitted bit was "1" as illustrated in Fig. 1.

A primary feature and advantage of systems based on direct detection is simplicity of implementation. However, the transmission bit-rate is not easily scaled. In such systems, an increase in bit rate is achieved through an increase in the time-division-multiplexed (TDM) rate and requires shorter transmission time slots and wider signal spectral widths. In the 1990s and 2000s, a significant amount of research was devoted to investigating the performance and practicality of >10 Gb/s transmission systems based on electrical and optical time division multiplexing (OTDM and ETDM) (Makovejs et al. 2008; Turkiewicz et al. 2005). Since an increase in the TDM rate also increases the spectral width, it does not necessarily enable a substantial increase in spectral efficiency and capacity per fiber. In addition, the move to 40 Gb/s intensity-modulated systems made evident a new range of significant intra-channel nonlinear effects that had to be managed. For several reasons, 10 Gb/s WDM systems were the last mass-adopted systems that employed direct detection; the success of 40 Gb/s direct-detection systems was limited and short-lived.

Another notable feature of 2.5–10 Gb/s direct-detection long-haul systems is that they required in-line optical dispersion compensation modules in order to realize received signals with low accumulated chromatic dispersion. These compensation modules were periodically deployed at amplification sites in the mid-stage of 2-stage EDFAs. The compensation technology was largely based on either highly negative dispersion compensating fiber (DCF), or dispersion compensating gratings (DCGs). Both solutions introduced an additional loss and therefore an additional amplifier. Thus, the use of low-dispersion transmission fibers in terrestrial systems was beneficial in terms of reducing the total number of dispersion compensation modules, dual-stage amplifiers, and total link loss.

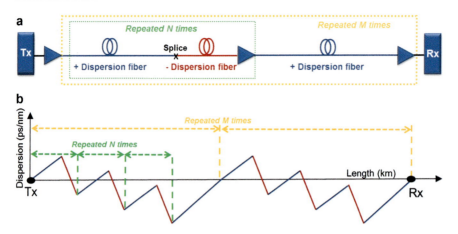

Fig. 2 Dispersion compensation in previous-generation submarine systems. (**a**) Schematic diagram; (**b**) Dispersion map

While both terrestrial and submarine optical fiber transmission systems obey similar rules of physics, a different approach to dispersion compensation was pursued in direct detection submarine systems operating at bit rates such as 10 Gb/s. One of the most popular approaches relied on splicing (and subsequently re-coating) positive and negative dispersion fibers within the same span to achieve dispersion management. In the type of configuration shown in Fig. 2, hybrid spans comprised of both positive and negative dispersion fibers had a small net negative dispersion. After N hybrid fiber spans, the next span was made from only positive dispersion fiber to bring the total dispersion back to near zero. The larger section of N hybrid spans followed by one positive dispersion span was repeated M times to form the whole submarine cable.

History of Fiber Evolution (2007 Onwards)

In the late 2000s, the first commercial 100 Gb/s systems with coherent detection were developed and became commercially available. The advent of coherent receiver technology revolutionized the way the industry approached the design of optical networks. Coherent detection allowed information to be encoded not only in amplitude, but also in phase and state of polarization, thereby increasing the number of bits per symbol period and the amount of information encoded per channel. For example, 100 Gb/s systems (total bit rate in the range of 112–128 Gb/s including forward error correction overhead) are now based on polarization-multiplexed quadrature phase shift keying (PM-QPSK). With TDM symbol rates of only 28–32 Gbaud, the use of QPSK modulation encodes 2 bits of information per symbol, and transmitting independent QPSK data on two orthogonal polarizations produces an overall bit rate multiplier of 4, resulting in bit rates of 112–128 Gb/s.

Such spectrally efficient increases of the overall bit rate were made possible through integration of digital signal processing (DSP) functions in coherent receivers to create digital coherent receivers.

Significant additional benefits of systems using digital coherent receivers include the ability to compensate for large amounts of chromatic dispersion and polarization mode dispersion in the receiver DSP, which has two important implications. First, the use of in-line dispersion compensation became superfluous and even undesirable from the standpoint of transmission performance. All-dispersive systems (sometimes called dispersion-uncompensated as compared to systems with periodic, in-line dispersion compensation) proved to be more resilient toward the impact of intra-channel and inter-channel nonlinear effects. In these systems, the accumulated chromatic dispersion is compensated digitally in the receiver DSP. Second, the ability to compensate for large amounts of polarization mode dispersion (PMD) in DSP (not previously possible with direct-detection systems) facilitates an upgrade in the TDM rate from 10 Gbaud to more than 25 Gbaud. In direct-detection systems with on-off keying modulation, upgrading to data rates higher than 10 Gb/s was largely limited by PMD (particularly pronounced for upgrades over legacy deployed optical fibers). Currently, 100 Gb/s coherent transmission systems now represent the workhorse of long-haul transmission systems deployed.

As will be shown shortly, coherent technology also modified the requirements for optical fibers with a distinct shift toward lower loss and larger effective area. Another important fiber attribute that has gained attention in coherent systems is the nonlinear refractive index n_2, which, along with effective area, determines the nonlinear tolerance of a fiber through the nonlinear coefficient $\gamma = \frac{2\pi}{\lambda} \frac{n_2}{A_{eff}}$. For example, in traditional Germania-doped fibers the nonlinear index n_2 is approximately 2.3×10^{-20} m^2/W, although the exact value will be dependent on Germania concentration in the core. Large effective area fibers typically have slightly lower n_2 due to lower Germania concentration in the core to achieve the required difference in refractive indices between the core and the cladding. The n_2 can be reduced to approximately 2.1×10^{-20} m^2/W (up to $\sim$10% reduction) by using silica-core fiber designs with Fluorine-doped cladding.

In systems with coherent detection, chromatic dispersion (CD) also contributes to system performance. All other things being equal, non-zero dispersion shifted fibers (NZ-DSF) with CD equal to $\sim$4 ps/nm/km at 1550 nm (ITU-T G.655 2009) incur a transmission penalty compared to standard single-mode fibers with CD equal to $\sim$16 ps/nm/km at 1550 nm (ITU-T G.652 2016). This is explored in more detail later. The performance advantage of higher dispersion fibers is partly because they enable a more rapid spreading of the signal, causing faster reduction in signal peak-to-mean ratio, and thereby improving tolerance toward nonlinear effects. The higher dispersion also promotes faster walkoff of channels with different wavelengths, decreasing interchannel nonlinear effects. A further increase in CD from $\sim$16 to $\sim$21 ps/nm/km (typical CD value for large effective area fibers) (ITU-T G.654 2016) will provide additional tolerance toward nonlinear effects, albeit the improvement will be less pronounced compared to an increase from 4 to 16 ps/nm/km.

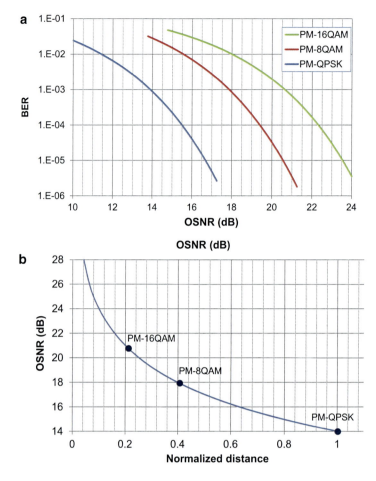

Fig. 3 (**a**) OSNR sensitivity of different modulation formats, (**b**) Normalized reach of each format

Overall, there seems to be a consensus within the industry that optical transport systems with coherent detection will dominate the landscape for the foreseeable future. The future bit-rates will continue to grow beyond 100 Gb/s through the use of even more spectrally efficient modulation (e.g., 8 or 16-state quadrature amplitude modulation) and probabilistic constellation shaping. As those spectrally efficient modulation formats have more stringent optical signal-to-noise ratio requirements, the importance of advanced fiber characteristics discussed previously will be even more pronounced for >100 Gb/s systems (particularly, for long-reach transmission). For example, consider Fig. 3a which shows the nominal OSNR sensitivity data (bit error rate BER vs. OSNR) for 32 Gbaud PM-QPSK, PM-8QAM, and PM-16QAM. The net data rate 150 Gb/s PM-8QAM signal requires about 4 dB higher OSNR than 100 Gb/s PM-QPSK, and 200 Gb/s PM-16QAM requires about 6.7 dB higher OSNR

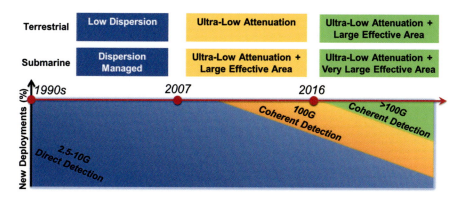

Fig. 4 Schematic illustration of the evolution of optical fiber transmission systems

than PM-QPSK. These higher OSNR requirements for higher spectral efficiency formats translate directly into shorter reach lengths for the same system. In Fig. 3b, the normalized reach lengths for the three formats are shown for a hypothetical system corresponding to the required OSNR values at a BER of about 1×10^{-3}. The reach lengths of PM-8QAM and PM-16QAM are reduced relative to PM-QPSK by about 60% and 80%, respectively. This reach reduction for higher spectral efficiency modulation formats illustrates the need for optical fiber improvements to help extend the possible reach of such formats to practical and cost-effective distances. The overall progression of technologies in terrestrial and submarine optical fiber transmission systems, along with key optical fiber characteristics, are summarized in Fig. 4.

Optical Fiber Designs for Long-Haul Transmission

To design an optical fiber for long-haul transmission, there are several factors to consider such as attenuation, mode field diameter or effective area, cutoff wavelength, and chromatic dispersion.

The total attenuation of an optical fiber is the sum of intrinsic and extrinsic attenuation. Figure 5 shows the major attenuation components of a generic silica-based optical fiber. Intrinsic attenuation is due to fundamental properties of glass materials used to construct the fiber core and cladding. Factors include Rayleigh scattering, and infrared and ultraviolet absorption tails in the transmission window. Extrinsic attenuation factors include absorption due to impurities such as transition metals and OH ions, small angle scattering (SAS) due to waveguide imperfections, and loss due to fiber bending effects.

The most important intrinsic attenuation factor is Rayleigh scattering loss. Rayleigh scattering loss (α_{RS}) is the sum of scattering due to density fluctuations (α_ρ) and scattering from concentration fluctuations (α_c). This is expressed as

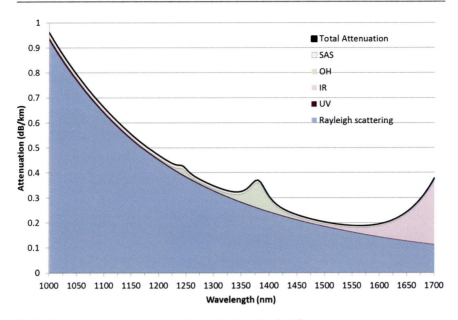

Fig. 5 Major attenuation components for a silica-based optical fiber

$$\alpha_{RS} = \alpha_\rho + \alpha_c, \qquad (1)$$

where α_ρ is given by

$$\alpha_\rho = \frac{8\pi^3}{3\lambda^4} n^8 p^2 \beta_T k_B T_f, \qquad (2)$$

T_f is fictive temperature, λ is incident wavelength, p is the photoelastic coefficient, n is refractive index, k_B is the Boltzmann constant, and β_T is isothermal compressibility (Maksimov et al. 2011). Of these, the most important parameter is glass fictive temperature, the temperature where glass structure is the same as that of a supercooled liquid. The concentration fluctuation scattering loss is proportional to the gradient of dopant concentration C as

$$\alpha_c \sim \left(\frac{\partial n}{\partial C}\right)^2 \langle \Delta C^2 \rangle T_f \qquad (3)$$

Because both the density and concentration fluctuation components of Rayleigh scattering are proportional to the fictive temperature, it is important to reduce T_f as much as possible to increase structural relaxation. To further reduce the concentration fluctuation, it is advantageous to reduce the dopant level in the core (Lines 1994; Kakiuchida et al. 2003).

1 Single-Mode Fibers for High Speed and Long-Haul Transmission

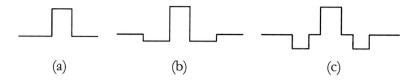

Fig. 6 Profile designs for large effective area fibers

The aforementioned impurities which contribute to extrinsic attenuation can be mostly addressed and eliminated in the chemical vapor deposition process utilized in fiber production. Scattering losses due to waveguide imperfections can be minimized by matching core and cladding glass viscosity and using advanced fiber-manufacturing technology. Extrinsic attenuation due to both macro- and micro-bending losses is primarily addressed in fiber design.

Optical fiber properties such as effective area, cutoff wavelength, chromatic dispersion, and bending loss are governed by refractive index profile design. As mentioned earlier, fiber chromatic dispersion was a very important fiber parameter for direct detection systems due to transmission limiting pulse broadening. However, with new developments in coherent detection and digital signal processing technologies, transmission impairments from fiber dispersion can be compensated digitally. Low dispersion is no longer a factor that needs to be considered in fiber design. In fact, high dispersion is desirable because it can reduce nonlinear effects. Due to the reduced impact of dispersion, the most influential fiber parameters are effective area, cable cutoff wavelength, and attenuation due to bending losses.

Figure 6 shows three general profile design types that can be used for fibers with large effective area and low attenuation. Figure 6a is a step index profile design. This simple design has two profile parameters: the relative refractive index change (core delta) and the core radius. To increase the effective area, the core radius can be increased but core delta must be reduced to keep the cable cutoff wavelength below the minimum wavelength of the application window, e.g., 1530 nm for 1550 nm window. Due to the cable cutoff wavelength specification limitations in international standards, the bend losses will tend to increase when the effective area gets larger. In general, there are system-based specifications for attenuation due to macrobending, e.g., less than 0.5 dB loss for 100 turns at a bend diameter of 60 mm. For a step index design with cable cutoff and macro-bend loss constraints, the maximum effective area that can be achieved is approximately 110 μm^2.

To further increase the effective area, a depressed cladding layer (low index trench) can be added, as shown in Fig. 6b and c. This suppresses macro-bend losses while keeping the cable cutoff wavelength below 1530 nm. For example, Fig. 7 shows measured bend loss as a function of effective area for design (b) at 1550 nm. For 30 mm bend diameter, the bend loss starts to increase rapidly when effective area is greater than 130 μm^2, while for 40 mm bend diameter, the effective area can be as large as 175 μm^2 and maintain minimal attenuation due to bending.

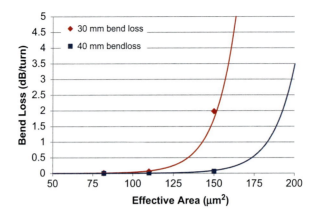

Fig. 7 Bend loss as a function of effective area for profile design (**b**) shown in Fig. 6

Table 1 ITU-T macrobend loss fiber specifications

	G.652.B	G.652.D	G.654.A	G.654.B	G.654.C	G.654.D	G.654.E
Macrobend loss at 1625 nm with 30 mm bend radius (dB/100 turns)	0.1	0.1	0.5	0.5	0.5	2.0	0.1

Table 1 shows the ITU-T specifications for maximum macrobend loss for G.652 and G.654-compliant fibers (ITU-T G.652 2016; ITU-T G.654 2016). The specifications are provided at 1625 nm. The maximum cable cutoff wavelength for the G.652 standards is 1260 nm, and these fibers are meant for terrestrial transmission systems which usually involve distances between splices of 2 and 8 km. Splices in terrestrial systems are usually done in splice trays that include a number of fiber turns or loops on each side of the splice. The G.654 standards A-D were designed for submarine fibers with larger effective areas for which the specification for maximum cable cutoff is 1530 nm. A key difference in splicing of submarine fibers compared to terrestrial fibers is that intra-span splices are done in a straight-through configuration (no loops or turns) in which the fiber is re-coated after splicing. The fibers are spliced in something similar to a splice tray configuration only at the repeater sites where it is spliced to fiber jumpers attached to the optical amplifiers. It is worth noting that besides the macrobend loss specifications for submarine fibers as defined by the ITU-T standards, companies that build submarine cables may have their own specifications that a fiber manufacturer must meet. The G.654.E standard recently adopted was also for a larger effective area fiber with cable cutoff of 1530 nm, but since this fiber is nominally intended for terrestrial system deployment, the macrobend loss specification is the same as the G.652 fiber specifications.

In addition to macro-bend induced attenuation, attenuation due to micro-bends is also a limiting factor for large effective area optical fibers (Bickham 2010). If the fiber-coating system currently used for standard single-mode fiber (G.652) is employed, it has been reported that the fiber effective area may only be increased to approximately 120 μm^2 due to micro-bend attenuation (Bigot-Astruc et al. 2008). Micro-bending is an attenuation increase caused by high frequency longitudinal perturbations to the waveguide. These perturbations couple power from the guided fundamental mode in the core to higher-order cladding modes that are lost to the outer medium. A phenomenological model introduced by Olshansky captures the importance of treating the glass and coating as a composite system by predicting that the micro-bend losses scale as

$$\gamma_{micro} \propto \frac{a^4}{b^6 \Delta^3} E^{3/2} \quad (4)$$

where γ_{micro} is attenuation due to micro-bending, a is the core radius, b is the cladding radius, Δ is the relative refractive index of the core, and E is the elastic modulus of the primary coating layer that surrounds the glass (Olshansky 1975). As mentioned earlier, core delta and core radius are determined by the desired effective area and the cutoff wavelength. These variables are not completely independent and together do not offer a significant path to manage micro-bending sensitivity. This leaves the inner primary modulus as the key factor for addressing increased micro-bend loss in large effective area fibers. Making the inner primary coating softer helps to cushion the glass from external perturbations and improve the micro-bending performance.

The role of primary coating modulus in mitigation of micro-bend induced attenuation has been demonstrated experimentally. In the experiment, fibers of different effective areas were made with two coatings which have higher and lower inner primary moduli (Coating A and B, respectively). Figure 8 shows measured fiber attenuation data for fiber wound under tension on a shipping spool, a condition that promotes microbending. For effective areas between 110 and 115 μm^2, the

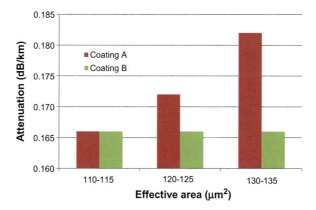

Fig. 8 Attenuation measured on shipping reels of fibers with different effective areas and different coatings

attenuation of fibers with the two coatings are the same. This shows that for fibers with an effective area less than 115 μm^2, intrinsic attenuation dominates. For fiber effective areas larger than 120 μm^2, the micro-bend loss starts to increase with Coating A, while Coating B minimizes the micro-bend loss to nearly zero for effective areas up to 135 μm^2. It is evident that to achieve ultra-low attenuation, a fiber with very large effective area will require a coating with an optimized inner primary modulus to protect against micro-bend induced attenuation. With optimized primary coating, fibers with low loss and effective area of about 150 μm^2 are now commercially available.

The effective area of the fundamental mode can be increased further by increasing the fiber cutoff wavelength beyond the operating wavelength. In this case, the fiber becomes a few mode fiber. A few mode fiber can be used as a quasi-single-mode (QSM) fiber by launching the light into the fundamental mode (Yaman et al. 2010). It has been shown that the effective area of the LP_{01} mode can be increased to over 200 μm^2 using this approach (Yaman et al. 2015). The large-effective-area nature of the LP_{01} mode may help extend the transmission distance considerably. However, an issue of multi-path interference (MPI) arises in QSM transmission that may limit the transmission distance. In general, mode coupling between the LP_{01} and LP_{11} modes is expected to be present in practical QSM systems and may induce MPI and therefore signal degradation. MPI must be compensated before the full benefit of the larger effective area of QSM can be realized. It has been shown that MPI can be compensated to a large degree by using digital signal processing techniques in the receiver (Sui et al. 2014).

Quantification of System Level Performance

There are many ways to assess and quantify overall system performance of different optical fibers for current high-speed coherent transceivers. Some of these metrics include bit error rate (BER) or alternatively Q-factor defined as 20 log (Q) where $Q = \sqrt{2} erfc^{-1} (2 \cdot BER)$, a fiber figure of merit (FOM) function, the generalized optical-signal-to-noise ratio (G-OSNR), and the reach length attainable at the forward error correction (FEC) threshold or with a defined Q-factor margin. It is important to mention that there are many factors involved in transmission performance beyond fiber type, including type of optical amplification, span length between optical amplifiers, signal modulation format, symbol rate, FEC overhead, and net coding gain. Here, we will mainly concentrate on the fiber characteristics and attributes that affect modern coherent system performance. We discuss the relative impact of the fiber traits in both conventional long-haul (LH) or ultra-long-haul (ULH) repeatered systems with transmission through a long chain of optical amplifiers, and specialized unrepeatered systems which consist of a long single span with no locally powered amplification provided within the span.

Long-Haul and Ultra-Long-Haul Transmission Systems

We begin by assuming that new coherent transmission systems are designed and deployed without any optical dispersion compensation anywhere in the line system. Thus, all optical amplifiers can be of a single stage configuration and we can assume all dispersion compensation is performed digitally in the digital signal processing (DSP) carried out in the coherent receiver. In principle, some dispersion compensation can also be performed digitally in the transmitter, and somewhat enhanced performance may be obtained in this case. In this context of coherent detection and for standard or Nyquist wavelength division multiplexing (WDM) transmission, a fiber figure of merit (FOM) has been developed as a means of comparing different fibers and their performance (Curri et al. 2013; Hirano et al. 2013; Makovejs et al. 2016a). The FOM is based on the Gaussian noise model of coherent transmission systems (Poggiolini 2012). A simplified version of the FOM is given below in Eq. 5, in which the FOM of a fiber under consideration is defined in a relative sense to a reference fiber.

$$FOM(dB) = \frac{2}{3} 10 \log \left[\frac{A_{eff}}{A_{eff,ref}} \frac{n_{2,ref}}{n_2} \right] - \frac{2}{3} \left(\alpha_{dB} - \alpha_{dB,ref} \right) L \\ - \frac{1}{3} 10 \log \left[\frac{L_{eff}}{L_{eff,ref}} \right] + \frac{1}{3} 10 \log \left[\frac{D}{D_{ref}} \right] \quad (5)$$

In Eq. 5, A_{eff} is fiber effective area [μm^2], n_2 is fiber nonlinear index of refraction [m^2/W], α_{dB} [dB/km] is fiber attenuation, D is chromatic dispersion [ps/nm/km], L is the span length between amplifiers [km], L_{eff} is the nonlinear effective length $L_{eff} = \frac{(1-e^{-\alpha L})}{\alpha} \approx \frac{1}{\alpha}$ [km], and α is the fiber attenuation in linear units [1/km]. While some forms of the FOM also include the effect of splice loss to a jumper, we have neglected that here, as we will address splice loss issues later. As it is defined here and based on the Gaussian noise model, the FOM represents the expected difference in 20log(Q) between the fiber under study and the reference fiber in the same system configuration (same link length, span length) if the optimal channel power is used in the system for each fiber. The optimal channel power occurs at the power level for which the noise power from amplified spontaneous emission (ASE) is equal to twice the noise power from nonlinear effects, and represents the channel power that promotes the highest Q-factor and best performance (Poggiolini et al. 2014).

As illustrated in Eq. 5, the main fiber parameters that affect overall system performance are attenuation, nonlinear tolerance as governed by effective area and nonlinear index, and chromatic dispersion. As mentioned earlier, attenuation and fiber effective area contribute the most to transmission performance. We therefore begin by temporarily ignoring dispersion effects (essentially assuming equal dispersion for the fiber under study and reference fiber) and examine the FOM as a function of only A_{eff} and attenuation, as shown in Fig. 9. The reference fiber

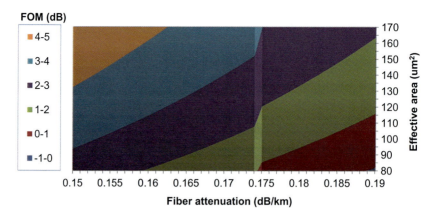

Fig. 9 Fiber FOM as function of attenuation and effective area for 100 km spans in a long-haul repeatered system

for this data has attenuation $\alpha = 0.19$ dB/km and effective area $A_{eff} = 82$ μm², representative of a generic G.652-compliant standard single-mode fiber. The span length for Fig. 9 is 100 km, typical for many terrestrial LH networks. The data shows clearly that FOM is increased by both reducing fiber attenuation and increasing fiber effective area. The reduction of attenuation from 0.19 to 0.155 dB/km results in about 2.3 dB FOM improvement, while increasing A_{eff} from 82 to 150 μm² yields almost 1.75 dB advantage for this span length. If the fiber has both 0.155 dB/km attenuation and 150 μm² effective area (as is the case for some state-of-the-art submarine system fibers), the FOM advantage is over 4 dB. The discontinuity between 0.174 and 0.175 dB/km reflects an assumption that attenuation values below 0.175 dB/km are achieved with silica core fibers with lower nonlinear index n_2 compared to Ge-doped fibers (Kim et al. 1994; Makovejs et al. 2016a). The n_2 values assumed here are 2.3×10^{-20} m²/W for Ge-doped fiber and 2.1×10^{-20} m²/W for silica core fiber.

The fiber figure of merit function (neglecting dispersion effect) can also be used to understand the relative effects of lower fiber attenuation and greater fiber effective area. The comparison is dependent on the system span length, as given in the second term of Eq. 5. Figure 10 illustrates the equivalent increase in effective area as a function of attenuation reduction. The equivalency is calculated by determining the same increase in FOM with respect to the nominal parameters of the reference fiber described above. This is done for four different span lengths of 120, 100, 80, and 60 km. It is evident that lowering the fiber attenuation has a larger effect for longer spans. For example, a reduction of attenuation (from 0.19 dB/km) of −0.04 dB/km is equivalent to increasing the effective area (above 82 um²) by approximately 160, 120, 85, and 60 um² for span lengths of 120, 100, 80, and 60 km, respectively.

To obtain a more comprehensive comparison of different optical fibers in terms of the FOM (including all relevant parameters), we evaluate fiber characteristics that are representative of commercially available optical fibers today. The fibers

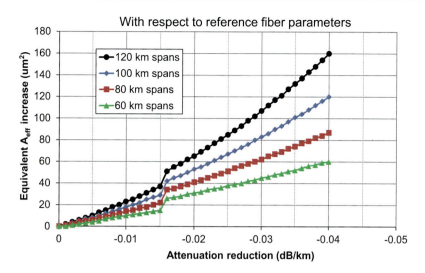

Fig. 10 Equivalent effective area increase as a function of attenuation reduction for different span lengths, as estimated from FOM

Table 2 Realistic optical fiber characteristics

	Fiber #1	Fiber #2	Fiber #3	Fiber #4	Fiber #5	Fiber #6
ITU standard description	Conventional G.652	G.652 with low loss	G.652 with ultra-low loss	G.654.E with ultra-low loss	G.654.B with ultra-low loss	G.654.D with ultra-low loss
Attenuation (dB/km)	0.19	0.183	0.162	0.168	0.154	0.154
Effective area (μm^2)	82	82	82	125	112	150
Dispersion at 1550 nm (ps/nm/km)	17	17	17	21	21	21
Nonlinear index n_2 (m^2/W)	2.3×10^{-20}	2.3×10^{-20}	2.1×10^{-20}	2.1×10^{-20}	2.1×10^{-20}	2.1×10^{-20}

considered are described in Table 2. Fiber #1 represents a generic G.652 fiber widely used in terrestrial networks. Fiber #2 is a similar fiber with slightly lower attenuation. Fiber #3 is another G.652 fiber made with a silica core that provides significantly lower attenuation. Fiber #4 is a silica core terrestrial fiber with a large effective area compliant with the ITU-T G.654.E standard. Fibers #5 and #6 are silica core submarine system fibers with large effective areas compliant with ITU-T G.654.B and ITU-T G.654.D standards, respectively.

The FOM was calculated according to Eq. 5 for Fibers #2–6, relative to the generic G.652 standard single-mode fiber represented by Fiber #1, over a range

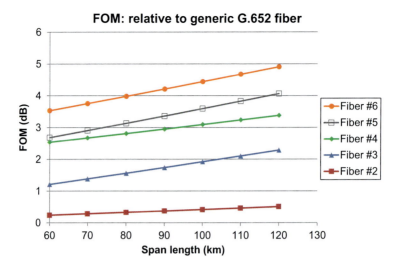

Fig. 11 FOM as a function of span length for five fibers relative to generic G.652-compliant fiber. Fiber descriptions: Fiber #1: conventional G.652, Fiber #2: G.652 with low loss, Fiber #3: G.652 with ultra-low loss, Fiber #4: G.654.E with ultra-low loss, Fiber #5: G.654.B with ultra-low loss, Fiber #6: G.654.D with ultra-low loss

of span lengths from 60 to 120 km. The results are shown in Fig. 11. As described earlier, the greater impact of lower attenuation with longer span lengths is illustrated by the larger slope of the FOM function with respect to span length for the various silica core fibers.

Finally, it is interesting to break out the various components of the FOM functions for each of the fibers at individual span lengths. This data is shown in Fig. 12 for 60 km and 100 km spans. The results confirm that greater benefit is derived from low attenuation with longer spans, while reducing nonlinear effects with larger effective area, lower n_2, or greater dispersion has the same impact for any span length. For 60 km spans, the large effective areas of Fibers #4–6 generally provide the leading FOM component, while for 100 km spans, the lower attenuation of those fibers has equal or greater impact on the FOM as effective area.

Raman Gain Considerations

While the FOM as constructed in Eq. 5 is independent of the type of optical amplifiers used in a system, it is also worthwhile to examine the nominal expected behavior and performance of the different fibers when used with Raman amplification. Raman amplification is becoming more widely accepted in terrestrial networks because it enables greater reach lengths, especially important for multilevel modulation formats such as PM-16QAM and data rates of 200 Gb/s and higher. In general, the effective noise figure of a distributed Raman amplifier decreases with higher

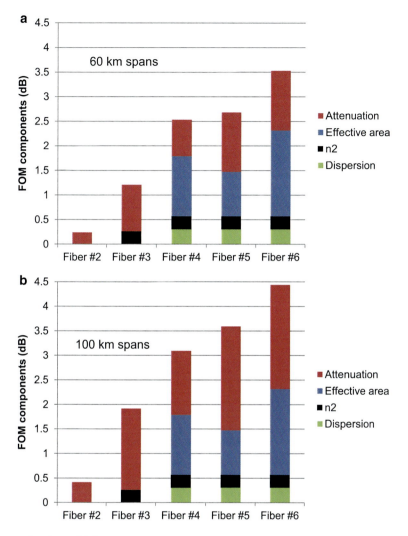

Fig. 12 Breakout of different FOM components for five fibers: (**a**) 60 km spans, (**b**) 100 km spans. Fiber descriptions: Fiber #1: conventional G.652, Fiber #2: G.652 with low loss, Fiber #3: G.652 with ultra-low loss, Fiber #4: G.654.E with ultra-low loss, Fiber #5: G.654.B with ultra-low loss, Fiber #6: G.654.D with ultra-low loss

Raman gain (Evans et al. 2004). Therefore, one aspect of interest with regard to optical fiber is the level of ON/OFF Raman gain achievable with a given level of Raman pump power.

For the following analysis, we examine the simplest case of the ideal ON/OFF Raman gain for the six optical fibers considered in the un-depleted pump approximation. In this approximation, the Raman gain can be calculated as

$$Gain(dB) = \frac{10}{\ln(10)} g_R P_{pump} L_{eff} \qquad (6)$$

where g_R is the fiber's Raman gain coefficient in units of (1/W/km), P_{pump} is the Raman pump power in W, and L_{eff} is the *effective length of the pump* laser at the pump wavelength. The pump effective length is defined as

$$L_{eff} = \frac{(1 - e^{-\alpha_p L})}{\alpha_p} \approx \frac{1}{\alpha_p} \qquad (7)$$

where α_p is the attenuation of the pump wavelength in linear units (1/km). As can be seen from Eq. 6, the Raman gain is affected by fiber characteristics in terms of the pump wavelength attenuation and the Raman gain coefficient g_R. The gain coefficient is essentially a function of the fiber material system and the index profile. In particular, since Raman amplification is a nonlinear effect, the gain coefficient is affected by the fiber effective area and the fiber nonlinear index of refraction. To compare the six different fibers in Table 2, we make the simplifying assumption that the gain coefficient of each fiber can be calculated by scaling the gain coefficient of Fiber #1 as

$$g_R = g_{R,\#1} \cdot \frac{n_2}{n_{2,\#1}} \cdot \frac{A_{eff,\#1}}{A_{eff}} \qquad (8)$$

By taking the gain coefficient of the generic standard single-mode fiber #1 as about 0.4 W^{-1} km^{-1}, and making a further simple assumption that the attenuation at a pump wavelength of about 1450 nm is 0.05 dB/km higher than the attenuation at 1550 nm for all fibers, it is possible to calculate idealized Raman ON/OFF gain values for the different fibers in an un-depleted pump approximation. The results of that calculation are shown in Fig. 13 for a span length of 100 km and a nominal pump power of 500 mW at 1450 nm. This comparison is simplified, as real systems will likely have more than one pump wavelength to produce relatively flat gain over a wide bandwidth for a large number of optical channels and the un-depleted pump approximation will likely be less accurate. However, this simple analysis is useful for understanding the relative Raman gains likely with the different optical fibers and the same level of pump power. In particular, it is evident that effective area has a significant influence, and fibers with larger effective areas will have smaller ON/OFF Raman gains due to smaller gain coefficients g_R.

As mentioned earlier, the effective noise figure of a Raman amplifier decreases with increasing Raman gain, so for the same pump power, the overall relative performance advantage of larger A_{eff} fibers will be slightly smaller than the prediction of FOM in Figs. 11 and 12 in hybrid Raman/EDFA systems in which the Raman gain is less than the total span loss. However, it is also important to realize that fibers with lower attenuation require less Raman gain, which can significantly lessen the pump power demand. An easy way to demonstrate this is to consider systems in which all amplification is obtained through distributed Raman amplifiers. Consider, for example, systems with 100 km spans for which all

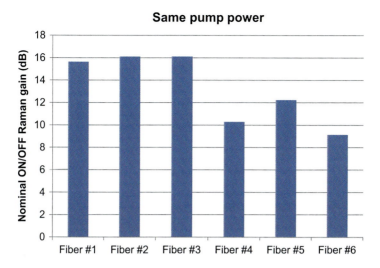

Fig. 13 Nominal Raman gain in dB for the same pump power and span length. Fiber descriptions: Fiber #1: conventional G.652, Fiber #2: G.652 with low loss, Fiber #3: G.652 with ultra-low loss, Fiber #4: G.654.E with ultra-low loss, Fiber #5: G.654.B with ultra-low loss, Fiber #6: G.654.D with ultra-low loss

fiber loss is to be compensated with distributed Raman amplification. Then Fiber #1 would require 19 dB of Raman gain (assuming no extra losses) while Fiber #4 would require 16.8 dB of Raman gain. We can use Eq. 6 to calculate relative estimates of the required pump powers needed to produce the gains that would fully compensate for the span losses. The results for all six fibers are shown in Fig. 14, normalized to the pump power required for Fiber #1. We see that even though Fibers #2 and #3 have nearly the same ON/OFF gain for the same pump power as Fiber #1, they require approximately 6% and 17% lower pump power to achieve full span loss compensation, respectively. Similarly, while Fiber #4 has nearly 5.5 dB lower Raman ON/OFF gain than Fiber #1 for the same 500 mW pump power, only about 34% more Raman pump power is required to compensate for the lower span loss of Fiber #4. In that case, under these idealized conditions, Fiber #1 would require 607 mW of pump power and Fiber #4 would require 816 mW of pump power.

Unrepeatered Span Transmission Systems

The previous analysis of fiber FOM and the relative effects of attenuation and effective area were in the context of conventional repeatered multispan systems with optical amplifiers compensating for the loss of each span. In contrast, an unrepeatered span system is significantly different and connects a transmitter and receiver pair over a single long span with no active equipment between the terminals. The single span length of these systems can often be up to several hundreds of km

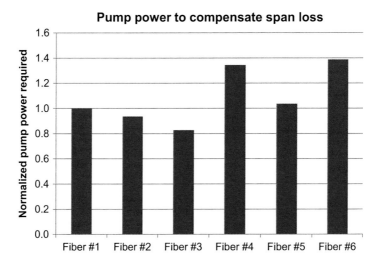

Fig. 14 Normalized pump power required for full span loss compensation with distributed Raman amplification. Fiber descriptions: Fiber #1: conventional G.652, Fiber #2: G.652 with low loss, Fiber #3: G.652 with ultra-low loss, Fiber #4: G.654.E with ultra-low loss, Fiber #5: G.654.B with ultra-low loss, Fiber #6: G.654.D with ultra-low loss

and are designed for both submarine and terrestrial networks. They may connect islands, an island to the mainland, coastal mainland points to each other in a festoon arrangement, or two terrestrial cities with forbidding and difficult terrain between them. In each case, the systems require the transmission of optical signals over a long distance without powered amplifiers between the terminals. A primary goal in the design of many unrepeatered systems is to maximize the reach so the distance between the desired terminal points can be achieved with the desired data rate. There have been many recent research examples of such systems with coherent transmission of 100 and 200 Gb/s channels (Downie et al. 2010; Mongardien et al. 2013).

We now examine the relative fiber characteristics that have the most influence on performance and reach in unrepeatered span systems. For repeatered multispan systems, we found that the fiber parameters contributing the most to transmission performance are attenuation and effective area. While dispersion and nonlinear index of refraction have some effect, the range of these parameters is smaller in real fibers and they thus have smaller impact. The same is true for unrepeatered span systems so we will focus again on attenuation and effective area. The fiber FOM given in Eq. 5 was derived in the context of repeatered multispan systems and may not be as directly relevant to unrepeatered span systems. However, we can make relatively simple estimations of the effects of fiber attenuation and effective area on the maximum unrepeatered span length (Downie et al. 2016).

To begin, we consider the role of nonlinear tolerance as given by fiber effective area. Similar to a conventional system, the channel launch power into the span will normally be optimized in order to get the best performance. The optimal power represents a balance between linear and nonlinear impairments. We recognize that higher optimal channel powers translate into higher optical signal-to-noise (OSNR) values at the receiver. Therefore, the difference in the received OSNR between a fiber under evaluation with effective area A_{eff} and a reference fiber with effective area $A_{eff,ref}$ and the same attenuation can be approximated by

$$\Delta OSNR(dB) = 10 \log \left[\frac{A_{eff}}{A_{eff,ref}} \right] \qquad (9)$$

Under the assumption that a given OSNR value is required at the receiver to produce the desired Q-factor value, the difference in attainable span length ΔL between the evaluated and reference fibers can be simply related to the common attenuation of the fibers as

$$\Delta L(km) = \frac{\Delta OSNR(dB)}{\alpha_{dB}} \qquad (10)$$

The results predicted by Eqs. 9 and 10 for a hypothetical reference fiber with effective area $A_{eff,\,ref} = 80$ μm^2 are shown in Fig. 15 for two different values of fiber attenuation. The actual predicted increase in reach ΔL is relatively small and less than 22 km, even for fiber effective areas as large as 170 μm^2 with very low attenuation of 0.15 dB/km. This simple analysis does not account for lower Raman gain that accompanies larger fiber effective area, which may serve to reduce the ΔOSNR and thus further reduce the reach increase compared to the reference fiber.

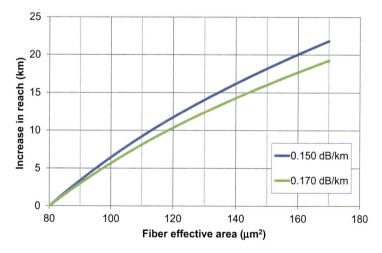

Fig. 15 Increase in the reach or span length of unrepeatered span systems as a function of fiber effective area relative to 80 μm^2

We next consider the effect of lower fiber attenuation on the reach of an unrepeatered span system. There are many different designs for these systems, including the use of forward and backward pumped Raman amplification, remote optically pumped amplifiers (ROPAs) in mid-span locations, higher-order Raman pumping, etc. (Chang et al. 2015; Huang et al. 2017). However, for a given type of system design, a commonly used means of quantifying the performance of the design is by the maximum span loss that can be tolerated for a given level of signal quality and capacity. To compare the reach L of a fiber with attenuation α_{dB} to the reach L_{ref} of a reference fiber with attenuation $\alpha_{dB,\,ref}$, we can equate the maximum loss accommodated by the system design as

$$Loss_{\max}(dB) = \alpha_{dB} \cdot L = \alpha_{dB,ref} \cdot L_{ref} \qquad (11)$$

It is then straightforward to calculate the difference in reach length $\Delta L = L - L_{ref}$ between the fiber under evaluation and the reference fiber as

$$\Delta L(km) = \frac{(\alpha_{dB,ref} - \alpha_{dB})}{\alpha_{dB}\alpha_{dB,ref}} \cdot Loss_{\max}(dB) \qquad (12)$$

Advanced long reach unrepeatered spans may have maximum loss values between 60 and 100 dB (Chang et al. 2015; Huang et al. 2017). The data in Fig. 16 represents the increase in reach length ΔL in km for fibers with attenuation values <0.19 dB/km, compared to a reference fiber with 0.19 dB/km attenuation. It is clear that lowering the attenuation has a significantly greater effect on increasing the reach or span length of an unrepeatered span system than increasing fiber effective area. By reducing the fiber attenuation to 0.15 dB/km, the reach can be increased between

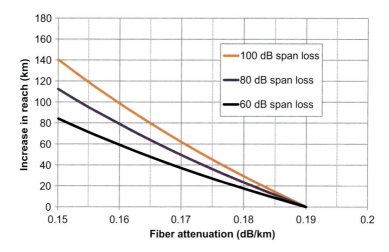

Fig. 16 Increase in the reach or span length of unrepeatered span systems as a function of fiber attenuation relative to 0.19 dB/km

about 85 and 140 km, depending on the maximum loss tolerated by the system design.

The previous analyses estimated unrepeatered span system reach increases for larger effective area and lower attenuation assuming that the span was homogeneously comprised of a single fiber type. In fact, many long unrepeatered spans may be designed and constructed with a hybrid fiber span configuration in which two or more fiber types are used in order to maximize performance and reach. As observed earlier, Raman gain varies in an inverse relationship with fiber effective area since it is a nonlinear effect. Since nearly all long unrepeatered span systems will employ distributed Raman amplification as a means to increase reach, span construction can be performance optimized by employing the largest effective area fiber at the beginning of a span where the signal power is highest to minimize signal nonlinearity, and a smaller effective area fiber where greater Raman gain is desired, such as at the back end of the span if backward pumped Raman amplification is used. Such designs take advantage of both large effective area fibers for increased channel launch power as well as enhanced Raman gain from fibers with somewhat smaller effective area to optimize the entire system (Downie et al. 2010, 2014; Puc et al. 2009).

Transmission System Modeling and Experiments

In this section, we briefly present some modeling and experimental results comparing various single-mode optical fibers for high-speed transmission systems. For the modeling data, we used the Gaussian noise formalism developed to predict transmission performance of dispersion uncompensated coherent systems with Nyquist WDM channels (Poggiolini 2012; Carena et al. 2014). The conventional repeatered system modeled had 100 km spans with EDFA-only amplification at the end of each span. The EDFA noise figure was 5 dB, there were 80 optical channels, the transmitter and receiver were assumed to be ideal, and there were no extra losses in the spans beyond the fiber loss. The channels were encoded with the polarization-multiplexed 16 quadrature amplitude modulation (PM-16QAM) modulation format, with a symbol rate of 32 Gbaud. The results are given in Fig. 17 as Q-factor (dB) vs. distance for all six fiber types described in Table 2. In the figure, the dashed line represents a Q-factor threshold of 6.25 dB for soft-decision forward error correction (SD-FEC). The absolute maximum reach for each fiber system can be considered as the distance at which the Q-factor meets the SD-FEC threshold line. Below this Q-factor level, the signal will not be error-free after error correction is applied. The second dashed line is at a Q-factor of 9.25 dB, representing a 3 dB margin above the SD-FEC threshold. Real commercial systems are deployed with some level of margin to allow system degradation over time due to factors such as future fiber repairs and component aging effects. Inspection shows that the reach lengths expected with the different fibers measured at either the SD-FEC threshold or the 3 dB margin level are consistent with the relative performances predicted by the fiber FOM examined earlier.

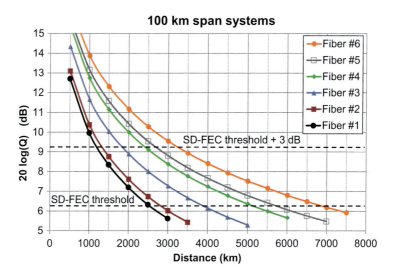

Fig. 17 Gaussian noise model transmission predictions of Q-factor vs. distance for systems with 100 km spans of different fiber types. Fiber descriptions: Fiber #1: conventional G.652, Fiber #2: G.652 with low loss, Fiber #3: G.652 with ultra-low loss, Fiber #4: G.654.E with ultra-low loss, Fiber #5: G.654.B with ultra-low loss, Fiber #6: G.654.D with ultra-low loss

For comparison, experimental transmission system measurements over commercially available fibers with similar properties to hypothetical fibers #1, #3, #5, and #6 were performed in a re-circulating loop configuration (Downie et al. 2016). A schematic illustration of the general experimental setup is shown in Fig. 18. In the experiments, 20 channels spaced by 50 GHz and modulated with 32 Gbaud PM-16QAM signals were transmitted over systems with 100 km spans of fiber. The loop length was 3 × 100 km. An EDFA was used to amplify the channels back up to the original launch power at the end of each fiber span.

Experimental results of Q-factor vs. transmission distance for a central channel in the 20 channel plan are shown in Fig. 19. These results were obtained by first determining the optimal channel launch power for each different fiber separately, and then transmitting at the optimal channel power and measuring Q-factor vs. distance. There are significant differences in the absolute reach lengths attained from the experimental transmission measurements compared to the model results. However, the relative performance of the various fibers was consistent between model and experiment, and clearly illustrated the longer reach afforded by fibers with lower attenuation and larger effective area. The relative reach lengths normalized to Fiber #1 are shown in Fig. 20 for several different Q margins above the SD-FEC threshold. In particular, the reach of Fiber #6 was more than twice that of Fiber #1, while Fiber #5 showed a reach advantage of about 80% over Fiber #1. There are several reasons why the experimental system results produced shorter reach lengths than the model. The model assumed an ideal transmitter and receiver while the real devices significantly depart from ideal performance, meaning that higher OSNR

1 Single-Mode Fibers for High Speed and Long-Haul Transmission

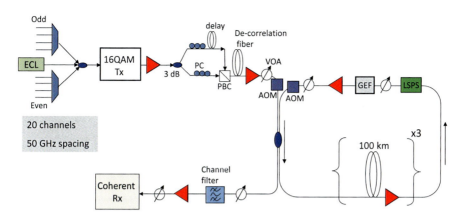

Fig. 18 Re-circulating loop experimental transmission system setup. PBC: polarization beam combiner, VOA: variable optical attenuator, AOM: acousto-optic modulator, GEF: gain equalization filter, LSPS: loop synchronous polarization scrambler

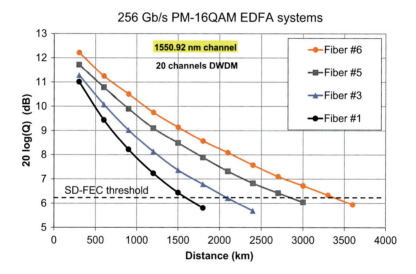

Fig. 19 Experimental transmission results for four different optical fibers in a re-circulating loop setup with 32 Gbaud PM-16QAM signals. Fiber descriptions: Fiber #1: conventional G.652, Fiber #3: G.652 with ultra-low loss, Fiber #5: G.654.B with ultra-low loss, Fiber #6: G.654.D with ultra-low loss

values are required to produce the same BER. Moreover, the re-circulating loop has losses from elements that are not present in the straight-line model system, and these extra losses tend to negatively impact systems with lower attenuation fiber to a greater extent. Another effect is that the noise figure of the EDFAs can be larger for the smaller gains required for the fiber spans with lower attenuation fiber.

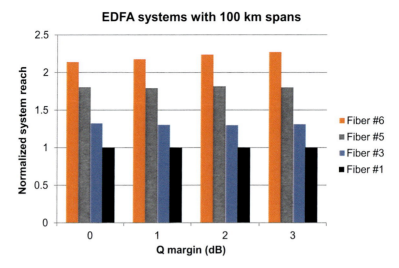

Fig. 20 Normalized reach lengths of experimental transmission for different Q margin levels above SD-FEC threshold. Fiber descriptions: Fiber #1: conventional G.652, Fiber #3: G.652 with ultra-low loss, Fiber #5: G.654.B with ultra-low loss, Fiber #6: G.654.D with ultra-low loss

Other Factors and Considerations

Splice Loss

In general, the splicing of standard ITU-T G.652-compliant fibers is a well-known process. Splice losses of <0.1 dB are considered typical, with average splice loss better than 0.03 dB in a controlled lab environment. The requirements on splice loss in the field may be slightly relaxed due to the fact that splicing may be performed in nonideal conditions for temperature, cleanliness, equipment maintenance, operator skill, etc., and in a confined space. However, the average splice loss guaranteed by the contractor should still be well within 0.1 dB. The frequency of the splicing will be determined by the length of the cable drum (typically between 2 and 8 km for terrestrial long-haul deployments) and the total number of splices depends on frequency and span length. It is also worth noting that the loss of the very first splice within the span (from the amplifier pigtail to the transmission fiber) can be compensated by increasing the amplifier output power by the respective amount without sacrificing fiber nonlinear transmission performance.

More generally, the splice loss between any two fibers will primarily depend on the mode field diameters of the two fibers being spliced. A theoretical estimation of such splice loss can be determined using Eq. 13, where $2W_1$ and $2W_2$ are mode field diameters (A_{eff} is proportional to $(2\,W)^2$) of the two fibers, d is the radial splice offset, and α_d is the splice loss (Ohashi et al. 1987).

$$\alpha_d = -10\, log \left[\left(\frac{2W_1 W_2}{W_1^2 + W_2^2} \right)^2 \times \exp \left(\frac{-2d^2}{W_1^2 + W_2^2} \right) \right] \quad (13)$$

The theoretically predicted splice loss of a fiber with conventional A_{eff} of 82 μm^2 to a fiber with an effective area 150 μm^2 (the largest A_{eff} currently available on the market) is around 0.3 dB. In practice, radial splice offset can vary from splice to splice, leading to variation of splice loss for the same two fiber types. In addition, the mode field diameter (or, A_{eff}) for a particular fiber type can vary, leading to further spread in splice loss values. The allowable variation in mode field diameter is determined by the ITU-T standards and is typically within ± 0.4 and ± 0.7 μm range for long-haul terrestrial and submarine fibers (ITU-T G.652 2016; ITU-T G.654 2016).

Figure 21a shows the ranges of measured difference in effective area for different fiber pair combinations. Each fiber type contains low, medium, and large effective areas from its production distribution. The average effective areas of Fibers #6, #1, and #5 were about 150, 82, and 112 μm^2, respectively. Naturally, the possible variation in A_{eff} is lowest when the same fibers are being spliced. When fibers with dissimilar A_{eff} are spliced, the possible range or variation increases (e.g., Fiber #6 with A_{eff} from upper distribution tail, and Fiber #5 with A_{eff} from lower distribution tail). As seen from Fig. 21b, larger variations in A_{eff} lead to larger variations in splice loss, and in the worst case scenario the splice loss may exceed 0.4 dB. It is worth noting, however, that splices between fibers with dissimilar effective areas are only likely to occur when transmission fiber (which could have effective area up to 150 μm^2) is spliced to the amplifier pigtail with significantly smaller A_{eff}. Techniques involving "bridge" fibers and tapering can be employed to reduce splice losses between fibers with dissimilar effective area (Makovejs et al. 2016b). Overall, the vast majority of splices are between like fiber types.

Practical Benefits of Ultra-Low Attenuation and Large Effective Area Fibers

The key performance benefits of low attenuation and large effective area optical fibers are higher OSNR and higher Q-factor. Depending on network goals and strategy, higher OSNR/Q-factor can promote and enable a number of secondary benefits as illustrated in Fig. 22.

First, higher OSNR/Q-factor enables an increase in distance between amplifiers while maintaining the same performance, thereby reducing the total number of amplifier huts and associated amplification equipment needed for a given optical fiber link. The reduction in number of amplifier huts lowers the amount of total power required for delivery to the amplifiers and also to provide temperature and humidity control for effective and reliable operation. This is particularly impactful for greenfield deployments, where there is no preexisting infrastructure that may prescribe the design rules. In addition, a greenfield deployment would necessitate

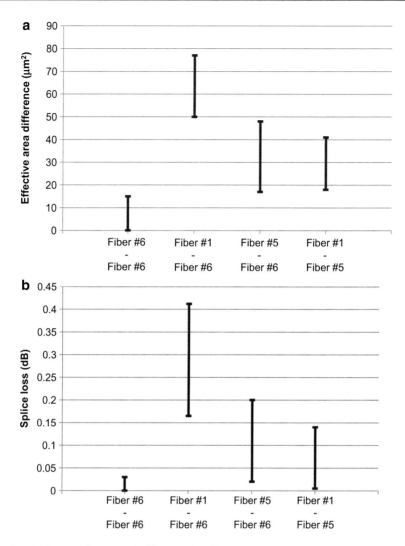

Fig. 21 (a) Range of measured difference in effective areas of two fibers under evaluation; (b) Range of measured splice losses for different pairs of fibers. Fiber descriptions: Fiber #1: conventional G.652, Fiber #5: G.654.B with ultra-low loss, Fiber #6: G.654.D with ultra-low loss

the expensive construction of new amplifier sites. Another difficulty is particularly pertinent in remote areas where access to electricity to power amplification sites may not be immediately available. The choice is then to either modify the existing power line by creating an extension to bring the power supply closer to the amplifier site, an expensive and lengthy process, or to use mobile diesel generators. The latter approach represents its own challenges. It is also expensive, particularly on the operational side, not always reliable, and requires frequent visits by a

Fig. 22 Examples of practical benefits enabled by ultra-low attenuation and large effective area fibers

technician to refill its supply. As typical long-haul deployments contain 48–288 fiber count cables, the reduction of amplification site count would result in the overall system reduction of 48–288 amplifiers per site eliminated. This can clearly lead to significant cost savings overall.

Similar benefits can be realized in submarine optical fiber links, where the removal of submarine repeaters can decrease the cable construction cost and overall submarine power requirements. The latter is particularly important for submarine optical cables that not only carry optical information in the fibers, but also electrical power in a copper conductor needed to operate the undersea optical amplifiers. Because electrical power is a precious and finite resource (current designs only allow to supply 12–15 kV from the power feed equipment from each end of the shore), any reduction in the amplifier power supply can be beneficial and enable significant cost savings.

An increase in Q-factor also enables longer reach, which we have defined as the distance at which the signal Q-factor drops to the minimum level allowed in the system design. For terrestrial applications, this is important given the evolution from opaque (signals regenerated at every node to add/drop traffic) to transparent networks (with optical add-drop multiplexers in which pass-through channels remain in the optical domain until they arrive at the final destination). Reach requirements in networks with transparent architecture are typically longer than in networks with opaque architecture. In optically transparent networks, optical fibers that enable long system reach can provide significant cost savings by reducing the number of expensive regenerators. Reach requirements can be even larger for networks with optical protection, in which geographically diverse protection routes are provisioned, typically longer than the main routes.

In addition to longer reach, higher margin over the FEC threshold can be achieved with an increase in Q-factor. This can improve network reliability and provide repair resilience. If a cable is cut (e.g., due to excavation/construction works or vandalism), the repair usually requires splicing a short fiber section to the two loose ends of the cable. This repair consisting of two splices per fiber will add insertion loss to the system. While the additional loss resulting from a single repair may not be significant, frequent cable cuts and repairs can eventually lead to a substantial increase in the overall cable loss potentially causing system failure. Fibers enabling higher initial signal Q-factor can therefore increase the total allowable number of repairs and increase cable longevity.

Finally, as the Internet traffic continues to grow with ~25% rate in fixed Internet networks and ~50% in mobile networks (Cisco VNI 2017), the demand for increased capacity in optical fiber cables is great. Traffic growth publically reported by Cloud providers at major conferences in even higher. As mentioned earlier, one way to increase capacity beyond 100 Gb/s is to adopt spectrally efficient modulation formats such as 8QAM, 16QAM, or even 64QAM, perhaps in combination with other advanced techniques, e.g., probabilistic constellation shaping. Ultra-low attenuation and large effective area fibers will be needed to support such transmission because of the more stringent OSNR requirements compared to well-established 100 Gb/s QPSK systems. This again is related to the ability of such fibers to provide longer reach before re-generation is needed as the intrinsic reach of higher modulation format and spectral efficiency signals diminishes due to greater OSNR requirements.

Figure 23 shows that on one hand, maximum reach can be increased when using fibers with advanced attenuation and effective area characteristics. On the other hand, for certain ranges of distance advanced fibers enable a transition to higher-density modulation formats, providing higher bit-rate per wavelength and ultimately higher capacity per fiber. We will refer to the distance range where such a transition

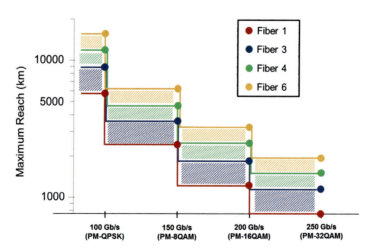

Fig. 23 Model results for maximum reach and bit-rate values for different fibers. Fiber descriptions: Fiber #1: conventional G.652, Fiber #3: G.652 with ultra-low loss, Fiber #4: G.654.E with ultra-low loss, Fiber #6: G.654.D with ultra-low loss. Fiber characteristics are described in Table 2

occurs as a capacity enhancement window. The modeling results in Fig. 23 are based on Gaussian noise formalism with Nyquist WDM channels (Poggiolini 2012; Carena et al. 2014), previously used in section "Transmission System Modeling and Experiments." The amplifier span length was chosen to be 100 km for consistency with the experimental transmission results described in section "Transmission System Modeling and Experiments." It is important to note that the results in Fig. 23 assume 9.25 dB Q-factor threshold (i.e., 6.25 dB FEC threshold +3 dB margin) to ensure that the distances are representative of what may be achieved in practice. The impact of splice and connector losses as well as the impact of implementation penalty were neglected. In practice, implementation penalty generally increases with the modulation-format density and varies from vendor to vendor. We expect that the distances in deployed networks will be somewhat shorter. The step in bit-rate was chosen to be 50 Gbs/s; an industry-accepted granularity for adaptive-rate transponders that can be achieved using well-established modulation formats (PM-QPSK, PM-8QAM, PM-16QAM, PM-32QAM). In the future, however, we may expect even smaller steps (e.g., 25 Gb/s), which could result via advanced transponder technologies such as probabilistic constellation shaping (Roberts et al. 2017).

The different shades represent the capacity enhancement windows, as defined previously. For example, the blue region shows the capacity enhancement window enabled when using Fiber 3, as compared to Fiber 1. The green represents an additional increase when using Fiber 4 compared to Fiber 3, and the orange shows an additional increase when using Fiber 6 compared to Fiber 4. In other words, Fiber 6 (with ultra-low attenuation of 0.154 dB/km and very large A_{eff} of 150 μm^2) can enable transitions to a higher-density modulation format over the widest range of distances. For example, Fiber 6 enables the use of 200 Gb/s bit-rate instead of 150 Gb/s over distances spanning from 1200 to 3200 km.

Potential Future Directions

It is highly likely that optical fiber technology will continue to evolve along with system technologies. Coherent transmission systems are now the dominant type of systems deployed, and this is expected to continue into the foreseeable future. While this is the case, fiber attenuation and effective area will remain the most important characteristics influencing system performance and reach in LH and ULH systems. Lowering fiber attenuation will continue to be a focus of fiber manufacturers (Makovejs et al. 2016a; Tamura et al. 2017). As a reaction to traffic growth, much recent research has been devoted to spatial division multiplexing technologies such as multicore fibers, as explored in another chapter in this book. Another potential fiber advancement that has garnered attention recently is quasi-single-mode fibers, an approach that could serve to further increase effective area and nonlinear tolerance (Downie et al. 2017).

As described earlier, fiber effective area is a function of the fiber index profile design, and larger effective area decreases the light intensity in the core and increases the nonlinear tolerance. This permits larger channel launch powers and higher OSNR values at the receiver. The largest effective area of commercially available optical fibers is currently about 150 μm^2 (Makovejs et al. 2016a). It becomes difficult to design single-mode fibers with effective areas larger than that due to practical considerations of bend loss, which tends to increase with effective area. One means of enabling larger effective area while minimizing bend loss is to relax the requirement of pure single-mode propagation in the fiber. This has been described as quasi-single-mode (QSM) fiber transmission. In this case, the effective area of the fundamental mode may be increased significantly by allowing the next higher order mode (usually LP_{11}) to also propagate, preferably with higher attenuation. Signals are launched only into the fundamental mode, but may couple back and forth from the fundamental to the higher-order mode during propagation in the fiber spans (Mlejnek et al. 2015). This mode coupling can then lead to multipath interference (MPI), the level of which is largely governed by mode coupling strength and differential mode attenuation.

In a sense, the MPI introduced by a fiber that supports a higher order mode during propagation represents a tradeoff of a new linear impairment in exchange for lowering nonlinear impairments due the larger effective area of the fundamental mode. However, it has been demonstrated that MPI can be effectively compensated by digital signal processing in a coherent receiver (Sui et al. 2014), and recently, a system employing quasi-single-mode fiber transmission demonstrated better overall performance than a system with purely single-mode transmission (Yaman et al. 2015). The use of hybrid fiber span designs in which each span is comprised of a combination of QSM fiber and single-mode fiber may be particularly effective. In this approach, an optimized length of QSM fiber is deployed in the first part of each span, followed by single-mode fiber in the latter part of the span (Downie et al. 2017). This type of design takes full advantage of the higher nonlinear tolerance of the QSM fiber in the portion of the span where the channel powers are highest, while minimizing MPI impairments (because the QSM fiber length is shorter than the full span length). It has also been found that signal modulation formats with multiple electrical subcarriers may be more tolerant to MPI than single carrier formats because of the lower baud rate of the subcarriers compared to a single carrier format with the same overall data rate (Yaman et al. 2015).

Conclusions

Over the years, optical fiber has proven itself to be the most scalable medium for information transmission. The advent of coherent detection and digital signal processing in the late 2000s opened up a new realm of opportunities to achieve large transmission capacity in a cost-effective way. Coherent technology is likely to remain the main technology in the foreseeable future. As a result, ultra-low attenuation and large effective area are now and will continue to be the most important fiber characteristics to enable superior transmission performance. These recent advancements provide the ability to achieve higher transmission capacity, which is particularly important for mobile networks and networks carrying data center related traffic. Alternatively, superior performance can translate to longer system reach. This is also very important, since recent trends show that networks are increasingly migrating from opaque to transparent architectures.

Fiber design and manufacturing is a complex and nuanced process. Achieving ultra-low levels of attenuation and large effective area depends on the appropriate refractive index profile, material composition, draw conditions, and established quality processes. Many fiber attributes are interconnected, such as effective area, dispersion, bend loss, and cut-off wavelength, and typically cannot be modified in isolation. Balancing these interdependencies is essential. Finally, fiber geometrical and mechanical characteristics must be managed to ensure compliance with International Telecommunications Union standards.

The performance of different optical fibers can be quantified with a number of theoretical and experimental approaches. Theoretical approaches vary in their complexities and may involve analytical calculations of optical signal-to-noise ratio and figure of merit, semi-analytical calculations of Q-factor based on enhanced Gaussian-noise model, or complete split-step propagation modeling. Experimental approaches involve either a straight-line experiment (preferred for a complete system characterization, but expensive) or re-circulating loop experiments. We used several of these modeling and experimental techniques to illustrate the relative performance and practical benefits of low attenuation and large effective area fibers in modern coherent systems. Our primary focus was on conventional long-haul and ultra-long-haul systems with optical amplifiers compensating for fiber loss at the end of each individual span, but we also explored the relative roles of attenuation and nonlinear tolerance in unrepeatered span systems. For conventional long-haul systems, the roles of lower fiber attenuation and larger effective area can be comparable in terms of increasing Q-factor or fiber FOM relative to standard single-mode fibers, depending on the span length. The slightly lower nonlinear index n_2 obtained with silica core fibers compared to Germania doped cores also plays a small role in increasing nonlinear tolerance, as does higher dispersion. For unrepeatered span systems, we showed that attenuation is the most important fiber attribute for increasing reach. We also briefly considered the effects on Raman amplification of different fibers, as this amplifier technology is becoming more widely used in modern coherent terrestrial systems, and is well established in unrepeatered span systems. While larger effective area fibers have lower Raman gain coefficients and therefore generally lower ON/OFF Raman gain compared to smaller effective area fibers for the same Raman pump power, it is important to consider the effect of fiber attenuation. In particular, lower fiber attenuation may serve to significantly reduce the larger pump power needed for fibers with larger effective area, because the required Raman gain to compensate for the total span loss is also significantly smaller.

References

S.R. Bickham, *Optical Fiber Communication Conference. Exposition and National Fiber Optic Engineers Conference*, paper OWA5 (2010)

M. Bigot-Astruc et al., *Proceedings of European Conference on Optical Communication*, paper Mo.4.B.1 (2008)

A. Carena, G. Bosco, V. Curri, Y. Jiang, P. Poggiolini, F. Forghieri, Opt. Express **22**, 16335 (2014)

D. Chang, P. Perrier, H. Fevrier, S. Makovejs, C. Towery, X. Jia, L. Deng, B. Li, Opt. Express **23**, 25028 (2015)

Cisco VNI, White Paper (2017)

L.G. Cohen, C. Lin, W.G. French, Electron. Lett. **15**, 334 (1979)

V. Curri, A. Carena, G. Bosco, P. Poggiolini, M. Hirano, Y. Yamanoto, F. Forghieri, *Optical Fiber Communication Conference and Exposition National Fiber Optic Engineers Conference*, paper OTh3G.2 (2013)

E. Desurvire, J.R. Simpson, P.C. Becker, Opt. Lett. **12**, 888 (1987)

J.D. Downie, J. Hurley, J. Cartledge, S. Ten, S. Bickham, S. Mishra, X. Zhu, A. Kobyakov, *Proceedings of European Conference on Optical Communication*, paper We.7.C.5 (2010)

J.D. Downie, J. Hurley, I. Roudas, D. Pikula, J.A. Garza-Alanis, Opt. Express **22**, 10256 (2014)

J.D. Downie, M.-J. Li, S. Makovejs, J. Opt. Commun. Netw. **8**, A1 (2016)

J.D. Downie, M. Mlejnek, I. Roudas, W.A. Wood, A. Zakharian, J.E. Hurley, S. Mishra, F. Yaman, S. Zhang, E. Ip, Y.-K. Huang, IEEE J. Sel. Top. Quantum Electron. **4400312**, 23 (2017)

A. Evans, A. Kobyakov, and M. Vasilyev, in Raman Amplifiers for Telecommunications 2: Sub-Systems and Systems, ed. By M. N. Islam (Springer-Verlag, New York/Berlin/Heidelberg, 2004), p. 388

M. Hirano, T. Haruna, Y. Tamura, S. Kawano, S. Ohnuko, Y. Yamamoto, Y. Koyano, T. Sasaki, Optical Fiber Communication Conference and Exposition, and National Fiber Optic Engineers Conference, paper PDP5A.7 (2013)

Y.K. Huang, E. Ip, Y. Aono, T. Tajima, S. Zhang, F. Yaman, Y. Inada, J.D. Downie, W. Wood, A. Zakharian, J. Hurley, S. Mishra, *Optical Fiber Communications Conference*, paper Th2A.59 (2017)

ITU-T G.652 standard (2016)

ITU-T G.654 standard (2016)

ITU-T G.655 standard (2009)

W.B. Joyce, R.W. Dixon, R.L. Hartman, Appl. Phys. Lett. **31**, 756 (1977)

H. Kakiuchida, E.H. Sekia, K. Saito, A.J. Ikushima, Jpn. J. Appl. Phys. **42**, 1526 (2003)

C.K. Kao, G.A. Hockham, Proc. IEE **133**, 1151 (1966)

F.P. Kapron, D.B. Keck, R.D. Maurer, Trunk Telecom. Guided Waves, IEE 148 (1970)

K.S. Kim, R.H. Stolen, W.A. Reed, K.W. Quoi, Opt. Lett. **19**, 257 (1994)

M.E. Lines, J. Non-Cryst. Solids **171**, 209 (1994)

Y. Liu, *Optical Fiber Communication Conference* (1997)

S. Makovejs, G. Gavioli, V. Mikhailov, R.I. Killey, P. Bayvel, Opt. Express **16**, 18725 (2008)

S. Makovejs, J.D. Downie, J.E. Hurley, J.S. Clark, I. Roudas, C.C. Roberts, H.B. Matthews, F. Palacios, D.A. Lewis, D.T. Smith, P.G. Diehl, J.J. Johnson, C.R. Towery, S.Y. Ten, J. Lightwave Tech. **34**, 114 (2016a)

S. Makovejs, A. Zakharian, J.D. Downie, J. Hurley, J. Clark, S. Ten, *SubOptic Conference*, paper TU1A.1 (2016b)

L. Maksimov, A. Anan'ev, V. Bogdanov, T. Markova, V. Rusan1, O. Yanush, *IOP Conference Series: Materials Science and Engineering* 25 (2011)

T. Miya, Y. Terunuma, T. Hosaka, T. Moyashita, Electron. Lett. **15**, 106 (1979)

M. Mlejnek, I. Roudas, J.D. Downie, N. Kaliteevskiy, K. Koreshkov, IEEE Photonics J. **7**, 7100116 (2015)

D. Mongardien, C. Bastide, B. Lavigne, S. Etienne, H. Bissessur, *Proceedings of European Conference on Optical Communication*, paper Tu.1.D.2 (2013)

M. Ohashi, N. Kuwaki, N. Uesugi, J. Lightwave Technol. **5**, 1676 (1987)

R. Olshansky, Appl. Opt. **14**, 20 (1975)

P. Poggiolini, J. Lightwave Technol. **30**, 3857 (2012)

P. Poggiolini, G. Bosco, A. Carena, V. Curri, Y. Jiang, F. Forghieri, J. Lightwave Technol. **32**, 694 (2014)

A. Puc, D. Chang, W. Pelouch, P. Perrier, D. Krishnappa, S. Burtsev, *Proceedings of European Conference on Optical Communication*, paper 6.4.2 (2009)

K. Roberts, Q. Zhuge, I. Monga, S. Gareau, C. Laperle, J. Opt. Commun. Netw. **9**, C12 (2017)

R.J. Sanferrare, AT&T Tech. J. **66**, 95 (1987)

Q. Sui, H.Y. Zhang, J.D. Downie, W.A. Wood, J. Hurley, S. Mishra, A.P.T. Lau, C. Lu, H.Y. Tam, P.K.A. Wai, *Optical Fiber Communications Conference*, paper M3C.5 (2014)

H. Taga, S. Yamamoto, M. Mochizuki, H. Wakabayashi, Trans. IEICE **E71**, 940 (1988)

Y. Tamura, H. Sakuma, K. Morita, M. Suzuki, Y. Yamamoto, K. Shimada, Y. Honma, K. Sohma, T. Fujii, T. Hasegawa, *Optical Fiber Communications Conference*, paper Th5D.1 (2017)

R.W. Tkach, A.R. Chraplyvy, F. Forghieri, A.H. Gnauck, R.M. Derosier, J. Lightwave Technol. **13**, 841 (1995)

J.P. Turkiewicz, E. Tangdiongga, G. Lehmann, H. Rohde, W. Schairer, Y.R. Zhou, E.S.R. Sikora, A. Lord, D.B. Payne, G.-D. Khoe, H. de Waardt, J. Lightwave Tech. **23**, 225 (2005)

A.R. Tynes, R.M. Derosier, Electron. Lett. **13**, 673 (1977)

F. Yaman, N. Bai, B. Zhu, T. Wang, G. Li, Opt. Express **18**, 13250 (2010)

F. Yaman, S. Zhang, Y.K. Huang, E. Ip, J.D. Downie, W.A. Wood, A. Zakharian, S. Mishra, J. Hurley, Y. Zhang, I.B. Djordjevic, M.F. Huang, E. Mateo, K. Nakamura, T. Inoue, Y. Inada, T. Ogata, *Optical Fiber Communications Conference*, paper Th5C.7 (2015)

Multimode Fibers for Data Centers

2

Xin Chen, Scott R. Bickham, John S. Abbott, J. Doug Coleman, and Ming-Jun Li

Contents

Introduction of Multimode Fibers	42
Basics of Multimode Optical Fibers	45
Limitation of VCSEL-MMF Transmission and Novel Solutions	61
Multimode Fiber for Long Wavelength Applications	74
Universal Fibers, a New Fiber Concept Bridging SM and MM Transmissions in Data Centers	79
Optical Trends in Data Centers and Concluding Remarks	87
References	96

Abstract

Data centers (DCs) have evolved rapidly to deliver higher data rates, higher density, and longer distances while staying as economical as possible. Multimode fiber (MMF) operated at 850 nm is the leading optical medium now used in DCs for distances up to 100–150 m, enabling utilization of vertical-cavity surface-emitting lasers (VCSELs) to provide low-cost optical connectivity compared to single-mode fiber solutions. However recent trends in DC drive the MMF-based systems toward several new fronts, including wavelength division multiplexing involving longer wavelengths (BiDi and SWDM), extended reach through engineered links, and variations in the core dimensions or operating wavelength of the MMF. In this chapter, the role of MMFs in DCs will be reviewed, beginning with a discussion of the fundamental aspects of light propagation, modal bandwidth and other fiber characteristics, and link models. Various approaches to address transmission limitations due to chromatic and

---

X. Chen (✉) · S. R. Bickham · J. S. Abbott · J. D. Coleman · M.-J. Li
Corning Incorporated, Corning, NY, USA
e-mail: ChenX2@Corning.com; bickhamsr@corning.com; abbottjs@corning.com; doug.coleman@corning.com; LiM@Corning.com

© Springer Nature Singapore Pte Ltd. 2019
G.-D. Peng (ed.), *Handbook of Optical Fibers*,
https://doi.org/10.1007/978-981-10-7087-7_68

modal dispersion are then summarized. One such approach is to operate the MMF at longer wavelengths to take advantage of the lower chromatic dispersion. The concept of a "universal" fiber that bridges the gap between the multimode and single-mode transmission is then introduced. Recent trends in DC are also reviewed, and one clear conclusion is that the role of MMF in DC is still evolving to meet the increased needs for scalability, density, data rate, and economic requirements.

Introduction of Multimode Fibers

Multimode optical fiber (MMF) is a type of optical fiber mostly used for communication over short distances, such as within a building, on a campus, or in a data center. Compared to single-mode fibers, MMF has a large core diameter and a high numerical aperture, and these allow the use of lower-cost light sources such as light-emitting diodes (LEDs) and vertical-cavity surface-emitting lasers (VCSELs) while enabling low-cost splices and connectors between fibers. However, compared to single-mode fiber, the modal bandwidth of MMF is lower due to modal dispersion associated with multiple modes propagating in the fiber.

MMF was introduced in early optical fiber networks in the 1970s with LEDs transceivers for both long-haul and short-reach applications. The first generation of MMF had a core diameter of 50 μm and a relative refractive index, or core delta, of 1%. In the early 1980s, single-mode semiconductor lasers were developed and enabled single-mode fiber applications in long-haul networks. However MMF was still preferred for short-reach applications because it offered low-cost solutions with LED light sources. However because the LED had a very large spot size, the 50 μm MMF could not fully capture the available power, and this produced an insertion loss penalty. As a result, in systems using 850 nm LEDs with 50 μm MMF, the transmission distance was limited to 1.2 km at 10 Mb/s. To overcome this limitation, a different MMF with a core diameter of 62.5 μm and a core delta of 2% was developed. The higher numerical aperture (NA) and larger core enabled more light to be coupled to the fiber from a LED source, and this combination supports up to 2 km transmission at 10 Mb/s and even faster data rates of 100 Mb/s in shorter-distance applications such as the "Fast Ethernet" standard of the early 1990s. However, in the mid-1990s, the developments of a 1 Gb/s optical Ethernet standard and low-cost VCSELs at 850 nm again favored the use of the 50 μm MMF due to its lower material dispersion and higher modal bandwidth. The coupling to 50 μm MMF was no longer an issue due to the smaller spot size and lower NA of VCSELs. For these reasons, 50 μm MMF has become the fiber of choice for 1 Gb/s and 10 Gb/s Ethernet applications. Over time, the VCSEL transceiver data rates have increased (Kuchta et al. 2012), and 10 Gb/s and 25 Gb/s are now commonly used with an even higher baud rate of 28 Gb/s used in 32G Fiber Channel. Until very recently, the only transmission modulation method has been the non-return-to-zero (NRZ) format. Pulse-amplitude-modulation (PAM4) format has now been proposed and is being built into new generation of transceivers which doubles the data rate based on a baud rate at 50% of the effective data rate. PAM4 enables evolving systems to double the amount of information (bits) sent within each symbol. For example, PAM4 doubles the symbol bandwidth from 1 bit/symbol to 2 bits/symbol, so with a 25 Gb/s baud rate, the data rate is doubled to 50 Gb/s. IEEE and Fiber Channel standards groups are now including PAM-4 for 50G/100G/200G/400G and 64G/256G data rates that use 32G–50G and higher optical lanes. When using PAM4 technology, transmission electronics need to include forward error correction (FEC) to offset the 4–6 dB optical penalty to ensure sufficiently robust bit error rate (BER) performance. FEC is also commonly used with NRZ modulation to enhance system performance.

Multimode fibers designed for short-distance communication have near-parabolic index profiles optimized to equalize the time of flight of the various modes. Differences in these times of flight produce the intermodal dispersion which limits the bandwidth of multimode fiber compared to single-mode fiber. The data rate can also be limited by other factors such attenuation and chromatic dispersion, as is also the case with single-mode fiber. The performance of an optical fiber link as measured by bit error rate or intersymbol interference depends on both the multimode fiber, which determines the mode delays and chromatic dispersion, and the VCSEL laser, which determines the relative power in the different modes and operating wavelength(s) and the line width, which interacts with the chromatic dispersion of the fiber.

There are several categories of MMF. For ease of interoperability, standard classes of MMF have been established by the IEC and given the designations OM1, OM2, OM3, OM4, and OM5. The earliest categories OM1 and OM2 were designed for use with LEDs near 1300 nm. As discussed in section "Basics of Multimode Optical Fibers," the bandwidths of these fibers were characterized using an "overfilled launch" mimicking an LED source. The newer categories OM3, OM4, and OM5 are "laser optimized" and have bandwidths characterized using a high-resolution differential mode delay (HRDMD) measurement to address the various mode power distributions that can arise with a VCSEL launch. OM3 and OM4 are designed for use at 10G and higher data rates used in data centers. Simply, VCSELs render the 850 nm serial solutions more economical when compared to a 1300 nm solution using long-wavelength lasers. High data rates in conjunction with the desired application distances support 50 μm laser-optimized MMF as the fiber type for VCSEL-based systems. The relatively smaller core size of the 50 μm MMF yields an inherently higher bandwidth capability than 62.5 μm MMF. The 10G standard requirement for 300 m maximum reach requires the use of 50 μm OM3 fiber with an effective modal bandwidth (EMB) of at least 2000 MHz·km. In 2012, the 10G standard was amended to include 50 μm OM4 fiber with an EMB of at least 4700 MHz·km to support distances up to 400 m. The IEEE 802.3ba 40/100G Ethernet standard ratified in June 2010 provides specific guidance for 40G and 100G transmission with multimode and single-mode fibers, as well as twinax copper cable. OM3 and OM4 are the only MMFs included in the standard to support 100 and 150 m distances, respectively. Since then, OM4 has become the default MMF for defining Ethernet and Fiber Channel distance

Table 1 Summary of standardized categories

Multimode optical fiber nomenclature and bandwidths							
Nomenclature reference: ISO/IEC 11801 and ANSI/TIA-568-C.3							
			Minimum modal bandwidth MHz.km				
Fiber type	Core diameter (μm)	Approximate year introduced	Overfilled launch bandwidth (OFL BW)			Effective bandwidth (laser launch EMB)	
			850n	1300 nm	953	850 nm	953 nm
OM1[a]	62.5	~1986–89	200	500	N/A	N/A	N/A
OM2[a]	50	~1981	500	500	N/A	N/A	N/A
OM3	50	~2002	1500	500	N/A	2000	N/A
OM4	50	2009	3500	500	N/A	4700	N/A
OM5	50	2016	3500	500	1850	4700	2470

[a]OM1 and OM2 are now in an informative annex for IEC11801 as grandfathered specifications

objectives for higher data rates. The Telecommunications Industry Association (TIA) initiated a work group in October 2014 to develop guidance for a wide band multimode fiber (WB MMF) 50 μm fiber standard to support short wavelength division multiplexing (SWDM) transmission. The TIA-492AAAE Standard was published in June 2016, and the IEC WB MMF standard was completed in 2017. ISO/IEC JTC 1/SC 25 approved the OM5 designation for inclusion into the ISO/IEC 11801-1 document that is slated to be published in 2017. TIA is expected to harmonize with the ISO/IEC 11801-1 document and implement OM5 usage in 2017. This OM5 MMF is a version of the OM4 fiber with the additional bandwidth requirement of EMB $\geq$ 2470 MHz·km at 953 nm. The WB MMF optical and mechanical attributes are compliant with OM4 50 μm specifications and include the additional specifications of a minimum EMB and maximum attenuation at 953 nm. WB MMF is intended for operation using VCSEL transceivers across the 846–953 nm wavelength range. To date, SWDM is a proprietary wavelength division multiplexing (WDM) technology that uses up to four wavelengths across the 850–940 nm range while operating with VCSEL transceivers. The required EMB is lower at 953 nm compared to 850 nm due to lower chromatic dispersion at 953 nm. The zero dispersion wavelength of MMF is near 1310 nm, and consequently the chromatic dispersion is lower at 953 nm than at 850 nm. Consequently the EMB requirement is lower at 953 nm than at 850 nm to achieve the same system performance (Table 1).

The remainder of the chapter consists of five sections. Section "Basics of Multimode Optical Fibers" covers the basics of MMF including light propagation, bandwidth characterization, and link performance. Section "Limitation of VCSEL-MMF Transmission and Novel Solutions" discusses the limitation of VCSEL-MMF transmissions around 850 nm and reviews several novel approaches of addressing reach limitations at high data rates. Section "Multimode Fiber for Long Wavelength Applications" explores the possibility of new MMFs operating at wavelengths other than 850 nm, including fiber design considerations and experimental demonstrations of performance at 1060 nm and 1310 nm. Section "Universal Fibers, a New Fiber Concept Bridging SM and MM Transmissions in Data Centers" presents another novel concept of MMF, called universal fiber, which can be used for both single-mode and multimode transmission. The final section, "Optical Trends in Data Centers and Concluding Remarks," provides an overview of the optical data center trends with emphasis on the implications of MMF and also presents some brief concluding remarks.

Basics of Multimode Optical Fibers

This section focuses on the specific use of MMF in data centers. Related discussion of optical fibers in general, including single-mode fibers or multimode fibers (including non-data-center topics), can be found in other Springer handbooks (Nouchi et al. 2012; Freude 2002).

The performance of an NRZ optical fiber link with VCSEL transceivers and MMF with a given modal bandwidth can be estimated using a link model described in sections "System Link Models," "Light Propagation, Characterization, and Link Performance," "MMF Characterization: Modal Bandwidth and DMD," and "Source Characterization (Encircled Flux)" will first explain the fiber and laser properties leading to the key parameters used in the model.

Light Propagation, Characterization, and Link Performance

Introduction and Nomenclature

To enable ease of use and interoperability, MMFs are standardized, with the fiber specification determined by TIA and/or IEC standards and the "link" specifications (which use the specified fiber) determined by IEEE 802.3 T11, Fiber Channel, and other groups. As discussed in section "Introduction of Multimode Fibers," the MMFs typically used in data centers are OM3 or OM4 fibers (as designated by IEC standards), with a 125 μm cladding diameter, a 50 μm core diameter, and an NA of 0.2. These MMFs have a graded index core surrounded by a cladding, as shown in Fig. 1a. The cores of these MMFs typically have parabolic or nearly parabolic shapes that are created by doping the SiO_2 host glass with GeO_2, which increases the index of refraction relative to the SiO_2 cladding. The variation of this refractive index with radius is the refractive index "profile."

In 2009, the first bend-insensitive multimode fiber (BI-MMF) was commercialized, and this fiber delivered enhanced macrobend performance while maintaining the high modal bandwidth needed for compliance with the OM3/OM4 standard. The BI-MMF achieves improved bend performance through the addition of a low index fluorine-doped "ring," or "trench," in the cladding, as shown in Fig. 1b. This trench creates a barrier that improves the confinement of the optical signal propagating in the core of the MMF. The location and dimensions of the trench are chosen to maintain OM3/OM4 standards-compliance and ensure sufficiently high modal bandwidth (Bickham et al. 2011).

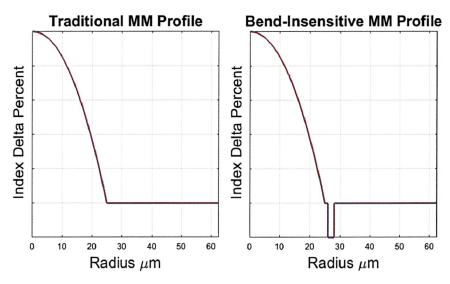

Fig. 1 (a) Legacy MMF with a 50 μm core diameter and an NA of 0.2, (b) BI-MMF fiber with a "trench" in the cladding

The refractive index profile of the legacy MMF is given by:

$$n^2(r) = n_1^2 [1 - 2\Delta(r/a)^\alpha], \quad 0 \leq r \leq a$$
$$n^2(r) = n_2^2, \quad r \geq a \quad (1)$$
$$\Delta = \frac{(n_1^2 - n_2^2)}{2n_1^2}.$$

where n_1 is the maximum refractive index of the core and n_2 is the refractive index of the cladding. α defines the curvature of the core and has an optimum value of $\alpha_{opt} = 2 - C\Delta - 2P$, where $C \approx 2$, $P = \frac{n_1}{N_1} \frac{\lambda}{\Delta} \frac{d\Delta}{d\lambda}$, $N_1 = n_1 - \frac{dn_1}{d\lambda}$, and λ is the wavelength. The exact value of coefficient C depends on the assumed modal power distribution and the metric that is being minimized. For example, $C \approx 12/5$ corresponds to minimizing the weighted RMS spread of the mode delays when the individual modes are equally weighted. The optimum α is 2 to first order in Δ.

For BI-MMF, the refractive index profile of the graded index core is the same as the first line of Eq. (1), but the cladding index is no longer constant and is now described by:

$$n^2(r) = n_T^2, \quad R_1 \leq r \leq R_2$$
$$n^2(r) = n_2^2, \quad R_2 \leq r \quad (2)$$

where R_1 and R_2 are the inside and outside radii of the trench, respectively, and n_2, the refractive index of the trench is less than the index of the outer cladding (i.e., $n_T < n_2$). Careful control of the location and dimensions of the trench enables high bandwidth while simultaneously yielding a significant improvement in macrobend performance (Bickham et al. 2011).

Light Propagation in Multimode Fibers; Basic Definition of Bandwidth

The index profile of MMF is optimized to reduce the spread in group delays of the modes. The way this spreading is measured and quantified is with the "modal bandwidth," which is determined by measuring a reference pulse $P_{in}(t)$ on a very short length and a output pulse $P_{out}(t)$ (with spreading) on the longer test length. The modal bandwidth is determined by taking the Fourier transform of the "input" and "output" pulses and determining a transfer function $H(f)$ from their ratio:

$$H(f) = \frac{H_{out}(f)}{H_{in}(f)} \tag{3}$$

The "3 dB bandwidth" is the frequency where $|H(f)| = 0.5$ and is used to calculate how the fiber will interact with the other components of the link (e.g., the rise and fall time of the laser pulse in an NRZ system and the temporal response of the receiver). In early data center applications, the data rate was low enough that the fiber was the limiting part of the link and determined the maximum possible link distance or reach. In systems operating at higher data rates ($\geq$40 Gb/s Ethernet), the receiver bandwidth and the rise/fall characteristics of the laser also contribute to the reach limitation. The IEEE Ethernet link model, which uses these parameters, will be introduced later in this section.

Theory: Scalar Wave Equation, Modes, and Mode Groups

The mathematical expression that describes the light propagation in the fiber and the connection between the index profile and the modes and mode delays is known as the scalar wave equation:

$$\left(\nabla^2 + k^2 n^2(r)\right) \Psi_{lm} = \beta_{lm}^2 \Psi_{lm}. \tag{4}$$

Here $n(r)$ is the refractive index profile and $k = 2\pi/\lambda$ is the wavenumber. The modal functions ψ_{lm} and propagation parameters β_{lm} are the eigenfunctions and eigenvalues of the equation, respectively. The scalar wave equation makes approximations to Maxwell's equations for electric field E and magnetic field H when the maximum index change Δn is small and when dn/dr can be neglected.

The guided modes and the propagation parameter β depend on both a radial mode number m and an azimuthal mode number l and have the form:

$$\Psi_{l,m}(r, \theta) = \Psi_{l,m}(r) \begin{Bmatrix} \cos l\theta \\ \sin l\theta \end{Bmatrix}. \tag{5}$$

There are pure "radial" modes with $l = 0$ and no angular dependence, compared to "azimuthal" or "spiral" modes with $l > 0$. If there is no angular dependence to the index profile (as in the ideal case), the *cosine* and *sine* "modes" are degenerate but need to be counted twice in considering the total number of modes. Finally, each individual mode has two polarization states which are also degenerate.

For an arbitrary index profile, the scalar wave equation can be solved numerically with the boundary condition that ψ_m decays exponentially in the constant index cladding. If the core of the MMF has a parabolic profile with $\alpha = 2$ (see Eq. 1), there is an analytical solution: the eigenfunctions are the classic "parabolic cylinder functions," or Gauss-Laguerre functions, with the radial dependence having the form of a Laguerre polynomial multiplied by a Gaussian function:

$$\Psi_{l,m}(r, \theta) = e^{-\rho/2} \rho^{1/2} L_m^l(\rho) \begin{Bmatrix} \cos l\theta \\ \sin l\theta \end{Bmatrix}$$
$$\rho = V(r/a)^2$$
$$V = n_1 k a \sqrt{2\Delta}$$
(6)

where L_m^l is the generalized Laguerre polynomial. For many of the modes (that do not extend very far into the cladding), the analytical solution given by Eq. (6) is a good approximation to the exact one.

The modal bandwidth of a MMF depends on the relative delays of the modes. The modal solutions of the MMF are thus characterized by the propagation parameter β_{lm}, and the corresponding group delays τ_{lm} are given by:

$$\tau_{lm} = \frac{1}{c} \frac{d\beta_{lm}}{dk}.$$
(7)

where c is the "velocity of light" in a vacuum. In a dispersive medium, pulses do not travel at "the speed of light" but at the so-called group velocity associated with $d\beta/dk$. Because the group velocity depends on $d\beta/dk$, the dependence of refractive index profile $n(r)$ on wavelength is needed to calculate the mode delays. Very small deviations of $n(r)$ from the optimal profile are enough to change the relative mode delays to shift the bandwidth from OM4 (≥ 4700 MHz·km) to OM3 (≥ 2000 MHz·km).

In cylindrical systems, such as an optical fiber, modes with the same $M = (2 m + l)$ have the same β (analogous to the electronic orbitals of an atom, which have multiple electrons in the same "shell"). These individual modes with the same β are said to be in the same "group." A second simplification observed in MMFs is that there is strong coupling between the individual modes in a group and these groups tend to arrive as a single pulse with a delay that is approximately the arithmetic average of the individual delays of the modes in the group (counting azimuthal modes twice for the sine and cosine components). For a standard OM3 or OM4 fiber with 18 (or sometimes 19) mode groups, the output pulse structure will show the signature of 18 groups rather than the 90 individual modes that are solutions of the scalar wave equation. Group "g" has in fact "g" modes, with the azimuthal modes counted twice. The "even" numbered groups have only azimuthal modes, while the "odd" numbered groups also have one radial mode with $l = 0$.

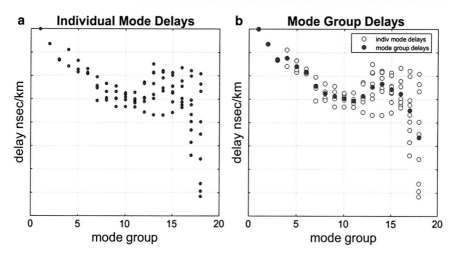

Fig. 2 (a) Individual mode delays for an example profile; (b) the mode group delays which are observed in the DMD (see section "MMF Characterization: Modal Bandwidth and DMD")

Figure 2 plots the delay versus the mode group number for a MMF for the (a) 90 individual modes and (b) 18 mode groups.

In standard or legacy MMF, the modal bandwidth is primarily determined by the α parameter, which determines the wavelength at which the peak bandwidth is located, and the accuracy in achieving the target α over the entire core region, which determines the magnitude of the peak bandwidth. For BI-MMF, the depth of the trench and the proximity of the trench to the graded index core must also be optimized to achieve high bandwidth. Figure 3a is a plot of the overfilled bandwidth as a function of the spacing between the core and the trench for two different trench depths. When the trench is far away from the core, it primarily impacts the bend loss of the outer mode groups, not their modal dispersion. The overfilled bandwidth is low in this regime because the outer mode groups tend to travel faster than the other mode groups in the fiber. As the trench moves closer to the core, the outer mode groups slow down and the bandwidth increases. For a given trench depth, there is an optimal spacing between the core and the trench at which the differential delays between the inner and outer mode groups in the fiber are minimized (Bickham et al. 2011). This optimal spacing depends on the core delta, the trench delta, and the trench width, so it is only shown generically in Fig. 3a. As the trench moves even closer to the core, these outer mode groups continue to slow down, and this again has a negative impact on the modal bandwidth because the outer mode groups now lag the other mode groups propagating in the fiber.

Advantages of BI-MMF

The enhanced bending performance of BI-MMF has been documented in various journal articles and was used to support the integration of a bend loss specification into the OM3/OM4 standard. Figure 3b shows the measured bend loss at 850 nm of

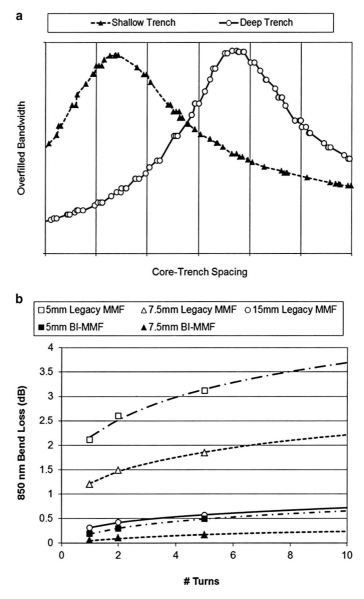

Fig. 3 (a) Dependence of the overfilled bandwidth on the core-trench spacing for two trench depths; (b) bend loss versus number of mandrel wraps for standard MMF and BI-MMF

a legacy MMF and BI-MMF as a function of the number of turns around mandrels with radii of 5, 7.5, and 15 mm. The data was generated using an encircled flux launch (EFL), which can be approximated by launching through a short length of input legacy MMF that incorporates one turn around a 25 mm diameter mandrel.

The measured data is plotted using the discrete symbols, while the modeled bend losses are plotted with curved lines. The BI-MMF clearly provides much better macrobend performance, with lower losses at a bend radius of 5 mm than legacy MMF exhibits at a much larger bend radius of 15 mm.

The data plotted in Fig. 3b exhibits the typical behavior in which the total macrobend loss increases approximately with the natural log of the number of turns. This behavior is a consequence of the discreteness of the mode groups. Even at relatively large bend radii, several of the outer mode groups in legacy MMF suffer leakage losses. However, when one of these mode groups is completely stripped out, it will no longer contribute to the bend loss with subsequent mandrel wraps. The incremental bend loss decreases, and this leads to a flattening of the measured bend loss versus the number of mandrel wraps. In contrast, only the outermost one or two mode groups in the BI-MMF become leaky when the fiber is bent, and the dependence of bend loss with the number of mandrel wraps is nearly linear because it requires several mandrel wraps to strip them out of the core.

It is interesting to note that when BI-MMF was first introduced, it was questioned whether the cables deployed in data centers are actually subjected to bends that are small enough to induce significant macrobend losses. While instances of this event occurring are probably rare, BI-MMF did enable innovative new designs with cable and hardware that resulted in a significant increase in the deployment density compared to legacy systems. For example, BI-MMF has been integrated into smaller, more flexible cables which have proven to be craft-friendly during installation because they tolerate more robust routing than products based on legacy MMF (Hurley and Cooke 2009). Bend-insensitive multimode fibers have also enabled smaller hardware components such as connectors with shorter boots, as shown in Fig. 4a. These shorter boots permit easier finger access when installing or removing the connectors and, when combined with smaller and more flexible cables, allow more compact routing of fiber guides in equipment racks. The advent of BI-MMF also enabled a doubling in the port density from the traditional density of 144 ports to 288 ports (or 576 fibers) within a single four-unit (4 U) housing, as shown in Fig. 4b. These innovations and others have made cable and hardware and equipment products incorporating BI-MMF indispensable in contemporary enterprise data center deployments.

MMF Characterization: Modal Bandwidth and DMD

Modal Bandwidth
The output pulse of a MMF, neglecting broadening due to chromatic dispersion, can be approximated as the sum of N_g pulses corresponding to the N_g mode groups, each with a modal power P_g corresponding to the total power in all modes in that group. Theoretically the pulse can be represented by a sum of delta functions, each with a relative delay corresponding to the group:

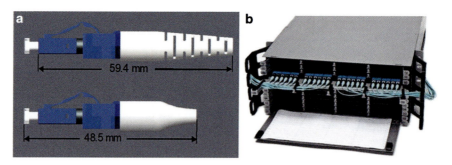

Fig. 4 (a) Comparison of connector boots designed for use with legacy MMF (top) and BI-MMF (bottom); (b) high density 288 port housing (Hurley and Cooke 2009)

$$P(t) = \sum_g P_g \delta\left(t - \Delta\tau_g\right)$$
$$\sum_g P_g = 1 \tag{8}$$
$$\Delta\tau_g = \tau_g - \sum_g P_g \tau_g$$

Note that the mode power distribution P_g is a function of the launch condition, and the delay $\Delta\tau_g$ relative to the centroid is a function of the fiber. A subtle point is that the absolute arrival time of the pulse (i.e., its centroid) is $\sum_g P_g \tau_g$, and this quantity depends on both the modal delays and power distribution.

To achieve low BER and low intersymbol interference (ISI) for high data rate systems largely limited by the transceiver and system penalties, it is important to ensure that little power resides outside the unit bit window. If the input pulse and output pulse are measured in nanoseconds, the 3 dB bandwidth of the fiber is defined as the frequency (in MHz or GHz) where the modulus of the transfer function drops to 50% of its peak value. For an ideal fiber with a "delta function" input and an output pulse given by Eq. (8), the transfer function $H(f)$ can be calculated with elementary properties of Fourier transforms:

$$H(f) = \sum_g P_g \exp\left(2\pi i f \Delta\tau_g\right)$$
$$|H(f)| = \sqrt{\left(\sum_g P_g \cos\left(2\pi f \Delta\tau_g\right)\right)^2 + \left(\sum_g P_g \sin\left(2\pi f \Delta\tau_g\right)\right)^2} \tag{9}$$

The low-frequency part of this curve with $|H(f)| > 0.5$ is the region of interest. If the transfer function is non-Gaussian, a modification of the 3 dB estimate is often used, for example, including extrapolation from the 1.5 dB value or Gaussian fitting. The "effective modal bandwidth" or EMB of a fiber depends on the mode delays and the mode power distribution. If the pulse and $|H(f)|$ are approximately

Gaussian, the bandwidth in GHz is approximately $0.187/\sigma$, where σ is the RMS pulse in nanoseconds.

A simplified model with two modes having powers P_1 and P_2 can show non-Gaussian behavior when P_1 and P_2 are not equal. The limit $P_2 = P_1$ results in the sum of two equal pulses (a so-called dual Dirac pulse), which is described by:

$$cP(t) = 0.5 \{\delta(t + \Delta\tau) + \delta(t - \Delta\tau)\}$$
$$\Delta\tau = \frac{\tau_2 - \tau_1}{2}$$
$$H(f) = \cos(2\pi f \Delta\tau) \qquad (10)$$
$$f_{3dB} = \frac{1}{6\Delta\tau} = \frac{1}{3(\tau_2 - \tau_1)}$$

Prior to the introduction of OM3 and OM4 MMF, bandwidth measurements used a prescribed launch. For example, the overfilled (OFL) bandwidth measurement used a mode conditioning patch cord to simulate a Lambertian source with equal power at all angles, putting equal power into all the individual modes. This approximated the launch of an LED which was the important source for 100 Mb/s systems. For 1 Gb/s systems, the modification of a restricted mode launch (RML) was introduced, using a special MMF with a nominal NA of 0.2 but with a core diameter of ~23 μm rather than 50 μm. This smaller core was designed to simulate in a repeatable way the launch from 1 GHz VCSELs. With the arrival of higher-frequency 10 Gb/s VCSELS in about 2001, a different approach for measuring bandwidth was needed, and this led to the development of the HRDMD technique to characterize the mode group delays.

Differential Mode Delay Measurement

The DMD measurement is made by launching a diffraction-limited spot into the MMF at different offset positions from the center. At each offset position x, the output pulse $F\_x(t)$ is recorded and the 2D array F_xt is analyzed. HRDMDs use the entire pulse at each offset position to generate as much information as possible about the differential mode delays in the MMF. The power P_m going into each individual mode can be theoretically calculated from overlap integrals.

Two parallel specifications of MMF using the HRDMD have been implemented in standards: the "minEMBc" approach and the "DMDmask" approach. Both are based on an extensive Monte Carlo simulation carried out by the TIA and IEEE standards group (Pepeljugoski et al. 2003a, b) using 5000 model fibers and 2000 model lasers, which resulted in specifications for both the MMF and the VCSEL. The "DMDmask" approach (DiGiovanni et al. 2002) is specified in TIA/EIA 492AAAC/D. Using Eq. 9, the leading and trailing edge of the DMD pulse structure can be estimated to ensure that $\tau_2 - \tau_1$ is small enough to ensure that any pulse is within the "bit window." This idea was extended to include realistic mode power distributions with the EMBc approach.

The "minEMBc" approach (Abbott et al. 2013) is also specified in TIA/EIA 492AAAC/D. The motivation was to construct a weighted sum of the HRDMD pulses to simulate the output pulse from a VCSEL launch. The "c" in EMBc' indicates that it is calculated from the DMD data, not measured with an actual source; the "min" in minEMBc indicates that what is reported is the minimum EMBc of a "test set" of sources. The minEMBc approach is different from the DMDmask approach in that it is modeling the system bandwidth, including both the fiber (i.e., DMD data) and laser (the "weights"). The advantages are that (a) it yields a bandwidth and is thus tied to the system link model, and (b) it is easier to modify the specification to encompass the new generations of lasers. EMBc also has a more direct link to BER and ISI and provides a method of predicting system performance for any particular fiber and particular laser. The EMBc metric will be discussed in more detail in section "The "DMD Weight Function" and the EMBc Metric" after the EF metric used to characterize the laser source is introduced. The "minEMBc" approach is the only method adopted by TIA for the new OM5 fiber (discussed in section "Performance of MMF with Different Peak Wavelength in WDM Based Transceivers").

Source Characterization (Encircled Flux)

Introduction
In the TIA Monte Carlo simulation, each of laser modes corresponding to the 2000 theoretical laser sources was generated with a model, and the coupling of these to the modes of a legacy MMF with a 50 μm core diameter and an NA of 0.2 was calculated. The launch was characterized by P_m, the power going into the individual modes, and $EF_{in}(r)$, the integral of the modal power versus radius for all modes (weighted by P_m, remembering to count once for purely radial modes and twice for azimuthal modes):

$$EF_{in}(r) = \int_0^r \sum_m P_m \Psi_m^2(r) r \, dr \qquad (11)$$

$EF_{in}(r)$ is normalized so that it goes to 1.00 as r goes to infinity, and it corresponds to the integral of a measured near field profile on a short length of fiber. An example of measured $EF(r)$ is given in Fig. 5a.

These measurements also show the specification developed for 10GbE, where $EF < 0.30$ at a radius of 4.5 μm and $EF > 0.86$ at a radius of 19.0 μm. Joint fiber and laser specifications were needed to describe both the laser (P_m or ultimately P_g) and the fiber (τ_g). A scatter plot of the fraction of power inside the 4.5 μm radius annulus versus the radius in microns which encircles 86% of the power is shown in Fig. 5b for the 2000 theoretical sources used in the TIA 10GbE Monte Carlo modeling (gray dots), the 10 theoretical 850 nm sources used in developing the EMBc specification (white circles), and actual VCSEL data measured subsequent to the introduction of the specification (blue dots). The red lines indicate the source specification. Note

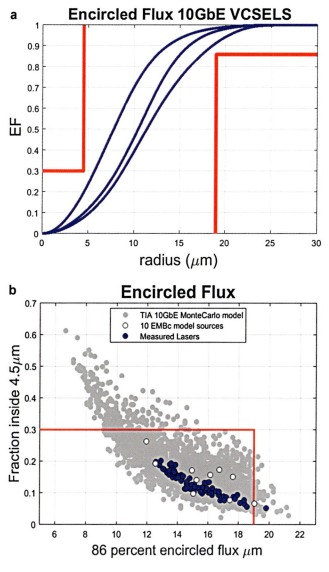

Fig. 5 (**a**) Encircled flux versus radius for three typical 10 Gb/s VCSELS, with the inner and outer specifications marked; (**b**) distribution of sources used to develop the 10GbE specification

that the blue dots corresponding to actual lasers measured after the specification was adopted correspond to a narrow portion of the theoretical distribution.

As noted in section "Theory: Scalar Wave Equation, Modes and Mode Groups," there is strong coupling between the degenerate individual modes in a mode group, and the analysis for the TIA 10GbE work made the assumption that there is full coupling within a group with no phase cancelation. Thus the power is divided evenly between all the modes in the group, and there is a single power P_g and a single-mode delay τ_g for the group. Although this is an approximation, it agrees well with observations of DMD measurements.

When this intra-mode group coupling is considered, it results in an implicit encircled flux $EF_{\text{fiber}}(r)$ of interest, which corresponds to:

$$EF_{\text{fiber}}(r) = \int_0^r \sum_g P_g \Psi_g^2(r) r dr$$
$$\Psi_g^2(r) = \sum_{m \in g} \Psi_m^2(r) \qquad (12)$$
$$P_g = \sum_{m \in g} P_m$$

Thus the source determines the modal power weighting P_g, and the fiber determines the mode group delay τ_g. The impulse response is given by Eq. 12, and the bandwidth can be calculated as described above. Alternatively, the impulse response (or bandwidth) can be incorporated into a system model, as discussed in section "System Link Models."

The "DMD Weight Function" and the EMBc Metric

The ISI of the link depends on the EMB calculated from P_g and τ_g. In the Monte Carlo model, the required ISI for the set of 40,000 modeled links is assumed to be acceptable if (a) there is an encircled flux spec for the source and (b) the EMB for a small selected set of fibers is high enough. This analysis led to the introduction of the ten sources used in developing the TIA EMBc standard, as shown as white circles in Fig. 5b.

The idea behind the EMB sources is to use a weighting of the DMD pulses $P_x(t)$ at different offset positions x, which when summed together will give a single pulse $P(t)$ representing the correct mode power distribution P_g. The MPD P_g can be determined by deconvolving an EF measurement and solving the equation:

$$I(r) = \sum_g P_g \Psi_g^2(r) \qquad (13)$$

Equation 13 can be set up as a nonnegative least squares problem. Once the modal powers P_g are known, the problem of approximating them with a weighted sum of DMD pulses involves a second least squares problem. In the DMD measurement, the power going into group g at offset x can be written as an overlap integral and reduced to a matrix P_{xg}. This matrix is used to solve for a DMD weighting function W_x which satisfies:

$$P_g = \sum_x W_x P_{xg}. \qquad (14)$$

Equation 14 again defines a least squares problem. Once the weights are determined, the pulse simulating the laser with a given encircled flux (and hence P_g) is given by the weighted sum of the DMD pulses $F_x(t)$ described by:

Fig. 6 EMB(850 nm) versus EMB(1300 nm) in GHz·km for the 5000 theoretical "fibers" in the Monte Carlo analysis of Pepeljugoski et al. (2003a, b), using a fixed mode power distribution (source 1946 = "Source five" of the ten min EMBc sources). The 1.5 dB frequency has been extrapolated to 3 dB by multiplying by 1.414

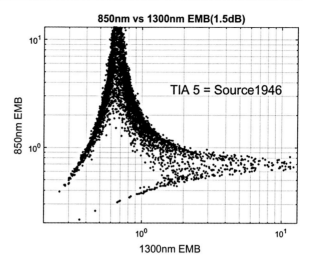

$$P(t) = \sum_x W_x F_x(t). \qquad (15)$$

The EMBc is then determined from the calculated output pulse in Eq. 15 and a reference pulse measured on a short length of fiber. In the 10GbE standard, fibers meeting the DMD mask were defined as having an effective modal bandwidth (EMB) of at least 2000 MHz·km. To generate an equivalent specification using EMBc, the EMB of a fiber is defined as 1.13 × minEMBc, where "minEMBc" is the minimum EMBc of the ten sources. For purposes of system modeling, the EMBc value can be useful since it includes the mode power distribution P_g of the laser and the mode delays τ_g of the fiber. Neither the DMDmask nor minEMBc represent any single laser, but rather are conservative estimates for a larger population.

As noted in section "Light Propagation, Characterization, and Link Performance," the optimum "alpha" depends on wavelength. When the 850 nm and 1300 nm bandwidths of a series of alpha profiles with different alphas are calculated and plotted, a characteristic "triangular" shape results. Figure 6 is a plot of the EMB at 1300 nm versus the EMB at 850 nm using the 5000 theoretical "fibers" in the TIA Monte Carlo study. The EMB values at 850 nm are based on source number 1946, which corresponds to source five of the ten minEMBc sources, (the white circle farthest to the right in Fig. 5b). The 1300 nm delays were calculated using the approximation in Pepeljugoski et al. (2003a, b), and the 1300 nm mode power distribution is calculated by interpolating source five to the correct number of mode groups. The plot in Fig. 6 represents the modal bandwidth without the chromatic dispersion of the laser, but it does incorporate approximately the effect of the parameter P discussed in section "Light Propagation, Characterization, and Link Performance." The high 850 nm BWs correspond to an alpha near 2.1 and the high 1300 nm BWs correspond to an alpha near 2.0. As the alpha value increases from ∼1.8 to ∼2.3, the (x,y) plot maps out a pattern that is approximately triangular.

Non-alpha errors give points inside the triangle, while points which appear to lie outside the triangle typically have non-Gaussian transfer functions.

System Link Models

Current multimode link models are based on the 1 Gb/s work by Nowell et al. (2000) and the book by Cunningham and Lane (1999). The system modeling for 10 Gb/s systems (Pepeljugoski et al. 2003a, b) was used for the 10 Gb Ethernet standard. In developing this standard, an Excel® spreadsheet (Hanson et al., http://www.ieee802.org/3/ae/public/index.html) incorporated the features of the 1 Gb/s model and extended them to 10 Gb/s. This 10GbE spreadsheet was used to develop the specifications (the specifications are "official," meaning that the link model is generally accepted but is not an "official" document). This basic spreadsheet continues to be updated in new projects in IEEE802.3 and Fiber Channel.

The system link models incorporate two constraints: (a) the ISI must lie below a targeted level, and (b) the total system power must be above a center budgeted level. These reflect two types of degradation of a pulse with length: (a) spreading due to dispersion effects (intermodal dispersion, material effects, etc.) and (b) attenuation of the pulse due loss of power (attenuation, connectors, etc.). Depending on the assumptions for the link, the key constraint can be either (a) or (b). The IEEE 10GbE model (Hanson et al., http://www.ieee802.org/3/ae/public/index.html) provides a portion of the spreadsheet for entering the many parameters, a portion for generating tables of output values, and plots of the eye diagram (giving the ISI) and power penalties affecting the budget. Figure 7 shows an example. The link model also captures sources of noise, the finite bandwidth of the receiver, the rise-time of the laser, and other effects limiting transmission.

The IEEE 10GbE model necessarily includes various simplifications. At higher bit rates, the method of accurately handling the jitter due to the laser becomes more complicated. The link models generally approximate the fiber impulse response by a Gaussian model, which misses subtleties arising from multiple pulses. They also typically approximate the spectral character of the laser by an RMS spectral width; however current 850 nm VCSELS have multiple modes and hence multiple wavelengths. The rise/fall parameters in the model are approximate values; in fact each laser mode has its own rise/fall structure (as well as its own wavelength). However, the spreadsheet continues to be widely used.

One application of the spreadsheet is to understand the sensitivity of the link to parameter variations. Figure 8 gives an example of varying a single parameter in the spreadsheet (the EMB) and determining the fiber length where there is either zero margin remaining in the power budget or the ISI has reached the value of 3.60 dB, which is used as a consensus upper limit. In this example, the 25GbE model (Petrilla 2013, http://www.ieee802.org/3/bm/public/may13/index.html) is always limited by the power budget, while the 10GbE model is limited by ISI at EMBs up to about 1776 MHz·km and by the power budget for higher bandwidths. This idea can be extended to varying two or more parameters, including a weighting based on probability, which allows Monte Carlo estimates as discussed below.

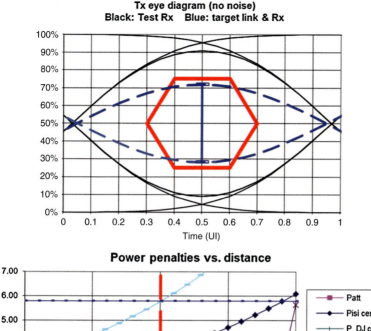

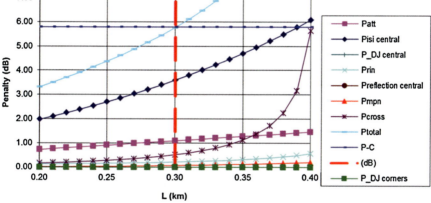

Fig. 7 An example calculation from the IEEE 10GbE spreadsheet model showing plots of the resulting eye diagram at system length (300 m) and the various power penalties as a function of link length. (Hanson et al., http://www.ieee802.org/3/ae/public/index.html)

The IEEE link models are used to develop specifications for link components to guarantee a targeted system reach, including the VCSEL-MMF transmission system used in data centers. The IEEE standard uses a "worst case analysis" which estimates the link distance when all components are at their specified values. However, in practice, each attribute in a real system forms a statistical distribution and only has the worst case value at the tail of the distribution. The parameters from different components are typically independent and therefore unlikely to be simultaneously worst case. For this reason the modeled system reach is generally acknowledged to be conservative. While these conditions ensure robustness, the weakness is that the system is overdesigned (creating unnecessary cost) or the

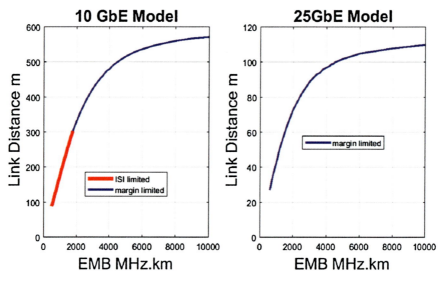

Fig. 8 Example calculations from IEEE 10GbE and 25GbE spreadsheet models showing the link distance at zero margin or the ISI limit as a function of fiber EMB

capability including supportable link length is underestimated. The example below looks at the system reach capability from a statistical point of view by introducing distributions for two of the key parameters (Chen et al. 2016c). This idea has been introduced in some IEEE projects since the early 2000s, particularly as the data rates increase and designing solutions becomes increasingly challenging. For example, the 10GbE development (Pepeljugoski et al. 2003a, b) sought a 1% risk in establishing the fiber EMB and VCSEL encircled flux targets.

For this example, version 3.1.16a of the 10GbE model (Hanson et al., http://www.ieee802.org/3/ae/public/index.html) and the "May 3, 2013" version of the 100GbE Example model (Petrilla 2013, http://www.ieee802.org/3/bm/public/may13/index.html) are used with estimated bandwidth and connector loss distributions. The bandwidth distribution from the 10GbE development is shown in Fig. 9. The fiber is specified by the minimum bandwidth value when tested against the ten sources in the EMBc procedure discussed above, so that when used with a random laser, the EMB is almost always higher and usually much higher (consistent with the assumed 1% risk).

The connector loss distribution is summarized in Fig. 10, using the standard grade (Grade B) connector loss from the IEC MMF connector insertion loss guidance. The tail of the distribution only exceeds the 1.5 dB total connector loss specification used in the IEEE link model when there are six connector pairs, which is not a common system configuration in DCs.

Using the modal bandwidth and connector loss distributions, the resulting reach distribution can be evaluated. The link model is evaluated for each combination of (BW, Loss), weighted by the relative probability of this pair of values occurring.

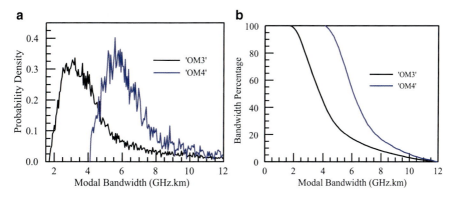

Fig. 9 (**a**) The probability density function of "OM3" and "OM4" fibers. (**b**) Percentage of "OM3" and "OM4" fibers exceeding a given bandwidth value

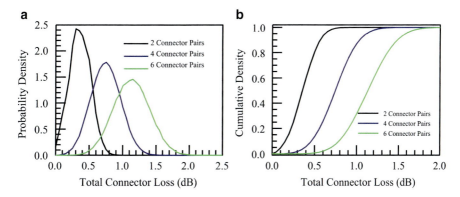

Fig. 10 (**a**) The probability density function and (**b**) cumulative density function for connector loss with two, four, and six connector pairs

Figure 11 breaks down the results by fiber type (OM3, OM4) and by the number of connector pairs (two, four, or six). For the majority of the cases, the system reach significantly exceeds the worst case scenario, illustrating the usefulness of this approach to quantify system robustness.

Limitation of VCSEL-MMF Transmission and Novel Solutions

In this section, the role of chromatic dispersion and modal dispersion in limiting the VCSEL-MMF system performance is discussed. These two factors are first reviewed in section "Limiting factors for VCSEL-MMF transmission" to illustrate how they can limit the system reach. For the remaining subsections, different approaches for mitigating the limitation are presented. Section "Chromatic Dispersion Compensation" discusses a novel approach to compensate the chromatic

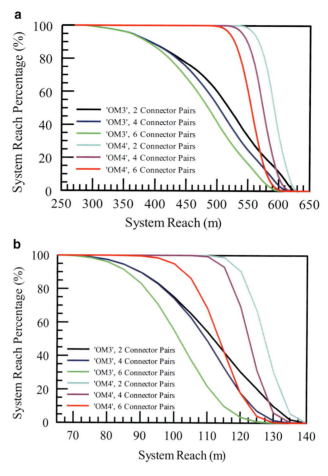

Fig. 11 System reach statistics using EMB and connector loss distributions and parameters in the (**a**) 10GbE link model and (**b**) 25GbE link model

dispersion effect, while section "Performance of MMF with Different Peak Wavelength in WDM Based Transceivers" illustrates how the choice of peak wavelength for OM4 can affect the VCSEL-MMF transmission for the SWDM application. Finally, section "Modal Dispersion Compensation for SWDM applications" summarizes a novel approach to convert OM4 fibers optimized for 850 nm for long reach at higher wavelengths up to 953 nm, using a multiplexer and demultiplexer divided into two wavelength bands.

Limiting Factors for VCSEL-MMF Transmission

Among many limiting factors for VCSEL-MMF transmission, two of the primary ones are chromatic dispersion and the modal dispersion at wavelengths away from peak or optimal wavelength for a given MMF.

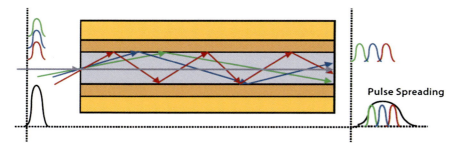

Fig. 12 Illustration of the chromatic dispersion effect in multimode fiber

As performance requirements for MMF increase, there have been questions about the need for a more advanced optical fiber type with even higher bandwidth than OM4. However transmitting in the 850 nm window at 25 Gb/s speeds and beyond, chromatic dispersion, not modal bandwidth, is the limiting factor. Chromatic dispersion describes the spread that occurs as a light pulse travels through an optical fiber (Fig. 12). Light pulses from VCSELs are not necessarily monochromatic but can contain different wavelength components. These different wavelength components can arrive later or earlier than others at the other end of the fiber, thereby effectively spreading the pulse.

Chromatic dispersion is measured in picoseconds (ps) of pulse spreading per nm of spectral width per kilometer of fiber length (ps/nm·km). The chromatic dispersion of MMF at 850 nm dispersion is typically about −95 ps/nm·km. In transmission systems, dispersion is generally limited to about 30% of the bit period. With a 10 Gb/s VCSEL, the bit period is 100 ps, which means the system can tolerate up to 30 ps of chromatic dispersion. At 10 Gb/s, this is not an issue; however at 25 Gb/s the bit period drops to 40 ps, and at 30% tolerance, only 12 ps of chromatic dispersion is allowed.

Because of the impairments due to chromatic dispersion, increasing the EMB to values significantly higher than the 4700 MHz.km value specified by the OM4 standard does not generally result in longer reach when operating at 850 nm. Table 2 provides an understanding of what the additional distances are when modeling 28 Gb/s, which is representative of 32G Fiber Channel. With the transceivers that are used for multimode fiber, the typical spectral widths fall between 0.45 nm and 0.65 nm. For an OM4 fiber with an EMB of 4700 MHz·km has modeled distance capabilities of 100 m at 0.65 nm and 130 m at 0.45 nm. Table 2 shows that increasing the bandwidth of the fiber only provides a small gain in distance capability. While it is possible to manufacture a 10 GHz·km multimode fiber, the higher bandwidth provides only 5 m additional reach at a spectral width of 0.65 nm and only 10 m more reach at 0.45 nm. It can also be seen from Table 2 that a multimode fiber with higher bandwidth is only beneficial if VCSEL manufacturers implement lower spectral widths on their VCSELs in the 850 nm region, but this is unlikely due to transceiver reliability concerns.

Table 2 Simulated maximum transmission length through optical fiber at 28 Gb/s

Modeled maximum transmission distance at 32 Gb/s (850 nm)			
Spectral width (nm)	4700 MHz·km	10 GHz·km	20 GHz·km
0.65[a]	90 m	100 m	100 m
0.45[a]	120 m	135 m	140 m
0.20	160 m	225 m	250 m
0.05	180 m	290 m	375 m

[a]Typical industry VCSEL spectral widths

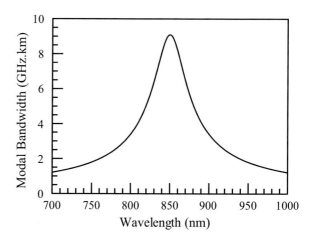

Fig. 13 Illustrative chart of modal bandwidth versus wavelength for MMF optimized for 850 nm

Due to material dispersion, the modal bandwidth of MMF decreases from the peak value when the wavelength moves away from the peak wavelength. Figure 13 illustrates the dependence of modal bandwidth on the wavelength for an MMF of OM4 quality optimized for 850 nm, in which case the peak or optimum bandwidth is located near 850 nm. For SWDM applications, the system reach is not only determined by the modal bandwidth and chromatic dispersion at 850 nm but also the values of these attributes at the longer wavelengths. The SWDM transceivers utilize four VCSELs at wavelengths near 850 nm, 880 nm, 910 nm, and 940 nm, while bi-directional (BiDi) transceivers utilize two wavelengths near 850 nm and 900 nm. Note that the rate that the modal bandwidth decreases from the peak value is largely determined by the material composition of the MMF core but also to some extent is affected by the refractive index profile.

Chromatic Dispersion Compensation

As noted insection "Limiting Factors for VCSEL-MMF Transmission," chromatic dispersion poses significant limitation for VCSEL-MMF transmission. Recently established standards are based on a high VCSEL spectrum line width of around 0.6 nm, which has the benefit of easing manufacturing difficulty and increased reliability. However the increased chromatic dispersion (CD) effects due to larger VCSEL spectral line widths also introduce a system performance penalty which

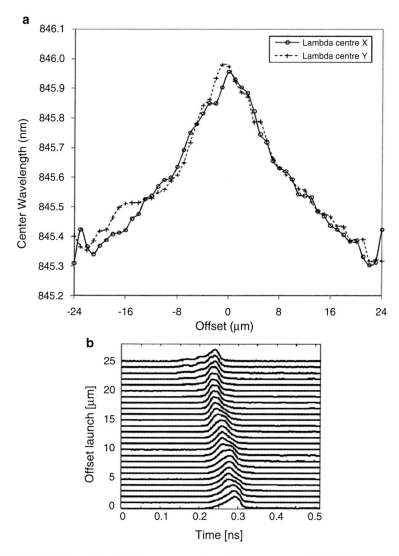

Fig. 14 (**a**) Radial variation in the center wavelength for a representative VCSEL transceiver coupled into a multimode fiber as reproduced from Fig. 5 of Pimpinella et al. (2011). (**b**) A fiber with left-tilted DMD characteristic from Fig. 1 of Molin et al. (2011) with length of 560 m

limits the system reach. Recent research has demonstrated a way to mitigate the CD effect (Gholami et al. 2009; Molin et al. 2011). It was observed that in some cases, the light coupled from the VCSEL into the MMF experiences a spatial wavelength dependence such as the one shown in (Pimpinella et al. 2011), which is reproduced below in Fig. 14a. The wavelengths of the VCSEL modes propagating closer to the edge of the core of the MMF have lower values than the modes at the center

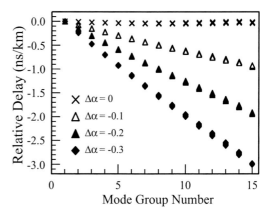

Fig. 15 Relative delay versus mode group number for different values of $\Delta\alpha$

of the core, with a wavelength shift $\Delta\lambda_c$. This shift leads to a time delay for the portion of the signal traveling through the center of the fiber core compared to the portion propagating near the edge of the core. The light at the edge of the core will therefore take a longer time to reach the end of the fiber than the light traveling near the center of the core. Normalized to the unit length of the fiber, the relative delay is $\Delta t = D \cdot \Delta\lambda_c$. It has been suggested that such time delay can be compensated if the MMF has a differential mode delay that has a left tilt, as shown by Fig. 14b. The hypothesis is that when a MMF with this left-tilted DMD is paired with a VCSEL exhibiting a spatial wavelength dependence, the chromatic dispersion observed in Fig. 14b will be compensated. In some cases (Gholami et al. 2009; Molin et al. 2011; Pimpinella et al. 2011), improved system performance was observed.

The approach used in Gholami et al. 2009, Molin et al. 2011, and Pimpinella et al. 2011 requires that the MMF used in the transmission have the left-tilted DMD characteristic. An alternative approach was proposed in Chen et al. (2013) without this requirement. A short MMF jumper that has the desired left-tilted DMD is incorporated into the MMF link so that the overall MMF link possesses the desired modal delay characteristics. This proposal also compensates the CD effect and therefore supports a longer system reach. Since a typical MMF link comprises several spans, this approach of adding one additional, short jumper to the existing network is cost-effective since the rest of the MMF link remains unchanged.

The optimum performance of an MMF is determined by the α value of the refractive index profile defined in section "Light Propagation, Characterization, and Link Performance." When α value is below optimum value α_0, the DMD shows a left tilt toward higher radial offsets of the MMF core. Figure 15 illustrates the relative modal delays (relative to that with lowest mode group number) versus mode group number for a typical MMF with 50 μm core and 1% delta over for several values of $\Delta\alpha = \alpha - \alpha_0$. As the mode group number increases, the relative delay scales linearly with the mode group number, and a more negative $\Delta\alpha$ yields a higher differential delay or stronger DMD tilt. Figure 16a shows the DMD output of 70 m of a MMF with $\Delta\alpha \sim -0.2$. The root mean square (RMS) centroid at each radial

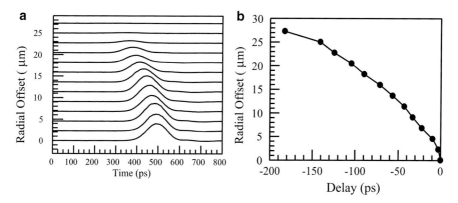

Fig. 16 (**a**) DMD plot of a 70 m MMF jumper; (**b**) the measured DMD centroid of the 70 m MMF jumper at different radial offsets

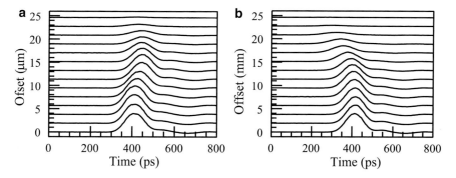

Fig. 17 (**a**) DMD output for 1 km MMF. (**b**) DMD output for combined MMFs

offset, which is referred to as the DMD centroid, is shown in Fig. 16b. This lower α MMF example clearly produces the desired left-tilted DMD.

Next, a conventional MMF is concatenated with the MMF jumper to create a left-tilted DMD plot for a MMF link. The primary fiber in the link is a 1 km length of conventional MMF that has a slight right tilt, as shown in Fig. 17a. The 70 m MMF jumper characterized in Fig. 16 was then concatenated with this primary MMF, yielding the DMD trace for the concatenated MMFs shown in Fig. 17b. The amount of left DMD tilt in the link can be controlled by adjusting the length of the MMF jumper.

A transmission experiment using the 25 Gb/s VCSEL and ROSA obtained from V-I-Systems was also conducted, yielding the results shown in Fig. 18. To illustrate the benefit of CD compensation, 11 m of a MMF jumper was concatenated with several lengths of OM4 MMF. The length of the OM4 MMF was chosen so that the differential delays of the modes propagating at different core radial positions balanced the delay due to CD. With the added CD compensation jumper, the system

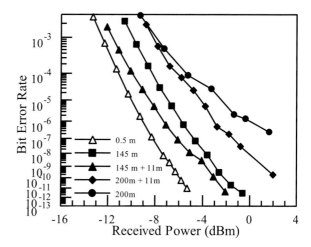

Fig. 18 Center wavelength versus radial position for the VCSEL-MMF coupling

performance was improved in links with both 145 m and 200 m OM4. Without the CD compensation, the system performance is far worse, as exhibited by the waterfall curve for the 200 m MMF without the 11 m jumper.

The chromatic dispersion compensation using either a MMF with a left-tilted DMD MMF or with the modal delay compensation jumper requires the VCSEL-MMF coupling to result in a specific spatial wavelength dependence in order for the compensation to work. While the efforts illustrate an interesting mechanism to improve the system performance, in practice the spatial wavelength dependence varies dramatically with the VCSEL alignment to MMF, and therefore it cannot be reliably used broadly for commercial VCSEL transceivers.

Performance of MMF with Different Peak Wavelength in WDM-Based Transceivers

As mentioned in section "Limiting Factors for VCSEL-MMF Transmission," WDM-based multimode transceivers that couple multiple wavelengths into a single MMF are being developed and deployed. Such transceivers include 40G/100G BiDi and 40G/100G SWDM. Recent studies have been conducted using a 40G BiDi transceiver with a QSFP form factor that transmits 20 Gb/s VCSEL light near 850 nm in one direction and near 900 nm in the other direction (Chen et al. 2016a). This BiDi transceiver uses a 4 × 10G electric connection that is compatible with a QSFP 40G SR4 interface. 40G BiDi supports distances of 100/150/200 m for OM3/OM4/OM5 MMF. Another emerging technology is short wavelength division multiplexing (SWDM) (Tatum et al. 2015), which utilizes VCSELs at four wavelengths (850/880/910/940 nm) operating at 10 Gb/s for 40G transmission or at 25 Gb/s for 100G transmission. While SWDM and BiDi can use existing MMFs such as OM3 and OM4, these transceivers are also expected to take advantage of OM5 MMF. For 40G SWDM, the specified transmission distances are 240 m/350 m/440 m, respectively, for OM3/OM4/OM5. For 100G

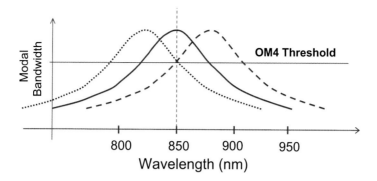

Fig. 19 The schematic illustration of OM4 with different peak wavelengths

SWDM, the expected transmission distances are 70 m/100 m/150 m, respectively, for OM3/OM4/OM5.

The OM4 standard was adopted several years ago as a high performance MMF with a data center focus. As discussed in section "Introduction of Multimode Fibers," OM4-grade MMF is defined to have an EMB of at least 4700 MHz·km at 850 nm. The modal bandwidth of MMF is wavelength dependent and is generally peaked at a certain wavelength λ_P, and falls off on either side, as shown in Fig. 13. OM4-grade MMF meets the specified EMB value at 850 nm, but this can occur when λ_P is located at wavelengths higher and lower than 850 nm. Figure 19 schematically illustrates the variation of the modal bandwidth for three hypothetical OM4 MMFs with different λ_P values. To meet the EMB requirement of 4700 MHz·km at 850 nm, the λ_P of OM4 fibers can be as low as about 810 nm and as high as about 890 nm. The OM4 example with the low λ_P value will yield a lower modal bandwidth at a longer wavelength (e.g., 900 nm) than the OM4 example with the high λ_P value. A recent experiment was conducted to evaluate the performance OM4 fibers with different in 40G BiDi transmission (Chen et al. 2016a). Two OM4 fibers with λ_P values that are near the lower and upper ends of the allowable range for OM4 fiber were used in the experiment. The two fibers are labeled as low-peak-wavelength OM4 (LPW-OM4) and high-peak-wavelength OM4 (HPW-OM4). The LPW-OM4 fiber represents roughly the lowest possible bandwidth that can be achieved by OM4 at 900 nm. On the other hand, the HPW-OM4 represents approximately the best performance that can be achieved by OM4 at 900 nm. This HPW-OM4 was verified to be an OM5 fiber that meets 2470 MHz·km EMB requirement at 953 nm.

The system reach for the OM4 MMF samples was evaluated to find out the longest distance over which the system can perform error-free at both 850 nm and 900 nm. Table 3 shows the system reach of the testing results. Six transceivers labeled "T1" to "T6" were used to explore the natural variations of the product characteristics. The testing was conducted at room temperature. For all six transceivers, the maximum system reach for the LPW-OM4 is 325 m, and the maximum system reach for the HPW-OM4 varies from 350 m to 390 m. in the testing shown in

Table 3 The maximum system reaches for LPW- and HPW-OM4 MMFs for 40G BiDi transceivers

Transceiver label	Max length using LPW-OM4 (m)	Max length using HPW-OM4 (m)
T1	325	350
T2	325	390
T3	325	350
T4	325	370
T5	325	370
T6	325	370

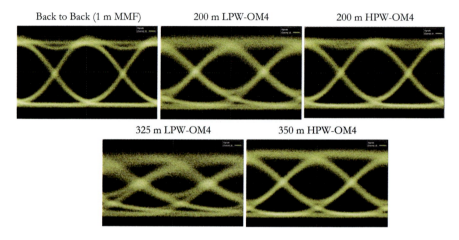

Fig. 20 Optical eye diagrams at 900 nm for several fiber configurations using transceiver "T4"

Fig. 20, it was found that the 900 nm transmission limits the maximum transmission length, which suggests that modal bandwidth at 900 nm may be a contributing factor.

The optical eye diagrams of one of the 900 nm VCSELs with different lengths and types of fiber were also measured. As shown in Fig. 21, at 200 m, the LPW-OM4 shows a slightly more degraded optical eye compared to the HPW-OM4 fiber but can still reach error-free performance easily. At lengths greater than 300 m, the HPW-OM4 fiber has a more open optical eye than the LPW-OM4 fiber.

In addition to the above testing using 40G BiDi transceivers, two 100G SWDM transceiver alpha samples from Finisar were obtained for further testing. The system reach capability using LPW-OM4 and HPW-OM4 obtained for these 100G SWDM transceivers is shown in Table 4. For 100G SWDM transceivers, LPW-OM4 performed bit error-free at 250 m and 200 m, while for HPW-OM4, the maximum transmission distance was 350 m and 250 m, respectively. Some performance variation is clearly observed with the different transceivers, but overall, HPW-OM4 performs better than LPW-OM4 due to the higher modal bandwidth at longer wavelengths.

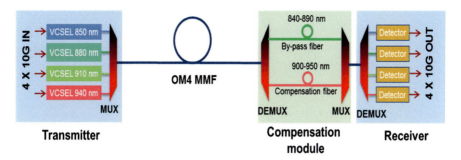

Fig. 21 SWDM system configuration using OM4 MMF with modal dispersion compensation

Table 4 The maximum system reaches for LPW- and HPW-OM4 MMFs for 100G SWDM

Transceiver label	Max length using LPW-OM4 (m)	Max length using HPW-OM4 (m)
100G-SWDM-1	250	350
100G-SWDM-2	200	250

Modal Dispersion Compensation for SWDM Applications

As shown above in section "Performance of MMF with Different Peak Wavelength in WDM Based Transceivers," OM4 fibers with higher λ_P values have higher modal bandwidth at wavelengths greater than 850 nm. OM5 is indeed a subcategory of OM4 with higher modal bandwidth at longer wavelengths to enable WDM transmission over longer system lengths. However, it is also desirable for installed systems with OM4 fiber to have an upgrade path to SWDM. This modal dispersion compensation approach also offers a solution to upgrade OM4 MMF systems to SWDM for extended reach transmission. A novel approach of using OM4 for SWDM application was recently proposed in Chen et al. (2017).

Figure 21 is a schematic of an SWDM system using OM4 MMF and a modal dispersion compensation module. The transmitter has four VCSELs with wavelengths that are multiplexed together and coupled into a single OM4 MMF. At the receiver, the four wavelengths are demultiplexed and sent to four detectors. For SWDM using OM5 MMF, the fiber is required to have minimal EMBs of 4700, 3888, 3087, and 2552 MHz·km at 850, 880, 910, and 940 nm, respectively. These EMBs cannot be guaranteed in a randomly selected OM4 MMF. For an OM4 MMF, the bandwidths of 850 and 880 nm are above the minimal EMBs, but bandwidths of 910 and 940 nm can be well below the minimal EMBs required by the OM5 standard.

Figure 22a shows the required bandwidths at the four wavelengths as a function of distance for a nominal OM4 MMF with a bandwidth of 4700 MHz·km at 850 nm. 40G SWDM transmission over a link distance of 600 m is used to illustrate the modal dispersion compensation approach using this OM4 MMF. For 600 m, the bandwidths for the four wavelengths are about 7.8, 6.5, 5.1, and 4.3 GHz based on the OM5 standard. It can be seen in Fig. 23a that all the four wavelength channels will work with the OM4 MMF up to about 350 m link distance. For link distance

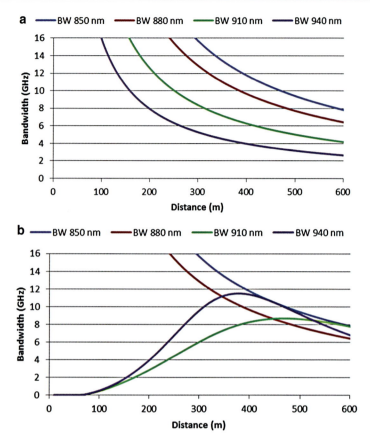

Fig. 22 Bandwidth of four wavelength channels as a function of transmission distance (**a**) OM4 fiber before and (**b**) OM4 fiber after compensation

between 350 and 600 m, the two short wavelength channels of 850 and 880 nm have bandwidths above the OM5 values, but the two long wavelength channels of 910 and 940 nm have bandwidths below the OM5 values.

To increase the bandwidths at the two long wavelengths, a modal dispersion compensation module is added to the transmission link as shown in Fig. 21. The compensation module includes a demultiplexer (DEMUX), modal dispersion compensation fibers, and a multiplexer (MUX). The input wavelength channels from a MMF are demultiplexed into two wavelength bands, 840–890 and 900–950 nm. The short wavelength band is sent through a bypass MMF fiber because no compensation is needed for the two channels of 850 and 880 nm. The long wavelength band is coupled to a modal dispersion compensation fiber (MDCF) with an optimal length to correct the differential modal delays of 910 and 940 nm channels to a functional level. As an example, 20 m of MDCF having a graded index profile with an alpha value of 1.55, a core delta of 1%, and a core diameter of 50 μm

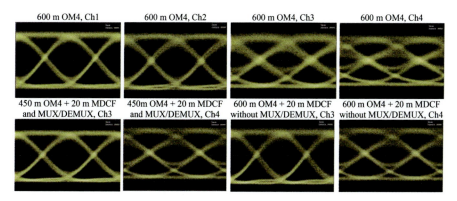

Fig. 23 Optical eye diagrams of Channels 1–4 with 600 m OM4 only, optical eye diagrams of Channels 3 and 4 with 450 m OM4 with 20 m MDCF connected to Channels 3 and 4 using MUX/DEMUX, and Channels 3 and 4 with 600 m OM4 with 20 m MDCF without MUX/DEMUX

was used in the compensation module. Figure 22b plots the bandwidths of the four channels after compensation was used. Figure 22b shows that the MDCF increases the bandwidths for all four wavelength channels above the values needed to enable system reach in the 350–600 m range.

SWDM transmission over OM4 fiber with modal dispersion compensation has also been demonstrated in system testing using a 40G SWDM transceiver. The transceiver transmitted four streams of 10 Gb/s data at wavelengths around 850, 880, 910, and 940 nm. The system performed error-free over OM4 fiber with lengths up to 350 m without compensation. For longer fiber lengths, modal dispersion compensation was needed to increase the bandwidths for the two long wavelength channels. For a 600 m link using the OM4 without a MDCF, Channels 1 and 2 with 850 and 880 nm wavelengths both transmitted bit error-free over 10 minutes, while Channels 3 and 4 with wavelengths of 910 and 940 nm had very high bit error values around 10^{-3}. This is understandable given that the modal bandwidths of the OM4 fiber at 910 and 940 nm drop significantly from the value at the peak wavelength. The optical eye diagrams in Fig. 23 show that the eyes for Channels 1 and 2 are wide open, but the eyes of Channels 3 and 4 are degraded due to the low modal bandwidth at the longer wavelengths.

To improve the performance of Channels 3 and 4, modal dispersion compensation was applied to the two channels. A MUX/DEMUX device was used to separate and combine the four wavelength channels centered at 850, 880, 910, and 940 nm. A MDCF was used for both Channels 3 and 4 at 910 and 940 nm. It was observed that with 450 m OM4 without MDCF, Channels 1 and 2 performed error-free over 10 min but with an unacceptably high BER of 2×10^{-9} for Channels 3 and 4. With 20 m of MDCF added to Channels 3 and 4, these two channels performed bit error-free over 10 min, and the corresponding eye diagrams were improved, as shown in Fig. 23. The 450 m base distance is shorter than target 600 m length due to insertion loss penalties for Channel 4 and the use of OM3 fiber pigtails in the

MUX/DEMUX devices. By omitting the MUX/DEMUX, the transmission distance reaches 600 m for Channels 1 and 2 without the MDCF and for Channels 3 and 4 with the MDCF. The experimental results suggest that if two wavelength-band MUX/DEMUX devices were used with optimized design and packaging to provide lower the insertion loss, 350–600 m transmission for the two long wavelength channels could be realized with a single MDCF.

Multimode Fiber for Long Wavelength Applications

Motivation for Long Wavelength MMF Systems

As noted previously, when the modal bandwidth of MMF reaches OM4 levels, with an EMB of 4700 MHz·km or greater at 850 nm, the chromatic dispersion becomes a limiting factor for high data rates and/or long reach links because of the large transceiver linewidth (around 0.6 nm) associated with VCSELs. The system reach is limited to about 100 m for a data rate of 25 Gb/s as defined by IEEE 802.3 bm. However proprietary 100G-SR4 transceivers that enable OM4 to transmit up to 300 m are available. Figure 24a is a plot of the CD and attenuation of a representative MMF over a range of wavelengths. The magnitude of the CD decreases from about −95 ps/nm/km at 850 nm to −35 ps/nm/km at 1060 nm, as indicated by the blue and green vertical lines, respectively. The chromatic dispersion crosses zero near 1300 nm, as indicated by the vertical red line.

To alleviate the deleterious impact of high CD, the data center industry is exploring different options for transceivers (Chen et al. 2015). Although most VCSEL-based transceivers operate at wavelengths near 850 nm, GaAs-based VCSEL technology, which is superior to the InP-based in terms of speed, efficiency, manufacturability, and cost-efficiency, can be extended to wavelengths near ~1100 nm using conventional compound semiconductors without compromising reliability (Suzuki et al. 2012). In addition, for high-density applications, the use of coarse wavelength division multiplexing (CWDM)) of long wavelength VCSELs has also been demonstrated (Tan et al. 2017). In this scheme, several sources with different wavelengths are multiplexed so that they co-propagate in the same MMF. Another wavelength range of interest is around 1310 nm, where silicon-photonics transceivers are being developed to take advantage of the absence of chromatic dispersion in single-mode and multimode fibers (Bickham et al. 2014).

The wavelength dependence of the CD and attenuation are inherent properties of a silica-based waveguide that advantage MMF-based transmission systems operating at wavelengths above 850 m. Another wavelength-dependent attribute of a given MMF design is a decrease in the number of guided mode groups with increasing wavelength, as shown in Fig. 24b. Fibers with fewer mode groups can potentially enable higher modal bandwidth, but the LP01 mode and outer mode groups carry a higher fraction of the optical power, so their modal delays need to be carefully controlled by optimizing the fiber design and manufacturing process.

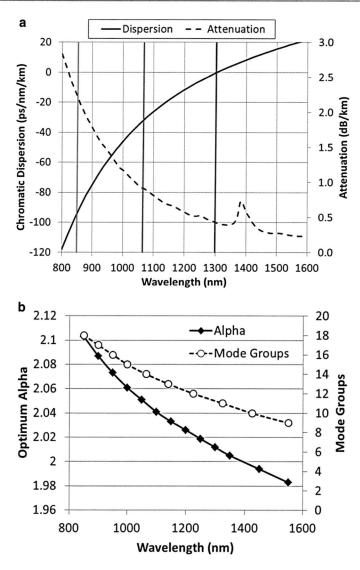

Fig. 24 (a) Chromatic dispersion and spectral attenuation of a representative MMF; (b) wavelength dependence of the optimum alpha and the number of guided mode groups

Design of MMF for Long Wavelength Transceivers

From a design perspective, the main characteristic of the refractive index profile that needs to be optimized to achieve high modal bandwidth at the target wavelength is the alpha parameter of the graded index core. Figure 24b shows that the optimum alpha decreases from a value of about 2.1 at 850 nm to about 2.0 at 1310 nm. A decrease of only 5% in the alpha value shifts the location of the bandwidth peak by

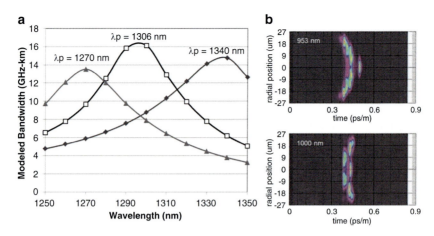

Fig. 25 (a) Modeled bandwidth versus wavelength for three perturbations of a MMF optimized for the 1300 nm window; (b) DMD traces of a prototype MMF with high modal bandwidth in the 990–1065 nm window

over 450 nm, and this sensitivity illustrates one of the challenges of fabricating a MMF with high modal bandwidth.

The λ_P value is primarily determined by the alpha of the graded index core, but other variations from the ideal refractive index profile can also influence the location of λ_P. In order for a MMF to have high modal bandwidth at the system wavelength, λ_{SYS}, the λ_P must be located within a few tens of nanometers of λ_{SYS}. However that condition is necessary but not sufficient. The maximum modal bandwidth must also be high enough to ensure that the specified modal bandwidth at λ_{SYS} is achieved, and this requires precise control of the refractive index profile during the manufacturing process. Figure 25a is a plot of the modal bandwidth versus wavelength for three modeled Monte Carlo perturbations of the refractive index profile of a MMF optimized for the 1300 nm window (Bickham et al. 2014). Each of the three-modeled fibers has a sufficiently high-peak bandwidth to yield modal bandwidths greater than 8 GHz·km at 1300 nm and greater than 4 GHz·km over the 1270–1330 nm window.

New long-wavelength CWDM transceivers are being designed that multiplex VCSELs operating at different wavelengths together and couple them into a single MMF. Researchers at Hewlett Packard Enterprises have prototyped a Tx module comprised of four independent bottom-emitting VCSEL arrays, each operating at a different wavelength (990, 1015, 1040, and 1065 nm) and each with integrated lenses fabricated directly onto the back of the GaAs substrate (Tan et al. 2017). This CWDM transceiver will require a special MMF with high modal bandwidth over the 990–1065 nm range. Figure 25b illustrates the 953 nm and 1000 nm DMD traces of a prototype fiber that satisfies this condition. The EMB value at 953 nm is 8190 MHz·km, and the left curvature indicates that λ_P is slightly higher than this wavelength. At 1000 nm, the spread in the modal delays has compressed compared to 953 nm, and the EMB value has increased to 9890 MHz·km, which is very close to the peak value for this fiber. The DMD trace was not recorded at 1065 nm, but it would exhibit some right curvature (opposite of the 953 nm DMD), and the EMB is expected to decrease to the 7500–8000 MHz·km range.

System Testing of MMF Optimized for 1060 nm

The first reports on VCSELs operating in the 1060 nm window came in 2006 from NEC, which used strained InGaAs/GaAs quantum wells (QWs) with doped barriers for high differential gain and reduced gain compression to produce a multimode VCSEL operating at 1090 nm with a modulation bandwidth of 20 GHz and transmission capacity of 25 Gb/s (Suzuki et al. 2006). Most of the recently reported work on these long wavelength VCSELs has come from Furukawa, which employed double intracavity contacts and a dielectric top distributed Bragg reflector (DBR) to make a low-resistance high-efficiency oxide-confined 1060 nm VCSEL with a modulation bandwidth of 20 GHz. This design enabled error-free transmission over 300 m of MMF optimized for high modal bandwidth at 1060 nm (Kise et al. 2014). HP demonstrated CWDM transmission at 25 Gb/s over 75 m of OM3 fiber using 990–1065 nm oxide-confined, bottom-emitting, and lens-integrated VCSELs with strain-compensated InGaAs/GaAsP QWs. They were able to extend the reach to 300 m at 990 and 1065 nm using a MMF optimized for high modal bandwidth in this wavelength range (Tan et al. 2017). Figure 26a illustrates the eye diagrams for the back-to-back (BtB) configuration and for error-free transmission over 100 m of 1060 nm-optimized MMF using a 1065 nm VCSEL on a probe station with a Arden mode controller to launch an expanded beam into the MMF. The VCSELs were equalized using an arbitrary waveform generator (AWG). The eye diagrams at the top were measured with an NRZ line rate of 25 Gb/s, while those at the bottom were measured with PAM4 modulation with an aggregate bit rate of 50 Gb/s.

Researchers at Chalmers University of Technology also recently demonstrated new high-speed oxide-confined 1060 nm single- and multimode VCSELs designed for extended reach VCSEL-MMF optical interconnects. The single-mode design promotes single-mode emission with a relatively large oxide aperture through weak transverse optical guiding and strong transverse confinement of optical gain and was shown to support data rates up to 50 Gb/s (Simpanen et al. 2017). Using a mode-selective launch, they demonstrated error-free transmission over 1000 m of 1060 nm-optimized MMF at bit rates up to 25 Gb/s, as shown in Fig. 26b. The λ_P of the MMF used in their experiment was approximately 1010 nm, and the peak EMB was 6500 MHz·km.

System Testing of MMF Optimized for 1310 nm

The ideal transmission wavelength in terms of chromatic dispersion is at 1310 nm, as shown in Fig. 24a. A MMF with a design similar to that shown in Fig. 1b was fabricated using the outside vapor deposition process and determined to have EMB value of 13 GHz·km at 1310 nm. One type of transceiver paired with this fiber in the experiments was a prototype SiPh transceiver comprised of an integrated optical module that includes hybrid silicon lasers, silicon modulators, photodetectors,

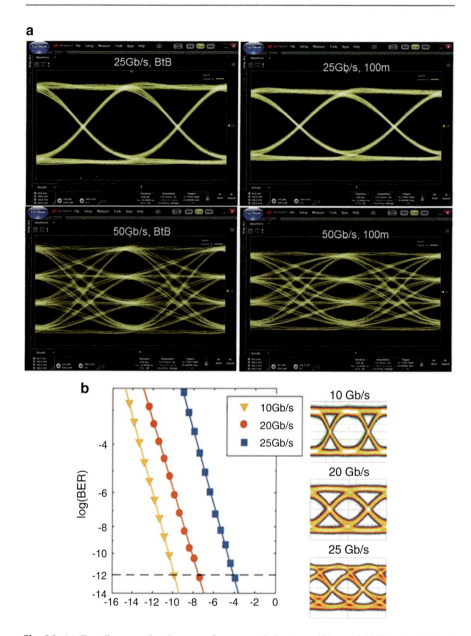

Fig. 26 (a) Eye diagrams showing error-free transmission over 100 m of 1060 nm-optimized MMF at 25 Gb/s (top) and at 50 Gb/s using PAM4 (bottom) (Data courtesy of Mike Tan, HP Enterprises); (b) transmission over 1000 m of 1060 nm-optimized MMF with PRBS-15 signals (Simpanen et al. 2017)

waveguides, and high-speed electronic circuitries. The performance of the SiPh module and the 1310 nm-optimized MMF was evaluated by conducting 25 Gb/s transmission experiments over 410 m and 820 m lengths (Chen et al. 2014). The waterfall curves (BER vs. received optical power curves) were first measured in the BtB condition and then with the 410 m and 820 m lengths of MMF inserted into the link. The optical power penalties at 10^{-12} BER for 410 m and 820 m of MMFs were 1.4 dB and 3.4 dB, respectively, and Fig. 27a shows that there is little degradation in the eye diagram after 420 m compared to the BtB configuration.

The second type of transceiver used in the system testing in this wavelength range was a single-mode 10 Gb/s VCSEL employing GaInNAs quantum wells, operating at a wavelength of 1280 nm. These transmission experiments were performed using an 8-fiber ribbon cable having a length of 410 m. By looping through three of the fibers in the ribbon cable, the total length was extended to 1230 m. The BER was measured for three different launch conditions from the single-mode VCSEL: (1) a center launch, (2) an 8 μm offset launch, and (3) a multimode launch obtained by expanding the VCSEL launch into one that approximately matches the encircled flux launch standard at 1300 nm (Bickham et al. 2014). The 8 μm offset launch condition is designed to evaluate the sensitivity to lateral misalignments such as those that occur when passive alignment is used in transceiver manufacturing. The waterfall curves in Fig. 27b illustrates that there is essentially no penalty compared with the back-to-back condition (BtB) with the center launch and less than 2 dB penalty with both the offset launch and the multimode launch.

Universal Fibers, a New Fiber Concept Bridging SM and MM Transmissions in Data Centers

Universal Fiber Concept and Benefits

Enterprise data centers primarily use OM3/OM4 multimode fiber for data transmission as most channel lengths are less than 100 m, and the trend looks to continue since multimode (MM) transmission remains a cost-effective solution covering the majority of transmission distances. On the other hand, single-mode fiber is used in data centers both at short distances, similar to MMF, and at distances above 100 m. Because SM transmission enjoys high system bandwidth and is capable of longer system reaches, mega- and hyper-data centers tend to predominantly adopt SM transmission. However, even in such large-scale data centers, there is still a large percentage of SM transmission for distances less than 100 m. For very short distances in enterprise data centers, single-mode fiber is used for the carrier interface to provide linkage to the router and FICON (i.e., Fiber Connection) mainframe for storage applications. For distances up to several hundred meters, both MM and SM transmission coexist, depending on the data rate and the type of data centers. While it is feasible to mix both multimode and single-mode fibers in data centers, a uniform type of optical fiber that can accommodate both types of transmissions is preferred to simplify cable management, transceiver/connectivity logistics and provide flexibility for future transceiver upgrades.

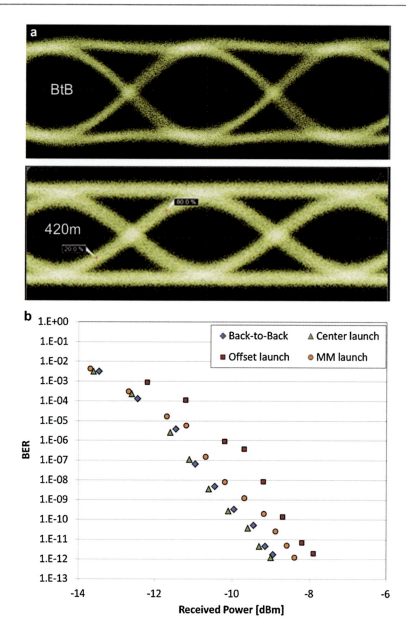

Fig. 27 (a) Eye diagrams in the BtB configuration and after transmission through 420 m of MMF with a 25 Gb/s SiPh transceiver operating at 1310 nm; (b) waterfall curves for the BtB configuration and with various launch conditions after transmission through 1230 m of MMF with a 10 Gb/s VCSEL operating at 1280 nm

Table 5 Comparison of cost and power consumption for multimode and single-mode transceivers. For both cost and power consumption, the values are shown relative to that of 10G MM transceivers

Fiber	Relative transceiver cost			Relative power consumption		
	10G	40G	100G	10G	40G	100G
MMF (OM1–OM4)	1	6	13	1	2	4
SMF	2	15–21	13–37	1.5	5	5.5

The lower cost of MM systems is driven mostly by the large difference in prices between SM and MM transceivers, as shown in Table 5. However, as speeds approach 100G and beyond, the difference in price is shrinking. In fact, at or beyond 100G, many data center operators are being advised and are expected to deploy single-mode optics due to the price and capability requirements. Another factor that has attracted increased attention is the power consumption in data centers. Data center operations are highly sensitive to cost and power consumption – in particular for mega- and hyper-data center operators. Single-mode systems consume more power, as also shown in Table 5. Therefore, operators who would like to build an infrastructure that is capable of supporting the needs for today, yet be flexible enough to also support the future, are faced with a difficult choice between two options. The first option is to deploy MMF today to take advantage of the lower cost of MM transceivers today but then rip out the cable infrastructure and replace with a SM version within the first or second upgrades. The second option is to deploy single-mode fiber today and pay more for the SM transceivers with each upgrade until the SM transceiver price becomes comparable to the MM option. Therefore, selecting an appropriate fiber is both a financial and technological decision with consequences for today as well as for the future. Multimode and single-mode fibers both have advantages and drawbacks. It is desirable to have a universal fiber (UF) that provides the freedom to choose low-cost MM transceivers for the current and future needs but is ready for upgrade to SM transceivers in the future, when MM technology becomes limited by distance and when SM transceiver prices become more favorable. This section discusses designs and feasibility of UF.

Fiber Designs

The key consideration of the UF design is for the fiber to be able to operate both for multimode transmission around from 850 nm to 950 nm and to be used for single-mode transmission around 1300 nm and 1550 nm. The core of the UF has a simple alpha profile similar to conventional MMFs as described by Eq. 1. In order to accommodate the single-mode transmission, it is necessary for the mode field diameter of the fundamental LP_{01} mode to approximately match that of a standard single-mode fiber, which is around 9.2 μm at 1310 nm.

In the design optimization, the LP_{01} MFD versus the core diameter was calculated for different core deltas at 1310 nm and 1550 nm. As shown in Fig. 28, when the core delta increases, the same LP_{01} MFD can be achieved with a larger core diameter. This allows increasing the core diameter for better VCSEL-MMF

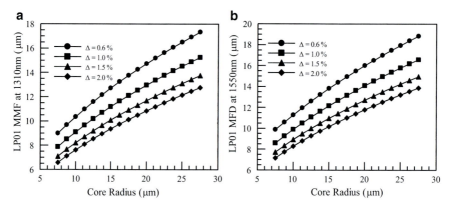

Fig. 28 The LP_{01} MFD for the UF at 1310 nm (**a**) and 1550 nm (**b**)

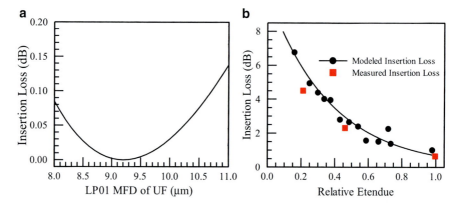

Fig. 29 (**a**) Insertion loss between standard single-mode fiber and UF as a function of MFD of UF; (**b**) insertion loss between a standard 50 μm MMF and UF as a function of relative etendue of UF

coupling around 850 nm, without causing MFD mismatch at 1310 nm. Figure 28 also shows that, when the MFD is matched for single-mode fiber at 1310 nm, it is also matched at 1550 nm. Note that the MFD of standard single-mode fiber is around 9.2 um at 1310 nm and around 10.4 μm at 1550 nm.

For single-mode operation, a mismatch in MFD will cause coupling loss which results in an insertion loss in the system. To understand how large the core diameter can be for a given core delta, the coupling loss due to MFD mismatch at 1310 nm is calculated. Figure 29a shows the insertion loss as a function of MFD of UF when coupled to a standard single-mode fiber. The MFD of the standard single-mode fiber is assumed to be 9.2 μm. Figure 29a indicates that the MFD of UF does not need to match perfectly to 9.2 μm to have a low insertion loss. For insertion losses below 0.1 dB, the MFD can be as large as 10.7 μm.

For multimode operation, both the core diameter and core delta (or core numerical aperture) of UF affect the insertion loss to a standard MMF. To calculate the insertion loss from a 50 μm standard MMF to a UF, numerical modeling was performed by evaluating the overlap integrals between different modes. UFs with different delta and core diameters were considered in the calculation. It was found that the fiber etendue, which is proportional to $NA^2 \times D^2$, could be used to gauge the insertion loss, where NA is the numerical aperture of the fiber core and D is the core diameter. Figure 2b shows modeled insertion loss as a function of the etendue of UF normalized to that of the 50 μm standard MMF. The MMF insertion loss of UF can be reduced by increasing the etendue. This was verified through measurements of insertion losses of 50 μm MMF coupled to three different UFs, as shown in Fig. 29b. The measured insertion losses are in good agreement with the modeling results.

System Level Testing and Verification for Major MM and SM Applications

This section reviews the system level testing results for UF with several 100G transceivers with QSFP form factors (Chen et al. 2016b, d). The optical transceivers used in data centers almost exclusively have the QSFP form factor for 40G and 100G. As shown in Table 6, both MM and SM transmissions were tested. Standards based and proprietary 40/100G MM and SM variants are commercially available. Standards-defined 40/100G variants include parallel (8F) transmission for MMF and duplex fiber transmission for SMF. Proprietary-defined 40/100G MM and SM variants are also available. 40/100G MM duplex fiber WDM transmission as well as 100G SM duplex WDM and parallel (8F) transmission options are now being deployed.

The transceivers tested included:

- 100G SR4: This is a VCSEL-based commercial transceiver with QSFP form factor operating at 25 Gb/s per lane in compliance with IEEE 802.3bm standard. It uses an 8-fiber MPO connector with four fibers to transmit light and four to receive light. The targeted distance is 100 m for OM4 and 70 m for OM3.
- 100G SWDM (proprietary): This VCSEL transceiver utilizes four wavelengths around 850 nm, 880 nm, 910 nm, and 940 nm, each operating at 25 Gb/s for an aggregate 100G data rate within a single fiber. The transceiver only utilizes two fibers with two LC connectors.
- 100G BiDi (proprietary): This is a VCSEL transceiver that utilizes duplex LC connectivity and two wavelengths (850 nm/900 nm) similar to the 40G BiDi

Table 6 Typical 100G QSFP form factor optical transceivers used in data center

	Connectivity form factor	
Transmission type	Duplex LC	Parallel optics
SM	100G CWDM4	100G PSM4
MM	100G SWDM, 100G BiDi	100G SR4/eSR4

Table 7 Demonstrated transmission distance for the five types of 100G transceivers using UF

Transceiver	Transmission distance
100G SR4	200 m
100G SWDM	150 m
100G BiDi	200 m
100G CWDM4	2700 m
100G PSM4	2000 m

transceiver. In addition to moving to higher baud rate, it also adopts PAM4 technology, which doubles the number of bits in serial data transmissions by increasing the number of levels of pulse-amplitude modulation, therefore doubling the data rate. The 100G BiDi transceiver combines the host card four 25G NRZ electrical lanes into two 50G wavelengths (850 nm/900 nm) using a PAM4 ASIC inside the module. On receiving the 50G PAM4 signals, the transceiver module PAM4 ASIC converts the 50G PAM4 signals back to 25G NRZ, to interface to the host card four 25G lane CAUI electrical interface. FEC technology is included in the module ASIC to address the PAM4 insertion loss and to provide system performance at equal to, or better than, 10^{-12} BER. The 100G BiDi transceiver has the QSFP form factor and supports up to 70/100/150 m over OM3/OM4/OM5 multimode fibers.

- 100G CWDM4 (proprietary): This is a 100G or 4 × 25G single-mode transceiver targeting DC applications with an expected system reach of up to 2 km. It operates at four wavelengths around 1300 nm. It utilizes duplex LC connectivity involving two fibers while having QSFP28 electric interface. By design, it is used with single-mode fibers.
- 100G PSM4: Another type of SM transmission that is enjoying wide use is parallel single-mode transmission. It is designed for use in 100 Gb Ethernet links and utilizes a parallel single-mode fiber infrastructure to support reach of up to 500 m. This transceiver is compliant with the QSFP28 MSA, PSM4 MSA, and applicable portions of IEEE P802.3bm. The transmission of 100G PSM4 is based on 1310 nm lasers. A 1550 nm SM parallel transceiver in accordance with the MSA was also evaluated.

Using the UF design reported in Chen et al. (2016b), system testing was conducted for all five types of 100G transceivers that cover both SM and MM transmissions with both duplex LC and MPO connectors. The transmission distances which show bit error-free performance are listed in the Table 7. Some of the results have previously been reported in Chen et al. (2016b), but new results are added here and updated with longer distances than previously reported for some transceiver types. Two system testing examples that show more details are also included.

Multimode Transmission with 100G eSR4 Transceiver
A prototype UF at 850 nm with one channel of a 100G eSR4 VCSEL-based transceiver was also tested. With this UF design, this fiber reaches bit error-free

Fig. 30 The BER versus received power for BtB and 150 m system configurations

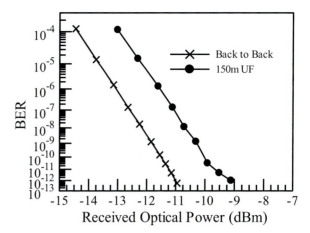

Fig. 31 Optical eye diagrams for BtB and 150 m system configurations

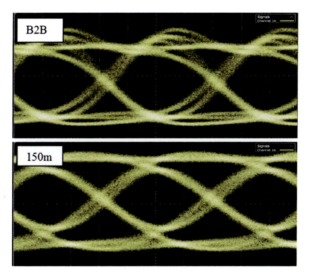

performance at −9 dBm at 150 m, as shown in Fig. 30. The optical eye diagrams are shown in Fig. 31 for both BtB and 150 m conditions. The eye opens widely even at 150 m. This indicates that the UF can have the potential to perform in extended reach regime longer than the 100 m limit defined by the standard using OM4. A Viavi optical network tester (ONT) was also used to conduct the system testing through a cable with MPO connectors using UF. With physical layer, forward error correction (FEC), PCS, and Mac/IP layers all turned on, error-free performance was demonstrated over 200 m of UF cable over multiple days.

Single-Mode Transmission Using 100G PSM4 at 1310 Nm and 1550 Nm
Two 100G PSM4 transceivers operated at wavelengths of 1310 nm and 1550 nm, respectively, were used for the system testing. Although it is common for 100G

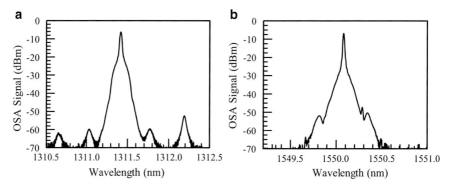

Fig. 32 Optical spectrum of (**a**) 1310 nm 100G PSM4 transceiver and (**b**) 1550 nm 100G PSM4 transceiver

PSM4 transceivers to be implemented in the 1310 nm window, the PSM4 can also utilize the 1550 nm wavelength window, which is optimal for merging with silicon-photonics technology at either 1310 nm or 1550 nm. PSM4 transmission can reach up to 2 km distance. The 1550 nm 100G PSM4 transceiver has full interoperability with most other industry 100G PSM4s based on 1310 nm wavelength, even though the wavelength is different, because it uses a wide band receiver that can detect both 1310 nm and 1550 nm wavelengths. The fundamental mode transmission of UF can be done at both 1310 nm and 1550 nm. The optical spectra of the two transceivers are shown in Fig. 32. The wavelengths of the peak power for the two lasers are very close to 1310 nm and 1550 nm, respectively.

The two 100G PSM4 transceivers use an angle-polished MPO connector. A commercial fan-out cable made from standard single-mode fiber with an angled MPO connector was inserted into the transceivers to gain access to individual transmitters or receivers. Figure 33 shows the BER versus received optical power curves for the (a) 1310 nm transceiver and (b) the 1550 nm transceiver. In the experiment, two conditions, BtB and with 2 km of UF, were tested. The 2 km distance well exceeds the specified distance of 500 m for using 100G PSM4 with standard single-mode fiber. For the 1310 nm transceiver, it was observed that BtB and transmission over 2 km of UF both have very robust performance. They reach bit error-free condition at received optical powers below −11 dBm, and the performance of the 2 km UF is very close to that of BtB condition. For 1550 nm, in the BtB condition, the system reaches bit error-free around −6.75 dBm, while for 2 km of UF, the error-free optical power is around −6 dBm. There is a slight power penalty due to the use of 2 km UF at 1550 nm. This is believed to be due to the fact that at 1550 nm, the higher chromatic dispersion of the fiber induces an additional impairment, but at 1310 nm there is minimal penalty since the chromatic dispersion is nearly zero. In addition to BER testing, an optical network tester (ONT) was used for additional testing for both 1310 nm and 1550 nm 100G PSM4 transceivers. It was observed that the link with 2 km of UF performed error-free for more than 10 h.

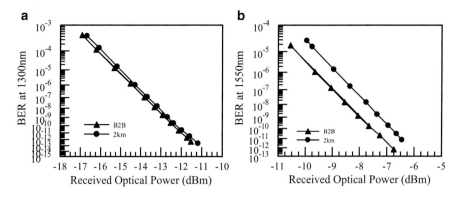

Fig. 33 The BER versus received optical power curves for (**a**) 1300 nm 100G PSM4 transceiver and (**b**) 1550 nm 100G PSM4 transceiver

Optical Trends in Data Centers and Concluding Remarks

In this final section, the optical trends for data centers are summarized in section "Optical Trends in the Data Center," and short concluding remarks are presented in section "Concluding Remarks."

Optical Trends in the Data Center

Just a few years ago, there was a mix of transmission media in data centers involving both copper and optical fibers. Today, the question is not whether optical fiber will be the primary connectivity medium in data centers but whether one should install multimode or single-mode fiber in data centers configured for operation at 10G and higher speeds. Optical connectivity offers the performance attributes sought by end users. These include:

Scalability: Network managers are looking for physical layer solutions that will not only support current speeds but will support migration to the emerging data rates that are expected from the Ethernet Alliance and Fiber Channel Industry Association speed roadmaps shown in Fig. 34 (Ethernet Alliance 2017). With what is known today about future distance considerations, it is possible to design physical layer solutions that can adequately support lifecycles of 15–20 years without having to reconfigure that physical layer solution significantly. Ethernet and Fiber Channel transmission standards develop guidance based on specific criteria that includes technical and commercial feasibility. A primary objective is to deliver economical solutions that meet distance objectives representative of deployed multimode fiber connectivity channel lengths. Trends have shown that as Ethernet data rates have increased from 10 to 40 to 100G and Fiber Channel data rates have increased for 8 to 16 to 32G, the 100 m channel distance represents approximately 95% of deployed OM3 and 90% of deployed OM4 channel lengths as shown in Fig. 35. In other words, for the vast majority of data center users, a 100 m channel distance is more than sufficient to meet their needs.

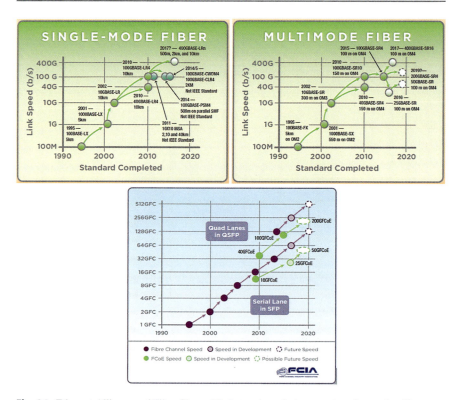

Fig. 34 Ethernet Alliance and Fiber Channel Industry Association speed roadmaps that illustrates the migration to the higher data rates

Another contribution to the higher multimode fiber percentage at 100 m has been the cannibalization of copper connectivity. For data centers operating at 10G and higher data rates, copper connectivity has been relegated to server connections where twinaxial copper cable is mostly used. Twinaxial copper cable is expected to continue as the primary server interconnect until 50G, where optical interconnects will be preferred.

High density: To help increase network efficiencies and drive down the overall cost per circuit, network managers are looking to install high-density solutions, not just in pathway spaces and in the vertical and horizontal management within the racks but also in the electronics. Because fiber ports have low power consumption and generate less heat, port densities can be much higher on line cards, making optical connectivity a more attractive option.

Reliability: Data center downtime can translate into thousands or millions of dollars of lost revenue. Data centers must stay up and running, and optical connectivity is the most robust solution available, mechanically as well as environmentally, to ensure long-term, reliable performance.

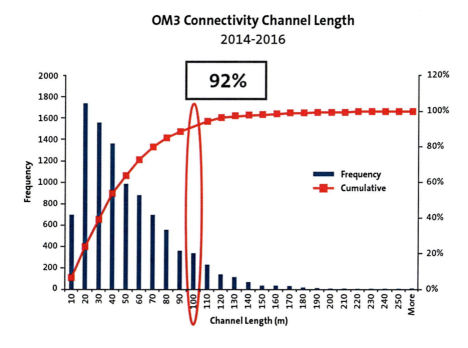

Fig. 35 Data center OM3 and OM4 channel length distribution. (Source: Corning)

Green: Companies today are considering how to minimize their overall environmental footprint. Optical connectivity uses fewer raw materials in the fabrication of the connectivity solutions, consumes less power, and optimizes cooling, as it has much higher density within the pathway and spaces.

Multimode Fiber Prevails in Traditional Enterprise Data Centers

The choice of optical fiber type is driven primarily by the transceivers and the physical configuration of the data center. In the traditional enterprise data centers, multimode fiber in conjunction with VCSELs is the dominant connectivity and transmission technology. Single-mode fiber is being considered for enterprise DC with distances from 150 m to 300 m and beyond but is mostly utilized in hyperscale data centers where the sheer size of the data center requires links of 500 m to 2 km or more. For the vast majority of data centers, however, the added reach of single-mode fiber is not necessary or cost-effective. This is good news for data center economics because VCSELs are easy to fabricate and test, and they are easy to package into the transceiver optical subassembly (TOSA) of the transceiver. VCSEL transmission on multimode optical fiber represents the lowest power and lowest cost currently available. In short, VCSELs offer a great value proposition for data center installations.

Figure 36 provides an estimate of the fiber types that ship into data centers. The amounts of OM3 and OM4 have significantly increased, with OM4 experiencing higher growth over the last couple of years. This is a result of OM4 being recommended in TIA and international standards and the fact that Fiber Channel and Ethernet transmission standards now use OM4 to set distance objectives for data rates $=>$ 32GFC Fiber Channel and $=>$ 40G Ethernet.

The demand for single-mode fiber in the data center remained flat until about 2013, as it was typically used in enterprise spaces for carrier interfaces, uplinks within colocation data centers, and some application where FICON is being used. The emergence of hyperscale computing changed that – Fig. 36 shows the overall growth of single-mode fiber has increased since 2013, with more single-mode being used in these hyperscale data centers.

Serial and Parallel Transmission Multimode Fiber Transmission

Ethernet multimode fiber transmission relied on duplex fiber serial transmission up to 10GBASE-SR. When the Ethernet standards 802.3ba task group first addressed 40G multimode fiber transmission, the 850 nm VCSEL modulation capability was unable to deliver a single 40G wavelength variant. Ethernet chose parallel optic transmission based on development simplicity and the ability to provide a reliable, low-cost solution. Parallel optics differs from traditional duplex fiber optic serial communication in that data is simultaneously transmitted and received over multiple optical fibers. 40GBASE-SR4 parallel optics requires eight OM3 or OM4 fibers with 10G transmission on each fiber: four fibers (four fibers × 10G/fiber) to transmit (Tx) and four fibers (four fibers × 10G/fiber) to receive (Rx). The 802.3ba standard specifies a maximum 100/150 m distance (OM3/OM4). Proprietary extended-reach 40GeSR4 parallel optic transceivers are now available to support distances up to

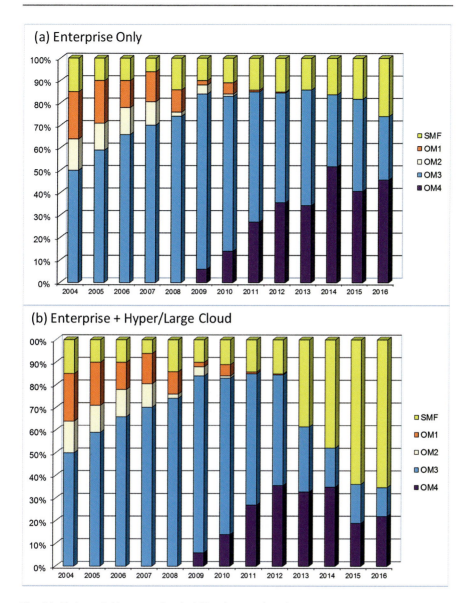

Fig. 36 Estimated shipments of optical fiber, by type, into data centers

300/400 m (OM3/OM4). With Corning MTP® connectivity, the extended-reach 40GeSR4 transceiver delivers up to 330/550 m (OM3/OM4).

Ethernet 802.3ba 100GBASE-SR10 also used 10G per fiber and required 20 OM3 or OM4 fibers (10 fibers × 10G/fiber Tx and 10 fibers × 10G/fiber Rx) but had limited industry demand. As VCSEL technology evolved, 25G modulation rates became available so there is now the Ethernet 802.3bm 100GBASE-SR4

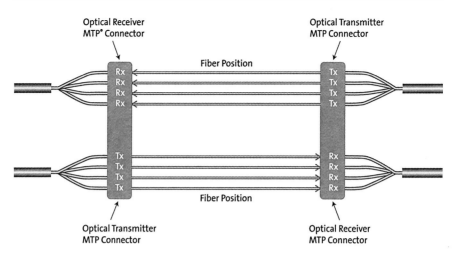

Fig. 37 Schematic for 40GBASE-SR4, 100GBASE-SR4, 200GBASE-SR4, and 400GBASE-SR4.2 parallel transmission

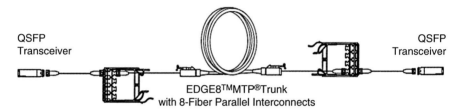

Fig. 38 EDGE8 MTP trunk cable with MTP adapter module connectivity for 40GBASE-SR4 and 100GBASE-SR4

transmission variant that has identical connectivity requirements as the 40GBASE-SR4. See Figs. 37 and 38. The 802.3bm standard specifies up to 70/100 m (OM3/OM4). The OM3/OM4 70/100 m distances are now expected to be the default distance objectives on future data rates such as 200/400G. Proprietary extended-reach 100GeSR4 transceivers are expected in the future to service up to 200/300 m (OM3/OM4). Base-8 multimode connectivity is the optimal choice for parallel optics as existing 40G/100G variants utilize and future 200G/400G variants are expected to utilize SR4 optics with eight fibers (e.g., 4F Tx and 4F Rx).

As discussed earlier, beyond 10G, the VCSELs were unable to support a serial 40G duplex fiber solution. The Ethernet 802.3ba task group evaluated WDM and parallel technologies for 40/100G and selected parallel optics for the standard as it would result in faster and more economical near-term 40/100G deployments. 10GBASE-SR is a widely deployed standards-based technology that uses OM3/OM4. 10GBASE-SR variant also includes OM1 and OM2 options but has minimal traction. 10G connectivity using EDGE8™ MTP® trunk cables with an MTP-to-LC universal polarity module and/or MTP-to-LC universal polarity harness provides four breakout duplex LC circuits per 8-fiber MTP connector. The duplex LC connector plugs into the 10GBASE-SR SFP transceivers. 10GBASE-SR deployments that utilize MTP trunk cables easily migrate to 40GBASE-SR4 parallel optics.

The migration of brownfield 10GBASE-SR to 40G in cabling environments without MTP connectivity in the backbone led transceiver manufacturers to offer proprietary 40G transceivers with a duplex multimode fiber interface. Two proprietary variants can be considered, 40G BiDi which is commercially available now and 40G SWDM which is expected to be commercially available sometime in 2017.

The 40G BiDi transceiver has two 20G optical channels that are bidirectionally transmitted and received over each fiber. This feature results in an aggregate bandwidth of 40 Gb/s with a duplex LC cable. The 40G BiDi transceiver is specified up to 100/150/200 m on OM3/OM4/OM5. With Corning EDGE™ and EDGE8 solutions, the 40G BiDi transceiver delivers up to 100/200/200 m on OM3/OM4/OM5 connectivity. 100G BiDi transceivers became commercially available in 2018 with distance specifications of 70/100/150 m for OM3/OM4/OM5. The 100G BiDi transceiver has two 50G PAM4 optical channels that are bidirectionally transmitted and received over each fiber.

The 40G SWDM transceiver multiplexes four 10G wavelengths that are codirectionally transmitted over each simplex fiber to provide 40G total bandwidth as illustrated in Fig. 39. The 40G SWDM transceiver is expected to support up to 240/350/440 m on OM3/OM4/OM5. The SWDM transceiver has been designed to primarily operate over OM3/OM4 fiber in existing brownfield data centers. 100G SWDM became commercially available in 2017 with distance specifications of 70/100/150 m for OM3/OM4/OM5. The 100G SWDM transceiver has four 25G optical channels that are unidirectionally transmitted and received over each fiber as illustrated in Fig. 39.

The Emergence of Base-8 Connectivity

With the current transmission standards and transceiver technology advancements, network switch vendors are all tooling up their 100GbE switch portfolios based on 25GbE electrical traces instead of 10GbE electrical traces. This reduces the fiber requirement for 100GbE from 20 to 8 fibers. In addition, 40GbE transceivers are currently available in two and eight fiber solutions for both single-mode/multimode links. At the same time, similar solutions are currently in the development phase for Ethernet 200GbE and 400GbE. Recently, Fiber Channel vendors on the SAN side of the DC are following a similar path at 128GFC by using SR4 parallel communications, utilizing an eight fiber MTP®. After much discussion with transceiver and switch vendors, the solutions today (40/100GbE) and in the future (200/400GbE) all converge on duplex and 8f parallel solutions with some interim solutions along the way.

With current transceivers and switch vendor's trends leading to 2F and 8F transceivers, there is a need for optimized solutions. Traditional MTP solutions are based on 12F connectors, which is not always divisible by eight. Based on this information, to simplify network design and operation, improve fiber utilization,

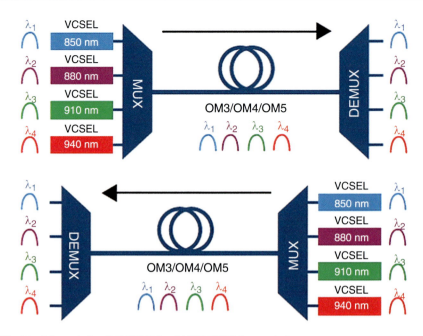

Fig. 39 Schematic for 40G SWDM and 100G SWDM transmission

and reduce costs and attenuation in an optical link, an 8-fiber-based infrastructure would provide you with the optimal solution.

A Base-8 infrastructure as illustrated in Fig. 40 consists of backbone trunks that have 8-fiber legs and modules that have four ports (8-fibers) and eight fiber harnesses. Since parallel connectivity uses 8-fibers out of the available 12-fibers in the connector, the issue arises with a Base-12 infrastructure to either leave the four middle fibers dark or use some type of conversion device. A conversion device can convert two 12-fiber links into three 8-fiber links. This allows all fibers to be utilized giving you three parallel links for each 24 fibers of installed Base-12 trunk cables. This is not necessary when a Base-8 infrastructure is installed. Migration from duplex to parallel links is easy without the added complexity since the infrastructure is already 8-fibers (each trunk cable has 8-fiber legs). The conversion is made easy by removing the modules used for duplex communication and replacing them with MTP adapter panels. MTP-MTP array jumpers would provide the link from the trunks to the QFSP+ transceivers. The installation of a Base-8 infrastructure allows for 100% fiber utilization when migrating to parallel links without the need of conversion modules or harnesses. Figure 40 illustrates Base-8 connectivity for SFP and QSFP transceivers that interfaces to the electronics.

Concluding Remarks

MMF has been used in data centers for several decades. Despite its initial use at very low data rates with LED type light sources and their inherent severe

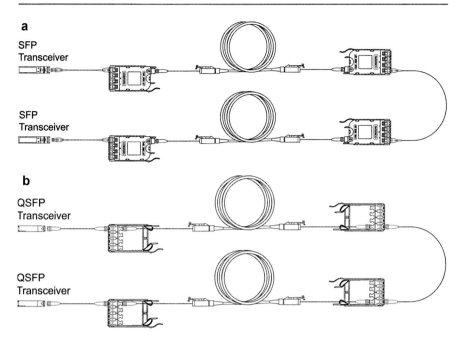

Fig. 40 (**a**) Schematic of Base-8 connectivity for SFP transceivers with cross-connect structured cabling; (**b**) schematic of Base-8 connectivity for QSFP transceivers with cross-connect structured cabling

limitation on modal bandwidth, MMF has evolved over the time to adapt to VCSEL-based laser sources and to improve the modal bandwidth dramatically. Now MMF connectivity is still the leading low-cost optical solution used in the data center for short-reach distances up to 100–150 m at data rates of 10/40/100G. The data center is still rapidly evolving along with the technology associated with optical fibers. There are multiple considerations and forces driving this evolution. With aggregate data rates migrating to 200/400G and beyond, MMF is facing new challenges. Higher data rates and higher density of data transmission require that MMF to have even higher bandwidth and to overcome chromatic dispersion limitations. The use of WDM transmission requires the MMF to have sufficient modal bandwidth at longer wavelength. On the other hand, hyperscale data center users are pushing toward the full adoption of single-mode fiber and single-mode transceivers. These forces are pushing the industry in different directions, and it is not clear-cut where the systems will evolve. However, MMF is expected to continue to be an important transmission medium to meet the system demands. As discussed in this chapter, several new MMF variants have been proposed to address or mitigate the limitations of VCSEL-MMF transmission. These new MMFs will be integral components of the system solutions that will enable data centers to deliver higher data rates, higher density, and longer distances.

References

J. S. Abbott, S. Bickham, P. Dainese, M-J Li, Fibers for short-distance applications, in *Optical Fiber Telecommunications VIA*, ed. by I. Kaminow, T. Li, A. Wilner (Elsevier, New York, 2013), p. 243

S. Bickham et al. Design and characterization of bend-insensitive multimode fiber, in *Proceedings of IWCS* 2011, (2011), pp. 154–159

S. Bickham et al., High bandwidth multimode fiber for the 1300 nm window, in *Paper 7–7, Proceedings of IWCS* 2014 (2014)

X. Chen, M.-J. Li, J.E. Hurley, K.A. Hoover, D.R. Powers, R.S. Vodhanel, Chromatic dispersion compensated 25Gb/s multimode VCSEL transmission around 850nm with a MMF jumper, in *OFC/NFOEC Technical Digest OW1B* (2013)

X. Chen et al. 25 Gb/s transmission over 820m of MMF using a multimode launch from an integrated silicon photonics transceiver. Opt. Express **22**(2), 2070–2077 (2014)

X. Chen et al. Long wavelength multimode fiber transmissions for high speed data center applications, in *Proceedings of OECC* (2015)

X. Chen, J.E. Hurley, S. Bickham, J. Abbott, B. Chow, D. Coleman, M.-J. Li, Evaluation of extended reach capability of 40G BiDi VCSEL-based WDM transmission over OM4 multimode fibers, in *Proceedings of SPIE* 9772, Broadband Access Communication Technologies X, 977206 (2016a)

X. Chen, J.E. Hurley, J. Stone, J.D. Downie, I. Roudas, D. Coleman, M. Li, Universal fiber for both short-reach VCSEL transmission at 850 nm and single-mode transmission at 1310 nm, in *Optical Fiber Communication Conference, OSA Technical Digest* (online) (Optical Society of America, 2016), paper Th4E.4 (2016b)

X. Chen, J. Abbott, D. Powers, D. Coleman, M.-J. Li, Statistical treatment of IEEE spreadsheet model for VCSEL-multimode fiber transmissions, in *OECC/PS2016* July 2016 Niigata, Japan Paper TuD4-3 (2016c)

X. Chen, J.E. Hurley, J.S. Stone, A.R. Zakharian, D. Coleman, M.-J. Li, Design of universal fiber with demonstration of full system reaches over 100G SR4, 40G sWDM, and 100G CWDM4 transceivers. Opt. Express **24**, 18492–18500 (2016d)

X. Chen, J.E. Hurley, D. Gui, Y. Li, J.S. Stone, M.-J. Li, Demonstration of SWDM transmission over OM4 multimode fiber with modal dispersion compensation, in *Optical Fiber Communication Conference, OSA Technical Digest* (Online) (Optical Society of America, 2017), Paper Tu2B.3 (2017)

D.G. Cunningham, W.G. Lane, *Gigabit Ethernet Networking* (Macmillan Technical Publishing, New York, 1999)

G. DiGiovanni, S. Das, L. Blyler, W. White, R. Boncek, S.G. Golowich, Chapter 2: design of optical fibers for communications systems. In: I. Kaminow, T. Li (eds.) Optical Fiber Telecommunications IVB, Academic Press, (New York, 2002)

Ethernet Alliance, 2017. www.ethernetalliance.org/wp-content/.../Ethernet-Roadmap-2sides-Final-5Mar.pdf

W. Freude, Vielmodenfasern, in *Optische Kommuniikationstechnik: Handbuch fur Wissenschaft und Industrie*, ed. by E. Voges K. Petermann (Springer, Berlin, 2002)

A. Gholami, D. Molin, P. Sillard, Compensation of chromatic dispersion by modal dispersion in MMF- and VCSEL- based gigabit Ethernet transmissions. Photonics Technol. Lett. **21**, 645–647 (2009)

D. Hanson, D. Cunningham, P. Dawe, D. Dolfi, 10Gb/s link budget spreadsheet (version 3.1.16a) Excel® spreadsheet on IEEE website. http://www.ieee802.org/3/ae/public/index.html

W. Hurley, T. Cooke, Bend-insensitive multimode fibers enable advanced cable performance, in *Paper 13-4, Proceedings of IWCS* (2009)

T. Kise et al. Development of 1060 nm 25-Gb/s VCSEL and demonstration of 300 m and 500 m system reach using MMFs and link optimized for 1060 nm, OFC 2014, paper Th4G (2014)

D.M. Kuchta, A.V. Rylyakov, C.L. Schow, J.E. Proesel, C. Baks, C. Kocot, L. Graham, R. Johnson, G. Landry, E. Shaw, A. MacInnes, J. Tatum, A 55Gb/s directly modulated 850 nm VCSEL-based optical link, in *IEEE Photonics Conference* 2012 *(IPC 2012)*. Post Deadline Paper PD 1.5 (2012)

D. Molin, M. Bigot-Astruc, P. Sillard, Chromatic dispersion compensated multimode fibers for data communications. ECOC 2011, paper Tu.3.C.3 (2011)

P. Nouchi, P. Sillard, D. Molin, Optical fibers, in *Fibre Optic Communication-Key Devices*, ed. by H. Venghaus N. Grote (Springer, Berlin, 2012)

M.C. Nowell, D.G. Cunningham, D.C. Hanson, L.G. Kazovsky, Evaluation of Gb/s laser based fibre LAN links: Review of the gigabit Ethernet model. Opt. Quant. Electron. **32**, 169 (2000)

P. Pepeljugoski, M.J. Hackert, J.S. Abbott, S.E. Swanson, S.E. Golowich, A.J. Ritger, P. Kolesar, Y.C. Chen, P. Pleunis, Development of system specification for laser-optimized 50μm multimode Fiber for multigigabit short-wavelength LANS. J. Lightwave Technol. **21**, 1256 (2003a)

P. Pepeljugoski, S.E. Golowich, A.J. Ritger, P. Kolesar, A. Ristekski, Modeling and simulation of next-generation multimode Fiber links. J. Lightwave Technol. **21**, 1242 (2003b)

J. Petrilla, Example MMF link model "ExampleMMF LinkModel 130503.xlsx" (2013). http://www.ieee802.org/3/bm/public/may13/index.html

R. Pimpinella, J. Castro, B. Kose, B. Lane, Dispersion compensated multimode fiber, in *Proceedings of the 60th IWCS Conference* (2011), p. 410

E. Simpanen et al. 1060 nm single and multimode VCSELs for up to 50 Gb/s modulation, to be published in *Proceedings of* 2017 *IEEE Photonics Conference* (2017)

N. Suzuki et al. 25 Gbit/s operation of InGaAs-based VCSELs. Electron. Lett. **42**, 975 (2006)

T. Suzuki et al. Reliability study of 1060nm 25Gbps VCSEL in terms of high speed modulation, in *Proceedings of SPIE* 8276, 827604-1-8 (2012)

M. Tan et al. Universal photonic interconnect for data centers, in *Proceedings of OFC* 2017, Paper Tu2B.4 (2017)

J.A. Tatum, D. Gazula, L.A. Graham, J.K. Guenter, R.H. Johnson, J. King, C. Kocot, G.D. Landry, I.E.E.E. Senior Member, I. Lyubomirsky, A.N. MacInnes, E.M. Shaw, K. Balemarthy, R. Shubochkin, D. Vaidya, M. Yan, F. Tang, VCSEL-based interconnects for current and future data centers. J. Lightwave Technol. **33**, 727 (2015)

Multi-core Fibers for Space Division Multiplexing

Tetsuya Hayashi

Contents

Introduction	100
Basics of the Coupled-Mode Theory for Optical Fibers	101
Coupled-Mode Theory for Orthogonal Modes	101
Coupled-Mode Theory for Non-orthogonal Modes	102
Reciprocity of the Mode Coupling Coefficient	103
Uncoupled Multi-core Fibers	103
Mode Coupling in Weakly Coupled MCF	104
Random Mode Coupling Due to Longitudinal Perturbations	106
Discrete Coupling Model and Statistical Distribution of the Crosstalk	108
Coupled Power Theory for Predicting the Statistical Mean of the Crosstalk	112
Crosstalk Suppression Strategy	120
Coupled Multi-core Fibers	125
Systematically Coupled Multi-core Fiber	125
Randomly Coupled Multi-core Fiber	128
Common Design Factors for Uncoupled and Coupled Multi-core Fibers	139
Excess Loss Due to the Power Coupling to the Coating	139
Cutoff Wavelength Variation Due to Surrounding Cores	139
Cladding Diameter	140
Conclusion	142
References	142

Abstract

Space division multiplexing (SDM) through an optical fiber is an attractive technology to cope with the "capacity crunch" in single-mode fiber transmission systems anticipated in the near future, and the multi-core fiber (MCF) has been

---

T. Hayashi (✉)
Optical Communications Laboratory, Sumitomo Electric Industries, Ltd., Yokohama, Kanagawa, Japan
e-mail: t-hayashi@sei.co.jp

© Springer Nature Singapore Pte Ltd. 2019
G.-D. Peng (ed.), *Handbook of Optical Fibers*,
https://doi.org/10.1007/978-981-10-7087-7_66

intensively researched by various groups in recent years. This chapter provides a basic understanding of the MCF for the SDM transmission. To understand the propagation and coupling behaviors in MCFs, the coupled-mode theory for the optical fibers is briefly reviewed. After that, the detailed characteristics of the propagation and coupling in uncoupled and coupled MCFs are discussed. Design factors common for these uncoupled and coupled MCFs are also described.

Introduction

In the optical fiber communication research field, the single-mode fiber (SMF) transmission systems have achieved capacities up to about 100 Tb/s per fiber by employing time-, wavelength-, polarization-division multiplexing and multilevel modulations (Qian et al. 2011; Sano et al. 2012). However, the transmission capacity of the single-core fiber is rapidly approaching its fundamental limit, and the current trends of traffic growth – increasing by a factor of 10 in 5 years – and system capacity growth will result in capacity crunch in the near future (Essiambre et al. 2010; Essiambre and Tkach 2012). In such a situation, spatial division multiplexing (SDM) is an attractive technology for further enlargement of the fiber capacity or spatial capacity – the capacity per cross-sectional area of the fiber – and for reducing the cost and energy per bit in transmission systems (Morioka 2009; Winzer 2015; Winzer and Neilson 2017).

Concepts of SDM optical fiber transmission are not so new (Inao et al. 1979a, b; Berdagué and Facq 1982). However, the active research and development on the SDM by various groups had to wait until the late 2000s, since other technologies – such as optical fiber ribbon, passive optical network, etc. – were able to fulfill the demands for the high-density and efficient optical network. In the late 2000s, the future capacity crunch became a reality, and the MCF has come to attract lots of attention again (Morioka 2009; Winzer 2015).

This chapter describes the MCF for the SDM transmission, which has multiple cores in one common cladding. To understand the propagation and coupling behaviors in MCFs, the coupled-mode theory can be the basis of the discussion. Therefore, in section "Basics of the Coupled-Mode Theory for Optical Fibers," the coupled-mode theory for the optical fibers is briefly reviewed before going into detail on the properties of the MCFs.

After that, the detailed characteristics of the propagation and coupling in MCFs are discussed. Since the MCF has many degrees of freedom in its design, various types of MCFs have been proposed. The MCF can be divided into uncoupled MCF – described in section "Uncoupled Multi-core Fibers" – and coupled MCF, described in section "Coupled Multi-core Fibers."

In section "Common Design Factors for Uncoupled and Coupled Multi-core Fibers," common design factors for the uncoupled and coupled MCFs are described.

Basics of the Coupled-Mode Theory for Optical Fibers

Coupled-Mode Theory for Orthogonal Modes

The coupled-mode equation for an optical fiber with parallel cores is expressed as

$$\frac{d}{dz}|E\rangle = -j\left(\boldsymbol{\beta} + \boldsymbol{\kappa}\right)|E\rangle, \tag{1}$$

where $|E\rangle = [E_1, E_2, \ldots, E_N]^T$ is the column vector of the electric fields E_n of the N core modes, $\boldsymbol{\beta} = \mathrm{diag}\,(\beta_1, \beta_2, \ldots, \beta_N)$ is $N \times N$ diagonal matrix which includes the propagation constants $\beta = (2\pi/\lambda)n_{\mathrm{eff}}$, λ is the wavelength in vacuum, n_{eff} is the effective refractive index, and $\boldsymbol{\kappa}$ is $N \times N$ matrix including the mode coupling coefficients κ_{nm} from Core m to Core n, which is expressed as

$$\kappa_{nm} \equiv \frac{\omega \varepsilon_0 \iint \left(n^2 - n_m^2\right) \mathbf{e}_n^* \cdot \mathbf{e}_m dx dy}{\iint \widehat{z} \cdot \left(\mathbf{e}_n^* \times \mathbf{h}_n + \mathbf{e}_n \times \mathbf{h}_n^*\right) dx dy}, \tag{2}$$

where ω is the angular frequency of the light in vacuum, ε_0 is the vacuum permittivity, n is the actual index profile, n_m is the refractive index profile (RIP) of Core m in the absence of the other cores, $\boldsymbol{e}$ is the normalized vector core mode profile of the electric field, $\boldsymbol{h}$ is the normalized vector core mode profile of the magnetic field, the superscript $*$ indicates the complex conjugate, and $\widehat{z}$ is the unit vector in the direction of z axis.

Since $|E\rangle$ can be expressed as

$$|E\rangle = \exp\left(-j\boldsymbol{\beta}z\right)|A\rangle, \tag{3}$$

where $|A\rangle = [A_1, A_2, \ldots, A_N]^T$ is the vector of the complex amplitudes A_n of E_n, Eq. 1 can be rewritten as

$$\frac{d}{dz}|A\rangle = -j\exp\left(j\boldsymbol{\beta}z\right)\boldsymbol{\kappa}\exp\left(-j\boldsymbol{\beta}z\right)|A\rangle, \tag{4}$$

A component of Eq. 4 is the well-known form of the coupled-mode equation and represented as

$$\frac{dA_n}{dz} = \sum_m -j\kappa_{nm} \exp\left[j\left(\beta_n' - \beta_m'\right)z\right] A_m. \tag{5}$$

where $\beta_n' = \beta_n + \kappa_{nn}$. When the core-to-core coupling is weak, $\kappa_{nn} \sim 0$ can be assumed.

These formulations assume the orthogonality of the modes and thus are called orthogonal coupled-mode theory.

Coupled-Mode Theory for Non-orthogonal Modes

The orthogonal coupled-mode theory is accurate enough for weakly coupled MCF with identical cores (homogeneous MCF). However, the orthogonal coupled-mode theory becomes inaccurate for strongly coupled MCF (SC-MCF) because the core modes are not the eigenmodes of the whole multi-core waveguide and not orthogonal to each other. To describe the mode coupling between strongly coupled parallel cores, non-orthogonal coupled-mode theory was developed (Streifer et al. 1987; Chuang 1987; Haus et al. 1987), which was formulated as

$$\mathbf{P}\frac{d}{dz}|E\rangle = -j\mathbf{H}|E\rangle, \tag{6}$$

where $\mathbf{P}$ is the power matrix associated with the core modes whose elements are expressed as

$$P_{nm} \equiv \frac{\iint \hat{z} \cdot \left(\mathbf{e}_n^* \times \mathbf{h}_m + \mathbf{e}_n \times \mathbf{h}_m^*\right) dxdy}{\iint \hat{z} \cdot \left(\mathbf{e}_n^* \times \mathbf{h}_n + \mathbf{e}_n \times \mathbf{h}_n^*\right) dxdy}. \tag{7}$$

and $\mathbf{H}$ is the overall Hermitian coupling matrix whose elements are represented as

$$H_{nm} = P_{nm}\beta_m + \kappa_{nm}. \tag{8}$$

The power matrix $\mathbf{P}$ has the diagonal elements equal to 1, and its off-diagonal elements are the cross power terms due to the mode non-orthogonality. When the modes are orthogonal to each other, $\mathbf{P}$ becomes an identity matrix, and Eq. 6 is reduced to Eq. 1. The propagation and coupling of the non-orthogonal modes can be solved by rewriting Eq. 6 as

$$\frac{d}{dz}|E\rangle = -j\mathbf{P}^{-1}\mathbf{H}|E\rangle = -j\left(\mathbf{B} + \mathbf{K}\right)|E\rangle, \tag{9}$$

where $\mathbf{B}$ is the diagonal matrix with the diagonal elements of $\mathbf{P}^{-1}\mathbf{H}$ and corresponds to β in the orthogonal coupled-mode theory and $\mathbf{K} = \mathbf{P}^{-1}\mathbf{H} - \mathbf{B}$ corresponds to κ in the orthogonal coupled-mode theory.

In the simple two core case, Eq. 9 can be rewritten as

$$\frac{d}{dz}|E\rangle = -j\begin{bmatrix} B_1 & K_{12} \\ K_{21} & B_2 \end{bmatrix}|E\rangle, \tag{10}$$

or

$$\frac{d}{dz}|A\rangle = -j \begin{bmatrix} 0 & K_{12}\exp[j(B_1-B_2)z] \\ K_{21}\exp[j(B_2-B_1)z] & 0 \end{bmatrix}|A\rangle, \quad (11)$$

where $B_n = \beta_n + \frac{\kappa_{nn}-P_{nm}\kappa_{mn}}{1-P_{nm}\kappa_{mn}}$ and $K_{nm} = \frac{\kappa_{nm}-P_{nm}\kappa_{mm}}{1-P_{nm}\kappa_{mn}}$.

Reciprocity of the Mode Coupling Coefficient

For power conservation in the multi-core system, the reciprocity of the coupling coefficients is necessary as

$$\kappa_{nm} = \kappa_{mn}^*, \quad (12)$$

in the orthogonal coupled-mode theory and as

$$\begin{aligned} T_{nm} &= T_{mn}^* \\ P_{nm}\beta_m + \kappa_{nm} &= P_{mn}^*\beta_n^* + \kappa_{mn}^*, \end{aligned} \quad (13)$$

in the non-orthogonal coupled-mode theory. Based on Eqs. 13 and 12 cannot hold unless $\beta_m = \beta_n$. In the weakly coupled case, the cross power can be negligibly low; thus Eqs. 6, 7, 8, 9, 10, and 11 can be reduced to the orthogonal coupled-mode theory, and the reciprocity assumption of Eq. 13 also can be reduced to Eq. 12. However, when the power confinements of dissimilar cores are very different as shown in Fig. 1, the mode coupling coefficient cannot be reciprocal, because the mode coupling coefficient is the overlap integral of the modes within the permittivity perturbation, as defined in Eq. 2. To cope with this problem, the mode coupling coefficient can be redefined as (Huang 1994; Koshiba et al. 2011)

$$\bar{\kappa}_{nm} = \bar{\kappa}_{mn} \equiv \frac{(\kappa_{nm}+\kappa_{mn})}{2}. \quad (14)$$

for calculating the mode coupling in a weakly coupled MCF using the orthogonal coupled-mode theory. In this chapter, the coupling coefficients κ in the equations are not barred explicitly, but the redefined mode coupling coefficient should be used for MCFs with dissimilar cores (heterogeneous MCFs).

Uncoupled Multi-core Fibers

In the uncoupled MCF (also referred to as weakly coupled MCF), the coupling between the cores is weak, and each core can be used as an individual spatial channel. Therefore the uncoupled MCF is basically compatible with conventional

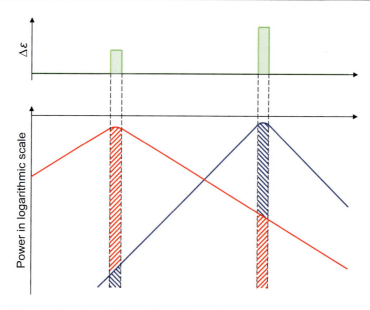

Fig. 1 Schematic illustration of the profiles of the permittivity and the core mode powers of two dissimilar cores (Reused fromHayashi (2013) with permission of the author)

transceivers for single-mode fibers. The characteristics of the inter-core crosstalk (XT) of the uncoupled MCF have been intensively studied for simultaneously improving the core density and suppressing XT. In this section, the coupling characteristics of the uncoupled MCF are described.

Mode Coupling in Weakly Coupled MCF

In the simplest case of two weakly coupled perfect cores – Core m and Core n – without any longitudinal perturbations, and only Core m is excited ($|A_m(0)| = 1$, $|A_n(0)| = 0$), the powers in Core n and Core m are derived as

$$\begin{cases} |A_m|^2 = 1 - F_{nm}\sin^2(q_{nm}z), \\ |A_n|^2 = F_{nm}\sin^2(q_{nm}z), \end{cases} \qquad (15)$$

$$q_{nm} = \sqrt{\kappa_{nm}^2 + \left(\frac{\beta_n - \beta_m}{2}\right)^2}, \qquad (16)$$

$$F_{nm} = \frac{\kappa_{nm}^2}{q_{nm}^2}, \qquad (17)$$

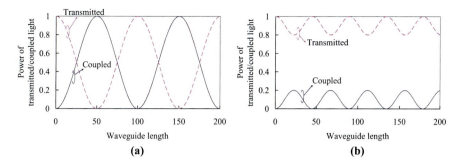

Fig. 2 Power conversions between coupled waveguides with the same coupling coefficients. (**a**) Propagation constants of the waveguides are the same. (**b**) Propagation constants of the waveguides are different (Reused from Hayashi et al. (2011a). ©2011 OSA)

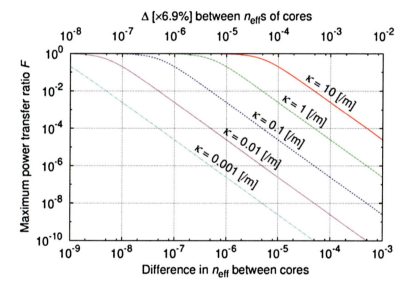

Fig. 3 The relationship between the difference in the effective indices between cores and the maximum power transfer ratio F at 1550 nm (Reused from Hayashi (2013) with permission of the author)

where $\pi/(2q)$ is the coupling length and F is the maximum power transfer ratio. In this case, the power in each core sinusoidally oscillates, and the propagation constant mismatch reduces F, as shown in Fig. 2. Figure 3 shows the relationship between the difference in the effective indices between cores and the maximum power transfer ratio F at 1550 nm. Indeed, F is suppressed to be less than 0.001 (-30 dB) by only a slight difference of 0.001 in n_{eff} between cores, even when $\kappa = 10$ [/m].

Random Mode Coupling Due to Longitudinal Perturbations

This type of systematic coupling between the core modes may be valid for very short optical fiber components, such as a fused fiber coupler. However, it is not the case for long MCFs, because the actual MCFs are longitudinally perturbed by bends, twists, and structure fluctuations. To cope with such perturbations, the coupled-mode equations have to be modified as

$$\frac{d}{dz}|E\rangle = -j\left(\boldsymbol{\beta}_{eq} + \boldsymbol{\kappa}\right)|E\rangle, \tag{18}$$

$$\boldsymbol{\beta}_{eq} = \boldsymbol{\beta}_c + \boldsymbol{\beta}_v, \quad \boldsymbol{\beta}_v = \boldsymbol{\beta}_b + \boldsymbol{\beta}_h, \tag{19}$$

where $\boldsymbol{\beta}_c = \mathrm{diag}(\beta_{c,1}, \beta_{c,2}, \ldots, \beta_{c,N})$ is $N \times N$ diagonal matrix which include unperturbed constant components of propagation constants $\beta_{c,n} = (2\pi/\lambda)n_{\mathrm{eff},c,n}$, $\boldsymbol{\beta}_v = \mathrm{diag}(\beta_{v,1}, \beta_{v,2}, \ldots, \beta_{v,N})$ includes the variable components of the propagation constants, $\boldsymbol{\beta}_b = \mathrm{diag}(\beta_{b,1}, \beta_{b,2}, \ldots, \beta_{b,N})$ is including low-spatial-frequency perturbation $\beta_{b,n}$ of the propagation constants which can be induced by the macrobends and twists of the fiber, and $\boldsymbol{\beta}_h = \mathrm{diag}(\beta_{h,1}, \beta_{h,2}, \ldots, \beta_{h,N})$ is including high-spatial-frequency perturbation $\beta_{h,n}$ of the propagation constants which can be induced by the longitudinal structural fluctuation and the microbends of fiber. In this chapter, the effective index corresponding to β_x is expressed as $n_{\mathrm{eff},x}$, like $\beta_{c,n} = (2\pi/\lambda)n_{\mathrm{eff},c,n}$. Figure 4 shows the schematics of perturbations on β – expressed in n_{eff}. As shown in Fig. 4a, b, the bend and the structure fluctuation can induce a slight change in β_v in one core, which can be the causes of β_h. Between two cores, as shown in Fig. 4c, the bend can induce relatively large β_v in a core by taking the other core as a reference, which can be the cause of β_b.

Now, the β_b-dominant case is considered – β_h is ignored. A bent fiber with the RIP $n(r, \theta)$ can be described as a corresponding straight fiber with the equivalent refractive index (Marcuse 1982):

$$n_{eq}(r, \theta) \approx n(r, \theta)\left(1 + \frac{r\cos\theta}{R_b}\right), \tag{20}$$

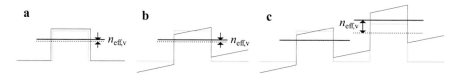

Fig. 4 (a) Slight fluctuations in the refractive index contrast and/or core diameter can induce n_{eff} fluctuation (corresponding to β_h). (**b, c**) Fiber bends can induce the perturbation on the RIPs, which can be regarded as a tilt of the RIP. The RIP tilt can induce $n_{\mathrm{eff,eq}}$ change (**b**) within a core and (**c**) between cores. In MCFs, β change between cores is a dominant perturbation from the bends. Narrow and thick lines represent RIP and effective index $n_{\mathrm{eff,eq}} = \beta_{eq}/(2\pi/\lambda)$

where (r, θ) is the local polar coordinate in the cross section of the fiber, $\theta = 0$ in radial direction of the bend, and R_b is the bending radius of the MCF. One can replace R_b with the effective bend radius $R_{b,\text{eff}}$ to account for photoelastic effect due to bend-induced stress. According to Eq. 20, the equivalent effective index $n_{\text{eff,eq},n}$ of Core n can be represented as

$$n_{\text{eff,eq},n} \approx n_{\text{eff},c,n}\left(1 + \frac{r_n \cos\theta_n}{R_b}\right), \quad (21)$$

and $\beta_{\text{eq},n}$ and $\beta_{b,n}$ can be represented as

$$\beta_{\text{eq},n} \approx \beta_{c,n} + \beta_{b,n} \approx \beta_{c,n} + \beta_{c,n}\frac{r_n \cos\theta_n}{R_b}. \quad (22)$$

Now, $|E\rangle$ can be expressed as

$$|E\rangle = \exp\left(-j\int_0^z \boldsymbol{\beta}_{\text{eq}} dz\right)|A\rangle, \quad (23)$$

thus Eq. 4 can be rewritten as

$$\frac{d}{dz}|A\rangle = -j\exp\left(j\int_0^z \boldsymbol{\beta}_{\text{eq}} dz\right)\boldsymbol{\kappa}\exp\left(-j\int_0^z \boldsymbol{\beta}_{\text{eq}} dz\right)|A\rangle. \quad (24)$$

A component of Eq. 24 is represented as

$$\frac{d}{dz}A_n = \sum_m -j\kappa_{nm}\exp\left[-j\int_0^z \left(\beta_{\text{eq},m} - \beta_{\text{eq},n}\right)dz\right]A_m. \quad (25)$$

When β_{eq} is constant, Eq. 25 is reduced to Eq. 5. However, as already mentioned above, β_b and β_h of actual MCF are not negligible and fluctuated along the longitudinal direction of the MCF, and the XT in the actual MCF cannot be predicted using the simple unperturbed coupled-mode equation in Eq. 5 (Takenaga et al. 2010, 2011a; Fini et al. 2010a, b; Hayashi et al. 2010).

For example, Fig. 6 shows an example of the longitudinal evolution of the coupled power or $|A_n|^2$ in a MCF, simulated using Eqs. 22 and 25. The MCF was bent at constant R_b and twisted continuously at a constant twist rate of two turns/m. Discrete-like dominant changes (bend-induced resonant couplings) are observed at every phase-matching point (PMP) where the difference in β_{eq} between the cores equals zero, and the XT changes in the other positions can be regarded just as local fluctuations. Even if the propagation constants are slightly different between the cores, such resonant coupling can occur in a small bending radius (Fini et al. 2010a, b; Hayashi et al. 2010). By using the local coordinate shown in Fig. 5, the threshold bending radius R_{pk}, whether the phase matching can be induced by the bend or not, is derived as

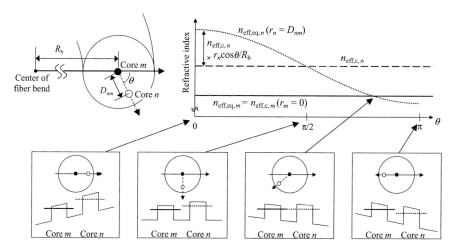

Fig. 5 Fiber bends and twists induce the variation of equivalent effective refractive index ($n_{\text{eff,eq}}$) from intrinsic (constant) effective index ($n_{\text{eff,c}}$). The center of Core m is taken as the origin of the local coordinate for simple description

$$|\beta_{c,n} - \beta_{c,m}| = \beta_{c,n} \frac{D_{nm}}{R_{\text{pk}}},$$

$$R_{\text{pk}} = \frac{D_{nm}\beta_{c,n}}{|\beta_{c,n} - \beta_{c,m}|} = \frac{D_{nm} n_{\text{eff,c},n}}{|n_{\text{eff,c},n} - n_{\text{eff,c},m}|}, \quad (26)$$

where D_{nm} is the core-to-core distance between Cores m and n or the distance between the centers of the cores. The dominant changes appear random, because the phase differences between Cores m and n are different for every PMP. The phase differences can easily fluctuate in practice by slight variations of the perturbations. Therefore, the XT can be understood as a practically stochastic parameter.

Discrete Coupling Model and Statistical Distribution of the Crosstalk

When the PMPs are discretized by the bends and twists, such as in the case of Fig. 6, the longitudinal evolution of the XT X_{nm} from Core m to Core n can be approximated by approximating the coupled-mode equation as the discrete changes (Hayashi et al. 2011a):

$$A_n(N_{\text{PM}}) \approx A_n(N_{\text{PM}} - 1) - j \chi_{nm}(N_{\text{PM}}) \exp\left[-j \varphi_{\text{rnd}}(N_{\text{PM}})\right] A_m(N_{\text{PM}} - 1)$$
$$\approx A_n(0) - j \sum_{l=1}^{N_{\text{PM}}} \chi_{nm}(l) \exp\left[-j \varphi_{\text{rnd}}(l)\right] A_m(l-1). \quad (27)$$

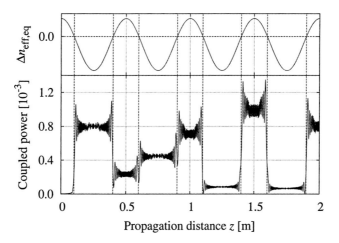

Fig. 6 An example of longitudinal evolution of coupled power in a bent and twisted MCF (Replotted from the data in Hayashi et al. (2010))

where $A_n(N_{PM})$ represents the complex amplitude of Core n after N_{PM}-th PMP, $\phi_{rnd}(N_{PM})$ is the phase difference between Cores m and n at N_{PM}-th PMP, and χ_{nm} is the coefficient for the discrete changes caused by the coupling from Core m to Core n. ϕ_{rnd} can be regarded as a random sequence, and ϕ_{rnd} varies with wavelength and with time variations of the bend, twist, and other perturbations. In the special case of the homogeneous MCF with constant bend and twist, χ_{nm} can be approximated as (Hayashi et al. 2011a)

$$\chi_{nm} \cong \sqrt{\frac{\kappa_{nm}^2}{\beta_c} \frac{R_b}{D_{nm}} \frac{2\pi}{\omega_{twist}}} \exp\left[-j\left(\frac{\beta_c D_{nm}}{\omega_{twist} R_b} - \frac{\pi}{4}\right)\right]. \tag{28}$$

When the XT is adequately low ($|A_n(N_{PM})| \ll 1$) and $A_m(0) = 1$, Eq. 27 is simplified as

$$A_n(N_{PM}) \approx -j \sum_{l=1}^{N_{PM}} \chi_{nm} \exp(-j\varphi_{rnd,l}), \tag{29}$$

and the XT X_{nm} can be approximated as $|A_n(N_{PM})|^2$.

Since $\Re[\chi_{nm}\exp(j\phi_{rnd})]$ and $\Im[\chi_{nm}\exp(j\phi_{rnd})]$ – the I–Q components of each discrete coupling – have the variance $\sigma_{2df,nm}^2$ of $|\chi_{nm}|^2/2$, probability density functions (pdf) of $\Re A_n(N)$ and $\Im A_n(N)$ converge to Gaussian distributions whose variance $\sigma_{2df,nm}^2$ is $N_{PM}|\chi_{nm}|^2/2$ if N_{PM} is adequately large, based on the central limit theorem. When assuming random polarization mode coupling, coupled power can be equally distributed to two polarization modes statistically. Therefore, pdf's of $\Re A_n(N_{PM})$'s and $\Im A_n(N_{PM})$'s of two polarization modes converge to the Gaussian distribution with the variance $\sigma_{\text{4df},nm}^2$ of $N_{\text{PM}}|\chi_{nm}|^2/4$. In this case, the XT X can be represented as a sum of powers of $\Re A_n(N_{\text{PM}})$'s and $\Im A_n(N_{\text{PM}})$'s of two polarization modes. Since the sum of powers of four normally distributed random numbers divided by their own variances is chi-squared distributed with 4 degrees of freedom (df):

$$f_{\chi^2,\text{4df}}(x) = \frac{x}{4}\exp\left(-\frac{x}{2}\right), \qquad (30)$$

$X_{nm}/\sigma_{\text{4df},nm}^2$ can be chi-square distributed with 4 df, and its statistical average $\mu_{X,nm}/\sigma_{\text{4df},nm}^2$ is 4. Therefore, the pdf and the statistical average $\mu_{X,nm}$ of X_{nm} can be obtained as

$$f_{X,\text{4df}}(X_{nm}) = f_{\chi^2,\text{4df}}\left(\frac{X_{nm}}{\sigma_{\text{4df},nm}^2}\right)\frac{d}{dX}\left(\frac{X_{nm}}{\sigma_{\text{4df},nm}^2}\right) = \frac{X_{nm}}{4\sigma_{\text{4df},nm}^4}\exp\left(-\frac{X_{nm}}{2\sigma_{\text{4df},nm}^2}\right), \qquad (31)$$

$$\mu_{X,nm} = 4\sigma_{\text{4df},nm}^2 = N_{\text{PM}}|\chi_{nm}|^2. \qquad (32)$$

In the case of the homogeneous MCF ($\Delta\beta_{c,nm} = 0$) with constant bend and twist, the mean of the statistical distribution of the XT from Core m to Core n can be derived as (Hayashi et al. 2011a)

$$\mu_{X,nm} \approx \frac{2\kappa_{nm}^2 R_b L}{\beta_{c,n} D_{nm}}, \qquad (33)$$

which corresponds to Eq. 60.

In actual uncoupled MCFs, the phase difference ϕ_{rnd} between cores varies with wavelength and time, as mentioned above. Figure 7 shows an example of the XT spectrum of the actual MCF and the probability distribution of the XT obtained from the spectrum. The measured XT distribution is well fitted by the theoretical fitting line of Eq. 31. The fringes in the XT spectrum are the interference fringes of the coupled lights from many PMPs; therefore, the inverse Fourier transform can extract the information of the skew (differential group delay between cores) accumulated between PMPs – the fringe period becomes shorter when the skew becomes larger. The relationship among the XT parameters, distribution, and time/wavelength is summarized in Fig. 8.

Now, the XT for a modulated signal can be considered based on the XT behavior, which is discussed above and summarized in Fig. 8. From the short-period fringes of the XT spectrum shown in Fig. 7a, and from the Gaussian distribution on the in-phase and quadrature (I–Q) plane, XT may behave as additive Gaussian-distributed noise; when the bandwidth of the signal light is adequately broad, the instantaneous frequency of the modulated signal light may rapidly vary with time over the broad bandwidth of the XT spectrum. In this case, the short-term average XT (STAXT, X averaged over a short term) equals to the statistical average of the

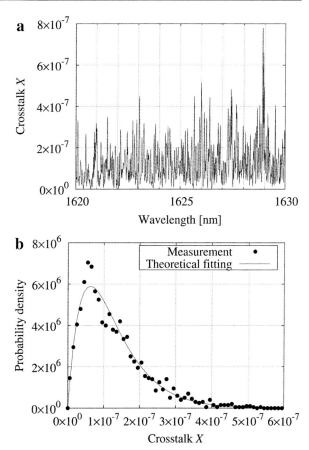

Fig. 7 Examples of (**a**) a crosstalk spectrum of a 17-km-long homogeneous MCF and (**b**) a probability distribution of the crosstalk obtained from the spectrum (Replotted from the data in Hayashi et al. (2011b, 2012))

XT, because the modulated signal light averages the wavelength dependence of X. When the bandwidth becomes narrower, the XT distribution on the I–Q plane can become more distorted from the Gaussian distribution. Moreover, if the bandwidth is adequately narrow, the XT may behave like a static coupling, since the XT spectrum variation is very gradual compared to the symbol rate in the order of picoseconds to nanoseconds. However, this static coupling may gradually vary with time, and the STAXT varies with time and obeys the chi-square distribution in the power – or the normal distribution in the I–Q plane – as shown in Fig. 8.

Rademacher et al. investigated such an XT behavior from the relationship between the variance ($\mathrm{Var_{XT}}$) of the STAXT and the modulated signal bandwidth for the on-off keying (OOK) case and the quadrature phase-shift keying (QPSK) case, as shown in Fig. 9, by numerically solving the extension of Eq. 29 for a skew and dual polarizations (Rademacher et al. 2017a):

$$A_n(L,\omega) \approx -j \sum_{l=1}^{N_{\mathrm{PM}}} \chi_{nm,l} \mathbf{R}_l \begin{bmatrix} \exp[-j\varphi_{\mathrm{rnd},l} - j\tau_{nm,l}\omega] \\ \exp[-j\varphi_{\mathrm{rnd},l} - j\tau_{nm,l}\omega] \end{bmatrix}. \tag{34}$$

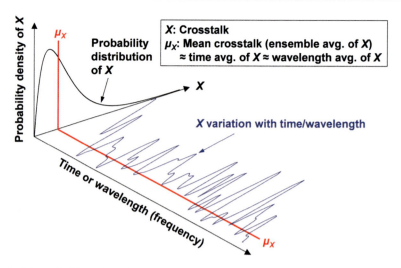

Fig. 8 Schematic illustration of the parameters related to the crosstalk (Reused from (Hayashi et al. 2014) ©2014 IEICE. Permission number: 17RA0069)

where $\mathbf{R}_l$ is a random unitary 2×2 matrix simulating random polarization rotations between the PMPs and τ_{nm} is the differential group delay between Core n and Core m at l-th PMP. In Rademacher et al. (2017a), τ_{nm} was assumed to be proportional to the fiber longitudinal position, and the linear birefringence was ignored. According to Fig. 9, the variance of the STAXT is high when the skew and symbol rate product is low and low when the skew and symbol rate product is high. In the case of QPSK modulation, Var_{XT} converges to zero when the skew and symbol rate product is high. However, in the case of OOK modulation, Var_{XT} never converge to zero even when the skew and symbol rate product is very high. This can be understood from the spectrum of the modulated signal lights, shown in the bottom of Fig. 9. The spectrum of the QPSK signal light has a flat-top-like shape; therefore, the XT of the signal light is well averaged over the signal band. On the other hand, the spectrum of the OOK signal light has a strong and sharp peak at the center of the spectrum, which comes from the carrier frequency of the OOK signal; therefore, the XT of the signal light is not well averaged over the signal band. So, the modulation formats with a strong carrier component – such as OOK, pulse amplitude modulation (PAM), etc. – require some margin from the statistical average XT for the XT suppression, because of the STAXT variation. These XT behaviors were also experimentally confirmed (Rademacher et al. 2017a).

Coupled Power Theory for Predicting the Statistical Mean of the Crosstalk

The mean XT can be considered using the coupled power theory. Since the perturbations β_b and β_s behave as random processes, coupling in the MCF is

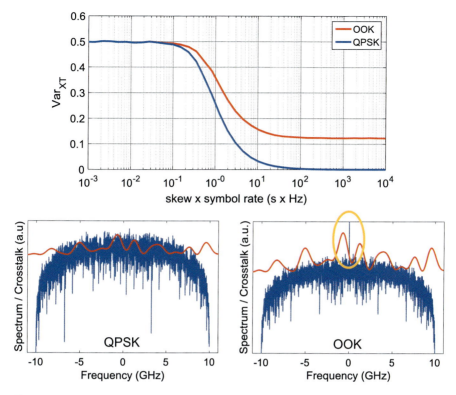

Fig. 9 (Top) The variance of the short-term averaged inter-core crosstalk of modulated signals (Reused from Rademacher et al. (2017a) ©OSA 2017), and (bottom) schematic spectra of signal and crosstalk for the OOK and QPSK modulations (Reused from the presentation material of Rademacher et al. (2017b) by courtesy of the authors)

described using power-coupled equations (Takenaga et al. 2010, 2011a):

$$\frac{dP_n}{dz} = \sum_{m \neq n} h_{nm} [P_m - P_n], \tag{35}$$

where P_n is the average power in Core n, h_{nm} is the power coupling coefficient from Core m to Core n, and z is the longitudinal position of the MCF. The XT – that is, average/mean XT, or statistical average/mean μ_X of XT X, to be exact – can be defined in some ways, but a widely used definition of the average XT from Core m to Core n is

$$\mu_{X,nm} \equiv \frac{P_{nm}}{P_{mm}}, \tag{36}$$

where P_{nm} is the output power of Core n when only Core m is excited. The average XT to Core n also can be defined as

$$\mu_{X,n} \equiv \frac{\sum_{m \neq n} P_{nm}}{P_{nn}}, \tag{37}$$

which may correspond to the XT-to-signal ratio. In case that μ_X is small enough, or $P_m - P_n \approx P_m|_{z=0}$ can be assumed in Eq. 35, Eq. 36 can be approximated as

$$\mu_{X,nm} \approx \int_0^L h_{nm} dz, \tag{38}$$

and Eq. 37 as

$$\mu_{X,n} \approx \int_0^L \sum_{m \neq n} h_{nm} dz, \tag{39}$$

where L is the fiber length. In case that h_{nm} is constant or averaged along the MCF, the XT accumulation can be regarded as linear accumulation ($+10$ dB/decade).

In the following subsections, the characteristics of the XT in the MCF are described by starting from the coupled-mode equation with perturbed propagation constants, based on the references. The equations in this chapter may be different from those in the references because of the differences of definitions of symbols and functions.

Local Power Coupling Coefficient Under High-Spatial-Frequency Perturbations Without Bend Radius Change and Fiber Twisting

At first, the power coupling coefficient under the effect of high-spatial-frequency perturbations – e.g., the structure fluctuation and the microbends – is discussed. The power coupling coefficient can be derived from the coupled-mode equation with longitudinally perturbed propagation constants. For the sake of the derivation simplicity, the low-spatial-frequency perturbations change the propagation constants gradually enough to regard that they are constant. Then, the coupled-mode equation can be expressed as

$$\frac{dA_n}{dz} = -j\kappa_{nm} \exp\left[-j\left(\beta_{c+b,m} - \beta_{c+b,n}\right) z - j \int_0^z (\beta_{h,m} - \beta_{h,n}) dz\right] A_m, \tag{40}$$

where

$$\beta_{c+b,n} = \beta_{c,n} + \beta_{b,n} = \beta_{c,n} \left(1 + \frac{r_n \cos \theta_n}{R_b}\right). \tag{41}$$

3 Multi-core Fibers for Space Division Multiplexing

Based on Eq. 40, in the case of low XT, the XT in amplitude within the fiber segment $[z_1, z_2]$ can be expressed as

$$\Delta x_{nm} = \frac{\Delta A_n}{A_m} \approx -j\kappa_{nm} \int_{z_1}^{z_2} \exp\left[-j\left(\beta_{c+b,m} - \beta_{c+b,n}\right)z\right] f(z)dz, \qquad (42)$$

by using

$$f(z) \equiv \exp\left[-j \int_0^z \left[\beta_{h,m}(z') - \beta_{h,n}(z')\right] dz'\right]. \qquad (43)$$

Accordingly, the mean XT increase in power within the segment $[z_1, z_2]$ can be expressed as

$$\begin{aligned}
\langle \Delta X_{nm} \rangle &= \left\langle |\Delta x_{nm}|^2 \right\rangle = \langle \Delta x_{nm} \Delta x_{nm}^* \rangle \\
&\approx \kappa_{nm}^2 \int_{z_1}^{z_2}\int_{z_1}^{z_2} \exp\left[-j\left(\beta_{c+b,m} - \beta_{c+1,n}\right)(z-z')\right] \langle f(z)f^*(z')\rangle dz dz' \\
&\approx \kappa_{nm}^2 \int_{z_1}^{z_2}\int_{z_1-z'}^{z_2-z'} \exp\left[j\Delta\beta_{c+b,nm}\zeta\right] \langle f(z'+\zeta)f^*(z')\rangle d\zeta dz' \\
&\approx \kappa_{nm}^2 \int_{z_1}^{z_2} dz' \int_{-\infty}^{\infty} R_{ff}(\zeta)\exp(j\Delta\beta_{c+b,nm}\zeta)d\zeta \\
&\approx \kappa_{nm}^2 \Delta z \int_{-\infty}^{\infty} R_{ff}(\zeta)\exp(j\Delta\beta_{c+b,nm}\zeta)d\zeta,
\end{aligned} \qquad (44)$$

where R_{ff} is the autocorrelation function (ACF) of $f(z)$, Δz is $z_2 - z_1$, and the correlation length l_c of R_{ff} is assumed to be adequately shorter than Δz. $R_{ff}(\zeta)$ can be understood as the correlation between the coupled and non-coupled lights that are propagated for the length of ζ after the coupling. For example, where $\zeta \gg l_c$, the coupled and non-coupled lights lose their coherency even if the lights are very coherent. Based on the Wiener-Khinchin theorem, the power spectrum density (PSD) is the Fourier transform of the ACF:

$$S_{ff}^{(\tilde{\nu})}(\tilde{\nu}) = \int_{-\infty}^{\infty} R_{ff}(\zeta)\exp(j2\pi\tilde{\nu}\zeta)d\zeta, \qquad (45)$$

where $\tilde{\nu} = n_{\text{eff}}/\lambda = \beta/(2\pi)$ represents the wave number (or spatial frequency) in the medium. Note that $\tilde{\nu}$, n_{eff}, and β have common subscripts, e.g., $\tilde{\nu}_{c+b} = n_{\text{eff},c+b}/\lambda = \beta_{c+b}/(2\pi)$. To describe the PSDs with respect to $\tilde{\nu}$ and β with common expressions, we would like to define the PSD with respect to β, whose total power is equivalent to Eq. 45. From Parseval's theorem, the average power of $f(z)$, or expected value of $|f(z)|^2$, is equivalent to the integral of the PSD over whole $\tilde{\nu}$, and the following equation holds between $f(z)$ and the PSDs of $f(z)$:

$$\int_{-\infty}^{\infty} S_{ff}^{(\tilde{\nu})}(\tilde{\nu}) \, d\tilde{\nu} = \int_{-\infty}^{\infty} S_{ff}^{(\tilde{\nu})}(\tilde{\nu}) \frac{d\tilde{\nu}}{d\beta} d\beta = \mathrm{E}\left[|f(z)|^2\right] = 1, \qquad (46)$$

where $\mathrm{E}[\cdot]$ represents the expected value. Therefore, in this paper, the PSD $S_{ff}^{(\beta)}(\beta)$ with the scale of the propagation constant β (the angular wave number in the medium) is defined as

$$S_{ff}^{(\beta)}(\beta) \equiv S_{ff}^{(\tilde{\nu})}(\tilde{\nu}) \frac{d\tilde{\nu}}{d\beta} = \frac{1}{2\pi} S_{ff}^{(\tilde{\nu})}(\tilde{\nu}) = \frac{1}{2\pi} \int_{-\infty}^{\infty} R_{ff}(\zeta) \exp(j\beta\zeta) \, d\zeta, \qquad (47)$$

From Eqs. 44 and 47, the power coupling coefficient can be expressed as

$$h_{nm} = \frac{\langle \Delta X_{nm} \rangle}{\Delta z} \approx \kappa_{nm}^2 S_{ff}^{(\tilde{\nu})} \left(\frac{\Delta n_{\text{eff},c+b,nm}}{\lambda}\right) = \kappa_{nm}^2 \left[2\pi S_{ff}^{(\beta)}(\Delta\beta_{c+b,nm})\right], \qquad (48)$$

It is not easy to assume a proper β_h, R_{ff}, and S_{ff} from theoretical consideration. However, Koshiba et al. reported that the mean XT of actual MCFs can be well predicted using Eqs. 35 and 48 by assuming the exponential ACF (Koshiba et al. 2011):

$$R_{ff}(\zeta) = \exp\left(-\frac{|\zeta|}{l_c}\right). \qquad (49)$$

The exponential ACF has been introduced to microbend loss analysis. Since the PSD of the exponential ACF is the Lorentzian distribution, the local power coupling coefficient is obtained from Eq. 48 as (Koshiba et al. 2012)

$$\begin{aligned} h_{nm}(z) &= \kappa_{nm}^2 \frac{1}{\pi} \frac{1/(2\pi l_{\text{cor}})}{1/(2\pi l_c)^2 + \left[\Delta n'_{\text{eff},c+b,nm}(z)/\lambda\right]^2} \\ &= \kappa_{nm}^2 2\pi \frac{1}{\pi} \frac{1/l_{\text{cor}}}{1/l_{\text{cor}}^2 + \left[\Delta\beta_{c+b,nm}(z)\right]^2}, \end{aligned} \qquad (50)$$

The mean XT μ_X estimated using the coupled power equation with the power coupling coefficient of Eq. 50 may be valid in the cases where changes of R_b and θ

are gradual enough compared to l_c, since $\Delta\beta_{c+b,nm}$ was assumed as a constant for the derivation.

Power Coupling Coefficient Averaged Over Fiber Twisting
By using Eq. 50, the power coupling coefficient can be averaged over fiber twisting. Now, $\Delta\beta_{c,nm}$ can be written as

$$\Delta\beta_{c+b,nm}(R_b, \theta_{nm}) = \Delta\beta_{c,nm} + \Delta\beta_{b,nm}(R_b, \theta_{nm}), \tag{51}$$

$$\Delta\beta_{b,nm}(R_b, \theta_{nm}) = \Delta\beta_{b,nm}^{dev}(R_b)\cos\theta_{nm}, \tag{52}$$

$$\Delta\beta_{b,nm}^{dev}(R_b) = \beta_{c,n}\frac{D_{nm}}{R_b}, \tag{53}$$

where θ_{nm} represents the angle between the radial direction of the bend and a line segment from Core m to Core n, $\Delta\beta_{b,nm}$ is the peak deviation of $\beta_{b,nm}$, and D_{nm} is the center-to-center distance between Core m and Core n.

Let $\Pr(\theta_{nm})$ and $\Pr(R_b)$ be the probability density functions of θ_{nm} and of R_b, respectively, along the MCF; by assuming that $\Pr(\theta_{nm})$ and $\Pr(R_b)$ are statistically independent, the twist of the MCF is gradual enough, and mean XT is adequately low; the mean XT $\mu_{X,nm}$ from Core m to Core n can be expressed as

$$\begin{aligned}\mu_{X,nm}(L) &\approx \int_0^L h_{nm}(z)dz \approx L\left[\frac{1}{L}\int_0^L h_{nm}(z)dz\right] \approx L\,\mathrm{E}[h_{nm}] \\ &\approx L\int_0^\infty \Pr(R_b)\Pr(\theta_{nm})h_{nm}(R_b,\theta_{nm})\,d\theta_{nm}dR_b,\end{aligned} \tag{54}$$

where the average power coupling coefficient over θ can be defined as

$$\overline{h}_{nm}(R_b) = \int_0^{2\pi} \Pr(\theta_{nm})h_{nm}(R_b,\theta_{nm})\,d\theta_{nm}, \tag{55}$$

and $\Pr(\theta_{nm}) = 1/(2\pi)$ can hold by assuming that the twist of the MCF is random enough and the MCF is adequately long. By substituting $\Delta\beta_{b,nm} = \Delta\beta_{b,nm}^{dev}\cos\theta_{nm}$ and using Eq. 48 and $\sin\theta_{nm} = \sin\left[\arccos\left(\Delta\beta_{b,nm}/\Delta\beta_{b,nm}^{dev}\right)\right] = \sqrt{1-\left(\Delta\beta_{b,nm}/\Delta\beta_{b,nm}^{dev}\right)^2}$, Eq. 55 can be rewritten as

$$\bar{h}(R_b) = \int_0^{2\pi} \frac{1}{2\pi} h(R_b, \theta) \, d\theta = \int_0^{2\pi} \frac{1}{2\pi} \kappa^2 2\pi S_{ff}^{(\beta)} [\Delta\beta_{c+b}(R_b, \theta)] \, d\theta$$

$$= 2\kappa^2 \int_0^{\pi} S_{ff}^{(\beta)} \left[\Delta\beta_c + \Delta\beta_b^{dev}(R_b) \cos\theta \right] d\theta$$

$$= 2\pi\kappa^2 \int_{-\Delta\beta_{b,nm}^{dev}}^{\Delta\beta_{b,nm}^{dev}} \frac{1}{\pi \sqrt{\left[\Delta\beta_b^{dev}(R_b)\right]^2 - \Delta\beta_b^2}} S_{ff}^{(\beta)}(\Delta\beta_c + \Delta\beta_b) \, d(\Delta\beta_b),$$

(56)

where S_{ff} is the Lorentzian distribution as shown in Eq. 50. By using the arcsine distribution,

$$\text{ASD}(a, b) = \begin{cases} \frac{1}{\pi \sqrt{a^2 - b^2}}, & |a| \geq |b|, \\ 0, & \text{otherwise,} \end{cases}$$

(57)

Equation 56 can be rewritten as

$$\bar{h}_{nm}(\Delta\beta_{c,nm}, R_b)$$

$$= \kappa_{nm}^2 2\pi \left[\text{ASD}\left(\Delta\beta_{b,nm}^{dev}, \Delta\beta_{b,nm}\right) * S_{ff}^{(\beta)}(\Delta\beta_{c,nm} + \Delta\beta_{b,nm}) \right]_{\Delta\beta_b} \quad (58)$$

$$= \kappa_{nm}^2 \left[\text{ASD}\left(\Delta\tilde{v}_{b,nm}^{dev}, \Delta\tilde{v}_{b,nm}\right) * S_{ff}^{(\tilde{v})}(\Delta\tilde{v}_{c,nm} + \Delta\tilde{v}_{b,nm}) \right]_{\Delta\tilde{v}_b},$$

where the expression of $(f * g)_x$ denotes the convolution of f and g with respect to x and the expression with respect to $\tilde{v}$ is also shown for comparison. Here, $\text{ASD}\left(\Delta\beta_{b,nm}^{dev}, \Delta\beta_{b,nm}\right)$ is the probability distribution of $\Delta\beta_b$. By regarding that PSD S_{ff} in Eq. 48 includes both the effects of the structure fluctuation and the macrobend, the convolution term in Eq. 58 may be understood as the PSD S_{ff} in Eq. 48.

Particularly where $|\Delta\beta_{c,nm}|$ and the bandwidth of $S_{ff}^{(\beta)}$ are adequately smaller than $\Delta\beta_{b,nm}^{dev}$, S_{ff} becomes a narrow delta-function-like distribution, and the convolution contains only a gradually varying part of $\Pr(\Delta\beta_{b,nm})$; therefore, Eq. 58 can be approximated as

$$\bar{h}_{nm}(\Delta\beta_{c,nm}, R_b)$$

$$\approx \kappa_{nm}^2 \left[2\pi \text{ASD}\left(\Delta\beta_{b,nm}^{dev}, \Delta\beta_{c,nm}\right) \right] = \kappa_{nm}^2 \frac{2}{\sqrt{(\beta_{c,n} D_{nm}/R_b)^2 - \Delta\beta_{c,nm}^2}} \quad (59)$$

$$\approx \kappa_{nm}^2 \text{ASD}\left(\Delta\tilde{v}_{b,nm}^{dev}, \Delta\tilde{v}_{c,nm}\right) = \kappa_{nm}^2 \frac{\lambda}{\pi \sqrt{(n_{\text{eff},c,n} D_{nm}/R_b)^2 - \Delta n_{\text{eff},c,nm}^2}},$$

which is also obtained from Eq. 48 by approximating the PSD $S_{ff}^{(\beta)}$ as the probability distribution of $\Delta\beta_b$ – ASD $\left(\Delta\beta_{b,nm}^{dev}, \Delta\beta_{b,nm}\right)$ – with constant R_b. In case of homogeneous MCFs ($\Delta\beta_{c,nm} = 0$), Eq. 59 is reduced to

$$\bar{h}_{nm}(R_b) \approx \kappa_{nm}^2 \frac{2R_b}{\beta_{c,n}D_{nm}} = \kappa_{nm}^2 \frac{\lambda R_b}{\pi n_{\text{eff},c,n}D_{nm}}, \tag{60}$$

which coincides with Eq. 33. The difference between Eqs. 59 and 60 is less than 0.1 dB when $\Delta\beta_c < 0.21\beta_c D/R_b$; therefore, Eq. 60 may also be used for estimating the XT of a bent heterogeneous MCF with small $\Delta\beta_c$.

The analytical solution of Eq. 58 was derived by assuming constant R_b and twist rate as (Koshiba et al. 2012)

$$\bar{h}_{nm}(\Delta\beta_{c,nm}, R_b) \approx \sqrt{2}\kappa_{nm}^2 l_{cor}\left[1/\sqrt{a\left(b + \sqrt{ac}\right)} + 1/\sqrt{c\left(b + \sqrt{ac}\right)}\right] \tag{61}$$

$$a = 1 + (\Delta\beta_{c,nm}l_c - B_{nm}l_c/R_b)^2 \approx 1 + l_c(\Delta\beta_{c,nm}l_c - \beta_{c,n}D_{nm}l_c/R_b)^2, \tag{62}$$

$$b = 1 + (\Delta\beta_{c,nm}l_c)^2 - (B_{nm}l_c/R_b)^2 \approx 1 + (\Delta\beta_{c,nm}l_c)^2 - (\beta_{c,n}D_{nm}l_c/R_b)^2, \tag{63}$$

$$c = 1 + (\Delta\beta_{c,nm}l_c + B_{nm}l_c/R_b)^2 \approx 1 + (\Delta\beta_{c,nm}l_c + \beta_{c,n}D_{nm}l_c/R_b)^2, \tag{64}$$

$$B_{nm} = \sqrt{(\beta_{c,n}x_n - \beta_{c,m}x_m)^2 + (\beta_{c,n}y_n - \beta_{c,m}y_m)^2}, \tag{65}$$

where B_{nm} can be approximated as $\beta_{c,n}D_{nm}$ if $\beta_{c,m}/\beta_{c,n} \simeq 1$.

Figure 10 shows comparisons between $\bar{h}$ calculated by using Eq. 58 and $\bar{h}$ calculated by using Eqs. 61, 62, 63, 64, and 65. Figure 10a, b shows the PSDs normalized with respect to the Lorentzian S_{ff} and to the arcsine distribution ASD($\Delta\beta_b^{dev}$, $\Delta\beta_c$), respectively. The Lorentzian and arcsine distributions represent the spectra of the perturbations induced by the structure fluctuation and by the macrobend, respectively. Solid lines represent $\bar{h}$ calculated by using Eq. 58, and dashed lines represent $\bar{h}$ calculated by using Eqs. 61, 62, 63, 64, and 65; however, the solid lines and the dashed lines are overlapped, and we can only see the solid lines. Accordingly, it was clearly confirmed that Eq. 58 is equivalent to the expression of $\bar{h}$ with Eqs. 61, 62, 63, 64, and 65, and it can be also said that the set of Eqs. 61, 62, 63, 64, and 65 is a closed-form solution of the convolution of the Lorentzian and the arcsine distribution. It is difficult to interpret physical meaning of Eqs. 61, 62, 63, 64, and 65 intuitively, but this closed-form expression is powerful and easy to estimate the mean XT.

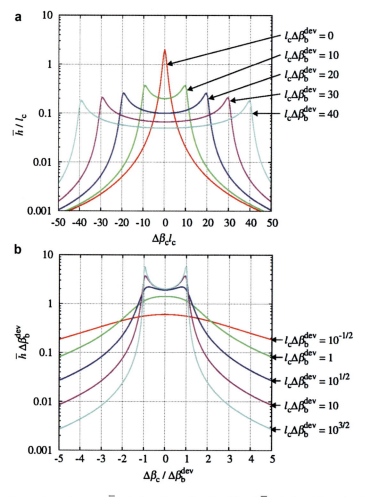

Fig. 10 Comparisons between $\bar{h}$ calculated by using Eq. 58 and $\bar{h}$ calculated by using Eqs. 61, 62, 63, 64, and 65. (a) $\bar{h}$ normalized with respect to the Lorentzian, (b) $\bar{h}$ normalized with respect to the arcsine distribution. Solid lines, $\bar{h}$ calculated by using Eq. 58; dashed lines, $\bar{h}$ calculated by using Eqs. 61, 62, 63, 64, and 65. The solid lines and the dashed lines are overlapped (Reused from Hayashi et al. (2013). ©2012 OSA)

Crosstalk Suppression Strategy

Based on the above descriptions, it can be understood that the XT is proportional to the power of the mode coupling coefficient and to the PSD of the perturbations, which can be intuitively explained as the amount of the phase matching. Accordingly, various XT suppression methods have been proposed and demonstrated by ways of reductions of these parameters. In this subsection, the methods for suppressing the mode coupling coefficient and the phase matching are described.

Suppression of the Mode Coupling Coefficient

The mode coupling coefficient can be suppressed by reducing the overlap of the modes, so it can be suppressed by confining the modes to the cores strongly and/or enlarging the core-to-core distance. Of course, the core-to-core distance should be shortened as much as possible, in respect to the core density. Accordingly, high confinement core design is important for the suppression of the mode coupling coefficient. Generally, there is a trade-off between strong light confinement and preferable optical characteristics, such as large A_{eff}, short cutoff wavelength, and long operating wavelength. High-index and small-diameter core structure is one of the options to achieve strong light confinement, but it degrades the effective area A_{eff} and increases the nonlinear noise. Specially designed RIPs – such as trench- or hole-assisted core structures, shown in Fig. 11 – can confine the mode strongly while preserving a large A_{eff}. Photonic crystal (photonic bandgap) structures (also shown in Fig. 11) can also be leveraged for the strong mode confinement.

Another approach is adopting quasi-single-mode cores. The quasi-single-mode core is technically a few-mode core; therefore, the confinement of the fundamental mode can be higher than a single-mode core. It was originally proposed for enhancing the effective area in the single-core fiber transmissions by using only the fundamental mode of a few-mode fiber while keeping the bending loss of the fundamental mode low enough (Sui et al. 2015). When it is used for MCFs, the XT between the fundamental modes can be suppressed. Since the intra-core modal XT is difficult to be suppressed at fiber splicing, the light that coupled to the higher-order modes must be attenuated before the next splicing point. Sasaki et al. (Sasaki et al. 2016, 2017) proposed to use 1 km cutoff wavelength instead

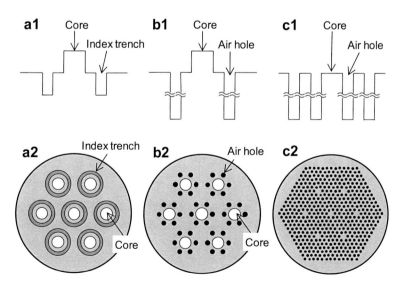

Fig. 11 Schematic examples of (1) RIPs and (2) cross sections of (**a**) trench-assisted MCF, (**b**) hole-assisted MCF, and (**c**) photonic bandgap MCF

of the conventional 22 m cutoff wavelength – i.e., the cable cutoff wavelength – because the typical minimum interval of the fiber splicing in metro trunk line optical fiber cable is said to be more than 1 km. As reported in Sui et al. (2015), the mode coupling during the propagation also induces the multipath interference (MPI) between the fundamental mode and higher-order propagation modes, even in weakly coupled few-mode fibers. So, digital signal processing (DSP) may be necessary to compensate the effect of the MPI. The power coupling from the fundamental mode to the higher-order modes can also induce the transmission loss degradation, which has to be cared for in fiber design.

Suppression of the Phase Matching

The phase-matching suppression methods can be categorized into some types according to the utilization of the perturbations. Here, four types of the suppression methods are explained in Items A–D. A schematic example of $\tilde{h}$ ($\Delta\beta_c$, R_b) in Eq. 58 for the Lorentzian S_{ff} is shown in Fig. 12 and will help with a better understanding.

Utilization of the Propagation Constant Mismatch

One is the method utilizing the propagation constant mismatch $\Delta\beta_c$ ($=(2\pi/\lambda)\Delta n_{\text{eff},c}$) to suppress the phase matching. The propagation constant mismatch $\Delta\beta_c$ which is larger than the maximum bend-induced perturbation $\Delta\beta_b^{\text{dev}}$ can prevent the bend-induced phase matching between the dissimilar cores (the non-phase-matching regions in Fig. 12). In other words, the bending radius of the MCF has to be managed to be adequately larger than R_{pk} – some margin is needed for avoiding the phase matching induced by the spectral broadening of S_{ff} due to the structural fluctuations. In the heterogeneous MCFs, it is preferred if the correlation length l_c of the structural fluctuation can be elongated, because the spectral broadening of S_{ff} can be narrowed and the PSD leakage in the non-phase-matching region can be suppressed, as shown in Figs. 10 and 12.

If most of a MCF is deployed in gentle-bend conditions, a slight difference in propagation constants or effective indices may be enough for the phase-matching suppression. If MCFs are deployed in more bend-challenged conditions, the large $\Delta n_{\text{eff},c}$ is required for avoiding the bend-induced phase matching. For example, to suppress R_{pk} lower than 10 cm, a large difference in core Δ and/or diameter is necessary for simple step-index cores. Therefore, as shown in Fig. 13, by employing various complex RIPs, R_{pk} can be suppressed while keeping the optical characteristics – especially A_{eff} – similar among the cores.

It should be noted that the early-stage research on the MCF in the 2000s had been conducted before the effects of various longitudinal perturbations were found. So, the design studies of heterogeneous MCFs in the early research stage may ignore such effects, and the actual XT of such MCFs might be degraded as reported in Hayashi et al. (2010). Also, the papers on heterogeneous MCF fabrication results may have reported the XT measured in the phase-matching region when the knowledge on the effects of the bends were not found or widely spread.

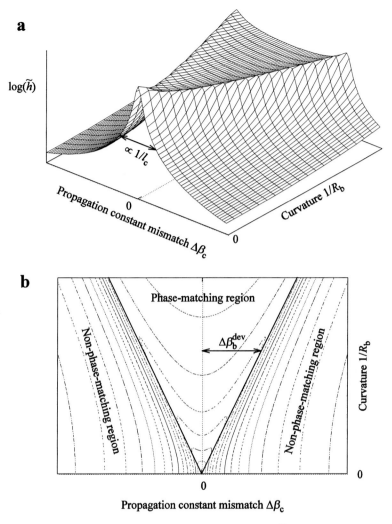

Fig. 12 A schematic example of the power coupling coefficient $\overline{h}$ averaged over the rotation direction, as a function of the propagation constant mismatch $\Delta\beta_c$ and the curvature $1/R_b$, in case that twist of a MCF is gradual and random enough. (**a**) A three-dimensional plot, (**b**) a contour map of $\log(\overline{h})$. Thick solid lines in (**b**) are the thresholds between the phase-matching region and the non-phase-matching region

Utilization of the Phase Mismatch Induced by Macrobends

The bend can also be utilized for the phase-matching suppression. As shown in Fig. 12, enlargement of the bend-induced perturbation – caused by the increase of the curvature or the decrease of the bending radius – can spread the PSD and suppress the XT even in case of a homogeneous MCF ($\Delta\beta_c = 0$); identical core structure is rather desirable for suppressing the PSD. The PSD changes gradually

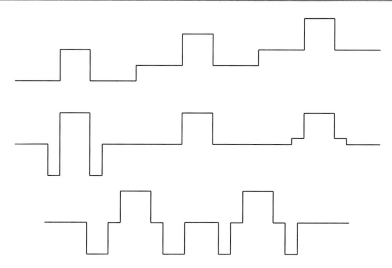

Fig. 13 Schematic RIPs of cores for enlarging propagation constant mismatch while suppressing the optical characteristic mismatch. (Upper) double-cladding structure where the RIP shapes are the same among the cores but the RIP values are offset among the cores, (middle) another type of double-cladding structure where a depressed/raised cladding modifies the propagation constant, and (bottom) trench-assisted type where the depth and width of the index trench and the separation from the core of the trench can modify the propagation constant of the core

with the bend radius, and there is no drastic PSD increase like that around R_{pk} in case of heterogeneous MCFs, since the PSD is suppressed in the phase-matching region. As already described, the average XT of a homogeneous MCF is proportional to the average bending radius. Therefore, if the average bending radius of the MCF is managed to be smaller than a certain value, or if the MCF is deployed in bend-challenged conditions, low XT can be achieved with identical cores. In actual fiber deployments, too small fiber bending radius would not be adopted from the point of view of macrobend loss and practical cable structure. Thus, of course, this method has to be employed with the mode coupling coefficient reduction like the other phase-matching reduction methods.

Utilization of the Phase Mismatch Induced by Fiber Structural Fluctuation and/or Microbends

As shown in Fig. 12, the power spectrum of the perturbations is broadened by the longitudinal structural fluctuations and/or the microbends. If a MCF has a very short correlation length l_c, the power spectrum spreads broadly over the propagation constant mismatch $\Delta \beta_c$, and thus the PSD may be suppressed even in case of an unbent homogeneous MCF ($\Delta \beta_c = 0$, $1/R_b = 0$). To the author's knowledge, the XT suppression by the structural fluctuation has not actually been observed yet, because the bend-induced perturbations are much larger than the fluctuation-induced perturbations in the measurement conditions. However, the structural fluctuation may work when the MCF is cabled and installed in very gently bent conditions. The correlation length l_c is different among various MCFs, and it is still under study what kinds of parameters determine the value of l_c and whether they are relating the propagation loss.

Coupled Multi-core Fibers

In the coupled MCF (also referred to as strongly coupled MCF, coupled core fiber, and coupled core MCF), the coupling between the cores is strong, and the one set of multiple coupled cores guides multiple spatial modes; therefore, each core cannot be used as an independent spatial channel. The coupled MCF can be further subdivided to systematically coupled MCF and randomly coupled MCF, depending on the coupling strength. These two subtypes of the coupled MCFs are often simply called "coupled MCF" or the like without distinction in much literature.

Systematically Coupled Multi-core Fiber

When the coupling between the cores is strong, the eigenmodes of the whole multi-core system can be supermodes, i.e., the superpositions of core modes, and stably propagate in the MCF without coupling. Thus, the coupling between the core modes is very systematic and deterministic. Such supermodes can be discussed with the core modes using the non-orthogonal coupled-mode theory. By defining $\mathbf{H}' = \mathbf{P}^{-1}\mathbf{H}$, Eq. 9 can be rewritten as

$$\frac{d}{dz}|E\rangle = -j\mathbf{H}'|E\rangle, \tag{66}$$

where $\mathbf{H}'$ is Hermitian for power conservation. Any Hermitian matrix can be diagonalized as

$$\mathbf{H}' = \mathbf{U}\mathbf{\Lambda}\mathbf{U}^*, \tag{67}$$

where $\mathbf{U}$ is a unitary matrix, the superscript * represents the conjugate transpose, and $\mathbf{\Lambda}$ is a diagonal matrix. The diagonal elements of $\mathbf{\Lambda}$ are the eigenvalues of $\mathbf{H}'$ and are the propagation constants of the eigenmodes. $\mathbf{U}$ is composed of the orthonormal eigenvectors of $\mathbf{H}'$ written as its column and converts the eigenmodes $\mathbf{E}_{\text{eig}}$ and the core modes $\mathbf{E}$ as

$$\begin{cases} |E\rangle = \mathbf{U}|E_{\text{eig}}\rangle, \\ |E_{\text{eig}}\rangle = \mathbf{U}^*|E\rangle. \end{cases} \tag{68}$$

The orthonormal eigenvectors are the eigenmodes represented as the superpositions of the core modes.

Using Eqs. 67 and 68, the coupled-mode equation of Eq. 66 can be reduced to

$$\frac{d}{dz}|E_{\text{eig}}\rangle = -j\Lambda |E_{\text{eig}}\rangle, \tag{69}$$

which is a simple propagation equation without coupling. So, there are no couplings between the supermodes in ideal cases. In actual cases, various longitudinal perturbations can induce mode coupling between the supermodes, but such a coupling can be suppressed by increasing the propagation constant mismatch between the supermodes.

To understand these equations in detail, a two-core case is considered for the sake of simplicity. Equation 66 can be rewritten as

$$\mathbf{H}' = \begin{bmatrix} B_1 & K \\ K & B_2 \end{bmatrix}, \tag{70}$$

$\mathbf{H}'$ can be eigendecomposed with the eigenvalues and the corresponding orthonormal eigenvectors:

$$\Lambda_1 = B_{\text{avg}} + \sqrt{\Delta^2 + K^2},$$

$$|u_1\rangle = \frac{1}{\sqrt{\left(\Delta + \sqrt{\Delta^2 + K^2}\right)^2 + K^2}} \begin{bmatrix} \Delta + \sqrt{\Delta^2 + K^2} \\ K \end{bmatrix}, \tag{71}$$

$$\Lambda_2 = B_{\text{avg}} - \sqrt{\Delta^2 + K^2},$$

$$|u_2\rangle = \frac{1}{\sqrt{\left(\Delta + \sqrt{\Delta^2 + K^2}\right)^2 + K^2}} \begin{bmatrix} -K \\ \Delta + \sqrt{\Delta^2 + K^2} \end{bmatrix}, \tag{72}$$

where $B_{\text{avg}} = (B_1 + B_2)/2$ and $\Delta = (B_1 - B_2)/2$. Here, the diagonal matrix Λ and the unitary matrix $\mathbf{U}$ can be chosen as

$$\Lambda = \begin{bmatrix} \Lambda_1 & 0 \\ 0 & \Lambda_2 \end{bmatrix}, \quad \mathbf{U} = [|u_1\rangle, |u_2\rangle] \tag{73}$$

According to Eqs. 71 and 72, when $\Delta \neq 0$, the eigenmodes are the superpositions of unequally excited core modes. When $\Delta = 0$ (identical cores, $B = B_1 = B_2$), Eqs. 71 and 72 are reduced to

$$\Lambda_1 = B + K, \quad |u_1\rangle = \frac{1}{\sqrt{2}} \begin{bmatrix} 1 \\ 1 \end{bmatrix}, \tag{74}$$

$$\Lambda_2 = B - K, \quad |u_2\rangle = \frac{1}{\sqrt{2}} \begin{bmatrix} -1 \\ 1 \end{bmatrix}, \tag{75}$$

Here, each supermode is the superposition of the two core modes with equal amplitude where even mode (Eq. 74) consists of the core modes with the same phase and the odd mode (Eq. 75) consists of the core modes with the opposite phases. The propagation constant mismatch $\Delta\Lambda$ between the supermodes is

$$\Delta\Lambda_{12} = \Lambda_1 - \Lambda_2 = 2K, \tag{76}$$

and increases when core-to-core coupling becomes strong. In general cases of more than two identical cores, Λ and $\Delta\Lambda$ can be represented as

$$\Lambda_i = B + \sum_n a_{i,n} K_n, \tag{77}$$

$$\Delta\Lambda_{ij} = \sum_n \left(a_{i,n} - a_{j,n}\right) K_n \tag{78}$$

where $a_{n,i}$ represents the coefficient of the coupling coefficient K_n, for i-th supermode.

By taking advantage of the orthogonality of the supermodes, Kokubun and Koshiba proposed uncoupled SDM transmission over the supermodes of linearly arranged ($1 \times n$) coupled cores (Kokubun and Koshiba 2009). In the linear layout, all the supermodes have different propagation constants, and there are no degenerated modes. So, when K is large enough, the coupling between the supermodes can be suppressed even with small longitudinal perturbations, and all the spatial modes can be transmitted without mode mixing. In the transmission experiments, the modal XT suppression at the mode multiplexing/demultiplexing and at fiber splicing/connection would be the challenge for the transmission without modal XT compensation, like the weakly coupled few-mode fiber transmissions.

When the XT compensation using the multiple-input-multiple-output (MIMO) digital signal processing (DSP) is necessary, low differential group delay (DGD) between the supermodes is important for suppressing the calculation complexity of the DSP. By differentiating Eq. 78 by the angular frequency ω, the DGD τ between the supermodes can be expressed as

$$\frac{\partial \tau}{\partial z} = \frac{\partial \Delta\Lambda_{ij}}{\partial \omega} = \sum_n \left(a_{i,n} - a_{j,n}\right) \frac{\partial K_n}{\partial \omega}. \tag{79}$$

Here, if $\partial K_n/\partial \omega = 0$ for all n, τ can be zero. When the coupled core system has many different K for many different core-to-core distances, it is difficult to suppress $\partial K_n/\partial \omega$ for all the combinations of the cores. However, in the cases of two cores and the three cores with equilateral layout, only one kind of K exists. Xia et al. reported that in such case $\partial K/\partial \omega$ can have zero value with an optimized core design and pitch (Xia et al. 2011). Though the DGD depends on the wavelength, they reported that the DGD less than 60 ps/km can be achieved over the entire C-band (1530–1565 nm) when $\Delta = 0.06\%$ and the normalized frequency V is 1.707 at 1550 nm.

Randomly Coupled Multi-core Fiber

When the core-to-core coupling is weaker than the systematically coupled MCF but not negligible like the uncoupled MCF, the strong and random mode mixing can occur in the MCF caused by the longitudinal perturbations on the MCF (Hayashi 2017). This type of the MCF is referred to as randomly coupled MCF (RC-MCF). The RC-MCF requires the MIMO DSP to undo the random mode mixing, but this random mixing provides some preferable features for the transmission (Ryf et al. 2017). The strong random mode mixing can suppress the accumulations of the DGD, the mode dependent loss/gain (MDL/MDG), and the nonlinear impairments. As already mentioned, the DGD suppression is a critical factor for reducing the calculation complexity in the MIMO DSP for the XT compensation. The RC-MCF can simultaneously realize low DGD and ultralow loss comparable to the lowest loss realized in the SMF, because the RC-MCF can suppress the DGD with simple step-index-type pure-silica cores thanks to the random coupling. In contrast, the single-core few-/multi-mode fiber (FMF/MMF) needs a GeO_2-doped precisely controlled graded-index-type core for the DGD suppression, and the mode coupling in the FMF/MMF is weak.

In this section, the mechanism of the random coupling is described by using the two-core case for the sake of simplicity. In addition, the DGD characteristics are also discussed.

Mechanism of Random Mode Coupling

The propagation and coupling of the orthogonal eigenmodes can be formulated as

$$\frac{d}{dz}\begin{bmatrix} E_e \\ E_o \end{bmatrix} = -j \begin{bmatrix} \Lambda_e & Q_{eo} \\ Q_{oe} & \Lambda_o \end{bmatrix} \begin{bmatrix} E_e \\ E_o \end{bmatrix}, \qquad (80)$$

where the subscripts "e" and "o" represent the even and odd modes, respectively, Λ represents the propagation constant of the eigenmode, and Q represents the coupling coefficient between the eigenmodes. By using the slowly varying envelope approximation,

$$\begin{bmatrix} E_e \\ E_o \end{bmatrix} = \exp\left(-j \int_0^z \Lambda \, dz \right) \begin{bmatrix} A_e \\ A_o \end{bmatrix}, \qquad (81)$$

Equation 80 can be rewritten as

$$\frac{dA_m}{dz} = -j Q_{mn} \exp\left[j \int_0^z (\Lambda_m - \Lambda_n) \, dz \right] A_n. \qquad (82)$$

where (m, n) can be (e, o) or (o, e). When $Q \gg 0$ and $\Lambda_e \sim \Lambda_o$, the mode propagation becomes nonadiabatic, and non-negligible coupling between the eigenmodes can occur (Sakamoto et al. 2016).

3 Multi-core Fibers for Space Division Multiplexing

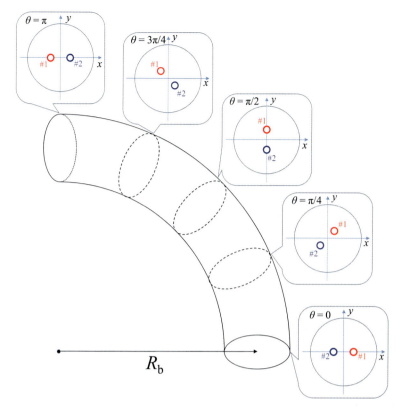

Fig. 14 Schematic illustration showing the coordinates relating to the fiber bend and twist. θ is the fiber rotation angle that is defined such that $\theta = 0$ when the direction from Core 2 to Core 1 is the radial direction of the bend. The direction of the x axis of the local coordinates on a fiber cross section is chosen to always be the radial direction of the fiber bend

When the longitudinal perturbation is dominantly induced by the bends and twists, the propagation constant mismatch between the eigenmodes can be expressed as

$$\Delta \Lambda = 2\sqrt{\left(\frac{\beta_c D}{2R}\cos\theta\right)^2 + \kappa^2}, \tag{83}$$

and the minimum of the propagation constant mismatch $\Delta \Lambda$ is 2κ. Therefore, the smaller core mode coupling coefficient κ induces the larger eigenmode coupling. Here, the coordinates are defined as shown in Fig. 14.

Next, the coupling coefficient Q is considered. Under the longitudinal perturbations to the fiber, the coupling between the orthogonal eigenmodes can be approximated as the butt coupling of short fiber segments where each segment is uniform without the perturbations. When the segment length is Δz and small enough, Eq. 80 can be approximated as

$$\begin{bmatrix} E_e(z+\Delta z) - E_e(z) \\ E_o(z+\Delta z) - E_o(z) \end{bmatrix} = \begin{bmatrix} -j\Lambda_e\Delta z & -jQ_{eo}\Delta z \\ -jQ_{oe}\Delta z & -j\Lambda_o\Delta z \end{bmatrix} \begin{bmatrix} E_e \\ E_o \end{bmatrix}. \tag{84}$$

where the diagonal elements represent the propagations (phase rotations) of the eigenmodes and the off-diagonal elements represent the butt coupling before or after the segment. Therefore, the mode coupling coefficient for one butt coupling is $-jQ_{mn}\Delta z$ and can be defined as

$$-jQ_{mn}\Delta z \equiv \frac{\iint \hat{z} \cdot \left[\mathbf{e}_m^*(z+\Delta z) \times \mathbf{h}_n(z) + \mathbf{e}_n(z) \times \mathbf{h}_m^*(z+\Delta z)\right] dxdy}{\iint \hat{z} \cdot \left(\mathbf{e}_m^* \times \mathbf{h}_m + \mathbf{e}_m \times \mathbf{h}_m^*\right) dxdy}. \tag{85}$$

By using normalized electric field profile e of a scalar mode, Eq. 85 can be simplified as

$$-jQ_{mn}\Delta z = \iint e_m(x,y,z+\Delta z) e_n(x,y,z) \, dxdy. \tag{86}$$

According to Eqs. 68, 71, 72, and 73, the electric field profiles of even and odd modes can be expressed by core modes e_1 (Core 1) and e_2 (Core 2) as

$$\begin{bmatrix} e_e(x,y,z) \\ e_o(x,y,z) \end{bmatrix} = \begin{bmatrix} a(z)e_1(x,y,z) + b(z)e_2(x,y,z) \\ -b(z)e_1(x,y,z) + a(z)e_2(x,y,z) \end{bmatrix}, \tag{87}$$

where

$$a = \frac{\Delta + \sqrt{\Delta^2 + K^2}}{\sqrt{\left(\Delta + \sqrt{\Delta^2 + K^2}\right)^2 + K^2}},$$

$$b = \frac{K}{\sqrt{\left(\Delta + \sqrt{\Delta^2 + K^2}\right)^2 + K^2}}, \quad a^2 + b^2 = 1 \tag{88}$$

By assuming the negligible cross power between the core modes ($\iint e_1 e_2 dxdy \simeq 0$), Eq. 86 can be further simplified by substituting Eq. 87 as

$$\begin{aligned} -jQ_{12}\Delta z &= \iint e_1(x,y,z+\Delta z) e_2(x,y,z) \, dxdy \\ &\approx -a(z+\Delta z)b(z) + b(z+\Delta z)a(z) \\ &\approx a(z)\left[b(z+\Delta z) - b(z)\right] - b(z)\left[a(z+\Delta z) - a(z)\right] \end{aligned} \tag{89}$$

By taking the limits of Eq. 89, as Δz approaches zero, Q_{12} can be derived as

$$Q_{12} \approx j\left(a\frac{\partial b}{\partial z} - b\frac{\partial a}{\partial z}\right). \tag{90}$$

By using the excitation ratio ρ of the core modes in the eigenmode, defined as

$$\rho \equiv \frac{a}{b} = \frac{\Delta}{K} + \sqrt{\left(\frac{\Delta}{K}\right)^2 + 1}, \tag{91}$$

Q_{12} can be also derived as

$$-jQ_{12}\Delta z \approx -b(z+\Delta z)b(z)\left[\frac{a(z+\Delta z)}{b(z+\Delta z)} - \frac{a(z)}{b(z)}\right] \tag{92}$$

$$Q_{12} \approx \frac{-j}{1+\rho^2}\frac{\partial \rho}{\partial z}. \tag{93}$$

Equations 90 and 93 are the same except for their notations. Equation 90 is a more symmetric notation to each core mode, but Eq. 93 can indicate that Δ/K is a key parameter for the coupling, and its analytical solution may be simple. According to Eqs. 90 and 93, when the mode field patterns change steeply, a large coupling between the eigenmodes occurs. In the MCF, the perturbations on Δ are the main sources of the perturbation on ρ, since the K can be regarded as constant. When the longitudinal perturbation is dominantly induced by the bends and twists, ρ can be expressed as

$$\rho = M\cos\theta + \sqrt{M^2\cos^2\theta + 1}, \tag{94}$$

where M is the maximum ratio of the bend-induced perturbation to the core coupling-induced perturbation on the propagation constant:

$$M = \frac{\beta_c D}{2R_b}/\kappa. \tag{95}$$

Now, B and K can be approximated as β and κ, respectively, because the orthogonal coupled-mode theory is accurate enough in the random coupling region. Based on Eqs. 93 and 94, $|Q|$ can be derived as

$$|Q_{12}| = \frac{\rho}{1+\rho^2}\frac{M\sin\theta}{\sqrt{M^2\cos^2\theta+1}}\frac{\partial\theta}{\partial z}. \tag{96}$$

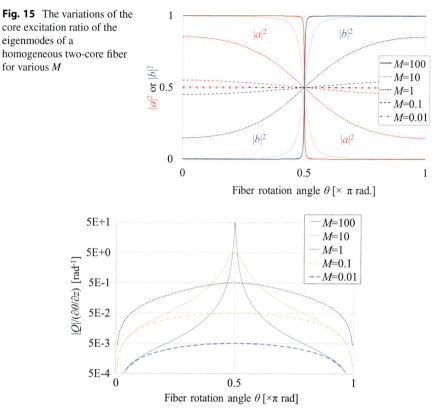

Fig. 15 The variations of the core excitation ratio of the eigenmodes of a homogeneous two-core fiber for various M

Fig. 16 The variations of the mode coupling coefficient between the eigenmodes of a homogeneous two-core fiber for various M. The coefficient Q represents the coupling per unit length and $Q/(\partial\theta/\partial z)$ the coupling per unit rotation angle

Here, the variations of the core excitation ratio ($|a|^2$ and $|b|^2$) of the eigenmodes are shown in Fig. 15, and the mode coupling coefficient between the eigenmodes is shown in Fig. 16, both for various M values. At $\theta = \pi/2$, the slopes of the changes of $|a|^2$ and $|b|^2$ are steepest, and $|Q|/(\partial\theta/\partial z)$ has the highest value of $M/2$. The shapes of $|Q|/(\partial\theta/\partial z)$ are similar among the cases of $M = 0.01, 0.1,$ and 1, because the variations of a and b along θ are similar to sinusoidal variations whose amplitudes depend on M. When M is 10 or 100, the variations of a and b become more localized around $\theta = \pi/2$. Therefore, the coupling coefficient not around $\theta = \pi/2$ becomes more suppressed.

Figure 17 shows the intensity evolutions – or mode couplings – of eigenmodes and core modes in a constantly bent and twisted homogeneous two-core fiber, which were calculated using Eqs. 82, 83, and 96 for the eigenmodes and Eqs. 19, 22, and 24 for the core modes; $\beta_h = 0$ was assumed. Only one eigenmode (even mode) or core mode (Core 1, see Fig. 14) is excited at $\theta = 0$. The intensity evolutions for the core pitches of 15 μm, 20 μm, and 30 μm were calculated. The core diameter was 9 μm, the refractive index contrast was 0.35%, the bend radius was 8 cm, and the twist

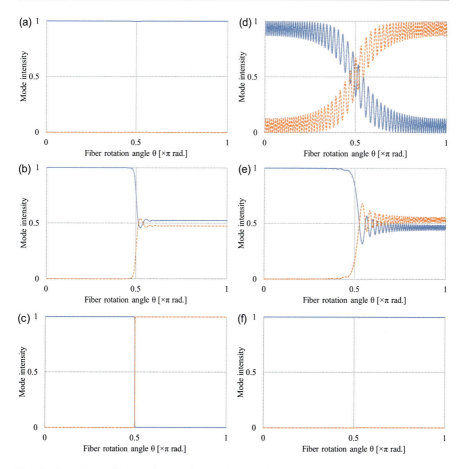

Fig. 17 Intensity evolutions of (**a–c**) eigenmodes and (**d–f**) core modes in a homogeneous two-core fiber with a core pitch of (**a, d**) 15 μm, (**b, e**) 20 μm, or (**c, f**) 30 μm, when only one (**a–c**) eigenmode or (**d–f**) core mode is excited at $\theta = 0$. The core diameter is 9 μm, the refractive index contrast is 0.35%, the bend radius is 8 cm, and the twist rate is 2π radians

rate was 2π rad/m. Here the M value becomes larger when the core pitch becomes larger. When the core pitch was 15 μm, the eigenmodes were almost uncoupled, and no significant intensity variation is observed. In contrast, the core modes couple significantly but it is a systematic coupling. The high-frequency oscillation of the intensity is due to the beating between the eigenmodes. The low-frequency intensity variations represent the transfer of the power from the excited mode to the other mode, which is because most of the power of the even mode is always guided in a core that is located at the outer side of the bend. When the core pitch was 20 μm, both the eigenmodes and the core modes experience a significant coupling around $\theta = \pi/2$ which can be regarded as a PMP. So, every time the lights pass a PMP, random power mixing occurs. When the core pitch was 30 μm, the core modes were almost uncoupled, and no significant intensity variation is observed. On the other hand, the eigenmodes exhibit a full power transfer from the even mode to

Table 1 Characteristics of multi-core fibers

Fiber type	Coupled MCF		Uncoupled MCF
	Systematically coupled MCF	Randomly coupled MCF	
Core pitch	Small	Medium	Large
Coupling between core modes	Strong		Weak
Coupling between eigenmodes	Systematic	Random	
	Weak	Strong	
Dominant source of DGD	Random		Systematic
	DGD between eigenmodes	Both may affect, but core mode DGD may be dominant	DGD between core modes
DGD accumulation	Linear proportion to the fiber length	Square root proportion to the fiber length	Linear proportion to the fiber length

the odd mode, and the power transfer is greatly localized at $\theta = \pi/2$. This can be interpreted as follows:

(i) At $\theta = 0$, the even mode is excited. The even mode is almost equivalent to the core mode of Core 1, because, when $\theta < \pi/2$, Δ is very large and the even mode is localized in Core 1 (outside the bend) and the odd mode in Core 2 (inside the bend).
(ii) At $\theta = \pi/2$, Δ becomes zero and the eigenmodes suddenly spread over the two cores. Therefore, the light in the outside core mode in $\theta < \pi/2$ couples to both even and odd modes at $\theta = \pi/2$. However, all power is still guided in Core 1, because the phases of the even and odd modes are the same in Core 1 and opposite in Core 2.
(iii) When $\theta > \pi/2$, Δ becomes very large. Therefore, each eigenmode is localized to each core again. However, the even mode is localized in Core 2 and the odd mode in Core 1, because Core 2 is outside the bend and Core 1 is inside the bend now.
(iv) Since all the power of the two eigenmodes is guided only in Core 1 at $\theta = \pi/2$, the light in Core 1 at $\theta = \pi/2$ couples to the eigenmode localized in Core 1 in $\theta > \pi/2$, which is the odd mode.

For better understanding, Table 1 summarizes the characteristics of MCFs.

Group Delay Spread in Randomly Coupled Multi-core Fiber

Discrete Coupling Model for Simple DGD Estimation for the Special Case

In terms of the coupling behaviors, the randomly coupled MCF and the uncoupled MCF have the random coupling between the core modes, and their difference is just the amount of the coupling. In addition, the random coupling between the core modes is localized around the PMPs. Therefore, the coupling in the randomly coupled MCF can also be approximated by the discrete coupling model. When the discrete coupling at every PMP is strong and random enough, the variance of the total GD $\tau_{g,tot}$ of the MCF is the sum of the variances of the local GD of each uncoupled segment between two adjacent PMPs as

$$\sigma^2_{\tau_{g,tot}} = \sum_{l=1}^{N_{PM}} \sigma^2_{\tau_{g,l}}, \qquad (97)$$

where σ_x^2 is the variance of x and $\tau_{g,l}$ is the local GD between the l-th uncoupled segments. The local GD can be numerically calculated using

$$\tau_{g,l} = \int_{z_{l-1}}^{z_l} \frac{\partial \tau_{g,eig}}{\partial z} dz, \qquad (98)$$

$$\frac{\partial \tau_{g,eig}}{\partial z} = \frac{\partial \Lambda}{\partial \omega} = \frac{\partial}{\partial \omega}\left(\beta_c \pm \sqrt{\left(\frac{\beta_c D}{2R_b}\cos\theta\right)^2 + \kappa^2}\right). \qquad (99)$$

In the special case where the bend-induced DGD is dominant, Eq. 99 can be simplified as

$$\frac{\partial \tau_{g,eig}}{\partial z} \approx \frac{\partial \beta_c}{\partial \omega} \pm \frac{\partial \beta_c}{\partial \omega}\frac{D}{2R_b}\cos\theta. \qquad (100)$$

When the constant twist with the twist rate of ω_{twist} [rad/unit length] is assumed, $\sigma^2_{\tau_{g,l}}$ between two adjacent PMPs can be derived as

$$\sigma^2_{\tau_{g,l}} \approx \left(\frac{\partial \beta_c}{\partial \omega}\frac{D_{nm}}{R_b}\frac{1}{\omega_{twist}}\right)^2, \qquad (101)$$

and N_{PM} can be approximated as

$$N_{PM} \approx \frac{\omega_{twist} L}{\pi}, \qquad (102)$$

Therefore, $\sigma^2_{\tau_{g,tot}}$ and $\sigma_{\tau_{g,tot}}$ can be expressed as

$$\sigma^2_{\tau_{g,tot}} = \sigma^2_{\tau_{g,l}} N_{PM} = \left(\frac{\partial \beta_c}{\partial \omega}\frac{D_{nm}}{R_b}\right)^2 \frac{L}{\omega_{twist}\pi}, \qquad (103)$$

$$\sigma_{\tau_{\text{g.tot}}} = \frac{\partial \beta_c}{\partial \omega} \frac{D_{nm}}{R_b} \sqrt{\frac{L}{\omega_{\text{twist}} \pi}}, \qquad (104)$$

Definition of Group Delays of Randomly Coupled Modes by "Principal Modes"
The above simplification cannot hold when the bend-induced DGD is not dominant or the core count is more than two. In this case, the DGD has to be numerically calculated. In the same manner as the GDs of randomly coupled polarizations are defined (Gordon and Kogelnik 2000), the GDs of randomly coupled modes can be defined (Fan and Kahn 2005), which is reviewed in this subsection.

When a lossless system is assumed for the sake of simplicity, an input state $|E_{\text{in}}\rangle$ and an output state $|E_{\text{out}}\rangle$ of the modes are related by the unitary transfer matrix $\mathbf{T}_{\text{tot}}$ of an RC-MCF as

$$|E_{\text{out}}\rangle = \mathbf{T}_{\text{tot}} |E_{\text{in}}\rangle \qquad (105)$$

The generalized (unitary) Jones matrix $\mathbf{J}_{\text{tot}}$ corresponding to $\mathbf{T}_{\text{tot}}$ can be defined as

$$\mathbf{T}_{\text{tot}} = \exp(-j\phi_0) \, \mathbf{J}_{\text{tot}}, \qquad (106)$$

where ϕ_0 is the common phase. The mean τ_0 of the GDs of the modes can be expressed as

$$\tau_0 = \frac{\partial \phi_0}{\partial \omega}. \qquad (107)$$

Since $\mathbf{J}_{\text{tot}}$ is unitary, the following relationship holds:

$$\mathbf{J}_{\text{tot}} \mathbf{J}_{\text{tot}}^* = \mathbf{I}, \qquad (108)$$

where $\mathbf{I}$ is the identity matrix.

By using Eqs. 105, 106, 107, and 108, the derivative of $|E_{\text{out}}\rangle$ with respect to ω can be expressed as

$$\frac{\partial |E_{\text{out}}\rangle}{\partial \omega} = -j \, (\tau_0 \mathbf{I} + \mathbf{G}) \, |E_{\text{out}}\rangle, \qquad (109)$$

$$\mathbf{G} = j \frac{\partial \mathbf{J}_{\text{tot}}}{\partial \omega} \mathbf{J}_{\text{tot}}^*, \qquad (110)$$

where the operator $\mathbf{G}$ is referred to as "group delay operator." All the eigenvalues of $\mathbf{G}$ are real, because $\mathbf{G}$ is Hermitian ($\mathbf{G} = \mathbf{G}^*$), which can be proved by differentiating Eq. 108 with respect to ω.

Now, substituting an eigenvector of $\mathbf{G}$ to $|E_{\text{out}}\rangle$ in Eq. 109 yields

$$\frac{\partial |E_{\text{out}}\rangle_n}{\partial \omega} = -j\left(\tau_0 \mathbf{I} + \mathbf{G}\right) |E_{\text{out}}\rangle_n = -j\left(\tau_0 + \tau_n\right) |E_{\text{out}}\rangle_n, \qquad (111)$$

where $|E_{\text{out}}\rangle_n$ and τ_n are an eigenvector and its corresponding eigenvalue of the group delay operator $\mathbf{G}$, respectively. From Eq. 111, $\tau_0 + \tau_n$ is the GD that is independent of ω – zero dispersion – to first order when the output state of the modes is an eigenvector $|E_{\text{out}}\rangle_n$. Therefore, every eigenvector of the group delay operator $\mathbf{G}$ can be understood as an output "principal mode" (Fan and Kahn 2005) in analogy to "principal state of polarization." The GD spread of a randomly coupled MCF can be evaluated by the standard deviation of the GDs of the principal modes as

$$\sigma_{\tau_{g,\text{tot}}} = \sqrt{\frac{1}{N} \left\langle \sum_{n=1}^{N} \tau_n^2 \right\rangle}. \qquad (112)$$

where N is the number of the modes. The standard deviation of the GDs of the principal modes corresponds to the standard deviation of the impulse response.

The generalized Jones matrix $\mathbf{J}_{\text{tot}}$ of the whole fiber can be calculated by modeling the fiber as a concatenation of short enough uniform fiber segments as

$$\mathbf{J}_{\text{tot}} = \prod_{k=1}^{N_{\text{seg}}} \mathbf{J}_k = \prod_{k=1}^{N_{\text{seg}}} \exp\left[-j \Delta \mathbf{H}_k \Delta z_k\right], \qquad (113)$$

where the subscript k represents a matrix of k-th segment, N_{seg} is the number of the segments, and $\Delta \mathbf{H}_k$ is defined as

$$\Delta \mathbf{H}_k = \mathbf{H}_k - \mathbf{I} \operatorname{tr}(\mathbf{H}_k)/N, \qquad (114)$$

so that the common phase rotation in $\mathbf{H}_k$ can be eliminated – $\operatorname{tr}(\mathbf{H}_k)$ represents the trace of $\mathbf{H}_k$, i.e., the sum of the diagonal elements of $\mathbf{H}_k$, and thus $\operatorname{tr}(\mathbf{H}_k)/N$ is the average of the diagonal elements of $\mathbf{H}_k$. Here, the mode basis of $\mathbf{H}_k$ can be either of the core modes or the eigenmodes. Basically, the calculation of Eq. 113 requires the very short segment to assure the longitudinal uniformity of each segment under various perturbations; therefore, the calculation complexity can be exhaustive. However, the method is basically applicable to the DGD calculation for any type of the MCFs.

When the discrete coupling model is valid and when constant bend radius and twist rate are assumed, the calculation complexity can be reduced by adopting the split-step method – where the coupling is assumed to discretely occur only

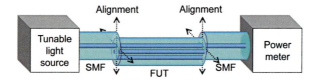

Fig. 18 The schematic experimental setup for observing the transmission spectrum of a core of the CC-MCF (FUT, fiber under test), for measuring the SMD in the similar way to the fixed analyzer method for PMD measurement (Reused from Hayashi et al. (2017) ©2016 IEEE)

at each PMP – for the GD spread estimation (Sakamoto et al. 2016). In the split-step method, a unitary rotation matrix $\mathbf{C}$ that represents the coupling at each PMP is calculated with the small segment method at first. Then, the generalized Jones matrix $\mathbf{J}_l$ between $(l-1)$-th and l-th PMPs can be calculated as

$$\mathbf{J}_l = \mathbf{C} \exp\left(-j \int_{z_{l-1}}^{z_l} \Delta \mathbf{B} dz\right), \qquad (115)$$

where

$$\Delta \mathbf{B} = \mathbf{B} - \operatorname{Itr}(\mathbf{B})/N \simeq \boldsymbol{\beta}_{\mathrm{eq}} - \operatorname{Itr}(\boldsymbol{\beta}_{\mathrm{eq}})/N. \qquad (116)$$

Spatial Mode Dispersion
The GD spread of randomly coupled MCFs can be measured using the method similar to the fixed analyzer technique with Fourier analysis for the polarization mode dispersion (PMD) measurement.

As pointed out by Sakamoto et al. (2016), the GD spread of randomly coupled MCFs can be measured using the method similar to the fixed analyzer technique – sometimes also referred to as the wavelength scanning technique/method – with Fourier analysis for the polarization mode dispersion (PMD) measurement (ITU-T G.650.2 2015). Figure 18 shows a measurement setup for a randomly coupled MCF. In the measurement, the transmission spectrum of one core is measured, the transmission spectrum is Fourier-transformed, and the probability distribution of the DGD was obtained as the result of the Fourier transform, since the interference fringe in the transmission spectrum contains the group delay information of various randomly coupled optical paths. Figure 19 shows an exemplary DGD distribution with Gaussian shape. After the PMD definition in the fixed analyzer technique (ITU-T G.650.2 2015), Sakamoto et al. defined the "spatial mode dispersion" as the standard deviation – square root of the second moment – of the DGD distribution. This DGD distribution is the autocorrelation function of the impulse response, as with the case of the PMD measurement (Gisin et al. 1994). Therefore, the SMD corresponds to $2\sigma_{\tau_{\mathrm{g,tot}}}$ – twice the standard deviation of the impulse response.

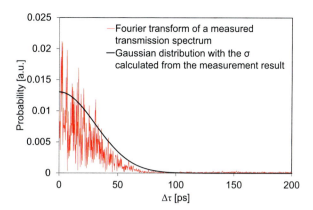

Fig. 19 An example of measured DGD distributions (Reused from Hayashi et al. (2017) ©2016 IEEE)

Common Design Factors for Uncoupled and Coupled Multi-core Fibers

Excess Loss Due to the Power Coupling to the Coating

When a core in an MCF is arranged too near the cladding-coating interface, the light in the core can couple to the coating, and the transmission loss of the core can degrade because the refractive index of the coating is higher than those of cores and cladding for suppressing the light propagation in the cladding modes. The excess loss due to the coupling to the coating is referred to as "excess loss," "coating-leakage loss," "leakage loss," "confinement loss," "tunneling loss," etc. The minimum distance between a core center and the cladding-coating interface – the so-called "outer cladding thickness (OCT)" – is an important parameter for the leakage loss to the coating. Figure 20 schematically illustrates the OCT in a cross section and in refractive index profile of an MCF and the mode field pattern of an outer core that is of interest in terms of the leakage loss. Qualitatively, the leakage loss can be understood as the mode coupling from the outer core mode to the leaky and lossy coating modes. So, as more fraction of the outer core mode field penetrates the coating, the more power couples from the outer core mode to the coating modes. To simultaneously achieve a small OCT and a low leakage loss, the strong light confinement into the outer core is preferred.

Cutoff Wavelength Variation Due to Surrounding Cores

Other than the XT, another important issue for realizing the short core pitch is the effect of surrounding cores on the cutoff wavelength of each core, especially for trench- or hole-assisted MCFs. When a core (surrounded core) is surrounded by other cores (surrounding cores), the trenches or holes for the surrounding cores can also serve as those for the surrounded cores (Takenaga et al. 2011b). Therefore, the

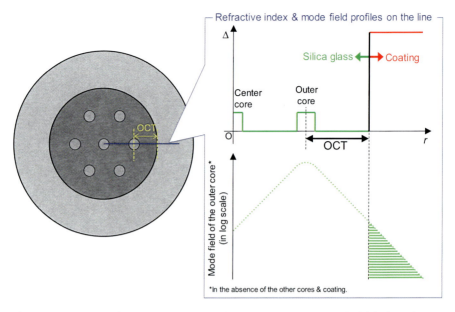

Fig. 20 Schematic profiles of the refractive index and an outer core mode field of a polymer-coated silica glass MCF

higher-order modes of the surrounded core become more confined, and the cutoff wavelength can be elongated. Sometimes, this cutoff wavelength variation can be the limiting factor for shortening the core pitch, rather than the XT. Figure 21 shows the relationship between the core pitch and how easily the higher-order modes can couple to the leaky cladding modes. As shown in the left figure, when the core pitch is long enough, there is adequate space for cladding mode propagation adjacent to the center core. Thus, the higher-order modes of the center core can easily couple to cladding modes. On the other hand, as shown in the right figure, when the core pitch is too short, the cladding space adjacent to the center (surrounded) core is limited, and not enough cladding modes can propagate; thus, the higher-order modes of the center core have to penetrate the outer (surrounding) cores for coupling to cladding modes.

Cladding Diameter

To pack many cores into one common cladding and/or reduce the fraction of unutilized areas (outer cladding and coating area) in fiber cross section, the cladding diameters of MCFs are often designed to be larger than that of the standard optical fiber, i.e., 125 μm. Ultrahigh-capacity transmission experiments have been conducted using various MCFs with a cladding diameter more than 200 μm

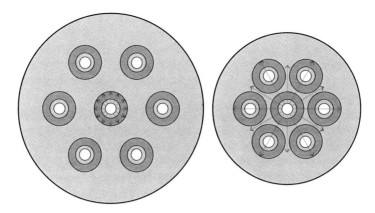

Fig. 21 Schematic showing the relationship between the core pitch and how easily the higher-order modes can couple to the leaky cladding modes by radial arrows

(Puttnam et al. 2015; Kobayashi et al. 2017; Soma et al. 2017). One concern for the large cladding diameter is the mechanical reliability degradation. Since the bend-induced stress on the glass fiber surface is in proportion to the glass diameter, the failure probability of a bent fiber can be degraded by a large cladding diameter. According to the power law theory (IEC/TR 62048 ed2.0 2011), the failure probability in the low-strength region – where failures are induced by the growth of subcritical cracks – is the same between different fibers if the ratio of the stress in use to the stress in a proof test is the same between those different fibers. Therefore, for example, 150 μm, 200 μm, 250 μm, and 300 μm claddings require 1.2 times, 1.6 times, 2 times, and 2.4 times higher proof stress levels compared to that of the standard fibers, respectively. On the other hand, the higher proof stress level cannot improve the intrinsic strength of the glass fiber even if there are no subcritical cracks on the fiber; thus the intrinsic glass strength may determine the fundamental upper limit of the cladding diameter. However, the practical upper limit of the cladding diameter may be lower than the fundamental one. For example, the proof tension is the product of the proof stress and the cladding diameter, which means 150 μm, 200 μm, 250 μm, and 300 μm claddings require 1.44 times, 2.56 times, 4 times, and 5.76 times higher proof tensions compared to that of the standard fibers, respectively. Too high tension forces would be unpractical for fiber mass production. In the recent research field, roughly around 250 μm is believed to be the practical upper limit of the cladding diameter. In addition, the fiber cost increase would be another issue because the length of a fiber drawn from a same-sized preform is inversely proportional to the square of the cladding diameter. To avoid such technical and economical hurdles, MCFs with a standard 125 μm cladding diameter have been actively proposed by various groups in recent years (Hayashi et al. 2015, 2016; Matsui et al. 2015; Gonda et al. 2016).

Conclusion

Mode/power coupling characteristics and other design factors of the uncoupled and coupled MCFs were described. Based on the knowledge described in the chapter, various MCFs have been proposed, and transmission experiments over those MCFs have been demonstrated.

The few-mode MCF (FM-MCF) – i.e., the MCF where each core guides multiple modes – was not described in this chapter, but its characteristics can be understood based on this chapter and the few-mode fiber chapter. The FM-MCF is suited to improve the number of the total spatial modes per fiber, and more than 100 spatial modes have been achieved. The highest fiber transmission capacity of 10.16 Pb/s was also achieved using an FM-MCF (6-mode 19-core fiber) over C + L band (1530–1625 nm) (Soma et al. 2017).

It should be noted that "uncoupled/coupled" or "weakly/strongly coupled" are very relativistic words and may be used in different definitions in other MCF articles. Also, one MCF can simultaneously be the uncoupled MCF for short-reach transmission and the randomly coupled MCF for long-haul transmission.

References

S. Berdagué, P. Facq, Mode division multiplexing in optical fibers. Appl. Opt. **21**, 1950–1955 (1982)

S.-L. Chuang, A coupled mode formulation by reciprocity and a variational principle. J. Lightw. Technol **5**, 5–15 (1987)

R.-J. Essiambre, R.W. Tkach, Capacity trends and limits of optical communication networks. Proc. IEEE **100**, 1035–1055 (2012)

R.-J. Essiambre, G. Kramer, P.J. Winzer, G.J. Foschini, B. Goebel, Capacity limits of optical fiber networks. J. Lightw. Technol. **28**, 662–701 (2010)

S. Fan, J.M. Kahn, Principal modes in multimode waveguides. Opt. Lett. **30**, 135–137 (2005)

J. M. Fini, T. F. Taunay, B. Zhu, M. F. Yan, Low cross-talk design of multi-core fibers, in *Conference on Lasers and Electro-Opticals (CLEO)*, Baltimore (Optical Society of America, Washington, DC, 2010a), p. CTuAA3

J.M. Fini, B. Zhu, T.F. Taunay, M.F. Yan, Statistics of crosstalk in bent multicore fibers. Opt. Express **18**, 15122–15129 (2010b)

N. Gisin, R. Passy, J.P.V. der Weid, Definitions and measurements of polarization mode dispersion: interferometric versus fixed analyzer methods. IEEE Photon. Technol. Lett. **6**, 730–732 (1994)

T. Gonda, K. Imamura, R. Sugizaki, Y. Kawaguchi, T. Tsuritani, 125 μm 5-core fibre with heterogeneous design suitable for migration from single-core system to multi-core system, in *European Conference on Optical Communication (ECOC)*, Dusseldorf (VDE, Frankfurt, 2016), pp. 547–549

J.P. Gordon, H. Kogelnik, PMD fundamentals: polarization mode dispersion in optical fibers. PNAS **97**, 4541–4550 (2000)

H.A. Haus, W.P. Huang, S. Kawakami, N.A. Whitaker, Coupled-mode theory of optical waveguides. J. Lightw. Technol. **5**, 16–23 (1987)

T. Hayashi, Multi-core fiber for high-capacity spatially-multiplexed transmission, Ph.D. Thesis, Hokkaido University (2013)

T. Hayashi, Coupled multicore fiber for space-division multiplexed transmission, in *Proceedings of SPIE, Vol 10130, Next–Generation Optical Communication: Components, Sub–Systems, and Systems VI* (IEEE, New York, 2017), p. 1013003

T. Hayashi, T. Nagashima, O. Shimakawa, T. Sasaki, E. Sasaoka, Crosstalk variation of multi-core fibre due to fibre bend, in *European Conference on Optical Communication (ECOC)*, Torino (IEEE, New York, 2010), p. We.8.F.6

T. Hayashi, T. Taru, O. Shimakawa, T. Sasaki, E. Sasaoka, Design and fabrication of ultra-low crosstalk and low-loss multi-core fiber. Opt. Express **19**, 16576–16592 (2011a)

T. Hayashi, T. Taru, O. Shimakawa, T. Sasaki, E. Sasaoka, Ultra-low-crosstalk multi-core fiber feasible to ultra-long-haul transmission, in *Optical Fiber Communication Conference (OFC)*, Los Angeles (Optical Society of America, Washington, DC, 2011b), p. PDPC2

T. Hayashi, T. Taru, O. Shimakawa, T. Sasaki, E. Sasaoka, Characterization of crosstalk in ultra-low-crosstalk multi-core fiber. J. Lightw. Technol. **30**, 583–589 (2012)

T. Hayashi, T. Sasaki, E. Sasaoka, K. Saitoh, M. Koshiba, Physical interpretation of intercore crosstalk in multicore fiber: effects of macrobend, structure fluctuation, and microbend. Opt. Express **21**, 5401–5412 (2013)

T. Hayashi, T. Sasaki, E. Sasaoka, Behavior of inter-core crosstalk as a noise and its effect on Q-factor in multi-core fiber. IEICE Trans. Commun. **E97.B**, 936–944 (2014)

T. Hayashi, T. Nakanishi, K. Hirashima, O. Shimakawa, F. Sato, K. Koyama, A. Furuya, Y. Murakami, T. Sasaki, 125-μm-cladding 8-core multi-core fiber realizing ultra-high-density cable suitable for O-band short-reach optical interconnects, in *Optical Fiber Communication Conference (OFC)*, Los Angeles (Optical Society of America, Washington, DC, 2015), p. Th5C.6

T. Hayashi, Y. Tamura, T. Hasegawa, T. Taru, 125-μm-cladding coupled multi-core fiber with ultra-low loss of 0.158 dB/km and record-low spatial mode dispersion of 6.1 ps/km$^{1/2}$, in *Optical Fiber Communication Conference (OFC)*, Anaheim (Optical Society of America, Washington, DC, 2016), p. Th5A.1

T. Hayashi, Y. Tamura, T. Hasegawa, T. Taru, Record-low spatial mode dispersion and ultra-low loss coupled multi-core fiber for ultra-long-haul transmission. J. Lightw. Technol. **35**, 450–457 (2017)

W.-P. Huang, Coupled-mode theory for optical waveguides: an overview. J. Opt. Soc. Am. A **11**, 963–983 (1994)

IEC/TR 62048 ed2.0, *Optical Fibres – Reliability – Power Law Theory* (International Electrotechnical Commission, Geneva, 2011)

S. Inao, T. Sato, S. Sentsui, T. Kuroha, Y. Nishimura, Multicore optical fiber, in *Optical Fiber Communication Conference (OFC)*, Washington, DC (Optical Society of America, Washington, DC, 1979a), p. WB1

S. Inao, T. Sato, H. Hondo, M. Ogai, S. Sentsui, A. Otake, K. Yoshizaki, K. Ishihara, N. Uchida, High density multicore-fiber cable, in *International Wire & Cable Symposium (IWCS)*, Fort Monmouth (1979b), pp. 370–384

ITU-T G.650.2, Definitions and test methods for statistical and non-linear related attributes of single-mode fibre and cable (2015)

T. Kobayashi, M. Nakamura, F. Hamaoka, K. Shibahara, T. Mizuno, A. Sano, H. Kawakami, A. Isoda, M. Nagatani, H. Yamazaki, Y. Miyamoto, Y. Amma, Y. Sasaki, K. Takenaga, K. Aikawa, K. Saitoh, Y. Jung, D.J. Richardson, K. Pulverer, M. Bohn, M. Nooruzzaman, T. Morioka, 1-Pb/s (32 SDM/46 WDM/768 Gb/s) C-band dense SDM transmission over 205.6-km of single-mode heterogeneous multi-core fiber using 96-Gbaud PDM-16QAM channels, in *Optical Fiber Communication Conference (OFC)*, Los Angeles (Optical Society of America, Washington, DC, 2017), p. Th5B.1

Y. Kokubun, M. Koshiba, Novel multi-core fibers for mode division multiplexing: proposal and design principle. IEICE Electron. Express **6**, 522–528 (2009)

M. Koshiba, K. Saitoh, K. Takenaga, S. Matsuo, Multi-core fiber design and analysis: coupled-mode theory and coupled-power theory. Opt. Express **19**, B102–B111 (2011)

M. Koshiba, K. Saitoh, K. Takenaga, S. Matsuo, Analytical expression of average power-coupling coefficients for estimating intercore crosstalk in multicore fibers. IEEE Photon. J. **4**(5), 1987–1995 (2012)

D. Marcuse, Influence of curvature on the losses of doubly clad fibers. Appl. Opt. **21**, 4208–4213 (1982)

T. Matsui, T. Sakamoto, Y. Goto, K. Saito, K. Nakajima, F. Yamamoto, T. Kurashima, Design of 125 μm cladding multi-core fiber with full-band compatibility to conventional single-mode fiber, in *European Conference on Optical Communication (ECOC)*, Valencia (IEEE, New York, 2015), p. We.1.4.5

T. Morioka, New generation optical infrastructure technologies: "EXAT initiative" towards 2020 and beyond, in *OptoElectronics and Communication Conference (OECC)*, Hong Kong (IEEE, New York, 2009), p. FT4

B. J. Puttnam, R. S. Luis, W. Klaus, J. Sakaguchi, J.-M. Delgado Mendinueta, Y. Awaji, N. Wada, Y. Tamura, T. Hayashi, M. Hirano, J. Marciante, 2.15 Pb/s transmission using a 22 core homogeneous single-mode multi-core fiber and wideband optical comb, in *European Conference on Optical Communication (ECOC)*, Valencia (IEEE, New York, 2015), p. PDP.3.1

D. Qian, M.-F. Huang, E. Ip, Y.-K. Huang, Y. Shao, J. Hu, T. Wang, 101.7-Tb/s (370×294-Gb/s) PDM-128QAM-OFDM transmission over 3×55-km SSMF using pilot-based phase noise mitigation. in *Optical Fiber Communication Conference (OFC)*, Los Angeles (Optical Society of America, Washington, DC, 2011), p. PDPB5

G. Rademacher, R.S. Luís, B.J. Puttnam, Y. Awaji, N. Wada, Crosstalk dynamics in multi-core fibers. Opt. Express **25**, 12020–12028 (2017a)

G. Rademacher, R. S. Luis, B. J. Puttnam, Y. Awaji, N. Wada, Crosstalk fluctuations in homogeneous multi-core fibers, in *Photonic Networks and Devices*, New Orleans (Optical Society of America, Washington, DC, 2017b), p. NeTu2B.4

R. Ryf, N. K. Fontaine, S. H. Chang, J. C. Alvarado, B. Huang, J. Antonio-Lopez, H. Chen, R.-J. Essiambre, E. Burrows, R. W. Tkach, R. Amezcua-Correa, T. Hayashi, Y. Tamura, T. Hasegawa, T. Taru, Long-haul transmission over multi-core fibers with coupled cores, in *European Conference on Optical Communication (ECOC)*, Gothenburg (IEEE, New York, 2017), p. M.2.E.1

T. Sakamoto, T. Mori, M. Wada, T. Yamamoto, F. Yamamoto, K. Nakajima, Fiber twisting and bending induced adiabatic/nonadiabatic super-mode transition in coupled multi-core fiber. J. Lightw. Technol. **34**, 1228–1237 (2016)

A. Sano, T. Kobayashi, S. Yamanaka, A. Matsuura, H. Kawakami, Y. Miyamoto, K. Ishihara, H. Masuda, 102.3-Tb/s (224 x 548-Gb/s) C- and extended L-band all-Raman transmission over 240 km using PDM-64QAM single carrier FDM with digital pilot tone, in *Optical Fiber Communication Conference (OFC)*, Los Angeles (Optical Society of America, Washington, DC, 2012), p. PDP5C.3

Y. Sasaki, R. Fukumoto, K. Takenaga, K. Aikawa, K. Saitoh, T. Morioka, Y. Miyamoto, Crosstalk-managed heterogeneous single-mode 32-core fibre, in *European Conference on Optical Communication (ECOC)*, Dusseldorf (VDE, Frankfurt, 2016), pp. 550–552

Y. Sasaki, K. Takenaga, K. Aikawa, Y. Miyamoto, T. Morioka, Single-mode 37-core fiber with a cladding diameter of 248 μm, in *Optical Fiber Communication Conference (OFC)*, Los Angeles (Optical Society of America, Washington, DC, 2017), p. Th1H.2

D. Soma, Y. Wakayama, S. Beppu, S. Sumita, T. Tsuritani, T. Hayashi, T. Nagashima, M. Suzuki, H. Takahashi, K. Igarashi, I. Morita, M. Suzuki, 10.16 Peta-bit/s dense SDM/WDM transmission over Low-DMD 6-mode 19-core fibre across C+L band, in *European Conference on Optical Communication (ECOC)*, Gothenburg (IEEE, New York, 2017), p. Th.PDP.A.1

W. Streifer, M. Osinski, A. Hardy, Reformulation of the coupled-mode theory of multiwaveguide systems. J. Lightw. Technol. **5**, 1–4 (1987)

Q. Sui, H. Zhang, J.D. Downie, W.A. Wood, J. Hurley, S. Mishra, A.P.T. Lau, C. Lu, H.-Y. Tam, P.K.A. Wai, Long-haul quasi-single-mode transmissions using few-mode fiber in presence of multi-path interference. Opt. Express **23**, 3156–3169 (2015)

K. Takenaga, S. Tanigawa, N. Guan, S. Matsuo, K. Saitoh, M. Koshiba, Reduction of crosstalk by quasi-homogeneous solid multi-core fiber, in *Optical Fiber Communication Conference (OFC)*, Los Angeles (Optical Society of America, Washington, DC, 2010), p. OWK7

K. Takenaga, Y. Arakawa, S. Tanigawa, N. Guan, S. Matsuo, K. Saitoh, M. Koshiba, An investigation on crosstalk in multi-core fibers by introducing random fluctuation along longitudinal direction. IEICE Trans. Commun. **E94.B**, 409–416 (2011a)

K. Takenaga, Y. Arakawa, S. Tanigawa, N. Guan, S. Matsuo, K. Saitoh, M. Koshiba, Reduction of crosstalk by trench-assisted multi-core fiber, in *Optical Fiber Communication Conference (OFC)*, Los Angeles (Optical Society of America, Washington, DC, 2011b), p. OWJ4

P.J. Winzer, Scaling optical fiber networks: challenges and solutions. Opt. Photonics News **29**, 28–35 (2015)

P.J. Winzer, D.T. Neilson, From scaling disparities to integrated parallelism: a decathlon for a decade. J. Lightw. Technol. (2017). https://doi.org/10.1109/JLT.2017.2662082

C. Xia, N. Bai, I. Ozdur, X. Zhou, G. Li, Supermodes for optical transmission. Opt. Express **19**, 16653–16664 (2011)

Optical Coherent Detection and Digital Signal Processing of Channel Impairments

4

Ezra Ip

Contents

Introduction	149
Transmitter	153
Mach-Zehnder Modulator	153
Signal Modulation	155
Analytical Baseband Model	155
Coherent Receiver	156
Optical-to-Electrical Downconversion	156
Single-Polarization Optical-to-Electrical Downconverter	157
Single-Sided Photodetection	159
Heterodyne Detection	160
Dual-Polarization Optical-to-Electrical Downconversion	160
I/Q Imbalance	161
Signal Conditioning Circuit	162
Analytical Baseband Model	163
Emulation of Other Detector Types	164
The Fiber Channel	165
Nonlinear Schrödinger Equation (NLSE)	165
Linear Time-Invariant (LTI) Model	166
Digital Signal Processing Preliminaries	168
Introduction	168
Sampling Rate Requirement	170
Discrete-Time Fourier Transform	172
Linear Equalization	173
Minimum Mean Square Error (MMSE) Equalizer	173
Equalizer Length Requirement	175
MMSE Performance	177
Frequency Domain Equalizer (FDE)	179
Adaptive Time-Domain Equalizer (TDE)	180

E. Ip (✉)
NEC Laboratories America, Princeton, NJ, USA
e-mail: ezra.ip@nec-labs.com

© Springer Nature Singapore Pte Ltd. 2019
G.-D. Peng (ed.), *Handbook of Optical Fibers*,
https://doi.org/10.1007/978-981-10-7087-7_54

Adaptive Frequency-Domain Equalizer (FDE)	183
Adaptive Multidelay Block Frequency-Domain Equalizer	184
Hybrid Equalizer Structure	185
Carrier Phase Estimation	186
Laser Phase Noise	186
Phase Estimation in the Absence of Data Modulation	187
Phase Estimation in the Presence of Data Modulation	189
Combining Laser Phase Noise Compensation with Linear Equalization	191
Nonlinear Compensation	193
Digital Backpropagation	193
Split-Step Fourier Method	195
Reduced Complexity Nonlinear Compensation Algorithms	198
Interchannel Nonlinear Compensation	199
Timing Recovery	202
Analog Timing Recovery	203
Digital Timing Recovery	203
Other Topics	206
Space-Division Multiplexing	206
Optical Performance Monitoring	209
Optical Sensing	210
Conclusions	210
References	212

Abstract

Optical transponders using coherent detection have been the mainstay in long-haul transmission since around 2010. By allowing the reconstruction of the optical electric field, coherent receivers achieve the following advantages:

- Increased spectral efficiency: Information can be encoded in all the available dimensions (polarization + quadrature) of an optical fiber.
- Improved optical power efficiency: Any arbitrary modulation format can be supported, allowing techniques like constellation shaping and probabilistic shaping to realize signal-to-noise performance closer to Shannon's limit.
- Digital signal processing compensation of channel impairments: Adaptive linear equalizers can be used to compensate linear impairments like chromatic dispersion and polarization mode dispersion. Fiber nonlinearity can also be mitigated using digital backpropagation, Volterra series, and other approaches. The combination of coherent detection and high-speed DSP enables a highly tunable receiver platform.
- Increased receiver sensitivity: It becomes possible to overcome shot-noise limit by increasing the power of the local oscillator, allowing performance that is ultimately limited by the optical signal-to-noise ratio of the link.
- Improved spectral management: Digital filters can be used to demultiplex an optical carrier of interest, allowing optical carriers to be spaced closer together, and facilitates "superchannel" transmission.

The first coherent systems to be widely deployed operated at 100-Gb/s using dual-polarization quadriphase shift keying (DP-QPSK). Subsequently, coherent detection has been used in 400-Gb/s superchannel transmission, with various contenders including 4 × 100-Gb/s DP-QPSK, 2 × 200-Gb/s DP-16QAM, and single-carrier 400-Gb/s DP-16QAM. As the cost and power consumption of coherent transponders have decreased over time, they have been pushed ever deeper into the network. Coherent transponders are now used in short-reach systems, and is even under consideration for intra-data center communications.

In this chapter, we review the theory of optical coherent detection, deriving mathematical models of the optical transmitter and receiver as well as an optical fiber link. We also review the most common digital signal processing operations that are performed in a coherent receiver, including linear equalization, optical phase noise compensation, and nonlinear compensation.

Introduction

In long-haul optical communications, it is generally desired to transmit the greatest data throughput over the longest distance, without signal regeneration if possible, at the lowest cost-per-bit. The capacity-distance product in b/s-km is an important metric in evaluating system performance. Total system capacity is the product of spectral efficiency (b/s/Hz) and the bandwidth (Hz) available for signal transmission. The usable bandwidth is constrained by the transparency region of glass and by the availability of cost-effective amplifiers. Dense wavelength-division multiplexing (DWDM) was an area of active research in the 1980s (Ishio et al. 1984; Brackett 1990). The invention of the erbium-doped fiber amplifier (EDFA) in the 1990s enabled repeaterless transmission over transcontinental and transoceanic distances (Desurvire 2002). Today, EDFAs facilitate long-haul optical fiber transmission in the C- (1530–1570 nm) and L- (1570–1610 nm) bands, giving a total utilizable bandwidth of around 10 THz.

Given the bandwidth constraint, maximizing data capacity requires maximizing spectral efficiency. In the absence of fiber nonlinearity, the capacity of a linear channel impaired only by additive white Gaussian noise (AWGN) was derived by Shannon and is given by Shannon (1948):

$$C = \log_2 (1 + \gamma_s) \quad (1)$$

where C is the ultimate spectral efficiency in b/s/Hz and γ_s is signal-to-noise ratio (SNR) in linear units. To approach Shannon's limit requires the signal to have Gaussian-distributed amplitude and the use of infinitely long forward error correction (FEC) codes (Mizuochi 2006).

In passband transmission, the sine and cosine functions are orthogonal, so in the complex representation of the signal, information can be encoded in both in-phase (I) and quadrature (Q) components (Proakis and Salehi 2001). A commonly used modulation format is quadrature amplitude modulation (QAM), where symbols are regularly spaced and arranged in a square or a cross on the complex plane. Due to the relative ease of generating such signals and its good power efficiency, QAM is commonly employed in wireless (Rappaport 1996), digital subscriber line (DSL) Sari et al. (1995), and optical systems (Ip et al. 2008). More advanced modulation formats such as iterative polar modulation (IPM) have also been proposed for optical transmission (Djordjevic et al. 2010). Inevitably, maximizing power efficiency requires utilizing all the degrees of freedom available for transmission. This necessitates a receiver capable of detecting both the signal's amplitude and its absolute phase.

The earliest fiber-optic systems were not particularly concerned with high spectral efficiency and used only a high-speed photodiode as a square-law detector of a signal's amplitude. This method of "direct detection" allowed a signal to be modulated only in positive amplitude, with on-off keying (OOK) being the simplest format with two amplitude levels, 0 (on) or 1 (off). Phase detection requires the interference of the signal of interest with a phase reference. The amplitude of the interference output depends on their relative phase, allowing phase to amplitude conversion. When the phase reference is a delayed copy of the signal, the detection method is known as "differential phase detection." Differential phase shift keying (DPSK) is a family of modulation formats where all the symbols have constant amplitude and information is encoded as a relative phase change between successive symbols. Differential binary phase shift keying (DBPSK) enjoys a 3 dB signal-to-noise ratio (SNR) advantage over OOK due to its constellation $\{-1, +1\}$ being bipolar instead of being unipolar $\{0, +1\}$. In the presence of fiber nonlinearity, DPSK enjoys further advantages over OOK when in-line dispersion compensation is employed, due to its constant amplitude (Liu 2004).

Coherent detection is a detection method where the received signal is interfered with an absolute phase reference. As both amplitude and phase are recovered, information can be encoded in both degrees of freedom, allowing the highest power efficiency. The absolute phase reference is usually provided by another laser, known as the local oscillator (LO), at the receiver. It is also possible to transmit a pilot tone alongside the signal of interest, however. The receiver can either extract the pilot tone by narrowband filtering and use it as an absolute phase reference; or the pilot can be placed on the orthogonal polarization from the signal, whereby a square-law detector achieves beating of the signal with the pilot tone (Miyazaki 2006). This method of detection is known as "self-coherent detection" and has certain advantages, such as improved tolerance to laser phase noise. Self-heterodyne orthogonal frequency-division multiplexing (OFDM) is another example of self-coherent detection (Armstrong 2009). Recently, the Kramers-Kronig receiver was proposed, where a signal's amplitude is detected using a square-law detector and, provided that the signal satisfies the condition of "minimum phase," its phase can be calculated from its amplitude by a Hilbert transform. However, this method also requires the transmission of a pilot (Mecozzi et al. 2016). Ultimately, self-coherent detection has worse performance than conventional coherent detection that uses a LO laser as phase reference. One reason is that noise accumulated during transmission causes the pilot tone to become an inaccurate phase reference. A second reason is that the presence of a pilot tone leads to enhanced nonlinear impairment due to cross-phase modulation (XPM). As a result, long-haul systems in commercial deployment today use the conventional approach. However, self-coherent detection can be useful short-reach and metro applications due to a simpler receiver.

Perhaps the most important factor that led to the resurgence in interest in coherent optical systems in the 2000s was the emergence of application-specific integrated circuits (ASICs) capable of performing digital signal processing (DSP) at gigahertz symbol rates. An optical channel is, to first degree, well modeled as a linear channel. If a coherent receiver can reconstruct of the signal's optical field, then channel impairments such as chromatic dispersion (CD) and polarization mode dispersion (PMD) can be efficiently compensated by DSP, achieving performance limited only by SNR. The flexibility afforded by DSP is a significant advantage. For example, before the advent of coherent systems, CD was traditionally compensated using in-line dispersion compensation fibers (DCF). System tolerance to residual dispersion is inversely proportional to the square of the symbol rate, because pulse broadening by CD is directly proportional to the signal's bandwidth, while the impact of the broadening is itself relative to the pulse duration, which is inversely proportional to the symbol rate. As a result, a 40-Gb/s DQPSK system (20-GHz baud rate) is four times as sensitive to residual dispersion as a 10-Gb/s OOK system, for example. To ensure low-power penalty, a transponder requires a tunable CD compensator to ensure that accumulated CD is as close to zero as possible prior to signal detection. But in an optical network, a signal may be dynamically routed due to link failure or change in network traffic. The use of tunable CD compensators can be cumbersome due to their limited tuning range and has typically required every span in an optical system to have in-line DCF roughly compensating the CD of the preceding span of transmission fiber. The use of in-line dispersion compensation leads to stronger nonlinear impairments, as signals of neighboring channels cannot decorrelate by the "walk-off" effect (Xie 2009). Furthermore, optical PMD compensation is difficult without coherent detection and usually limited to compensating first-order PMD only (Ozeki et al. 1994). When channel impairments are compensated in DSP, a system becomes vastly more flexible. Changing the amount of CD to be compensated requires no more than loading a different set of coefficients into a linear equalizer. Furthermore, adaptive algorithms enable rapid time-varying impairments such as PMD to be tracked at a speed comparable with the symbol rate, which is the order of magnitude faster than that achievable using a mechanical device. Linear equalization in DSP can further compensate such impairments as non-flatness in the channel's frequency response due to optical and electrical devices. A linear equalizer can adaptively converge to a solution that guarantees minimum mean square error (MMSE). Realizing such a compensator in analog hardware would be impossible. A coherent receiver employing DSP impairment compensation allows reuse of the same front-end hardware (the optical-to-electrical downconverter is discussed in section "Coherent Receiver") for every transponder, irrespective of the link, or the signal's parameters such as its modulation format, symbol rate, FEC code rate, and other parameters. Indeed, commercially available transponders today already allow not only flexible modulation format and tunable CD compensation, but entire functionalities can be turned on or off to enable dynamic flexible trade-off between performance and power consumption (Ishida et al. 2016).

The combination of coherent detection and DSP has also enabled new spectrally efficient multiplexing schemes (Huang et al. 2012). In conventional DWDM systems, optical channels are assigned on a fixed frequency grid, with the typical spacing between channels being 50 GHz and 100 GHz. A 100-Gb/s DWDM system on a fixed grid is shown in Fig. 1a. Due to soft-decision FEC and other overhead, the symbol rate is usually around 32 Gbaud, giving a raw data rate of around 128 Gb/s. But a 32-Gbaud signal with appropriate pulse shaping should only occupy a bandwidth slightly more than 32 GHz. When a 50 GHz grid is used, the gap between neighboring channels is wasted, reducing potential spectral efficiency. Recently, spectral "superchannels" have been used to realize line rates of 400 Gb/s or more per transponder. Figure 1b shows a 400-Gb/s superchannel realized using 4×100-Gb/s subcarriers which do not have to be at 50 GHz spacing. At the transmitter, the subcarriers are passively combined, while at the receiver, the signal can be passively split. A tunable LO at the receiver can be tuned to the appropriate frequency to downconvert the spectral region of interest for further processing. Since DSP can realize low-pass filters with near-rectangular frequency response much sharper than achievable using analog optics or electronics, the subcarriers can be placed much closer together than is possible without coherent detection. As an example, the subcarriers in a 4×100-Gb/s system may be spaced at 37.5 GHz, with the resulting superchannel occupying only 150 GHz of bandwidth, compared with 200 GHz using conventional WDM. The use of coherent receiver in conjunction with reconfigurable optical add/drop multiplexers (ROADMs) based on liquid crystal on silicon (LCos) having granularity of a few GHz has enabled nearly gapless transmission over the entire C- and L-bands and large increases in spectral efficiency.

Coherent detection was first introduced for optical communication in the 1980s (Okoshi and Kikuchi 1981; Okoshi 1982; Malyon 1984; Yamamoto and Kimura 1981). The performance of homodyne versus heterodyne receivers limited by quantum noise was studied in Shapiro (1985), while phase-locking requirements

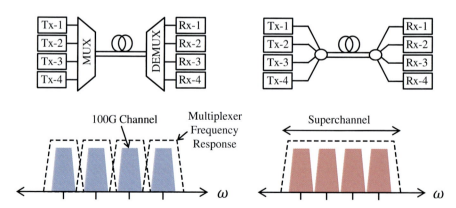

Fig. 1 Channel multiplexing using (**a**) conventional WDM with optical MUX/DEMUX and (**b**) superchannels with passive splitters

for optical homodyne receivers were studied in Kazovsky (1986). Dual-polarization coherent detection was proposed in Glance (1987). The potential for chromatic dispersion compensation in the electronic domain was studied in Iwashita (1990). An early demonstration of coherent binary phase shift keying (BPSK) was reported in Kahn et al. (1990), while quadriphase shift keying (QPSK) detection soon followed (Kahn et al. 1992). In these early experiments, laser phase noise was a key limitation due to the larger carrier-linewidth-to-symbol-rate ratio compared to wireless systems, and the early experiments used optical phase-locked loops (OPLL) which have significant feedback delays.

Following a hiatus in activity, coherent detection made a comeback in the mid-2000s. The potential of coherent QPSK for long-haul transmission was identified (Tsukamoto et al. 2005; Ly-Gagnon et al. 2006). Feedforward carrier phase recovery was proposed in Taylor (2005). Digital signal processing algorithms and subsystems for coherent optical receivers were studied at length in Ip et al. (2008) and Savory (2010). The first commercial 40G and 100G coherent transponders were reported in Roberts (2007) and Kline (n.d.). The late-2000s saw steady progress in the improvement of raw spectral efficiency as well as spectral efficiency-distance product. As of today, the largest constellation size demonstrated for optical transmission is 2048-QAM achieving a spectral efficiency of 15.3 b/s/Hz (Beppu et al. 2015). The total net data rate achievable over single-mode fiber (SMF) using C-+L-bands is just over 100 Tb/s (Qian et al. 2011; Sano et al. 2012). For ultra-long-haul submarine transmission exceeding transatlantic distance, 70.4 Tb/s was recently demonstrated in Cai et al. (2017). All of these results were made possible by coherent detection, with heavy use of DSP at both transmitter and receiver performing such operations as spectral shaping, nonlinearity compensation, and advanced FECs.

This work will be organized as follows: In sections "Transmitter," "Coherent Receiver," and "The Fiber Channel," we will review the constituent parts of a coherent optical transmission system, from the transmitter to the channel model to the receiver. We will derive the analytical linear baseband model of the overall optical system, which is valid at least in the lower power regime. The following sections "Digital Signal Processing Preliminaries," "Linear Equalization," "Carrier Phase Estimation," "Nonlinear Compensation," and "Timing Recovery" will focus on digital signal processing algorithms. We will review commonly used algorithms for linear equalization, laser phase noise compensation, nonlinear compensation, and timing recovery. Section "Other Topics" will also review other recent topics relevant to the field of coherent detection.

Transmitter

Mach-Zehnder Modulator

Coherent optical transceivers today typically use external modulators based on lithium niobate ($LiNbO_3$) for electrical-to-optical (E/O) upconversion. These modulators make use of the electro-optic effect, where the refractive index of

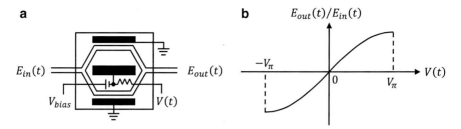

Fig. 2 (a) Mach-Zehnder modulator and its (b) amplitude characteristic

the waveguide material changes with applied voltage, resulting in phase modulation (Wooten et al. 2000). In a Mach-Zehnder (MZ) interferometer modulator (Fig. 2), the input optical wave is split into two arms, each undergoing phase modulation before they are interfered. The amplitude of the output wave depends on the relative phase shift between the two interfering arms. Compared with other modulator types, the MZ modulator has the advantages of high bandwidth, high extinction ratio, and zero chirp. These advantages have made MZ modulators ubiquitous in modern optical transceivers. It can be shown that the amplitude versus drive voltage characteristic of a MZ modulator is given by:

$$\frac{E_{out}(t)}{E_{in}(t)} = \sin\left(\frac{\pi}{2} \frac{(V(t) \otimes h_{mod}(t) + V_{bias})}{V_\pi} + \phi_{bias}\right) \qquad (2)$$

where $V(t)$ is the signal-bearing alternating current (AC) drive voltage, V_{bias} is a direct current (DC) bias voltage, ϕ_{bias} is the phase bias of the modulator, and V_π is the DC voltage required to swing the output power ratio $|E_{out}/E_{in}|^2$ from 0 to 1. Owing to bandwidth limitation of the electrical drive circuitry as well as mismatch between the group velocities of the optical wave and electrical drive signals, a MZ modulator will typically exhibit a low-pass filter response, which can be modeled by $h_{mod}(t)$.

A dual-quadrature in-phase (I) and quadrature (Q) modulator can be realized using two MZ modulators placed in parallel with an additional phase modulator (PM) on one arm to ensure that the two modulator outputs have relative phase shift of 90° before their fields are combined (Fig. 3). By further stacking two I/Q modulators in parallel with appropriate polarization optics, a dual-polarization MZ I/Q modulator can be realized. This modulator structure enables signal modulation on all the degrees of freedom available in single-mode fiber and is the most widely used modulator architecture in contemporary optical transceivers.

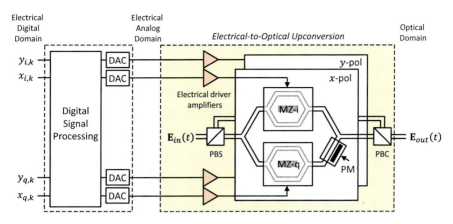

Fig. 3 Digital transmitter with dual-polarization Mach-Zehnder modulator

Signal Modulation

The MZ modulator characteristic in Eq. 2 is a nonlinear function (Fig. 2b). When transmitting higher-order modulation formats such as square-shaped QAM, the nonlinear characteristic will cause signal distortion. It is possible to restrict the amplitude of the drive signal to use only a small portion of the MZ characteristic. For example, setting $\max(V(t)) < 0.25 V_{\pi, ac}$ will reduce nonlinear distortion to only $\sim 2.5\%$. Using small amplitude signals will lead to poor SNR, however, due to noise in the electrical drive signal and finite modulator extinction ratio. If the electrical signal has sufficient amplitude to drive the MZ modulator into saturation, the small fluctuations can be suppressed.

Recently, "digital" transmitters have enabled the same flexibility at the transmitter as in a digital coherent receiver. The use of transmitter-side DSP in conjunction with digital-to-analog converter (DAC) generation of the drive waveform allows (i) the generation of arbitrary modulation format, (ii) shaping of the signal spectrum, (iii) linear pre-compensation of the frequency responses of the modulator and driver amplifiers, (iv) pre-compensation of the nonlinear characteristic of the MZ, and even (v) pre-compensation of link impairments such as transmitter-side digital backpropagation (see section "Nonlinear Compensation").

Analytical Baseband Model

A digital transmitter is shown in Fig. 3. In comparison with an analog optical transmitter represented by the E/O upconverter, a digital transmitter has an additional

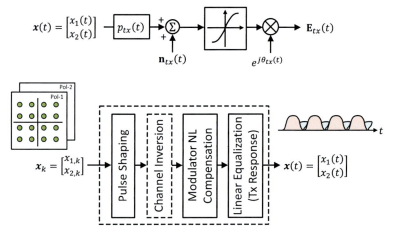

Fig. 4 (a) Analytical baseband model of the electrical-to-optical upconverter and (b) digital pre-equalization using transmitter-side DSP

front-end comprising a digital signal processor followed by parallel DACs that generates the analog I and Q drive signals for the two polarizations.

The digital transmitter can be described by the analytical baseband model shown in Fig. 4a. The drive signals form a complex-valued 2×1 vector $x(t) = \begin{bmatrix} x_{1i}(t) + jx_{1q}(t) \\ x_{2i}(t) + jx_{2q}(t) \end{bmatrix}^T$. Assuming the frequency responses of the modulator and DAC are the same for all four drive signals, it can be modeled as a low-pass impulse response $p_{tx}(t)$. The DACs add quantization noise, while the driver amplifiers add thermal noise. The total noises added by the transmitter can be modeled as an equivalent noise vector $\mathbf{n}_{tx}(t)$. The noisy low-pass filtered signal is passed through the MZ nonlinear characteristic, followed by modulation onto the optical carrier characterized by phase noise $\theta_{tx}(t)$. Apart from AWGN and phase noise, all other impairments of the electrical-to-optical upconverter are static and invertible. It is therefore possible to compensate them using the DSP architecture shown in Fig. 4b, where each impairment is inverted in the reverse order that they appear in Fig. 4a.

Coherent Receiver

Optical-to-Electrical Downconversion

Coherent detection is a detection method where the receiver computes decision variables based on recovery of the full electric field. This detection method enables information to be encoded in both amplitude and phase or, equivalently, in both the I and Q components of an optical carrier.

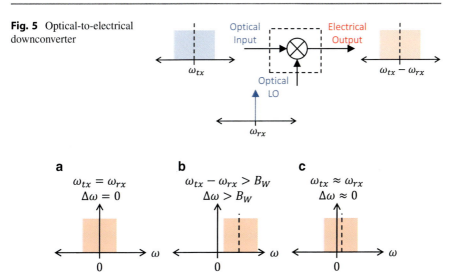

Fig. 5 Optical-to-electrical downconverter

Fig. 6 (**a**) Homodyne detection, (**b**) heterodyne detection, and (**c**) intradyne detection

The front end of a coherent receiver is an optical-to-electrical downconverter, as shown in Fig. 5. The input signal, modulated around a center frequency of ω_{tx}, is mixed with the receiver's local oscillator (LO) laser at a frequency of ω_{rx} serving as an absolute phase reference. The output electrical signal is centered around $\omega_{tx} - \omega_{rx}$. The case where $\omega_{tx} = \omega_{rx}$ is known as *homodyne detection* (Fig. 6a), whereas the case where $\omega_{tx} \neq \omega_{rx}$ is known as *heterodyne detection* (Fig. 6b). In heterodyne detection, the highest frequency of the electrical output signal is higher than in homodyne detection and is normally considered a disadvantage as it requires higher bandwidth components. Homodyne detection requires feedback to keep the LO locked to the signal's center frequency. Delay in the feedback circuit will lead to poor phase tracking performance (Barry and Kahn 1992). For these reasons, coherent receivers today typically use *intradyne detection* (Fig. 6c), where the intermediate frequency $\Delta\omega = \omega_{tx} - \omega_{rx} \approx 0$. This configuration allows the benefit of using of a free-running LO laser, and the bandwidth of the electrical output signal is only comparable to homodyne detection.

Single-Polarization Optical-to-Electrical Downconverter

An optical-to-electrical (O/E) downcoverter for single polarization can be constructed using a network of 3-dB couplers, a phase shifter, and photodiodes as shown in Fig. 7. Let the electric fields (E-field) of the input signal and LO be $E_s(t)$ and $E_{LO}(t)$, respectively. Since the Jones matrix of each 3-dB coupler is $\frac{1}{\sqrt{2}}\begin{bmatrix} 1 & j \\ j & 1 \end{bmatrix}$, the E-field incidents on the four photodiodes are:

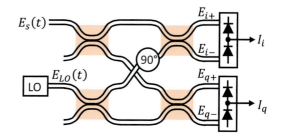

Fig. 7 Single-polarization optical-to-electrical downconverter constructed using four 3-dB couplers and four photodiodes

$$E_{i+} = \frac{1}{2}(E_s + E_{LO}) \tag{3}$$

$$E_{i-} = \frac{1}{2}(jE_s - jE_{LO}) \tag{4}$$

$$E_{q+} = \frac{1}{2}(jE_s - E_{LO}) \tag{5}$$

$$E_{q-} = \frac{1}{2}(-E_s + jE_{LO}) \tag{6}$$

Let the responsivities of the photodiodes be R_{i+}, R_{i-}, R_{q+} and R_{q-}, respectively. The output photocurrents are:

$$\begin{aligned} I_i &= \left(R_{i+}|E_{i+}|^2 - R_{i-}|E_{i-}|^2\right) + I_{i,sh} \\ &= \frac{R_{i+} - R_{i-}}{4}\left(|E_s|^2 - |E_{LO}|^2\right) + \frac{R_{i+} + R_{i-}}{2}\text{Re}\left(E_s E_{LO}^*\right) + I_{i,sh} \end{aligned} \tag{7}$$

$$I_q = \frac{R_{q+} - R_{q-}}{4}\left(|E_s|^2 - |E_{LO}|^2\right) + \frac{R_{q+} + R_{q-}}{2}\text{Im}\left(E_s E_{LO}^*\right) + I_{q,sh} \tag{8}$$

The spectra of the various terms in Eqs. 7 and 8 are shown in Fig. 8. $|E_s|^2$ and $|E_{LO}|^2$ are signal-signal and LO-LO beat terms and are interferences to the detection of the desired signal-LO beat terms that are proportional to the real and imaginary parts of $E_s E_{LO}^*$. The final terms $I_{i,\,sh}$ and $I_{q,\,sh}$ are shot noise produced by the photodiodes. As the amplitude of the LO laser is constant, LO-LO beating is a constant offset that can be remove by a DC blocker. Both signal-signal and LO-LO beat terms are weighted by the mismatch between the responsivities of the + and − photodiodes. As LO power is typically 20 dB higher than the signal power, and balanced photodiodes typically have common mode rejection ratios $CMRR = 2|R_+ - R_-|/(R_+ + R_-)$ greater than 25 dB, the interference terms are normally below the noise level and have negligible impact on signal detection.

The frequency downconversion property of the device in Fig. 7 is apparent from the signal-LO beat term $E_s E_{LO}^*$. Writing the modulated signal as $E_s(t) = E_{s,bb}(t)e^{j(w_{tx}t + \phi_{tx})}$, where ω_{tx} is the frequency of the transmitter's laser and

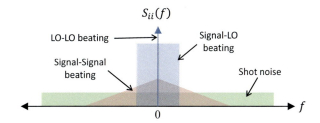

Fig. 8 Spectrum of the photocurrent output in an optical-to-electrical downconverter using balanced photodiodes

$E_{s,bb}(t)$ is baseband signal. Let $E_{LO}(t) = \sqrt{P_{LO}}e^{j(w_{rx}t+\phi_{rx})}$, where ω_{rx} and P_{LO} are the center frequency and power of the LO laser. We have $E_s E_{LO}^* = \sqrt{P_{LO}} E_{s,bb}(t) e^{j(\Delta\omega t + \Delta\phi)}$, which is the baseband signal modulated by the intermediate frequency $\Delta\omega = \omega_{tx} - \omega_{rx}$ as shown in Fig. 6.

To analyze the performance of a coherent receiver, we first note that shot noises $I_{i,sh}$ and $I_{q,sh}$ are AWGN terms with two-sided power spectral density (psd) (in units of $A^2 s/rad$) given by:

$$S_{I_i I_i}(\omega) = \frac{1}{2\pi} q \overline{I}_i \qquad (9)$$

$$S_{I_q I_q}(\omega) = \frac{1}{2\pi} q \overline{I}_q \qquad (10)$$

where $q = 1.6 \times 10^{-19}$ coulombs is the electronic charge and $\overline{I}_i \approx \left(\frac{R_{i+}+R_{i-}}{4}\right) P_{LO}$ and $\overline{I}_q \approx \left(\frac{R_{q+}+R_{q-}}{4}\right) P_{LO}$ are sums of the root-mean-square (RMS) photocurrents of each balanced photodiode pair. At high LO-to-signal power ratio, these terms are approximately proportional to LO power only.

Single-Sided Photodetection

The equations from Eqs. 7, 8, 9, and 10 can also describe single-sided detection by setting R_{i-} and R_{q-} to 0. In this case, signal-signal interference will not be canceled, and the desired signal-LO beat terms $\frac{R_{i+}}{2} Im(E_s E_{LO}^*)$ and $\frac{R_{q+}}{2} Re(E_s E_{LO}^*)$ will have only half the amplitude of balanced detection. The LO-LO beat term $|E_{LO}|^2$ can be removed by a DC blocker, while the signal-signal beat terms $\frac{R_{i+}}{4}|E_s|^2$ and $\frac{R_{q+}}{4}|E_s|^2$ can be made insignificant by increasing the LO-to-signal power ratio $\sqrt{P_{LO}/P_s}$. In terms of noise performance, halving the signal's amplitude means reducing signal *power* by a factor of four. By contrast, shot noise variance is only halved. If shot noise is the dominant noise mechanism – which is true in short-reach applications without in-line optical amplifiers, single-ended photodetection achieves only half the SNR of balanced photodetection. In long-haul systems, however, the received signal has typically passed through optical preamplifiers before O/E downconversion. Amplified spontaneous emission (ASE) from in-line optical amplifiers is normally the dominant noise source. Hence, in practice, single-ended detection will only degrade performance negligibly provided that $\sqrt{P_{LO}/P_s}$ is large enough so that the signal-to-interference ratio (SIR) from signal-signal beating is far below the noise level.

Heterodyne Detection

If $\Delta\omega = \omega_{tx} - \omega_{rx}$ is greater than the single-sided bandwidth of the signal (Fig. 6b), it is possible to drop the quadrature portion (bottom half) of the optical-to-electrical downconverter shown in Fig. 7. In this case, the photocurrent is:

$$I_i \propto Re\left(E_s E_{LO}^*\right) = \sqrt{P_{LO}} E_{s,bb} \cos\left(\Delta\omega t + \Delta\phi\right) \quad (11)$$

The spectrum of Eq. 11 is shown in Fig. 9. Due to the cosine, the signal-LO beat term has a mirror image in the negative frequencies. If the input signal has ASE noise from in-line amplifiers, care should be taken to filter out the noise in the "image band" before downconversion, as failure to do so will result in image-band noise folding back into the signa and a corresponding 3 dB degradation in SNR (Fig. 9b). Although the use of heterodyne detection halves the number of couplers and photodiodes required, the photocurrent will have a maximum frequency of $\Delta\omega + B_W \geq 2B_W$. This means that the photodiodes and all electrical components downstream, including transimpedance amplifiers (TIA) and digital-to-analog converters (DAC), will need at least twice the bandwidth required by homodyne or intradyne detection.

Dual-Polarization Optical-to-Electrical Downconversion

Since a single-mode fiber supports two polarization modes, a dual-polarization O/E downconverter is required to fully reconstruct the optical E-field. This can be accomplished using the structure shown in Fig. 10, where a LO laser is aligned at 45° relative to the reference x- and y-polarizations of the optical-to-electrical

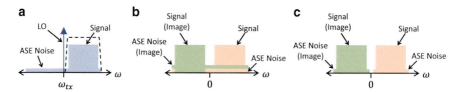

Fig. 9 (a) Optical spectrum of signal and LO with optional filtering (dotted) before O/E downconversion; electrical spectrum of O/E downconverter output (b) without and (c) with image-band filtering

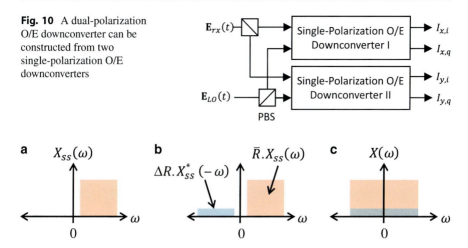

Fig. 10 A dual-polarization O/E downconverter can be constructed from two single-polarization O/E downconverters

Fig. 11 (a) Spectrum of a single side-band signal, (b) distortion arising from I/Q imbalance, and (c) I/Q balance is an in-band distortion for a dual side-band signal

downconverter. Polarization splitters are inserted in both the signal and LO paths so that each polarization component of the signal beats with its own LO in its dedicated O/E downconverter. The analog output photocurrents are the I and Q components of the two polarizations.

I/Q Imbalance

When a difference arises between the gains seen by the I and Q channels, the resulting signal distortion is known as I/Q imbalance. The impact of I/Q imbalance is most easily appreciated by considering a single side-band signal $x_{ss}(t)$ shown in Fig. 11, whose Fourier transform $X_{ss}(\omega)$ is non-zero only in the positive frequencies. Suppose the I and Q photodiode pairs are perfectly balanced, but $R_i \neq R_q$. Neglecting shot noise, the downconverted signal is:

$$I(t) = R_i Re(x_{ss}(t)) + j\, R_q Im(x_{ss}(t))$$
$$= \tfrac{R_i+R_q}{2}[Re(x_{ss}(t)) + j Im(x_{ss}(t))] + \tfrac{R_i-R_q}{2}[Re(x_{ss}(t)) - j Im(x_{ss}(t))]$$
$$= \tfrac{R_i+R_q}{2} x_{ss}(t) + \tfrac{R_i-R_q}{2} x_{ss}^*(t) \tag{12}$$

The first term reproduces the desired signal, while the second term is I/Q imbalance distortion and has a Fourier transform given by $x_{ss}^*(t) \overset{\mathcal{F}}{\leftrightarrow} X_{ss}^*(-\omega)$, so its frequency components are a mirror image of $x_{ss}(t)$. In this instance, because the signal is single-side band, the I/Q imbalance distortion $x_{ss}^*(t)$ falls outside of the bandwidth of the desired signal, so $x_{ss}(t)$ can be removed by filtering. For conventional double-side-band signals, however, I/Q imbalance will be an interference. The signal-to-interference ratio (SIR) is proportional to $(|\Delta R|/R)^2$.

If R_i and R_q were frequency-independent constants, it would be possible to eliminate I/Q distortion by rebalancing the amplitudes of the I and Q signals in DSP before complex addition. In practice, however, $R_i(\omega)$ and $R_q(\omega)$ are functions of frequency, arising from the difference between the frequency responses of the photodiodes, TIAs, and ADCs seen by each channel. Although $R_i(\omega)$ and $R_q(\omega)$ are static and can be digitally pre-compensated (Fig. 4 in section "Transmitter"), in practice, there is a limit to how precisely $R_i(\omega)$ and $R_q(\omega)$ can be measured. Even a small impedance mismatch in the downstream electrical circuit can result in reflections, which manifests as ripples in the frequency response or, equivalently, a long tail in the impulse response which requires many DSP taps to compensate. Note that I/Q imbalance also similarly arises at the transmitter from the difference between the frequency responses of the DACs, driver amplifiers, and modulator response for the I and Q channels. When a digital transmitter is used, these can be pre-compensated in DSP, with the same limitations as receiver-side I/Q imbalance compensation.

Signal Conditioning Circuit

Following O/E downconversion, each output photocurrent in Fig. 10 would typically drive a transimpedance amplifier (TIA), followed by sampling by an analog-to-digital converter (ADC) (Fig. 11). The ADC can be modeled as a low-pass (anti-aliasing) filter (LPF) followed by a sample-and-hold circuit with a sampling interval of T. The sampled signal is then quantized (Fig. 12), and the outputs s_k are then ready to be processed by DSP. On occasion, there may be other elements between the TIA and the ADC such as DC blockers, which are high-pass filters (HPF) with very low cutoff frequency. The signal conditioning circuit introduces two additional noise sources: thermal noise of the TIA and quantization noise of the ADC.

Thermal noise can be referenced to the input of the TIA and is an AWGN term with two-sided psd (in units of $V^2 s/rad$) of:

$$S_{vv}(\omega) = \frac{1}{\pi} k_B T_{amp} R_L \qquad (13)$$

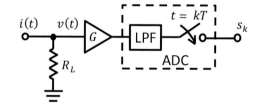

Fig. 12 Typical signal conditioning circuit in a coherent receiver comprising a transimpedance amplifier followed by an analog-to-digital converter (ADC)

where $k_B = 1.38 \times 10^{-23}$ J/K is Boltzmann's constant, T_{amp} is the "noise temperature" of the amplifier, and R_L is the load resistance, typically 50 Ω.

The ADC quantizes the analog signal to a fixed N_b bits at a quantization interval Δ. The range of signal amplitudes that can be represented is $-2^{N_b-1}\Delta \leq V \leq +\left(2^{N_b-1} - 1\right)\Delta$. If the amplitude of the ADC input falls outside of this range, the signal will be clipped. Clipping distortion is a highly nonlinear phenomenon. In long-haul transmission without in-line dispersion compensation, the amplitude distribution of I and Q in each polarization will be Gaussian. The quantization interval Δ can be selected to produce a predetermined clipping probability. Choosing a larger value of Δ will reduce clipping probability. However, the quantization process also adds quantization noise which is uniformly distributed between $-\Delta/2$ and $+\Delta/2$. As the variance of quantization noise is $\Delta^2/12$, there is a trade-off between minimizing clipping probability and minimizing quantization noise.

Analytical Baseband Model

Based on the foregoing discussion of the various elements, a coherent receiver can be represented using the analytical baseband model shown in Fig. 13. The O/E downconverter has impulse response $p_{rx,1}(t)$ and adds shot noise. The TIA has impulse response $p_{rx,2}(t)$ and adds thermal noise. Finally, the ADC has impulse response $p_{rx,3}(t)$, followed by the sampler, which adds quantization noise. We have assumed the frequency responses of all four I/Q tributaries to be identical. It is possible to combine all the impulse responses into a total impulse response of the receiver $p_{rx}(t)$, and refer all the noise sources to the input of the sampler as a single equivalent noise $\mathbf{n}_{rx}(t)$ as shown in Fig. 13.

In normal operation, shot noise should be much greater than either thermal noise or quantization noise, as according to Eqs. 7 and 8, the signal term $I = I_i + jI_q = R_p E_s E_{LO}^*$ can be increased by simply increasing the LO power, which is limited only by the power limitation of the photodiodes. The SNR of a shot noise limited

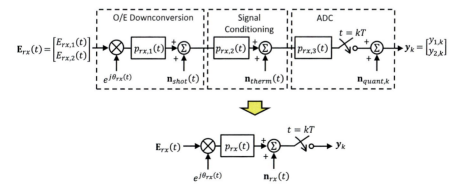

Fig. 13 Analytical baseband model of a coherent receiver

coherent receiver can be inferred from Eqs. 7, 8, 9, and 10. Assuming balanced photodetection, all photodiodes have responsivity of R_p, and the signal has baud rate $R_s = 1/T_s$; shot noise limited SNR is given by:

$$SNR_{shot} = \frac{R_p^2 P_{LO} P_s}{q R_p P_{LO} R_s} = \frac{R_p P_s T_s}{q} = \eta \bar{n}_s \qquad (14)$$

where $\eta = R_p h\nu/q$ is the quantum efficiency of the photodiodes, and $\bar{n}_s = P_s T_s/h\nu$ is the number of photons per symbol incident at the input of the O/E downconverter. Thus, Eq. 14 shows that shot noise limited SNR is equal to the number of *detected* photons per symbol (Ip et al. 2008; Agrawal 2002).

In optical systems without in-line amplifiers and at low power, Eq. 14 is usually a good model of a coherent system's performance. In long-haul systems where noise from in-line amplifiers dominates, shot noise creates a bit error rate (BER) floor, as shown in Fig. 14. The BER floor is the best achievable performance even in the absence of any other impairment in the optical fiber link.

Emulation of Other Detector Types

Since a dual-polarization O/E downcoverter recovers the complete optical E-field, it can emulate any signal detection method via digital signal processing in the electronic domain. Figure 15 and Eqs. 15, 16, 17, 18, and 19 illustrate how direct detection, differential detection, and Stokes parameters detection can be emulated:

$$A_k = |x_k|^2 + |y_k|^2 \qquad (15)$$

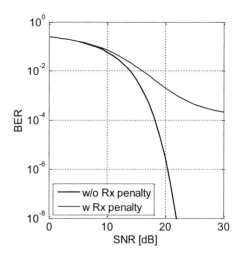

Fig. 14 Illustrating the impact of receiver noise causing a "noise floor" on the performance of 16-QAM. Receiver noise was set to an equivalent SNR of 18 dB

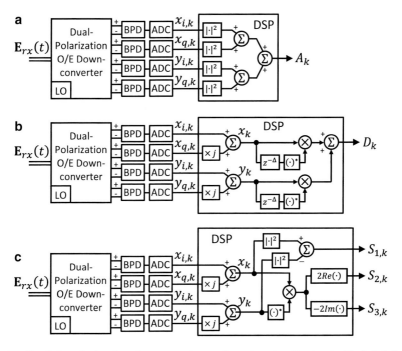

Fig. 15 Emulation of other detection types using a coherent receiver front-end with digital signal processing. (**a**) Direct detection, (**b**) differential detection, and (**c**) Stokes parameters detection

$$D_k = x_k x_{k-\Delta}^* + y_k y_{k-\Delta}^* \qquad (16)$$

$$S_{1,k} = |x_k|^2 - |y_k|^2 \qquad (17)$$

$$S_{2,k} = 2Re\{x_k y_k^*\} \qquad (18)$$

$$S_{3,k} = -2Im\{x_k y_k^*\} \qquad (19)$$

The Fiber Channel

Nonlinear Schrödinger Equation (NLSE)

Signal propagation in single-mode fiber (SMF) is described by a dual-polarization nonlinear Schrödinger equation (NLSE) (Evangelides et al. 1992):

$$\begin{aligned}\frac{\partial \mathbf{E}}{\partial z} &= \left(-\frac{1}{2}\alpha + \boldsymbol{\beta}_1\frac{\partial}{\partial t} - j\frac{1}{2!}\boldsymbol{\beta}_2\frac{\partial^2}{\partial t^2} - \frac{1}{3!}\boldsymbol{\beta}_3\frac{\partial^3}{\partial t^3}\right)\mathbf{E} + j\gamma\left[|\mathbf{E}|^2\mathbf{I} - \frac{1}{3}\left(\mathbf{E}^H\sigma_3\mathbf{E}\right)\sigma_3\right]\mathbf{E}\\
&\approx \left(-\frac{1}{2}\alpha + \boldsymbol{\beta}_1\frac{\partial}{\partial t} - j\frac{1}{2!}\boldsymbol{\beta}_2\frac{\partial^2}{\partial t^2} - \frac{1}{3!}\boldsymbol{\beta}_3\frac{\partial^3}{\partial t^3}\right)\mathbf{E} + j\frac{8}{9}\gamma|\mathbf{E}|^2\mathbf{E}\\
&= \left(\hat{\mathbf{D}} + \hat{\mathbf{N}}\right)\mathbf{E}\end{aligned} \qquad (20)$$

where $\hat{\mathbf{D}}$ and $\hat{\mathbf{N}}$ are the linear and nonlinear operators and $\mathbf{E}(z,t) = [E_1(z,t)\ E_2(z,t)]^T$ is the Jones representation of the optical E-field. In the second line of Eq. 20, we have assumed the length scale of polarization rotation during transmission to be much shorter than the nonlinear length scale. This enables the nonlinear operator to be approximated by a Manakov equation (Marcuse et al. 1997). The 2×2 matrices α, $\boldsymbol{\beta}_1$, $\boldsymbol{\beta}_2$, and $\boldsymbol{\beta}_3$ are the fiber's loss, group velocity, dispersion, and dispersion slope, respectively, and allows arbitrary polarization dependence of each parameter to be described. $\sigma_3 = \begin{bmatrix} 0 & -j \\ j & 0 \end{bmatrix}$ is a Pauli spin matrix (Gordon and Kogelnik 2000). Normally, α, $\boldsymbol{\beta}_2$, and $\boldsymbol{\beta}_3$ have negligible polarization dependence, so they can be replaced by scalars. A short section of fiber can also be modeled as a first-order PMD section. By choosing a coordinate system for Eq. 20 which propagates at the mean group velocity between the fiber section's fast and slow axes, the group velocity tensor reduces to:

$$\boldsymbol{\beta}_1(z) = \boldsymbol{\beta}_1 = \mathbf{R}(\theta,\phi)\left(\frac{\delta}{2}\sigma_1\right)\mathbf{R}^H(\theta,\phi) \qquad (21)$$

where

$$\mathbf{R}(\theta,\phi) = \begin{bmatrix} \cos\theta\cos\phi - j\sin\theta\sin\phi & -\sin\theta\cos\phi + j\cos\theta\sin\phi \\ \sin\theta\cos\phi + j\cos\theta\sin\phi & \cos\theta\cos\phi + j\sin\theta\sin\phi \end{bmatrix} \qquad (22)$$

is a rotation matrix state that can represent any point on the Poincaré sphere in Stokes space (Gordon and Kogelnik 2000). The columns of $\mathbf{R}(\theta,\phi)$ are the Jones vectors of the principal states of polarization (PSPs) for the fiber section, while δ is the differential group delay (DGD) per unit length.

Linear Time-Invariant (LTI) Model

When signal power is low, an optical fiber can be modeled as an LTI channel. Consider the integration of Eq. 20 with the nonlinear term set to zero ($\hat{\mathbf{N}} = 0$). We have:

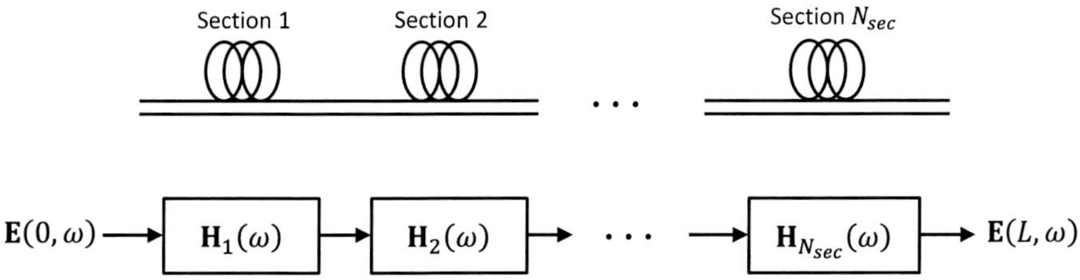

4 Optical Coherent Detection and Digital Signal Processing of Channel Impairments

Fig. 16 A single-mode fiber span can be modeled as the concatenation of first-order PMD sections

$$\mathbf{E}(L,\omega) = \exp\left(-\frac{\alpha}{2}L\right) \exp\left(j\left(\frac{\beta_2\omega^2}{2!} + \frac{\beta_3\omega^3}{3!}\right)L\right)$$
$$\mathbf{R}(\theta,\phi) \exp\left(j\frac{\omega\Delta\tau}{2}\sigma_1\right) \mathbf{R}^H(\theta,\phi) \, \mathbf{E}(0,\omega) = \mathbf{H}_{sec}(\omega) \, \mathbf{E}(0,\omega) \quad (23)$$

where $\mathbf{H}_{sec}(\omega)$ is the frequency response of the fiber section, $\mathbf{E}(0,\omega)$ and $\mathbf{E}(L,\omega)$ are the Fourier transforms of the E-field at the section's input and output, and $\Delta\tau = L\delta$ is the section's DGD. Higher-order PMD can be modeled by concatenating first-order PMD sections, with the PSP and DGD of each section being random variables (Lee et al. 2003) (Fig. 16). This allows the frequency response of a single-mode fiber span to be generalized as a left-sided product of the frequency responses of first-order PMD sections (Fig. 16):

$$\mathbf{H}^{(span)}(\omega) = \mathbf{H}_{N_{sec}}(\omega) \cdots \mathbf{H}_1(\omega) \quad (24)$$

A long-haul transmission system will typically comprise multiple concatenated spans of fibers. The loss of each fiber span is also compensated by in-line amplifiers, which are usually either EDFAs and/or Raman amplifiers (Fig. 17a). EDFAs have additive spontaneous emission (ASE) noise, while Raman amplifiers add spontaneous Raman scattering noise. In both cases, the noise electric field produced by an optical amplifier, referred to its input, can be modeled as an AWGN process with two-sided psd (in units of *W/Hz*) of:

$$\mathbf{S}_{E_n E_n}(\omega) = \frac{1}{4\pi} G\hbar\omega_{tx} (F-1) \mathbf{I} \quad (25)$$

where $\mathbf{I}$ is the 2×2 identity matrix, $\hbar\omega_{tx}$ is the energy per photon at the carrier frequency, G is the gain of the amplifier, and F is the amplifier's noise figure. For EDFAs, it can be shown that $F = 1 + n_{sp}\frac{G-1}{G}$, where n_{sp} is the spontaneous emission factor related to the inversion level of the gain medium (Haus 1998).

An LTI model of the long-haul transmission system is shown in Fig. 17b. The gains of each in-line amplifier have been subsumed into the frequency response of the preceding span of fiber, so $\mathbf{h}^{(i)}(t)$ is the frequency response of span i, and

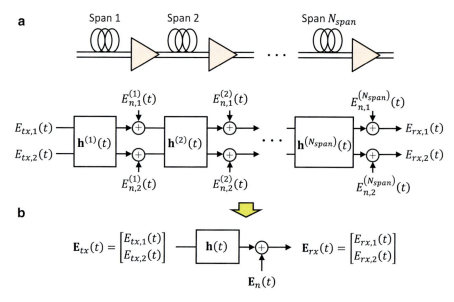

Fig. 17 (a) Long-haul transmission system with in-line amplification and (b) linear time-invariant (LTI) model

$E_n^{(i)}(t)$ is the noise added by the span. The E-fields at the input and output of the transmission system are then related by:

$$\mathbf{E}_{out}(t) = \mathbf{h}(t) \otimes \mathbf{E}_{in}(t) + \mathbf{E}_n(t) \qquad (26)$$

$$\mathbf{E}_{out}(\omega) = \mathbf{H}(\omega)\mathbf{E}_{in}(\omega) + \mathbf{E}_n(\omega) \qquad (27)$$

where $\mathbf{h}(t) \overset{\mathcal{F}}{\leftrightarrow} \mathbf{H}(\omega) = \prod_{i=1}^{N_{span}} \mathbf{H}^{(i)}(\omega)$ is the impulse/frequency response of the channel and $\mathbf{E}_n(t) \overset{\mathcal{F}}{\leftrightarrow} \mathbf{E}_n(\omega)$ is the equivalent noise added by all the amplifiers, referred to the channel output.

Digital Signal Processing Preliminaries

Introduction

The main advantage of coherent detection is that the recovery of the optical E-field enables compensation of channel impairments using DSP in the electronic domain. In sections "Transmitter," "Coherent Receiver," and "The Fiber Channel," we derived analytical baseband models for the transmitter, the fiber channel, and the

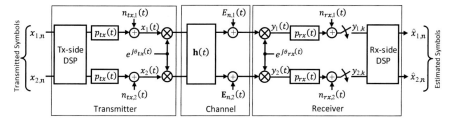

Fig. 18 Analytical baseband model of a complete coherent optical transmission system

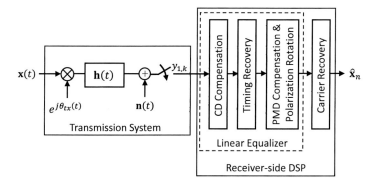

Fig. 19 Rx-side DSP compensation of an LTI channel and laser phase noise

receiver. Figure 18 shows the concatenation of these elements forming the complete optical system.

The transmitter response $p_{tx}(t)$ is normally static and can be measured and pre-compensated by a linear equalizer at the transmitter. Provided that the receiver's phase noise $\theta_{rx}(t)$ evolves "slowly" relative to the total memory length of the channel, it can be moved before the channel and combined with the transmitter's phase noise $\theta_{tx}(t)$ into a total phase noise process $\theta(t) = \theta_{tx}(t) + \theta_{rx}(t)$. The receiver response $p_{rx}(t)$ can then be subsumed into $\mathbf{h}(t)$, while the total equivalent noise $\mathbf{n}(t)$ added by the system can be referred to the channel output. Figure 19 thus shows the simplified LTI model of the system in Fig. 18, with $\theta(t)$ being a time-varying phase rotation of the modulated signal. This system can be inverted via the receiver-side operations shown on the right-hand side of Fig. 19. After sampling by the ADC at a synchronous rate, the residual CD, PMD, and group delay of the channel are compensated by a linear equalizer, followed by compensation of $\theta(t)$ by carrier recovery.

In submarine systems without in-line CD compensation, the channel's memory may be up to 100 ns long. The receiver's phase noise $\theta_{rx}(t)$ may not necessarily evolve non-negligibly over the channel's memory. The convolution of $\exp(j\theta_{rx}(t))$ with the equalizer's response will lead to "equalization-enhanced phase noise" (EEPN) which degrades system performance (Shieh and Ho 2008).

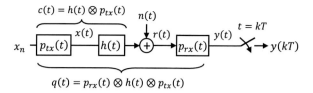

Fig. 20 Canonical model of a simple scalar channel

Sampling Rate Requirement

Consider the canonical model of a scalar channel shown in Fig. 20, where complex-valued symbols x_n are modulated by the transmitter's impulse response (pulse shape) $p_{tx}(t)$. The scalar LTI channel has an impulse response of $h(t)$ and adds noise of $n(t)$, with $c(t) = h(t) \otimes p_{tx}(t)$ being the received pulse shape. The receiver has impulse response $p_{rx}(t)$ followed by sampling at a rate of $R = 1/T$. We can write the sampled signal as:

$$y_k \triangleq y(kT) = \sum_n x_n q(kT - nT_s) + n'(kT) \tag{28}$$

where $q(t) = p_{rx}(t) \otimes h(t) \otimes p_{tx}(t)$ is total system impulse response and $n'(t) = p(t) \otimes n(t)$ is the noise after receiver filtering.

It is a well-known result in digital communications that the optimum receiver is a "matched filter" $p_{rx}(t) = c^*(-t)$ followed by a symbol-rate sampler ($T = T_s$) (Proakis and Salehi 2001). This receiver is optimum because it maximizes the SNR at the sampling instances. In addition, if $q(t) \overset{\mathcal{F}}{\leftrightarrow} Q(\omega)$ satisfies the Nyquist criterion:

$$\frac{1}{T_s} \sum_{k=-\infty}^{+\infty} Q\left(\omega - \frac{2\pi k}{T_s}\right) = 1, \tag{29}$$

then $q(t)$ will be zero at all the sampling instances except for the symbol of interest, resulting in a system impulse response that is free of intersymbol interference (ISI):

$$q(nT_s) = \begin{cases} 1 & n = 0 \\ 0 & otherwise \end{cases} \tag{30}$$

The absence of ISI allows low-complexity symbol-by-symbol detection without loss of performance. If ISI is non-zero, it will be necessary to use maximum-likelihood sequence estimation (MLSE) whose complexity scales exponentially with the ISI memory length (Proakis and Salehi 2001).

For systems where the signal does not experience frequency-selective fading, root-raised cosine (RRC) pulses are commonly used, since after a matched filter $p_{rx}(t) = p_{tx}^*(-t) = p_{rrc}^*(-t)$, the resultant impulse response $q(t)$ is a raised cosine (RC) function whose Fourier transform satisfies Eq. 29:

$$p_{rrc}(t) = \begin{cases} \frac{1}{T_s}\left(1+\beta\left(\frac{4}{\pi}-1\right)\right), & t=0 \\ \frac{\beta}{\sqrt{2T_s}}\left[\left(1+\frac{2}{\pi}\right)\sin\left(\frac{\pi}{4\beta}\right)+\left(1-\frac{2}{\pi}\right)\cos\left(\frac{\pi}{4\beta}\right)\right], & t=\pm\frac{T_s}{4\beta} \\ \frac{1}{\pi}\dfrac{\sin\left(\pi(1-\beta)\frac{t}{T_s}\right)+4\beta\frac{t}{T_s}\cos\left(\pi(1+\beta)\frac{t}{T_s}\right)}{t\left[1-\left(4\beta\frac{t}{T_s}\right)^2\right]}, & \text{otherwise} \end{cases} \quad (31)$$

$$P_{rrc}(\omega) = \begin{cases} 1, & |\omega| \leq \pi\frac{1-\beta}{T_s} \\ \sqrt{\frac{1}{2}\left[1+\cos\left(\frac{T_s}{2\beta}\left[|\omega|-\pi\frac{1-\beta}{T_s}\right]\right)\right]}, & \pi\frac{1-\beta}{T_s} \leq |\omega| \leq \pi\frac{1+\beta}{T_s} \\ 0, & \text{otherwise} \end{cases} \quad (32)$$

Symbol-rate sampling has disadvantages, however. Firstly, the exact channel impulse response $h(t)$ may not be known in advance and may be time-varying. Secondly, even if $h(t)$ is known, it may not be possible to synthesize the analog matched filter $p_{rx}(t) = c^*(-t)$ accurately. It is far easier to synthesize $p_{rx}(t)$ in DSP, where arbitrary filter shapes with arbitrarily sharp roll-off can be implemented. Digital filters can also adapt quickly to a time-varying channel.

In order for the sampled signal y_k to represent the analog signal $y(t)$ with no loss of information, it is necessary to sample at a rate above the "Nyquist frequency." Consider writing the sampled signal as the multiplication of $y(t)$ with a pulse train:

$$y_\delta(t) = y(t) \cdot \sum_{n=-\infty}^{+\infty} \delta(t-nT) \quad (33)$$

where $\delta(t)$ is the Dirac delta function. Since $\sum_{n=-\infty}^{+\infty}\delta(t-nT) \overset{\mathcal{F}}{\leftrightarrow} (1/T)\sum_{k=-\infty}^{+\infty}\delta(\omega-2\pi k/T)$, the Fourier transform of Eq. 33 is:

$$Y_\delta(\omega) = \frac{1}{T}\sum_{k=-\infty}^{+\infty} Y\left(\omega-\frac{2\pi k}{T}\right) \quad (34)$$

It is observed that sampling in time corresponds to replication in frequency, with the frequency replicas being separated by $2\pi/T$ (Fig. 21). Provided that the highest frequency component of $y(t)$ is less than π/T, the replicas will not overlap, and the sampled signal has all the essential information of $y(t)$. Sampling below the Nyquist frequency will result in "aliasing," which is the partial overlap of the frequency replicas. In the absence of aliasing, $y(t)$ can be reconstructed from $y_\delta(t)$ using a rectangular filter between $-\pi/T \leq \omega \leq \pi/T$. A digital filter $w_\delta(t) \overset{\mathcal{F}}{\leftrightarrow} W_\delta(\omega)$ which operates over $-\pi/T \leq \omega \leq \pi/T$ will perform the same operation as the equivalent analog filter $w(t)$. The design of a linear equalizer is covered in section "Linear Equalization."

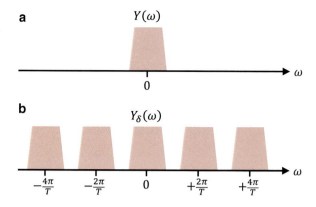

Fig. 21 Sampling in time corresponds to replication in frequency

Discrete-Time Fourier Transform

Consider an alternative method of writing the sampled signal in Eq. 33:

$$y_\delta(t) = y(t) \cdot \sum_{n=-\infty}^{+\infty} \delta(t - nT) = \sum_{n=-\infty}^{+\infty} y_n \delta(t - nT) \qquad (35)$$

where $y_n \triangleq y(nT)$ are the samples of the continuous-time signal. The Fourier transform of Eq. 35 is:

$$Y_\delta(\omega) = \int_{-\infty}^{+\infty} \sum_{n=-\infty}^{+\infty} y_n \delta(t - nT) \, e^{-j\omega t} \, dt = \sum_{n=-\infty}^{+\infty} y_n e^{-j\omega nT} \qquad (36)$$

The right-hand side of Eq. 36 is referred to as the discrete-time Fourier transform (DTFT) of y_n. Since Eqs. 34 and 36 are both Fourier transforms of the same signal, they have exactly the same value for all ω. The DTFT is thus no more than the Fourier transform of a continuous-time signal sampled by a pulse train with period T. The DTFT and its inverse are given by the following pair:

$$Y\left(e^{j\omega T}\right) = \sum_{n=-\infty}^{+\infty} y_n e^{-j\omega nT} \qquad (37)$$

$$y_n = \frac{T}{2\pi} \int_{-\pi/T}^{+\pi/T} Y\left(e^{j\omega T}\right) e^{j\omega nT} \, d\omega \qquad (38)$$

In addition to the DTFT, it is also common to design digital filters using the z-transform, which is the digital equivalent of the Laplace transform for continuous signals. The z-transform and its inverse are given by the following pair (Jury 1964):

$$Y(z) = \sum_{n=-\infty}^{+\infty} y_n z^n \tag{39}$$

$$y_n = \frac{1}{2\pi j} \oint_C Y(z) z^{n-1} \, dz \tag{40}$$

Linear Equalization

Minimum Mean Square Error (MMSE) Equalizer

For a static LTI channel impaired only by AWGN, the optimum linear equalizer is the "Wiener filter" which minimizes the mean squared error (MSE) between the equalizer output and the signal of interest. Consider again the baseband model of a dual-polarization transmission system (Fig. 22).

In the absence of laser phase noise and nonlinear effects, the signal just before the sampler is:

$$\mathbf{y}(t) = \mathbf{q}(t) \otimes \mathbf{x}(t) + \mathbf{n}'(t) \tag{41}$$

where $\mathbf{x}(t)$ is the baseband analog transmitted signal, $\mathbf{q}(t) = p_{rx}(t) \otimes \mathbf{h}(t) \otimes p_{tx}(t)$ is the total impulse response matrix of the system, and $\mathbf{n}'(t) = p_{rx}(t) \otimes \mathbf{n}(t)$ is the noise after receiver filtering, with $\mathbf{n}(t)$ being an AWGN process with two-sided psd of N_0 in each polarization. For single-mode fiber impaired by CD and first-order PMD, the channel matrix is:

$$\mathbf{h}(t) = \exp\left(-j\frac{1}{2}\int_0^L \beta_2(z)\, dz\, \omega^2\right) \mathbf{R}_1^{-1} \mathbf{D} \mathbf{R}_2 \tag{42}$$

where $\int_0^L \beta_2(z)\, dz$ is the accumulated dispersion of the link, $\mathbf{R}_1$ and $\mathbf{R}_2$ are unitary matrices that rotate the reference coordinates to the principle state of polarization

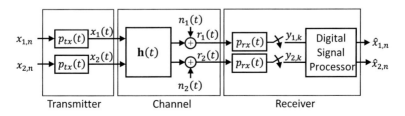

Fig. 22 Baseband model of a dual-polarization digital coherent receiver

(PSP) of the channel, and $\mathbf{D} = diag\left(e^{j\omega\tau_{DGD}/2}, e^{-j\omega\tau_{DGD}/2}\right)$ is the diagonal matrix of the link's DGD.

We assume $\mathbf{y}(t)$ is synchronously sampled at a rate of $R = (M/K)R_s$, where R_s is the symbol rate and M and K are integers. The two polarization components of the sampled signal are:

$$y_{i,k} \triangleq y_i(kT) = \sum_n \sum_{j=1}^{2} x_{j,n} q_{ij}(kT - nT_2) + n'_i(kT) \tag{43}$$

with $q_{ij}(t) = p_{rx}(t) \otimes h_{ij}(t) \otimes p_{tx}(t)$ and $n'_i(t) = p_{rx}(t) \otimes n_i(t)$.

The sampled signal is passed through a linear equalizer $\mathbf{w}$ whose outputs are estimates of the transmitted symbols:

$$\hat{\mathbf{x}}_k = \begin{bmatrix} \hat{x}_{1,k} \\ \hat{x}_{2,k} \end{bmatrix} = \begin{bmatrix} \mathbf{w}_{11}^T & \mathbf{w}_{21}^T \\ \mathbf{w}_{12}^T & \mathbf{w}_{22}^T \end{bmatrix} \begin{bmatrix} \mathbf{y}_{1,k} \\ \mathbf{y}_{2,k} \end{bmatrix} = \mathbf{w}^T \mathbf{y}_k \tag{44}$$

where $\mathbf{w}_{ij}^T = \begin{bmatrix} w_{ij,-L} & w_{ij,-L+1} & \cdots & w_{ij,L} \end{bmatrix}^T$ are column vectors of length $N = 2L + 1$, and $\mathbf{y}_{i,k}^T = \begin{bmatrix} y_{i,\lfloor \frac{kM}{K} \rfloor + L} & y_{i,\lfloor \frac{kM}{K} \rfloor + L-1} & \cdots & y_{i,\lfloor \frac{kM}{K} \rfloor - L} \end{bmatrix}^T$ are vectors of the N samples closest to symbol k, with $\lfloor x \rfloor$ denoting the largest integer less than or equal to x. Let $\boldsymbol{\varepsilon}_k = \mathbf{x}_k - \hat{\mathbf{x}}_k$ be the error vector. The filter which minimizes the mean square error (MSE) $tr\left(E\left[\boldsymbol{\varepsilon}_k^* \boldsymbol{\varepsilon}_k^T\right]\right)$ is given by:

$$\mathbf{w}_{opt}^T = \mathbf{A}^{-1} \boldsymbol{\alpha} \tag{45}$$

where $\mathbf{A} = E\left[\mathbf{y}_k^* \mathbf{y}_k^T\right]$ is the autocorrelation matrix of the samples and $\boldsymbol{\alpha} = E\left[\mathbf{y}_k^* \mathbf{x}_k^T\right]$ is the cross-correlation between the samples and the symbol of interest. The matrices $\mathbf{A}$ and $\boldsymbol{\alpha}$ can be partitioned as:

$$\mathbf{A} = \begin{bmatrix} \mathbf{A}_{11} & \mathbf{A}_{12} \\ \mathbf{A}_{21} & \mathbf{A}_{22} \end{bmatrix} \tag{46}$$

$$\boldsymbol{\alpha} = \begin{bmatrix} \boldsymbol{\alpha}_{11} & \boldsymbol{\alpha}_{12} \\ \boldsymbol{\alpha}_{21} & \boldsymbol{\alpha}_{22} \end{bmatrix} \tag{47}$$

The sub-matrices $\mathbf{A}_{ij} = E\left[\mathbf{y}_{i,k}^* \mathbf{y}_{j,k}^T\right]$ are of size $N \times N$, and the column vectors $\boldsymbol{\alpha}_{ij} = E\left[\mathbf{y}_{i,k}^* \mathbf{x}_{j,k}^T\right]$ are of length N. Let $A_{ij, lm}$ be the element at the intersection of the l-th row and m-th column of sub-matrix $\mathbf{A}_{ij}$. Similarly, let $\alpha_{ij,l}$ be the element at the l-th row of $\boldsymbol{\alpha}_{ij}$. Assuming the transmitted symbols are independent and identically distributed (i.i.d), i.e., they satisfy $E\left[x_{i,n_1} x_{j,n_2}^*\right] = P_x \delta_{ij n_1 n_2}$, where P_x is the signal power and δ is the Kronecker delta, we have:

$$A_{ij,lm} = P_x \sum_n \sum_{s=1}^{2} q_{is}^* \left(\left\lfloor \frac{kM}{K} \right\rfloor T - lT - nT_s \right) q_{js} \left(\left\lfloor \frac{kM}{K} \right\rfloor T - lT - nT_s \right)$$

$$+ N_0 \delta_{ij} \int_{-\infty}^{\infty} p_{rx}^*(t - lT) \, p_{rx}(t - mT) \, dt, \quad for -L \leq l, m \leq L \tag{48}$$

and

$$\alpha_{ij,l} = P_x q_{ij}^* \left(\left\lfloor \frac{kM}{K} \right\rfloor T - lT - kT_s \right), \quad for -L \leq l \leq L \tag{49}$$

It is observed that Eqs. 48 and 49 can have K different values depending on mod(kM, k). This is because when fractional oversampling is used ($K \neq 1$), the sampling instants relative to the symbol clock can take on K different patterns. Figure 23 illustrates an example for the case of 3/2-sampling. Odd symbols are sampled at their peaks, while even symbols have their peak straddling two samples. The equalizer thus has two different sets of coefficients for even and odd values of k. For a rate M/K system where M and K are relatively prime, the K different sets of coefficients for $\mathbf{w}_{opt}$ can be obtained by evaluating Eqs. 48 and 49 and all possible values of $mod(kM, k)$ and then computing Eq. 45.

Although fractional oversampling can reduce algorithmic complexity, 2× oversampling is still preferred in commercial systems. This is due to simpler symbol clock recovery. Moreover, in an adaptive implementation, an error signal for $\mathbf{w}_{opt}$ can only be generated every K symbols, so integer oversampling ($K = 1$) allows faster changes to be tracked.

Equalizer Length Requirement

A key parameter of the linear equalizer is its length $N = 2L + 1$. In order to compensate the channel accurately, NT needs to span the memory length of the channel. An example of a MMSE equalizer is shown in Fig. 24 assuming transmission over 1000 km of standard single-mode fiber (SSMF) with $D = 17$ ps/nm/km or $\beta_2 = 2.17 \times 10^{-26}$ s^2/m and without in-line CD compensation. Polarization effects are ignored in order to focus on a scalar channel with CD only. The signal uses RRC pulses with 1% roll-off at a symbol rate of 32 GHz, and $M/K = 2\times$

Fig. 23 In rate 3/2 oversampling, odd symbols are sampled at their peaks, whereas even symbols have their peaks midway between two samples

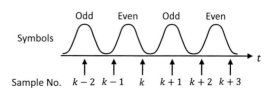

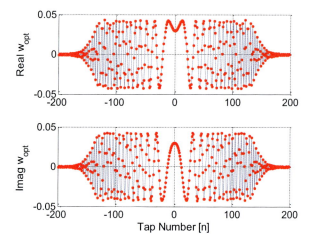

Fig. 24 Real and imaginary parts of the MMSE equalizer for a channel with 17,000 ps/nm of uncompensated CD. The signal uses RRC pulses with 1% roll-off at a symbol rate of 32 GHz, and 2× oversampling is assumed

oversampling ($1/T = 64$ GHz) is assumed. It is observed that the amplitude of the filter taps is non-negligible over a duration $2\pi|\beta_2 L|_{uncomp} R_s^2 (M/K) = 280$ taps. This is because $2\pi|\beta_2 L|_{uncomp} R_s$ is the group delay spread between the fastest and slowest frequency components of the signal, and the group delay spread is sampled at a rate of $R = (M/K)R_s$.

In a fiber channel, uncompensated CD is usually the greatest contributor to channel memory length. By contrast, PMD scales only as the square root of distance and is usually only a few symbols long. It is normally possible to use a filter length setting of $N = N_{CD} + N_{PMD} \approx N_{CD}$, where (Ip and Kahn 2007a):

$$N_{CD} \approx 2\pi|\beta_2 L|_{uncomp} R_s^2 (M/K) \qquad (50)$$

$$N_{PMD} \approx |\tau_{PMD}| R_s^2 (M/K) \qquad (51)$$

where $|\tau_{PMD}|$ is the DGD of the fiber channel. As $|\tau_{PMD}|$ is a Maxwellian distributed variable, setting Eq. 51 to around three times the mean DGD $\langle|\tau_{PMD}|\rangle = D_{PMD}\sqrt{L}$, where D_{PMD} is the PMD parameter of the fiber, is usually sufficient. Using an equalizer with fewer taps than the channel memory length would result in uncompensated ISI, requiring MLSE rather than symbol-by-symbol detection. Since CD and PMD are invertible using a linear equalizer whose complexity scales either linearly (time-domain implementation) or to the square root of memory length (frequency-domain implementation), a linear equalizer is preferable to MLSE.

MMSE Performance

We can analyze the performance of the MMSE equalizer by considering the DTFT of Eq. 41:

$$\mathbf{Y}\left(e^{j\omega T}\right) = \mathbf{Q}\left(e^{j\omega T}\right)\mathbf{X}\left(e^{j\omega T}\right) + \mathbf{N}'\left(e^{j\omega T}\right) \tag{52}$$

For convenience, we will drop the bracketed $e^{j\omega T}$ from here onwards. The error between the transmitted symbols and the equalizer output is:

$$\boldsymbol{\varepsilon} = \mathbf{X} - \mathbf{W}^T\mathbf{Y} = \left(\mathbf{I} - \mathbf{W}^T\mathbf{Q}\right)\mathbf{X} - \mathbf{W}^T\mathbf{N}' \tag{53}$$

Assuming the symbols are i.i.d. satisfying $E\left[x_{i,n_1} x^*_{j,n_2}\right] = P_x \delta_{ij} \delta_{n_1 n_2}$, and the noises in the two polarizations are uncorrelated satisfying $E[n_i(t) n_j(t-\tau)] = N_0 \delta_{ij} \delta(\tau)$, the MSE matrix is given by:

$$E\left[\boldsymbol{\varepsilon}\boldsymbol{\varepsilon}^H\right] = P_x\left(\mathbf{I} - \mathbf{W}^T\mathbf{Q}\right)\left(\mathbf{I} - \mathbf{W}^T\mathbf{Q}\right)^H - S_{N'N'}\mathbf{W}^T\mathbf{W}^* \tag{54}$$

where $S_{N'N'}$ is the psd of channel noise after receiver filtering:

$$S_{N'N'}\left(e^{j\omega T}\right) = N_0 \frac{1}{T^2} \sum_{m=-\infty}^{\infty} \left| P\left(\omega - \frac{2\pi m}{T}\right) \right|^2 \tag{55}$$

We can decompose the system frequency response $\mathbf{Q}$ in terms of its singular-value decomposition (SVD) $\mathbf{Q} = \mathbf{U}\mathbf{S}\mathbf{V}^H$. Then:

$$\begin{aligned} E\left[\boldsymbol{\varepsilon}\boldsymbol{\varepsilon}^H\right] &= P_x\left(\mathbf{I} - \mathbf{W}^T\mathbf{U}\mathbf{S}\mathbf{V}^H - \mathbf{V}\mathbf{S}\mathbf{U}^H\mathbf{W}^*\right) \mp \mathbf{W}^T\mathbf{U}\left(P_x\mathbf{S}^2 + S_{N'N'}\mathbf{I}\right)\mathbf{U}^H\mathbf{W}^* \\ &= \left(\mathbf{W}^T\mathbf{U} - P_x\mathbf{V}\mathbf{S}(P_x\mathbf{S}^2 + S_{N'N'}\mathbf{I})^{-1}\right)\left(P_x\mathbf{S}^2 + S_{N'N'}\mathbf{I}\right) \\ &\quad \left(\mathbf{W}^T\mathbf{U} - P_x\mathbf{V}\mathbf{S}(P_x\mathbf{S}^2 + S_{N'N'}\mathbf{I})^{-1}\right)^H \\ &\quad + \left(P_x - P_x^2\mathbf{V}\mathbf{S}(P_x\mathbf{S}^2 + S_{N'N'}\mathbf{I})^{-1}\mathbf{S}\mathbf{V}^H\right) \end{aligned} \tag{56}$$

The optimum equalizer minimizes MSE, which is given by $\mathrm{tr}(E[\boldsymbol{\varepsilon}\boldsymbol{\varepsilon}^H])$, and is therefore also equal to the sum of the eigenvalues of $E[\boldsymbol{\varepsilon}\boldsymbol{\varepsilon}^H]$. As $\left(P_x\mathbf{S}^2 + S_{N'N'}\mathbf{I}\right)$ is positive definite, MSE is minimized by setting the first term of Eq. 56 to zero. The frequency-domain representation of the MMSE filter is therefore:

$$\mathbf{W}^T_{opt} = P_x\mathbf{V}\mathbf{S}\left(P_x\mathbf{S}^2 + S_{N'N'}\mathbf{I}\right)^{-1}\mathbf{U}^H \tag{57}$$

An alternative expression for the equalizer coefficients in Eq. 45 can be obtained by taking the inverse DTFT of Eq. 57:

$$\mathbf{w}_{opt,n} = \frac{T}{2\pi} \int_{-\pi/T}^{\pi/T} \mathbf{W}_{opt}\left(e^{j\omega T}\right) e^{j\omega nT}\, d\omega, \quad for -L \leq n \leq L \qquad (58)$$

When $\mathbf{W} = \mathbf{W}_{opt}$ is used, Eq. 56 reduces to $E\left[\boldsymbol{\varepsilon}\boldsymbol{\varepsilon}^H\right] = P_x - P_x^2 \mathbf{V} S (P_x \mathbf{S}^2 + S_{N'N'}\mathbf{I})^{-1} \mathbf{S}\mathbf{V}^H$. As $\mathbf{V}$ is unitary, and $\mathbf{S} = \mathrm{diag}\,(s_1, s_2)$ is a diagonal matrix of the singular values of $\mathbf{Q}$, we have:

$$tr\left(E\left[\boldsymbol{\varepsilon}\boldsymbol{\varepsilon}^H\right]\right) = \sum_{i=1}^{2} \frac{P_x S_{N'N'}\left(e^{j\omega T}\right)}{P_x s_i^2\left(e^{j\omega T}\right) + S_{N'N'}\left(e^{j\omega T}\right)} \qquad (59)$$

We note that the singular values $s_i(e^{j\omega T})$ may be frequency-dependent. By Parseval's theorem, minimum mean squared error (MMSE) is given by:

$$\begin{aligned} MSE_{min} &= \sum_{i=1}^{2} E\left[|\varepsilon_{i,k}|^2\right] \\ &= \frac{T}{2\pi} \int_{-\infty}^{\infty} \sum_{i=1}^{2} \frac{P_x S_{N'N'}\left(e^{j\omega T}\right)}{P_x s_i^2\left(e^{j\omega T}\right) + S_{N'N'}\left(e^{j\omega T}\right)}\, d\omega \end{aligned} \qquad (60)$$

As an illustration that the frequency-domain expression of the MMSE equalizer in Eq. 57 is indeed optimal, consider the system in Fig. 22 where $\mathbf{Q}(e^{j\omega T}) = P_{tx}(e^{j\omega T}) P_{rx}(e^{j\omega T}) \mathbf{H}(e^{j\omega T})$. As $\mathbf{H}$ in Eq. 42 has unit singular values, $s_i^2\left(e^{j\omega T}\right) = \left|P_{tx}\left(e^{j\omega T}\right) P_{rx}\left(e^{j\omega T}\right)\right|^2$ for both polarizations $i = 1, 2$. Equation 54 is therefore equal to:

$$\mathbf{W}_{opt}^T\left(e^{j\omega T}\right) = \frac{P_x Q\left(e^{j\omega T}\right)}{P_x |Q\left(e^{j\omega T}\right)|^2 + S_{N'N'}\left(e^{j\omega T}\right)} \mathbf{H}^H\left(e^{j\omega T}\right) \qquad (61)$$

We note that:

$$\mathbf{H}^H\left(e^{j\omega T}\right) = \exp\left(j\frac{1}{2}\int_0^L \beta_2(z)\, dz\omega^2\right) \mathbf{R}_2^{-1} \begin{bmatrix} e^{-j\omega\tau/2} & 0 \\ 0 & e^{j\omega\tau/2} \end{bmatrix} \mathbf{R}_1 \qquad (62)$$

is indeed the filter response that exactly inverts CD and first-order PMD of the channel.

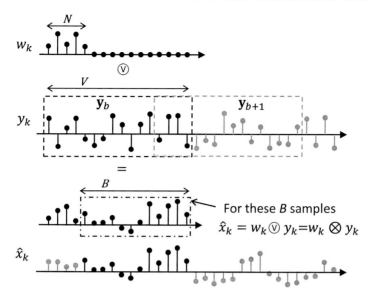

Fig. 25 Frequency-domain implementation of linear convolution using the overlap-and-save method

Frequency Domain Equalizer (FDE)

We have so far considered a time-domain implementation of the linear equalizer. For large equalizer length N, the filtering operation is much more efficiently implemented in the frequency domain using the overlap-and-add or overlap-and-save methods (Oppenheim and Schafer 1999). The overlap-and-save method is illustrated for a scalar example in Fig. 25. The equalizer input is partitioned into overlapping blocks $\mathbf{y}_b = \begin{bmatrix} y_{bb-N+1} & \cdots & y_{(b+1)B-1} \end{bmatrix}^T$ of length $V = N + B - 1 = 2^v$ which is an integer power of 2. The equalizer $\mathbf{w}$ of length N is padded with $B - 1$ zeros so it is also of length V. A circular convolution between $\mathbf{y}_b$ and $\mathbf{w}$ is performed by taking the fast Fourier transform (FFT) of the two vectors, multiplying them together, and then taking the inverse fast Fourier transform (IFFT). Since the final B output samples of the circular convolution are the same as the linear convolution between $\mathbf{y}_b$ and $\mathbf{w}$, these samples are saved. The window is then advanced by B samples to obtain $\mathbf{y}_{b+1}$, and the process is repeated to compute the next block of output samples.

A frequency-domain implementation of a linear equalization has lower algorithmic complexity than time-domain implementation because the FFT and IFFT can be efficiently implemented via a butterfly structure (Oppenheim and Schafer 1999). Provided that V is a power of two, a V-point FFT and IFFT requires only $(V/2)\log_2 V$ complex multiplications, while multiplying FFT$\{\mathbf{w}\}$ with FFT$\{\mathbf{y}_b\}$ together requires V complex multiplications. For dual-polarization signals, the matrix multiplication operation in Eq. 44 requires an FFT for each input polarization, an IFFT for each output polarization, and $4V$ complex multiplications in the frequency domain. Since B output symbols are computed every block, the algorithmic complexity of an FDE in complex multiplications per symbol is:

$$\min_{B:N+B-1=2^v} \left(\frac{1}{B} (N+B-1)(2\log_2(N+B-1)+4) \right) R_s \qquad (63)$$

For a given equalizer length N, there exists an optimal block size B that minimizes Eq. 63. For ease of implementation, however, V is commonly set to $2N$. The complexity per symbol in Eq. 63 scales as $O(\log_2 N)$. By contrast, $4N$ complex multiplications are required to compute Eq. 44 in the conventional manner, so the complexity of a time-domain equalizer scales as $O(N)$. A disadvantage of a frequency-domain implementation is higher latency, since an output sample is not available until the entire block has been processed. It can be shown that the FDE has a latency of $2\log_2 V + 1$ complex multiplications and additions, most of which is due to the FFT and IFFT operations.

Adaptive Time-Domain Equalizer (TDE)

In practice, the channel is time-varying due to mechanical vibration, changes in temperature, and other environmental conditions. In particular, lightning strikes can cause the state of polarization to change as rapidly as a few hundred krad/s (Krummrich et al. 2016). Hence, the linear equalizer needs to be adaptive. Adaptive equalizers have been studied in Oppenheim and Schafer (1999), Qureshi (1985), Gitlin and Weinstein (1981), and Haykin (2002).

A canonical model of an adaptive TDE is shown in Fig. 26. The received samples $\mathbf{y}_k$ are passed through the TDE whose output $\hat{\mathbf{x}}_k$ is subtracted from the desired signal $\mathbf{d}_k$ to obtain the error for that symbol:

$$\boldsymbol{\varepsilon}_k = \mathbf{d}_k - \hat{\mathbf{x}}_k \qquad (64)$$

The adaptive TDE seeks to minimize a cost function J. The gradient estimation block uses $\boldsymbol{\varepsilon}_k$ and $\mathbf{y}_k$ to compute an adjustment vector $\Delta \mathbf{w}_k$ for the equalizer that should lead to a reduction of J at the next iteration. Depending on the adaptive algorithm used, the desired signal $\mathbf{d}_k$ can either be the training symbols $\mathbf{x}_k$ or the equalizer output passed through a nonlinear function $\mathcal{G}\{\hat{\mathbf{x}}_k\}$. Table 1 shows the cost functions and desired signals for some commonly used adaptive algorithms (Ip and Kahn 2009).

If a system periodically inserts training symbols within data symbols, it is possible to use decision-aided (DA) training. The error $\boldsymbol{\varepsilon}_k = \mathbf{x}_k - \hat{\mathbf{x}}_k$ is then the difference between the known training symbols and the equalizer output. The use of training symbols reduces throughput, however. During normal transmission, the receiver can operate in decision-directed (DD) mode where symbol decisions made on the equalizer output are assumed to be correct and the error signal $\boldsymbol{\varepsilon}_k = [\hat{\mathbf{x}}_k]_D - \hat{\mathbf{x}}_k$ is the difference between the decision and the equalizer output. Provided

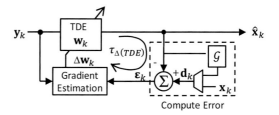

Fig. 26 Adaptive time-domain equalizer

the system is operating at a low symbol-error rate, DD operation is analogous to DA. However, a low symbol-error rate is not guaranteed, particularly at system initialization or after a system interruption.

In the absence of training symbols, "blind" adaptive algorithms can be used for initial training of the TDE coefficients. Once the symbol-error rate becomes sufficiently low, the system can switch to DD mode. Blind adaptive algorithms exploit known properties of the transmitted signal. In phase shift keying (PSK) constellations, for example, the constellation points lie on a circle of constant amplitude, allowing the use of radius-directed adaptive algorithms. Godard introduced a family of algorithms based on minimizing the cost function $J = E\left[\left\||\hat{x}_k|^p - R^{(p)}\right\|^2\right]$, where p is a positive integer and $R^{(p)} = E[|x_k|^{2p}]/E[|x_k|^p]$ is the target radius (Godard 1980). The case $p = 2$ is the most commonly used and is known as the "constant modulus algorithm" (CMA) (Leven et al. 2008). The CMA can be generalized for dual-polarization signals using the cost function $J_i = E\left[\left\||\hat{x}_{i,k}|^p - R_i^{(p)}\right\|^2\right]$ in each polarization.

Since $R_i^{(p)} = E\left[|x_{i,k}|^{2p}\right]/E\left[|x_{i,k}|^p\right]$ can be computed for any arbitrary constellation, the Godard algorithm can also for nonconstant modulus constellations such as QAM, albeit with loss of performance when used for densely packed constellations. Algorithms that are better suited for higher-order QAM have also been studied. They include the multi-modulus algorithm (MMA) (Yang et al. 2002) and the reduced-constellation algorithm (RCA) (Abrar 2004).

Coefficient Update
In an adaptive equalizer, the coefficients are adjusted according to:

$$\mathbf{w}^{(m+1)} = \mathbf{w}^{(m)} - \mu \nabla_\mathbf{w} J \tag{65}$$

where μ is the step size. We have defined the desired signals $\mathbf{d}_k$ in Table 1 so that:

$$\nabla_\mathbf{w} J = -2E\left[\mathbf{y}_k^* \boldsymbol{\varepsilon}_k^T\right] = -2E\left[\mathbf{y}_k^*(\mathbf{d}_k - \hat{\mathbf{x}}_k)^T\right] \tag{66}$$

When the equalizer is at its optimum setting, $\nabla_\mathbf{w} J = 0$, so the error vector is orthogonal to the vector of the received signal. For non-integer oversampling, recall that the equalizer has K sets of coefficients for each of the possible sampling phases.

Table 1 Cost functions and desired signals of commonly used adaptive algorithms

	Cost function (J)	Desired signal ($\mathbf{d}_k$)
Decision-aided (DA)	$E\left[\|\mathbf{x}_k - \hat{\mathbf{x}}_k\|^2\right]$	$\mathbf{x}_k$
Decision-directed (DD)	$E\left[\|[\hat{\mathbf{x}}_k]_D - \hat{\mathbf{x}}_k\|^2\right]$	$[\hat{\mathbf{x}}_k]_D$
Constant modulus algorithm (CMA)	$E\left[\left\||\hat{\mathbf{x}}_k|^2 - \mathbf{R}^{(2)}\right\|^2\right]$	$\begin{bmatrix} \hat{x}_{1,k}\left(1 + R_1^{(2)} - \|\hat{x}_{1,k}\|^2\right) \\ \hat{x}_{2,k}\left(1 + R_2^{(2)} - \|\hat{x}_{2,k}\|^2\right) \end{bmatrix}$
Godard	$E\left[\left\||\hat{\mathbf{x}}_k|^p - \mathbf{R}^{(p)}\right\|^2\right]$	$\begin{bmatrix} \hat{x}_{1,k}\left(1 + R_1^{(2)}\|\hat{x}_{1,k}\|^{p-2} - \|\hat{x}_{1,k}\|^{2p-2}\right) \\ \hat{x}_{2,k}\left(1 + R_2^{(2)}\|\hat{x}_{2,k}\|^{p-2} - \|\hat{x}_{2,k}\|^{2p-2}\right) \end{bmatrix}$

The superscript in Eq. 65 denotes the m-th update of that equalizer. E.g., when an oversampling rate of 3/2 is used, the two equalizers for the odd and the even symbols are adapted independently. By dissociating the superscript m from the symbol number k, it is assumed that the equalizer coefficients may not be updated every symbol. This may be especially true in optical systems where processing latency is much greater than a symbol period, precluding coefficient update on a per-symbol basis.

When the equalizer coefficients are adapted using the exact gradient function in Eq. 66, the coefficient update method is known as "steepest descent," and the rate of convergence is dependent on the step size and to the spread of eigenvalues of the autocorrelation matrix $\mathbf{A} = E\left[\mathbf{y}_k \mathbf{y}_k^H\right]$. Using a larger step size will lead to faster convergence but will result in larger "misadjustment" noise (Haykin 2002), and the coefficients may even diverge from the optimum solution. For the DA algorithm, the stability criterion requires that $0 < \mu < 1/\lambda_{\max}$, where $\lambda_{\max}$ is the largest eigenvalue of $\mathbf{A}$. By contrast, the time constant for convergence $N_{\text{iter}} = 1/(\mu \lambda_{\min})$ is limited by the smallest eigenvalue $\lambda_{\min}$ of $\mathbf{A}$ corresponding to the slowest mode.

Since $\mathbf{A}$ is an autocorrelation matrix, its eigenvalues are related to the amplitude of the spectrum of $\mathbf{y}_k$. As CD and PMD are unitary transformations, the frequency response of the channel is flat. Provided the signal has not experienced excessive spectral clipping by optical filters during transmission, the spread of eigenvalues should be small, and we expect $\lambda_{\max} \approx \lambda_{\min} \approx P_{rx}$.

Least Mean Square (LMS) Algorithm

In practice, the exact gradient function in Eq. 63 is difficult to estimate. Many symbols will be required to gain sufficient knowledge of $\nabla_\mathbf{W} J$ before an update is possible. Such a delay may lead to unacceptable adaptation speed. The most commonly used least mean square (LMS) adaptive algorithm uses a single-shot estimator for the gradient function:

$$\nabla_\mathbf{W} J = -2 \mathbf{y}_k^* \boldsymbol{\varepsilon}_k^T \qquad (67)$$

where error $\boldsymbol{\varepsilon}_k$ is as defined in Eq. 64 in conjunction with Table 1. Since $\nabla_\mathbf{W} J$ is available immediately after processing a symbol, the equalizer can adapt as rapidly as processing latency allows. The time constant is therefore N_iter times the latency of the feedback loop (Fig. 26), which we denote as $\tau_{\Delta(\text{TDE})}$. The cost of computing the gradient in Eq. 67 is $4N$ complex multiplications. If the equalizer is adapted every symbol, the computational cost of adaptation is 100% more than the linear equalizer itself.

Recursive Least Squares (RLS) Algorithm

There exists a class of recursive least squares (RLS) algorithms which achieves faster convergence than LMS by learning the inverse of the signals' autocorrelation matrix and incorporating this knowledge to gradient estimation, therefore eliminating the eigenvalue spread problem. The RLS algorithm can be found in Haykin (2002), while a modified RLS algorithm suitable for use with CMA was studied in Chen et al. (2004). RLS suffers from high complexity due to the extra computations needed to estimate the inverse autocorrelation matrix.

Singularity Problem

"Single-mode" fiber in reality supports two polarization modes. When a signal has passed through many in-line elements, each having finite polarization-dependent loss (PDL), it is possible for the channel to become near degenerate, meaning it has large eigenvalue spread or, equivalently, a large condition number. For such channels, the rows of the equalizer $\mathbf{w}$ may converge to the same solution, leading to the same polarization (mode) being extracted. This is known as the "singularity problem." Whether an equalizer converges to singularity depends on the initial value used for $\mathbf{w}$ during training.

It is possible to modify the error function to enforce nondegeneracy of the rows of $\mathbf{w}$ during adaptation. Independent component analysis (ICA) is a class of blind source separation algorithms which can extract statistically independent non-Gaussian signals (Zhang et al. 2008). A two-stage singularity-free CMA algorithm has also been studied in Xie and Chandrasekhar (2010). For channel with very strong mode-dependent loss (especially in few-mode fibers to be discussed in section "Timing Recovery"), successive interference cancellation (SIC) is a robust blind source separation method, where at each iteration, a new signal component is extracted, which is then convolved with the estimated channel response and is then subtracted from the received signal. Removal of the stronger signals makes it easier for the receiver to recover weaker signals from the residue at the next iterations. SIC has been studied in the context of few-mode fibers in Chen et al. (2012).

Adaptive Frequency-Domain Equalizer (FDE)

To reduce algorithmic complexity, the adaptive TDE in Fig. 26 can be turned into an adaptive FDE by inserting FFTs and IFFTs, as shown in Fig. 27. Since the FDE coefficients are frequency-domain values, whereas both the sampled signal $\mathbf{y}_k$ and

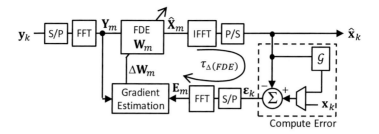

Fig. 27 Adaptive frequency-domain equalizer

the errors ε_k are time-domain values, they must be transformed to the frequency domain prior to calculating an adjustment for the equalizer.

Adaptive FDEs have been studied in Shynk (1992), while its application to coherent optical systems was studied in Kuschnerov et al. (2009) and Bai and Li (2012). The operations performed by the adaptive FDE are completely analogous to those performed by an adaptive TDE, so the error signals for the DA, DD, and blind algorithms can be calculated using Table 1 (Haykin 2002).

In terms of complexity, the gradient estimation block in the frequency-domain LMS (FLMS) algorithm requires an additional FFT and IFFT, due to the need to apply a gradient constraint in time (Haykin 2002; Kuschnerov et al. 2009). Total complexity is dominated by the five Fourier transforms. If the equalizer is adapted after every block of symbols processed, complexity per symbol is increased by 150% compared to a nonadaptive FDE which has only one FFT/IFFT pair. It is possible to use a computationally simpler "unconstrained gradient" algorithm which removes the need for an additional FFT/IFFT pair in the gradient estimation block. This results in only a 50% increase in complexity over nonadaptive FDE. However, this approach has worse performance than conventional FLMS (Mansour and Gray 1982).

The disadvantage of an adaptive FDE is feedback delay, due to the presence of FFTs and IFFTs in the feedback loop as shown in Fig. 27. As the coefficients cannot be updated faster than latency of the DSP operations within the loop, an adaptive FDE has a time constant of $N_{\text{iter}}\tau_{\Delta(\text{FDE})}$, which is much longer than the time constant of $N_{\text{iter}}\tau_{\Delta(\text{TDE})}$ for an adaptive TDE without FFT and IFFTs.

Adaptive Multidelay Block Frequency-Domain Equalizer

It is possible to trade off fast adaptation of a TDE against the low computational cost of an FDE by using a multidelay block frequency domain (MB-FDE) equalizer (Soo and Pang 1990). The MB-FDE parallelizes the input signal into block of size $2N/B_K$, where B_K is the number of processing blocks (Fig. 28). Each block contains an adaptive FDE. As the FFT and IFFT sizes are $2N/B_K$, the number of butterfly stages in each FFT is reduced by $\log_2 B_K$, lowering latency. However, computational

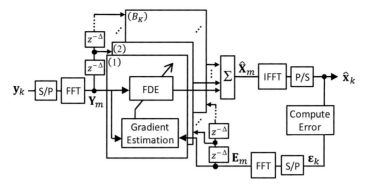

Fig. 28 Adaptive multidelay block frequency domain equalizer

Table 2 Scaling of latency and computational cost with equalizer length for a TDE, FDE, and MB-FDE

	Processing requirement (multiplications per symbol)	Feedback latency (# of multiplications)
Time-domain equalizer (TDE)	$O(N)$	$O(1)$
Frequency-domain equalizer (FDE)	$O(\log_2 N)$	$O(\log_2 N)$
Multidelay block FDE (MDF)	$O\left(\log_2 \frac{N}{B}\right) + O(B)$	$O\left(\log_2 \frac{N}{B}\right) + O(1)$

cost is increased as the FFT and IFFT operations have to be performed at B_K times the rate of a full FDE. The FDE and TDE can be considered limiting cases of an MB-FDE when the number of blocks are $B_K = 1$ and $B_K = 2N$, respectively. Table 2 summarizes the latency and computational requirement of various adaptive equalizer structures.

Hybrid Equalizer Structure

In a typical optical fiber channel, CD is a static impairment, whereas PMD is time-varying. It is therefore possible to use a hybrid equalizer structure shown in Fig. 29 (Savory 2008), which uses a computationally efficient nonadaptive FDE for CD compensation, followed by a low-latency adaptive TDE which allows fast polarization tracking. The hybrid equalizer is the most widely used equalizer structure in commercial real-time coherent receivers. The length of the FDE depends on the amount of residual channel CD that needs to be compensated. From Eq. 50, it is observed that each equalizer tap compensates $1/2\pi R_s^2$ (M/K) of residual CD. At a baud rate of 32 GHz and 2× oversampling, this corresponds to 61 ps/nm per tap at an optical carrier wavelength of 1550 nm. A submarine system with 10,000 km of

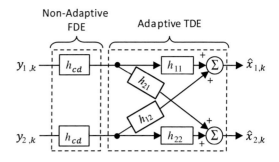

Fig. 29 Hybrid FDE-TDE structure for compensating CD and PMD in a fiber channel

SSMF and no in-line CD compensation will require an FDE length of 2787, while a metro system comprising 80 km of SSMF will only require an FDE length of 23. The vast discrepancy in potential equalizer sizes has resulted in a wide diversity of coherent receiver products aimed at different markets. A transceiver for short haul will use a smaller FDE to reduce power consumption, whereas a transceiver for ultra-long-haul transmission will have an FDE that can compensate channel memory lengths thousands of taps long. In a software-defined network (SDN), a higher layer controller can specify to a transponder the approximate residual CD to be compensated by the FDE. These initial coefficients are then loaded from memory. If the initial estimate of residual CD is inaccurate, any leftover CD after the FDE will be compensated by the adaptive TDE, which will normally have between 7 and 15 taps, which is sufficient to combat both PMD and small amounts of residual CD.

Carrier Phase Estimation

Laser Phase Noise

The electric field of a laser is not perfectly monochromatic. This is due to the origin of a laser's output being spontaneous emission noise being amplified inside a resonant cavity. The output phase of a laser is therefore "noisy." Coherent optical communication systems typically use external cavity semiconductor lasers (ECL), whose phase noise is usually well modeled as a Wiener process:

$$\theta(t) = 2\pi \int_{-\infty}^{dt'} v(t) \, dT' \qquad (68)$$

Where frequency noise $v(t)$ is a Gaussian process with the autocorrelation function:

$$E\left[v(t)v(t-\tau)\right] = \frac{\Delta v}{2\pi} \delta(\tau) \qquad (69)$$

4 Optical Coherent Detection and Digital Signal Processing of Channel Impairments

It can be shown that the output of a laser $l(t) = \sqrt{P}e^{j\theta(t)}$ will have Lorentzian spectrum given by:

$$S_{ll}(f) \sim \frac{1}{f^2 + \left(\frac{\Delta\nu}{2}\right)^2} \tag{70}$$

The phase noise property of a laser is thus characterized by its linewidth $\Delta\nu$, which is simply the two-sided 3-dB bandwidth of its spectrum. In the absence of all other impairments, laser phase noise will cause the signal constellation to rotate in a time-varying manner according to the instantaneous phase $\theta(t)$. It is possible to subsume this time-varying constellation rotation as part of a time-varying channel. However, the linewidth of an ECL is typically on the order of tens of kHz, which is normally much faster than polarization changes. It is generally desirable to separate linear equalization from phase noise compensation, as the former can use a smaller "step size" to allow lower misadjustment noise, while the latter needs to use a larger "step size" to track a faster-varying process.

Phase Estimation in the Absence of Data Modulation

We first consider the simplest carrier phase estimation scenario where the signal is a laser tone in a single polarization without data modulation. The only impairment we consider other than phase noise is AWGN. A system model for this scenario is shown in Fig. 30a. It is assumed that carrier phase estimation is performed at the symbol rate $1/T_s$. The true laser phase θ_k is the integration of a Gaussian noise process Ω_k which has zero mean and a variance of $\sigma_\Omega^2 = 2\pi\Delta\nu T_s$. Due to the presence of AWGN n_k, which has zero mean and variance σ_n^2, the receiver is only able to measure a phase of $\psi_k = \theta_k + n'_k$ (Fig. 30b), with the variance of n'_k being $\sigma_{n'}^2 = \sigma_n^2/2$, since only the component of noise orthogonal to the signal phasor will affect carrier phase estimation. As θ_k is correlated in time, it is possible to pass ψ_k through a digital filter W_ψ whose output $\hat{\theta}_k$ is the receiver's best estimate of θ_k. Writing all quantities in terms of their bilateral z-transforms (Oppenheim and Schafer 1999):

$$X(z) = \sum_{k=-\infty}^{\infty} x_k z^{-k} \tag{71}$$

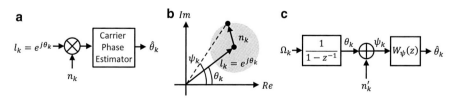

Fig. 30 (**a**) Carrier phase estimation for an unmodulated tone, (**b**) phasor diagram, and (**c**) linear model

it can be shown that the W_ψ which minimizes the mean square error $E\left[\left|\theta_k - \hat{\theta}_k\right|^2\right]$ is given by (Ip and Kahn 2007b):

$$W(z) = \frac{E\left[\theta(z)\psi\left(z^{-1}\right)\right]}{E\left[\psi(z)\psi\left(z^{-1}\right)\right]} = \frac{rz^{-1}}{-1 + (2+r)z^{-1} - z^{-2}} \qquad (72)$$

where $r = \sigma_\Omega^2/\sigma_{n'}^2 > 0$. This filter has two poles that are reciprocals of each other:

$$z_1, z_2 = \left(1 + \frac{r}{2}\right) \pm \sqrt{\left(1 + \frac{r}{2}\right)^2 - 1} \qquad (73)$$

with z_1 inside the unit circle mapping to a causal sequence and z_2 outside the unit circle mapping to an anti-causal sequence. The inverse z-transform of Eq. 73 is:

$$w_{\psi,k} = \frac{\alpha\, r}{1 - \alpha^2}\alpha^{|k|} \qquad (74)$$

where $\alpha = \left(1 + \frac{r}{2}\right) - \sqrt{\left(1 + \frac{r}{2}\right)^2 - 1}$. The filter in Eq. 74 is a two-sided decaying exponential symmetric about the origin. The exponential decay arises due to phase estimates at times $k - l$ far from the symbol k of interest being progressively less accurate estimators of θ_k, so they are given decreasing emphasis in computing $\hat{\theta}_k$. The symmetry of the filter is due to the accuracy of the estimator ψ_{k-l} being dependent only on $|l|$. In the limit of low phase noise compared with AWGN ($\sigma_\Omega^2 \ll \sigma_{n'}^2$), the decay rate of the filter is slow ($\alpha \to 1$) to exploit the longer temporal coherence of θ_k. Conversely, in the limit of high phase noise compared with AWGN ($\sigma_\Omega^2 \gg \sigma_{n'}^2$), the filter has a fast decay rate as ψ_{k-l} rapidly becomes inaccurate estimators of θ_k. Finally, $\sum_{k=-\infty}^{\infty} w_{\psi,k} = 1$, so Eq. 74 is an unbiased estimator.

Since it is impractical to implement a two-sided exponential with infinite tails, Eq. 74 can be approximated by an infinite impulse response (IIR) filter in parallel with a finite impulse response (FIR) filter as shown in Fig. 31. The performance of the MMSE estimator has been studied in Ip and Kahn (2007b).

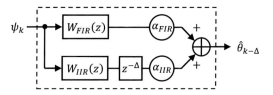

Fig. 31 Emulating the ideal two-sided exponential filter with an IIR filter in parallel with an FIR filter

Phase Estimation in the Presence of Data Modulation

While the carrier phase estimation algorithm in section "Phase Estimation in the Absence of Data Modulation" is sufficient for a system which transmits a pilot tone alongside the signal, most systems in operation today do not use any pilot tones. The phase measurement ψ_k at symbol k must therefore be extracted from the modulated signal. Like linear equalization, this can be achieved by the use of training symbols (decision-aided), symbol decisions (decision-directed), or by exploiting known properties of the modulated signal (nondecision-aided).

Decision-aided phase estimation is not common, as the carrier phase will be unknown to the receiver between training symbols. Since laser phase evolves much faster than the channel impulse response, large numbers of training symbols will have to be inserted at regular intervals, reducing the rate of useful data throughput.

Nondecision-aided (NDA) and decision-aided (DD) phase estimation can be accomplished using the feedforward structure shown in Fig. 32a. In the upper path, an instantaneous phase estimate ψ_k is first obtained from the signal. ψ_k and is then passed through a filter W_ψ (with delay Δ) to produce an MMSE estimate $\hat{\theta}_k$ of the carrier phase. This phase is then used to de-rotate the signal in the bottom path, which has been appropriately delayed. Compared with feedforward carrier estimation described for a pilot tone in section "Phase Estimation in the Absence of Data Modulation," the only additional component is the instantaneous phase estimator.

NDA phase estimation is commonly used for M-ary PSK, as the M-fold rotational symmetry of the constellation means that raising the signal to be the M-th power removes data modulation. This method of phase estimation is also commonly referred to as the Viterbi-Viterbi algorithm (Viterbi and Viterbi 1983). Let the set of possible transmitted symbols be $x_m = e^{j2\pi m/M}$, $m \in [0, M-1]$. Raising the received signal to the M-th power yields:

$$y_k^M = \left(x_k e^{j\theta_k} + n_k\right)^M = e^{jM\theta_k} + m_k \tag{75}$$

where $e^{jM\theta_k}$ is the desired term dependent on θ_k, and $m_k = \sum_{p=1}^{M} \binom{M}{p} \left(x_k e^{j\theta_k}\right)^{M-p} n_k^p$ is a summation of signal-noise beat terms contributing a noise vector to the phasor $e^{jM\theta_k}$. The instantaneous phase can be found from the angle of y_k^M and divided by $1/M$ (Fig. 32b):

$$\psi_k = \frac{1}{M} \arg\left(y_k^M\right) \approx \theta_k + n_k' \tag{76}$$

where Mn_k' is the angular projection of m_k. At high SNR (γ_s), it can be shown that n_k' is approximately i.i.d. Gaussian with zero mean and variance (Ip and Kahn 2007b):

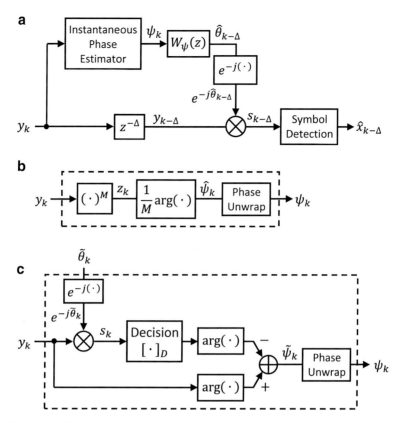

Fig. 32 (a) Feedforward carrier phase estimation and compensation. The instantaneous phase estimator can be (b) nondecision aided (NDA) or (c) decision-directed (DD)

$$\sigma_{n'}^2 = \eta(M, \gamma_s) \frac{1}{\gamma_s} \qquad (77)$$

where $\eta(M, \gamma) = \frac{1}{2M^2} \sum_{p=1}^{M} \binom{M}{p}^2 p! \gamma_s^{-p+1}$.

For non-(D)PSK constellations, ψ_k can be estimated using a DD instantaneous phase estimator shown in Fig. 32c. A major difference between the NDA and the DD structures is that DD requires an initial estimate of the carrier phase $\tilde{\theta}_k$. In the DD algorithm, the received signal y_k is first de-rotated by $\tilde{\theta}_k$. Provided $\tilde{\theta}_k$ is not significantly different than the real carrier phase θ_k, the de-rotated constellation will be properly aligned and can be detected by a decision device:

$$\hat{x}_k = \left[y_k e^{-j\tilde{\theta}_k} \right]_D \qquad (78)$$

At high SNR, the decision is correct ($\hat{x}_k = x_k$) with high probability. This allows the instantaneous carrier phase to be estimated:

$$\psi_k = \arg(y_k) - \arg(\hat{x}_k) = \theta_k + n'_k \tag{79}$$

where n'_k is the angular projection of noise n_k in the direction orthogonal to the rotated signal $x_k e^{j\theta_k}$. It can be shown that at high SNR, n'_k are approximately i.i.d. and Gaussian with zero mean and variance:

$$\sigma_{N'}^2 = \eta \frac{1}{\gamma_s} \tag{80}$$

Where $\eta = \frac{1}{2} E\left[|x|^2\right] E\left[\frac{1}{|x|^2}\right]$ is equal to half times a "constellation penalty."

Phase Unwrapping

The instantaneous phase produced the NDA and DD phase estimators, as well as the pilot tone phase estimator covered in section "Phase Estimation in the Absence of Data Modulation," which need to be phase unwrapped before passing through the MMSE filter W_ψ. This is a consequence of the arg(·) function output being limited to $-\pi$ and π, as angles differing by integer multiples of 2π are indistinguishable. In the absence of phase unwrapping, the NDA phase estimator output is constrained between $-\pi/M$ and π/M, whereas the DD output is constrained between $-\pi$ and π. Since θ_k is a Wiener process, its value is unconstrained. Phase unwrapping has been studied in Taylor (2004) and involves adding integer multiples of $2\pi/M$ (or 2π) to the output of the NDA (or DD) phase estimator to ensure that the magnitude of the phase difference between adjacent symbols is less than π/M (or π). Phase unwrapping can cause "cycle slipping" and is a highly nonlinear phenomenon (Meyr et al. 1997).

Combining Laser Phase Noise Compensation with Linear Equalization

In presence of both laser phase noise and an LTI channel, a system that uses receiver-side impairment compensation has the model shown in Fig. 33. The phase of the transmitter and receiver local oscillator lasers is denoted as $\theta_{tx}(t)$ and $\theta_{rx}(t)$, respectively. The receiver can invert the impairments in reverse order as shown by the operations in the dotted box in Fig. 33. For clarity, the function blocks in the channel inverter are shown as continuous-time functions. Provided the receiver signal is sampled above the Nyquist criterion (section "Sampling Rate Requirement"), the equivalent operations can be performed in DSP.

Since multiplication by laser phase noise is non-commutable with dispersion, the DSP blocks cannot be reordered without loss of performance. Moreover, as $\theta_{tx}(t)$, $\theta_{rx}(t)$, and $\mathbf{h}(t)$ are time-varying, the DSP compensators $\hat{\theta}_{tx}(t)$, $\hat{\theta}_{rx}(t)$, and

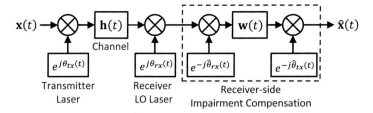

Fig. 33 Receiver-side impairment compensation taking into account laser phase noise and an LTI channel

$\mathbf{w}(t)$ must be adaptive. Previously, we saw in linear equalization in section "Linear Equalization" that adaptation requires the receiver to feedback the error between the output of each DSP block and the desired output. This is also true for laser phase noise compensation. For the Tx phase noise compensator, the desired output is the transmitted symbols. Since the linear equalizer precedes the Tx phase compensator, the desired output is $\mathbf{d}_k e^{-j\hat{\theta}_{tx,k}}$, where $\mathbf{d}_k$ is given in Table 1 for the various adaptive linear equalizer algorithms. The error signal is therefore $\boldsymbol{\varepsilon}_k e^{-j\hat{\theta}_{tx,k}}$. By extension, the error for each block is obtained by backpropagating (The principle of backpropagating the error to adapt a nonlinear system has been studied in the field of neural networks (Haykin 1998). The usage of backpropagation here should not be confused with the nonlinear compensation technique of "backward propagation" introduced in section "Nonlinear Compensation") the error $\boldsymbol{\varepsilon}_k$ through all the other blocks downstream from it, as shown in Fig. 34a.

It is observed that the Rx phase noise compensator is a digital LO controlled by a phase-locked loop (PLL), where the feedback delay is equal to the sum of the latencies of all the other DSP blocks, including the linear equalizers in the impairment compensation and error feedback paths. This feedback delay is functionally analogous to the physical delay of a hardware PLL. Delay reduces linewidth tolerance. It was shown in Barry and Kahn (1992) that for large feedback delays τ_Δ, the linewidth requirement scales as $2\pi \Delta v \tau_\Delta$. Most of the delay in τ_Δ comes from the group delays of $\mathbf{w}$ and $\mathbf{w}^{-1}$, as these have to be causal filters. The linewidth requirement for the Rx local oscillator laser will therefore scale as $2\pi \Delta v_{rx} N_{cd} T_s$, where N_{cd} is the memory length of the channel due to CD given in Eq. 50. By contrast, the linewidth requirement for the Tx laser scales as $2\pi \Delta v_{tx} T_s$, as the feedback path does not enclose the filters $\mathbf{w}$ and $\mathbf{w}^{-1}$. Hence for a receiver that uses feedback adaptation only, as shown in Fig. 34a, the linewidth requirement on the receiver LO is much more severe than the Tx laser.

Laser phase noise tolerance can be increased by using feedforward adaptation. Consider Fig. 34b, in which the operations in Fig. 34a are labeled "Iteration #1" and where the error feedback path is labeled as the "parameter update" circuit. A second copy of the received signal is delayed by the feedback latency. The receiver uses the parameters obtained in the initial adaptation (Iteration #1) to process the received signal a second time (Iteration #2) to produce a better estimate of the transmitted

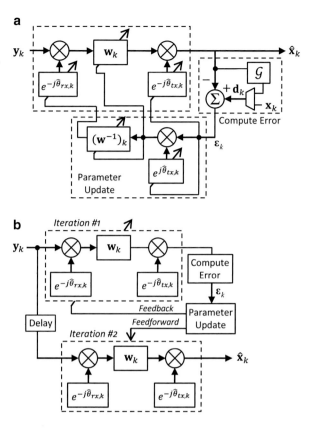

Fig. 34 Compensation of phase noise and LTI impairments using (**a**) feedback adaptation only and (**b**) using both feedback and feedforward adaptation

signal. This additional processing represents "feedforward" operation. It is possible to increase the number of iterations of impairment compensation + error estimation + parameter update to obtain asymptotically better performance.

Nonlinear Compensation

Digital Backpropagation

Signal propagation in fiber is ultimately distorted by a combination of both chromatic dispersion and Kerr nonlinearity, as described by the NLSE introduced in section "The Fiber Channel." A typical characteristic of transmission performance versus signal power is shown in Fig. 35. The vertical axis denotes signal quality (Q_{SNR}), which we define as signal power divided by the total variance of "interference" arising from noise and nonlinearity. At low launch power, optical fiber is well modeled as an LTI channel, so a 1 dB increase in launch power results in a 1 dB improvement in signal quality. As signal power is increased, the nonlinear term in the NLSE will begin to dominate.

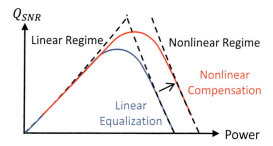

Fig. 35 Signal quality versus power characteristic with and without nonlinear compensation

For fibers having "large" CD (say, larger than 2 ps/nm/km) and where in-line CD compensation is not used, it has been shown that nonlinear distortion is well approximated as a Gaussian noise (Poggiolini 2012; Poggiolini et al. 2014; Mecozzi and Essiambre 2012; Savory 2013), whose variance grows as the cube of signal power. The cubic scaling is due to the Kerr effect being a third-order nonlinearity. Assuming nonlinearity is not compensated, Q_{SNR} will have the form:

$$Q_{SNR} = \frac{P}{N + \eta P^3} \quad (81)$$

where P is signal power, N is total noise power added by the link, and η is a parameter that depends on the fibers propagated through. From Eq. 81, it is observed that Q_{SNR} will decrease by 2 dB for every 1 dB increase in launch power at high powers.

Traditionally, nonlinear impairments in DWDM systems are categorized into the following types (Agrawal 2002):

1. Self-phase modulation (SPM): Nonlinearity arising in a channel as a result of its own intensity
2. Cross-phase modulation (XPM): Nonlinearity arising in a channel as a result of the intensity of other channels in a WDM system
3. Four-wave mixing (FWM): Nonlinearity arising from three interacting fields at different wavelengths producing a nonlinear field at a fourth wavelength within the bandwidth of the channel of interest.
4. Nonlinear phase noise (NLPN): Nonlinear interaction between signal and noise (Gordon and Mollenauer 1990).

The first three types of nonlinear impairments are deterministic given the full electric field of the WDM signal. In the absence of noise, these nonlinearities can be completely compensated as the NLSE is an invertible equation. It is therefore possible to simultaneously compensate CD and nonlinearity by solving the inverse NLSE:

$$\frac{\partial \mathbf{E}}{\partial z} = -\left(\hat{\mathbf{D}} + \hat{\mathbf{N}}\right)\mathbf{E} \quad (82)$$

where $\hat{\mathbf{D}}$ and $\hat{\mathbf{N}}$ are the linear and nonlinear operators defined in Eq. 20. The operation performed by Eq. 82 is analogous to passing the received signal $\mathbf{E}(L, t)$ through a fictitious fiber with opposite signs of loss, dispersion, and nonlinearity, yielding an estimate of the transmitted signal $\mathbf{E}(0, t)$. This method of nonlinear compensation by solving the inverse NLSE is commonly referred to as "backward propagation" or "backpropagation" (BP). When implemented in DSP, it is known as digital backpropagation (DBP).

In the context of Eq. 82, linear equalization of CD and PMD, whether performed optically using DCF or digitally using a linear equalizer, can be viewed as a "simplified DBP" taking only into account the linear operator. Even mid-span phase conjugation can be viewed as a form of BP, as inverting the signal phase at midpoint causes the second half of the transmission medium to become a backpropagation on the first half of the link (Chowdhury and Essiambre 2004).

DBP was first proposed as a transmitter-side electronic pre-compensation algorithm in Essiambre et al. (2006) (Fig. 36a), since before the advent of a dual-polarization coherent receiver, the complex-valued electric field is only available at the transmitter. The transmitter's DSP solves the inverse NLSE to find the modulator drive signal necessary so that the received optical waveform is free of both linear and nonlinear distortions after propagation. When the dual-polarization coherent receiver became available, receiver-side DBP was made possible. Receiver-side DBP was studied in Mateo et al. (2008) and Ip and Kahn (2008) (Fig. 36b). It is also possible to split the DBP operation between the transmitter and receiver (Fig. 36c). Since noise added by in-line amplifiers causes the actual waveform in forward transmission to be different to that digitally backpropagated, splitting the DBP operation 50/50 between the transmitter and receiver yields the best performance. When nonlinear compensation is employed, the DSP operations performed at the transmitter and receiver are shown in Fig. 37. Since BP is only concerned with the electric field, it is a universal algorithm that operates independently of the modulation format. Moreover, as DBP simultaneously compensates both LTI impairments and Kerr nonlinearity, the only additional DSP operation required at the receiver is laser phase noise compensation. In practice, DBP can only take into account the linear impairments known to the receiver (and/or) transmitter. Polarization rotation and PMD are generally unknown. Therefore, a short linear equalizer after DBP is usually required prior to carrier recovery.

Split-Step Fourier Method

In DBP, the inverse NLSE in Eq. 82 is normally solved using the split-step Fourier method (SSFM) (Agrawal 2001), which divides the fiber channel into sections (Fig. 38). To backpropagate the signal through a section of fiber from $z+h$ to z, it is possible to use a non-iterative, asymmetric SSFM (NA-SSFM) (Fig. 39a):

$$\mathbf{E}(z, t) \approx \exp\left(-h\hat{\mathbf{N}}(z+h)\right) \exp\left(-h\hat{\mathbf{D}}\right) \mathbf{E}(z+h, t) \tag{83}$$

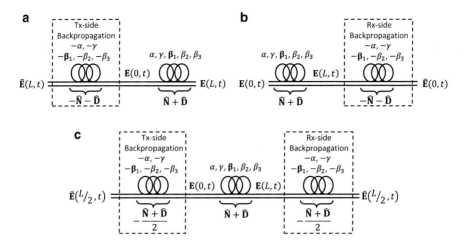

Fig. 36 Digital backpropagation performed at the (**a**) transmitter side, (**b**) receiver side, and (**c**) splitting between transmitter and receiver sides

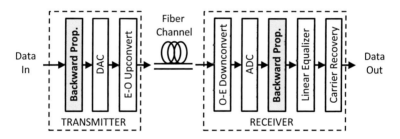

Fig. 37 Nonlinear compensation transmitter and receiver

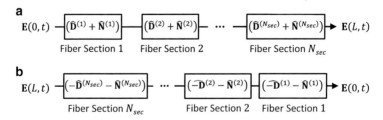

Fig. 38 (**a**) Forward and (**b**) backward propagation for a fiber channel modeled as a concatenation of N_{sec} fiber sections

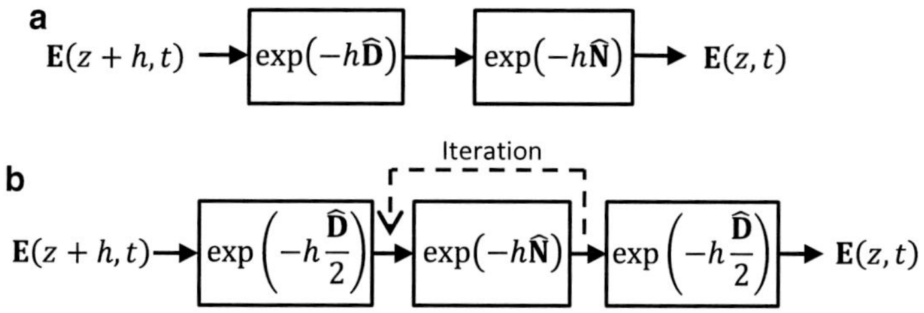

```
                    a
              E(z + h, t) → [exp(-hD̂)] → [exp(-hN̂)] → E(z, t)

                                    Iteration
                                    ┌ - - - - ┐
              b                     ↓         │
                         ┌──────┐ ┌────────┐ ┌──────┐
              E(z + h, t)→│exp(-hD̂/2)│→│exp(-hN̂)│→│exp(-hD̂/2)│→ E(z, t)
                         └──────┘ └────────┘ └──────┘
```

Fig. 39 Backpropagation through one section of fiber using the (**a**) non-iterative asymmetric SSFM and (**b**) iterative symmetric SSFM

Alternatively, a more accurate iterative, symmetric SSFM (IS-SSFM) can also be used (Fig. 39b):

$$\mathbf{E}(z,t) \approx \exp\left(-\frac{h}{2}\hat{\mathbf{D}}\right) \exp\left(-h\frac{\hat{\mathbf{N}}(z+h) + \hat{\mathbf{N}}(z)}{2}\right) \exp\left(-\frac{h}{2}\hat{\mathbf{D}}\right) \mathbf{E}(z+h,t) \tag{84}$$

The latter algorithm is "symmetric" because the linear operator is split equally on each side of the integral of the nonlinear operator. In Eq. 84, this integral has been approximated using the trapezoidal rule (Agrawal 2001). Since $\hat{\mathbf{N}}(z)$ depends on $\mathbf{E}(z,t)$, whose value is initially unknown, an iterative algorithm is required to solve Eq. 84. In the NA-SSFM, the need for iteration is removed by approximating the integral of $\hat{\mathbf{N}}(z)$ using a one-sided rectangle. The IS-SSFM is more accurate than the NA-SSFM but is more computationally expensive.

The linear operator is most efficiently computed in the frequency domain as an all-pass filter with polynomial phase:

$$\mathcal{F}\left\{\exp\left(-h\hat{\mathbf{D}}\right) \mathbf{E}(z',t)\right\} = \exp\left(-h \mathcal{F}\left\{\hat{\mathbf{D}}\right\}\right) \mathbf{E}(z',\omega)$$
$$= e^{\alpha h/2} e^{-j(\beta_2 \omega^2/2! + \beta_3 \omega^3/3!)h} \mathbf{R}(\theta,\phi) e^{-j\omega(\Delta\tau/2)\sigma_1} \mathbf{R}^H(\theta,\phi) \mathbf{E}(z',\omega) \tag{85}$$

where $\mathbf{R}(\theta,\phi)$ was defined in Eq. 22.

The nonlinear operator is most easily evaluated in the time-domain as a phase shift proportional to the instantaneous signal power in both polarizations:

$$\exp\left(-h\hat{\mathbf{N}}\right) \mathbf{E}(z',t) = \exp\left(-j\frac{8}{9}\gamma \left(\|\mathbf{E}(z',t)\|^2\right) h\right) \mathbf{E}(z',t) \tag{86}$$

Solving the SSFM requires the use of FFTs and IFFTs to switch between frequency and time, which accounts for the algorithm's high computational cost.

Step-Size Requirement

To compute the NLSE accurately, the step size h is normally chosen so that the phase accumulated by the linear and nonlinear step (in time or frequency) is less than $\Delta \zeta$:

$$h \ll \Delta \zeta \min \left\{ \frac{2}{\beta_2 (\pi R_s)^2}, \frac{9}{8\gamma \left(\mathbf{E}(z)^2\right)} \right\} \quad (87)$$

Here, we have assumed the use of the Nyquist pulse shaping so that the highest frequency component of the signal is $R_s/2$. In fibers having non-zero dispersion and at typical power levels, the nonlinear phase accumulated at each step is typically much smaller than the phase accumulated by CD.

Other methods of step-size setting have also been considered in the context of transmission system simulations. For example, a method which keeps the normalized simulation error $\Delta \zeta$ fixed at each step was suggested by Zhang and Hayee (2008):

$$h = \left[\frac{\Delta \xi}{\gamma \|\mathbf{E}(z)\|^2 \beta_2 R_s^2} \right]^{1/3} \quad (88)$$

Since signal power changes along a fiber during propagation, Eq. 88 allows the use of nonconstant step size to reduce the number of steps required.

Reduced Complexity Nonlinear Compensation Algorithms

The high computational complexity of digital backpropagation has been a major impediment to its widespread implementation in commercial systems. Consider typical step sizes required for standard single-mode fiber (SSMF) with $D = 17$ ps/nm/km, $\gamma = 0.0013$ and a baud rate $R_S = 32$ GHz typical of 100G systems. Using Zhang and Hayee (2008) with $\Delta \xi = 10^{-3}$, we have $h = 3.4$ km. However, since the signal will be noisy, it is not necessary to solve the NLSE with higher accuracy than the expected signal fidelity at the receiver. In Ip and Kahn (2008), it was proposed to perform DBP at a resolution of one step per span. It is possible to increase the step size still further by using a "filtered DBP" approach as shown in Fig. 40, where the signal amplitude is low-passed filtered before it is used for phase de-rotation:

$$\exp\left(-h\hat{\mathbf{N}}\right) \mathbf{E}\left(z',t\right) = \exp\left(-j\frac{8}{9}\gamma\left(\|\mathbf{E}(z',t)\|^2 \otimes h_{lpf}(t)\right)h\right) \mathbf{E}\left(z',t\right) \quad (89)$$

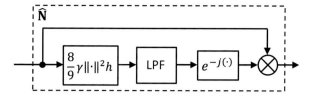

Fig. 40 Nonlinear operator in filtered DBP enables the use of larger step sizes

This method relies on the fact that higher frequencies are dispersed by CD so that for large step size, the nonlinear electric fields contributed by the high frequencies become averaged. Filtered DBP was studied in Li et al. (2011) and Du and Lowery (2010), where it was found that in systems without in-line CD compensation using DCF, the step size can be increased to around four spans without significant loss of performance compared with using one step per span. Even with this improvement, the complexity of DBP remains much higher than linear equalization, which is effectively one stage of DBP without any nonlinear phase compensation. A submarine link with 100 spans of fiber will require roughly 25 times the complexity of linear equalization, as complexity is dominated by the FFT/IFFT operations, whose complexity only scales logarithmically with total accumulated CD per step.

DBP algorithms based on the Volterra series expansion of the NLSE have also been studied in the frequency domain in Weidenfeld et al. (2010) and Ip et al. (2011) and in the time domain by Yan et al. (2011). All reduced complexity algorithms ultimately come with a trade-off between performance versus algorithmic complexity. Improved performance can be obtained at the expense of complexity using smaller step sizes or retaining more terms in the Volterra series. Ultimately, even with the use of arbitrarily small step size and high complexity, published experimental results have shown that performance improvement expected using single-channel nonlinear compensation (NLC) has only been on the order of 1–1.5 dB. This limitation is due to a variety of factors, including XPM from neighboring channels which cannot be compensated in single-channel NLC and PMD rendering it impossible to solve the exact inverse NLSE since PMD fluctuations within the link cannot be easily measured.

Complexity reduction may ultimately be most effective in "dispersion-managed" system that uses DCF for in-line CD compensation after each fiber span, as this returns the signal's amplitude profile to roughly the same shape after every span, save for the cumulative effect of noise and nonlinearity. This allows the use of "folded DBP" which treats the entire link as a single equivalent span with N_{span} times the nonlinearity, reducing the computation requirement of DBP to only a few times that of linear equalization (Fischer et al. 2009; Zhu and Li 2011).

Interchannel Nonlinear Compensation

In WDM transmission over most fiber types that have non-zero dispersion, XPM is the dominant nonlinear effect. This limits the performance improvement achievable by single-channel DBP which can only compensate self-phase modulation SPM. It is possible to compensate interchannel nonlinearties by downconverting the WDM signal using a bank of parallel O/E downconverters driven by local oscillators at wavelengths of ω_1 to ω_{N_ω} as shown in Fig. 41. The LO lasers may be individual lasers or may be tones from a frequency comb. Reconstruction of the electric fields $\mathbf{E}_n(t)$ around each LO enables joint processing of their electric fields to remove their mutual nonlinear interaction.

The most obvious interchannel nonlinear compensation method is to solve the inverse NLSE for the full WDM electric field. This method requires the bank of LO lasers to be phase locked, which can be provided by a frequency comb with a repetition rate of $\Delta\omega$. After determining the constant phase offsets θ_n between the local oscillators, it is possible to reconstruct:

$$\mathbf{E}(t) = \sum_{n=1}^{N_\omega} \mathbf{E}_n(t) e^{j(n\Delta\omega t + \theta_n)} \tag{90}$$

The inverse NLSE in Eq. 82 is then performed on this full WDM electric field. The step-size requirement is determined as per Eq. 87 or 88, where signal power and bandwidth are those of the WDM signal. This "total-field NLSE" (T-NLSE) method fully compensates XPM and FWM but has very high complexity.

In WDM systems where the bandwidth per channel $2\pi B_W$ is much less than their frequency spacing ω_{sp}, some computational savings can be obtained by solving a "coupled NLSE" (NLSE) between the $\mathbf{E}_n(t)$'s, rather than the inverse NLSE for the full WDM electric field. Consider the nonlinear operator in Eq. 20, where we expand the WDM electric field in terms of its constituent channels as $\mathbf{E}(t) = \sum_n \mathbf{E}_n(t) e^{j\omega_n t}$, and where $\mathbf{E}_n = \begin{bmatrix} E_{x,n} & E_{y,n} \end{bmatrix}^T$ are the polarization components of each channel:

$$\hat{\mathbf{N}}.\mathbf{E}(t) = j\frac{8}{9}\gamma |\mathbf{E}(t)|^2 \mathbf{E}(t) = j\frac{8}{9}\gamma \sum_{l,m,n} \left(E_{x,l}(t) E_{x,m}^*(t) \right.$$
$$\left. + E_{y,l}(t) E_{y,m}^*(t) \right) \mathbf{E}_n(t) e^{j(\omega_l - \omega_m + \omega_n)t} \tag{91}$$

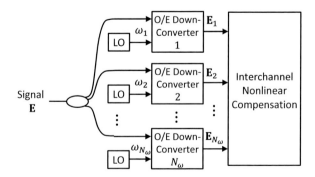

Fig. 41 Receiver front end for interchannel nonlinear compensation in a WDM system

4 Optical Coherent Detection and Digital Signal Processing of Channel Impairments

Assuming the WDM channels are evenly spaced in frequency, the nonlinear electric field for the k-th channel depends on the summation terms in Eq. 91 that satisfy $k = l - m + n$:

$$\hat{\mathbf{N}}_k \cdot \begin{bmatrix} E_{x,k}(t) \\ E_{y,k}(t) \end{bmatrix} = j\frac{8}{9}\gamma \begin{bmatrix} \sum_{l-m+n=k} \left(E_{x,l}(t)E^*_{x,m}(t) + E_{y,l}(t)E^*_{y,m}(t) \right) E_{y,n}(t) \\ \sum_{l-m+n=k} \left(E_{x,l}(t)E^*_{x,m}(t) + E_{y,l}(t)E^*_{y,m}(t) \right) E_{y,n}(t) \end{bmatrix} \quad (92)$$

The term where $l = m = n = k$ is SPM, while the terms where either ($n = k$ and $l = m \neq k$) or ($l = k$ and $m = n \neq k$) are XPM. All the other terms are FWM and are negligible in fibers with appreciable CD (say greater than 2 ps/nm/km). Discarding the FWM terms, we are left with:

$$\hat{\mathbf{N}}_k \cdot \begin{bmatrix} E_{x,k} \\ E_{y,k} \end{bmatrix} \approx j\frac{8}{9}\gamma \begin{bmatrix} \left(\sum_{m \neq k} \left(2|E_{x,m}|^2 + |E_{y,m}|^2 \right) + |E_{x,k}|^2 + |E_{y,k}|^2 \right) E_{x,k} + \left(\sum_{m \neq k} E_{x,m} E^*_{y,m} \right) E_{y,k} \\ \left(\sum_{m \neq k} \left(2|E_{y,m}|^2 + |E_{x,m}|^2 \right) + |E_{y,k}|^2 + |E_{x,k}|^2 \right) E_{y,k} + \left(\sum_{m \neq k} E_{y,m} E^*_{x,m} \right) E_{x,k} \end{bmatrix}$$

$$= j\frac{8}{9}\gamma \begin{bmatrix} C_{xx} & C_{xy} \\ C_{yx} & C_{yy} \end{bmatrix} \begin{bmatrix} E_{x,k} \\ E_{y,k} \end{bmatrix} \quad (93)$$

Thus, $\hat{\mathbf{N}}_k$ can be represented as a matrix $j\frac{8}{9}\gamma\, \mathbf{C}_k$. The diagonal elements of $\mathbf{C}_k$ generate nonlinear fields in the same polarization as the signal $E_{x,k}$ and $E_{y,k}$, while the off-diagonal elements of $\mathbf{C}_k$ generate nonlinear fields in the orthogonal polarization as $E_{x,k}$ and $E_{y,k}$. The nonlinear fields associated with these off-diagonal terms are known as cross-polarization modulation (XPolM). We also observe that $\mathbf{C}_k$ has terms $|E_{x,m}|^2$ and $|E_{y,m}|^2$ that depend on the electric fields of neighboring channels $m \neq k$. Hence, the NLSEs for the different channels couple via the nonlinear operator.

Since $\mathbf{C}_k$ is Hermitian, it has an eigenvalue decomposition $\mathbf{C}_k = \mathbf{V}_k \mathbf{\Lambda}_k \mathbf{V}_k^H$. Integration of Eq. 93 yields:

$$\exp\left(-h\hat{\mathbf{N}}_k\right) \cdot \mathbf{E}_k = \mathbf{V}_k \exp\left(-h\mathbf{\Lambda}_k\right) \mathbf{V}_k^H \mathbf{E}_k \quad (94)$$

For the linear operator, if we ignore PMD and polarization dependence, we have:

$$\hat{\mathbf{D}} \cdot \mathbf{E}(t) = \left(-\frac{1}{2}\alpha + \beta_1 \frac{\partial}{\partial t} - j\frac{1}{2!}\beta_2 \frac{\partial^2}{\partial t^2} - \frac{1}{3!}\beta_3 \frac{\partial^3}{\partial t^3} \right) \sum_n \mathbf{E}_n(t) e^{j\omega_n t} \quad (95)$$

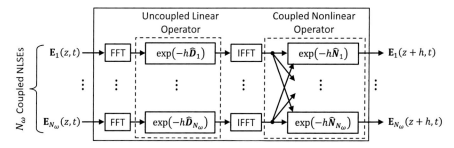

Fig. 42 Illustrating one stage of the C-NLSE algorithm for multichannel digital backpropagation

Taking the Fourier transform of the above equation and integrating over step size h, we have the following propagator for k-th channel:

$$\exp\left(-h\hat{\mathbf{D}}_k\right).\mathbf{E_k}(\omega) = \exp\left(\frac{1}{2}\alpha h - j\left(\beta_{0,k} + \beta_{1,k}\omega + \frac{1}{2!}\beta_{2,k}\omega^2 \right.\right.$$
$$\left.\left. + \frac{1}{3!}\beta_{3,k}\omega^3\right)h\right)\begin{bmatrix} E_{x,k} \\ E_{y,k} \end{bmatrix} \quad (96)$$

where $\beta_{3,k} = \beta_3$, $\beta_{2,k} = \beta_2 + \beta_3\omega_k$, $\beta_{1,k} = \beta_1 + \beta_2\omega_k + \frac{1}{2}\beta_3\omega_k^2$ and $\beta_{0,k} = \beta_1\omega_k + \frac{1}{2}\beta_2\omega_k^2 + \frac{1}{6}\beta_3\omega_k^3$ are the coefficients of the propagation constant for the k-th channel. Since the different channels have different values of $\beta_{1,k}$, they "walk off" due to dispersion.

The N_ω parallel equations (94) and (96) form the coupled NLSE (C-NLSE) for the WDM channels. These equations can be numerically solved using the architecture shown in Fig. 42. The advantage of the C-NLSE over the T-NLSE is that the linear operator operates on the electric field of one channel with bandwidth $2\pi B_W$, rather than the whole WDM channel with total bandwidth of $N_\omega\omega_{sp}$. This enables the use of much larger step size. On the other hand, the C-NLSE involves solving N_ω parallel equations. Most of the algorithmic complexity comes from taking FFTs and IFFTs for each coupled NLSE. Provided $2\pi B_W \ll \omega_{sp}$, the C-NLSE will have lower computational complexity than the T-NLSE. In the limit where $2\pi B_W \approx \omega_{sp}$, however, their complexities are the same.

Timing Recovery

In the previous sections, we had assumed that the receiver's sampling clock was synchronous with the symbol clock. In practice, this condition is not guaranteed, as the receiver's RF clock is at a different physical location, so it is independent of the transmitter's clock. To guarantee synchronous sampling, "timing recovery" is required.

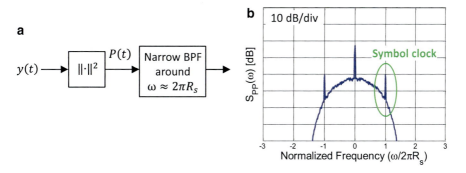

Fig. 43 (a) Analog timing recovery by taking the amplitude of the input signal and applying narrowband filtering around the clock tone and (b) spectrum of $P(t)$

Analog Timing Recovery

Timing recovery can be performed in the analog domain. An example setup is shown in Fig. 43a. The electrical input signal is passed through an amplitude squarer followed by narrow bandpass filtering (BPF) around the symbol clock frequency. Figure 43b shows the spectrum of $P(t)$, assuming the input signal $y(t) = \sum_n y_n q(t - nT_s)$ is modulated by QPSK symbols and $q(t)$ is a nonreturn-to-zero (NRZ) pulse shape. The spectrum of $P(t)$ has tones at $\omega = 2\pi R_S$ that can be extracted by the BPF. A digital implementation of square timing recovery has also been studied in Oerder and Meyer (1988).

A squarer timing recovery circuit has the advantage of simple implementation. However, performance is highly dependent on the carrier-to-interference ratio (CIR) of the extracted clock tone. A low CIR will lead to timing jitter causing performance degradation. The CIR of the extracted clock tone is highly dependent on the pulse shape $q(t)$. When raised cosine or root-raised cosine pulses are used, the clock tones disappear. Even with the use of NRZ pulses, if $q(t)$ has passed through a channel with chromatic dispersion, the power at the clock tone will reduce.

Digital Timing Recovery

The use of a digital coherent receiver enables timing recovery in the digital domain. Digital timing recovery was studied in Gardner (1993) and Erup et al. (1993). A canonical model for a digital timing recovery circuit is shown in Fig. 44. The input signal $x(t)$, which can be any of the analog output tributaries of the O/E downconverter in Fig. 10, is sampled by an ADC that is triggered by the receiver's analog sampling clock, whose frequency $\tilde{R} = 1/\tilde{T}$ which is not synchronized with the transmitter's symbol clock $R_s = 1/T_s$, i.e., $\tilde{R}$ and R_S, is not related by a simple integer fraction. The timing recovery circuit has an interpolation filter that takes the

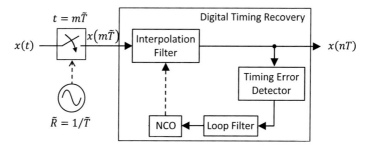

Fig. 44 Canonical model of a digital timing recovery circuit

asynchronous samples $x(m\tilde{T})$ and calculates synchronous samples $x(nT)$, where $T = (K/M)T_S$ as per section "Linear Equalization":

$$x(nT) = \sum_{m=-\infty}^{+\infty} x(m\tilde{T}) h_I(nT - m\tilde{T}) \qquad (97)$$

The ideal interpolation filter is the sinc(·) function:

$$h_I(t) = \frac{1}{\tilde{T}} \text{sinc}\left(\frac{t}{\tilde{T}}\right) \overset{\mathcal{F}}{\leftrightarrow} \text{rect}\left(\frac{\omega\tilde{T}}{2\pi}\right) = H_I(\omega) \qquad (98)$$

In practice, the interpolation filter is only the first digital filter stage and without adding noise, is invariably followed by an adaptive linear equalizer. Hence, $h_I(t)$ can be any low-pass function so long as $H_I(\omega)$ is always non-zero over the bandwidth of the input signal $x(t)$. The amplitude and phase response of $h_I(t)$ will be corrected by an adaptive equalizer downstream. The main criterion for $h_I(t)$ is low complexity. Hence rectangular or triangular functions in time are often used.

The interpolation filter is triggered by a numerically controlled oscillator (NCO), which is kept synchronized to the symbol clock by a feedback loop comprising a timing error detector followed by a loop filter. The timing relationship between the ADC samples and the interpolator outputs is shown in Fig. 45. We have:

$$nT = (m_n + \mu_n)\tilde{T} \qquad (99)$$

where $0 \leq \mu_n < 1$ is the fractional interval that either increases or decreases with time index n depending on whether the synchronous clock frequency is lower or higher than the ADC clock frequency. From Eq. 99, it can be deduced that $m_n = \lfloor nT/\tilde{T} \rfloor$ and $\mu_n = nT/\tilde{T} - m_n$. The NCO is thus governed by two parameters: the NCO clock frequency $\Omega = 2\pi(\tilde{T}/T)$ and its instantaneous phase Φ_n. The NCO supplies m_n and μ_n that control the interpolator. It can be shown that $\Phi_n = -\mu_n/\Omega$.

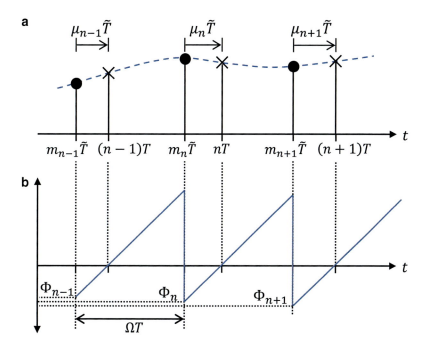

Fig. 45 (a) Illustrating the relationship between ADC samples and the interpolated values and their respective timing instances and (b) NCO phase relation

The timing recovery circuit can generate a synchronous clock by passing $\cos(\Phi_n)$ to a DAC (Auer 1990).

The timing error detector estimates instantaneous offset between the ideal sampling instants and the actual sampling instants. For BPSK/QPSK signals, Gardner introduced an early-late algorithm that calculates (Gardner 1986):

$$u_\tau = x(nT)\left[x((n+1)T) - x((n-1)T)\right] \quad (100)$$

where $x(t)$ is a bipolar signal from any of the analog outputs of the O/E downconverter in Fig. 10. The samples at $t = nT$ are those closest to the maximum eye opening of the symbols. As the Gardner algorithm assumes an oversampling of two ($T = T_s/2$), the timing recovery circuit can be conveniently placed between the FDE and the TDE in a hybrid equalizer, since the FDE output is naturally at 2× oversampling, while the TDE downsamples its output-to-symbol rate (Fig. 46).

The presence of residual uncompensated dispersion, PMD, or the use of larger signal constellation may degrade the performance of the Gardner timing error detector. However, other alternative methods exist for determining timing error. Consider that timing error is compensated by the adaptive TDE, whose output is kept synchronized to the optimal sampling instant by an adaptive algorithm. It is possible to take the Fourier transform of the TDE coefficients and estimate the group

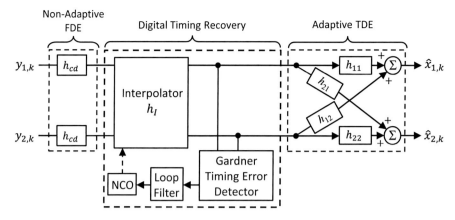

Fig. 46 Digital timing recovery operating at 2× oversampling using the Gardner timing error detector can be placed between the FDE and TDE in a hybrid equalization circuit

Fig. 47 Analytical model of a the timing recovery circuit

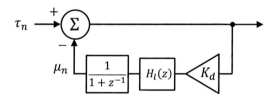

delay of the equalizer from its phase response. Since RF clocks are relatively stable, timing jitter is a slow process, so the group delay of the equalizer does not even have to be calculated at every symbol.

For small delay values, the timing error detector will produce an output of $K_d \tau_\epsilon$ that is approximately linear with timing error τ_ϵ, where K_d is the gain of the detector. The output (u_τ in case of the Gardner algorithm) drives a loop filter with transfer function $H_l(z)$ which controls the frequency of the NCO. A complete analytical model of the timing recovery circuit is shown in Fig. 47. A second-order phase-locked loop (PLL) has proven to be a popular choice for the timing recovery circuit (Kushnerov et al. 2009).

Other Topics

Space-Division Multiplexing

The combination of dense wavelength-division multiplexing (DWDM) and high spectral efficiency transmission enabled by coherent detection has pushed the capacity limit of single-mode fiber to around 100 Tb/s using the C-+L-bands (Qian et al. 2011; Sano et al. 2012), at a spectral efficiency of over 10 b/s/Hz. Ultimately, even in the absence of fiber nonlinearity, the spectral efficiency of a linear channel

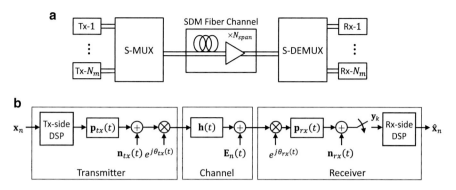

Fig. 48 (a) Coherent optical transmission system based on SDM fiber and (b) analytical model

only grows logarithmically with SNR. It is power inefficient to continue increasing total capacity by using increasingly large spectral efficiency. For a given fixed total power P_{tot}, that is available for transmission, and a fixed noise psd N_0 after transmission and N_{ch} parallel channels of bandwidth B_w available for use (these can be spectral or spatial channels), Shannon's capacity is:

$$C = N_{ch} B_W \log_2 \left(1 + \frac{P_{tot}}{N_{ch} B_W N_0}\right) \quad (101)$$

The capacity in Eq. 101 is an increasing function of N_{ch}. In the limit as $N_{ch} \to \infty$, the ultimate capacity using an infinite number of parallel channels is $C_\infty \to (P_{tot}/N_0)\log_2(e)$.

Dense wavelength-division multiplexing (DWDM) is a realization of spectral parallel transmission. However, as the usable bandwidth of single-mode fiber is already exhausted, space-division multiplexing has emerged as a proposed solution for increasing total system capacity beyond that achievable in single-mode fiber. Among the proposed candidate solutions are multi-core fibers (MCF); few-mode fibers (FMF), which support the propagation of many orthogonal modes (but fewer than that in traditional multimode fibers (MMF) with 50 or 62.5 μm core diameter); and strongly coupled multi-core fibers (SC-MCF). An SDM fiber is an optical transmission medium which supports the propagation of N_m spatial channels where $N_{ch} > 2$ polarization modes available in "single-"mode fiber.

All of the DSP techniques we have studied so far can be generalized for SDM fibers. Consider the generalized coherent receiver front-end shown in Fig. 48. Compared with the transmitter and receiver front-end shown in Figs. 3 and 10 for single-mode fiber, the beam splitter which acted as a spatial multiplexer for polarization modes has been replaced with generalized spatial multiplexer (S-MUX) and demultiplexer (S-DEMUX) that can access all N_m modes available in an SDM fiber. For FMFs, S-MUX and S-DEMUX have been successfully demonstrated using a variety of technologies, including the use of phase plates (Ryf et al. 2011), spot couplers (Ryf et al. 2012), integrated mode couplers (Koonen et al. 2012), photonic lanterns (Fontaine et al. 2013; Birks et al. 2015), and multiplane lightwave conversion (Labroille et al. 2016). In particular, mode-selective photonic lanterns and multiplane lightwave converters have low insertion loss and high modal extinction.

The linear equalization and carrier phase estimation techniques that were introduced in sections "Linear Equalization" and "Carrier Phase Estimation" can be directly generalized for SDM fibers. For instance, the linear equalizer input in Eq. 44 will now comprise of N_m vectors stacked on top each other: $\mathbf{y}_k = \begin{bmatrix} \mathbf{y}_{1,k}^T \cdots \mathbf{y}_{N_m,k}^T \end{bmatrix}^T$, with the vector for each mode still being $y_{i,k}^T = \begin{bmatrix} y_{i,\frac{kM}{K}+L} \; y_{i,\frac{kM}{K}+L-1} \cdots y_{i,\frac{kM}{K}-L} \end{bmatrix}^T$. The equalizer matrix $\mathbf{w}$ will likewise comprise of $N_m \times N_m$ compartments. The MMSE "Wiener" solution is still given by Eq. 42, while the performance of the equalizer can still be analyzed as per section "MMSE Performance."

One key distinction for linear equalization in FMF is that the channel's memory length will likely be much longer, as it is equal to the difference between the group delays of the fastest and slowest propagating modes of the fiber. The differential modal group delay (DMGD) of FMFs has typically ranged from tens of ps/km for DMGD-optimized FMFs supporting only the two lowest-order mode groups (Ip et al. 2014), to ns/km for FMFs supporting many mode groups without DMGD optimization (Fontaine et al. 2015). The channel memory length due to DMGD will be significantly greater than that of CD. If the modes do not couple during transmission, corresponding to a sparse channel matrix, linear equalization can be performed per mode group. However, modes do couple during transmission. Long-distance FMF transmission experiments to date have required the use of full multiple-input multiple-output (MIMO) equalization.

But even DMGD is not necessarily a limiting factor for DSP, as the computational complexity of an FDE scales only logarithmically with channel memory length. Moreover, it is possible to reduce channel memory length using techniques such as strong mode coupling (Ho and Kahn 2011). In terms of DSP complexity, the true limiting factor is that the complexity of full MIMO scales linearly with N_m. This is because the MIMO equalizer is ultimately a matrix multiplication operation at each frequency of the signal. Matrix multiplication requires $N_m \times N_m$ multiplications for every output vector of N_m symbols. The increase in complexity with N_m, along with other considerations, will limit the practical size of N_m for FMF or SC-MCF transmission, even as DSP continues to become more powerful (Ip et al. 2015).

Another important feature in SDM transmission is system MDL, which is generally much larger than polarization-dependent loss (PDL) in single-mode fiber systems. MDL in an FMF transmission system can arise from the S-MUX, S-DEMUX, few-mode EDFAs, filters, switches, and the fiber itself, as well as other elements. MDL is more likely in optical components where the optical electric field has to couple from one fiber to another, possibly with different refractive index geometry. From a system's perspective, MDL causes the channel matrix to have large singular values spread. In reported experiments, MDL as large as 20 dB is not uncommon, depending on the transmission distance and the number of modes supported by the fiber. This has key implications for linear equalization, as blind training algorithms become highly susceptible to the singularity problem, which was already evident for PDL in SMF (section "Adaptive Time-Domain Equalizer (TDE)"). Although ICA algorithms have been studied (Zhao et al. 2015; Yan and Hu 2016; He et al. 2016) and implemented at low baud rates and transmission distances, most offline experiments to date using FMF or SC-MCF have resorted to training symbols for initial adaptation of the equalizer coefficients.

In terms of carrier phase estimation, SDM fibers may offer computational savings when all the spatial channels are modulated by the same transmitter and receiver LO lasers. Since all spatial channels have the same phase noise, it is possible to perform carrier phase estimation on only one of the spatial channels or a subset thereof (e.g., the spatial channel with the best SNR) and then using this phase estimate to derotating the carrier phase on all the other spatial channels. Such a carrier recovery scheme for FMF was studied in van Uden et al. (2013) and for MCF in Mendinueta et al. (2013).

Optical Performance Monitoring

The recovery of the optical electric field and the use of DSP impairment compensation on that field allow unique performance monitoring techniques. Consider the MMSE equalizer matrix that was derived in Eq. 62, which is the solution that an adaptive equalizer will converge to in steady state. The equalizer matrix contains vital information about the channel (Hauske et al. 2009). For example, it is possible to perform quadratic curve fitting on the phase response of the adaptive TDE to learn the residual CD not already compensated by the FDE. This allows precise estimation of the CD of the channel. A singular-value decomposition (SVD) can also be performed on the channel matrix $\mathbf{H}(\omega) = \mathbf{U}(\omega)\mathbf{S}(\omega)\mathbf{V}^H(\omega)$. The difference in amplitude between the singular values on the diagonal of $\mathbf{S}(\omega)$ gives the system PDL (or MDL), while the difference in the slopes of their phase responses will reveal the system PMD. Furthermore, by monitoring of $\mathbf{H}(\omega)$ over time, it is possible to determine the rate at which the channel is changing. Since the channel matrix is expected to change slowly relative to the symbol rate, the calculations on $\mathbf{H}(\omega)$ can be performed by the transponder at a commensurate lower rate.

Further channel information can be gained by inspection of the quality of the equalized symbols. Assuming the impact of phase noise to be negligible, and there is no ISI or aliasing arising from sub-Nyquist sampling, the sources of distortion after equalization are noise (from both the channel and receiver) and nonlinearity. The impact of nonlinearity may be factored out either by knowing that the transmitted power is sufficiently low such that the system is in the linear regime. Otherwise, if the system is operating at an optimum power level, nonlinearity imposes a fixed penalty on the quality factor of the equalized constellation. Receiver noise is generally static and can be measured during transponder characterization and then factored out to determine the total noise added by the channel. In-band OSNR monitoring is an important feature of coherent transponders and has been studied in

Dong et al. (2012). Finally, the carrier recovery process produced an estimate of the instantaneous phase which is a Wiener process. Taking the spectrum of frequency noise (the derivative of the carrier phase) can reveal the combined linewidths of the transmitter and receiver LO lasers, as per Eq. 70.

Ultimately, channel information learned from a coherent transponder will be useful to higher layers in a software-defined network (SDN) and can greatly assist in channel provisioning and routing algorithms by the higher layers (Paolucci et al. 2012).

Optical Sensing

The benefits of optical coherent detection can also been applied to other fields. Optical sensing is one area that has been revolutionized by the information that can be gleaned from the recovery of the optical electric field. One example is distributed vibration sensing based on coherent detection of the Rayleigh backscatter. As acoustic vibration results in localized change in the fiber's refractive index, which manifests as a localized phase fluctuation, a coherent receiver can be used to perform phase optical time-domain reflectometry (OTDR), where DSP algorithms can be used to remove noise, set the spatial resolution, and perform averaging (Lu et al. 2010). Another example is distributed temperature sensing based on coherent detection of the spontaneous Brillouin backscatter. In this case, the coherent receiver serves as an ultrahigh-resolution optical spectrum analyzer (OSA): DSP is used to calculate the reflected Brillouin spectrum at a high frequency resolution, and curve fitting is then performed to estimate the instantaneous Brillouin frequency shift (Alahbabi et al. 2005).

In summary, all of these proposed methods have benefited from coherent detection in exactly the same manner as telecommunications, namely, recovery of the full optical electric field and arbitrary manipulation of that field by digital signal processing, which allows accurate determination of the desired channel parameters.

Conclusions

Optical coherent receivers have been made possible by the development of hardware components such as wide-bandwidth balanced photodetectors, low-linewidth tunable lasers, dual-polarization dual-quadrature Mach-Zehnder modulators, and dual-polarization optical hybrids. A major benefit of coherent detection is that the recovery of all the degrees of freedom available in single-mode fiber enables transmission at the highest spectral and power efficiency possible. An even greater advantage comes from powerful digital signal processing algorithms that can be employed to compensate channel impairments. It was the arrival of high-speed analog-to-digital converters (ADCs), digital-to-analog converts (DACs), and digital signal processors (DSP) that has made digital coherent receivers the optical receiver of choice for long-haul optical transmission. With the passage of time, and the resulting reduction in cost, digital coherent receivers have been pushed into metro and even access networks.

In this work, we have provided an introduction to coherent detection and the basic channel impairment compensation algorithms that are ubiquitous to modern digital coherent receivers. The key element in coherent detection is the optical-to-electrical downconverter that enables the recovery of the in-phase and quadrature electric fields in the two optical signal polarizations. We covered the theory of the operation of an O/E downcoverter and analyzed its noise performance. We also similarly derived linear models for the dual-polarization optical transmitter and channel to derive an analytical baseband model of the full coherent optical system.

In terms of channel impairment compensation, the most important algorithms are linear equalization of the chromatic dispersion (CD), polarization mode dispersion (PMD), and other linear time-invariant (LTI) impairments and optical carrier recovery to compensate laser phase noise. The most common equalizer structure is a hybrid equalizer which uses a computationally efficient frequency-domain equalizer (FDE) to compensate CD, as it is normally a fixed impairment that changes only slowly with time. The fixed FDE is followed by an adaptive TDE for compensating fast-varying impairments such as polarization rotation and PMD. A blind adaptive algorithm such as the constant modulus algorithm (CMA) is normally used to perform initial training of the TDE coefficients. Once initial convergence is achieved, adaptation is switched to a more accurate decision-directed (DD) algorithm to track the time-varying channel. The principles of nondecision-directed (NDA) and DD training also apply to laser phase noise compensation. We have covered the Viterbi-Viterbi (VV) algorithm as well as DD algorithms. We have also provided mathematical analysis of both linear equalization and laser phase noise compensation.

With continued improvement in DSP, it is possible to use higher-complexity algorithms to also compensate for fiber nonlinearity. We have seen that signal propagation in optical fiber is governed by a nonlinear Schrödinger equation (NLSE) that is fully invertible by digital backpropagation (DBP) in the absence of noise. There has been much effort in reducing the complexity of DBP to allow its practical implementation. These efforts have resulted in such algorithms as filtered DBP and Volterra series implementations. Nonlinear compensation involving only a few stages of DBP has already been demonstrated as a commercial product. Ultimately, there exists a trade-off between algorithmic complexity and performance. Most published works on nonlinear compensation have found only small performance gain of around 1 dB.

The combination of dense wavelength-division multiplexing (DWDM) and high spectral efficiency transmission enabled by coherent detection pushed the capacity limit of single-mode fiber to around 100 Tb/s using the C-+L-bands. Further improvements will require the use of parallel spatial channels. Space-division multiplexing (SDM) has received much recent research interest as an attempt to circumvent the nonlinear Shannon capacity for single-mode fiber. One proposal is to use few-mode fiber (FMF), whose orthogonal modes are merely a generalization of the two polarization modes of an SMF. Consequently, with the use of appropriate spatial multiplexers and demultiplexers to access the FMF modes, such as photonic lanterns with low insertion loss and low mode-dependent loss (MDL), it is possible to use the same DSP algorithms to compensate the impairments of a FMF channel. However, mode coupling will require the use of multiple-input multiple-output (MIMO) DSP, whose complexity per bit is expected to increase with the number of modes supported by the transmission fiber. This limitation has not yet been overcome. Ultimately, it is cost-per-bit that is the most important criterion in a telecommunications system. The poor scaling of DSP complexity may render FMF potentially useful only for unamplified, single-span short-reach systems. However, SDM using either multi-core fibers (MCF), or simply parallel SMF, will ultimately be required in long-haul transmission to facilitate increasing network demand.

Nevertheless, coherent detection is now a mature technology that has proven its value in cost-per-bit in single-mode fiber transmission. Future coherent systems may support additional technologies for improved performance monitoring and may even include real-time optical sensing to determine distributed temperature and/or strain along a transmission cable.

References

S. Abrar, Compact constellation algorithm for blind equalization of QAM signals, in *INCC 2004* (Lahore, 2004), pp. 170–174

G.P. Agrawal, *Nonlinear Fiber Optics*, 3rd edn. (Academic, San Diego, 2001)

G.P. Agrawal, *Fiber-Optic Communication Systems*, 3rd edn. (Wiley, New York, 2002)

M.N. Alahbabi, Y.T. Cho, T.P. Newson, 150-km range distributed temperature sensor based on coherent detection of spontaneous Brillouin backscatter and in-line Raman amplification. J. Opt. Soc. Am. B **22**(6), 1321–1324 (2005)

J. Armstrong, OFDM for optical communications. J. Lightwave Technol. **27**(3), 189–204 (2009)

E. Auer, An advanced variable data rate modem for Intelsat IDR/IBS services, in *2nd International Workshop Digital Signal Processing Techniques Applied to Space Communications*. Paper 1–3 (Turin, 1990)

N. Bai, G. Li, Adaptive frequency-domain equalization for mode-division multiplexed transmission. IEEE Photon. Technol. Lett. **24**(21), 1918–1921 (2012)

J.R. Barry, J.M. Kahn, Carrier synchronization for homodyne and heterodyne detection of optical quadriphase-shift keying. J. Lightwave Technol. **10**(12), 1939–1951 (1992)

S. Beppu, K. Kasai, M. Yoshida, M. Nakazawa, 2048 QAM (66 Gbit/s) single-carrier coherent optical transmission over 150 km with a potential SE of 15.3 bit/s/Hz. Opt. Express **23**(4), 4960–4969 (2015)

T.A. Birks, I. Gris-Sanchez, S. Yerolatsitis, S.G. Leon-Saval, R.R. Thomson, The photonic lantern. Adv. Opt. Photon. **7**(2), 107–167 (2015)

C.A. Brackett, Dense wavelength division multiplexing networks: Principles and applications. IEEE J. Sel. Areas Commun. **8**(6), 948–964 (1990)

J.-X. Cai, H.G. Batshon, M.V. Mazurczyk, O.V. SInkin, D. Wang, M. Paskov, W. Patterson, C.R. Davidson, P. Corbett, G. WOlter, T. Hammon, M. Bolshtyansky, D. Foursa, A. Pilipetskii, 70.4-Tb/s capacity over 7,600 km in C+L-band using coded modulation with hybrid constellation shaping and nonlinearity compensation, in *Optical Fiber Conference (OFC)*, Paper Th5B.2 (Los Angeles, 2017)

Y. Chen, T. Le-Ngoc, B. Champagne, C. Xu, Recursive least squares constant modulus algorithm for blind adaptive array. IEEE Trans. Signal Process **52**(5), 1452–1456 (2004)

X. Chen, J. Ye, Y. Xiao, A. Li, J. He, Q. Hu, W. Shieh, Equalization of two-mode fiber based MIMO signals with larger receiver sets. Opt. Express **20**(26), B413–B418 (2012)

A. Chowdhury, R.J. Essiambre, Optical phase conjugation and pseudolinear transmission. Opt. Lett. **29**(10), 1105–1107 (2004)

E. Desurvire, *Erbium-Doped Fiber Amplifiers, Principles and Applications* (Wiley, New York, 2002)

I.B Djordjevic, H.G. Batshon, L. Xu T. Wang, Coded polarization-multiplexed iterative polar modulation (PM-IPM) for beyond 400 Gb/s serial optical transmission, in *Optical Fiber Conference (OFC)*, Paper OMK2 (San Diego, 2010)

Z. Dong, A.P.T. Lau, C. Lu, OSNR monitoring for QPSK and 16QAM systems in presence of fiber nonlinearities for digital coherent receivers. Opt. Express **20**(17), 19520–19534 (2012)

L.B. Du, A.J. Lowery, Improved single channel backpropagation for intra-channel fiber nonlinearity compensation in long-haul optical communication systems. Opt. Express **18**(16), 17075–17088 (2010)

L. Erup, F.M. Gardner, R.A. Harris, Interpolation in digital modems – part II: implementation and performance. IEEE Trans. Commun. **41**(6), 998–1008 (1993)

R.-J. Essiambre, P.J. Winzer, X.Q. Qang, W. Lee, C.A. White, E.C. Burrows, Electronic predistortion and fiber nonlinearity. IEEE Photon. Technol. Lett. **18**(17), 1804–1806 (2006)

S.G. Evangelides, L.F. Mollenauer, J.P. Gordon, N.S. Bergano, Polarization multiplexing with solitons. J. Lightwave Technol. **10**(1), 28–35 (1992)

J.K. Fischer, C.-A. Bunge, K. Petermann, Equivalent single-span model for dispersion-managed fiber-optics transmission systems. J. Lightwave Technol. **27**(16), 3425–3432 (2009)

N.K. Fontaine, S.G. Leon-Saval, R. Ryf, J.R. Salazar Gil, B. ERcan, J. Bland-Hawthorn, Mode-selective dissimilar fiber photonic-lantern spatial multiplexers for few-mode fiber, in *European Conference on Optical Communications (ECOC)*, Paper PD1.C.3 (London, 2013)

N.K. Fontaine, R. Ryf, H. Chen, A.V. Benitez, B. Guan, R. Scott, B. Ercan, S.J.B. Yoo, L.E. Grüner-Nielsen, Y. Sun, R. Lingle, E. Antonio-Lopez R. Amezcua-Correa, 30×30 MIMO transmission over 15 spatial modes, in *Optical Fiber Communication Conference (OFC)*, Paper Th5C.1 (Los Angeles, 2015)

F.M. Gardner, A BPSK/QPSK timing-error detector for sampled receivers. IEEE Trans. Commun. **COM-34**(5), 423–429 (1986)

F.M. Gardner, Interpolation in digital modems – part I: fundamental. IEEE Trans. Commun. **41**(3), 501–507 (1993)

R.D. Gitlin, S.B. Weinstein, Fractionally spaced equalization: an improved digital transversal equalizer. Bell Syst. Technol. J. **60**(2), 275–296 (1981)

B. Glance, Polarization independent coherent optical receiver. J. Lightwave Technol. **LT**(2), 274–276 (1987)

D.N. Godard, Self- recovering equalization and carrier tracking in two-dimensional data communication systems. IEEE Trans. Commun. **COM-28**, 1867–1875 (1980)

J.P. Gordon, H. Kogelnik, PMD fundamentals: Polarization mode dispersion in optical fibers. Proc. Natl. Acad. Sci. **97**(9), 4541–4550 (2000)

J.P. Gordon, L.F. Mollenauer, Phase noise in photonic communications systems using linear amplifiers. Opt. Lett. **15**(23), 1351–1353 (1990)

H.A. Haus, The noise figure of optical amplifiers. IEEE Photon. Technol. Lett. **10**(11), 1602–1604 (1998)

F.N. Hauske, M. Kushnerov, B. Spinnler, B. Lankl, Optical performance monitoring in digital coherent receivers. J. Lightwave Technol. **27**(16), 3623–3631 (2009)

S. Haykin, *Neural Networks: A Comprehensive Foundation*, 2nd edn. (Prentice Hall, Upper Saddle River, 1998)

S. Haykin, *Adaptive Filter Theory*, 4th edn. (Prentice Hall, Upper Saddle River, 2002)

Z. He, X. Li, M. Luo, R. Hu, C. Li, Y. Qiu, S. Fu, Q. Yang, S. Yu, Independent components analysis based channel equalization for 6×6 MIMO-OFDM transmission over few-mode fiber. Opt. Express **24**(9), 9209–9217 (2016)

K.-P. Ho, J.M. Kahn, Statics of group delays in multimode fiber with strong mode coupling. J. Lightwave Technol. **29**(21), 3119–3128 (2011)

Y.-K. Huang, E. Ip, P.N. Ji, Y. Shao, T. Wang, Y. Aono, Y. Yano, T. Tajima, Terabit/s optical superchannel with flexible modulation format for dynamic distance/route transmission, in *Optical Fiber Conference (OFC)*, Paper OM3.H.4 (Los Angeles, 2012)

E. Ip, J.M. Kahn, Digital equalization of chromatic dispersion and polarization mode dispersion. J. Lightwave Technol. **25**(8), 2033–2043 (2007a)

E. Ip, J.M. Kahn, Feedforward carrier recovery for coherent optical communications. J. Lightwave Technol. **25**(9), 2675–2692 (2007b)

E. Ip, J.M. Kahn, Compensation of dispersion and nonlinear impairments using digital backpropagation. J. Lightwave Technol. **26**(20), 3416–3425 (2008)

E. Ip, J.M. Kahn, Fiber impairment compensation using coherent detection and digital signal processing. J. Lightwave Technol. **28**(4), 502–519 (2009)

E. Ip, A.P.T. Lau, D.J.F. Barros, J.M. Kahn, Coherent detection in optical fiber systems. Opt. Express **16**(2), 753–791 (2008)

E. Ip, N. Bai, T. Wang, Complexity versus performance tradeoff in fiber nonlinearity compensation using frequency-shaped, multi-subband backpropagation, in *Optical Fiber Communication Conference (OFC)*, Paper OThF4 (Los Angeles, 2011)

E. Ip, M.-J. Li, K. Bennett, Y.-K. Huang, A. Tanaka, A. Korolev, K. Koreshkov, W. Wood, E. Mateo, J. Hu, Y. Yano, 146λ×6×19-Gbaud wavelength-and mode-division multiplexed transmission over 10×50-km spans of few-mode fiber with a gain-equalized few-mode EDFA. J. Lightwave Technol. **32**(4), 790–797 (2014)

E. Ip, G. Milione, Y.-K. Huang, T. Wang, Space division multiplexing for optical networks, in *Asia Communications and Photonics Conference (ACP)*, Paper ASu5.D.1 (Hong Kong, 2015)

O. Ishida, K. Takei, E. Yamazaki, Power efficient DSP implementation for 100G- and beyond multi-haul coherent fiber-optic communications, in *Optical Fiber Conference (OFC)*, Paper W3G.3 (Anaheim, 2016)

H. Ishio, J. Minowa, K. Nosu, Review and status of wavelength-division multiplexing technology and its applications. J. Lightwave Technol. **LT-2**(4), 448–463 (1984)

K. Iwashita, Chromatic dispersion compensation in coherent optical communications. J. Lightwave Technol. **8**(3), 367–375 (1990)

E.I. Jury, *Theory and Application of the z-Transform Method* (Wiley, New York, 1964)

J.M. Kahn, A.H. Gnauck, J.J. Veselka, S.K. Korotky, B.L. Kasper, 4-Gb/s PSK homodyne transmission system using phase-locked semiconductor lasers. IEEE Photon. Technol. Lett. **2**(4), 285–287 (1990)

J.M. Kahn, A.M. Porter, U. Padan, Heterodyne detection of 310-Mb/s quadriphase-shift keying using fourth-power optical phase-locked loop. IEEE Photon. Technol. Lett. **4**(12), 1397–1399 (1992)

L.G. Kazovsky, Balanced phase-locked loops for optical homodyne receivers: performance analysis, design considerations, and laser linewidth requirements. J. Lightwave Technol. **LT-4**, 182–195 (1986)

R. Kline, Verizon and Nortel complete the first 100G commercial deployment, http://www.ovum.com/news/euronews.asp?id=8315

A.M.J. Koonen, H. Chen, H.P.A. van den Boom, O. Raz, Silicon photonic integrated mode multiplexer and demultiplexer. IEEE Photon. Technol. Lett. **24**(21), 1961–1964 (2012)

P.M. Krummrich, D. Ronnenberg, W. Schairer, D. Wienold, F. Jenau, M. Herrmann, Demanding response time requirements on coherent receivers due to fast polarization rotations caused by lightning events. Opt. Express **24**(11), 12442–12457 (2016)

M. Kuschnerov, F.N. Hauske, K. Piyawanno, B. Spinnler, A. Napoli, B. Lankl, Adaptive chromatic dispersion equalization for non-dispersion managed coherent systems, in *Proceedings of the Optical Fiber Communication Conference (OFC 2009)*, Paper OMT1 (San Diego, 2009)

M. Kushnerov, F.N. Hauske, K. Piyawanno, B. Spinnler, M.S. Alfiad, A. Napoli, B. Lankl, DSP for coherent single-carrier receivers. J. Lightwave Technol. **27**(16), 3614–3622 (2009)

G. Labroille, P. Jian, N. Barré, B. Denolle, J.-F. Morizur, Mode selective 10-mode multiplexer based on multi-plane light conversion, in *Optical Fiber Communication Conference (OFC)*, Paper Th3E.5 (Anaheim, 2016)

J.H. Lee, M.S. Kim, Y.C. Chung, Statistical PMD emulator using variable DGD elements. IEEE Photon. Technol. Lett. **15**(1), 54–56 (2003)

A. Leven, N. Kaneda Y.-K. Chen, A real-time CMA-based 10 Gb/s polarization demultiplexing coherent receiver implemented in an FPGA, in *OFC 2008*, Paper OTuO2 (San Diego, 2008)

L. Li, Z. Tao, L. Dou, W. Yan, S. Oda, T. Tanimura, T. Hoshida, J. C. Rasmussen, Implementation efficient nonlinear equalizer based on correlated digital backpropagation, in *Optical Fiber Communication Conference (OFC)*, Paper OWW3 (Los Angeles, 2011)

X. Liu, Nonlinear effects in phase shift keyed transmission, in *Optical Fiber Conference (OFC)*, Paper ThM4 (Los Angeles, 2004)

Y. Lu, T. Zhu, L. Chen, X. Bao, Distributed vibration sensor based on coherent detection of phase OTDR. J. Lightwave Technol. **28**(22), 3243–3249 (2010)

D.-S. Ly-Gagnon, S. Tsukamoto, K. Katoh, K. Kikuchi, Coherent detection of optical quadrature phase-shift keying signals with carrier phase estimation. IEEE J. Lightwave Technol. **24**(1), 12–21 (2006)

D.M. Malyon, Digital fiber transmission using optical homodyne detection. Electron. Lett. **20**, 281–283 (1984)

D. Mansour, A.H. Gray, Unconstrained frequency-domain adaptive filter. IEEE Trans. Acoust. Speech Signal. Process. **ASSP-30**(5), 726–734 (1982)

D. Marcuse, C.R. Menyuk, P.K.A. Wai, Application of the Manakov-PMD equation to studies of signal propagation in optical fibers with randomly varying birefringence. J. Lightwave Technol. **15**(9), 1735–1746 (1997)

E. Mateo, L. Zhu, G. Li, Impact of XPM and FWM on the digital implementation of impairment compensation for WDM transmission using backward propagation. Opt. Express **16**(20), 16124–16137 (2008)

A. Mecozzi, R. Essiambre, Nonlinear Shannon limit in pseudolinear coherent systems. J. Lightwave Technol. **30**(12), 2011–2024 (2012)

A. Mecozzi, C. Antonelli, M. Shtaif, Kramer-Kronig coherent receiver. Optical **3**(11), 1220–1227 (2016)

J.M.D. Mendinueta, B.J. Puttnam, J. Sakaguchi, R.S. Lus, W. Klaus, Y. Awaji, N. Wada, A. Kanno, T. Kawanishi, Investigation of receiver DSP carrier phase estimation rate for self-homodyne space-division multiplexing communication systems, in *Optical Fiber Communication Conference (OFC)*, Paper JTh2A.48 (Anaheim, 2013)

H. Meyr, M. Moeneclaey, S. Fechtel, *Digital Communication Receivers* (Wiley, New York, 1997)

T. Miyazaki, Linewidth-tolerant QPSK homodyne transmission using a polarization-multiplexed pilot carrier. IEEE Photon. Technol. Lett. **18**(2), 388–390 (2006)

T. Mizuochi, Recent progress in forward error correction and its interplay with transmission imparments. IEEE J. Sel. Top. Quantum Electron. **12**(4), 544–554 (2006)

M. Oerder, H. Meyer, Digital filter and square timing recovery. IEEE Trans. Commun. **36**(5), 605–612 (1988)

T. Okoshi, Heterodyne and coherent optical fiber communications: Recent progress. IEEE Trans. Microw. Theory Tech. **MIT-30**(8), 1138–1148 (1982)

T. Okoshi, K. Kikuchi, Heterodyne type optical fiber communications. IEEE J. Opt. Commun. **2**(3), 82–88 (1981)

A.V. Oppenheim, R.W. Schafer, *Discrete-Time Signal Processing* (Prentice Hall, Upper Saddle River, 1999)

T. Ozeki, M. Yoshimura, T. Kudo, H. Lbe, Polarization-mode dispersion equalization experiment using a variable equalizing optical circuit controlled by a pulse-waveform-comparison algorithm, in *Optical Fiber Conference (OFC)*, Paper TuN4 (San Jose, 1994)

F. Paolucci, F. Cugini, N. Hussain, F. Fresi, L. Poti, OpenFlow-based flexible optical networks with enhanced monitoring functionalities, in *European Conference on Optical Communications (ECOC)*, Paper Tu1.D.5 (Amsterdam, 2012)

P. Poggiolini, The GN model of non-linear propagation in uncompensated coherent optical systems. J. Lightwave Technol. **30**(24), 3857–3879 (2012)

P. Poggiolini, G. Bosco, A. Carena, V. Curri, Y. Jiang, F. Forghieri, The GN-model of non-linear propagation and its applications. J. Lightwave Technol. **32**(4), 694–721 (2014)

J. Proakis, M. Salehi, *Digital Communications*, 4th edn. (McGraw-Hill, New York, 2001)

D. Qian, M.-F. Huang, E. Ip, Y.-K. Huang, Y. Shao, J. Hu, T. Wang, 101.7-Tb/s (370×294-Gb/s) PDM-128QAM-OFDM transmission over 3×55-km SSMF using pilot-based phase noise mitigation, in *Optical Fiber Conference (OFC)*, Paper PDPB5 (Los Angeles, 2011)

S.U.H. Qureshi, Adaptive equalization. Proc. IEEE **73**(9), 1349–1387 (1985)

T. Rappaport, *Wireless Communications: Principles and Practice* (Prentice-Hall, Englewood Cliffs, 1996)

K. Roberts, 40 Gb/s optical systems with electronic signal processing, in *Proceedings of the Lasers and Electro-Optics Society (LEOS)*, Paper WFF4 (Hualien, 2007)

R. Ryf, S. Randel, A.H. Gnauck, C. Bolle, R.-J. Essiambre, P. Winzer, D.W. Peckham, A. McCurdy, R. Lingle, Space-division multiplexing over 10 km of three-mode fiber using coherent 6×6 MIMO processing, in *Optical Fiber Communication Conference (OFC)*, Paper PDPB10 (Los Angeles, 2011)

R. Ryf, N. Fontaine, R.-J. Essiambre, Spot-based mode couplers for mode-multiplexed transmission in few-mode fiber. IEEE Photon. Technol. Lett. **24**(21), 1973–1976 (2012)

A. Sano, T. Kobayashi, S. Yamanaka, A. Matsuura, H. Kawakami, Y. Miyamoto, K. Ishihara, H. Masuda, 102.3-Tb/s (224×548-Gb/s) C- and extended L-band all-Raman transmission over 240 km using PDM-64QAM single carrier FDM with digital pilot tone, in *Optical Fiber Conference (OFC)*, Paper PDP5C3 (Los Angeles, 2012)

H. Sari, G. Karam, I. Jeanclaude, Transmission techniques for digital terrestrial TV broadcasting. IEEE Commun. Mag. **33**, 100–109 (1995)

S.J. Savory, Digital filters for coherent optical receivers. Opt. Express **16**(2), 804–817 (2008)

S. Savory, Digital coherent optical receivers: Algorithms and subsystems. IEEE J. Sel. Top. Quantum Electron. **16**(5), 1164–1179 (2010)

S. Savory, Approximations for the nonlinear self-channel interference of channels with rectangular spectra. IEEE Photon. Technol. Lett. **25**(10), 961–964 (2013)

C.E. Shannon, A mathematical theory of communication. Bell Syst. Tech. J. **27**, 379–423 (1948)

J. Shapiro, Quantum noise and excess noise in optical homodyne and heterodyne receivers. IEEE J. Quantum Electron. **QE-21**(3), 237–250 (1985)

W. Shieh, K.-P. Ho, Equalization-enhanced phase noise for coherent-detection systems using electronic digital signal processing. Opt. Express **16**(20), 15718–15727 (2008)

J.J. Shynk, Frequency-domain multirate adaptive filtering. IEEE Signal Process. Mag. **9**(1), 14–37 (1992)

J.S. Soo, K.K. Pang, Multidelay block frequency domain adaptive filter. IEEE Trans. Acoust. Speech Signal. Process. **38**(2), 373–376 (1990)

M.G. Taylor, Coherent detection method using DSP for demodulation of signal and subsequent equalization of propagation impairments. IEEE Photon. Technol. Lett. **16**(2), 674–676 (2004)

M.G. Taylor, Accurate digital phase estimation process for coherent detection using a parallel digital processor, in *European Conference on Optical Communications (ECOC)*, Paper Tu4.2.6 (Glasgow, 2005)

S. Tsukamoto, D.-S. Ly-Gagnon, K. Katoh, K. Kikuchi, Coherent demodulation of 40-Gbit/s polarization-multiplexed QPSK signals with 16-GHz spacing after 200-km transmission, in *Optical Fiber Conference (OFC)*, Paper PDP29 (Anaheim, 2005)

R.G.H. van Uden, C.M. Okonkwo, V.A.J.M. Sleiffer, M. Kushnerov, H. de Waardt, A.M.J. Koonen, Single DPLL joint carrier phase compensation for few-mode fiber transmission. IEEE Photon. Technol. Lett. **25**(14), 1381–1384 (2013)

A.J. Viterbi, A.M. Viterbi, Nonlinear estimation of PSK-modulated carrier phase with applications to burst digital transmission. IEEE Trans. Inf. Theory **IT-29**(4), 543–551 (1983)

R. Weidenfeld, M. Nazarathy, R. Noe, I. Shpantzer, Volterra nonlinear compensation of 100G coherent OFDM with baud-rate ADC, tolerable complexity and low intra-channel FWM/XPM error propagation, in *Optical Fiber Communication Conference (OFC)*, Paper OTuE3 (San Diego, 2010)

E.L. Wooten, K.M. Kissa, A. Yi-Yan, E. Murphy, D.A. Lafiaw, P.F. Hallemeier, D. Maack, D.V. Attanasio, D.J. Fritz, G.J. McBrien, D.E. Bossi, A review of lithium niobate modulators for fiber-optics communications systems. IEEE J. Sel. Top. Quantum Electron. **6**(1), 69–82 (2000)

C. Xie, Interchannel nonlinearities in coherent polarization-division multiplexed quadrature phase-shift-keying systems. IEEE Photon. Technol. Lett. **21**(5), 274–276 (2009)

C. Xie, S. Chandrasekhar, Two-stage constant modulus algorithm equalizer for singularity free operation and optical performance monitoring in optical coherent receiver, in *Optical Fiber Communication Conference (OFC)*, Paper OMK3 (San Diego, 2010)

Y. Yamamoto, T. Kimura, Coherent optical fiber transmission systems. IEEE J. Quantum Electron. **QE-17**, 919–935 (1981)

L. Yan, G. Hu, Demultiplexing in mode-division multiplexing system using multichannel blind deconvolution. IEEE Photon. J. **8**(2), 1–11 (2016)

W. Yan, Z. Tao, L. Dou, L. Li, S. Oda, T. Tanimura, T. Hoshida, J.C. Rasmussen, Low complexity digital perturbation back-propagation, in *European Conference on Optical Communications (ECOC)*, Paper Tu3.A.2 (Geneva, 2011)

J. Yang, J.-J. Werner, G.A. Dumont, The multimodulus blind equalization and its generalized algorithms. J. Sel. Areas Commun. **20**(5), 997–1015 (2002)

Q. Zhang, M.I. Hayee, Symmetrized split-step Fourier scheme to control global simulation accuracy in fiber-optic communication systems. J. Lightwave Technol. **26**(2), 302–316 (2008)

H. Zhang, Z. Tao, L. Liu, S. Oda, T. Hoshida, J. C. Rasmussen, Polarization multiplexing based on independent component analysis in optical coherent receivers, in *European Conference on Optical Communications (ECOC)*, Paper Mo.3.D.5 (Brussels, 2008)

L. Zhao, G. Hu, L. Yan, H. Wang, L. Li, Mode demultiplexing based on frequency-domain independent component analysis. IEEE Photon. Technol. Lett. **27**(2), 185–188 (2015)

L. Zhu, G. Li, Folded digital backward propagation for dispersion managed fiber-optic transmission. Opt. Express **19**(7), 5953–5959 (2011)

Part II
Solitons and Nonlinear Waves in Optical Fibers

5 A Brief History of Fiber-Optic Soliton Transmission

Fedor Mitschke

Contents

Introduction	222
Prehistory of Fiber Solitons: From an Idea to a First Experiment	222
Nonlinear Waves	222
Nonlinear Optics	224
Optical Fiber Technology	224
Toward Experimental Proof of Principle	225
Some Facts About Fibers	227
Some Facts About the NLSE and Its Soliton Solution	228
The Soliton Solution	230
Deviations from the Exact Solution	231
From Lab Curiosity to Commercial Deployment	232
The Soliton Laser	232
The Raman Shift	233
Soliton Interaction	234
Dark Solitons	235
Coding Formats for Optical Telecommunications	235
Generalized NLSE	236
Optical Amplifiers	238
Gordon-Haus Jitter	239
Four-Wave Mixing	242
Dispersion Managed Solitons	243
Commercial Soliton Systems	245
Fiber Solitons in the Twenty-First Century	246
Telecommunications and Limits to Growth	246
Soliton Molecules	248
Soliton Structures on a Background	250
Supercontinuum Generation	251
Rogue Waves	254

F. Mitschke (✉)
Institut für Physik, Universität Rostock, Rostock, Germany
e-mail: fedor.mitschke@uni-rostock.de

© Springer Nature Singapore Pte Ltd. 2019
G.-D. Peng (ed.), *Handbook of Optical Fibers*,
https://doi.org/10.1007/978-981-10-7087-7_71

Fiber Lasers.. 254
Beyond the Nonlinear Schrödinger Ansatz................................... 256
Conclusion... 257
References... 258

Abstract

This is a review of fiber-optic solitons, with an emphasis on their use for data transmission. The historic account begins with the first concept of nonlinear waves in the nineteenth century. In the 1980s satisfactory fibers became available, and research into fiber-optic solitons took off. Around the turn of the millennium, maturity of soliton transmission for commercial deployments was reached. Since then, various extensions and generalizations of the soliton concept have been found and investigated, and some of them have led to interesting new applications. This review outlines these developments and briefly touches upon current concepts and problems.

Keywords

Optical fiber · Soliton · Nonlinear Schrödinger equation

Introduction

Solitons are the paradigm of nonlinear fiber optics.

Nonlinear waves of that name exist in different fields of science, but fiber-optic solitons were the first type to prove their real-world usefulness. Optical telecommunications, after driving much soliton research, was the application for which they moved out of the lab and entered the competitive commercial world. This report aims to sketch the historical development which led to fiber-optic solitons, all the way from an observation of a peculiar water wave in the nineteenth century, through efforts to refine both mathematical and experimental techniques, via suggestions that they might exist in optical fibers, and on to commercial application around the turn of the millennium, plus some further developments until today.

Three quite different subject areas needed to come together and merge until the topic of nonlinear fiber optics could come into existence, and the study of fiber-optic solitons came into its own: an understanding of nonlinear waves, the evolution of nonlinear optics, and a certain degree of maturity of optical fiber technology.

Prehistory of Fiber Solitons: From an Idea to a First Experiment

Nonlinear Waves

A keen observer started the discussion of nonlinear waves; from there, it took 146 years until the first fiber-optic soliton was reported. The development in between is outlined very briefly in this section.

Scottish civil engineer John Scott Russell was involved with building the Union Canal, a waterway which was used to haul coal from Glasgow to Edinburgh. Ever an attentive observer, he noticed a peculiar phenomenon in 1834. His words have been repeated so often (see, e.g., Mitschke 2016) that it suffices here to reiterate his key point: A large *solitary heap of water* (his wording), once excited, propagated down the canal for a long distance without appreciable diminution of speed, or change of shape. This is the origin of the term *solitary wave*.

Today what once was a bustling waterway has banks overgrown with greenery, and the scenery is quaint. There is only a short section with sharp boundaries where the canal crosses a motorway near Hermiston by bridge (with the road below), and indeed in this section, a reenactment was undertaken in 1995 (Soliton wave receives crowd of admirers 1995).

The relevance of Russell's observation was not recognized, and indeed questioned by many, for a long time. Several luminaries of the nineteenth century, like Airy and Stokes, disputed that Russell's observation described a new type of wave. In the 1870s, both Lord Rayleigh (John William Strutt) in England and Joseph Valentin Boussinesq in France independently attempted to find a theory to Russell's observations, and in 1872, Boussinesq's work confirmed that Russell was right. In 1895, the mathematician Diederik Johannes Korteweg and his former PhD student Gustav de Vries in the Netherlands came up with a wave equation which is now named KdV equation after their names. (There is some dispute whether Boussinesq deserves priority de Jager 2011). The KdV equation can reproduce what Russell had observed: it does have soliton solutions, but unfortunately sufficient mathematical tools were not yet developed at the time. Therefore, the matter lay dormant again for many years. Meanwhile nonlinear waves began to receive some interest in other fields. In solid-state physics, Jakow Iljitsch Frenkel and Tatyana Abramovna Kontorova found a related model ca. 1937 which also allows soliton solutions. Again, with mathematical tools insufficient at the time, the matter rested for decades.

It was probably the advent of electronic computing that finally propelled the discussion of nonlinear waves to new insights. The surprising result of Fermi, Pasta, and Ulam in 1955 (Fermi et al. 1955) about a recurrence in an ensemble of oscillators with nonlinearity is a well-known example and generated renewed interest in nonlinear wave phenomena. In 1965, Martin Kruskal and Norman Zabusky numerically studied the KdV equation and made a remarkable observation (Kruskal and Zabusky 1965): Solitary waves could arise as a sequence of pulse-like spikes with relative velocities to each other, so that they could undergo collisions. The counterintuitive observation was that pulses passed through each other with no change of form or speed – unexpected in a nonlinear setting! This prompted the authors to call these special solitary waves *solitons*, a name intentionally reminiscent of particles. (Other particle-like properties, like interaction forces, would be found much later; see below.) On the mathematical side, Clifford Gardner, John Green, Kruskal, and Robert Mitsuru Miura in 1967 introduced the inverse scattering theory (Gardner et al. 1967), a technique which allows to find analytic solutions of integrable nonlinear wave equations. Finally, the mathematical toolbox was well equipped to treat solitons.

At about the same time, another yet one other integrable nonlinear wave equation began to be mentioned. The nonlinear Schrödinger equation (NLSE) derives its name from a formal similarity to the Schrödinger equation of quantum physics fame. It can describe water surface waves in certain circumstances but derives its relevance from the fact that it applies to several other physical situations, too. The NLSE saw its first use in optics in the context of self-focusing of beams, even though Raymond Y. Chiao, Elsa M. Garmire, and Charles Hard Townes did not have a name for this equation in their 1964 paper (Chiao et al. 1964). It should be noted that simultaneously Vladimir Il'ich Talanov in Russia pursued similar ideas and in 1967 presented both the NLSE and a soliton solution (Litvak and Talanov 1967) without using either name. The suggestion for the equation's name was made by Tosiya Taniuti and Nobuo Yajima in (1969).

Nonlinear Optics

Nonlinear optics was an arcane subject before the advent of the laser. As early as 1929, Maria Goeppert-Mayer wrote her PhD thesis (advised by Max Born) on two-photon transitions. She submitted it to the university of Göttingen in February 1930 and as an article to *Annalen der Physik* in November of that year where it appeared a few months later (Maria Göppert-Mayer 1931). In that paper she wrote on page 284: "Andererseits wirkt aber die quadratische Abhängigkeit von der Lichtdichte ungünstig, so dass zur Beobachtung sicher große Lichtintensitäten erforderlich sind." (On the other hand, the square dependence on light power is unfavorable so that for an observation surely large light intensities are required.) She had no way of knowing that the shortcoming of light sources available in her day would be relieved by the invention of the laser 30 years later.

A few other nonlinear phenomena were also discussed prior to the advent of the laser. Saturated absorption was observed by Gilbert N. Lewis, David Lipkin, and Theodore T. Magel in 1941 (Lewis et al. 1941). Brillouin scattering was discussed by Immanuel Lazarevich Fabelinskii in 1965 (Fabelinskii 1965). His concept of optical pumping in the 1950s earned Alfred Kastler (Kastler 1957) the Nobel Prize in 1966. But the true beginning of nonlinear optics is usually considered to have taken place when Peter Franken and collaborators performed an experiment just 1 year after Maiman's first laser (Franken et al. 1961). They irradiated a crystal with laser light and demonstrated the generation of an optical "overtone," the second harmonic. From there on, the field "exploded" as numerous other researchers joined the field. Frequency conversion of all kinds (sum, difference, second or third harmonic, etc.) is now routinely used in countless laboratories and in many commercial products.

Optical Fiber Technology

Glass, like many other materials, can be spun into thin strands or fibers. Cotton candy is entangled fiber of sugar.

It seems possible to guide light through glass fibers by way of total internal reflection; however, attempts at doing so remained futile for a long time for two reasons: (1) Once the surface of the fiber touched anything that would perturb the index difference to the outside, light was lost. (2) The power loss of the best glass around the middle of the twentieth century stood at ca. 1 dB/m. With these impediments, it seemed like a given that all light was lost after a few meters and nobody considered light guiding over longer distances (Hecht et al. 1999).

Charles Kuen Kao and George Hockham suggested in 1966 (Kao and Hockham 1966) that the power loss was not truly inherent in the glass material, but rather was dominated by chemical impurities. They suggested that improved technology of glassmaking would allow glasses with loss of about 20 dB/km. Several researchers jumped at it, and soon that mark had been reached. It was also understood that the remaining dominant contribution to loss was not absorption but Rayleigh scattering. Its characteristic λ^{-4} wavelength dependence promised much lower loss at longer wavelengths, and that promise held true. At the global loss minimum at a wavelength around 1.5 µm (where infrared absorption starts to kick in), a loss just below 0.2 dB/km has been routinely obtained for many years now. Kao's bold insight earned him a Nobel Prize in 1990 (http://www.nobelprize.org/nobel_prizes/physics/laureates/2009/kao-facts.html).

Once optical fiber became a mass-produced item, the prize per meter came down considerably and is now comparable to that of copper cable – or much lower if one compares similar data-carrying capacity. After successful field trials in the 1970s, fibers were installed more and more often as replacements or upgrades of existing cables or in new installations. In 1988, the first fiber-optic transatlantic cable took up service and vastly improved audio quality while reducing cost.

Toward Experimental Proof of Principle

In 1972, Vladimir Evgen'evich Zakharov and Aleksei B. Shabat in the Soviet Union applied inverse scattering to the NLSE (without calling it such) (Zakharov and Shabat 1972). They found the soliton solution, and they used that word. They also discussed what is now known as higher-order solitons for which they use the somewhat unfortunate description as "bound states of solitons." This is correct only in the sense that it was meant to draw a distinction to the case of a group of solitons of different frequencies which they also studied; those, unsurprisingly in a dispersive medium, split rapidly. It is incorrect insofar as the binding energy between the constituent solitons in higher-order solitons is zero, and these structures split up after infinitesimal perturbation.

The next important step forward was the work by Junkichi Satsuma and Nobuo Yajima in Japan in 1974 (Satsuma and Yajima 1974) which treated the initial value problem of the Zakharov-Shabat soliton solution. They found an explicit solution for the first higher-order soliton, the $N = 2$ soliton (for more details, see section "Some Facts About the NLSE and Its Soliton Solution" below). Incidentally, they repeated the moniker "bound state" for it. They also found that a wave with a shape that deviates from the prescription of the soliton solution propagates with some oscillation of its shape, but the oscillation is damped, and eventually one sees the emerging soliton.

In 1973, Akira Hasegawa and Frederick Tappert at Bell Labs suggested (Hasegawa and Tappert 1973b) that the NLSE describes the propagation of picosecond light pulses in optical fiber in the presence of nonlinearity, i.e., of fiber-optic solitons. At the time, fibers as optical waveguides were still quite novel, and transmission of short pulses was hampered by the fiber's group velocity dispersion. It was a bold proposal to use soliton pulses, i.e., exploit nonlinearity, to overcome this difficulty. They presented numerical simulations demonstrating the self-stabilizing action of the nonlinearity, and the robustness of solitons in the presence of perturbations and noise, and even of mild loss.

An experimental test of these predictions had to wait for several years. The state of technology in the 1970s was that on the one hand the required laser sources (color center lasers) were under development; on the other hand, fibers were still too lossy in the wavelength regime of anomalous dispersion. On both counts, there was progress toward the end of that decade, so that in 1980 Linn Frederick Mollenauer, Roger Hall Stolen, and James Power Gordon at Bell Labs could experimentally verify the existence of solitons (Mollenauer et al. 1980). They used a color center laser providing pulses of 7 ps width and a 700 m long optical fiber. Dispersion broadened out these pulses at low power, but when power was increased, there was indeed the characteristic restoration of the pulse shape and width: and it occurred just at the theoretically expected power level. For even higher powers, further narrowing and pulse splitting was indicative of the transition to an $N = 2$ soliton.

This perfect confirmation of theoretical predictions was the final starting shot which prompted many other researchers to join this line of research soon thereafter. With a view to long-distance transmission, Hasegawa (1984b) considered Raman amplification to balance the loss. Mollenauer discussed design examples for transoceanic distances with vastly improved transmission capacity (Mollenauer 1986), which sounded fantastic at the time (while in hindsight, they were quite moderate). Telecom companies like AT&T (his employer), however, were reluctant because they were worried about how to "fill" existing capacity. At the time, telecommunication was about telephone calls; it was not anticipated that the introduction of the Internet would lead to an explosion of demand years later. But there was yet another reason that his proposal was initially greeted with much skepticism.

The concept of solitons for telecommunication really implies a radical departure from existing concepts. It amounts to an acceptance that the transmission channel is nonlinear.

Linear distortion, like dispersion, is known well in classical electronic telecommunications engineering. In fact, a misguided attempt to overcome it led to catastrophic damage of the very first transatlantic electric cable in 1858 because back then it was not understood that dispersive distortion can be undone by an inversely dispersive element (The Atlantic Cable website 2011). But nonlinearity is a different matter. To be sure, electronic engineers know nonlinear components well: Diodes and transistors have highly nonlinear, in fact exponential, voltage-current relations. Of all electronic circuits, only passive RLC circuits are linear (self-heating of resistors, saturation of magnetization in cores of inductors, etc. can be neglected for the weak signals typically encountered). The art of electronics largely consists of ways to achieve linearity in amplifiers, etc.: in part through nonlinearity cancellation by using, e.g., differential or push-pull amplifier designs, but mostly through negative feedback. All the time, however, the transmission medium – whether cable or radio wave – is always considered linear.

The only exception actually proves the rule: A strong amplitude-modulated radio wave can, in suitable conditions, affect ionospheric electron density. A process known as ionospheric cross modulation can then transfer its modulation onto other radio waves. The annoying result is that a radio listener simultaneously hears two programs superimposed (Benson 1964). The phenomenon has been dubbed the *Luxemburg effect* after the first station for which this interference was noted (Tellegen 1933). Radio Luxemburg emitted extremely high power in the heart of Europe; it started operations in the 1930s and was discontinued in the 1980s.

Nonlinearity in the transmission channel basically has two consequences. First, it creates distortion for a single data stream; however, the soliton paradigm stipulates that this can actually be beneficial. Second, and more profoundly, the superposition principle – the hallmark of linear systems – does no longer hold. Simultaneous transmission of several independent data streams, without which no real-world installation makes sense, would suffer from mixing and cross talk which surely would wreak havoc on data integrity. In short, with regard to fibers, many were afraid of the Luxemburg effect on steroids.

Such was the state of affairs in the mid-1980s. Before we continue the timeline, we need to familiarize the reader with some specifics.

Some Facts About Fibers

Optical fibers are made from fused silica, i.e., silicon dioxide in glassy form. This material is chemically inert, mechanically quite strong, and not expensive. Most importantly, it can be produced with very low optical loss.

The simplest structure of an optical fiber would be a cylindrical body of glass to guide light by total internal reflection at its surface. However, the surface is exposed to the environment, and the guided light is easily lost if something touches the surface that can couple the light out. Therefore, the simplest *realistic* fiber structure is known as step index fiber: In the center, there is the light-guiding core; it is surrounded by the cladding. By the use of dopants to the glass, the index of the core is slightly raised (occasionally, the cladding index has been lowered). The waveguiding property is then dominated by the core-cladding interface which is decoupled from the environment.

Several factors contribute to the optical power loss. Electronic transitions in the ultraviolet part of the spectrum create absorption; its long-wavelength tail reaches across the visible into the near infrared. Similarly, molecular vibrations in the mid-infrared have a short-wavelength tail which extends all the way to the visible.

Both cross near $\lambda = 1.5\,\mu\text{m}$. On top of this, there is Raleigh scattering due to the statistical nature of the glass structure; loss from this factor rolls off toward long wavelengths as λ^{-4}. Around the intersection of the two absorption tails, this is the dominant contribution to fiber loss. As Kao and Hockham had suggested (Kao and Hockham 1966), glass must be made with a high degree of chemical purity, or else the loss would be dominated by impurity absorption. Meanwhile fiber manufacturers have mastered the art of purifying materials to such high degree that in the best fibers, absorption loss is swamped by Raleigh scattering and can be ignored. At the wavelength of lowest loss around $\lambda = 1.55\,\mu\text{m}$, a loss of $0.2\,\text{dB/km}$ is routinely reached. Note that this corresponds to a half-power distance of 15 km; this makes optical fiber the best light-transmitting solid-state material in existence.

A wave equation describing light propagation in fibers must consider the proper boundary conditions at the core-cladding interface. One obtains a discrete set of modes, i.e., patterns of field distribution across the cross section of the fiber (see, e.g., Mitschke 2016). As the problem is cylindrically symmetric, it is unsurprising that size and shape of the modes involve Bessel functions. A dimensionless number determines which modes can propagate; it combines both refractive indices (n_K for the core and n_M for the cladding) and the core radius a with the wavelength of the light λ:

$$V = \frac{2\pi}{\lambda} a \sqrt{n_K^2 - n_M^2}. \tag{1}$$

Analysis reveals that for $V \leq 2.4048$, only a single mode can propagate (the numerical value comes from the first zero of the J_0 Bessel function). This is called the fundamental mode; it plays a dominant role in optical telecommunication. It has a bell-shaped field distribution which is specified by Bessel J_0 in the center, with some part of the field stretching into the cladding where it decays radially as given by a Bessel K_0 function. The remainder of this report will deal almost exclusively with single-mode fibers. Given the realistically possible amount of dopants, the core-cladding index difference cannot be more than about one percent and is usually less, so that as a rule of thumb, the core diameter is about six wavelengths for a single-mode fiber.

We note in passing that in a circularly symmetric fiber, all modes exist in two orthogonal polarization states so that, in a very strict interpretation, the name "single-mode fiber" is a bit of a misnomer. Nonetheless, it is universally accepted, and when it comes to counting the number of modes existing in some fiber, it is necessary to specify whether the number includes all degeneracies or not.

Some Facts About the NLSE and Its Soliton Solution

The leading causes of change to pulse shapes in a fiber are group velocity dispersion and the Kerr nonlinearity. The former, by itself, typically broadens pulses temporally; the latter creates a dynamic phase shift across the pulse known as *self-phase modulation* which by itself broadens the pulse spectrum (Stolen 1978). The nonlinear Schrödinger equation combines both influences. It is the relevant propagation equation for light pulses in optical fiber when loss can be neglected, when the dispersion can be represented by a single constant (no higher-order corrections), and when no further nonlinear effects need to be considered (but see section "Generalized NLSE"). We briefly sketch some facts; for more details, see e.g., Mitschke (2016) and Agrawal (2013). The NLSE is often written as

$$i\frac{\partial}{\partial z}A - \frac{\beta_2}{2}\frac{\partial^2}{\partial t^2}A + \gamma|A|^2 A = 0 \quad (2)$$

where we maintain physical quantities for ease of interpretation. It describes the envelope of the pulse amplitude $A(z,t)$; the optical frequency does not appear here. The envelope evolves in time t and distance along the fiber z and is referred to a frame of reference moving with the group velocity at the optical (carrier) frequency. β_2 is the coefficient of group velocity dispersion and γ the coefficient of nonlinearity.

In fused silica, group velocity dispersion changes sign in the near-infrared near 1.3 μm. At wavelengths longer than this, $\beta_2 < 0$; this is known as anomalous dispersion. Conversely, at shorter wavelengths (and all the way to the visible), there is normal dispersion. Typical values at 1.5 μm are $\beta_2 = -25 \cdot 10^{-27}$ s^2 m^{-1}. For wide bandwidth signals or for operating wavelengths close to the zero dispersion wavelength, it may be important to consider higher-order corrections to the dispersion.

The coefficient of nonlinearity follows from the fact that the refractive index is slightly dependent on light intensity. Typical values are $\gamma = 10^{-3}$ W^{-1} m^{-1}; this number is only mildly wavelength-dependent.

The NLSE has a continuous wave solution:

$$A = \sqrt{P_0}\, e^{i\gamma P_0 z}. \quad (3)$$

It turns out that this solution is stable for normal, but unstable for anomalous dispersion. In the latter case, that means that a perturbation of the constant field amplitude at some Fourier component will grow with a rate that depends on the frequency of the perturbation. This is known as *modulation instability* (Agrawal 2013) and is important to understand when it comes to optical signal transmission (Anderson and Lisak 1984). A first experimental demonstration was given in Tai et al. (1986); a review of the wider topic can be found in Zakharov, and Ostrovsky (2009). If white noise perturbs the continuous wave, there is always a Fourier component at the frequency of maximum gain which is at the carrier frequency plus/minus

$$\omega = \sqrt{\frac{2\gamma P_0}{|\beta_2|}} \quad (4)$$

so that the constant power gives way to a periodic ripple. The further evolution is interesting in itself; we will return to it below (see section "Soliton Structures on a Background").

If a pulse of light is launched into the fiber at normal dispersion, the interplay of self-phase modulation and dispersion can lead to the phenomenon of optical wave breaking when frequency-shifted light from the edges of the pulse overtakes unshifted light farther out in the tails (Tomlinson et al. 1985). If, however, the same is done in the anomalous dispersion regime, the fate of the pulse is much more interesting.

The Soliton Solution

For anomalous dispersion, the NLSE has a stable pulse-like solution known as the fundamental soliton. It takes the form

$$A(z,t) = \sqrt{\frac{\beta_2}{\gamma} \frac{1}{T_0}} \operatorname{sech}\left(\frac{t - t_c}{T_0}\right) \exp\left(i \frac{\gamma P_0 z}{2} + i\varphi_c\right). \tag{5}$$

where the soliton is centered at t_c in the comoving frame of reference and has a phase offset φ_c. In the following, we will assume both to be zero. T_0 is the pulse duration. The prefactor of the sech term in Eq. (5) sets the peak amplitude and can be written as

$$\sqrt{\frac{|\beta_2|}{\gamma} \frac{1}{T_0}} = A_{\max} = \sqrt{P_0}. \tag{6}$$

It is a convenient convention that amplitudes are scaled so they can be written as square root of power. Obviously amplitude and timing are coupled for this type of solution, so that always

$$P_0 T_0^2 = \frac{|\beta_2|}{\gamma}. \tag{7}$$

The LHS combines both relevant pulse parameters, the RHS the fiber parameters. Note that by integration of Eq. (5), one immediately obtains the soliton energy as

$$E_{\text{sol}} = 2 P_0 T_0. \tag{8}$$

The quantity $P_0 T_0^2$ in Eq. (7) therefore has the dimension of a time integral of energy which is known as action in classical mechanics; for want of a better name, we call Eq. (7) the *constraint of constant action*.

As we will use them below, we cite here the length scales of the situation commonly defined as

$$L_D = \frac{T_0^2}{|\beta_2|} \quad \text{dispersion length} \tag{9}$$

$$L_{NL} = \frac{1}{\gamma P_0} \quad \text{nonlinear length} \tag{10}$$

A physical understanding of the soliton is that linear and nonlinear effects balance each other out when Eq. (7) is fulfilled. That is equivalent to saying that $L_D = L_{NL}$. The power-dependent phase in Eq. (5) rotates through 2π over a distance

$$z_{2\pi} = 4\pi L_D = 4\pi L_{NL}, \tag{11}$$

but the pulse shape in terms of power profile is invariant.

Deviations from the Exact Solution

To quantify a deviation from the exact soliton solution, we can use the soliton order $N \geq 0$, $N \in \mathbb{R}$. N can be inserted as a multiplier on the RHS of Eq. (5); then, Eq. (7) is modified into

$$P_0 T_0^2 = N^2 \frac{|\beta_2|}{\gamma} \tag{12}$$

(Note that some authors use different terminology. For example, our N is called $A = N + \alpha$ with $N \in \mathbb{N}$ in Satsuma and Yajima 1974). For the soliton proper, $N = 1$; a pulse with non-integer N cannot be a pure soliton. As was shown by inverse scattering theory by Zakharov and Shabat (1972), in that case, one obtains a soliton plus a nonsolitonic remainder which propagates as a linear (purely dispersive) wave known in the soliton literature as radiation.

For $N > 3/2$ a second soliton is generated. At integer $N = 2$, a two-soliton compound without radiation is obtained; an explicit expression was given in Satsuma and Yajima (1974). It can be understood as a beating between both solitons; the power profile repeats itself after $z_0 = z_{2\pi}/8$. Beginning at $N = 5/2$, a third soliton is generated, etc. The integer closest to N indicates the number of solitons, and the deviation from the integer value sets the radiative part.

This is illustrated in Fig. 1 in which energy is normalized to soliton units E_{sol} (Eq. 8). The total pulse energy scales as N^2 and is shown in the top part as a dashed parabola. A first soliton is created at $N = 1/2$; upon further increase of N, its energy rises linearly. The difference between the straight soliton line and the total-energy parabola is shown in the bottom part on an inverted and expanded scale. The first soliton is created when radiation exceeds $1/4 \, E_{sol}$, goes through zero at $N = 1$, and then rises again. At $N = 3/2$, there is enough surplus energy available to create second soliton. Note that the "capture range" for generating a soliton, $1/2 \leq N \leq 3/2$, in technical terms extends from -6 dB to $+3.5$ dB relative to the target value. As the

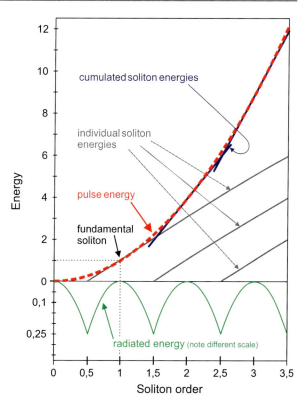

Fig. 1 Soliton and radiation energies as a function of soliton order N. Scales normalized to units of fundamental soliton (black dotted markers). Individual soliton energies (gray) add up to a cumulative value (blue); this approximates the total pulse energy (red dashed). The difference is shown on an expanded, inverted scale in the lower part (green) (After Mitschke 2016)

soliton solution is a stable solution, a suddenly perturbed soliton can readjust within the bounds of the capture range.

The process of generation of further solitons repeats at all half-integer N values. The cumulated solitonic energies approximate the parabola and are tangent to it at integer values of N; in those places, there is no radiation.

The NLSE is integrable; it is therefore not immediately applicable to realistic fibers which always have some loss. Fortunately, for weak loss, the soliton concept remains applicable; see section "Generalized NLSE" below.

From Lab Curiosity to Commercial Deployment

The Soliton Laser

After the initial demonstration of the soliton principle, Mollenauer realized that a reliable pulse source suitable for further studies was required. At the time, color center lasers were an obvious choice to access the wavelength region around 1.5 μm, and he conceived the idea to couple a fiber resonator to the main laser resonator, so that the fiber could assist in pulse shaping. The original idea was that the invariant shape of the fundamental soliton would force the laser to generate pulses of just that shape. Coupling two resonators required interferometric stability, though. The device introduced in a first report of a *soliton laser* (Mollenauer and Stolen 1984) was not stable enough for practical use but provided proof of principle. A stabilization scheme introduced in Mitschke and Mollenauer (1986a) solved the practicality issue and greatly facilitated further studies. Among other things, it was possible to generate pulses of a mere 50 fs duration with this laser; by using post-compression by exciting an $N = 2$ soliton in an external fiber of length $z_0/2$, this was reduced to 19 fs or less than four cycles (Mitschke and Mollenauer 1987a). These pulses of 1986 have been the shortest infrared light pulses for many years.

It was also demonstrated in Mitschke and Mollenauer (1986a) that the inner working of the soliton laser did not rely on a fundamental soliton; rather, some pulse shape variation went on in the fiber, and the pulse was closer to an $N = 2$ soliton. Many laboratories copied the concept including the stabilization scheme and applied it to lasers other than the color center type. Names like "coupled cavity mode locking" (Barr and Hughes 1989; Zhu 1989) or "interferential pulse mode locking" (Morin and Piché 1989) were used, but the term *additive pulse mode locking* (Ippen et al. 1989) ("APM") found the widest acceptance. With time, it became clear (Ippen et al. 1989; Blow and Nelson 1988; Ouellette and Piché 1988; Wang 1992; Krausz and Brabec 1991) that the nonlinear phase shift in the fiber pulses could interfere with the nearly unchirped pulses in the main cavity such as to create constructive interference at pulse center and destructive interference in the wings, so that upon every round trip, the pulse was shortened. This went on until dispersive effects limited further pulse compression. The scheme turned out particularly useful for Nd:YAG lasers, a laboratory workhorse which before, unfortunately, could only be made to produce short pulses by active mode locking with ca. 100 ps wide pulses; in APM operation, the full gain bandwidth could be exploited, and the pulse duration was about one order of magnitude shorter (Liu et al. 1990; Mitschke et al. 1993).

The Raman Shift

Studies of the propagation of light pulses in fiber benefitted from the soliton laser as a reliable pulse source. During such investigations the spectrum of the fiber output was checked with a scanning Fabry-Perot interferometer. On Dec. 4, 1985, this author noticed a peculiar effect: While some spectral power remained at the laser frequency, the soliton itself was downshifted considerably. The precise amount depended on power but could reach several THz (recall that a wavelength of 1.5 μm corresponds to a carrier frequency of 200 THz). An explanation was found in cooperation with J. P. Gordon; two papers with the experimental report (Mitschke and Mollenauer 1986b) and the theoretical explanation (Gordon 1986) appeared in the same issue of Optics Letters. The Raman gain in fibers, which had been studied a few years earlier by R. H. Stolen et al. (Stolen et al. 1984), peaks at a frequency offset at about 13 THz but is nonzero all the way down to zero offset. Therefore, a differential gain arises between the high-frequency and the low-frequency side of the pulse spectrum. As a result, the "center of mass" of the pulse spectrum gradually shifts downward in frequency as the pulse propagates in the fiber, a process called *soliton self-frequency shift*.

As it turned out, just before this discovery, researchers in Russia had come very close to finding the same thing. Dianov et al. propagated powerful pulses through a fiber and observed a spectral wing on the low-frequency side which they attributed to Raman scattering (Dianov et al. 1985).

Evaluation of the amount of shift led Gordon to the conclusion that it scales with the inverse fourth power of the pulse width. Hence, for 5 ps pulses, there would be a shift of its own spectral width (ca. 60 GHz) after 1000 km of fiber – a very mild effect. In contrast, for 50 fs pulses, a similar shift by the spectral width (in that case ca. 6 THz) would occur after a distance of just 1 m. It should be pointed out that such shift applies to *any* light pulse, not just a soliton: but the soliton would hang together as a unit, whereas a non-soliton might suffer from dispersion.

There was an immediate and profound conclusion from this discovery: It would never be possible to increase transmission rates in optical telecommunications by just scaling down the pulse width and ramp up the clock rate correspondingly. Integrity of the signal would seriously suffer for pulses shorter than a few picoseconds. One would have to find other ways of exploiting the entire spectral range over which fiber is highly transparent (ca. 30 THz). Indeed, up to this day, all subsequent schemes make use of wavelength division multiplexing (see section "Coding Formats for Optical Telecommunications") with moderate clock rates in each channel, so that pulses shorter than the said limit were never used.

On the other hand, the Raman effect would turn out to have its beneficiary side, too. Fiber loss over long distance makes it necessary to amplify signals, and the Raman effect is useful for that. The suggestion was originally Hasegawa's (1984b), but Mollenauer and collaborators performed first successful tests (Mollenauer and Smith 1988) (see below in section "Optical Amplifiers").

Soliton Interaction

One consequence of the fact that the channel is nonlinear is that signals can mutually couple, or interact. In 1981, Karpman and Solov'ev (Karpman and Solov'ev 1981) had found this in a study of pulse pairs with a perturbation ansatz; Gordon in 1983 (Gordon 1983) started from the exact two-soliton expression. In either case, the prediction was that in a pair of equal solitons, an effective "interaction force" exists which decays exponentially with separation of the pulses and varies sinusoidally with their relative phase. Thus there would be either attraction or repulsion, depending on the phase being equal or opposite, respectively. This prediction was tested experimentally in Mitschke and Mollenauer (1987b). A Mach-Zehnder interferometer served to split pulses from the soliton laser into pairs, with adjustable separation and relative phase when the arm lengths were fine-tuned. The prediction was confirmed quite well in the repulsive case. In the attractive case, it held up only until the first close encounter of the two pulses; once they had collided, they also experience repulsion, and the pair split. Soon thereafter, this was explained as a Raman shift at the collision moment which upset their phases (Kodama and Nozaki 1987).

Again, there was a consequence for applications of light pulses – indeed of any type, not just solitons – for telecommunication. Signaling light pulses require sufficient separating distance between them such that the interaction cannot appreciably shift pulse positions which would give rise to detection errors.

Dark Solitons

In the wavelength regime near $\lambda = 1.5\,\mu\text{m}$, optical fibers have their lowest loss. At the same time, in this regime group, velocity dispersion is anomalous. Therefore, most attention has been given to solutions of the NLSE for $\beta_2 < 0$. On the other hand, there is a solution for normal dispersion which has quite interesting properties in its own right.

The solution basically has a $\tanh(t)$, rather than $\text{sech}(t)$ shape. Considering the power profile (square of amplitude), and recalling that $\tanh^2(t) = 1 - \text{sech}^2(t)$, it appears as a dark pulse on an infinitely extended bright background. This shape has given rise to the moniker *dark soliton*. A defining feature is the phase jump at the center where the power dips to zero. (As it turns out, there is an entire family of dark solitons of varying degrees of darkness and also of sharpness of the phase change Agrawal 2013; the quoted type would then be called the black among a family of gray solitons.)

This solution seems to have been derived first in Zakharov and Shabat (1973); the moniker *dark pulse* appeared shortly thereafter (Hasegawa and Tappert 1973a). Beyond numerical studies, at least two groups pursued this phenomenon experimentally in 1987 (Emplit et al. 1987; Krökel et al. 1988). It was later argued that dark solitons would be more robust in the presence of perturbations (Kivshar et al. 1994). However, their generation is technically much more difficult than that of bright pulses, and the bright background demands a lot of optical power. With these complications, dark solitons have received much less research than bright solitons.

Coding Formats for Optical Telecommunications

In optical telecommunications, throughout the 1990s, growing demand prompted the installation of several transoceanic cables, each one with larger data-carrying capacity than its predecessor. The late 1980s had seen rising popularity of telefax; during the 1990s, the first Internet services began to appear. Recall that the Internet was introduced in 1989, the World Wide Web started in 1991, and Yahoo was founded in 1994 and Google in 1996.

In that timeframe, there was some discussion as to what the best coding formats are; however, at the time, basically all coding was done in a binary format. In what became known as *on-off keying*, the presence of light in a particular time slot (clock period) signaled a logical "1," whereas its absence represented a "0." The light pulse could either have the same duration as the time slot, i.e., fill its duration entirely: then, a sequence of several "1s" would be represented by light power that remained on. This was called NRZ, as in *no return to zero*. If on the other hand the light pulse was shorter than its slot, power would always return to zero; this was known as RZ (*return to zero*). The relative advantages of NRZ vs. RZ were discussed for some time. RZ uses less power and makes clock regeneration easier; NRZ uses less optical bandwidth.

The other distinction of coding methods arises from the fact that no single data source can ever produce data rates approaching what the fiber can handle. Therefore, multiple data streams are combined into one or *multiplexed*. One can multiplex in the time domain, by interleaving n bit streams to a combined stream with n times higher clock rate. One can alternatively multiplex in the spectral domain, by having a grid of optical carrier waves each of which is modulated with one data stream. The former is called *time domain multiplexing* or TDM and the latter *wavelength division multiplexing* or WDM. It should be no surprise that a given amount of data covers the same optical bandwidth, whether coded in TDM or in WDM or any combination. However, it is technologically challenging to push high-frequency electronic circuitry to ever-higher clock rates. On the other hand, it is economically disadvantageous to have very many WDM channels as each requires its own laser, etc. Therefore a compromise must be made: Data are multiplexed by TDM to whichever rate is feasible, and then several such signals are combined optically by WDM. The transition between both has moved over the years from a few GHz to about 100 GHz today.

Generalized NLSE

The NLSE has been derived with some approximations. Generally it describes pulse propagation in fibers quite well, but in some situations, it is not sufficiently accurate unless certain modifications to the equation are made. We already mentioned the Raman effect; in this section, we introduce a few more. The generalized NLSE can take the form

$$i\frac{\partial}{\partial z}A - \frac{\beta_2}{2}\frac{\partial^2}{\partial t^2}A - \underbrace{i\frac{\beta_3}{6}\frac{\partial^3}{\partial t^3}A + \frac{\beta_4}{24}\frac{\partial^4}{\partial t^4}A \ldots}_{\text{higher-order dispersion}} + \gamma|A|^2 A$$

$$+ \underbrace{i\frac{\gamma}{\omega_0}\frac{\partial}{\partial t}\left(|A|^2 A\right)}_{\text{self steepening}} - \underbrace{T_R \gamma A\frac{\partial}{\partial t}|A|^2}_{\text{Raman}} + \underbrace{i\frac{\alpha}{2}A}_{\text{loss}} = 0. \qquad (13)$$

which must almost always be treated numerically.

The term labeled "loss" contains Beer's coefficient α and describes linear power loss. This term alone can upset the equilibrium of dispersive and nonlinear influences as expressed by Eq. (7) or equivalently by the condition $L_D = L_{NL}$. However, the equilibrium is maintained at least to good approximation as long as the loss is sufficiently weak, and the pulse is still stabilized by nonlinearity. In such case, one can apply perturbation analysis (Kaup 1976; Karpman and Maslov 1977; Hasegawa and Kodama 1981; Kivshar and Malomed 1989); it is consistently found that the soliton smoothly rearranges its duration and peak power while its energy slowly decays. When the pulse energy is attenuated with propagation distance z according to $E_z = E_0 \exp(-\alpha z)$, the simultaneous solution of Eq. (7) and the modified Eq. (8) written in the form

$$\frac{|\beta_2|}{\gamma} = P_z T_z^2 = P_0 T_0^2 \tag{14}$$

$$\frac{E(z)}{2} = P_z T_z = P_0 T_0 \, e^{-\alpha z} \tag{15}$$

yield that

$$P_z = P_0 \, e^{-2\alpha z} \quad \text{and} \quad T_z = T_0 \, e^{+\alpha z}. \tag{16}$$

The peak power droops while the duration widens. This conclusion has been confirmed in numerous numerical studies (Blow and Doran 1985) and experiments (Mollenauer et al. 1985; Smith and Mollenauer 1989), but it must be emphasized that it is valid only for the "adiabatic" case in which power loss is so slow that Eq. (7) always remains a good approximation. Outside this realm, there are deviations (Kubota and Nakazawa 1990).

Terms labeled "higher-order dispersion" constitute a series expansion of the fiber's dispersion curve around the center operating wavelength. The dispersion curve is determined by several factors, some of which arise from the properties of the glass and some from modifications due to the waveguiding geometries. Within limits, one can tailor fibers to have somewhat different dispersion values at a given wavelength. In the derivation of the NLSE, it is assumed that the value of the dispersion taken at the carrier frequency is representative for the entire bandwidth of the pulses to be described. This amounts to saying that the dispersion curve is represented by a single value, $\beta_2(\omega_0)$. Clearly, once the pulses have appreciable bandwidth, one will want to consider also the slope of the dispersion with frequency, i.e., $\beta_3 = \partial \beta_2 / \partial \omega$. Speaking more generally, the dispersion curve can be written as a series expansion at a carrier frequency; all terms with $\beta_i, i > 2$ are referred to as *higher-order dispersion* terms. A typical situation where higher-order dispersion becomes important is close to the point where $\beta_2 = 0$. Research shows that the impact on solitons is relatively benign (Wai et al. 1986). Another relevant case is when one considers the shortest possible pulses which of course have very wide bandwidth.

The "self-steepening" term (Anderson and Lisak 1983) leads to an asymmetry in the pulse shape, and it can produce shifts in spectral and temporal positions even in the absence of the Raman term. It can also cause higher-order solitons to disintegrate (Golovchenko et al. 1985). The Raman effect has been discussed in section "The Raman Shift"; the corresponding term is shown here in a simplified version which contains T_R as a characteristic Raman time. More accurate models have been given (Stolen et al. 1989; Mamyshev and Chernikov 1990; Lin and Agrawal 2006).

Beyond these terms, there is one more issue, and that regards the state of polarization of the light. One typically considers a fiber to be circularly symmetric. But then, each mode exists in two orthogonal states of polarization but fully identical in all other respects. Strictly speaking, a single-mode fiber is somewhat of a misnomer because the fundamental mode is twofold degenerate. In reality, though, fibers are never perfectly circularly symmetric; even a slight bend of the fiber introduces an amount of birefringence which is difficult to predict and usually not constant in time. Light signals which are launched with a well-defined state of polarization soon acquire the state of "scrambled" polarization. The scrambling distance is usually very short distances in comparison to the length of a typical communication link. Adequate modeling for short fibers would require two coupled equations of the NLSE type, but with mutual cross-phase modulation terms (Manakov 1974; Menyuk 1989). Quite often, the two coupled pulses remain locked, or trapped, to each other (Menyuk 1987; Wai et al. 1991; Hasegawa 2004). If they do not (as is the case in linear systems), the stochastic "random walk" nature of the evolution of the state of polarization leads to a timing difference which grows as the square root of distance. It is of the order of $50\,\mathrm{fs}/\sqrt{\mathrm{km}}$ and began to pose a problem once data rates above 10 Gbit/s over long distances were introduced after the turn of the century (Iannone et al. 1993; Xie et al. 2003). The upside of this issue is that while the state of polarization is unpredictably shifted, still two orthogonal states typically remain orthogonal, to good approximation. This opens a chance for polarization multiplexing (Evangelides et al. 1992) which can enhance the data-carrying capacity by a factor of two in the best case.

We note in passing that when the nonlinear increase of the index exceeds the index difference from birefringence, a new type of instability is expected (Winful 1985; Matera and Wabnitz 1986).

Optical Amplifiers

In the first optical transmission schemes, it became clear that fiber loss poses a serious problem for distances beyond roughly 100 km. Signal attenuation by some 20 dB makes it necessary to restore the power level by amplification. This was initially done in *repeaters*, boxes which were periodically inserted in long-haul lines. They contained photodetectors to convert the signal to an electronic format, circuitry to amplify and regenerate, and lasers to convert back to an optical format. Considering the intended use on the ocean floor, they had to be constructed to perform maintenance-free and error-free for the cable's life expectancy. This in itself is more than a trivial challenge, but when later the number of WDM channels was increased, this technology became increasingly cumbersome. It was inflexible with respect to modifications of TDM data rates or WDM carrier frequencies; nothing could be done once the fiber was laid.

Engineers at the time dreamed of all-optical repeaters. They would employ optical amplifiers without any need of conversion to an electronic format and back. A major advantage, besides reduced complexity, would be that such devices are "transparent" to the optical data format; a fiber line can be upgraded later without having to modify the hardware of the repeaters.

The most popular optical amplifier is based on a piece of specialty fiber, which is doped with Er ions, and is spliced into the main fiber. These were introduced in 1987 by two groups practically simultaneously (Desurvire et al. 1987; Mears et al. 1987). When suitable pump light is supplied from a cw laser diode included in the amplifier package, the fiber amplifies signals with quite good fidelity. Within a couple of meters of Er-doped fiber, several tens of dB of gain can be had, virtually free of polarization dependence, with practically no saturation-induced cross talk by virtue of the extremely long upper-state lifetime (11 ms) in the erbium system. For more details, see Desurvire (1994) and Becker et al. (1999). Laboratory experiments like Mollenauer et al. (1990) were performed to study possible detrimental effects that might occur over very long distances.

As a next step toward practicality, WDM needed to be considered; it introduces a few additional complications. The amplifier gain is never spectrally flat so that different WDM channels unavoidably experience slightly different gain. After some propagation distance, they can differ vastly in their signal power, so that large error rates can occur. Frequency filters in a soliton system, however, stabilize the power to a value proportional to the dispersion at each channel's center frequency; as a result, the power distribution across the gain band varies only mildly (Mamyshev and Mollenauer 1996). Also, different WDM channels will have different group velocities due to the fiber's dispersion curve. Therefore, signals in these channels can and will collide. A study of a long line with WDM and lumped amplifiers showed that detrimental effects from such collisions can be avoided by observation of some simple design rules (Mollenauer et al. 1991), and further tests showed the feasibility of error-free soliton transmission over up to 15,000 km (Mollenauer et al. 1992a).

An alternative to Er-doped amplifying fibers is the use of Raman amplification. The transmission fiber itself is the amplifying medium (Fig. 2). This arrangement comes with the advantage of not being restricted to the spectral window over which Er ions can provide gain. It was demonstrated in Mollenauer and Smith (1988) and became increasingly popular when in later years more and more optical bandwidth was used (Bromage 2004).

Gordon-Haus Jitter

Optical amplifiers invariably introduce a certain amount of noise to a signal stream due to spontaneous emission (Heffner 1962). This noise, when superimposed with

Fig. 2 Raman amplifiers require no specialty fibers. Pump lasers provide cw light energy some of which the Raman process transfers to enhance the signal pulses. One often configures for counterpropagating pump to obtain a more uniform gain

the signal, can stochastically alter pulse amplitude, phase, and center frequency. When the data stream consists of solitons, the amplitudes are fairly stable due to the self-correcting capability of the soliton principle. However, fluctuations of the center frequency, in a dispersive medium, translate into fluctuations of group velocity. That, in turn, causes a random walk of the pulse's timing, and as a result, its arrival time will also fluctuate. This undesired effect grows with third power of distance. The phenomenon, first realized in 1986 by James P. Gordon and Hermann A. Haus (Gordon et al. 1986), has become known as the *Gordon-Haus jitter*. When in a laboratory experiment soliton transmission was demonstrated over more than 10,000 km, it was also confirmed that the Gordon-Haus effect exists in the predicted amount (Mollenauer et al. 1990).

Such research showed that understanding of soliton physics was scientifically sound. Nevertheless, engineers tasked with planning the next transoceanic cables were concerned that the soliton concept might not yet be fully mature for a technical application. Therefore, in fiber-optic cables of the 1990s, the soliton concept was not adopted.

Researchers kept working on the case, though. They found that the Gordon-Haus jitter can be kept in check by inserting frequency comb filters into the fiber which have transmission peaks at the WDM channel center frequencies and roll-off for offset frequencies. Such filters would nudge solitons back to center frequency in the event that they had been shifted (Mecozzi et al. 1991; Kodama and Hasegawa 1992); see Fig. 3. Fabry-Perot filters automatically possess the comb-like periodic response that is required to filter an entire WDM data mix with a single filter. Some researchers went beyond this and suggested that fancier filters like Butterworth type (Mecozzi 1995) would be even better; such filter was used, e.g., in Suzuki et al. (1994).

Even with filters, the buildup of spontaneous emission noise in a very long fiber with several amplifiers still was a concern. The issue was addressed by the sliding filter concept (Mollenauer et al. 1992b) which is a truly elegant idea. In a sequence of filters, not all have the same center frequency; rather, the center frequencies are staggered (Fig. 4). This means that for linear waves including spontaneous emission noise, the fiber line is opaque because any wavelength which can pass through one

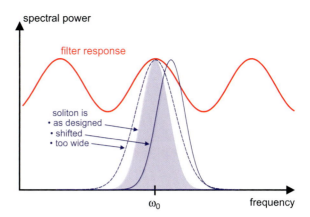

Fig. 3 Soliton spectrum (blue area) and etalon filter response (red). If the soliton is shifted (blue), it receives an asymmetric attenuation which moves it back. If its energy is too high, it temporally compresses; then the spectrum widens (dashed). The filter then causes more attenuation, counteracting the energy offset

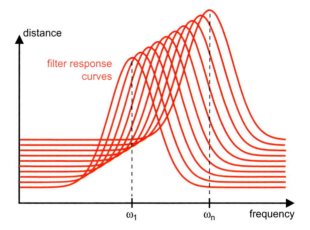

Fig. 4 Successive filters in the transmission line have staggered center frequencies: ω_1 for the first, ω_n for the n-th filter. The line is opaque for linear waves at all frequencies

of the filters will be blocked by some of the other filters. Solitons, in contrast, can readjust after perturbation; they just gradually move their center frequency in accordance to what the string of filters demands. In effect, what such a sliding filter accomplishes is to discriminate linear vs. nonlinear waves.

In the early 1990s, erbium fiber amplifiers were well understood, and several laboratories performed long-distance propagation tests, usually in recirculating loops. In a much heralded result by a group at NTT in Japan, bit patterns were correctly transmitted with solitons over 1 million km (Nakazawa et al. 1991) and in a subsequent refinement over as much as 180 million km (Nakazawa et al. 1993) – which really means over *unlimited distance*. The experiment involved multiple round trips in an ≈500 km fiber loop with Er amplifiers inserted every ≈50 km. Most importantly, a modulator impressed the clock rate onto the circulating signals. This way solitons got periodically reshaped, their interaction was suppressed, and the Gordon-Haus jitter was removed. However, skeptics questioned whether the setup really is a simulator for long-distance transmission or whether it rather constitutes an actively mode-locked laser – the authors themselves concede that similarity – and measures coherence time in the laser.

Four-Wave Mixing

In the presence of the Kerr nonlinearity, the refractive index modulation caused by *one* wave is also felt by others. In this way, cross talk can be induced between WDM channels. Consider two carrier waves with frequencies ω_A and $\omega_B = \omega_A + \Delta\omega$ copropagating in a fiber. This is a special case of the more general framework of four-wave mixing where three Fourier components, in the presence of a third-order nonlinearity, create a fourth. In our case, we find that frequency mixing can give rise to $2\omega_A$ and $2\omega_B$ (frequency doubling) and $\omega_A + \omega_B$ (sum frequency generation); these are of no concern as the new frequencies are an octave away. It can also give rise to $2\omega_A - \omega_B$ and $2\omega_B - \omega_A$. Close to the two original frequencies, these are two new ones, placed by $\Delta\omega$ above and below (Maeda et al. 1990), as schematically shown in Fig. 5. The new frequency components can then, in turn, couple energy to further frequency components, all spaced with $\Delta\omega$ in frequency.

We conclude that four-wave mixing couples frequencies together which sit on an equally spaced frequency grid. That is bad news since it means a maximum of nuisance on a regular WDM frequency grid. By international standardization, the wavelength grid for WDM in optical fibers is defined by the ITU grid (ITU recommendation ITU-T 2012) (ITU stands for International Telecommunication Union, a United Nations agency). It is referred to a frequency of 193,100 GHz with uniform steps of 100 GHz both up and down so that the range from 191,700 to 196,100 GHz is covered. It has later been expanded further, and subdivisions at integer multiples of 50 or 25 GHz have also been used.

A measure to eliminate the four-wave mixing-induced cross talk is the introduction of *dispersion management*, as will be discussed in the next section.

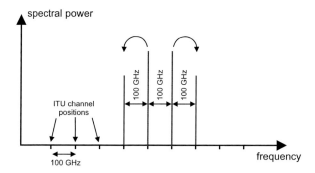

Fig. 5 Four-wave mixing couples wavelength channels of a uniformly spaced channel grid, like the ITU grid with its $\Delta\omega = 2\pi \cdot 100$ GHz spacing. This cross talk is, of course, highly undesired

Dispersion Managed Solitons

A uniformly spaced frequency grid for WDM channels was adopted, but from the viewpoint of avoiding cross talk, it is the worst choice. The saving grace comes from *phase matching* or, in this case, mismatching. In nonlinear optics, it is well understood that for frequency conversion, both the generating and the generated wave must propagate with the same propagation constant, so that their relative phase is constant over the interaction length. It is the relative phase which determines the direction of energy flow, much as is the case for coupled oscillators. In our present context, we wish to suppress, rather than enhance, the generation of new frequencies; this implies that we need as much phase mismatch as possible. Mismatch is large when adjacent wavelength channels have very different group velocities, in other words if there is strong dispersion. On the other hand, low dispersion is desirable in order to keep pulse energies low.

These conflicting objectives can be reconciled by introducing high *local* dispersion and simultaneously low *path-average* dispersion in a fiber with alternating sign of the dispersion coefficient. One makes use of the fact that a fiber's dispersion curve can be tailored, within certain limits, by introducing a structure more involved than the step index fiber. Adding several radial zones of various refractive index makes it possible to shift the dispersion, taken at some operation frequency, to be normal or anomalous. If segments of fibers are concatenated such that pieces of positive and negative β_2 alternate, the path-average dispersion $\bar{\beta}$ can be brought to any desired value close to or at zero, whereas local dispersion is much higher. Such an arrangement, shown in Fig. 6, is called *dispersion management*; the dispersion map is described by two values of β_2 designated by β^+ and β^-, respectively, and by the spatial period $L_{\text{map}} = L^+ + L^-$ with which the alternating pattern is repeated (Chraplyvy et al. 1993). For an extensive review of dispersion management, see Turitsyn et al. (2012).

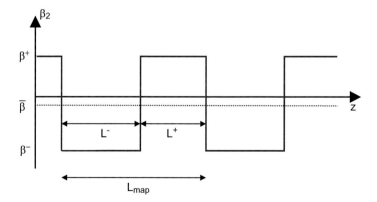

Fig. 6 Dispersion-managed fiber consists of alternatingly normally and anomalously dispersive fiber segments. The dispersion parameter β_2 alternates between values of β^+ and β^-; the lengths of segments are L^+ and L^-, respectively. The dispersion map period is $L_{\text{map}} = L^+ + L^-$

It soon turned out that the approach of perfectly nulling the path-average dispersion is not really successful. On the one hand, such null can be obtained at a single wavelength only. On the other, the fiber has nonlinearity after all; it turns out that a small remaining anomalous dispersion is helpful. To understand this, one should consider that over most of the distance, the pulses are dispersively broadened; only in those places where their width contracts does the nonlinearity give a contribution to the dynamics. That is a small effect, but it is not zero. The concept was dubbed the *chirped RZ format* (Mu et al. 2002).

Solitons are solutions of the NLSE at anomalous dispersion. It is therefore not immediately clear whether solitons or anything like them can exist in dispersion-managed fibers with their alternating dispersion. After finding that the average dynamics of pulse propagation in a dispersion-managed fiber can be approximated by the NLSE (Gabitov and Turitsyn 1996), further research led several groups almost simultaneously to the insight that a type of pulses exists which bears similarity to solitons in that they are stabilized through nonlinearity (Smith et al. 1996; Nijhof et al. 1997; Chen 1998; Turytsin and Shapiro 1998; Kutz and Evangelides 1998; Grigoryan and Menyuk 1998). Their shape periodically "breathes" over the period of L_{map}, but when sampled once per dispersion period, it is maintained, i.e., *stable in the stroboscopic sense*. It resembles a Gaussian more closely than the sech of conventional solitons (Turitsyn et al. 2003) but may have "wiggles" in the slopes (Turitsyn et al. 2003; Lushnikov 2004). These soliton-like pulses in dispersion-managed (DM) fibers became known as DM solitons (Turitsyn et al. 2012). Chirped RZ can also be understood as a DM soliton format.

It was found that DM solitons can exist near the zero-dispersion wavelength or even a slight bit into the regime of path-average normal dispersion (Nijhof et al. 1997; Chen 1998; Turytsin and Shapiro 1998; Kutz and Evangelides 1998; Grigoryan and Menyuk 1998). Correspondingly, dark solitons can also exist on either side of the zero dispersion wavelength (Stratmann et al. 2001). This raises the question whether dark and bright solitons can coexist under the right circumstances. Such thinking led to the discovery of soliton molecules in Stratmann et al. (2005) and will be discussed further in section "Soliton Molecules" below.

It turned out that beyond four-wave mixing suppression, dispersion-managed fiber presents other benefits: the Gordon-Haus jitter is also reduced (Suzuki et al. 1995). On the other hand, there was concern whether DM solitons would survive collisions in a WDM system as easily as standard solitons. After all, in a collision in a DM system, both signals do not just slide past each other: with the back-and-forth of the dispersion, both signals will repeatedly move through each other to and fro. Even when a single encounter gives rise to only a small displacement in time or frequency, the repetition can accumulate a much larger net effect. Therefore, unacceptable jitter can be produced (McKinstrie 2002; Mamyshev and Mollenauer 1999). Countermeasures are frequency-guiding filters or certain group delay compensators; the latter were used for tests over up to 20,000 km (Mollenauer et al. 2003).

A peculiar feature of DM solitons is that in the stroboscopic representation, there can be slow shape oscillations. Slow means that the spatial period is many dispersion periods. This phenomenon has been traced to deviations between the launched pulse and the precise shape of the soliton (Hartwig 2010). Note that in contrast to standard solitons of the NLSE, there is no closed expression for the precise shape of the DM soliton. The slow oscillations are, in a sense, the DM counterpart of the beat between soliton and radiation in standard solitons.

Commercial Soliton Systems

Around the turn of the millenium, the soliton concept had been tested in field experiments in the USA, Japan, and Europe (Andrekson 1999) and was found to be mature enough for an introduction to the commercial market. In France, former French telecom researcher Thierry Georges in 1999 founded a company named Algety in order to bring soliton technology to market. His company soon became part of the US company Corvis which in turn was purchased by Broadwing. For the same purpose, in 2000, Nick Doran in England founded a company named Solstis as an independent subunit of British equipment manufacturer Marconi. Marconi Solstis in 2001 formally introduced its product (Marconi Introduces Soliton-Based Ultra Long Haul System at SuperComm 2001) which was chosen for an installation across Australia, from Perth to Adelaide (Amcom IP1 (TM) Chooses Marconi's Soliton-based Technology for Ultra Long Haul Network: World's Longest Overland Optical Transmission Without Regeneration 2002; Forysiak 2003; Solitons go the distance in ultralong-haul DWDM 2003; http://www.electronicsweekly.com/news/archived/resources-archived/marconi-start-up-to-target-solitons-2000-06/). This all-optical network operated with picosecond soliton pulses in standard fiber. It had amplifiers spaced ca. 90 km over its total length of 287 km length; every second amplifier site had a dispersion-compensating element. As much of the distance was located in a desert, amplifiers could be solar-powered. After thorough testing in 2002, the system went operational in early 2003 (The world's first terabit transcontinental optical communications system exploiting dispersion managed solitons 2014). At the time, it was the world's longest commercially deployed unregenerated 10 GBit/s DWDM terrestrial transmission system; successful field trials of the same technology over 5745 km were reported in Pratt et al. (2003). In 2001, Lucent Technologies, one of the major equipment providers for the telecom industry, introduced a new DWDM system called Lambda Extreme. Designed for long-haul use, it works with dispersion-managed soliton transmission and with Raman amplification (Alcatel-Lucent 2011). The following year, this technology was deployed between Tampa and Miami (both Florida), and trials took place elsewhere, too (www.opticalkeyhole.com/eventtext.asp?ID=23702&pd=3/20/2002&bhcp=1).

Several sales of soliton-based systems have taken place over the next few years, as one can learn through private communication. However, vendors and buyers alike do not publicly disclose any details. Clearly, in a market where a very small number of equipment providers serve an industry of the order of 100 potential customers, there is no incentive for either to publicly advertise the inner workings of the technology or much less write scientific papers about them. Also, current data sheets do not go into technical detail (see, e.g., a data sheet about a soliton system The world's first terabit transcontinental optical communications system exploiting dispersion managed solitons 2014 which does not mention this fact at all http://archive.ericsson.net/service/internet/picov/get?DocNo=28701-FGC1010609&Lang=EN&HighestFree=Y%20%282010%29). Available information about commercial soliton systems is therefore spotty.

Fiber Solitons in the Twenty-First Century

Telecommunications and Limits to Growth

Around the turn of the century, transmission rates in commercial fiber networks had been increased to 10 Gbit/s, corresponding to time slots of 100 ps duration for each bit. Efforts to introduce 40 Gbit/s met with difficulties due to polarization mode dispersion. Compensators were developed, and that took some time; not before ca. 2011 did the industry introduce 100 Gbit/s technology.

In the meantime, another more fundamental problem arose. The growth of demand for data-carrying capacity kept driving further developments. When traffic in the 1980s consisted mostly of telephony, in the meantime, customers routinely used the Internet to download data and increasingly demanded streaming video. These services require much higher data rates, and global traffic kept rising exponentially, with an extra boost in the 1990s with the introduction of massive WDM, and a slowing in the years after the dotcom crisis (Korotky 2013).

Technology must find ways to keep up with demand. It is therefore useful to consider whether the data-carrying capacity of optical fibers is subject to some fundamental limit. A good starting point for such consideration is Shannon's celebrated theorem(Shannon 1948):

$$C = B \log_2 \left(1 + \frac{S}{N}\right). \tag{17}$$

This defines the *channel capacity* C which is the maximum data rate for which error-free transmission is possible. C is proportional to B, the available bandwidth, i.e., the spectral width of the fiber's transparency window around $\lambda = 1.5\,\mu\text{m}$ corresponding to $\nu = 200\,\text{THz}$. Depending on how much loss one is willing to accept, one can put the number at 30 THz or maybe 50 THz. Shannon was discussing a radio-frequency amplitude-modulated system in which the argument of the logarithm, with signal power S and noise power N, describes the number of distinguishable levels.

For binary systems, as in conventional on-off keying, there are only two symbols: mark and space. Then the capacity has the same numerical value as the bandwidth and comes out as 30...50 Gbit/s. In the first years of the new millennium, it became apparent that this limit was soon to be reached. Massive use of WDM approached the point where the entire available bandwidth was filled with data. Nonlinear corrections to Shannon's result were discussed (Mitra and Stark 2001; Kahn and Ho 2001), but it became clear that a bottleneck would arise. The discussion about the limit of growth began in 2002 (Mitschke 2002). In 2009, Andrew Chraplyvy of Alcatel-Lucent Bell Labs coined the term "capacity crunch" (Chraplyvy 2009), and the matter was widely discussed (Richardson 2010; Tkach 2010; Essiambre and Tkach 2012).

As the bandwidth is an inherent property of the fiber, it cannot be increased. Polarization multiplexing provides another factor-of-two at best. The only option to increase C is then to forgo the binary coding, i.e., use codings "beyond binary." A key factor contributing to further data rate increase was that a method from radio engineering known as QAM (quadrature amplitude modulation) was adopted. QAM is a combination of amplitude-and-phase modulation. It allows to transmit several bits per time slot; the number is limited again by considerations of signal-to-noise ratio. This was combined with sophisticated coding schemes, complex error compensation techniques, and a few more technical features we have no room here to discuss, but which are outlined in Li (2009). In all these schemes, nonlinearity was avoided by keeping the power low. Basically, the conventional thinking was to treat the fiber as a linear conduit. It goes without saying that low launch power is not the best proposition to guarantee good signal-to-noise ratio at the receiver. Meanwhile, society's demands on data-carrying capability keep growing relentlessly, driven mostly by data-hungry services using the Internet. After email came streaming video, and the next entrant may be Internet connection of all kinds of appliances and objects (the Internet of things). Barring an unlikely end to this growth, new ways must be found to meet demand in a couple of years from now.

The soliton format fell out of grace in the context of beyond-binary coding as it was considered to provide only on-off keying. However, this assumption has meanwhile been shown not to be true; see section "Soliton Molecules" below.

Currently, it seems impossible to lay more fiber to keep up with demand. Fiber manufacturing plants already operate as fast as they can, and fiber is being deployed at an amazing rate of currently $\approx$420 million km annually and growing (Mack 2017), corresponding to 13 km/s or almost 40 times the speed of sound. Perhaps more fundamentally, to duplicate the same technology keeps the cost per bit constant, that is, not sustainable in the long run. Therefore, researchers look out for further degrees of freedom that might be used in coding. There is now a discussion of *space division multiplexing* by using fibers which either admit more than one mode (Mushid et al. 2008) or which have multiple cores (Saitoh et al. 2016). With this approach, a remarkable transmission rate of just above 1 Pb/s was obtained in lab tests by two groups in 2012 (Takara et al. 2012; Qian et al. 2012). An obvious technical challenge is the issue of cross talk between modes or cores, but again more fundamentally, one may doubt whether proposals to insert non-compatible specialty fibers into the existing vast worldwide fiber network are realistic.

A suggestion for a quite unconventional coding of information was originally proposed by Hasegawa in 1993 (Hasegawa Nyu 1993). It is based on the fact that in integrable systems, the Zakharov-Shabat eigenvalues of solitons, which determine energy and velocity, are preserved quantities. The argument is that use of eigenvalues as information carriers would provide more robustness than pulse shapes, etc. The required computer processing power to extract eigenvalues from a received signal is no longer a problem now, and several researchers have taken up discussion of this subject recently (Prilepsky et al. 2014; Hari et al. 2014; Kamalian et al. 2016a, b; Yousefi and Kschischang 2014a, b).

With all these efforts, plus a few more for which this presentation does not have room, it is still not clear how the ever-growing amount of data that society wants to transmit can be handled in a couple of years. A full-blown capacity crunch seems to be coming our way after all (Ellis et al. 2016). Current conferences deal with this pressing issue (Capacity Crunch: When, Where and What Can Be Done? 2017). Some new thinking is required; maybe the strict adherence to linear formats, i.e., avoidance of nonlinearity, is not sustainable. It turns out that inherently nonlinear techniques like soliton transmission can also be used for more than binary coding formats, as the following section shows.

Soliton Molecules

It was explained in section "Dispersion Managed Solitons" that the range of existence of dispersion-managed solitons extends slightly into the path-average normal dispersion regime, whereas that of dark DM solitons extends slightly into the path-average anomalous dispersion regime. This implies that in a certain range around the zero dispersion wavelength, a DM fiber can support either. In the course of investigations into a possible coexistence, the group of the present author discovered in numerical work that a certain combination of pulses is remarkably stable in DM fibers. The particular structure can be described alternatively as a dark soliton in a background so narrowed down that on either side there is just a bright soliton or as a pair of bright solitons with a phase jump in between. The bright antiphase pulses have a certain preferred separation; if the separation is perturbed, pulses return to that equilibrium position, and there are relaxation oscillations in the process. All this is quite reminiscent of the behavior of two nuclei in a diatomic molecule and led to the moniker *soliton molecule*. This result was not published until it had been experimentally verified (Stratmann et al. 2005). At the time of discovery, the authors of Stratmann et al. (2005) were not aware that around the same time, the same phenomenon, with different terminology like bi-soliton or antisymmetric soliton, was also noted in theoretical work (Parè 1999; Maruta et al. 2002; Ablowitz et al. 2002; Feng and Malomed 2004).

Note that in conventional (constant dispersion) fibers, equal-energy soliton pairs would be subject to interaction (see section "Soliton Interaction"); they cannot exist with constant temporal separation. Therefore, apparently similar phenomena in constant-dispersion fibers as discussed early on in Uzunov et al. (1993) and Akhmediev et al. (1994) and later in Al Khawaja (2010) and Al Khawaja and Boudjemâa (2012) are qualitatively different. In DM fibers, the situation is such that stable bound states of equal-energy DM solitons exist. The favored temporal separation pertaining to a stable equilibrium results from attracting and repelling

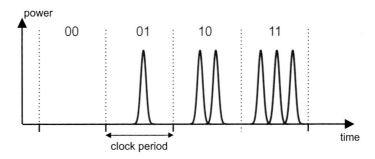

Fig. 7 Four different symbols can be used to encode two bits of information per clock period

forces which are different in DM fibers due to the ever-changing chirp. The binding mechanism therefore relies on the phase dynamics; this was shown in detail in Hause et al. (2007) and Hause et al. (2008).

Some numerical indications hinted at the possibility of soliton molecules of more than two bright pulses (Maruta et al. 2002; Ablowitz et al. 2002; Diaz-Otero and Chamorro-Posada 2012; Maruta and Yoshika 2009). And indeed, it was shown both in theory and experiment (Rohrmann et al. 2012, 2013) that a three-pulse molecule exists. The significance of this result is that with its inclusion, there are four different "letters of the alphabet": a three-molecule, a two-molecule, a single soliton, or no light. Each of these four different symbols could be transmitted stably through a DM fiber. This makes it possible to transmit two bits of information in one clock period (Fig. 7) and demonstrates that soliton-based transmission can go beyond binary. In a practical transmission scheme, there is no need, of course, for the fancy detection schemes used in the lab experiment: The two-soliton molecule contains very nearly twice, the three-soliton molecule very nearly three times the energy of the single soliton so that common detectors combined with thresholding (as is routinely used in QAM transmission) fully suffice to decode the data (Rohrmann et al. 2013).

There were indications that the equilibrium might not be unique (Gabitov et al. 2007; Shkarayev and Stepanov 2009); this led to reevaluation of the binding mechanism. It turned out that certain higher-order equilibria should exist, and this could be experimentally corroborated (Hause and Mitschke 2013).

Other researchers joined the study of soliton molecules: there are investigations of their stability (Boudjemâa and Khawaja 2013) and dynamics (Alamoudi et al. 2014) as well as applications (Johnson et al. 2011). The formation process of soliton molecules was very recently addressed in Herink et al. (2017). Incidentally, comparable soliton compounds exist in fiber lasers; see section "Fiber Lasers" below.

The use of a solitonic format for data transmission holds some promise as regards the signal-to-noise ratio issue over the long haul. Nonlinearity has been taken into account from the outset and does not have to be avoided; solitons with their somewhat higher power than "linear" signals do not suffer nearly as much from detection noise, etc. For all that is known, the soliton molecule format can be combined with WDM and with both polarization and phase multiplexing. Of course it cannot be combined with amplitude multiplexing. For short distance, some advanced linear coding formats may be superior, but with reduced noise problems and some extra robustness due to the solitonic format, long-haul transmission may well be served best by the soliton molecule format.

Soliton Structures on a Background

Section "Some Facts About the NLSE and Its Soliton Solution" mentioned that the cw solution of the NLSE is unstable to perturbations in the anomalous dispersion regime. We return to this issue here because within the last decade or so, a full understanding of the further fate of the destabilized wave was obtained. The irony is that the mathematical background had been fully worked out as early as 1986 (Akhmediev and Korneev 1986). In the meantime, occasionally conjectures about the relation of the modulation instability and soliton generation were discussed, but it took about 20 years to set the record straight.

Consider this solution to the NLSE which has become known as the Akhmediev breather:

$$A(Z,T) = \sqrt{P_0}\, \frac{(1-4a)\cosh(bZ) + ib\sinh(bZ) + \sqrt{2a}\cos(\omega T)}{\sqrt{2a}\cos(\omega T) - \cosh(bZ)} \exp(iZ). \tag{18}$$

The parameters are $0 < a < 1/2$ and $b = \sqrt{8a - 16a^2}$. Position is conveniently normalized as $Z = z/L_{\mathrm{NL}}$. The modulation frequency is $\omega = \omega_c\sqrt{1-2a}$, and

$$\omega_c = \pm\sqrt{\frac{4\gamma P_0}{|\beta_2|}} \tag{19}$$

is the highest frequency at which gain can occur.

Inspection of Eq. (18) reveals that time appears as oscillation of the form of $\cos(\omega T)$, whereas position appears in hyperbolic functions of the argument bZ. For large bZ, the hyperbolic function terms swamp the oscillation terms; the maximum oscillation amplitude therefore occurs at $Z = 0$, and for $Z \to \pm\infty$, it vanishes. Therefore, Eq. (18) describes the evolution of a continuous wave starting at $|A(Z,T)|^2 = P_0$ at $Z \to -\infty$. The modulation amplitude grows exponentially, saturates at $Z = 0$, and decays again. For $Z \to +\infty$, the initial $|A(Z,T)|^2 = P_0$ is recovered: a "life cycle" of growth and decay (Fig. 8).

In a situation where a continuous wave is sent into the fiber, the cw power will be slightly perturbed, e.g., by quantum noise. Such fluctuation will dominantly excite the frequency pertaining to the gain maximum. It occurs at $a = 1/4$ where it takes a value of

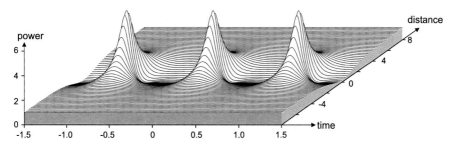

Fig. 8 Evolution of an Akhmediev breather at $a = 1/4$

$$g_{max} = 2\gamma P_0 \quad \text{at} \quad \omega = \frac{\omega_c}{\sqrt{2}} \qquad (20)$$

as in Eq (4). This special case of Eq. (18) reproduces what had been known all along about the gain and its frequency dependence. Figure 8 gives an illustration. It may seem natural to assume that this mechanism is a way to produce an entire comb of solitons at once (Hasegawa 1984a). However, such thinking turns out to be misguided: there is no one-to-one correspondence between the peaks of the Akhmediev breather at $Z = 0$ and solitons (Mahnke and Mitschke 2012, 2013).

First experiments to demonstrate the Akhmediev breather in optical fibers were described in Dudley et al. (2009) and in more detail in Hammani et al. (2011). The process starts either from random perturbation (Dudley et al. 2009) or by suitable modulation as suggested in Hasegawa (1984a) and done experimentally in Hammani et al. (2011). Perturbations, including those by higher-order dispersion, can lead to a recurrence of the breather (Mussot et al. 2013; Chin et al. 2015) which has been compared to the famous Fermi-Pasta-Ulam recurrence phenomenon (Fermi et al. 1955).

If $a \to 1/2$, the distance between the peaks of the comb diverges, and as a result, one obtains a structure which is localized in both time and space: an isolated, singular extremum (see Fig. 9). It is known as the Peregrine soliton, after Peregrine (1983) who obtained an exact but simplified expression for it. We point out that the moniker "soliton" appears here for historical reasons; there is no similarity to the standard NLSE solitons other than that this structure is also a solution of the NLSE. Discussion of the Peregrine soliton was revived with an experimental demonstration in Kibler et al. (2010). Similarly, the Kuznetsov-Ma soliton is a further member of the family of NLSE solutions described as a modulation on top of a constant background and was studied in Kibler et al. (2012).

Supercontinuum Generation

A remarkable development recently involving solitons is the generation of optical supercontinuum. The term designates spectrally broadband light (implying short

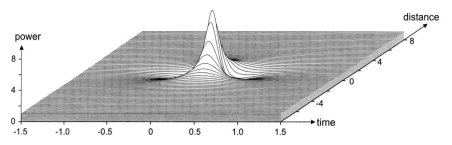

Fig. 9 Evolution of a Peregrine soliton

temporal coherence) generated from laser light after passing through a length of fiber. Light emerging from a single mode fiber has perfect spatial coherence and is easily focused. These properties make it an ideal light source for applications like coherence tomography where the resolution of the generated image is determined by spatial resolution in the transverse and temporal coherence length in the longitudinal direction (Froehly and Meteau 2012). White light from thermal sources is limited in its spectral power density by Planck's radiation law and accessible temperatures; moreover, it loses most of its power when one imposes spatial coherence with an aperture. Laser-generated supercontinuum suffers from no such limitations.

In most cases (but not always, see Swiderski and Maciejewska 2012), special fibers are used for supercontinuum generation known as microstructured, photonic crystal, or holey fibers. Their design differs radically from conventional fibers: A number of voids – hollow channels – run the entire length of the fiber, parallel to the core and surrounding it (Russell 2003). In the type most frequently used for supercontinuum generation, the holes are arranged in a geometrical pattern, at the center of which one hole is missing. That missing hole position takes the role of the light-guiding core, and the surrounding hole pattern acts as the cladding zone. The difference between core index and effective cladding index is much larger than can be obtained in ordinary fibers, and so the light confinement is stronger. That allows to make the core smaller, and consequently the nonlinearity coefficient γ (Eq. (2)) is larger.

The formation of supercontinuum proceeds, broadly speaking, as follows: A powerful light signal is launched into a highly nonlinear optical fiber. If the light is in the form of pulses which are at least many picosecond wide, one will initially see the development of spectral sidebands by modulation instability. In principle, the evolution follows that of the Akhmediev breather: There is sideband gain as in Eq. (20), and that gain is the dominant factor that determines the fiber length until strong sidebands have formed. However, there are numerous perturbations from higher-order dispersion, the Raman effect, etc., so that the pulse comb will not develop to such clean form as in Fig. 8 but rather will disintegrate into a number of solitons as in Mahnke and Mitschke (2012).

If on the other hand the launched pulses are shorter, i.e., in the femtosecond range, a different process is at work. The pulses will disintegrate, too, but by what is called *soliton fission* (Husakou and Herrmann 2001), i.e., the breakup of high-order N-solitons into individual fundamental solitons. The fiber length for fission may be estimated as $(1/\sqrt{2})\, L_D/N$ (Chen and Kelley 2002); for femtosecond pulses, this is shorter than the typical growth distance for modulation instability. For example, the octave-spanning spectra used in Nobel-winning metrology were generated in this way (Jones et al. 2000; Holzwarth et al. 2000).

In either case, several solitons of various frequencies, durations, and powers are created very soon, along with some radiation. Subsequently, all kinds of interactions between these components will take place: The Raman effect will shift all pulses toward the low-frequency side of the spectrum at a rate dependent on their power. There are coupling mechanisms between spectral components from four-wave mixing, and there is phase matching between solitons and radiation in the presence of higher-order dispersion (Roy et al. 2009), etc.

Once the spectrum begins to widen, increasingly the dispersion curve across the relevant spectral range dominates much of the further evolution. For example, qualitatively new phenomena arise through higher-order dispersion (Akhmediev and Karlsson 1995). If, for example, a soliton is Raman-shifted toward a zero dispersion wavelength, it will not cross that point; instead, some of its energy will leak into the normally dispersive region, and subsequently that dispersive wave and what is left of the soliton remain coupled together (Skryabin et al. 2003). Phase-matching conditions can be modified very much when the fiber's dispersion curve is designed to have two or even more zero-dispersion wavelengths (Zhao et al. 2015). Cross-phase modulation may couple spectral components in regimes of similar absolute value, but different sign, of dispersion (Gorbach and Skryabin 2007). It is hardly disputed that dispersion is the most relevant design parameter (Tse et al. 2006) for the final spectral shape of the supercontinuum.

There is no fixed universal definition of what exactly constitutes a supercontinuum, but a spectral width of at least an octave seems to be an acceptable criterion. One of the first experiments in fibers already obtained two octaves (Ranka et al. 2000). Beyond spectral width, flatness, and power density, another important characteristic of supercontinuum is the stability of the source. When it comes to tomographic imaging, fluctuations can thwart the noise reduction usually obtained from temporal averaging. In other words, some long-term contribution to the coherence is desirable. Coherence in supercontinua has been investigated, e.g., in Genty et al. (2011).

We have described that in most cases, intense laser pulses are used to start supercontinuum generation and should point out that it can be generated successfully even with cw irradiation (Cumberland et al. 2008; Kudlinski and Mussot 2008; Travers et al. 2008). An extensive review of supercontinuum generation was given in Dudley et al. (2006). An application to the generation of a multicarrier signal for optical telecommunications was suggested in (Mori et al. 2003).

Rogue Waves

The history of solitons began with water waves, and so it may be fitting that a much-discussed paper in 2007 (Solli et al. 2007) suggested a connection between optical supercontinuum and a certain remarkable type of water waves. Mariners have talked for centuries about "freak waves," "monster waves," or "rogue waves" which would appear without warning, have gigantic height, and be able to sink a ship in moments. For a long time, such reports were considered as tall stories until a first objective recording by an instrument aboard an oil-drilling rig became known (Taylor 2005). Research soon showed that rogue waves are rare but not *exceedingly rare*.

In Solli et al. (2007), the usual setup for the generation of optical supercontinuum was modified insofar as through filtering only the long-wavelength end of the spectrum was considered. A statistical analysis showed that every once in a while some pulse appeared that was exceptionally large – an *optical rogue wave*. This may be explained by the fact that the strongest pulses in the supercontinuum get Raman-shifted the most and eventually appear at the long-wavelength end of the spectrum. The argument was that there might be similarities to the situation on the ocean.

This prompted further research (Akhmediev et al. 2009a, b; Bonatto et al. 2011; Ankiewicz et al. 2013), and a commonly accepted definition of a "rogue wave" became necessary. While such definition exists in oceanography, in optics an informal consensus was formed that three criteria are required (Akhmediev and Pelinovsky 2010): (1) The wave appears suddenly and disappears without a trace, (2) its peak amplitude is at least twice the "significant" wave height (average of the largest third), and (3) a histogram of crest heights has a "fat tail," meaning that the probability of very high crests far exceeds standard statistics.

Subsequently, various mechanisms have been proposed for rogue wave generation: collision of solitons (Erkintalo et al. 2010) or of Akhmediev breathers (Akhmediev et al. 2009b), Peregrine solitons (Shrira and Geogjaev 2010), and more (Armaroli et al. 2015). However, some doubts remain (Birkholz et al. 2016; Fedele et al. 2016). A main difference between the fiber and the ocean surface is that the former has one and the latter two spatial dimensions; that can make a lot of difference. It is probably fair to say that there may be no such thing as a single explanation for all rogue wave phenomena.

Fiber Lasers

As pointed out in section "Optical Amplifiers," fibers can be doped with laser-active materials and thus become amplifiers when provided suitable pump light. Several dopants are available for gain in different wavelength regimes. Neodymium, for example, is useful for a wavelength at 1064 nm. As this report is about solitons, only fibers doped with erbium ions concern us here; they have been discussed in breadth

(Desurvire 1994; Becker et al. 1999). Closing such a fiber onto itself provides a feedback mechanism which can generate lasing; addition of a mode-locking mechanism can ensure pulsing which may take the form of a soliton travelling around the cavity (Duling 1991a). Such mechanisms employed APM (section "The Soliton Laser") with a nonlinear fiber loop mirror (Duling 1991b) or nonlinear polarization rotation (Matsas et al. 1992) or with saturable absorbers (De Souza et al. 1993). However, there is the fundamental conundrum that the output power available from a soliton fiber laser is quite limited to the order of 10 picojoules. The soliton energy is set by fiber parameters and pulse width as specified in Eqs. (7) and (8), and only a fraction of the intracavity power can be coupled out. If pump power is increased, the individual soliton does not become more energetic; rather, multiple pulses will circulate in the cavity. While it is true that in view of the typically long cavities of fiber lasers, repetition rates higher than the fundamental cavity round-trip frequency (*harmonic mode locking*) can be welcome, multiple pulsing often comes with fluctuating relative timing. Additional means were used to stabilize the pulse train (Harvey and Mollenauer 1993).

An additional complication is that a circulating soliton periodically transits various resonator components with different dispersion. This constitutes a periodic perturbation; in the presence of some linear radiation, this leads to phase matching between radiation and soliton at certain frequency offsets from the soliton center frequency. In those positions, sidebands appear which compromise the pulse shape; they are known as Kelly sidebands (Kelly 1992).

The first concept to solve the dilemma was to reduce the nonlinear impact on the pulse by temporally stretching it in much of the cavity; the idea of a *stretched pulse fiber laser* (Tamura et al. 1993) is similar to the concept of dispersion management and yielded one order of magnitude higher-output pulse energy. Where high power is not a prime consideration, this concept in combination with harmonic mode locking can be quite successful (He et al. 2016).

A new idea, introduced in 2004 (Ilday et al. 2004), was to forgo solitons entirely and build a laser using parabolic pulses, a "self-similar" type of pulse in the sense that its power can be scaled without change of width or shape. These self-similar pulses are described by an asymptotic solution of the NLSE (Fermann et al. 2000) and are also called *similaritons* (Finot et al. 2004). They exist in normal dispersion, have a linear chirp (which facilitates external pulse compression), and are not susceptible to optical wave breaking (section "Some Facts About the NLSE and Its Soliton Solution"). This makes them successful in fiber lasers (Renninger et al. 2010).

But even when lasers are operated in a regime where there are solitons, they are subject to an alternation of gain and loss. This situation is clearly beyond the realm of the NLSE and is much better modeled by a cubic-quintic complex Ginzburg-Landau equation (Soto-Crespo et al. 1996). The solitons of that equation are subject to a twofold equilibrium: dispersion vs. nonlinearity like ordinary solitons and additionally gain vs. loss (Grelu and Akhmediev 2012). Just like NLSE solitons, *dissipative solitons* are found in a large variety of contexts throughout physics and beyond (Akhmediev and Ankiewicz 2008).

In fiber lasers, one can find a rich phenomenology of dynamic effects including bifurcations and chaos (Soto-Crespo et al. 2004), formation of bound states of solitons (Grelu and Soto-Crespo 2008), and complex soliton ensemble dynamics (Chouli and Grelu 2010). For good overview of the field of fiber lasers, see Okhotnikov (2012).

Beyond the Nonlinear Schrödinger Ansatz

The NLSE was originally derived with picosecond light pulses in mind; the relative spectral bandwidth of such pulses is small enough that there is a well-defined center (or carrier) frequency and that in most cases the dispersion can be assumed to be constant across the spectral width. However, pulses in the low femtosecond regime have wide relative bandwidth, and higher-order corrections to the dispersion as in section "Generalized NLSE" are mandatory. This is all the more true when it comes to optical supercontinuum which covers wide parts of the spectrum, sometimes several octaves, and within that width, there may be one or even more zero-dispersion wavelengths.

For the NLSE, typically a series expansion of the dispersion around some central reference frequency is used, as indicated in Eq. (13). Taylor series expansions have disadvantages when used over wide intervals: they diverge outside the interval and cannot reproduce poles. Some effort has been made to improve modeling of such cases by use of a rational expression of the dispersion curve, rather than an polynomial, known as the *Padé approximation*. It can reproduce resonances, gives a much better fit to the realistic dispersion curve, and was shown to make a difference for few-cycle pulses (Amiranashvili et al. 2010). However, even that may be not good enough in some challenging cases. In the derivation of the NLSE, the carrier frequency is removed as a consequence of the *slowly varying envelope approximation* (SVEA) which stipulates that during one cycle of the field oscillation, the change in the envelope is small. This approximation comes to the limit of its validity in some experimental situations.

In Husakou and Herrmann (2001), a propagation equation was given that does not require SVEA and Taylor expansion of dispersion; it is called *forward Maxwell equation*. Further research led to the following insight: Both positive and negative frequencies arise naturally from Fourier transforms in pulse propagation problems, but in the framework of SVEA, the negative frequencies are ignored. This is at variance with situations in radio engineering where the *analytic signal* is considered; it contains frequencies of either sign. The need to do the same in optics was emphasized in Amiranashvili and Demircan (2011). A model has been proposed that takes all combinations of forward/backward propagation of positive/negative frequencies into account (Conforti et al. 2013). In comparison to the more conventional purely positive-frequency models like Schrödinger and Ginzburg-Landau equations, the new model also has solitons, but beyond that, it captures certain mixing processes which are absent in previous models. Some experimental data seem to hint at just such processes (Rubino et al. 2012); further evidence will hopefully settle the issue.

Conclusion

The harmonic wave is the paradigm of linear waves.

When in the course of historical development linear waves were increasingly better understood (Jean-Baptiste Joseph Fourier contributed a milestone in 1822), the attention shifted to situations of practical interest which build on harmonic waves but go beyond that simplified concept. No real-world example of linear waves – be it about musical sound or radio transmission – relies on pure harmonic waves.

The soliton is the paradigm of nonlinear waves in just the same sense. It seems to be a general rule that once the understanding of a concept has matured, and proficiency in its use has been reached, few applications use the pure concept but rather some generalization of it. From optical supercontinuum to optical rogue waves to possible future nonlinear transmission schemes, not the pure NLSE soliton serves as a model but rather its more complex generalizations.

In the specific application of optical telecommunications, the conventional thinking is that nonlinearity in the channel is a nuisance. Concepts from radio engineering are easily applied to fibers as long as the optical power is kept low enough so as to keep nonlinear effects almost negligible. On the other hand, nonlinearity can be outright beneficial – for a nice example, see Turitsyn (2012) – and may get embraced eventually, after all.

The story of solitary waves is almost 200 years old. The particular embodiment of fiber-optic solitons was proposed 44 years ago and first verified 37 years ago. This review took a look back in time, to outline how solitons have come a long way from a lab curiosity to a living reality. In the process of both increased understanding and advanced application, the pure and simple idealization gave way to messier realities, away from integrability. But the studies involved in this progression have carried rich fruit. The nonlinear Schrödinger equation is now a starting point to describe the dynamics of Bose-Einstein condensates, plasma waves, and more. Both bright and dark solitons, in various physical situations, have been described in detail and are now quite well understood. The optical soliton in a fiber is a convenient test bed for many such concepts.

Meanwhile, new applications have been adopted for commercial products: Optical supercontinuum sources can be purchased off the shelf for applications in metrology and coherence tomography. All told, our concept of solitons has matured, as this review attempted to explain. That does not mean, however, that their story has come to its end. Optical telecommunications, largely performed in the linear regime so far, currently runs into severe limitations, and there is a desperate search for new ideas to cope with growing demand. Maybe nonlinear transmission schemes involving solitons and their generalizations will see fresh appreciation. So far, all reports claiming the end of soliton history have turned out to be premature.

References

M.J. Ablowitz, T. Hirooka, T. Inoue, Higher-order asymptotic analysis of dispersion managed transmission systems: solutions and their characteristics. J. Opt. Soc. Am. B **19**, 2876 (2002)

G.P. Agrawal, *Nonlinear Fiber Optics*, 5th edn. (Academic Press, Amsterdam, 2013)

N. Akhmediev, A. Ankiewicz (eds.), *Dissipative Solitons: From Optics to Biology and Medicine*. Lecture Notes in Physics, vol. 751 (Springer, Berlin, 2008)

N. Akhmediev, M. Karlsson, Cherenkov radiation emitted by solitons in optical fibers. Phys. Rev. A **51**, 2602 (1995)

N.N. Akhmediev, V.I. Korneev, Modulation instability and periodic solutions of the nonlinear Schrödinger equation. Theor. Math. Phys. **69**, 1089 (1986)

N. Akhmediev, E. Pelinovsky, Introductory remarks on "discussion & debate: rogue waves – towards a unifying concept?" Eur. Phys. J. Spec. Top. **185**, 1 (2010)

N.N. Akhmediev, G. Town, S. Wabnitz, Soliton coding based on shape invariant interacting soliton packets: the three-soliton case. Opt. Commun. **104**, 385 (1994)

N. Akhmediev, A. Ankiewicz, M. Taki, Waves that appear from nowhere and disappear without a trace. Phys. Lett. A **373**, 675 (2009a)

N. Akhmediev, J.M. Soto-Crespo, A. Ankiewicz, Extreme waves that appear from nowhere: on the nature of rogue waves. Phys. Lett. A **373**, 2137 (2009b)

U. Al Khawaja, Stability and dynamics of two-soliton molecules. Phys. Rev. E **81**, 056603 (2010)

U. Al Khawaja, A. Boudjemâa, Binding energy of soliton molecules in time-dependent harmonic potential and nonlinear interaction. Phys. Rev. E **86**, 036606 (2012)

S.M. Alamoudi, U. Al Khawaja, B.B. Baizakov, Averaged dynamics of soliton molecules in dispersion-managed optical fibers. Phys. Rev. A **89**, 053817 (2014)

Alcatel-Lucent, 1625 Lambda Extreme Transport Brochure (2011). http://www.telecomnetworks.ru/datadocs/doc_1587tu.pdf

Amcom IP1 (TM) Chooses Marconi's Soliton-Based Technology for Ultra Long Haul Network: World's Longest Overland Optical Transmission Without Regeneration, Marconi Corporation, press release 19 Mar 2002. Archived at: http://www.prnewswire.com/news-releases/amcom-ip1tm-chooses-marconis-soliton-based-technology-for-ultra-long-haul-network-76538107.html

S. Amiranashvili, A. Demircan, Ultrashort optical pulse propagation in terms of analytic signal. Adv. Opt. Technol. **2011**, 989515 (2011)

S. Amiranashvili, U. Bandelow, A. Mielke, Padé approximant for refractive index and nonlocal envelope equations. Opt. Commun. **283**, 480 (2010)

D. Anderson, M. Lisak, Nonlinear asymmetric self-phase modulation and self-steepening of pulses in long optical waveguides. Phys. Rev. A **27**, 1393 (1983)

D. Anderson, M. Lisak, Modulational instability of coherent optical-fiber transmission signals. Opt. Lett. **9**, 468 (1984)

P. Andrekson, Solitons are near commercialization. LEOS Newslett. **13**, 3 (1999)

A. Ankiewicz, J.M. Soto-Crespo, A. Chowdhury, N. Akhmediev, Rogue waves in optical fibers in presence of third-order dispersion, self-steepening, and self-frequency shift. J. Opt. Soc. Am. B **30**, 87 (2013)

A. Armaroli, C. Conti, F. Biancalana, Rogue solitons in optical fibers: a dynamical process in a complex energy landscape. Optica **2**, 497 (2015)

J.R.M. Barr, D.W. Hughes, Coupled cavity modelocking of a Nd:YAG laser using second-harmonic generation. Appl. Phys. B **49**, 323 (1989)

P.C. Becker, N.A. Olsson, J.R. Simpson, *Erbium-Doped Fiber Amplifiers: Fundamentals and Technology* (Academic Press, San Diego, 1999)

R.F. Benson, A discussion of the theory of ionospheric cross modulation. Radio Sci. J. Res. NBS/USNC-URSI **68D**, 10 (1964)

S. Birkholz, C. Brée, I. Veselić, A. Demircan, G. Steinmeyer, Ocean rogue waves and their phase space dynamics in the limit of a linear interference model. Sci. Rep. **6**, 35207 (2016)

K.J. Blow, N.J. Doran, The asymptotic dispersion of soliton pulses in lossy fibres. Opt. Commun. **52**, 367 (1985)

K.J. Blow, B.P. Nelson, Improved mode locking of an F-center laser with a nonlinear nonsoliton external cavity. Opt. Lett. **13**, 1026 (1988)

C. Bonatto, M. Feyereisen, S. Barland, M. Giudici, C. Masoller, Deterministic optical rogue waves. Phys. Rev. Lett. **107**, 053901 (2011)

A. Boudjemâa, U. Al Khawaja, Stability of N-soliton molecules in dispersion-managed optical fibers. Phys. Rev. A **88**, 045801 (2013)

J. Bromage, Raman amplification for fiber communications systems. J. Lightwave Technol. **22**, 79 (2004)

Capacity Crunch: When, Where and What Can Be Done? Announcement for a Workshop at Optical Fiber Conference (2017). http://www.ofcconference.org/en-us/home/program-speakers/workshops/capacity-crunch/

Y. Chen, H.A. Haus, Dispersion-managed solitons with net positive dispersion. Opt. Lett. **23**, 1013 (1998)

C.-M. Chen, P.L. Kelley, Nonlinear pulse compression in optical fibers: scaling laws and numerical analysis. J. Opt. Soc. Am. B **19**, 1961 (2002)

R.Y. Chiao, E. Garmire, C.H. Townes, Self-trapping of optical beams. Phys. Rev. Lett. **13**, 479 (1964)

S.A. Chin, O.A. Ashour, M.R. Beli, Anatomy of the Akhmediev breather: cascading instability, first formation time, and Fermi-Pasta-Ulam recurrence. Phys. Rev. E **92**, 063202 (2015)

S. Chouli, P. Grelu, Soliton rains in a fiber laser: an experimental study. Phys. Rev. A **81**, 063829 (2010)

A. Chraplyvy, The coming capacity crunch, Plenary Paper 1.0.2, in *Proceedings of 35th European Conference on Optical Communications ECOC* (2009)

A.R. Chraplyvy, A.H. Gnauck, R.W. Tkach, R.M. Derosier, 8 × 10 Gbit/s transmission through 280 km of dispersion-managed fiber. IEEE Photon. Technol. Lett. **5**, 1233 (1993)

M. Conforti, A. Marini, T.X. Tran, D. Faccio, F. Biancalana, Interaction between optical fields and their conjugates in nonlinear media. Opt. Express **21**, 31239 (2013)

B.A. Cumberland, J.C. Travers, S.V. Popov, J.R. Taylor, Toward visible CW-pumped supercontinua. Opt. Lett. **33**, 2122 (2008)

E.M. de Jager, *On the Origin of the Korteweg-de Vries Equation* (2011). https://arxiv.org/pdf/math/0602661v1.pdf

E.A. De Souza, C.E. Soccolich, W. Pleibel, R.H. Stolen, J.R. Simpson, D.J. DiGiovanni, Saturable absorber modelocked polarization maintaining Erbium-Doped fiber laser. Electron. Lett. **29**, 447 (1993)

E. Desurvire, *Erbium-Doped Fiber Amplifiers: Principles and Applications* (Wiley, Hoboken, 1994)

E. Desurvire, J.R. Simpson, P.C. Becker, High-gain erbium-doped traveling-wave fiber amplifier. Opt. Lett. **12**, 888 (1987)

E.M. Dianov, A.Y. Karasik, P.V. Mamyshev, A.M. Prokhorov, V.N. Serkin, M.F. Stel'makh, A.A. Fomichev, Stimulated-Raman conversion of multisoliton pulses in quartz optical fibers. Pis'ma Zh. Eksp. Teor. Fiz. **41**, 242 (1985); JETP Lett. **41**, 294 (1985)

F.J. Diaz-Otero, P. Chamorro-Posada, Propagation properties of strongly dispersion-managed soliton trains. Opt. Commun. **285**, 162 (2012)

J.M. Dudley, G. Genty, S. Coen, Supercontinuum generation in photonic crystal fiber. Rev. Mod. Phys. **78**, 1135 (2006)

J.M. Dudley, G. Genty, F. Dias, B. Kibler, N. Akhmediev, Modulation instability, Akhmediev Breathers and continuous wave supercontinuum generation. Opt. Express **17**, 21497 (2009)

I.N. Duling III, All-fiber ring soliton laser mode locked with a nonlinear mirror. Opt. Lett. **16**, 539 (1991a)

I.N. Duling III, Subpicosecond all-fiber erbium laser. Electron. Lett. **27**, 544 (1991b)

A.D. Ellis, N. MacSuibhne, D. Saad, D.N. Payne, Communication networks beyond the capacity crunch. Phil. Trans. R. Soc. A **374**, 20150191 (2016)

P. Emplit, J.P. Hamaide, F. Reynaud, C. Froehly, A. Barthelemy, Picosecond steps and dark pulses through nonlinear single mode fibers. Opt. Commun. **62**, 374 (1987)

M. Erkintalo, G. Genty, J.M. Dudley, Giant dispersive wave generation through soliton collision. Opt. Lett. **35**, 658 (2010)

R.-J. Essiambre, R.W. Tkach, Capacity trends and limits of optical communication networks. Proc. IEEE **100**, 1035 (2012)

S.G. Evangelides, L.F. Mollenauer, J.P. Gordon, N.S. Bergano, Polarization multiplexing with solitons. J. Lightwave Technol. **10**, 28 (1992)

I.L. Fabelinskii, Molecular scattering of light (Nauka Press, Moscow, 1965); transl. (Plenum Press, New York, 1968)

F. Fedele, J. Brennan, S. Ponce de León, J. Dudley, F. Dias, Real world ocean rogue waves explained without the modulational instability. Sci. Rep. **6**, 27715 (2016)

B.-F. Feng, B.A. Malomed, Antisymmetric solitons and their interactions in strongly dispersion-managed fiber-optic systems. Opt. Commun. **229**, 173 (2004)

E. Fermi, J. Pasta, S. Ulam, Studies of Non Linear Problems. Document LA-1940 (May 1955) Los Alamos Report LA-1940 (1955) Reprinted in E. Fermi, Collected Papers Vol. II (ed. E. Segrè) 978, University of Chicago Press, 1965

M.E. Fermann, V.I. Kruglov, B.C. Thomsen, J.M. Dudley, J.D. Harvey, Self-similar propagation and amplification of parabolic pulses in optical fibers. Phys. Rev. Lett. **84**, 6010 (2000)

C. Finot, G. Millot, J.M. Dudley, Asymptotic characteristics of parabolic similariton pulses in optical fiber amplifiers. Opt. Lett. **29**, 2533 (2004)

W. Forysiak, *Dispersion Managed Solitons and Real-World Link Installations*, paper ThI2, Lasers and Electro-Optics Society LEOS 2003. The 16th Annual Meeting of the IEEE (2003)

P.A. Franken, A.E. Hill, C.W. Peters, G. Weinreich, Generation of optical harmonics. Phys. Rev. Lett. **7**, 118 (1961)

L. Froehly, J. Meteau, Supercontinuum sources in optical coherence tomography: a state of the art and the application to scan-free time domain correlation techniques and depth dependant dispersion compensation. Opt. Fiber Technol. **18**, 411 (2012)

I. Gabitov, S.K. Turitsyn, Averaged pulse dynamics in a cascaded transmission system with passive dispersion compensation. Opt. Lett. **21**, 327 (1996)

I. Gabitov, R. Indik, L. Mollenauer, M. Shkarayev, M. Stepanov, P.M. Lushnikov, Twin families of bisolitons in dispersion-managed systems. Opt. Lett. **32**, 605 (2007)

C.S. Gardner, J.M. Greene, M.D. Kruskal, R.M. Miura, Method for solving the Korteweg-de Vries equation. Phys. Rev. Lett. **19**, 1095 (1967)

G. Genty, M. Surakka, J. Tutunen, A.T. Friberg, Complete characterization of supercontinuum coherence. J. Opt. Soc. Am. B **28**, 2301 (2011)

E.A. Golovchenko, E.M. Dianov, A.M. Prokhorov, V.N. Serkin, Decay of optical solitons, JETP Lett. **42**, 87 (1985)

M. Göppert-Mayer, Über Elementarakte mit zwei Quantensprüngen. Annalen der Physik **401**, 273 (1931)

A. Gorbach, D. Skryabin, *Gravity-Like Potential Traps Light and Stretches Optical Supercontinuum*, paper NThB5, Nonlinear Photonics 2007, Quebec (The Optical Society of America, Washington, DC, 2007)

J.P. Gordon, Interaction forces among solitons in optical fibers. Opt. Lett. **8**, 596 (1983)

J.P. Gordon, Theory of the soliton self frequency shift. Opt. Lett. **11**, 662 (1986)

J.P. Gordon, H.A. Haus, Random walk of coherently amplified solitons in optical fiber transmission. Opt. Lett. **11**, 665 (1986)

P. Grelu, N. Akhmediev, Dissipative solitons for mode-locked lasers. Nat. Photon. **6**, 84 (2012)

P. Grelu, J.M. Soto-Crespo, Temporal soliton "molecules" in mode-locked lasers: collisions, pulsations, and vibrations, in Akhmediev and Ankiewicz (2008), p. 137

V.S. Grigoryan, C.R. Menyuk, Dispersion-managed solitons at normal average dispersion. Opt. Lett. **23**, 609 (1998)

K. Hammani, B. Wetzel, B. Kibler, J. Fatome, C. Finot, G. Millot, N. Akhmediev, J.M. Dudley, Spectral dynamics of modulation instability described using Akhmediev breather theory. Opt. Lett. **36**, 2140 (2011)

S. Hari, F. Kschischang, M. Yousefi, Multi-eigenvalue Communication Via the Nonlinear Fourier Transform, in *27th Biennial Symposium on Communications (QBSC)* (2014)

H. Hartwig, M. Böhm, A. Hause, F. Mitschke, Slow oscillations of dispersion managed solitons. Phys. Rev. A **81**, 033810 (2010)

G.T. Harvey, L.F. Mollenauer, Harmonically mode-locked fiber ring laser with an internal Fabry-Perot stabilizer for soliton transmission. Opt. Lett. **18**, 107 (1993)

A. Hasegawa, Generation of a train of soliton pulses by induced modulational instability in optical fibers. Opt. Lett. **9**, 288 (1984a)

A. Hasegawa, Numerical study of optical soliton transmission amplified periodically by the stimulated Raman effect. Appl. Opt. **23**, 3302 (1984b)

A. Hasegawa, Effect of polarization mode dispersion in optical soliton transmission in fibers. Phys. D **188**, 241 (2004)

A. Hasegawa, Y. Kodama, Signal transmission by optical solitons in monomode fiber. Proc. IEEE **69**, 1145 (1981)

A. Hasegawa, T. Nyu, Eigenvalue communication. J. Lightwave Technol. **11**, 395 (1993)

A. Hasegawa, F. Tappert, Transmission of stationary nonlinear optical pulses in dispersive dielectric fibers. II. Normal dispersion. Appl. Phys. Lett. **23**, 171 (1973a)

A. Hasegawa, F. Tappert, Transmission of stationary nonlinear optical pulses in dispersive dielectric fibers. I. Anomalous dispersion. Appl. Phys. Lett. **23**, 142 (1973b)

A. Hause, F. Mitschke, Higher-order equilibria of temporal soliton molecules in dispersion-managed fibers. Phys. Rev. A **88**, 063843 (2013)

A. Hause, H. Hartwig, B. Seifert, H. Stolz, M. Böhm, F. Mitschke, Phase structure of soliton molecules. Phys. Rev. A **75**, 063836 (2007)

A. Hause, H. Hartwig, M. Böhm, F. Mitschke, Binding mechanism of temporal soliton molecules. Phys. Rev. A **78**, 063817 (2008)

W. He, M. Pang, C.R. Menyuk, P.S.J. Russell, Sub-100-fs 1.87 GHz mode-locked fiber laser using stretched-soliton effects. Optica **3**, 1366 (2016)

J. Hecht, *City of Light. The Story of Fiber Optics* (Oxford University Press, Oxford, 1999)

H. Heffner, The fundamental noise limit of linear amplifiers. Proc. IRE **50**, 1604 (1962)

G. Herink, F. Kurtz, B. Jalali, D.R. Solli, C. Ropers, Real-time spectral interferometry probes the internal dynamics of femtosecond soliton molecules. Science **356**, 50 (2017)

R. Holzwarth, T. Udem, T.W. Hänsch, J.C. Knight, W.J. Wadsworth, P.S.J. Russell, Optical frequency synthesizer for precision spectroscopy. Phys. Rev. Lett. **85**, 2264 (2000)

A.V. Husakou, J. Herrmann, Supercontinuum generation of higher-order solitons by fission in photonic crystal fibers. Phys. Rev. Lett. **87**, 203901 (2001)

E. Iannone, F. Matera, A. Galtarossa, G. Gianello, M. Schiano, Effect of polarization dispersion on the performance of IM-DD communication systems. IEEE Photon. Technol. Lett. **5**, 1247 (1993)

F.Ö. Ilday, J.R. Buckley, W.G. Clark, F.W. Wise, Self-similar evolution of parabolic pulses in a laser. Phys. Rev. Lett. **92**, 213902 (2004)

E.P. Ippen, H.A. Haus, L.Y. Liu, Additive pulse mode locking. J. Opt. Soc. Am. B **6**, 1736 (1989)

ITU recommendation ITU-T G.694.1 (2012). http://www.itu.int/rec/T-REC-G.694.1-201202-I/en

S. Johnson, S. Pau, F. Küppers, Experimental demonstration of optical retiming using temporal soliton molecules. J. Lightwave Technol. **29**, 3493 (2011)

D.J. Jones, S.A. Diddams, J.K. Ranka, A. Stentz, R.S. Windeler, J. Hall, S.T. Cundiff, Carrier-envelope phase control of femtosecond mode-locked lasers and direct optical frequency synthesis. Science **288**, 635 (2000)

J.M. Kahn, K.-P. Ho, A bottleneck for optical fibres. Nature **411**, 1007 (2001)

M. Kamalian, J.E. Prilepski, S.T. Le, S.K. Turitsyn, Periodic nonlinear Fourier transform for fiber-optic communications, Part I: theory and numerical methods. Opt. Express **24**, 18353 (2016a)

M. Kamalian, J.E. Prilepski, S.T. Le, S.K. Turitsyn, Periodic nonlinear Fourier transform for fiber-optic communications, Part II: eigenvalue communication. Opt. Express **24**, 18370 (2016b)

K.C. Kao, G.A. Hockham, Dielectric-fibre surface waveguides for optical frequencies. Proc. IEE **113**, 1151 (1966)

V.I. Karpman, E.M. Maslov, Perturbation theory for solitons. Sov. Phys. JETP **46(2)**, 281 (1977)

V.I. Karpman, V.V. Solov'ev, A perturbational approach to the two soliton systems. Physica **3D**, 487 (1981)

A. Kastler, Optical methods of atomic orientation and of magnetic resonance. J. Opt. Soc. **47**, 460 (1957)

D.J. Kaup, A perturbation expansion for the Zakharov-Shabat inverse scattering transform. SIAM J. Appl. Math. **31**, 121 (1976)

S.M. Kelly, Characteristic sideband instability of periodically amplified average soliton. Electron. Lett. **28**, 806 (1992)

B. Kibler, J. Fatome, C. Finot, F. Dias, G. Genty, N. Akhmediev, J.M. Dudley, The Peregrine soliton in nonlinear fiber optics. Nat. Phys. **6**, 790 (2010)

B. Kibler, J. Fatome, C. Finot, G. Millot, G. Genty, B. Wetzel, N. Akhmediev, F. Dias, J.M. Dudley, Observation of the Kuznetsov-Ma soliton dynamics in optical fibre. Sci. Rep. **2**, 463 (2012)

Y.S. Kivshar, B.A. Malomed, Dynamics of solitons in nearly integrable systems. Rev. Mod. Phys. **61**, 763 (1989)

Y.S. Kivshar, M. Haelterman, P. Emplit, J.-P. Hamaide, Gordon-Haus effect on dark solitons. Opt. Lett. **19**, 19 (1994)

Y. Kodama, A. Hasegawa, Generation of asymptotically stable optical solitons and suppression of the Gordon-Haus effect. Opt. Lett. **17**, 31 (1992)

Y. Kodama, K. Nozaki, Soliton interaction in optical fibers. Opt. Lett. **12**, 1038 (1987)

S.K. Korotky, Semi-empirical description and pojections of internet traffic trends using a hyperbolic compound annual growth rate. Bell Labs Tech. J. **18**, 5 (2013)

F. Krausz, T. Brabec, C. Spielmann, Self-starting passive mode locking. Opt. Lett. **16**, 235 (1991)

D. Krökel, N.J. Halas, G. Giuliani, D. Grischkowsky, Dark-pulse propagation in optical fibers. Phys. Rev. Lett. **60**, 29 (1988)

M.D. Kruskal, N.J. Zabusky, Interaction of "solitons" in a collisionless plasma and the recurrence of initial states. Phys. Rev. Lett. **15**, 240 (1965)

H. Kubota, M. Nakazawa, Long-distance optical soliton transmission with lumped amplifiers. IEEE J. Quantum Electron. **26**, 692 (1990)

A. Kudlinski, A. Mussot, Visible CW-pumped supercontinuum. Opt. Lett. **33**, 2407 (2008)

J.N. Kutz, S.G. Evangelides, Dispersion-managed breathers with average normal dispersion. Opt. Lett. **23**, 685 (1998)

G.N. Lewis, D. Lipkin, T.T. Magel, Reversible photochemical processes in rigid media. A study of the phosphorescent state. J. Am. Chem. Soc. **63**, 3005 (1941)

G. Li, Recent advances in coherent optical communication. Adv. Opt. Photon. **1**, 279 (2009)

Q. Lin, G.P. Agrawal, Raman response function for silica fibers. Opt. Lett. **31**, 3086 (2006)

A.G. Litvak, V.I. Talanov, A parabolic equation for calculating the fields in dispersive nonlinear media. Izvestiya VUZ. Radiofizika **10**, 539 (1967); Radiophysics quant. Electron. **10**, 296 (1967)

L.Y. Liu, J.M. Huxley, E.P. Ippen, H.A. Haus, Self-starting additive-pulse mode locking of a Nd:YAG laser. Opt. Lett. **15**, 553 (1990)

P.M. Lushnikov, Oscillating tails of a dispersion-managed soliton. J. Opt. Soc. Am. B **21**, 1913 (2004)

R. Mack, The global fiber optics market: running faster than 40,000 km per hour, in *Proceedings of 64th International Wire & Cable Symposium (IWCS) 2015 paper 1.1*; for current figure R. Mack cited after H. Hogan, *For optical fiber, more bandwidth looms*, Photonics Spectra Mar 2017, p. 36

M. Maeda, W.B. Sessa, W.I. Way, A. Yi-Yan, L. Curtis, R. Spicer, R.I. Laming, The effect of four-wave mixing in fibers on optical frequency-division multiplexed systems. J. Lightwave Technol. **8**, 1402 (1990)

C. Mahnke, F. Mitschke, Possibility of an Akhmediev breather decaying into solitons. Phys. Rev. A **82**, 033808 (2012)

C. Mahnke, F. Mitschke, Ultrashort light pulses generated from modulation instability: background removal and soliton content. Appl. Phys. B **116**, 15 (2013)

P.V. Mamyshev, S.V. Chernikov, Ultrashort-pulse propagation in optical fibers. Opt. Lett. **15**, 1076 (1990)

P.V. Mamyshev, L.F. Mollenauer, Wavelength-division-multiplexing channel energy self-equalization in a soliton transmission line by guiding filters. Opt. Lett. **21**, 1658 (1996)

P.V. Mamyshev, L.F. Mollenauer, Soliton collisions in wavelength-division–multiplexed dispersion-managed systems. Opt. Lett. **24**, 448 (1999)

S.V. Manakov, On the theory of two-dimensional stationary self-focusing of electromagnetic waves. Sov. Phys. JETP **38**, 248 (1974)

Marconi Introduces Soliton-Based Ultra Long Haul System at SuperComm 2001: 1.6 Terabit/s Solution Eliminates Need for High-Cost Regenerators, Marconi Corporation, press Release 5 June (2001). Archived at: https://www.thefreelibrary.com/Marconi+Introduces+Soliton-Based+Ultra+Long+Haul+System+At+Supercomm...-a075272418

A. Maruta, Y. Yoshika, Family of multi-hump solitons propagating in dispersion-managed optical fiber transmission system and their existent parameter ranges. Eur. Phys. J. Spec. Top. **173**, 139 (2009)

A. Maruta, T. Inoue, Y. Nonaka, Y. Yoshika, Bisoliton propagating in dispersion-managed system and its application to high-speed and long-haul optical transmission. IEEE J. Sel. Top. Quantum Electron. **8**, 640 (2002)

F. Matera, S. Wabnitz, Nonlinear polarization evolution and instability in a twisted birefringent fiber. Opt. Lett. **11**, 467 (1986)

V.J. Matsas, T.P. Newson, D.J. Richardson, D.J. Payne, Selfstarting passively mode-locked fiber ring soliton laser exploiting nonlinear polarization rotation. Electron. Lett. **28**, 1391 (1992)

J. McEntee, *Solitons go the distance in ultralong-haul DWDM*, FibreSystems Europe, Jan 2003, p. 19

C.J. McKinstrie, Frequency shifts caused by collisions between pulses in dispersion-managed systems. Opt. Commun. **205**, 123 (2002)

R.J. Mears, L. Reekie, I.M. Jauncey, D.N. Payne, Low noise Erbium-doped fibre amplifier operating at 1.54 μm. Electron. Lett. **23**, 1026 (1987)

A. Mecozzi, Soliton transmission control by Butterworth filters. Opt. Lett. **20**, 1859 (1995)

A. Mecozzi, J.D. Moores, H.A. Haus, Y. Lai, Soliton transmission control. Opt. Lett. **16**, 1841 (1991)

C.R. Menyuk, Stability of solitons in birefringent optical fibers I: equal propagation amplitudes. Opt. Lett. **12**, 614 (1987)

C.R. Menyuk, Pulse propagation in an elliptically birefringent Kerr medium. IEEE J. Quantum Electron. **25**, 2674 (1989)

P.P. Mitra, J.B. Stark, Nonlinear limits to the information capacity of optical fibre communications. Nature **411**, 1027 (2001)

F. Mitschke, *Optische Signalübertragung – Kommt das Ende des Wachstums?* Plenary talk (in German). Symposium on Photonics, Spring Meeting of DPG. Verhandlg. DPG (VI) **37**, SYIP II (2002)

F. Mitschke, *Fiber Optics. Physics and Technology*, 2nd edn. (Springer, Berlin, 2016)

F. Mitschke, L.F. Mollenauer, Stabilizing the soliton laser. IEEE J. Quantum Electron. **22**, 2242 (1986a)

F.M. Mitschke, L.F. Mollenauer, Discovery of the soliton self frequency shift. Opt. Lett. **11**, 659 (1986b)

F. Mitschke, L.F. Mollenauer, Ultrashort pulses from the soliton laser. Opt. Lett. **12**, 407 (1987a)

F. Mitschke, L.F. Mollenauer, Experimental observation of interaction forces between solitons in optical fibers. Opt. Lett. **12**, 355 (1987b)

F.M. Mitschke, G. Steinmeyer, M. Ostermeyer, U. Morgner, H. Welling, Additive pulse mode-locked Nd-YAG laser: an experimental account. Appl. Phys. B **56**, 335 (1993)

L.F. Mollenauer, The future of fiber communications: solitons in an all-optical system. Opt. News **12**, 42 (1986)

L.F. Mollenauer, K. Smith, Demonstration of soliton transmission over more than 4000 km in fiber with loss periodically compensated by Raman gain. Opt. Lett. **13**, 675 (1988)

L.F. Mollenauer, R.H. Stolen, The soliton laser. Opt. Lett. **9**, 13 (1984)

L.F. Mollenauer, R.H. Stolen, J.P. Gordon, Experimental observation of picosecond pulse narrowing and solitons in optical fibers. Phys. Rev. Lett. **45,** 1095 (1980)

L.F. Mollenauer, R.H. Stolen, M.N. Islam, Experimental demonstration of soliton propagation in long fibers: loss compensated by Raman gain. Opt. Lett. **10**, 229 (1985)

L.F. Mollenauer, M.J. Neubelt, S.G. Evangelides, J.P. Gordon, J.R. Simpson, L.G. Cohen, Experimental study of soliton transmission over more than 10,000 km in dispersion-shifted fiber. Opt. Lett. **15**, 1203 (1990)

L.F. Mollenauer, S.G. Evangelides, J.P. Gordon, Wavelength division multiplexing with solitons in ultra-long distance transmission using lumped amplifiers. J. Lightwave Technol. **9**, 362 (1991)

L.F. Mollenauer, E. Lichtman, G.T. Harvey, M.J. Neubelt, B.M. Nyman, Demonstration of error-free soliton transmission over more than 15,000 km at 5 Gbits/s, single-channel, and over more than 11,000 km at 10 Gbits/s in a two-channel WDM. Electron. Lett. **28**, 792 (1992a)

L.F. Mollenauer, J.P. Gordon, S.G. Evangelides, The sliding-frequency guiding filter: an improved form of soliton jitter control. Opt. Lett. **17**, 1575 (1992b)

L.F. Mollenauer, A. Grant, X. Liu, X. Wei, Ch. Xie, I. Kang, Experimental test of dense wavelength-division multiplexing using novel, periodic-group-delay-complemented dispersion compensation and dispersion-managed solitons. Opt. Lett. **28**, 2043 (2003)

K. Mori, K. Sato, H. Takara, T. Ohara, Supercontinuum light source generating 50 GHz spaced optical ITU grid seamlessly over S-, C- and L-bands. Electron. Lett. **39**, 544 (2003)

M. Morin, M. Piché, Interferential mode locking: Gaussian pulse analysis. Opt. Lett. **14**, 1119 (1989)

R.-M. Mu, T. Yu, V.S. Grigoryan, C.R. Menyuk, Dynamics of the chirped return-to-zero modulation format. J. Lightwave Technol. **20**, 47 (2002)

S. Mushid, B. Grossman, P. Narakorn, Spatial domain multiplexing: a new dimension in fiber optic multiplexing. Opt. Laser Technol. **40**, 1030 (2008)

A. Mussot, A. Kudlinski, M. Droques, P. Szriftgiser, N. Akhmediev, Appearances and disappearances of Fermi Pasta Ulam recurrence in nonlinear fiber optics, in *Conference on Lasers and Electro-Optics Europe and International Quantum Electronics Conference (CLEO EUROPE/IQEC)* (2013)

M. Nakazawa, E. Yamada, H. Kubota, K. Suzuki, 10 Gbit/s soliton data transmission over one million kilometers. Electron. Lett. **27**, 1270 (1991)

M. Nakazawa, K. Suzuki, E. Yamada, H. Kubota, Y. Kimura, M. Takaya, Experimental demonstration of soliton data transmission over unlimited distances with soliton control in time and frequency domains. Electron. Lett. **29**, 729 (1993)

J.H.B. Nijhof, N.J. Doran, W. Forysiak, F.M. Knox, Stable soliton-like propagation in dispersion managed systems with net anomalous, zero and normal dispersion. Electron. Lett. **33**, 1726 (1997)

O.G. Okhotnikov (ed.), *Fiber Lasers* (Wiley-VCH, Weinheim, 2012)

F. Ouellette, M. Piché, Ultrashort pulse reshaping with a nonlinear FabryPerot cavity matched to a train of short pulses. J. Opt. Soc. Am. B **5**, 1228 (1988)

C. Parè, P.-A. Bèlanger, Antisymmetric soliton in a dispersion-managed system. Opt. Commun. **168**, 103 (1999)

D.H. Peregrine, Water waves, nonlinear Schrödinger equations and their solutions. J. Aust. Math. Soc. B **25**, 16 (1983)

A.R. Pratt, P. Harper, S.B. Alleston, P. Bontemps, B. Charbonnier, W. Forysiak, L. Gleeson, D.S. Govan, G.L. Jones, D. Nesset, J.H.B. Nijhof, I.D. Phillips, M.F.C. Stephens, A.P. Walsh,

T. Widdowson, N.J. Doran, 5745 km DWDM transcontinental field trial using 10 Gbit/s dispersion managed solitons and dynamic gain equalization, paper PD26-1, in *Optical Fiber Communications Conference OFC* (2003)

J.E. Prilepsky, S.A. Derevyanko, K.J. Blow, I. Gabitov, S.K. Turitsyn, Nonlinear inverse synthesis and eigenvalue division multiplexing in optical fiber channels. Phys. Rev. Lett. **113**, 013901 (2014)

D. Qian, E. Ip, M.-F. Huang, M.-J. Li, A. Dogariu, S. Zhang, Y. Shao, Y.-K. Huang, Y. Zhang, X. Cheng, Y. Tian, P. Ji, A. Collier, Y. Geng, J. Linares, C. Montero, V. Moreno, X. Prieto, T. Wang, 1.05 Pb/s transmission with 109 b/s/Hz spectral efficiency using hybrid single- and few-mode cores, FW6C.3, in *Frontiers in Optics/Laser Science Conference (FiO/LS) XXVIII* (2012)

J.K. Ranka, R.S. Windeler, A.J. Stentz, Visible continuum generation in air-silica microstructure optical fibers with anomalous dispersion at 800 nm. Opt. Lett. **25**, 25 (2000)

D.J. Richardson, Filling the light pipe. Science **30**, 329 (2010)

W.H. Renninger, A. Chong, F.W. Wise, Self-similar pulse evolution in an all-normal-dispersion laser. Phys. Rev. A **82**, 021805 (2010)

P. Rohrmann, A. Hause, F. Mitschke, Solitons beyond binary: possibility of fibre-optic transmission of two bits per clock period. Sci. Rep. **2**, 866 (2012)

P. Rohrmann, A. Hause, F. Mitschke, Two-soliton and three-soliton molecules in optical fibers. Phys. Rev. A **87**, 043834 (2013)

S. Roy, S.K. Bhadra, G.P. Agrawal, Perturbation of higher-order solitons by fourth-order dispersion in optical fibers. Opt. Commun. **282**, 3798 (2009)

E. Rubino, J. McLenaghan, S.C. Kehr, F. Belgiorno, D. Townsend, S. Rohr, C.E. Kuklewicz, U. Leonhardt, F. König, D. Faccio, Negative-frequency resonant radiation. Phys. Rev. Lett. **108**, 253901 (2012)

P. Russell, Photonic crystal fibers. Science **299**, 358 (2003)

K. Saitoh, S. Matsuo, Multicore fiber technology. J. Lightwave Technol. **34**, 55 (2016)

J. Satsuma, N. Yajima, Initial value problem of one-dimensional self-modulation of nonlinear waves in dispersive media. Suppl. Progr. Theor. Phys. **55**, 284 (1974)

C.E. Shannon, A mathematical theory of communication. Bell Syst. Tech. J. **27**, 379, 623 (1948)

M. Shkarayev, M.G. Stepanov, New bisoliton solutions in dispersion managed systems. Phys. D **238**, 840 (2009)

V.I. Shrira, V.V. Geogjaev, What makes the Peregrine soliton so special as a prototype of freak waves? J. Eng. Math. **67**, 11 (2010)

D.V. Skryabin, F. Luan, J.C. Knight, P.S.J. Russell, Soliton self-frequency shift cancellation in photonic crystal fibers. Science **301**, 1705 (2003)

K. Smith, L.F. Mollenauer, Experimental onservation of adiabatic compression and expansion of soliton pulses over long fiber paths. Opt. Lett. **14**, 751 (1989)

N.J. Smith, F.M. Knox, N.J. Doran, K.J. Blow, I. Bennion, Enhanced power solitons in optical fibres with periodic dispersion management. Electron. Lett. **32**, 54 (1996)

Soliton wave receives crowd of admirers. Nature **376**, 373 (1995)

D.R. Solli, C. Ropers, P. Koonath, B. Jalali, Optical rogue waves. Nature **450**, 1054 (2007)

J.M. Soto-Crespo, N.N. Akhmediev, V.V. Afanasjev, Stability of the pulselike solutions of the quintic complex Ginzburg-Landau equation. J. Opt. Soc. Am. B **13**, 1439 (1996)

J.M. Soto-Crespo, M. Grapinet, P. Grelu, N. Akhmediev, Bifurcations and multiple-period soliton pulsations in a passively mode-locked fiber laser. Phys. Rev. E **70**, 066612 (2004)

R.H. Stolen, C. Lin, Self-phase-modulation in silica optical fibers. Phys. Rev. A **17**, 1448 (1978)

R.H. Stolen, C. Lee, R.K. Jain, Development of the stimulated Raman spectrum in single-mode silica fibers. J. Opt. Soc. Am. B **1**, 652 (1984)

R.H. Stolen, J.P. Gordon, W.J. Tomlinson, H.A. Haus, Raman response function of silica-core fibers. J. Opt. Soc. Am. B **6**, 1159 (1989)

M. Stratmann, M. Böhm, F. Mitschke, Stable propagation of dark solitons in dispersion maps of either sign of path-average dispersion. Electron. Lett. **37**, 1182 (2001)

M. Stratmann, T. Pagel, F. Mitschke, Experimental observation of temporal soliton molecules. Phys. Rev. Lett. **95**, 143902 (2005)

M. Suzuki, N. Edagawa, H. Taga, H. Tanaka, S. Yamamoto, S. Akiba, Feasibility demonstration of 20 Gbit/s single channel soliton transmission over 11,500 km using alternating amplitude solitons. Electron. Lett. **30**, 1083 (1994)

M. Suzuki, I. Morita, N. Edagawa, S. Yamamoto, H. Taga, A. Akiba, Reduction of Gordon-Haus timing jitter by periodic dispersion compensation in soliton transmission. Electron. Lett. **31**, 2027 (1995)

J. Swiderski, M. Maciejewska, Watt-level, all-fiber supercontinuum source based on telecom-grade fiber components. Appl. Phys. B **109**, 177 (2012)

K. Tai, A. Hasegawa, A. Tomita, Observation of modulational instability in optical fiber. Phys. Rev. Lett. **56**, 135 (1986)

H. Takara, A. Sano, T. Kobayashi, H. Kubota, H. Kawakami, A. Matsuura, Y. Miyamoto, Y. Abe, H. Ono, K. Shikama, Y. Goto, K. Tsujikawa, Y. Sasaki, I Ishida, K. Takenaga, S. Matsuo, K. Saitoh, M. Koshiba, T. Morioka, 1.01-Pb/s (12 SDM/222 WDM/456 Gb/s) Crosstalk-Managed Transmission with 91.4-b/s/Hz Aggregate Spectral Efficiency, in *European Conference on Optical Communication (ECOC)*, Th 3 C. 1 (2012)

K. Tamura, E.P. Ippen, H.A. Haus, L.E. Nelson, 77-fs pulse generation from a stretched-pulse mode-locked all-fiber ring laser. Opt. Lett. **18**, 1080 (1993)

T. Taniuti, N. Yajima, Perturbation method for a nonlinear wave modulation. I. J. Math. Phys. **10**, 1369 (1969)

P. Taylor, The Shape of the Draupner Wave of 1st January, Department of Engineering Science, University of Oxford. http://www.icms.org.uk/archive/meetings/2005/roguewaves/presentations/Taylor.pdf

B.D.H. Tellegen, Interaction between Radio-Waves? Nature **131**, 840 (1933)

The Atlantic Cable website, *History of the Atlantic Cable & Undersea Communications: Cyrus Field* (2011). http://atlantic-cable.com//Field/

The world's first terabit transcontinental optical communications system exploiting dispersion managed solitons, Impact case study REF3b, Research Excellence Framework (2014). www.impact.ref.ac.uk/casestudies2/refservice.svc/GetCaseStudyPDF/37025

R.W. Tkach, Scaling optical communications for the next decade and beyond. Bell Labs Techn. J. **14**, 3 (2010)

W.J. Tomlinson, R.H. Stolen, A.M. Johnson, Optical wave breaking of pulses in nonlinear optical fiber. Opt. Lett. **10**, 457 (1985)

J.C. Travers, A.B. Rulkov, B.A. Cumberland, S.V. Popov, J.R. Taylor, Visible supercontinuum generation in photonic crystal fibers with a 400 W continuous wave fiber laser. Opt. Express **16**, 14435 (2008)

M.L.V. Tse, P. Horak, N.G.R. Broderick, J.H.V. Price, J.R. Hayes, D.J. Richardson, Supercontinuum generation at 1.06 µm in holey fibers with dispersion flattened profiles. Opt. Express **14**, 4445 (2006)

S.K. Turitsyn, Nonlinear communication channels with capacity above the linear Shannon limit. Opt. Lett. **37**, 3600 (2012)

S.K. Turytsin, E.G. Shapiro, Dispersion-managed solitons in optical amplifier transmission systems with zero average dispersion. Opt. Lett. **23**, 682 (1998)

S. Turitsyn, E. Shapiro, S. Medvedev, M.P. Fedoruk, V. Mezentsev, Physics and mathematics of dispersion managed optical solitons. C. R. Phys. **4**, 145 (2003)

S.K. Turitsyn, B.G. Bale, M.P. Fedoruk, Dispersion-managed solitons in fibre systems and lasers. Phys. Rep. **521**, 135 (2012)

I.M. Uzunov, V.D. Stoev, T.I. Tzoleva, Influence of the initial phase difference between pulses on the N-soliton interaction in trains of unequal solitons in optical fibers. Opt. Commun. **97**, 307 (1993)

P.K.A. Wai, C.R. Menyuk, Y.C. Lee, H.H. Chen, Nonlinear pulse propagation in the neighborhood of the zero-dispersion wavelength of monomode optical fibers. Opt. Lett. **11**, 464 (1986)

P.K. Wai, C.R. Menyuk, H.H. Chen, Stability of solitons in randomly varying birefringent fibers. Opt. Lett. **16**, 1231 (1991)

J. Wang, Analysis of passive additive-pulse mode locking with eigenmode theory. IEEE J. Quantum Electron. **28**, 562 (1992)

H.G. Winful, Polarization instabilities in birefringent nonlinear media: application to fiber-optic devices. Opt. Lett. **11**, 33 (1985)

C. Xie, L. Möller, H. Haunstein, S. Hunsche, Comparison of system tolerance to polarization-mode dispersion between different modulation formats. IEEE Photon. Technol. Lett. **15**, 1168 (2003)

M.I. Yousefi, F.R. Kschischang, Information transmission using the nonlinear Fourier transform, Part I: mathematical tools. IEEE Trans. Inf. Theory **60**, 4312 (2014a)

M.I. Yousefi, F.R. Kschischang, Information transmission using the nonlinear Fourier transform, Part II: numerical methods. IEEE Trans. Inf. Theory **60**, 4329 (2014b)

V.E. Zakharov, L.A. Ostrovsky, Modulation instability: the beginning. Phys. D **238**, 540 (2009)

V.E. Zakharov, A.B. Shabat, Exact theory of two-dimensional self-focusing and one-dimensional self-modulation of waves in nonlinear media. Sov. Phys. JETP **34**, 62 (1972)

V.E. Zakharov, A.B. Shabat, Interaction between solitons in a stable medium. Zh. Eksp. Teor. Fiz **64**, 1627 (1973); Sov. Phys.-JETP **37**, 823 (1973)

X. Zhao, X. Liu, S. Wang, W. Wang, Y. Han, Z. Liu, S. Li, L. Hou, Numerical calculation of phase-matching properties in photonic crystal fibers with three and four zero-dispersion wavelengths. Opt. Express **23**, 27899 (2015)

X. Zhu, P.N. Kean, W. Sibbett, Coupled-cavity mode locking of a KCl:Tl laser using an Erbium-doped optical fiber. Opt. Lett. **14**, 1192 (1989)

Perturbations of Solitons in Optical Fibers

Theodoros P. Horikis and Dimitrios J. Frantzeskakis

Contents

Introduction	270
Physical Model and Nonlinear Schrödinger (NLS) Equation	271
Bright and Dark Solitons and the Effect of Perturbations	274
Bright Solitons	274
Dark Solitons	275
Solitons Under Perturbations	277
Bright Solitons Under Perturbations	278
Perturbation Theory for Dark Solitons	280
The Background	280
The Soliton and the Shelf	282
Adiabatic Dynamics	284
Dark Solitons Under Perturbations	288
Beyond the Adiabatic Theory: Soliton Radiation	292
Summary and Conclusions	296
References	297

Abstract

Bright and dark solitons, namely, decaying localized pulses and dips off of a continuous-wave background, were predicted to occur in optical fibers more than 40 years ago. Since then, they have been extensively studied in theory and in experiments as they were proposed for applications in optical fiber communications. In the ideal case, optical fiber solitons are described by the completely

---

T. P. Horikis (✉)
Department of Mathematics, University of Ioannina, Ioannina, Greece
e-mail: horikis@uoi.gr

D. J. Frantzeskakis
Department of Physics, National and Kapodistrian University of Athens, Athens, Greece
e-mail: dfrantz@phys.uoa.gr

integrable nonlinear Schrödinger (NLS) equation. In practice, however, solitons in real optical fibers evolve under the presence of various perturbations, such as linear loss, third-order dispersion, stimulated Raman scattering, and so on. Therefore, the study of solitons under perturbations is a particularly relevant and interesting problem. Here, we review soliton propagation (for both bright and dark solitons) in optical fibers. We provide a general framework, relying on the adiabatic perturbation theory for solitons, and then apply our methodology to a rather general higher-order NLS model. Special emphasis is given to the case of dark solitons, because – especially in the presence of dissipation – perturbation theory should take into account an additional generic feature of the propagation: a "shelf." This is a linear wave that develops around the dark soliton as a result of the perturbations and accompanies the soliton during its propagation. Effects beyond the adiabatic perturbation theory, namely, the perturbation-induced emission of radiation and its influence on soliton interactions, are also discussed.

Introduction

Solitons, namely, localized in time envelopes of light waves in fibers, which propagate undistorted for long distances, were predicted to occur in the seminal works (Hasegawa and Tappert 1973a, b). In particular, it was shown that in the anomalous dispersion regime, bright solitons – namely, localized, decaying sech-shaped pulses – can be formed (Hasegawa and Tappert 1973a). On the other hand, in the normal dispersion regime, bright solitons do not exist; nevertheless, a continuous wave (cw) – i.e., a constant amplitude wave – is modulationally stable, and thus, solitons may exist only as "holes." Indeed, dark solitons – namely, intensity dips off of the cw background with a phase jump at the density minimum – are supported in optical fibers in this case (Hasegawa and Tappert 1973b). Both bright and dark solitons were observed in experiments (Mollenauer et al. 1980; Emplit et al. 1987; Krökel et al. 1988) and, since then, have been studied extensively (Hasegawa and Kodama 1995; Agrawal 2007) and have been proposed for applications in the field of optical fiber communications (Hasegawa and Kodama 1995; Agrawal 2001, 2007; Haus et al. 1996; Mollenauer et al. 1997).

As shown in Hasegawa and Tappert (1973a, b), solitons in optical fibers can be described by two different versions of the nonlinear Schrödinger (NLS) equation: bright (dark) solitons are governed by the so-called focusing (defocusing) NLS model, supplemented with vanishing (nonvanishing) boundary conditions at infinity. As was shown in Zakharov and Shabat (1972, 1973), both versions of the NLS equation are completely integrable by means of the inverse scattering transform (IST) (Ablowitz and Segur 1981). However, in real-life optical fibers used for transmission, important effects are present which are not modelled in the original NLS system. For instance, fiber loss rate per dispersion distance, although small enough not to allow soliton formation, is always present. Including this effect in the model leads to a dissipative version of the NLS, which incorporates a linear gain term. In practice, the loss is compensated for by optical amplifiers, and a pertinent gain term should also be added in the NLS model (see, e.g., Hasegawa and Kodama 1995; Agrawal 2007). Furthermore, in the case of pulse widths in the sub-picosecond regime, which are important for ultrahigh bit rates (Agrawal 2001), the NLS should also be modified to incorporate higher-order terms describing relevant effects. The latter include both linear and nonlinear ones, such as third-order dispersion, or self-steepening and stimulated Raman scattering, respectively. The relevant higher-order NLS model was rigorously derived, via the reductive perturbation method (Taniuti 1974), in the works (Kodama 1985; Kodama and Hasegawa 1987; Potasek 1989).

According to the above discussion, it becomes obvious that – in practice – solitons always evolve in optical fibers in the presence of a variety of perturbations. Thus, the theoretical analysis of this problem, namely, the study of the dynamics of solitons under perturbations, is quite relevant and interesting – also from the viewpoint of applications to optical communications and fiber lasers. Below, we present perturbation methods that can be used for the analytical study of this problem, both for bright and dark solitons. We focus on the adiabatic perturbation theory for solitons, based on the use of the underlying conserved quantities of the problem; we also touch upon effects beyond the adiabatic approximation, namely, the emission of radiation caused by the perturbations. Our considerations and results are rather general; nevertheless, for the sake of clarity, certain applications of our analytical approximations to a higher-order NLS model with gain and loss are also provided. Results of direct numerical simulations are also presented, corroborating our analytical findings.

The chapter is structured as follows. First, we present the NLS model and introduce its bright and dark soliton solutions (of the unperturbed problem) and their properties. We then study the problem of dynamics of bright and dark solitons under perturbations in more detail. First we study the case of bright solitons and then the case of dark solitons. Emphasis is given to the latter case, which is technically more involved due to the dissipation-induced emergence of a "shelf," namely, a linear wave which develops around the dark soliton. We also briefly discuss perturbation-induced emission of radiation and its effect on soliton interactions. We finally present a summary of our results and conclusions.

Physical Model and Nonlinear Schrödinger (NLS) Equation

The existence of solitons in optical fibers was first predicted in the pioneering works Hasegawa and Tappert (1973a, b), upon deriving the nonlinear evolution equation – which is the NLS equation – for the complex electric field envelope (i.e., the slowly varying Fourier amplitude). Actually, the NLS model is derived upon retaining the lowest-order (second-order) dispersion, i.e., the so-called group velocity dispersion (GVD), and nonlinearity, which – for a glass fiber – is cubic and originates from the Kerr effect.

The easiest way to derive the NLS model is to Taylor expand the wave number $k(\omega, |\mathbf{E}|^2)$ around the carrier frequency ω_0, the carrier wave number k_0, and electric field intensity $|\mathbf{E}|^2$, namely:

$$k - k_0 \approx k'(\omega_0)(\omega - \omega_0) + \frac{1}{2}k''(\omega_0)(\omega - \omega_0)^2 + \frac{\partial k}{\partial |\mathbf{E}|^2}|\mathbf{E}|^2, \tag{1}$$

where $k' = 1/v_g$ is the group velocity and $k'' = \partial^2/\partial\omega^2$ is the GVD. Then, replacing $k - k_0$ and $\omega - \omega_0$ by $i\partial_z$ and $-i\partial_t$, respectively, and acting the resulting operator on the electric field envelope, one obtains the equation:

$$i(\mathbf{E}_z + k'\mathbf{E}_t) - \frac{1}{2}k''\mathbf{E}_{tt} + \frac{\partial k}{\partial |\mathbf{E}|^2}|\mathbf{E}|^2\mathbf{E} = 0, \tag{2}$$

where subscripts denote partial derivatives. To connect the above NLS model with the nonlinear fiber optics context, it is noted that the refractive index $n(k, \omega, |\mathbf{E}|^2)$ of a glass fiber exhibiting the Kerr nonlinearity is given by:

$$n \equiv \frac{kc}{\omega} = n_0(\omega) + n_2|\mathbf{E}|^2, \tag{3}$$

where $n_0(\omega)$ is the linear frequency-dependent part of the fiber's refractive index, $n_2 \approx 1.3 \times 10^{-22}$ (m/V)^2 is the Kerr coefficient, and c is the velocity of light in vacuum. Then, the coefficients appearing in Eq. (2) can be approximated as:

$$k' \approx \frac{n_0(\omega_0)}{c}, \quad k'' \approx \frac{2}{c}\frac{\partial n_0}{\partial \omega_0}, \quad \frac{\partial k}{\partial |\mathbf{E}|^2} \approx \frac{\omega_0 n_2}{c}. \tag{4}$$

Taking into regard the above approximations and using a retarded reference frame so that $t \to t - k'z$, Eq. (2) becomes:

$$i\mathbf{E}_z - \frac{1}{2}k''\mathbf{E}_{tt} + g|\mathbf{E}|^2\mathbf{E} = 0, \tag{5}$$

where $g = \omega_0 n_2/c$.

The above equation can further be elaborated upon decomposing the electric field as $\mathbf{E} \propto \mathbf{e} F(\mathbf{r}_\perp) u(z, t)$, where $F(\mathbf{r}_\perp)$ accounts for the modal distribution of the fiber, $u(z, t)$ is the time-dependent longitudinal electric field envelope, and $\mathbf{e}$ is a polarization vector (Hasegawa and Kodama 1995; Agrawal 2007; Kivshar and Agrawal 2003). In this case, it can be shown that the nonlinearity coefficient is redefined as $g \to g/A_{\text{eff}}$, where $A_{\text{eff}} = \left(\int_{\mathbb{R}^2} |F(\mathbf{r}_\perp)|^2 d\mathbf{r}_\perp\right)^2 / \int_{\mathbb{R}^2} |F(\mathbf{r}_\perp)|^4 d\mathbf{r}_\perp$ is the effective core area. Then, measuring length, time, and the field intensity $|u|^2$ in units of the dispersion length $L_D = t_0^2/|k''|$, initial pulse width t_0, and L_D/g, respectively, one obtains the following normalized NLS equation:

$$iu_z - \frac{s}{2}u_{tt} + |u|^2 u = 0, \tag{6}$$

where $s \equiv \mathrm{sgn}(k'')$, and $s = -1$ ($s = +1$) corresponds to the anomalous (normal) dispersion regime with $k'' < 0$ ($k'' > 0$).

In addition, to take into regard the effect of (linear) fiber loss or gain – which is assumed to be on the order of GVD and Kerr nonlinearity – one may also introduce an imaginary part to the linear part of the refractive index. The relevant parameter, say α, characterizes the absorption or gain coefficient, which sets the coefficient of a linear loss/gain term which should be incorporated in the NLS, namely:

$$\mathrm{i} u_z - \frac{s}{2} u_{tt} + |u|^2 u = \mathrm{i}\gamma u, \tag{7}$$

where $\gamma = \alpha L_D$, and $\gamma < 0$ ($\gamma > 0$) corresponds to linear loss (gain).

On the other hand, while the NLS Eq. (6) [or (7)] can successfully describe propagation of picosecond pulses in optical fibers, it fails when the pulse width is further decreased. Indeed, in the case of ultrashort pulses (of widths in the sub-picosecond regime), the relevant spectral width becomes comparable to the carrier frequency ω_0 and, thus, the NLS has to be complemented with higher-order terms. This way, employing the reductive perturbation method (RPM) (Taniuti 1974), a higher-order NLS model was derived (Kodama 1985; Kodama and Hasegawa 1987; Potasek 1989), which incorporates third-order linear dispersion and nonlinear dispersion terms, namely:

$$\mathrm{i} u_z - \frac{s}{2} u_{tt} + |u|^2 u = \mathrm{i}\gamma u + \mathrm{i}\delta |u|^2 u + \mathrm{i}\beta u_{ttt} + \mathrm{i}\mu (|u|^2 u)_t + (\mathrm{i}\nu + \sigma_R) u (|u|^2)_t. \tag{8}$$

The left-hand side of Eq. (8) corresponds to the usual NLS Eq. (6); the right-hand side, apart from accounting for the presence of linear gain/loss (of strength γ) discussed above, also includes a nonlinear gain/loss term (of strength δ). Note that in the case of $\delta < 0$, the relevant term describes two-photon absorption process, which, in fact, appears as a by-product of enhanced nonlinearity (Silberberg 1990). On the other hand, in the case of $\delta > 0$, the relevant term describes nonlinear gain, which is included to counterbalance the effects on the dark soliton's background (see below) from the linear gain/loss term (Ikeda et al. 1995).

Furthermore, Eq. (8) incorporates third-order dispersion and self-steepening effects (characterized by the coefficients β and μ, respectively), as well as other conservative and dissipative higher-order effects (of strength ν and σ_R, respectively), with the latter corresponding to the stimulated Raman scattering (SRS) effect. Physically speaking, the origin of the SRS effect is the non-instantaneous, delayed response of the fiber nonlinearity and is known to be one of the dominant effects for very short optical pulses (Stolen et al. 1989).

Variants of Eq. (8), stemming from the omission of various terms – with the idea of investigating the role of a specific higher-order effect on soliton dynamics – have been studied extensively in the literature (see detailed discussion and references in the books Hasegawa and Kodama 1995; Agrawal 2007; Kivshar and Agrawal 2003). Nevertheless, there are only few works (see, e.g., Horikis and Frantzeskakis 2013, for dark solitons) that have been devoted to the study of the combined role of *all* effects, i.e., of all terms appearing in the right-hand side of Eq. (8).

Motivated by the above discussion, we will first discuss the soliton solutions of the unperturbed NLS Eq. (6) and then consider the problem of the dynamics of bright and dark solitons under perturbations. We will focus on the specific perturbation appearing in the right-hand side of Eq. (8) and analyze the soliton dynamics in this context.

Bright and Dark Solitons and the Effect of Perturbations

The NLS, Eq. (6), is a well-known completely integrable system for both types of dispersion, i.e., for the anomalous dispersion ($s = -1$) (Zakharov and Shabat 1972) and for the normal one ($s = +1$) (Zakharov and Shabat 1973). Thus, being an infinite-dimensional Hamiltonian dynamical system, the NLS possesses an infinite number of conserved quantities (integrals of motion), with the lowest-order ones being the energy:

$$E = \int_{-\infty}^{-\infty} |u|^2 dt, \tag{9}$$

the momentum (alias "mean frequency"):

$$P = \frac{i}{2} \int_{-\infty}^{-\infty} (u\bar{u}_t - \bar{u}u_t) \, dt, \tag{10}$$

where overbar denotes complex conjugate, and the Hamiltonian:

$$H = \frac{1}{2} \int_{-\infty}^{-\infty} \left(s|u_t|^2 + |u|^4 \right) dt. \tag{11}$$

Furthermore, it is known (Zakharov and Shabat 1972, 1973) that the type of soliton solutions of the NLS depends on the parameter s. The two types of solitons, bright and dark, for the anomalous and normal dispersion regimes, respectively, are appended below.

Bright Solitons

In the anomalous dispersion regime ($s = -1$), the NLS equation supplemented with vanishing conditions at infinity, i.e., $|u| \to 0$ as $t \to \infty$, possesses a *bright* soliton solution of the following form (Zakharov and Shabat 1972):

$$u_{bs}(z,t) = \eta \operatorname{sech}[\eta(t + \kappa z)] \exp\left[-i\kappa t + \frac{i}{2}(\eta^2 - \kappa^2)z\right], \tag{12}$$

where η is the amplitude and inverse temporal width of the soliton, while κ is the frequency shift and velocity (in the (t, z)-plane).

Introducing the solution (12) into the integrals of motion E, P, and H, it is readily found that:

$$E = 2\eta, \quad P = 2\eta\kappa, \quad H = \eta\kappa^2 - \frac{1}{3}\eta^3. \tag{13}$$

These equations imply that the bright soliton behaves as a classical particle with effective mass M_{bs}, momentum P_{bs}, and Hamiltonian (total energy) H_{bs}, respectively, given by $M_{bs} = 2\eta$, $P_{bs} = M\kappa$, and $H_{bs} = \frac{1}{2}M\kappa^2 - \frac{1}{24}M^3$ (recall that the soliton velocity is k). Notice that in the equation for the energy, the first and second terms in the right-hand side are, respectively, the kinetic energy and the binding energy of the quasiparticles associated with the soliton (Kaup 1976). Differentiating the soliton energy and momentum over the soliton velocity, the following relation is found: $\partial H_{bs}/\partial P_{bs} = \kappa$, which underscores the particle-like nature of the bright soliton.

Dark Solitons

In the normal dispersion regime ($s = +1$), the NLS equation supplemented with nonvanishing conditions at infinity, i.e., $|u| \rightarrow u_\infty \neq 0$ as $t \rightarrow \infty$ (where u_∞ is an arbitrary real constant), admits a dark soliton solution. The latter, can be considered as an excitation of the cw solution $u = u_\infty \exp(i u_\infty^2 z)$. The dark soliton may be expressed as (Zakharov and Shabat 1973):

$$u_{ds}(z, t) = u_\infty (\cos\varphi \tanh\xi + i \sin\varphi) \exp(i u_\infty^2 z), \tag{14}$$

where $\xi \equiv u_\infty \cos\varphi (t - \upsilon z)$ (with t_0 being the time central position of the soliton), while the remaining parameters υ and φ are connected through the relation $\upsilon = u_\infty \sin\varphi$. Here, φ is the so-called soliton phase angle or, simply, the phase shift of the dark soliton ($|\varphi| < \pi/2$), which describes the *darkness* of the soliton through the relation, $|u|^2 = 1 - \cos^2\varphi \operatorname{sech}^2\xi$; this way, the limiting cases $\varphi = 0$ and $\varphi \neq 0$ correspond to the so-called *black* and *gray* solitons, respectively. The amplitude and velocity of the dark soliton are given by $\cos\varphi$ and $\sin\varphi$, respectively; thus, the black soliton, $u = u_\infty \tanh(u_\infty t) \exp(i u_\infty^2 t)$, is a stationary dark soliton ($\varphi = 0$), while the gray soliton is a moving one.

In the limiting case of $\cos\varphi \ll 1$, corresponding to a small-amplitude gray soliton, the soliton velocity is close to the so-called speed of sound C (i.e., $\upsilon \sim C = u_\infty$ in our units). The speed of sound is, therefore, the maximum possible velocity of a dark soliton which, generally, always travels with a velocity less than the speed of sound. In the same limit, the small-amplitude dark soliton can be well approximated by the soliton solution of an effective Korteweg-de Vries (KdV) equation (Tsuzuki 1971). The formal reduction of the NLS to the

KdV equation (Zakharov and Kuznetsov 1986) has been particularly useful in studies of dark solitons under perturbations – cf., e.g., Kivshar (1990), Kivshar and Afanasjev (1991), and Frantzeskakis (1996), as well as the review Kivshar and Luther-Davies (1998).

As the integrals of motion of the NLS equation refer to *both* the background and the dark soliton, a new definition of the integrals is necessary so as to add the contribution of the background (Uzunov and Gerdjikov 1993; Barashenkov and Panova 1993; Kivshar and Yang 1994) and avoid any diverging behavior. As such, the integrals of motion for the dark soliton read:

$$E_{ds} = \int_{\infty}^{+\infty} \left(u_\infty^2 - |u|^2\right) dt, \tag{15}$$

$$P_{ds} = \frac{i}{2} \int_{-\infty}^{-\infty} (u\bar{u}_t - \bar{u}u_t) \, dt - 2u_\infty \varphi, \tag{16}$$

$$H_{ds} = \frac{1}{2} \int_{-\infty}^{+\infty} \left[|u_t|^2 + (u_\infty^2 - |u|^2)^2\right] dt. \tag{17}$$

Introducing, as before, the solution (14) into the above integrals of motion, we find:

$$E_{ds} = 2u_\infty \cos \varphi, \tag{18}$$

$$P_{ds} = -2v(C^2 - v^2)^{1/2} + 2C^2 \tan^{-1}\left[\frac{(C^2 - v^2)^{1/2}}{v}\right], \tag{19}$$

$$H_{ds} = \frac{4}{3}(C^2 - v^2)^{3/2}. \tag{20}$$

Differentiating H_{ds} and P_{ds} over the soliton velocity v yields $\partial H_{ds}/\partial P_{ds} = v$, which shows that, similarly to the bright soliton, the dark soliton effectively behaves like a classical particle. We note in passing that the particle properties of bright and dark solitons have been extensively analyzed – and exploited in various studies – in the physics of atomic Bose-Einstein condensates (BECs) (Carretero-González et al. 2008; Frantzeskakis 2010).

For our considerations below, it is also useful to introduce still another conserved quantity, namely, the mean time central position of the soliton(s) – alias "soliton center" – which is given by:

$$R_{bs} = \int_{-\infty}^{+\infty} t|u|^2 dt, \quad R_{ds} = \int_{-\infty}^{+\infty} t\left(u_\infty^2 - |u|^2\right) dt, \tag{21}$$

for the bright and dark solitons, respectively. At this point, it is relevant to note the following. The bright soliton solution (12) has two independent parameters: the amplitude η and velocity κ. The same holds for the dark soliton solution (14) as well, which has one free parameter for the background, u_∞, and one for the soliton, φ. In fact, in both types of solutions, there is also a freedom in selecting the initial time central position of the solitons [cf. Eq. (21)], and phase, which were set equal to zero in Eqs. (12) and (14). Here, it should be recalled that, being completely integrable, the NLS Eq. (6) has infinitely many symmetries, including translational, gauge, and Galilean invariances.

Solitons Under Perturbations

From the discussion in the previous section, it becomes clear that – generally – bright and dark solitons evolve in optical fibers in the presence of perturbations. In the case of Eq. (8), the latter can be distinguished, depending on weather conserves or not the Hamiltonian, as follows: (a) Hamiltonian perturbations, i.e., the third-order dispersion $\propto \beta$ and the higher-order nonlinear terms $\propto \mu, \nu$, and (b) dissipative perturbations, i.e., the linear/nonlinear loss/gain terms $\propto \gamma$ and $\propto \delta$, respectively, as well as the SRS term $\propto \sigma_R$.

When the perturbations are sufficiently small, it is possible to apply a perturbation technique in order to analyze the problem. Several such techniques have been developed in the past, both for bright (Karpman and Maslov 1977; Kaup and Newell 1978a; Karpman 1979; Kivshar and Malomed 1989) and dark (Uzunov and Gerdjikov 1993; Kivshar and Yang 1994; Konotop and Vekslerchik 1994; Kivshar and Krolikowski 1995; Chen et al. 1999; Lashkin 2004; Ablowitz et al. 2011a) solitons; pertinent perturbation approaches include the perturbed inverse scattering method, the variational approach (alias Lagrangian method), the Lie transform method, and others (see books Hasegawa and Kodama 1995; Kivshar and Agrawal 2003 and references therein). Among these techniques, a particularly convenient method to study the soliton dynamics is the adiabatic perturbation method. According to this approach, an adiabatic relation is the balance between nonlinearity and dispersion, so that (amplitude) × (width) = const. In other words, it is assumed that – in the presence of the perturbations – the functional form of the soliton remains unchanged, but the soliton parameters change (slowly) as the soliton travels along the optical fiber. Thus, the soliton parameters are treated as unknown functions of z, and their evolution is determined by the evolution of the conserved quantities of the NLS.

Below, our aim is to study the following perturbed NLS equation:

$$iu_z - \frac{s}{2}u_{tt} + |u|^2 u = \varepsilon F[u], \tag{22}$$

where $0 < \varepsilon \ll 1$ is a formal small parameter and $F[u]$ is a general functional perturbation. We will first discuss general aspects of this problem for both bright and dark solitons and then focus on the particular perturbation suggested in Eq. (8), namely, for $F[u]$ given by:

$$F[u] = i\gamma u + i\delta |u|^2 u + i\beta u_{ttt} + i\mu(|u|^2 u)_t + (i\nu + \sigma_R)u(|u|^2)_t. \tag{23}$$

Bright Solitons Under Perturbations

We start with the case of the anomalous dispersion regime ($s = -1$), i.e., the case of bright solitons. First, using Eq. (22) and its complex conjugate, one may arrive at the following equations for the conserved quantities:

$$\frac{dE}{dz} = \varepsilon \int_{-\infty}^{+\infty} \left(\bar{u}F + u\bar{F} \right) dt, \tag{24}$$

$$\frac{dP}{dz} = \varepsilon i \int_{-\infty}^{+\infty} \left(\bar{u}_t F - u_t \bar{F} \right) dt, \tag{25}$$

$$\frac{dH}{dz} = 2\varepsilon \int_{-\infty}^{+\infty} \left[\left(\frac{1}{2}\bar{u}_{tt} + |u|^2 \bar{u} \right) F + \left(\frac{1}{2}u_{tt} + |u|^2 u \right) \bar{F} \right] dt. \tag{26}$$

As discussed above, for sufficiently small perturbation, the form of the soliton solution $u_{\mathrm{bs}}(z,t)$ may be assumed to have the following rather general form, where all its parameters are allowed to vary in z as:

$$u_{\mathrm{bs}}(z,t) = \eta(z)\,\mathrm{sech}[\eta(z)(t - t_0(z))] \exp\left[-i\kappa(z)t + i\phi(z)\right], \tag{27}$$

where the soliton's amplitude (and inverse width) η, the time central position t_0, the frequency κ, and phase ϕ are unknown functions of z that have to be determined. Notice that, in the absence of the perturbation, t_0 and ϕ are constants which are obtained through the following equations:

$$\frac{dt_0}{dz} = -\kappa, \quad \frac{d\phi}{dz} = \frac{1}{2}\left(\eta^2 - \kappa^2\right). \tag{28}$$

Substituting the soliton (27) into Eqs. (25), (26) [and (21)], we obtain a set of four ordinary differential equations (ODEs) for the four unknown soliton parameters:

$$\frac{d\eta}{dz} = -\mathrm{Im} \int_{-\infty}^{\infty} F[u]\bar{u}\,dt, \tag{29}$$

$$\frac{d\kappa}{dz} = \mathrm{Re} \int_{-\infty}^{\infty} F[u]\tanh[\eta(t - t_0)]\bar{u}\,dt, \tag{30}$$

$$\frac{dt_0}{dz} = -\kappa - \frac{1}{\eta^2}\mathrm{Im} \int_{-\infty}^{\infty} F[u](t - t_0)\bar{u}\,dt, \tag{31}$$

$$\frac{d\phi}{dz} = \frac{1}{2}(\eta^2 - \kappa^2) + t_0\frac{d\kappa}{dz} - \mathrm{Re} \int_{-\infty}^{\infty} F[u]\left\{ \frac{1}{\eta} - (t - t_0)\tanh[\eta(t - t_0)] \right\}\bar{u}\,dt, \tag{32}$$

where Re and Im stand for the real and imaginary parts, respectively.

Before proceeding with the analysis of the full problem, where the perturbation $F[u]$ is given by Eq. (23), it is useful to study at first a simple example. Thus, we consider the NLS Eq. (7), with the linear loss/gain term assumed to be a small perturbation, i.e., $F[u] = i\gamma u$, with $\gamma \ll 1$. Then, substituting this form of $F[u]$ into Eqs. (29), (30), (31), (32) and performing the integrations, we find that the soliton frequency κ and time central position t_0 remain unaffected of the perturbation, while the soliton amplitude η and phase ϕ evolve along the fiber length due to the presence of the fiber loss/gain as follows:

$$\eta(z) = \exp(2\gamma z), \quad \phi(z) = \phi(0) - \frac{1}{8\gamma}[1 - \exp(4\gamma z)], \qquad (33)$$

where it was assumed that $\eta(0) = 1$ and $\kappa(0) = t_0(0) = 0$ (and thus $\kappa(z) = t_0(z) = 0$ for all z). Thus, in the presence of loss, $\gamma < 0$ (gain, $\gamma > 0$), the soliton amplitude decreases (increases), while its width increases (decreases), i.e., the soliton broadens (is compressed) along the fiber.

Let us now return to the full problem and study the effect of the perturbation (23) on the dynamics of bright solitons. Following the same procedure, i.e., substituting Eq. (23) into Eqs. (29), (30), (31), (32) and performing the integrations, we find that the soliton parameters evolve according to the following system:

$$\frac{d\eta}{dz} = \frac{2}{3}\eta(3\gamma + 2\delta\eta^2), \qquad (34)$$

$$\frac{d\kappa}{dz} = -\frac{8}{15}\sigma_R \eta^4, \qquad (35)$$

$$\frac{dt_0}{dz} = -\kappa + \frac{1}{3}(3\beta - 3\mu - 2\nu)\eta + 3\beta\kappa^2\eta^2, \qquad (36)$$

$$\frac{d\phi}{dz} = -\kappa\left[(\mu - 3\beta)\eta^2 + \beta(\eta^2 - 2)q^2\right] + \frac{1}{2}(\eta^2 - \kappa^2) - \frac{8}{15}\sigma_R t_0 \eta^4. \qquad (37)$$

Although our findings in this case are more complicated, it is still possible to arrive at simple analytical results. Indeed, first we observe that Eq. (34) can be solved analytically to provide the functional form of $\eta(z)$, which is found to be:

$$\eta^2(z) = \frac{3K\gamma e^{4\gamma z}}{1 - 2K\delta e^{4\gamma z}}, \quad K = \frac{\eta^2(0)}{3\gamma + 2\delta\eta^2(0)}. \qquad (38)$$

Then, the frequency $\kappa(z)$ can be obtained from Eq. (35) by simply integrating the above expression for η. Finally, having found $\eta(z)$ and $\kappa(z)$, integration of Eqs. (36) and (37) yield, respectively, the functional forms of $t_0(z)$ and $\phi(z)$.

In Fig. 1 we show a typical evolution for a bright soliton under Eq. (22) with $\gamma = -\delta = \beta = \mu = \nu = \sigma_R = 0.01$. In this figure, the top panel depicts the complete evolution of the soliton and its center (also drawn in a white line the result

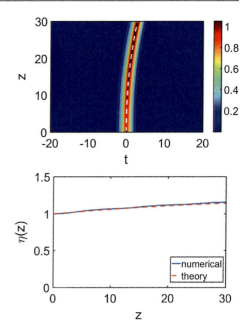

Fig. 1 Top panel: The contour plot of the evolution of a unit amplitude bright soliton (i.e., $\eta(0) = 1$) under Eq. (22) with $\gamma = -\delta = \beta = \mu = \nu = \sigma_R = 0.01$. The white dashed line is the soliton's center t_0 as predicted by the perturbation theory. Bottom panel: The evolution of the amplitude corresponding to the top panel and the prediction of the theory

of the perturbation theory), while in the bottom panel, we show a direct comparison between the numerically found amplitude and the theoretically predicted from the perturbation theory.

Perturbation Theory for Dark Solitons

Next, we consider the case of the normal dispersion regime ($s = +1$), i.e., the case of dark solitons. As we will see, the fact that the dark soliton is a more complicated object than the bright one – since it is composed by both the background and the pulse on top of it – makes the problem more difficult. For this reason, we first present a general perturbation theory for dark solitons and, in the next section, we will apply the derived results to Eq. (8).

The Background

To begin our considerations, we first rewrite Eq. (22) in a slightly different form, using capital letter U for the unknown electric field envelope (the reason will become obvious shortly), namely:

$$iU_z - \frac{1}{2}U_{tt} + |U|^2 U = \varepsilon F[U]. \tag{39}$$

Further, as is relevant in this case, we will assume nonvanishing boundary conditions at infinity, i.e., $|U| \not\to 0$ as $t \to \pm\infty$. The effect the perturbation has on the behavior of the solution at infinity is independent of any local phenomena, such as localized pulses which do not decay at infinity, i.e., dark solitons. In the case of a cw background, which is relevant to perturbation problems with dark solitons, we have $U_{tt} \to 0$ as $t \to \pm\infty$, and the evolution of the background at either end, $U \to U^{\pm}(z)$, is given by the equation:

$$i\frac{d}{dz}U^{\pm} + |U^{\pm}|^2 U^{\pm} = \varepsilon F[U^{\pm}]. \qquad (40)$$

We write $U^{\pm}(z) = u^{\pm}(z)e^{i\phi^{\pm}(z)}$, where $u^{\pm}(z) > 0$ and $\phi^{\pm}(z)$ are both real functions of z. Then, the imaginary and real parts of Eq. (40), respectively, read:

$$\frac{d}{dz}u^{\pm} = \varepsilon \operatorname{Im}\left[F[u^{\pm}e^{i\phi^{\pm}}]e^{-i\phi^{\pm}}\right], \qquad (41a)$$

$$\frac{d}{dz}\phi^{\pm} = (u^{\pm})^2 - \varepsilon \operatorname{Re}\left[F[u^{\pm}e^{i\phi^{\pm}}]e^{-i\phi^{\pm}}\right]/u^{\pm}. \qquad (41b)$$

The above equations completely describe the adiabatic evolution of the background under the influence of the perturbation $F[u]$. Although this is true for all choices of perturbation, we will further restrict ourselves to perturbations which maintain the phase symmetry of Eq. (39); i.e., $F[U(z,t)e^{i\theta}] = F[U(z,t)]e^{i\theta}$. As we show next, this is a sufficient condition to keep the magnitude of the background equal on either side and a property of most commonly considered perturbations, including the one in Eq. (23).

We now assume that at $z = 0$, $u^+(0) = u^-(0)$; then, since $u^{\pm}(z)$ satisfy the same equation, the evolution is the same for all z. Hence, $u^+(z) = u^-(z) \equiv u_\infty(z)$. While this restriction is convenient, the essentials of the method presented here apply in general. The equations for the background evolution (41) can now be further reduced by considering the phase difference $\Delta\phi_\infty(z) = \phi^+(z) - \phi^-(z)$ which is the parameter related to the depth of a dark soliton (see below); here $\phi^{\pm}(z)$ represents the phase as $t \to \pm\infty$, respectively:

$$\frac{d}{dz}u_\infty = \varepsilon \operatorname{Im}[F[u_\infty]], \qquad (42a)$$

$$\frac{d}{dz}\Delta\phi_\infty = 0. \qquad (42b)$$

Thus, while the magnitude of the background evolves adiabatically, the phase difference remains unaffected by the perturbation.

The Soliton and the Shelf

Let us now focus on the evolution of a dark soliton under perturbation. To simplify the calculations, introduce the transformation:

$$U = u \exp\left[\int_0^z u_\infty(s)^2 ds\right], \tag{43}$$

so Eq. (39) becomes:

$$iu_z - \frac{1}{2}u_{tt} + (|u|^2 - u_\infty^2)u = \varepsilon F[u]. \tag{44}$$

It is now convenient to express the dark soliton solution of the unperturbed equation in a slightly different form than the one in Eq. (14), namely:

$$u_s(t,z) = \{A + iB\tanh[B(t - Az - t_0)]\}\exp(i\sigma_0), \tag{45}$$

where the parameters of the soliton "core" A, B, t_0, and σ_0, denoting the velocity, amplitude, center, and phase of the soliton, are all real. Furthermore, the background amplitude is $(A^2 + B^2)^{1/2} = u_\infty$, and the phase difference across the soliton is $2\tan^{-1}(B/A)$, with $A \neq 0$. When the soliton velocity $A = 0$, Eq. (45) describes a black soliton, which has a phase difference of π, as was also explained above.

Below, we employ the method of multiple scales by introducing a slow scale variable $Z = \varepsilon z$, with the parameters A, B, t_0, and σ_0 being functions of Z. In addition, we seek the solution of Eq. (44) in the form of the following perturbation expansion:

$$u = u_0(Z, z, t) + \varepsilon u_1(Z, z, t) + O(\varepsilon^2) \tag{46}$$

The order $O(1)$ approximation $u_0(Z, z, t)$ should satisfy the slowly varying boundaries from Eqs. (42), which means that two of the parameters are already pinned down, namely, $A(Z) = u_\infty(Z)\cos(\Delta\phi_\infty/2)$ and $B(Z) = u_\infty(Z)\sin(\Delta\phi_\infty/2)$, and we take $\sigma_0(0) = 0$.

Now we consider the general case of a dark soliton with velocity $A(Z)$; also recall $A^2(Z) + B^2(Z) = u_\infty^2(Z)$. Let $u = q\exp(i\phi)$, where $q > 0$ and ϕ are real functions of z and t, and introduce moving frame of reference $T = t - \int_0^z A(\varepsilon s)ds - t_0$ and $\zeta = z$. This way, Eq. (44), with $u = u(\zeta, T, Z)$, becomes:

$$iu_\zeta - iAu_T - \frac{1}{2}u_{TT} + (|u|^2 - u_\infty^2)u = \varepsilon F[u] \tag{47}$$

Then, using $u = q\exp(i\phi)$, we obtain:

$$i(q_\zeta + i\phi_\zeta q) - iA(q_T + i\phi_T q) - \frac{1}{2}[q_{TT} + i2\phi_T q_T + (i\phi_{TT} - \phi_T^2)q]$$
$$+ q^3 - u_\infty^2 q = \varepsilon F[q, \phi]. \tag{48}$$

6 Perturbations of Solitons in Optical Fibers

The imaginary and real parts of the above equation, respectively, read:

$$q_\zeta = Aq_T + \frac{1}{2}(2\phi_T q_T + q\phi_{TT}) + \varepsilon \operatorname{Im}[F[q,\phi]], \tag{49a}$$

$$\phi_\zeta q = A\phi_T q - \frac{1}{2}(q_{TT} - \phi_T^2 q) + (|q|^2 - u_\infty^2)q - \varepsilon \operatorname{Re}[F[q,\phi]]. \tag{49b}$$

We now write Eq. (49b) in terms of the slow evolution variable $\zeta = \varepsilon Z$ and series expansions $q = q_0 + \varepsilon q_1 + O(\varepsilon^2)$ and $\phi = \phi_0 + \varepsilon \phi_1 + O(\varepsilon^2)$. At $O(1)$ the equations are satisfied by the soliton solution (45). On the other hand, at $O(\varepsilon)$ we have:

$$q_{1\zeta} = Aq_{1T} + \frac{1}{2}[2(\phi_{0T}q_{1T} + \phi_{1T}q_{0T}) + \phi_{1TT}q_0 + \phi_{0TT}q_1] + \operatorname{Im}[F[u_0]] - q_{0Z}, \tag{50}$$

$$\phi_{1\zeta}q_0 = -\phi_{0\zeta}q_1 + A(\phi_{0T}q_1 + \phi_{1T}q_0) - \frac{1}{2}(q_{1TT} - \phi_{0T}^2 q_1 - 2\phi_{0T}q_0\phi_{1T})$$
$$+ 3q_0^2 q_1 - u_\infty^2 q_1 - \operatorname{Re}[F[u_0]] - \phi_{0Z}q_0, \tag{51}$$

where $u_0 = q_0 \exp(i\phi_0)$. We look for stationary solutions at $O(\varepsilon)$:

$$Aq_{1T} + \frac{1}{2}\{2(\phi_{0T}q_{1T} + \phi_{1T}q_{0T}) + \phi_{1TT}q_0 + \phi_{0TT}q_1\} + \operatorname{Im}[F[u_0]] - q_{0Z} = 0, \tag{52a}$$

$$A(\phi_{0T}q_1 + \phi_{1T}q_0) - \frac{1}{2}(q_{1TT} - \phi_{0T}^2 q_1 - 2\phi_{0T}q_0\phi_{1T}) + 3q_0^2 q_1 - u_\infty^2 q_1$$
$$- \operatorname{Re}[F[u_0]] - \phi_{0Z}q_0 = 0, \tag{52b}$$

where

$$q_{0Z} = \frac{1}{2}\left(AA_Z + BB_Z \tanh^2(x)\right)q_0^{-1} + q_{0T}\left(\frac{B_Z}{B} - t_{0Z}\right), \tag{53a}$$

$$\phi_{0Z} = (AB_Z - BA_Z)\tanh(x)q_0^{-2} + \phi_{0T}\left(\frac{B_Z}{B} - t_{0Z}\right) + \sigma_{0Z}. \tag{53b}$$

Consider, now, Eq. (52a) in the limit $T \to \pm\infty$; using $q_0 \to u_\infty$ and $u_{\infty Z} = \operatorname{Im}F[u_\infty]$, we obtain:

$$Aq_{1T}^\pm + \frac{u_\infty}{2}\phi_{1TT}^\pm = 0. \tag{54}$$

We assume that q_1 tends to a constant with respect to t, i.e., $q_{1T} \to 0$ as $t \to \pm\infty$. As a result, $\phi_{1TT} \to 0$. Then q_1 and ϕ_{1T} both tend asymptotically to constants as

$t \to \pm\infty$, which corresponds to a *shelf* developing around the soliton (Ablowitz et al. 2011a). Substituting ϕ_{0T} into Eq. (52b), and in the limit $T \to \pm\infty$, we get:

$$A\phi_{1T}^{\pm} + 2u_\infty q_1^{\pm} = -\text{Re}\,[F[u_\infty]]/u_\infty \pm \frac{(AB_Z - BA_Z)}{u_\infty^2} + \sigma_{0Z} \tag{55}$$

We define $\Delta\phi_0$ by:

$$\Delta\phi_0 = 2\tan^{-1}\left(\frac{B}{A}\right), \tag{56a}$$

the phase change across the core soliton, as in the unperturbed case. This is consistent with the soliton parameters A and B being expressed in terms of background magnitude, u_∞, and phase change, $\Delta\phi_0$:

$$A = u_\infty \cos\left(\frac{\Delta\phi_0}{2}\right), \quad B = u_\infty \sin\left(\frac{\Delta\phi_0}{2}\right), \tag{56b}$$

again as in the unperturbed case. Using $\phi_Z^{\pm} = -\text{Re}\,[F[u_\infty]]/u_\infty$ and substituting Eq. (56b) into Eq. (55), we find:

$$A\phi_{1T}^{\pm} + 2u_\infty q_1^{\pm} = \phi_Z^{\pm} \pm \frac{\Delta\phi_{0Z}}{2} + \sigma_{0Z}. \tag{57}$$

Adiabatic Dynamics

Next, following a procedure reminiscent to the one used for the bright solitons, we will now employ the renormalized conservation equations to determine the slow evolution of the soliton parameters A and σ_{0Z}, as well as the shelf parameters $q_1^{\pm}$ and $\phi_{1t}^{\pm}$. The evolution of the renormalized integrals of the dark soliton read:

$$\frac{dH_{ds}}{dz} = \varepsilon \left(E \frac{d}{dZ} u_\infty^2 + 2\text{Re} \int_{-\infty}^{\infty} F[u]\bar{u}_z dt \right), \tag{58a}$$

$$\frac{dE_{ds}}{dz} = 2\varepsilon\,\text{Im} \int_{-\infty}^{\infty} F[u_\infty]u_\infty - F[u]\bar{u}\,dt, \tag{58b}$$

$$\frac{dP_{ds}}{dz} = 2\varepsilon\,\text{Re} \int_{-\infty}^{\infty} F[u]\bar{u}_t\,dt \tag{58c}$$

$$\frac{dR_{ds}}{dz} = -P + 2\varepsilon\,\text{Im} \int_{-\infty}^{\infty} t\,(F[u_\infty]u_\infty - F[u]\bar{u})\,dt. \tag{58d}$$

However, more work is required in order to find t_0. Note that if we find A, then $B = (u_\infty^2 - A^2)^{1/2}$. The edge of the shelf still propagates with velocity $V(Z) = u_\infty(Z)$;

6 Perturbations of Solitons in Optical Fibers

however, the speed may now vary in z. In terms of the moving frame of reference, the boundaries of the shelf are:

$$S_L(\zeta) = -\int_0^\zeta [u_\infty(\varepsilon s) + A(\varepsilon s)]\,ds, \tag{59a}$$

$$S_R(\zeta) = +\int_0^\zeta [u_\infty(\varepsilon s) - A(\varepsilon s)]\,ds, \tag{59b}$$

where S_L and S_R denote the positions (in T) of the left and right boundaries of the shelf, respectively, at ζ. Note that $A \leq u_\infty$ for all Z; thus, the soliton cannot overtake the shelf. The inner region consists of the soliton core and the shelf expanding around it, while the outer region consists of the infinite boundary conditions characterized by Eqs. (42).

We begin with the evolution equation for the Hamiltonian (58a), which reads:

$$\frac{d}{d\zeta}\int_{-\infty}^\infty \left[\frac{1}{2}|u_t|^2 + \frac{1}{2}(u_\infty^2 - |u|^2)^2\right] dt = \varepsilon\, (u_\infty^2)_Z \int_{-\infty}^\infty [u_\infty^2 - |u|^2]\,dt$$
$$+ 2\varepsilon\,\mathrm{Re}\int_{-\infty}^\infty F[u]\bar{u}_\zeta\,dt. \tag{60}$$

Substituting $u = (q_0 + \varepsilon q_1)\exp[i(\phi_0 + \varepsilon\phi_1)]$ and changing variables to the moving frame of reference, we have up to $O(\varepsilon)$:

$$\frac{d}{d\zeta}\int_{-\infty}^\infty [(q_{0T}^2 + \phi_{0T}^2 q_0^2) + (u_\infty^2 - q_0^2)^2]\,dT = 2\varepsilon\,(u_\infty^2)_Z \int_{-\infty}^\infty (u_\infty^2 - q_0^2)\,dT$$
$$- 4\varepsilon\,\mathrm{Re}\int_{-\infty}^\infty F[u_0]A\bar{u}_{0T}\,dT, \tag{61}$$

where, both here and later on, $u_0 = q_0\exp(i\phi_0)$. The Hamiltonian is unique among the evolution Eqs. (58) in that the contribution of the shelf appears only at $O(\varepsilon^2)$ or higher, and to $O(\varepsilon)$ may be ignored. We now put in the soliton form to get:

$$2B^2 B_Z = (u_\infty^2)_Z B - A\,\mathrm{Re}\int_{-\infty}^\infty F[u_0]\bar{u}_{0T}\,dT. \tag{62}$$

Taking a derivative with respect to Z of the equation $u_\infty^2 = A^2 + B^2$, we get:

$$(u_\infty^2)_Z = 2AA_Z + 2BB_Z, \tag{63}$$

which can be used to consolidate Eq. (62) to the form:

$$2BA_Z = \mathrm{Re}\int_{-\infty}^\infty F[u_0]\bar{u}_{0T}\,dT. \tag{64}$$

The evolution equations for energy (58b) and momentum (58c) both remain the same after transforming to the moving frame of reference:

$$\frac{d}{d\zeta} \int_{-\infty}^{\infty} (u_{\infty}^2 - |u|^2)\, dT = 2\varepsilon \mathrm{Im} \int_{-\infty}^{\infty} (F[u_{\infty}]u_{\infty} - F[u]\bar{u})\, dT, \tag{65}$$

$$\frac{d}{d\zeta} \mathrm{Im} \int_{-\infty}^{\infty} u\bar{u}_T\, dT = 2\varepsilon \mathrm{Re} \int_{-\infty}^{\infty} F[u]\bar{u}_T\, dT. \tag{66}$$

The inner region over which q_1 and ϕ_1 are relevant is $T \in [S_L(\zeta), S_R(\zeta)]$, and outside this region $q_1 = \phi_{1T} = 0$. At $O(1)$ the equations are obviously satisfied, while at $O(\varepsilon)$ we have:

$$B_Z - \frac{d}{d\zeta} \int_{S_L(\zeta)}^{S_R(\zeta)} q_0 q_1\, dT = \mathrm{Im} \int_{-\infty}^{\infty} (F[u_{\infty}]u_{\infty} - F[u_0]\bar{u}_0)\, dT, \tag{67a}$$

$$-2(AB)_Z - \frac{d}{d\zeta} \int_{S_L(\zeta)}^{S_R(\zeta)} \left(2\phi_{0T} q_0 q_1 + \phi_{1T} q_0^2\right) dT = 2\mathrm{Re} \int_{-\infty}^{\infty} F[u_0]\bar{u}_{0T}\, dT. \tag{67b}$$

Since the integrands on the left-hand side are not functions of ζ, we can apply the fundamental theorem of calculus to arrive at:

$$B_Z - u_{\infty}\left[(u_{\infty} - A)q_1^+ + (u_{\infty} + A)q_1^-\right] = \mathrm{Im} \int_{-\infty}^{\infty} (F[u_{\infty}]u_{\infty} - F[u_0]\bar{u}_0)\, dT, \tag{68a}$$

$$2(AB)_Z + u_{\infty}^2\left[(u_{\infty} - A)\phi_{1T}^+ + (u_{\infty} + A)\phi_{1T}^-\right] = -2\mathrm{Re} \int_{-\infty}^{\infty} F[u_0]\bar{u}_{0T}\, dT. \tag{68b}$$

We are left now with the evolution of the center of energy:

$$\frac{d}{d\zeta} \int_{-\infty}^{\infty} t(u_{\infty}^2 - |u|^2)\, dt = -\mathrm{Im} \int_{-\infty}^{\infty} u\bar{u}_t\, dt + 2\varepsilon \mathrm{Im} \int_{-\infty}^{\infty} t\, (F[u_{\infty}]u_{\infty} - F[u]\bar{u})\, dt, \tag{69}$$

which, after transforming to the moving frame of reference, becomes:

$$\frac{d}{d\zeta} \int_{-\infty}^{\infty} \left(T + \int_0^{\zeta} A + t_0\right)(u_{\infty}^2 - |u|^2)\, dt$$
$$= -\mathrm{Im} \int_{-\infty}^{\infty} u\bar{u}_T\, dT + 2\varepsilon \mathrm{Im} \int_{-\infty}^{\infty} \left(T + \int_0^{\zeta} A + t_0\right)(F[u_{\infty}]u_{\infty} - F[u]\bar{u})\, dT. \tag{70}$$

6 Perturbations of Solitons in Optical Fibers

After rearranging some terms, we have:

$$\frac{d}{d\zeta} \int_{-\infty}^{\infty} T(u_\infty^2 - |u|^2) dT \tag{71a}$$

$$+ \left(\int_0^\zeta A + t_0 \right) \left[\frac{d}{d\zeta} \int_{-\infty}^{\infty} (u_\infty^2 - |u|^2) dT - \varepsilon 2 \text{Im} \int_{-\infty}^{\infty} (F[u_\infty]u_\infty - F[u]\bar{u}) dT \right] \tag{71b}$$

$$+ A \int_{-\infty}^{\infty} (u_\infty^2 - |u|^2) dT + \text{Im} \int_{-\infty}^{\infty} u\bar{u}_T dT \tag{71c}$$

$$= -\varepsilon t_{0Z} \int_{-\infty}^{\infty} (u_\infty^2 - |u|^2) dT + 2\varepsilon \text{Im} \int_{-\infty}^{\infty} T(F[u_\infty]u_\infty - F[u]\bar{u}) dT. \tag{71d}$$

The terms on line (71a) yield:

$$\frac{d}{d\zeta} \int_{-\infty}^{\infty} T(u_\infty^2 - |u|^2) dT = -2 \left[S_R(u_\infty - A) q_1^+ + S_L(u_\infty + A) q_1^- \right] u_\infty. \tag{72}$$

The terms on line (71b) reproduce the energy equation (65) and cancel out. The terms on line (71c) are calculated up to $O(\varepsilon)$ using the previous results by integrating the energy and momentum equation (68):

$$E(\zeta) = 2B - 2 \left[S_R(Z) q_1^+ - S_L(Z) q_1^- \right] u_\infty + \varepsilon E_1(Z) + O(\varepsilon^2), \tag{73}$$

$$I(\zeta) = -2AB - u_\infty^2 \left[S_R(Z) \phi_{1t}^+ - S_R(Z) \phi_{1t}^- \right] + \varepsilon I_1(Z) + O(\varepsilon^2), \tag{74}$$

where it is noticed that $d/d\zeta = \varepsilon d/dZ$, while S_R and S_L are $O(1/\varepsilon)$ in terms of Z.

Now, putting everything together in terms of slow evolution variable $Z = \varepsilon \zeta$ we get from (71):

$$\varepsilon 2B t_{0Z} = 2\varepsilon \text{Im} \int_{-\infty}^{\infty} T\left(F[u_\infty]u_\infty - F[u_0]u_0^* \right) dT \varepsilon + AE_1(Z) + \varepsilon I_1(Z)$$

$$+ \{ 2u_\infty \left[S_R(u_\infty - A) q_1^+ + S_L(u_\infty + A) q_1^- \right] + 2u_\infty A \left[S_R q_1^+ - S_L q_1^- \right]$$

$$+ u_\infty^2 \left[S_R \phi_{1t}^+ - S_L \phi_{1t}^- \right] \}. \tag{75}$$

After some cancelations, this breaks into $O(1)$ terms:

$$2 \left[S_R q_1^+ + S_L q_1^- \right] + \left[S_R \phi_{1T}^+ - S_L \phi_{1T}^- \right] = 0, \tag{76}$$

and $O(\varepsilon)$ terms which include t_{0Z}, while higher-order energy and momentum terms have not been determined. The six Eqs. (57), (64), (68a), (68b), and (76) can now be used to solve for the set of six parameters $q_1^\pm$, $\phi_{1t}^\pm (= \phi_{1T}^\pm)$, A, and σ_0:

$$2BA_z = \text{Re}\left\{\int_{-\infty}^\infty F[u_s](u_s^*)_t \, dt\right\}, \tag{77}$$

$$Bt_{0z} = \text{Im}\left\{\int_{-\infty}^\infty t(F[u_\infty]u_\infty - F[u_s]u_s^*) \, dt\right\}, \tag{78}$$

$$u_\infty \sigma_{0z} = \text{Im}\left\{\int_{-\infty}^\infty (F[u_\infty]u_\infty - F[u_s]u_s^*) \, dt\right\} + \text{Re}\{F[u_\infty]\}, \tag{79}$$

$$BB_z = u_\infty u_{\infty z} - AA_z \tag{80}$$

$$u_\infty^2 \Delta\phi_{0z} = 2AB_z - 2BA_z \tag{81}$$

$$q_1^\pm = \frac{1}{2}\frac{\sigma_{0z} \pm \Delta\phi_{0z}}{u_\infty \mp A}, \quad \phi_{1t}^\pm = \mp 2q_1^\pm. \tag{82}$$

The above equations may now be solved from top to bottom.

Dark Solitons Under Perturbations

Our analysis starts with the dynamics of the soliton background. Assuming that $u(z, t \to \infty) = u_0(z)$, we derive from Eq. (6) the equation:

$$iu_{0z} - |u_0|^2 u_0 = i\gamma u_0 + i\delta|u_0|^2 u_0. \tag{83}$$

Then, employing the polar decomposition $u_0 = u_\infty(z)\exp(i\theta(z))$, we obtain:

$$u'_\infty = (\gamma + \delta u_\infty^2)u_\infty, \tag{84}$$

and $\theta' = u_\infty^2$, where primes denote differentiation with respect to z. Notice that the role of the term of strength δ is now more obvious: a nontrivial equilibrium (constant solution) exists if $\gamma\delta < 0$, which is $u_\infty^2 = -\gamma/\delta$. We focus here on these solutions, i.e., solutions that tend to stabilize the soliton, by keeping its parameters constant. Employing the results of the previous section, we find that the rest of soliton parameters evolve according to the following equations:

$$\begin{aligned}A' &= \frac{4}{15}\sigma_R A^4 + \frac{2}{3}\delta A^3 - \frac{8}{15}\sigma_R u_\infty^2 A^2 \\ &\quad + \left(\gamma + \frac{\delta}{3}u_\infty^2\right)A + \frac{4}{15}\sigma_R u_\infty^4,\end{aligned} \tag{85}$$

$$t'_0 = \left(2\beta - \mu - \frac{2\nu}{3}\right) A^2 - \left(2\beta + 2\mu + \frac{4\nu}{3}\right) u_\infty^2, \tag{86}$$

$$\sigma'_0 = \frac{B_z}{u_\infty} - \frac{2B}{3u_\infty} \left(3\gamma + 4u_\infty^2 \delta + 2\delta A^2\right). \tag{87}$$

These equations show that while the evolution of the soliton center, described by the equation $T'_0 = A + t'_0$, is affected by all parameters of Eq. (6) [directly or indirectly from $A(z)$], the solution parameters, i.e., the background, soliton core, and shelf, only depend on γ, δ, and σ_R. Thus, soliton stabilization can be targeted accordingly.

We now proceed by presenting numerical results for Eq. (6), which will be compared to the above analytical predictions. We use the parameter values $\beta = \mu = \nu = 0.01$ and vary γ, δ, and σ_R in order to investigate stabilization of the background and the soliton parameters. A typical result of the simulations, for $\gamma = -0.025$, $\delta = 0.02$, and $\sigma_R = 0.02$, with initial condition a unit-amplitude black soliton, is shown in Fig. 2. In this figure, top panel depicts the evolution of the soliton and the emergence of the shelf, propagating with a speed $u_\infty(z)$, so that its edge is $\int_0^z u_\infty(s)ds$. Additionally, bottom panel shows the evolution of the background amplitude and soliton velocity.

Let us now return to the system of Eqs. (84) and (85). Stable fixed points of this system correspond to stable solitons travelling on top of a constant background with a constant speed. Below, we identify two such solitons, namely, a gray and

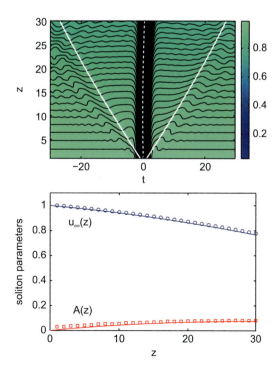

Fig. 2 Top panel: contour plot showing the evolution of a unit-amplitude black soliton; analytical predictions are depicted by the solid (white) lines, for the propagation of the shelf's edge, and the dotted (white) line, for the evolution of the soliton center. Bottom panel: evolution of $u_\infty(z)$ and $A(z)$; solid lines (circles/squares) correspond to analytical (numerical) results

a black one, supported in the presence ($\sigma_R \neq 0$) and in the absence ($\sigma_R = 0$) of the SRS effect, respectively. In both cases, the background assumes the same form: $u_\infty^2 = -\gamma/\delta$ [cf. Eq. (84)] for $\gamma\delta < 0$, i.e., for linear loss and nonlinear gain, or vice versa.

We start with the case $\sigma_R \neq 0$. Substituting $u_\infty^2 = -\gamma/\delta$ in Eq. (85) and seeking stationary solutions for the soliton velocity, we arrive at a fourth-order algebraic equation for A. We find that there exists only one root of this equation, which does not violate the fundamental relationship $A^2 + B^2 = u_\infty^2$; this root is: $A = (-5\delta^2 + \sqrt{25\delta^4 - 16\gamma\delta\sigma_R^2})/(4\delta\sigma_R)$. Thus, a stable soliton exists for:

$$u_\infty^2 = -\frac{\gamma}{\delta}, \quad A = \frac{-5\delta^2 + \sqrt{25\delta^4 - 16\gamma\delta\sigma_R^2}}{4\delta\sigma_R}. \tag{88}$$

Note that since $\gamma\delta < 0$, the quantity under the square root is always positive. Taking $-\gamma = \delta = \sigma_R = 0.02$, the soliton propagation for the above initial data is shown in the top panel of Fig. 3. Evidently, the soliton is characterized by a stable evolution (the background is fixed, and the center moves with constant speed), while suppression of the shelf is also observed. In fact, the calculated shelf is found to be of order $O(10^{-3})$, i.e., an order of magnitude less that that of the perturbation.

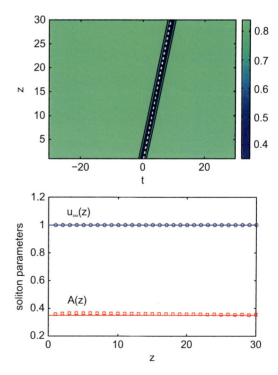

Fig. 3 The panels are similar to those in Fig. 2, but now for the stable gray soliton, characterized by the parameters given in Eq. (88)

The bottom panel of Fig. 3 also illustrates the background amplitude u_∞ and soliton velocity $A(z)$.

Next, we consider the situation where the SRS effect is absent, i.e., $\sigma_R = 0$. In this case, Eqs. (84) and (85) lead to the following equations for the background and soliton velocity:

$$u_\infty^2 = -\frac{\gamma}{\delta}, \quad A' = \frac{2}{3}(\delta A^2 - \gamma)A, \tag{89}$$

Obviously, the above equation for the velocity depicts a stationary solution $A = 0$ (recall that $\gamma\delta < 0$) that corresponds to a black soliton. Hence, when SRS is absent (which would result in a frequency downshift causing the soliton to move), a stable black soliton can exist, as seen in Fig. 4. The evolution of $u_\infty(z)$ and $A(z)$ follows a qualitatively similar behavior to that observed in the bottom panel of Fig. 3 and, hence, is not shown.

Here, it is important to mention that while these cases for u_∞ and A stabilize the soliton, this does not mean that the shelf is no longer present. The shelf is always present in the solution of the perturbed NLS equation, even if not noticeable due to its magnitude in the numerical simulations. In any case, it does not affect the single soliton propagation, but it will affect soliton interactions; nevertheless, addressing this problem is beyond the scope of this work. Notice, also, suppression of the shelf will result in the destabilization of the soliton.

We also briefly consider the case where gain/loss terms are absent, i.e., $\gamma = \delta = 0$. In this particular case, the dark soliton dynamics is merely driven by the SRS effect. Indeed, now the evolution of the background and soliton velocity is described by the following equations:

$$u'_\infty = 0, \quad A' = \frac{4}{15}\sigma_R(A^2 - u_\infty^2)^2. \tag{90}$$

Since $A^2 \neq u_\infty^2$, the right-hand side of the second equation is always positive and, thus, the minimum of the soliton density is always ascending and will finally

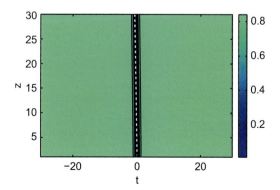

Fig. 4 Similar to the top panel of Fig. 2, but for the stable black soliton (for $\sigma_R = 0$), characterized by the parameters given in Eq. (89), for $-\gamma = \delta = 0.02$

be merged to the stationary background. Thus, obviously, no stable pulse (in the sense of stationary parameters) exists. We note that (90) recover the results obtained in Uzunov and Gerdjikov (1993) and Kivshar and Yang (1994), while numerical simulations illustrating the soliton evolution in this case provided results similar to those reported in Kivshar and Yang (1994) (and, hence, not repeated here).

Finally, it is important to make the following comment. The focus here was on the dynamics stemming from stationary nontrivial solutions of Eq. (84). However, this equation has the following general solution:

$$u_\infty^2(z) = \frac{\gamma u_0^2 e^{2\gamma z}}{\gamma + \delta u_0^2 - \delta u_0^2 e^{2\gamma z}}, \qquad (91)$$

where $u_0 = u_\infty(0)$. This suggests that there is a (finite) distance, $z_\star$, for which the background exhibits blowup. Indeed, the denominator is zero when:

$$z_\star = \frac{1}{2\gamma} \ln\left(1 + \frac{\gamma}{\delta u_0^2}\right). \qquad (92)$$

The unexpected feature here, which was avoided above by choosing $|\gamma| = \delta |u_0|^2$, is that the addition of the term $\delta |u|^2$ that counters the effects of the linear term may result in a blowup of the background in finite z, even when the other soliton parameters remain finite. Also, Eq. (91) shows that an equilibrium can also be reached in finite z when the denominator is a multiple of the numerator. While this will stabilize the background, it does not mean necessarily that it stabilizes the other parameters.

Beyond the Adiabatic Theory: Soliton Radiation

In the previous sections, we employed the adiabatic approximation to study perturbations of either bright or dark solitons in optical fibers. We have shown that when perturbations – such as damping, higher-order dispersion, higher-order nonlinearity, etc. – come into play, solitons do not propagate as undistorted objects but, instead, their physical properties (shape, velocity, center, etc.) over the propagation distance. Furthermore, during soliton evolution, energy radiation can additionally be excited, which can affect the soliton dynamics in nontrivial ways.

To motivate a short discussion on the effect of radiation that we append below, it is relevant to mention at this point the following. Soliton propagation in non-ideal optical fibers has attracted much attention, particularly in view of the relevance to high band-pass optical communication systems. In that regard, it should be recalled that there is a strong link between phase fluctuation and changes in the soliton velocity, resulting in jitter in the pulse train. One important source of phase fluctuation is background radiation present in the fiber, which accompanies the soliton pulses. This radiation may result either from perturbing influences in the fiber (such as the ones studied in the previous sections) or be present as part of the input pulse injected into the fiber (i.e., in the case where the initial condition is a perturbed soliton).

To be more specific, let us consider the anomalous dispersion regime and rewrite the perturbed NLS, Eq. (22), in the following form:

$$iu_t + su_{xx} - 2|u|^2 u = iF, \qquad (93)$$

where, as before, $s = -1$ ($s = +1$) corresponds to the anomalous (normal) dispersion regime. The perturbation iF in the NLS modifies the soliton solution in two distinct ways:

(i) The soliton parameters, which were constants of the motion in the unperturbed case, now vary with distance down the fiber. If the perturbation is small, this change is adiabatic and can be studied in the framework of the adiabatic perturbation theory exposed in the previous sections.
(ii) The perturbation is responsible also for the generation of a background radiation field, which is superimposed on the soliton pulse. Depending on the nature of the perturbation, this can exhibit quite complicated resonance features (see, e.g., Wai et al. 1986; Elgin 1992).

To incorporate the effect of radiation in the context of the perturbed NLS Eq. (93), one may assume that the field $u(x, t)$ departs only a little from being a pure soliton; then, one may seek solutions of Eq. (93) of the form (see, e.g., Kivshar and Malomed 1989):

$$u(z, t) = u_s(z, t) + \delta u(z, t), \qquad (94)$$

where $u_s(z, t)$ has the functional form of the (bright or dark) soliton solution [cf. Eqs. (12) and (14)], but with the soliton parameters depending on the propagation distance – as in the previous sections – and $\delta u(z, t)$, with $\delta u(0, t) = 0$, is the radiation emitted by the soliton.

Assume that the amplitude of the perturbing radiation field δu is much smaller than unity and its energy is much smaller than that of the soliton. Substituting Eq. (94) into the NLS and keeping first-order terms, one obtains a linear propagation equation for the radiation field, namely:

$$i\delta u_z + s\delta u_{tt} + 4|u_s|^2 \delta u + 2u_s^2 \delta u^* = iF, \qquad (95)$$

where, to simplify our notation below, star is used to denote complex conjugate. Here, even in the simpler case where a soliton evolves in an ideal fiber (i.e., $F = 0$), a difficulty arises when an attempt is made to solve the above equation using standard techniques. For instance, if a Fourier transformation with respect to the t variable is used, the presence of u_s results in convolution terms which prove intractable to further analysis.

The expression $u = u_s + \delta u$ is apparently not unique. Small changes in the parameters of u_s can be offset by modifications to δu, leaving the total u unchanged. However, consider a field that has this form at some t. If this field is permitted to propagate sufficiently far, some unique soliton will emerge, along with a small dispersed field of the order of δu that has left the vicinity of the soliton. This unique soliton, extrapolated back to the location of the perturbation, is called the *emergent soliton* (Gordon 1992). The emergent soliton is the same as u_s only if δu is purely dispersive; in other words, radiation is actually a dispersive field.

Two cases are usually considered: the first case considers an ideal fiber in the sense that subsequent pulse evolution is described by the NLS equation, with the pulse input to the fiber consisting of soliton plus radiation; in the second case, the fiber is not ideal, and pulse evolution is described by a perturbed form of the NLS.

The study of some of the properties of the radiation field, either accompanying the soliton pulse or generated by it, relies on inverse scattering theory (Ablowitz and Segur 1981). In the case of the anomalous dispersion regime, i.e., for bright solitons, particular solutions to a scattering problem associated with the latter – the Jost function solutions – are combined bilinearly to produce a complete basis of continuum states, onto which the radiation field is projected; this process resembles the situation when a simple function, say $x(t)$, is projected onto the kernel $\exp(i\omega t)$ in standard Fourier analysis. The manner in which the radiation field is projected onto these states and, more importantly, the reconstruction of this radiation field from a knowledge of these continuum modes together with a minimal set of "scattering data" is briefly outlined below.

In the case of dispersive perturbations, one may follow Gordon (1992); Elgin (1993) and introduce the *associate field* formalism. This approach provides results consistent with the ones obtained from a perturbation expansion for the Zakharov-Shabat inverse scattering transform (Kaup 1978b) (see also Haus 1997). According to the associate field formalism, the required form for $\delta u(z, t)$ reads:

$$\delta u(z,t) = -f_{tt} + 2\gamma f_t - \gamma^2 f + u_s^2 f^*, \qquad (96)$$

where $\gamma = 2\eta \tanh(2\eta t)$, while f evolves according to

$$\mathrm{i} f_z = f_{tt} - \frac{\mathrm{i}}{4\pi} \int_{-\infty}^{+\infty} \frac{\exp(-2\mathrm{i}\xi t)}{\xi^2 + \eta^2} \int_{-\infty}^{+\infty} (F^*\phi_1 \bar{\psi}_1 + F\phi_2 \bar{\psi}_2)^* \, \mathrm{d}t \, \mathrm{d}\xi, \qquad (97)$$

where ϕ_i, $\bar{\phi}_i$, ψ, and $\bar{\psi}_i$ are the components of Jost functions associated with the direct scattering problem of the NLS (Ablowitz and Segur 1981). Obviously, when $F = 0$, the associate field evolves according to:

$$\mathrm{i} f_z = f_{tt}. \qquad (98)$$

It then follows that the radiation field δq is again described by Eq. (96); however, in this case $\delta u(0,t)$ is a nonzero function of t, else $\delta u(z,t) \equiv 0$ due to the homogeneous character of Eq. (98).

Let us now briefly discuss the case of the normal dispersion regime, i.e., radiation for dark solitons. In this case, the situation is different. The scattering problem for the NLS equation with normal dispersion is different, and the closure of the relative Jost functions has not been shown. In addition, because these Jost functions differ from the focusing case, the direct analogue of the associate field has not been found, to date. Nevertheless, in some cases, e.g., in the case of the defocusing NLS perturbed by the third-order dispersion, it is possible to show the following (Karpman 1993; Afanasjev et al. 1996). Under the action of the third-order dispersion, radiation in the form of an oscillating tail is formed from the right of the soliton, and its front propagates with group velocity v_g, which is different from the dark soliton velocity v and the speed of sound c. Performing an asymptotic analysis, it is possible to show that, for sufficiently small coefficient β of the third-order dispersion term, the amplitude A of the tail depends on β exponentially for fixed v, namely (Karpman 1993; Afanasjev et al. 1996):

$$A \sim \exp\left[-\frac{\pi}{u_\infty \beta \sqrt{1 - v^2/c^2}}\right], \tag{99}$$

where u_∞ is the background amplitude. This dependence of A on β is similar to that for bright solitons, where it was found that the tail amplitude is exponentially small in the third-order dispersion coefficient, namely, $A \sim \exp(-1/\beta)$, and can be calculated by asymptotic analysis "beyond all orders" (Wai et al. 1986). These results are also consistent with the soliton robustness hypothesis (Menyuk 1993). Notice that the presence of the factor $\sqrt{1 - v^2/c^2}$ in the exponent in Eq. (99) shows that the radiation amplitude becomes exponentially small for any fixed β in the limit $v \to c$, i.e., for small-amplitude dark solitons (Kivshar and Afanasjev 1991; Frantzeskakis 1996).

It is also important to note that radiation also affects the interactions between solitons, a fact that is quite important for applications both in optical communications and in fiber lasers. For instance, an important effect that was predicted is that some perturbations tend to attenuate the interaction between bright solitons and may even lead to formation of *bound states* (BSs). This way, weakly overlapping solitons in the dissipatively perturbed NLS equation form a set of BSs, as was predicted in theory (Malomed 1991, 1992) and observed in numerical experiments (Agrawal 1991). In Malomed (1993), it was demonstrated that additional dissipative terms, e.g., a term describing the stimulated Raman scattering in a nonlinear optical fiber, destroy all the BSs provided the corresponding coefficient exceeds a certain critical value. On the other hand, in the absence of dissipation, and in the framework of the NLS equation with the third-order dispersion, it was demonstrated that two solitons – or a whole array of solitons – may form BSs, interacting with each other via emitted radiation (Malomed 1993). It is important to note that tightly BS solitons have been experimentally observed in nonlinear polarization rotation, figure-of-eight, carbon nanotubes, and graphene-based mode-locked fiber lasers (Akhmediev et al. 1998; Grelu et al. 2002; Seong and Kim 2002; Tang et al. 2005; Zhao et al.; Wu et al. 2011; Li et al. 2012; Gui et al. 2013; Tsatourian et al. 2013).

It is also relevant to mention that the interactions between dark solitons in the presence of perturbations have also been studied but less extensively. For instance, in Afanasjev et al. (1998), a perturbed defocusing NLS equation with a nonlinear saturable gain – actually a complex quintic Ginzburg-Landau equation – was studied. The focus of this work was on the formation and stability of two dark-soliton BSs, and it was found that such BSs do exist and appear to be fully stable in the quintic equation. A related dissipative equation – the so-called power-energy saturating model – was analyzed in Ablowitz et al. (2011b, 2013), and the effect of perturbations in the interactions between dark solitons was numerically studied. Notice that theoretical results reported in Afanasjev et al. (1998) and Ablowitz et al. (2011b, 2013) were in quantitative agreement with experimental and numerical results of the recent work (Guo et al. 2016), where the controlled generation of bright or dark solitons in a fiber laser was studied.

Summary and Conclusions

In this chapter, we have reviewed the dynamics of bright and dark solitons in optical fibers. Our exposition started with the introduction of the nonlinear Schrödinger (NLS) equation, which models soliton propagation in the ideal case (perfect fiber). Soliton solutions of the latter were introduced, and their fundamental properties were briefly discussed. We then focused on more realistic situations, where the NLS was modified to include extra terms accounting for various effects arising in real optical fibers and optical fiber links. These effects include loss, gain (for the compensation of loss), as well as third-order dispersion, self-steepening, and stimulated Raman scattering, which are particularly relevant for sub-picosecond pulses.

We have thus focused on the problem of studying the evolution of solitons under these perturbations. To analytically deal with this problem, we have adopted the rather general framework of the adiabatic perturbation theory. According to this approach, the functional form of the solitons does not change, but their parameters (e.g., amplitude, width, velocity, etc.) change along the fiber – i.e., they become unknown functions of the propagation distance.

Our analysis was performed independently for the cases of the bright and dark solitons. In the case of bright solitons, where the NLS model is supplemented with vanishing boundary conditions at infinity, the problem was rather straightforward to study. Indeed, using the evolution of the conserved quantities, it was possible to derive a set of ordinary differential equations (ODEs) for the soliton parameters that is even solvable analytically. In fact, once the evolution of the bright soliton amplitude was determined, all other parameters were found by simple integrations. The numerical results were shown to be in excellent agreement with the analytical predictions.

The study of the dynamics of dark solitons under perturbations proved to be more demanding. The problems in this case arise from the fact that the dark soliton is a more complicated object than its bright counterpart, because it is composed by a continuous-wave (cw) background and the soliton core. Thus, in this case, it was necessary to consider separately the dynamics of the dark soliton's constituents. In the case of the background, we have found that dissipative perturbations do have a nontrivial effect: boundary conditions in the vicinity of the soliton core and at infinity become unequal, which suggests the emergence of a "shelf," i.e., a linear wave that develops around the soliton core, and thus bridging the relevant difference. To account for this effect, it was necessary to develop a multiscale moving boundary layer theory, which accurately captures the dynamics of the soliton background, core, and shelf.

We have also briefly discussed effects that cannot be captured by the adiabatic perturbation theory. Such an effect is the perturbation-induced emission of radiation, which is particularly important for applications in optical communications and in the context of fiber lasers. We have thus presented a general framework and reviewed some analytical results for the case of bright solitons; pertinent results for dark solitons are not available, to the best of our knowledge, so far. Nevertheless, we also presented results for dark solitons, which can be obtained, in some cases (e.g., the NLS perturbed by the third-order dispersion), by means of proper asymptotic analysis. We also briefly described the effect of radiation on soliton interactions. We have thus explained that some perturbations tend to attenuate the interaction between bright solitons and may lead to the formation of soliton bound states; this effect was also observed in experiments with fiber lasers.

References

M.J. Ablowitz, H. Segur, *Solitons and the Inverse Scattering Transform* (SIAM, Philadelphia, 1981)
M.J. Ablowitz, S.D. Nixon, T.P. Horikis, D.J. Frantzeskakis, Proc. R. Soc. A **2133**, 2597 (2011a)
M.J. Ablowitz, S.D. Nixon, T.P. Horikis, D.J. Frantzeskakis, Opt. Lett. **36**, 793 (2011b)
M.J. Ablowitz, S.D. Nixon, T.P. Horikis, D.J. Frantzeskakis, J. Phys. A: Math. Theor. **46**, 095201 (2013)
V.V. Afanasjev, Yu.S. Kivshar, C.R. Menyuk, Opt. Lett. **21**, 1975 (1996)
V.V. Afanasjev, P.L. Chu, B.A. Malomed, Phys. Rev. E **57**, 1088 (1998)
G.P. Agrawal, Phys. Rev. A **44**, 7493 (1991)
G.P. Agrawal, *Applications of Nonlinear Fiber Optics* (Academic, San Diego, 2001)
G.P. Agrawal, *Nonlinear Fiber Optics* (Academic, London, 2007)
N.N. Akhmediev, A. Ankiewicz, J.M. Soto-Crespo, J. Opt. Soc. Am. B **15**, 515 (1998)
I.V. Barashenkov, E.Y. Panova, Physica D **69**, 114 (1993)
R. Carretero-González, D.J. Frantzeskakis, P.G. Kevrekidis, Nonlinearity **21**, R139 (2008)

X.J. Chen, Z.D. Chen, N.N. Huang, J. Phys. A: Math. Gen. **31**, 6929 (1998); N.N. Huang, S. Chi, X.J. Chen, J. Phys. A: Math. Gen. **32**, 3939 (1999)
J.N. Elgin, Opt. Lett. **17**, 1409 (1992)
J.N. Elgin, Phys. Rev. A **47**, 4331 (1993)
E. Emplit, J.E. Hamaide, E. Reynaud, G. Froehly, A. Barthelemy, Opt. Commun. **62**, 374 (1987)
D.J. Frantzeskakis, J. Phys. A: Math. Gen. **29**, 3631 (1996)
D.J. Frantzeskakis, J. Phys. A: Math. Theor. **43**, 213001 (2010)
J.P. Gordon, J. Opt. Soc. Am. B **9**, 91 (1992)
Ph. Grelu, F. Belhache, F. Gutty, J.M. Soto-Crespo, Opt. Lett. **27**, 966 (2002)
L.L. Gui, X.S. Xiao, C. X. Yang, J. Opt. Soc. Am. B **30**, 158 (2013)
J. Guo et al., IEEE Photonics J. **8**, 1 (2016)
A. Hasegawa, Y. Kodama, *Solitons in Optical Communications* (Clarendon Press, Oxford, 1995)
A. Hasegawa, F. Tappert, Appl. Phys. Lett. **23**, 142 (1973a)
A. Hasegawa, F. Tappert, Appl. Phys. Lett. **23**, 171 (1973b)
H.A. Haus, W.S. Wong, Rev. Mod. Phys. **68**, 423 (1996)
H.A. Haus, W.S. Wong, F.I. Khatri, J. Opt. Soc. Am. B **14**, 304 (1997)
T.P. Horikis, D.J. Frantzeskakis, Opt. Lett. **38**, 5098 (2013)
H. Ikeda, M. Matsumoto, A. Hasegawa, Opt. Lett. **20**, 1113 (1995)
V.I. Karpman, E.M. Maslov, Sov. Phys. JETP **46**, 281 (1977)
V.I. Karpman, Sov. Phys. JETP **50**, 58 (1979); Phys. Scr. **20**, 462 (1979)
V.I. Karpman, Phys. Lett. A **181**, 211 (1993)
D.J. Kaup, J. Math. Phys. **16**, 2036 (1976)
D.J. Kaup, A.C. Newell, Proc. R. Soc. Lond. Ser. A **361**, 413 (1978a)
D.J. Kaup, SIAM J. Appl. Math. **31**, 121 (1978b)
Yu.S. Kivshar, B.A. Malomed, Rev. Mod. Phys. **61**, 761 (1989)
Yu.S. Kivshar, Phys. Rev. A **42**, 1757 (1990)
Yu.S. Kivshar, V.V. Afanasjev, Opt. Lett. **16**, 285 (1991); Phys. Rev. A **44**, R1446 (1991)
Yu.S. Kivshar, X. Yang, Phys. Rev. E **49**, 1657 (1994)
Yu.S. Kivshar, W. Krolikowski, Opt. Commun. **114**, 353 (1995)
Yu.S. Kivshar, B. Luther-Davies, Phys. Rep. **298**, 81 (1998)
Yu.S. Kivshar, G.P. Agrawal, *Optical Solitons: From Fibers to Photonic Crystals* (Academic, San Diego, 2003)
Y. Kodama, J. Stat. Phys. **39**, 597 (1985)
Y. Kodama, A. Hasegawa, IEEE J. Quantum Electron. **23**, 510 (1987)
V.V. Konotop, V.E. Vekslerchik, Phys. Rev. E **49**, 2397 (1994)
D. Krökel, N.J. Halas, G. Giuliani, D. Grischkowsky, Phys. Rev. Lett. **60**, 29 (1988)
V.M. Lashkin, Phys. Rev. E **70**, 066620 (2004)
X.L. Li et al., Laser Phys. **22**, 774 (2012)
B.A. Malomed, Phys. Rev. A **44**, 6954 (1991)
B.A. Malomed, Phys. Rev. A **45**, R8321 (1992)
B.A. Malomed, Phys. Rev. E **47**, 2874 (1993)
C.R. Menyuk, J. Opt. Soc. Am. B **10**, 1585 (1993)
L.E. Mollenauer, R.H. Stolen, J.E. Gordon, Phys. Rev. Lett. **45**, 1095 (1980)
L.E. Mollenauer, J.E. Gordon, E.V. Mamyshev, *Optical Fiber Telecommunications III*, chap. 12, ed. by I.E. Kaminow, T.L. Koch (Academic, San Diego, 1997)
M.J. Potasek, J. Appl. Phys. **65**, 941 (1989)
N.H. Seong, D.Y. Kim, Opt. Lett. **27**, 1321 (2002)
Y. Silberberg, Opt. Lett. **15**, 1005 (1990)
R.H. Stolen, J.P. Gordon, W.J. Tomlinson, H.A. Haus, J. Opt. Soc. Am. B **6**, 1159 (1989)
T. Taniuti, Suppl. Prog. Theor. Phys. **55**, 1 (1974)
D.Y. Tang, B. Zhao, L.M. Zhao, H.Y. Tam, Phys. Rev. E **72**, 016616 (2005)
V. Tsatourian et al., Sci. Rep. **3**, 3154 (2013)
T. Tsuzuki, J. Low Temp. Phys. **4**, 441 (1971)
I.M. Uzunov, V.S. Gerdjikov, Phys. Rev. A **47**, 1582 (1993)

P.K.A. Wai, C.R. Menyuk, Y.C. Lee, H.H. Chen, Opt. Lett. **11**, 464 (1986)
X. Wu et al., Opt. Commun. **284**, 3615 (2011)
V.E. Zakharov, A.B. Shabat, Sov. Phys. JETP. **34**, 62 (1972)
V.E. Zakharov, A.B. Shabat, Sov. Phys. JETP. **37**, 823 (1973)
V.E. Zakharov, E.A. Kuznetsov, Physica D **18**, 455 (1986)
B. Zhao et al., Phys. Rev. E **70**, 067602 (2004)

Emission of Dispersive Waves from Solitons in Axially Varying Optical Fibers

A. Kudlinski, A. Mussot, Matteo Conforti, and D. V. Skryabin

Contents

Introduction	302
Emission of a Dispersive Wave from a Soliton	303
Fundamental Soliton	303
Dispersive Wave	304
Generation of Dispersive Waves from Solitons in Axially Varying Optical Fibers	306
Axially Varying Optical Fibers	306
Emission of Multiple Dispersive Waves Along the Fiber	307
Cascading of Dispersive Waves	309
Transformation of a Dispersive Wave into a Fundamental Soliton	310
Emission of Polychromatic Dispersive Waves	311
Generation of a Dispersive Wave Continuum	313
Conclusion and Perspectives	314
References	315

Abstract

The possibility of tailoring the guidance properties of optical fibers along the same direction as the evolution of the optical field allows to explore new directions in nonlinear fiber optics. The new degree of freedom offered by axially varying optical fibers enables to revisit well-established nonlinear phenomena

---

A. Kudlinski (✉)· A. Mussot · M. Conforti
CNRS, UMR 8523 – PhLAM – Physique des Lasers Atomes et Molécules, University of Lille, Lille, France
e-mail: alexandre.kudlinski@univ-lille1.fr; arnaud.mussot@univ-lille1.fr; matteo.conforti@univ-lille1.fr

D. V. Skryabin
Department of Nanophotonics and Metamaterials, ITMO University, St Petersburg, Russia

Department of Physics, University of Bath, Bath, UK
e-mail: d.v.skryabin@bath.ac.uk

© Springer Nature Singapore Pte Ltd. 2019
G.-D. Peng (ed.), *Handbook of Optical Fibers*,
https://doi.org/10.1007/978-981-10-7087-7_10

and even to discover novel short pulse nonlinear dynamics. Here we study the impact of meter-scale longitudinal variations of group-velocity dispersion on the propagation of bright solitons and on their associated dispersive waves. We show that the longitudinal tailoring of fiber properties allows to observe experimentally unique dispersive wave dynamics, such as the emission of cascaded, multiple, or polychromatic dispersive waves.

Introduction

Since its discovery in the frame of the Korteweg-de Vries equation by Zabusky and Kruskal (1965) and Gardner et al. (1967), the concept of solitons has been extended to many other systems described by integrable equations (Ablowitz et al. 1973), including the nonlinear Schrödinger equation (NLSE) (Zakharov and Shabat 1972) widely used to study nonlinear pulse propagation in optical fibers (Mollenauer et al. 1980; Hasegawa and Matsumoto 2002; Dudley et al. 2006). However, in real-world fibers, intrinsic higher-order dispersive and nonlinear effects break the integrability of the NLSE and therefore perturb the invariant propagation of solitons (Kivshar and Malomed 1989). Fundamental solitons, on the contrary to higher-order ones, are very stable in optical fibers, and they usually survive these perturbations, by continuously adapting their temporal and spectral shapes (Elgin 1993). The robustness of fundamental solitons has been exploited in the context of dispersion-managed optical communications in which the periodic evolution of losses and gain due to fiber attenuation and amplifiers is compensated by a periodic arrangement of dispersion and/or nonlinearity (Turitsyn et al. 2012). The so-called dispersion-managed solitons are propagating over kilometer-long systems, and they are usually relatively broad (in the picosecond duration scale), making them weakly altered by higher-order dispersion. In fact, in some cases, third-order dispersion might even help to stabilize them (Hizanidis et al. 1998). In the case of much narrower solitons (in the hundreds of femtosecond time scale), higher-order dispersion plays a much more significant role. In the case of a perturbation due to third-order dispersion, for example, a short soliton propagating near the zero-dispersion wavelength (ZDW) loses energy into a dispersive wave (also called resonant or Cherenkov radiation) across the ZDW (Wai et al. 1986, 1987) and experiences a spectral recoil in the opposite spectral direction in order to conserve the overall energy (Akhmediev and Karlsson 1995). This very well-known process has been studied extensively from theoretical and experimental points of views (Wai et al. 1986, 1987; Akhmediev and Karlsson 1995; Cristiani et al. 2004; Erkintalo et al. 2012; Webb et al. 2013; Conforti and Trillo 2013), in particular in the context of supercontinuum generation in which it plays a crucial role in the early dynamics (Dudley et al. 2006; Skryabin and Gorbach 2010). Following their emission, dispersive waves can also collide with solitons in the presence of Raman effect, leading to their nonlinear interaction (Yulin et al. 2004; Efimov et al. 2005; Skryabin and Yulin 2005). Understanding this nonlinear wave mixing process has been a key in extending supercontinuum sources toward the blue/ultraviolet spectral region

(Skryabin and Gorbach 2010; Gorbach et al. 2006; Gorbach and Skryabin 2007) and also toward long wavelengths in some specific cases (Chapman et al. 2010).

In this chapter, we study the process of dispersive wave emission from a soliton in various axially varying optical fibers. We show that the longitudinal variation of guiding properties allows to observe experimentally new and unique dynamics regarding dispersive waves. The first section introduces the basics of dispersive wave emission from a fundamental soliton, and the second one focuses on the peculiarities of this process in axially varying fibers.

Emission of a Dispersive Wave from a Soliton

In this section, we will introduce the basic concepts of fundamental temporal soliton propagation in optical fiber with second-order dispersion (or group-velocity dispersion (GVD)) and Kerr nonlinearity, as well as radiation of dispersive wave in the presence of third-order dispersion.

Fundamental Soliton

The nonlinear propagation of light in dispersive and nonlinear optical fibers is described by the nonlinear Schrödinger equation (NLSE)

$$\frac{\partial A}{\partial z} = -i\frac{\beta_2}{2}\frac{\partial^2 A}{\partial t^2} + i\gamma |A|^2 A \qquad (1)$$

Here, t is the retarded time in the frame traveling at the group velocity v_g of the input pulse. $A(z, t)$ is the envelope of the electric field, β_2 is the fiber GVD coefficient, and γ is the Kerr nonlinear parameter defined using the standard definition from Agrawal (2012). The fundamental soliton is an analytical solution of Eq. 1 (Zakharov and Shabat 1972) when the second-order dispersion coefficient is negative (anomalous GVD) and the nonlinear parameter γ is positive. Its amplitude takes the form

$$A(z,t) = \sqrt{P_0}\,\text{sech}\left(\frac{t}{T_0}\right) \qquad (2)$$

where P_0 is the soliton peak power and T_0 its duration. The soliton is termed *fundamental* when the following condition is fulfilled:

$$P_0 = \frac{|\beta_2|}{\gamma T_0^2} \qquad (3)$$

Figure 1 shows the results of a numerical integration of Eq. 1 in a standard telecommunication single-mode fiber. The second-order dispersion coefficient is $\beta_2 = -6.05 \times 10^{-28}$ s^2/m and the nonlinear parameter is $\gamma = 2.08$ W^{-1}.km^{-1}.

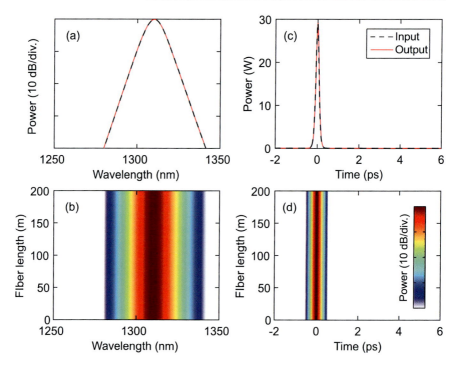

Fig. 1 Simulation of soliton propagation using Eq. 1, i.e., neglecting third-order dispersion. (**a**) Input (black dashed line) and output (red solid line) spectra in a standard telecommunication single-mode fiber of 200 m length. (**b**) Spectral dynamics along propagation, with the same colorscale as in (**d**). (**c**) Corresponding input (black dashed line) and output (red solid line) temporal profiles. (**d**) Temporal dynamics along propagation. Simulation parameters are given in the text

The input pulse is a fundamental soliton centered at 1310 nm with a duration T_0 of 100 fs and a peak power of 29.08 W (in accordance with Eq. 3). Figure 1a–d shows that the input pulse propagates without any spectral or temporal modification along the fiber, respectively. This is the most striking feature of temporal fundamental solitons, and this is the main reason why they have attracted so much interest both from fundamental and applicative point of views (Hasegawa and Matsumoto 2002; Taylor et al. 1996; Mollenauer and Smith 1988).

Dispersive Wave

In practice, however, one cannot always neglect the fiber third-order dispersion. In particular, it plays a major role when the soliton is located near the zero-dispersion wavelength (ZDW) of the fiber (Wai et al. 1986, 1987), i.e., when the soliton spectrum overlaps the normal GVD region located across the ZDW. In this case, Eq. 1 has to be rewritten in order to take into account the fiber third-order dispersion. It takes the form

$$\frac{\partial A}{\partial z} = -i\frac{\beta_2}{2}\frac{\partial^2 A}{\partial^2 t} + \frac{\beta_3}{6}\frac{\partial^3 A}{\partial^3 t} + i\gamma |A|^2 A \quad (4)$$

where β_3 is the third-order dispersion coefficient of the fiber. The third-order dispersion drastically modifies the soliton dynamics as compared to the case in which it is neglected. This is illustrated in Fig. 2 where we have simulated the propagation of the same input pulse as above using Eq. 4. We have considered a β_3 value of 6.9×10^{-41} s^3/m, the other fiber parameters being the same ones as above. The fiber ZDW (represented by the black dotted line in Fig 2a, b) is 1302 nm. The output spectrum obtained after 200 m (red solid line in Fig. 2a) strongly differs from the input one (black dashed line). First, it exhibits a strong peak around 1280 nm corresponding to a dispersive wave. Second, the soliton spectrum is distorted, and its central wavelength has redshifted from 1310 to 1313 nm. This is the spectral recoil accompanying the emission of the dispersive wave. The spectral dynamics displayed in Fig. 2b shows that the spectrum initially broadens toward the short-wavelength region and that the dispersive wave starts to form at about 50 m of propagation. In the time domain (Fig. 2c), it appears at the trailing edge of the soliton (red solid line) as a broad and relatively distorted pulse (see inset). Since the soliton has experienced a spectral recoil, it has slightly decelerated as compared to the fiber input. It has also lost some energy, which has been radiated into the dispersive wave. This can be further observed in the temporal dynamics plot of Fig. 2d, where we can see the slow deceleration of the soliton as well as the continuous radiation of a broad dispersive wave along propagation.

The frequency at which the dispersive wave is emitted can be predicted from phase-matching arguments between the two radiations as (Akhmediev and Karlsson 1995)

$$\frac{\beta_2}{2}\Omega^2 + \frac{\beta_3}{6}\Omega^3 = \frac{\gamma P_0}{2} \quad (5)$$

where $\Omega = \omega_S - \omega_{DW}$ is the frequency separation between the soliton at ω_S and the dispersive wave at ω_{DW}. Solving Eq. 5 with the parameters of our study gives a dispersive wave wavelength of 1283 nm, in excellent agreement with the numerical simulation results of Fig 2a, b. The process of dispersive wave emission from a soliton is very well-known and explained as long as optical fibers are uniform in length: when the soliton spectrum crosses the ZDW, the part of soliton energy located in the normal GVD region is radiated into a dispersive wave. This causes a spectral recoil of the soliton which thus moves away from the ZDW so that the emission process becomes less and less efficient and cannot occur again. Consequently, there is a unique dispersive wave which is generated and appears as a sharp spectral peak whose frequency does not change with propagation, as observed from Fig. 2b. However, the process can be radically different in optical fibers which are not uniform as a function of length, i.e., in axially varying fibers. This will be the focus of the remaining of this chapter.

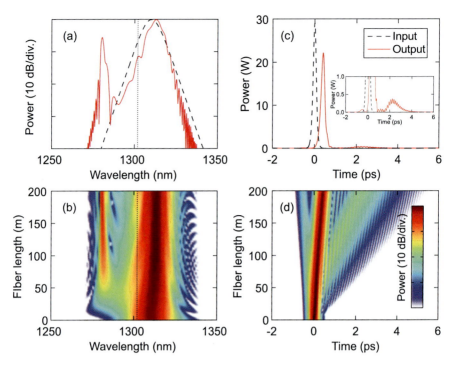

Fig. 2 Simulation of soliton propagation using Eq. 4, i.e., taking third-order dispersion into account. (**a**) Input (black dashed line) and output (red solid line) spectra in a standard telecommunication single-mode fiber of 200 m length. (**b**) Spectral dynamics along propagation, with the same colorscale as in (**d**). (**c**) Corresponding input (black dashed line) and output (red solid line) temporal profiles. (**d**) Temporal dynamics along propagation. Simulation parameters are given in the text. The black dotted lines in (**a**) and (**b**) represent the fiber ZDW

Generation of Dispersive Waves from Solitons in Axially Varying Optical Fibers

Axially Varying Optical Fibers

For the fabrication of axially varying optical fibers, the longitudinal evolution of the fiber diameter is controlled by adjusting the evolution of drawing speed with time (which is related to fiber length) using a servo-control system. This process is ruled by the conservation of glass mass between the preform and the fiber:

$$d_{\text{Fiber}} = d_{\text{Preform}} \sqrt{\frac{V_{\text{Preform}}}{V_{\text{Fiber}}}} \qquad (6)$$

where $d_{\text{Preform, Fiber}}$ are, respectively, the preform and fiber outer diameter and $V_{\text{Preform, Fiber}}$ are, respectively, the preform feed into the furnace and the drawing

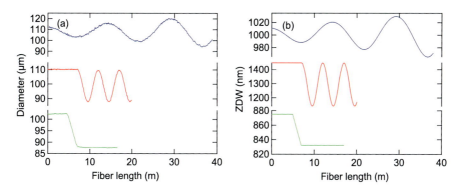

Fig. 3 (a) Evolution of the outer diameter as a function of fiber length for three different axially varying fibers measured during the fiber fabrication process. (b) Corresponding calculated ZDW. For the second fiber (red case), the ZDW plotted in (b) corresponds to the second one (located at long wavelengths)

capstan speed. In our process, d_{Preform} and V_{Preform} are fixed, and V_{Fiber} is adjusted with a desired $f(z)$ function (where z is the longitudinal space coordinate along the fiber), which results in a modulation of the fiber outer diameter $d_{\text{Fiber}}(z)$ and thus of the overall fiber structure. This results in a modulation of the mode(s) propagation constant(s) and thus of all guiding properties.

Figure 3a shows three examples of axially varying fibers used hereafter. These curves show the evolution of the outer diameter recorded during the fiber drawing process. Figure 3b shows the corresponding calculated ZDW. The example represented in red corresponds to a fiber with two ZDWs, and the plot in Fig. 3b corresponds to the second ZDW. In the range of diameter variations investigated here, the ZDW approximately follows a linear dependence with fiber diameter.

Emission of Multiple Dispersive Waves Along the Fiber

As recalled above, in uniform fibers, the soliton experiences a redshift (spectral recoil) during the generation of a dispersive wave. This redshift is usually strongly reinforced by the Raman-induced soliton self-frequency shift. As a consequence, the soliton moves far away from the ZDW, so that the dispersive wave emission process cannot occur again. We will first show an example in which the use of a suitable axially varying fiber allows for a single soliton to emit multiple dispersive waves along the fiber. Indeed, this can occur if the fiber ZDW moves along propagation so that it "hits" the soliton during its redshift following the emission of a dispersive wave (Arteaga-Sierra et al. 2014; Billet et al. 2014). In this case, the overlap between the soliton spectrum and the normal GVD region can become large enough to initiate the emission of a new dispersive wave. This is illustrated with numerical simulations and experiments in Fig. 4a, b, respectively. For these numerical simulations, the stimulated Raman scattering term has been added to

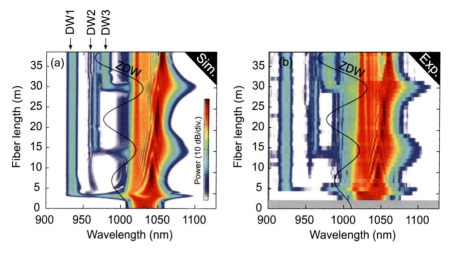

Fig. 4 (**a**) Numerical simulations and (**b**) experimental results in the spectral domain versus fiber length showing the emission of three distinct dispersive waves from a single soliton in an axially varying fiber. The evolution of the ZDW with fiber length is represented by black solid lines

Eq. 4, which now writes

$$\frac{\partial A}{\partial z} = -i\frac{\beta_2}{2}\frac{\partial^2 A}{\partial^2 t} + \frac{\beta_3}{6}\frac{\partial^3 A}{\partial^3 t} + i\gamma\left(A\int R(t')|A(t-t')|^2 dt'\right) \quad (7)$$

where $R(t) = (1 - f_R)\delta(t) + f_R h_R(t)$ includes both Kerr and Raman effects, where $h_R(t)$ corresponds to the Raman response function ($f_R = 0.18$) taken from Hollenbeck and Cantrell (2002).

We consider a fiber with a ZDW which evolves along the fiber (blue curve in Fig. 3a, b), following the profile represented by the black solid lines in Fig. 4a, b. The ZDW is 1011 nm at the fiber input and reaches 1021 and 1030 nm at 14 and 29 m, respectively. Simulations and experiments are performed with an input pulse of 410 fs full width at half maximum (FWHM) duration centered around 1030 nm and a peak power of 46 W. In order to be consistent with experiments, we consider a slight chirp with a chirp parameter of +3.7. All parameters can be found in Billet et al. (2014). The simulation in Fig. 4a shows the emission of a first dispersive wave around 925 nm at a length of 5 m, in a similar fashion to what happens in uniform fibers (see Fig. 2). The soliton experiences a redshift due to the combined action of spectral recoil and Raman effect. The ZDW increases to 1021 nm at 1 m, which brings it closer to the soliton and therefore enhances the spectral overlap between the soliton and the normal GVD region. This initiates the emission of a second dispersive wave around 960 nm at this location in the fiber. The same process occurs a third time around 29 m, resulting in the emission of a third dispersive wave. Experiments reported in Fig. 4b are in excellent agreement with simulations and demonstrate the process of multiple dispersive wave emission from a single soliton.

Cascading of Dispersive Waves

The process of dispersive wave emission can also occur in a slightly different scenario in uniform fibers: a soliton experiencing a strong Raman-induced self-frequency shift can hit the second ZDW of a fiber (located at long wavelengths), which cancels the redshift and generates a dispersive wave across the ZDW, at longer wavelengths (Skryabin et al. 2003; Biancalana et al. 2004). Here we will study this process in an axially varying fiber (red curve in Fig. 3a, b) with two ZDWs separated by a region of anomalous dispersion in which a fundamental soliton is excited. The evolution of the second ZDW (the long wavelength one) is represented by the white solid lines in Fig. 5. It is constant over the first 7 m and then oscillates as a function of length. We consider input pulses of 340 fs FWHM duration centered around 1030 nm, with a chirp parameter of +1.5 and a peak power of 75 W. Figure 5a, b shows, respectively, the numerical simulations performed with Eq. 7 and experimental results.

The input short pulse excites a fundamental soliton which initially experiences Raman-induced soliton self-frequency shift. Around 9 m, the ZDW has reached 1140 nm and is very close to the soliton so that a dispersive wave (labeled DW1 in Fig. 5a) is emitted across the ZDW, in the normal GVD region. At this point, the process is very similar to the one known in uniform fibers. The real novelty comes slightly after 10 m, when DW1 crosses the ZDW (which is increasing again

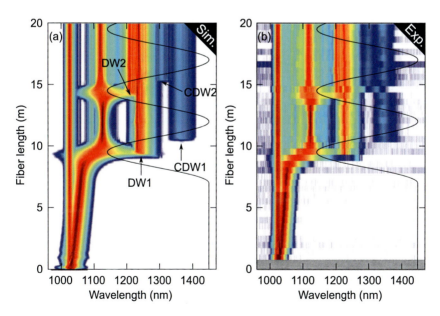

Fig. 5 (**a**) Numerical simulations and (**b**) experimental results in the spectral domain versus fiber length showing the emission of two cascaded dispersive waves in an axially varying fiber with two ZDWs. The evolution of the second ZDW (located at long wavelengths) with fiber length is represented by black solid lines

at this point): as soon as DW1 crosses the ZDW, a new radiation (labeled CDW1) is generated at even longer wavelengths, around 1350 nm. A careful analysis of this process reveals that the continuously evolving GVD prevents the dispersive wave to strongly spread out in time (as expected in uniform fibers, see Fig. 2d) and allows to keep it relatively localized in time as a short pulse (Bendahmane et al. 2014). As a consequence, when crossing the ZDW, this pulse can emit another dispersive wave which we term cascaded dispersive wave (CDW1), in analogy with cascaded four-wave mixing processes.

The exact same process occurs again farther in the fiber, around 15 m. The ZDW decreases again very close to the remaining of the soliton located slightly above 1100 nm so that a dispersive wave, DW2, is emitted. When DW2 crosses the ZDW slightly after, a cascaded dispersive wave, CDW2, is emitted. Experiments reported in Fig. 5b show again excellent agreement with numerical results and provide the evidence for the process of cascaded dispersive wave generation. We might also note that, similarly to the previous section, multiple dispersive waves (DW1 and DW2) have been observed from a single soliton, around the second ZDW.

Transformation of a Dispersive Wave into a Fundamental Soliton

As mentioned above and deeply analyzed in Bendahmane et al. (2014), the cascaded dispersive wave process is due to the fact that the dispersive wave initially generated experiences a GVD that varies with length, which allows it to remain temporally localized as a pulse to initiate the generation of a cascaded dispersive wave. Here, we will study this process more into details. For that, we consider a soliton launched in the vicinity of the ZDW so that it emits a dispersive wave in the normal GVD region, similarly to the usual case illustrated in Fig. 2. Once the dispersive wave is emitted, the fiber parameters are changed so that the GVD at the dispersive wave wavelength becomes anomalous (green curve in Fig. 3a, b) and the evolution of the radiation is studied. The results are summarized in Fig. 6. The input pulse is transform-limited with a 140 fs FWHM duration. It is centered around 881 nm and has a peak power of 42 W. The spectral evolution versus fiber length (simulations in Fig. 6a and experiments in Fig. 6c) shows the emission of a dispersive wave around 840 nm but does not highlight much change after the GVD change occurring between 5 and 7 m. The simulated time domain simulation of Fig. 6b provides much more information about the dynamics. The soliton decelerates due to the combined effects of spectral recoil and Raman-induced self-frequency shift, and the dispersive wave, which starts spreading out, is located slightly behind it, as expected. Once it enters the tapered region (materialized by the two dashed lines), the dispersive wave accelerates due to changing dispersion and crosses the soliton. At the same time, it recompresses and further propagated as a short pulse in the remaining of the fiber, recalling the behavior of a soliton. In fact, further theoretical analysis reveals that the dispersive wave has indeed been transformed into a fundamental soliton (Braud et al. 2016) and propagates as such in the remaining anomalous dispersion region. Experimental autocorrelation measurements reported in Fig. 6d confirm this

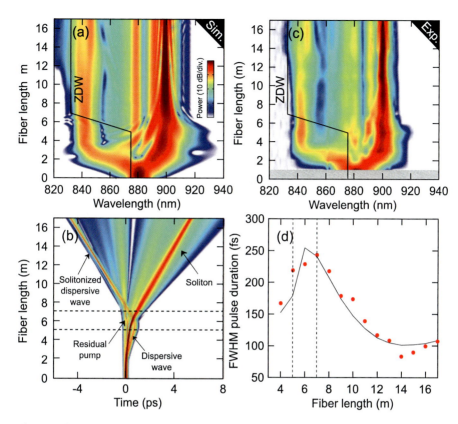

Fig. 6 (**a, b**) Numerical simulations and (**c, d**) experimental results showing the transformation of a dispersive wave into a fundamental soliton. (**a**) and (**b**) correspond, respectively, to the simulated spectral and temporal evolutions. (**c**) Measured spectral evolution versus fiber length. (**d**) Measured pulse duration around 840 nm (red full circles) and simulated one (black solid line). Black solid lines in (**a**) and (**c**) depict the ZDW. Dashed lines in (**b**) and (**d**) depict the limits of the tapered section

behavior: the pulse duration initially increases and then decreases after the tapered region, where they are perfectly fitted by square hyperbolic secant functions (Braud et al. 2016). These results show that a dispersive wave emitted from a soliton can be itself transformed into a fundamental soliton by carefully varying the longitudinal evolution of dispersion.

Emission of Polychromatic Dispersive Waves

In the previous section, we have shown that a dispersive pulse initially traveling in normal GVD can become transformed into a fundamental soliton when entering a region with anomalous dispersion. Here we will study a reversed situation in which a fundamental soliton propagating in anomalous dispersion enters a region with normal dispersion. Fundamentally, the pulse cannot be a soliton anymore, and it is expected to linearly disperse. We will see that it can excite a so-called polychromatic dispersive wave (Milin et al. 2012; Kudlinski et al. 2015).

Here, a fundamental soliton is excited by launching a transform-limited pulse with a 130 fs FWHM duration around 950 nm, with a peak power of 110 W. The fiber is uniform over the first 4.5 m, and then it is tapered down so that both ZDWs (depicted by solid lines in Fig. 7a, c) decrease until they join each other at 6 m. After this point, the fiber has all normal GVD all over the spectral range of interest here.

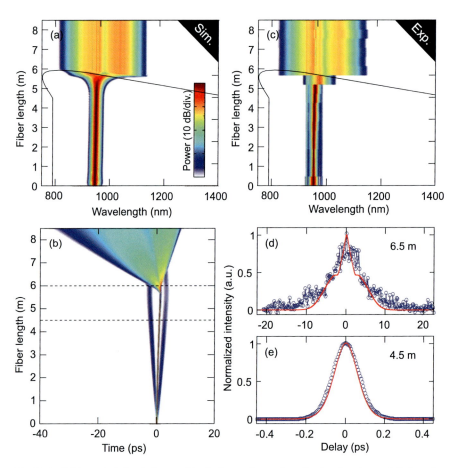

Fig. 7 (**a**,**b**) Numerical simulations and (**c**, **d**, **e**) experimental results showing the annihilation of a fundamental soliton into a polychromatic dispersive wave. (**a**) and (**b**) correspond, respectively, to the simulated spectral and temporal evolutions. (**c**) Measured spectral evolution versus fiber length. (**d**, **e**) Measured autocorrelation trace (open blue circles) and simulated ones (red solid line) at fiber length of (**e**) 4.5 m and (**d**) 6.5 m. Black solid lines in (**a**) and (**c**) depict the ZDWs. Dashed lines in (**b**) depict the limits of the tapered section

The spectral dynamics, displayed in Fig. 7a, c for simulations and experiments, respectively, show the initial propagation of a soliton until 5.5 m, where it crosses the second ZDW and thus enters the normal GVD region. At this point, a spectacular spectral broadening occurs, indicating that the pulse experiences a nonlinear. In fact, as long as the soliton spectrum starts to cross the ZDW, the emission of a dispersive wave is initiated. Since the second ZDW keeps decreasing and crossing the soliton, there is a continuous emission of radiation into the dispersive wave. Because the GVD properties of the fiber continuously changes, the phase-matching relation (5) linking the soliton and the dispersive wave continuously changes too, resulting in the generation of a broad radiation termed polychromatic dispersive wave (Milin et al. 2012; Kudlinski et al. 2015).

The process is as spectacular as in the time domain plot of Fig. 7b, where it can be seen that the significant broadening occurring at around 6 m is accompanied by a strong temporal broadening, which is consistent with the generation of a dispersive, strongly chirped pulse. This was confirmed experimentally by recording autocorrelation traces before and after the tapered section. At 4.5 m (Fig. 7e), the pulse has a duration of 113 fs (blue markers) and is well fitted by a square hyperbolic secant function, in good agreement with the simulation result (red solid line), which confirms its solitonic nature. But soon after the tapered section, at 6.5 m, the autocorrelation trace is much longer and has a much distorted profile, again in agreement with simulations. At this point, there is no clue of the presence of a soliton, which has therefore totally been annihilated into a polychromatic dispersive wave.

Generation of a Dispersive Wave Continuum

In this section, we will investigate a complex scenario in which multiple solitons and dispersive waves are generated in an axially varying fiber. This results in the generation of a 500 nm supercontinuum exclusively composed of dispersive waves.

For that, we use the same configuration as in section "Cascading of Dispersive Waves" except that we increase the peak power of the input pulse to 380 W. This exceeds the required power to form a fundamental soliton given by Eq. 3. In this case, the input pulse excites a higher-order soliton which immediately breaks up into several fundamental solitons (Beaud et al. 1987). Figure 8a, b shows, respectively, the simulated and experimental spectral dynamics recorded in this case. Three main solitons, labeled S_1 to S_3, form here. The first one, S_1, is also the most powerful one (Lucek and Blow 1992) and therefore experiences the most efficient Raman-induced frequency shift. It rapidly hits the second ZDW (in black solid line) so that the frequency shift stops and a strong dispersive wave (labeled DW_1) is emitted around 1600 nm. The soliton is frequency-locked around 1400 nm and hits the ZDW which starts to decrease at around 7 m. At this point, it emits a polychromatic dispersive wave (labeled PDW_1) following the process described in section "Emission of Polychromatic Dispersive Waves." The second soliton (S_2) experiences a less significant Raman-induced self-frequency shift than S_1 and hits

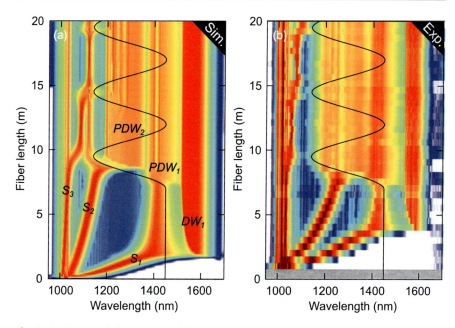

Fig. 8 (a) Numerical simulations and (b) experimental results in the spectral domain versus fiber length showing the generation of a dispersive wave continuum in an axially varying fiber with two ZDWs. The evolution of the second ZDW (located at long wavelengths) with fiber length is represented by black solid lines

the ZDW around 8 m, where it is already decreasing. Therefore, it emits a broad polychromatic dispersive wave (labeled PDW_2). Soliton S_3 remains relatively far from the ZDW all over the propagation and thus does not contribute significantly to the emission of dispersive waves.

Finally, the output spectrum between 1200 and 1600 nm is exclusively composed of dispersive waves and polychromatic dispersive waves generated from the two most powerful solitons.

Conclusion and Perspectives

The perturbation of a fundamental soliton by third-order dispersion in optical fibers causes the emission of a resonant dispersive wave across the zero-dispersion point. In axially varying fibers, the guiding properties can be tailored as a function of propagation distance so that the ZDW continuously evolves as the soliton propagates. This allows to observe specific dispersive wave dynamics, such as the emission of cascaded, multiple, or polychromatic dispersive waves.

Such axially varying fibers can also be used to control the properties of the soliton itself, such as harnessing its Raman-induced redshift (Judge et al. 2009; Bendahmane et al. 2013) or even inducing a blueshift (Stark et al. 2011).

In periodic fibers, multiple quasi-phase-matched dispersive waves can also be observed thanks to the periodicity (Conforti et al. 2015, 2016), following a completely different mechanism than the one presented here (Conforti et al. 2015). More generally, this work illustrates remarkable robustness of fundamental solitons against various types of perturbations and the extremely rich nonlinear dynamics that these perturbations can induce.

References

M.J. Ablowitz, D.J. Kaup, A.C. Newell, H. Segur, Phys. Rev. Lett. **31**(2), 125 (1973). https://link.aps.org/doi/10.1103/PhysRevLett.31.125. https://journals-aps-org.buproxy.univ-lille1.fr/prl/pdf/10.1103/PhysRevLett.31.125

G. Agrawal, *Nonlinear Fiber Optics*, 5th edn. (Academic Press, Amsterdam, 2012)

N. Akhmediev, M. Karlsson, Phys. Rev. A **51**(3), 2602 (1995). http://link.aps.org/doi/10.1103/PhysRevA.51.2602

F.R. Arteaga-Sierra, C. Milin, I. Torres-Gmez, M. Torres-Cisneros, A. Ferrando, A. Dvila, Opt. Express **22**(3), 2451 (2014). https://doi.org/10.1364/OE.22.002451. http://www.opticsexpress.org/abstract.cfm?URI=oe-22-3-2451

P. Beaud, W. Hodel, B. Zysset, H. Weber, IEEE J. Quantum Electron **23**(11), 1938 (1987). https://doi.org/10.1109/JQE.1987.1073262

A. Bendahmane, O. Vanvincq, A. Mussot, A. Kudlinski, Opt. Lett. **38**(17), 3390 (2013). https://doi.org/10.1364/OL.38.003390. http://ol.osa.org/abstract.cfm?URI=ol-38-17-3390

A. Bendahmane, F. Braud, M. Conforti, B. Barviau, A. Mussot, A. Kudlinski, Optica **1**(4), 243 (2014). https://doi.org/10.1364/OPTICA.1.000243. https://www.osapublishing.org/optica/abstract.cfm?uri=optica-1-4-243

F. Biancalana, D.V. Skryabin, A.V. Yulin, Phys. Rev. E **70**(1) (2004). http://web-support@bath.ac.uk. http://opus.bath.ac.uk/8984/

M. Billet, F. Braud, A. Bendahmane, M. Conforti, A. Mussot, A. Kudlinski, Opt. Express **22**(21), 25673 (2014). https://doi.org/10.1364/OE.22.025673. http://www.opticsexpress.org/abstract.cfm?URI=oe-22-21-25673

F. Braud, M. Conforti, A. Cassez, A. Mussot, A. Kudlinski, Opt. Lett. OL **41**(7), 1412 (2016). https://doi.org/10.1364/OL.41.001412. https://www.osapublishing.org/abstract.cfm?uri=ol-41-7-1412

B.H. Chapman, J.C. Travers, S.V. Popov, A. Mussot, A. Kudlinski, Opt. Express **18**(24), 24729 (2010). https://doi.org/10.1364/OE.18.024729. http://www.opticsexpress.org/abstract.cfm?URI=oe-18-24-24729

M. Conforti, S. Trillo, Opt. Lett. **38**(19), 3815 (2013). https://doi.org/10.1364/OL.38.003815. http://ol.osa.org/abstract.cfm?URI=ol-38-19-3815

M. Conforti, S. Trillo, A. Mussot, A. Kudlinski, Sci. Rep. **5**, srep09433 (2015). https://doi.org/10.1038/srep09433. https://www.nature.com/articles/srep09433

M. Conforti, S. Trillo, A. Kudlinski, A. Mussot, IEEE Photon. Technol. Lett. **28**(7), 740 (2016). https://doi.org/10.1109/LPT.2015.2507190

I. Cristiani, R. Tediosi, L. Tartara, V. Degiorgio, Opt. Express **12**(1), 124 (2004). https://doi.org/10.1364/OPEX.12.000124. http://www.opticsexpress.org/abstract.cfm?URI=oe-12-1-124

J.M. Dudley, G. Genty, S. Coen, Rev. Mod. Phys. **78**(4), 1135 (2006). http://link.aps.org/doi/10.1103/RevModPhys.78.1135

A. Efimov, A.V. Yulin, D.V. Skryabin, J.C. Knight, N. Joly, F.G. Omenetto, A.J. Taylor, P. Russell, Phys. Rev. Lett. **95**(21), 213902 (2005). http://link.aps.org/doi/10.1103/PhysRevLett.95.213902

J.N. Elgin, Phys. Rev. A **47**(5), 4331 (1993). http://link.aps.org/doi/10.1103/PhysRevA.47.4331

M. Erkintalo, Y.Q. Xu, S.G. Murdoch, J.M. Dudley, G. Genty, Phys. Rev. Lett. **109**(22), 223904 (2012). http://link.aps.org/doi/10.1103/PhysRevLett.109.223904

C.S. Gardner, J.M. Greene, M.D. Kruskal, R.M. Miura, Phys. Rev. Lett. **19**(19), 1095 (1967). https://link.aps.org/doi/10.1103/PhysRevLett.19.1095. https://journals-aps-org.buproxy.univ-lille1.fr/prl/pdf/10.1103/PhysRevLett.19.1095

A.V. Gorbach, D.V. Skryabin, Nat. Photon **1**(11), 653 (2007). https://doi.org/10.1038/nphoton.2007.202. http://www.nature.com/nphoton/journal/v1/n11/abs/nphoton.2007.202.html

A.V. Gorbach, D.V. Skryabin, J.M. Stone, J.C. Knight, Opt. Express **14**(21), 9854 (2006). https://doi.org/10.1364/OE.14.009854. http://www.opticsexpress.org/abstract.cfm?URI=oe-14-21-9854

A. Hasegawa, M. Matsumoto, *Optical Solitons in Fibers*, 3rd edn., revised and enlarged edition 2003 edn. (Springer, Berlin/Heidelberg; GmbH & Co. K, Berlin/New York, 2002)

K. Hizanidis, B.A. Malomed, H.E. Nistazakis, D.J. Frantzeskakis, Pure Appl. Opt. **7**(4), L57 (1998). http://stacks.iop.org/0963-9659/7/i=4/a=003

D. Hollenbeck, C.D. Cantrell, J. Opt. Soc. Am. B **19**(12), 2886 (2002). https://doi.org/10.1364/JOSAB.19.002886. http://josab.osa.org/abstract.cfm?URI=josab-19-12-2886

A.C. Judge, O. Bang, B.J. Eggleton, B.T. Kuhlmey, E.C. Mgi, R. Pant, C.M. de Sterke, J. Opt. Soc. Am. B **26**(11), 2064 (2009). https://doi.org/10.1364/JOSAB.26.002064. http://josab.osa.org/abstract.cfm?URI=josab-26-11-2064

Y.S. Kivshar, B.A. Malomed, Rev. Mod. Phys. **61**(4), 763 (1989). https://link.aps.org/doi/10.1103/RevModPhys.61.763. https://journals-aps-org.buproxy.univ-lille1.fr/rmp/pdf/10.1103/RevModPhys.61.763

A. Kudlinski, S.F. Wang, A. Mussot, M. Conforti, Opt. Lett. **40**(9), 2142 (2015). https://doi.org/10.1364/OL.40.002142. https://www.osapublishing.org/ol/abstract.cfm?uri=ol-40-9-2142

J.K. Lucek, K.J. Blow, Phys. Rev. A **45**(9), 6666 (1992). https://link.aps.org/doi/10.1103/PhysRevA.45.6666

C. Milin, A. Ferrando, D.V. Skryabin, J. Opt. Soc. Am. B **29**(4), 589 (2012). https://doi.org/10.1364/JOSAB.29.000589. http://josab.osa.org/abstract.cfm?URI=josab-29-4-589

L.F. Mollenauer, K. Smith, Opt. Lett. **13**(8), 675 (1988). https://doi.org/10.1364/OL.13.000675. https://www.osapublishing.org/ol/abstract.cfm?uri=ol-13-8-675

L.F. Mollenauer, R.H. Stolen, J.P. Gordon, Phys. Rev. Lett. **45**(13), 1095 (1980). http://link.aps.org/doi/10.1103/PhysRevLett.45.1095

D.V. Skryabin, A.V. Gorbach, Rev. Mod. Phys. **82**(2), 1287 (2010). http://link.aps.org/doi/10.1103/RevModPhys.82.1287

D.V. Skryabin, A.V. Yulin, Phys. Rev. E **72**(1), 016619 (2005). http://link.aps.org/doi/10.1103/PhysRevE.72.016619

D.V. Skryabin, F. Luan, J.C. Knight, P.S.J. Russell, Science **301**(5640), 1705 (2003). https://doi.org/10.1126/science.1088516. http://www.sciencemag.org/content/301/5640/1705

S.P. Stark, A. Podlipensky, P.S.J. Russell, Phys. Rev. Lett. **106**(8), 083903 (2011). http://link.aps.org/doi/10.1103/PhysRevLett.106.083903

J.R. Taylor, R. Arguello, J. Roger, Opt. Eng. **35**(8), 2437 (1996). http://doi.org/10.1117/1.600816

S.K. Turitsyn, B.G. Bale, M.P. Fedoruk, Phys. Rep. **521**(4), 135 (2012). https://doi.org/10.1016/j.physrep.2012.09.004. http://www.sciencedirect.com/science/article/pii/S0370157312002657

P.K.A. Wai, C.R. Menyuk, Y.C. Lee, H.H. Chen, Opt. Lett. **11**(7), 464 (1986). https://doi.org/10.1364/OL.11.000464. http://ol.osa.org/abstract.cfm?URI=ol-11-7-464

P.K.A. Wai, C.R. Menyuk, H.H. Chen, Y.C. Lee, Opt. Lett. **12**(8), 628 (1987). https://doi.org/10.1364/OL.12.000628. http://ol.osa.org/abstract.cfm?URI=ol-12-8-628

K.E. Webb, Y.Q. Xu, M. Erkintalo, S.G. Murdoch, Opt. Lett. **38**(2), 151 (2013). https://doi.org/10.1364/OL.38.000151. http://ol.osa.org/abstract.cfm?URI=ol-38-2-151

A.V. Yulin, D.V. Skryabin, P.S.J. Russell, Opt. Lett. **29**(20), 2411 (2004). https://doi.org/10.1364/OL.29.002411. http://ol.osa.org/abstract.cfm?URI=ol-29-20-2411

N.J. Zabusky, M.D. Kruskal, Phys. Rev. Lett. **15**(6), 240 (1965). https://doi.org/10.1103/PhysRevLett.15.240

V.E. Zakharov, A.B. Shabat, Sovi. J. Exp. Theor. Phys. **34**, 62 (1972). http://adsabs.harvard.edu/abs/1972JETP...34...62Z

Nonlinear Waves in Multimode Fibers

8

I. S. Chekhovskoy, O. S. Sidelnikov, A. A. Reduyk, A. M. Rubenchik,
O. V. Shtyrina, M. P. Fedoruk, S. K. Turitsyn, E. A. Zlobina,
S. I. Kablukov, S. A. Babin, K. Krupa, V. Couderc, A. Tonello,
A. Barthélémy, G. Millot, and S. Wabnitz

Contents

Introduction	319
Spatiotemporal Pulse Shaping in Multicore Fibers	321
Pulse Propagation in Multicore Fibers	322
Pulse Compression and Combining	324
Nonlinear Pulses in Multimode Fibers for Spatial-Division Multiplexing	328
Spatial-Division Multiplexing	328
Nonlinear Propagation in Multimode Fibers	329
The Influence of Nonlinear Effects on the Propagation of Optical Signals	331
Raman Cleanup Effect and Raman Lasing in Multimode Graded-Index Fibers	334
Experimental Observations and Theoretical Models of Raman Cleanup Effect	336
Raman Cleanup Effect in Raman Fiber Amplifiers and Lasers	340
GRIN Fiber Raman Lasers Directly Pumped by Multimode Laser Diodes	342
Combined Action of Raman Beam Cleanup and Mode-Selecting FBGs in GRIN Fiber Raman Lasers	346
Kerr Beam Self-Cleaning	353
Theoretical Models of Spatiotemporal Dynamics	355

I. S. Chekhovskoy · O. V. Shtyrina · M. P. Fedoruk
Novosibirsk State University, Novosibirsk, Russia

Institute of Computational Technologies SB RAS, Novosibirsk, Russia

O. S. Sidelnikov · A. A. Reduyk
Novosibirsk State University, Novosibirsk, Russia

A. M. Rubenchik
Lawrence Livermore National Laboratory, Livermore, CA, USA
e-mail: rubenchik@att.net

S. K. Turitsyn
Novosibirsk State University, Novosibirsk, Russia

Aston Institute of Photonic Technologies, Aston University, Birmingham, UK
e-mail: s.k.turitsyn@aston.ac.uk

© Springer Nature Singapore Pte Ltd. 2019
G.-D. Peng (ed.), *Handbook of Optical Fibers*,
https://doi.org/10.1007/978-981-10-7087-7_15

Kerr Beam Cleanup in GRIN MMF	358
Kerr Beam Cleanup in Step-Index Active MMF with Loss or Gain	363
Self-Cleaning in a MMF Laser Cavity	366
References	368

Abstract

We overview recent advances in the field of nonlinear guided wave propagation in multimode fibers. It is only in recent years that the study of nonlinear optics in multimode fibers has experienced a revival of research interest. Nonlinear arrays of linearly coupled multicore fibers permit spatiotemporal reshaping and coherent combining of ultrashort optical pulses. Spatial-division multiplexing is an emerging technology for increasing the capacity of optical communication links, and the presence of nonlinear mode coupling requires a careful analysis. Multimode nonlinear fibers have a strong potential for the implementation of a new class of high-power fiber lasers. We describe experiments of Raman beam cleanup that permit to implement quasi-single-mode Raman fiber laser with multimode pumps. Next we discuss the recently discovered effect of Kerr beam self-cleaning, whereby a speckled signal at the output of a multimode fiber may evolve toward a bell-shaped transverse profile.

E. A. Zlobina · S. I. Kablukov
Institute of Automation and Electrometry SB RAS, Novosibirsk, Russia
e-mail: kab@iae.nsk.su

S. A. Babin
Institute of Automation and Electrometry SB RAS, Novosibirsk, Russia

Novosibirsk State University, Novosibirsk, Russia
e-mail: babin@iae.nsk.su

K. Krupa
Department of Information Engineering, University of Brescia, Brescia, Italy
e-mail: katarzyna.krupa@unibs.it

V. Couderc · A. Tonello · A. Barthélémy
XLIM, UMR CNRS 7252, Université de Limoges, Limoges, France
e-mail: vincent.couderc@xlim.fr; alessandro.tonello@xlim.fr; alain.barthelemy@xlim.fr

G. Millot
ICB, UMR CNRS 6303, Université de Bourgogne, Dijon, France
e-mail: guy.millot@u-bourgogne.fr

S. Wabnitz (✉)
Novosibirsk State University, Novosibirsk, Russia

Department of Information Engineering, University of Brescia, Brescia, Italy

National Institute of Optics INO-CNR, Brescia, Italy
e-mail: stefan.wabnitz@unibs.it; stefano.wabnitz@ing.unibs.it

8 Nonlinear Waves in Multimode Fibers

Introduction

Multimode optical fibers (MMFs) have been investigated since the early 1980s, for their capacity of transmitting information over multiple spatial channels. However the presence of relatively large modal dispersion has largely prevented the use of MMFs for high-bit-rate transmissions over long-distance fiber-optic links. As a result, the long-distance fiber-optic infrastructure that has been built across the globe by operators and service providers over the past 30 years is virtually entirely based on single-mode optical fibers (SMFs). In recent years, there has been a revival of optical coherent communication systems, based on the technological development of fast, real-time digital signal processing techniques and hardware. This enables the digital compensation, in the electrical domain, of a large accumulated dispersion in the optical channel. Hence the technique of spatial-division multiplexing (SDM) using MMFs has received a renewed research and technological interest, as it may permit to overcome the capacity limitations of single-mode fibers for the next generation of fiber-optic networks. Guided modes of an ideal, longitudinally invariant optical fiber are orthogonal to each other, so that they can provide an alphabet for transmitting information. SDM may also be based on multicore optical fibers. Whenever the cores are sufficiently spaced apart, so that they do not interact via evanescent field overlap, a multicore fiber represents a simple way to spatially multiplex several independent information channels within the same optical fiber.

On the other hand, nonlinear fiber optics has also been extensively studied both theoretically and experimentally since the 1980s. It is now a mature field of technology, and it has led to many technological breakthroughs such as ultrashort pulse fiber lasers, ultrafast optical signal processing devices, supercontinuum sources, and self-referenced optical frequency combs. However, basically all of these developments have involved single-mode optical fibers.

Remarkably, it is not only until the last few years that nonlinear multimode fiber optics has started to receive the attention of researchers in a few leading groups. These initial studies have revealed a wealth of complex wave propagation phenomena, owing to the nonlinear coupling of spatial, temporal, and spectral degrees of freedom of the optical field. Besides their obvious interest from the viewpoint of basic science, nonlinear optical effects in multimode fibers hold the promise of several key technological advances, especially in the aforementioned fields of optical data communications, optical signal processing, high-power and brightness supercontinuum sources, and fiber lasers.

In this chapter, we overview recent research progress among some of the most promising directions in the field of nonlinear multimode fiber optics. We consider first in section "Introduction" pulse propagation in arrays of linearly coupled cores in a multicore fiber with either circular or hexagonal symmetry. The resulting set of coupled nonlinear evolution equations provides a discrete-continuous analogous of the two- or three-dimensional nonlinear Schrödinger equation. Based on this analogy, it is shown that it is possible to control the spatiotemporal propagation of light pulses injected in the array of cores, in a way to combine their intensity into a single core at the fiber output.

Next we investigate in section "Nonlinear Pulses in Multimode Fibers for Spatial-Division Multiplexing" nonlinear propagation of pulses in SDM transmission systems based on multimode fiber. We consider two essential cases of practical interest: weak- and strong-coupling regimes, in which the nonlinear propagation of optical signals is described by generalization of the so-called Manakov equations. The influence of nonlinear effects on the propagation of signals in a step-index multimode fiber is studied in the case of weak and strong coupling. These coupling regimes are compared with each other. We also investigate the impact of the number of propagating modes on the strength of nonlinear effects in multimode fibers.

If sections "Introduction" and "Nonlinear Pulses in Multimode Fibers for Spatial-Division Multiplexing" are dedicated to describe the mode coupling and reshaping of short optical pulses, in sections "Raman Cleanup Effect and Raman Lasing in Multimode Graded-Index Fibers" and "Kerr Beam Self-Cleaning," we describe essentially spatial nonlinear mode-coupling effects, activated either by stimulated Raman scattering or by the intensity-dependent refractive index of the fiber, known as Kerr effect. In both cases, we consider the highly multimode excitation of MMFs, which in linear conditions leads to a seemingly irregular spatial scrambling of the transverse intensity pattern at the fiber output owing to multimode interference. The presence of the Raman effect, on the other hand, permits under the condition of sufficiently high-power continuous wave (CW) excitation, to generate an essentially single-mode beam (typically close to the fundamental mode of the MMF) in the Stokes-shifted sidebands. As discussed in section "Raman Cleanup Effect and Raman Lasing in Multimode Graded-Index Fibers," this effect is known as Raman beam cleanup, and it has found important applications to the development of high-power, single-mode Raman fiber lasers pumped by multimode diodes.

We conclude the chapter with the overview in section "Kerr Beam Self-Cleaning" of recent experimental studies describing the discovery of Kerr beam self-cleaning in multimode optical fibers. In these experiments, using pulses with durations ranging from the nanosecond to the 100 femtosecond regime and propagating in the normal dispersion regime of a few meters long MMFs, it has been observed that, above a certain threshold power, the initially highly multimode beam evolves toward a bell-shaped transverse profile (sitting on a multimode, relatively low-power background), with a cross section very close to that of the fundamental mode of the fiber. The mechanism for this self-cleaning of light is the optical Kerr effect, as it is proved by the fact that the threshold power is inversely proportional to the fiber length. Thus beam self-cleaning is analogous to Raman beam cleanup, but it occurs also for the pump beam itself and not only for the Stokes-shifted wave. Initial demonstrations of this effect have involved graded-index, virtually lossless fibers. However it has also been reported that, in the presence of strong loss or gain, self-cleaning also occurs in a nearly step-index active ytterbium-doped MMF. Kerr self-cleaning has significant potential applications to high-power and ultrashort pulse fiber lasers, and initial experimental results of intracavity Kerr beam cleaning are presented.

Spatiotemporal Pulse Shaping in Multicore Fibers

Pulse shaping is commonly used in various fields of photonics and electronics; it is a process of changing the waveform of pulses for their better application. A lot of linear and nonlinear technics of pulse shaping in time and frequency domains were proposed in the last decades (Weiner 2000; da Silva et al. 1993; Boscolo et al. 2008; Andresen et al. 2011).

Multicore fibers (MCFs) represent a set of waveguides located under common cladding. Along with multimode fibers, they offer the great possibility of space-division multiplexing (SDM) for high-capacity optical communications. They are now used for a great variety of applications in various areas of photonics (Minardi et al. 2010; Gasulla and Capmany 2012; Eilenberger et al. 2013; Tünnermann and Shirakawa 2015). Multicore fibers attract a lot of attention in the fields of optical communications (Richardson et al. 2013; Saitoh and Matsuo 2016; Igarashi et al. 2014; van Uden et al. 2014) and fiber lasers (Cheo et al. 2001; Ramirez et al. 2015; Hadzievski et al. 2015).

MCFs, as an amplification media, are also attractive for linear beam combining applications in ultrafast laser systems. The increased effective area of MCF allows combining high-energy pulses with conserving a compactness of amplification scheme. In this approach, each core acts as an independent amplifying channel. In Lhermite et al. (2010), authors reported the co-phase combination of 49 beams at the output of a MCF. Experimental demonstration of pulse combining with the efficiency of 49% in the far field using a 7-core hexagonal MCF was presented in Ramirez et al. (2015). As mentioned in that paper, the deviation of residual phase differences, group delay differences among cores, and intensity variations decreased the efficiency, which theoretically can amount to approximately 76%. This discrepancy in amplification of each core is likely to result from MCF inhomogeneity. On the other hand, the cores in the MCF are coupled due to their proximity, reducing the phase mismatching among the cores.

Recent studies have shown the possibility of using nonlinear effects in multicore fibers based on wave collapse for effective combination and compression, which offers new opportunities for nonlinear pulse shaping (Chekhovskoy et al. 2016). Initially, the idea to use the collapse for pulse compression in fiber arrays was proposed more than 20 years ago (Aceves et al. 1995); however, to build the fiber arrays is a technological challenge. MCF is an example of a fiber array with a specific distribution of a relatively low total number of cores, where the proposed ideas can be implemented. This type of nonlinear combining is substantially different from currently popular schemes of linear beam combining (Fan 2005) and can be advantageous for some other energy transfer and delivery applications.

Here we discuss some latest results in nonlinear pulse combination and compression in multicore fiber and compare the effectiveness of ring and hexagonal 19-core fibers (Fig. 1a, d). The increase in the number of neighbors enhances the nonlinear effects and can make the compression and combination more robust and efficient. So the hexagonal structure should demonstrate improvements in performance as compared with ring core configurations. Massive numerical simulations were

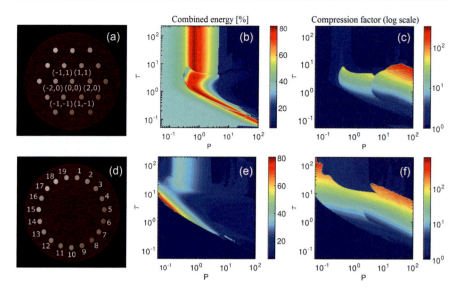

Fig. 1 Profiles of the 19-core MCF with hexagonal geometry (**a**) and the 19-core ring MCF (**d**). The dependence of the pulse compression factor (ratio of the initial width to the final one) (**b, e**) and the percentage of total energy combined in the central core at the first local maximum of the peak power of input pulse (**c, f**) on the parameters P and τ in logarithmic scale of input Gaussian pulses for the hexagonal MCF and for the ring MCF

performed to determine the conditions of most efficient coherent combination and compression of the pulses injected into considered MCFs. We also discuss the influence of pulse phase perturbations and relative pulse delays on the combining scheme.

Pulse Propagation in Multicore Fibers

The governing equations for a ring and square core configurations can be obtained in a similar way, so we consider the mathematical model on the example only of hexagonal multicore fibers. The electromagnetic field of optical pulses propagating along a N-core MCF with hexagonal core distribution can be well approximated by a superposition of modes

$$E(x, y, z, t) = \sum_{n,m} A_{n,m}(z,t) F_{n,m}(x - x_{n,m}, y - y_{n,m}) e^{i(\beta_{n,m}z - \omega t)} + cc, \quad (1)$$

where F gives the spatial mode structure and $A_{n,m}$ is the complex envelope of the electromagnetic field in core number (n, m), $\beta_{n,m}$ is the propagation constant, and the "cc" denotes "complex-conjugate." In the limit of a weak-coupling approximation, the dynamic of the field envelope $A_{n,m}$ of hexagonal MCF can be described using the discrete-continuous nonlinear Schrödinger equation (DCNLSE) (Mumtaz et al. 2013; Aceves et al. 2015; Chekhovskoy et al. 2016):

$$i\frac{\partial A_{n,m}}{\partial z} = \frac{\beta_2}{2}\frac{\partial^2 A_{n,m}}{\partial t^2} - \sum_{(k,l)\neq(n,m)} C_{n,m,k,l} A_{k,l} - \gamma |A_{n,m}|^2 A_{n,m}, \quad (2)$$

where β_2 is the group velocity dispersion parameter and γ is nonlinear Kerr parameter; both are the same for all cores for simplicity. Using variables substitution $A_{n,m} = \exp(i6z')\sqrt{C/\gamma}U_{n,m}$ with $z' = z/L$, $L = 1/C$, where C is equal for all neighboring cores coupling coefficient (other couplings may be neglected) and $t' = t/T$, $T^2 = -\beta_2/(2C)$, the linear couplings between envelopes in neighboring cores can be represented as a discrete analog of second derivatives with respect to time (omitting the primes):

$$i\frac{\partial U_{n,m}}{\partial z} + \frac{\partial^2 U_{n,m}}{\partial t^2} + (CU)_{n,m} + |U_{n,m}|^2 U_{n,m} = 0, \quad (3)$$

where $(CU)_{n,m} = U_{n-1,m-1} + U_{n+1,m-1} + U_{n-2,m} + U_{n+2,m} + U_{n-1,m+1} + U_{n+1,m+1} - 6U_{n,m}$. This equation conserves the total energy

$$E = \sum_{n,m} \int_{-\infty}^{\infty} |U_{n,m}(z,t)|^2 dt \quad (4)$$

and the Hamiltonian

$$H = \sum_{n,m} \int_{-\infty}^{\infty} \left[\left|\frac{\partial U_{n,m}}{\partial t}\right|^2 - (CU)_{n,m} U_{n,m}^* - \frac{|U_{n,m}|^4}{2} \right] dt. \quad (5)$$

For the ring core configurations, the governing equations take a form

$$i\frac{\partial U_k}{\partial z} + \frac{\partial^2 U_k}{\partial t^2} + (U_{k+1} - 2U_k + U_{k-1}) + |U_k|^2 U_k = 0. \quad (6)$$

With a large number of cores and smooth pulse intensity distribution across the cores, the continuous limit of the discrete models (6) and (3) in the form of the well-known nonlinear Schrödinger equation (NLSE) can be used for a qualitative understanding of the system evolution. In particular, the pulse evolution in the ring MCF can be approximately described by continuous 2D NLSE for the function $U(k, t, z)$, considering the index k as a continuous variable:

$$i\frac{\partial U}{\partial z} + \frac{\partial^2 U}{\partial t^2} + \frac{\partial^2 U}{\partial k^2} + |U|^2 U = 0. \quad (7)$$

Equation (7) is equivalent to the NLSE that describes the self-focusing of light in various nonlinear media. The continuous analog can be used for insight into discrete system evolution. In the conventional theory of self-focusing governed by the NLSE, the initial wave distribution collapses into a singularity when the Hamiltonian H is negative or when the beam power exceeds the critical value P_{cr}. This value depends on the beam shape and is minimal for the Townes mode. In our case, the role of power is played by the total energy injected into the MCF $E = \int dt\, dn |U_n|^2$. When the input energy exceeds the critical value E_0, (for the Townes beam, $E_0 = E_{cr} = 4\pi$, in terms of the dimensional variables $E_{cr} = 4\pi \sqrt{-C\beta_2/(2\gamma^2)}$), making $H < 0$, the intensity distribution is self-compressed over k and t. We can expect that the injected MCF pulses distributed over the cores with smooth maxima will be focused into a few cores around the maxima with simultaneous pulse compression. When the energy is concentrated into a few cores, the discreteness of the cores arrests further compression. When the input energy $E \gg 4\pi$, the distribution breaks into a few collapsing clusters with $E \approx 4\pi$ (similar to filamentation in the continuum limit). In every cluster, the compression and combining take place, but the location of the peaks is difficult to predict, and this situation is not practical for the goals of beam combining or pulse compression.

The continuous version of Eq. (3) takes the form of 3D NLSE:

$$i\frac{\partial U}{\partial z} + \frac{\partial^2 U}{\partial t^2} + \frac{\partial^2 U}{\partial k^2} + \frac{\partial^2 U}{\partial l^2} + |U|^2 U = 0. \tag{8}$$

An increase in the number of neighbors enhances the nonlinear interaction and makes collapse possible even for positive values of H (Kuznetsov et al. 1995). In this case, we can expect that the injected MCF pulses will be focused into a few central cores with simultaneous pulse compression. Nonlinear systems described by Eq. (8) have stronger collapsing features, and an MCF with 2D configuration of cores could demonstrate better compression results, which will be shown later. In the 3D situation, the collapse is "weak" (see definitions in Kuznetsov et al. 1995), and the energy involved in the compression processes decreases (Zakharov and Kuznetsov 2012). In this case, the optimal compression and combining are reached in a transient regime, and the choice of parameters is not universal. In the next subsection, we will present the results of the modeling of compression and combining in both situations and will examine the selection of the optimal parameters.

Pulse Compression and Combining

For 19-core hexagonal MCF, we simulated the dynamic of Gaussian pulses

$$U_{n,m}(t) = \sqrt{P} \exp(-t^2/\tau^2) \tag{9}$$

injected in all cores of this fiber to define values of amplitudes P and widths τ providing the most efficient pulse combining and pulse compression. The dependencies of the basic compression parameters are obtained for the range of parameters $P \in [0.05; 1000]$, $\tau \in [0.05; 230]$. We tracked the first local maximum of the peak power of the pulses propagating along the MCF. This event corresponds to the pulse compression for some domain of parameters P, τ. The pulse dynamic was modeled by the system (3). For numerical simulations we propose to use the generalization of the split-step Fourier method with the Padé approximation or finite-difference compact schemes (see Chekhovskoy et al. 2017).

We compared the domains with efficient combining and compression for a 19-core hexagonal MCF and for a 19-core ring MCF (Fig. 1). Recently it has been shown that initial modulation of initial Gaussian pulses improves the combining scheme stability for the ring MCF if the phases of injected pulses are perturbed. Therefore now we use pulses in the form

$$U_k(t) = \sqrt{P} \exp\left(-t^2/\tau^2\right)(1 + 0.3\cos(2\pi k/N)), \tag{10}$$

where N is the number of cores, $k = 1, \ldots, N$, with a smooth distribution of the peak power with a maximum in N-th core.

The maps of combining and compression performance for the MCFs under consideration are presented in Fig. 1. Figure 1b, e shows the dependence of the total energy percentage of all pulses combined at the first maximum of the peak power of the pulse in central core (combining efficiency). Due to the symmetry of the problem, both combining and compression take place in the N-th core for the ring MCF and in the (0,0)-core for the hexagonal fiber. We see that the optimal conditions for combining and compression in the case of hexagonal fiber are very different from the results for the ring core distribution. Efficient combining takes place in a much broader range of parameters in the vertical band, insensitive to the pulse duration. The efficiency of combining is comparable with the ring MCF but much less sensitive to variation of the initial parameters. Using ring 19-core fiber, it is possible to get combined pulse having about 80% of total injected in fiber energy at the distance along fiber $z = 65.9$. The maximum combining efficiency for a 19-core hexagonal MCF equals 80.9% (Fig. 2); the combined pulse can be obtained at the distance $z = 2.07$. In contrast to the ring core configurations, a wide region in the plane of parameters (P, τ) exists, where a part of the energy in the central core at the compression moment exceeds 70% of the initial energy E. The presence of this region allows us to obtain well-compressed pulses having most of the total energy E, which is of great practical importance. However, in this case a substantial part of the energy goes into the wings rather than into the central peak of the compressed pulse.

Figure 1c, f presents the dependence of the compression factor of a pulse (ratio of the initial full width at half maximum (FWHM) to the final one). The blue area corresponds to the pairs of the parameters P and τ, for which there is no pulse compression or the initial pulses (10) break into clusters as a result of the modulation instability. It is interesting that isolines of pulse compression factor in Fig. 1c, f

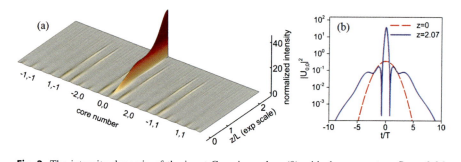

Fig. 2 The intensity dynamic of the input Gaussian pulses (9) with the parameters $P = 0.36$ and $\tau = 1.69$ injected in all cores of a hexagonal 19-core MCF (best combining) (**a**). The input intensity distribution in the central core (dashed red line) and the distribution at compression point (solid blue line) in logarithmic scale (**b**). 80.9% of total input energy E is combined in the core (0,0). The pulse width (FWHM) is reduced by 7.3×. The peak power increases in 103.4×

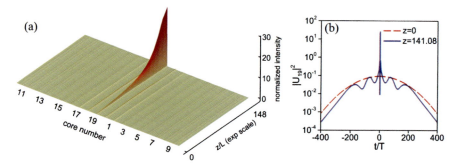

Fig. 3 The intensity dynamic of the input Gaussian pulses (10) with the parameters $P = 0.0545$ and $\tau = 184.0$ injected in all cores of the ring 19-core MCF (best compression) (**a**). The input intensity distribution in the 19th core (dashed red line) and the distribution at compression point (solid blue line) are shown in logarithmic scale (**b**). 10.9% of total input energy E is combined in the 19th core. The pulse width (FWHM) is reduced by 720.4×. The peak power increases in 314.8×

correlate with isolines of the ratio of the dispersion length $L_D = \tau^2/|\beta_2|$ and the nonlinear length $L_{NL} = 1/(\gamma P)$. The zone of optimal compression for the 19-core ring MCF is the stripe narrowing toward high total energies and is confined by the level $L_D/L_{NL} \approx 3000$. If $L_D/L_{NL} > 3000$, a large nonlinearity destroys a smooth pulse shape before the compression point. It is worth noting that the effective compression and effective combining cannot be reachable simultaneously. The maximal compression was obtained by the initial pulses with $L_D/L_{NL} \approx 3000$ and approximately equals 720, the compressed pulse has a smooth profile (see Fig. 3). The distance to the compression point is equal to 141.08 in this case.

The optimal parameters for the compression using hexagonal MCF differ from the results for the ring fiber. Using the 19-core hexagonal fiber, a maximal pulse compression up to 256 times can be achieved. In contrast to the combining pattern, at the point of maximal compression, a significant fraction of energy is left in the neighbor cores. The peak power increases greatly at the compression point. The best compression occurs in the case of a high-power input pulse. Too much nonlinearity ($L_D/L_{NL} \approx 4000$) destroys the pulse shape before the compression point. On the other hand, it is difficult to define the optimal compression case in the presence of high nonlinearity. The pulse compression factor close to its maximal value can be obtained for different pairs of parameters P, τ, and the distance to these compression points decreases with growth of the parameter P. The compression distance is sufficiently small for the 19-core hexagonal MCF ($z \leq 0.5$).

Thus, the stronger nonlinear interaction for a hexagonal MCF does not increase the efficiency of compression (combining) but greatly reduces the required distance and increases the process robustness. Therefore, the hexagonal MCFs have an advantage of being used as a base of the optical pulse compressor device.

It is critically important to control the relative phases of all injected pulses when using the linear pulse combining. In the case of nonlinear combining scheme, the phase-matching requirements can be relaxed. The results of a stability analysis for the first considered regime giving the best combining for the 19-core hexagonal MCF are presented in Fig. 4. We modeled the phase perturbations as uniformly distributed on the segment $[-\delta_p; \delta_p]$ with a random function C_{δ_p}, so the initial pulses were

$$\tilde{U}_{n,m}(t) = U_{n,m}(t) \exp[-iC_{\delta_p}]. \tag{11}$$

The parameter δ_p varied from 0 to π. All computed values were averaged on 1000 launches for every value of δ_p. The calculations showed that the combining scheme is stable for $\delta_p \in [0; \pi/5]$. If $\delta_p > \pi/5$, the pulse compression can occur in cores other than the central core (Fig. 4a). The distance to the compression point and the standard deviations from the mean value are shown in Fig. 4b. The limit values of temporal pulse delays that do not affect the combining and compression process were also determined. The random pulse delays were modeled as uniformly distributed on the segment $[-\delta_t; \delta_t]$ with the random function C_{δ_t}, making perturbed initial pulses

$$\tilde{U}_{n,m}(t) = U_{n,m}(t - C_{\delta_t}). \tag{12}$$

The simulations showed the stability of the combining scheme for $\delta_t \in [0; \tau]$, where $\tau = 1.8$ is the width of the injected pulses (see Fig. 4c, d).

Summing up, an effective nonlinear scheme for combining and compressing pulses using a 19-core ring and hexagonal MCF can be realized. In comparison with a ring geometry, the enhanced nonlinear interaction in the hexagonal fiber does not result in more efficient compression but makes the system shorter and, therefore, more practical. The limits of stability of this scheme to random phase perturbations of the input pulses were found.

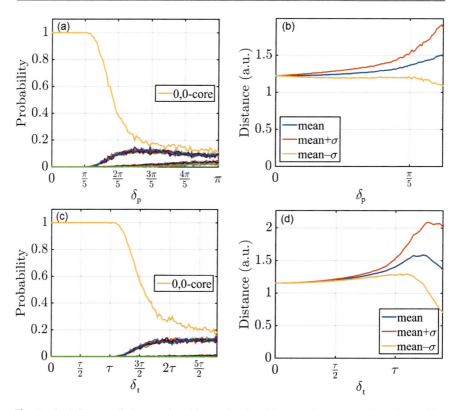

Fig. 4 The influence of phases mismatches and pulse delays on the compression scheme. The dependence of the probability of compression among cores of the 19-core hexagonal MCF on the phase-mismatch parameter δ_p (**a**) and on the pulse time delay parameter δ_t (**b**) for initial Gaussian pulses with amplitude $P = 0.53$ and width $\tau = 1.8$. The corresponding mean value and standard deviation of the distance to the peak power maximum (**b, d**) averaged on 1000 launches

Nonlinear Pulses in Multimode Fibers for Spatial-Division Multiplexing

Spatial-Division Multiplexing

Fiber-optic communication lines are by far the most effective information systems for transmitting large amounts of data over long distances at high speed. In modern systems based on a standard single-mode fiber (SSMF), all the available degrees of freedom (time, frequency, phase, and polarization) can be used to modulate and multiplex signals. Different technologies were called upon to achieve a data transfer rate of over 100 Tb/s for SSMF systems (Qian et al. 2011), but further increase in the capacity of such systems is difficult due to various limitations (Essiambre et al. 2010). At present, there is an imbalance between the growth of global information traffic, which is about 40% per year, and the growth of the total transmission capacity of modern fiber-optic links, which is about 20% a year. And if no new technology providing a significant increase in the data transfer rate is developed in the near future, we may face the problem of the traffic volume exceeding the capabilities of modern data transmission systems.

The spatial-division multiplexing (SDM) is currently considered as a promising technological pathway for further increasing the transmission capacity of optical networks (Winzer and Foschini 2011). Such systems require the development of fibers that support propagation in several spatial modes or cores. These fibers include multimode fibers (MMF), multicore fibers (MCF), and hollow-core fiber (HCF) (Essiambre et al. 2013). Due to a number of advantages (cost compared to MCF and HCF, ease of installation, integrability in most of existing data transmission systems), multimode communication systems are by far the most promising for increasing the transmission capacity of optical networks. Multimode fibers are fibers with one core of a sufficiently large diameter to support more than one spatial mode. The number of spatial modes supported by MMF grows rapidly depending on the core diameter and may number in the hundreds.

The use of multimode fibers allows significantly increasing the capacity of optical networks by simultaneously transmitting signals over different fiber modes. However, new effects influencing the transmitted signals appear at a simultaneous use of several modes, namely, linear mode coupling, differential mode delay (DMD) (Ho and Kahn 2011), and inter-mode nonlinear effects (Rademacher et al. 2012). Furthermore, the nonlinear effects are one of the major factors that limit the capacity of data transmission systems (Ellis et al. 2010). Therefore, for a successful use of multimode fibers as a method of increasing the data transfer rate, it is necessary to investigate and understand each of these effects.

Nonlinear Propagation in Multimode Fibers

The total electric field in a multimode fiber can be written as a sum over M of different spatial modes in the frequency domain (Mumtaz et al. 2013):

$$\tilde{E}(x, y, z, \omega) = \sum_{m}^{M} e^{i\beta_m(\omega)z} \tilde{A}_m(z, \omega) F_m(x, y)/\sqrt{N_m}, \qquad (13)$$

where $\tilde{A}_m(z, \omega) = [\tilde{A}_{mx}(z, \omega), \tilde{A}_{my}(z, \omega)]^T$ is the Fourier transform of the slowly varying envelope of the field in the time domain of the mth mode. This expression includes the slowly varying amplitudes of both polarization components of the spatial mode m with the spatial distribution $F_m(x, y)$ and the propagation constant $\beta_m(\omega)$. The normalization constant N_m represents the power carried by the mth mode and can be expressed as $N_m = \frac{1}{2}\varepsilon_0 \bar{n}_{\text{eff}} c I_m$, where $I_m = \bar{n}_m/\bar{n}_{\text{eff}} \int \int F_m^2(x, y) dx dy$, ε_0 is the vacuum permittivity, $\bar{n}_{eff}$ is the effective index of the fundamental mode, and $\bar{n}_m$ is the effective index of the mode m.

Here we consider two essential cases of propagation in multimode fibers, which are of practical interest: weak- and strong-coupling regimes. In the weak-coupling regime, the coupling between different spatial modes is weak compared to the coupling between two polarization components of one spatial mode. In addition, in this case the fiber length is not much longer than the correlation length (Kahn et al. 2012). In the strong-coupling regime, both types of coupling are of the same order, and the fiber length far exceeds the correlation length. In practice, some spatial modes of the multimode fiber can be weakly coupled, while others can be coupled more strongly.

The equations of nonlinear propagation of optical signals in multimode fibers are obtained from the Maxwell's equations using the approach of Kolesik and Moloney (2004). For more details of deriving the equations of propagation, see Mumtaz et al. (2013), Poletti and Horak (2008), and Mecozzi et al. (2012).

In this case, the nonlinear propagation of signals in multimode fibers in the weak-coupling regime is described by the following Manakov equation (Mumtaz et al. 2013):

$$\frac{\partial A_p}{\partial z} + \beta_{1p}\frac{\partial A_p}{\partial t} + i\frac{\beta_{2p}}{2}\frac{\partial^2 A_p}{\partial t^2} = i\gamma\left(f_{pppp}\frac{8}{9}|A_p|^2 + \sum_{m \neq p} f_{mmpp}\frac{4}{3}|A_m|^2\right)A_p, \tag{14}$$

where $A_p(z,t)$ is the slowly varying envelope of the pth mode in the time domain; β_{1p} and β_{2p} are the inverse group velocity and the group velocity dispersion of the pth spatial mode, respectively; and $\gamma = \omega_0 n_2/cA_{\text{eff}}$ is the nonlinear parameter of fiber, where n_2 is the nonlinear refractive index for glass, A_{eff} is the effective area of the fundamental mode at the central frequency ω_0, and f_{lmnp} are the coefficients of nonlinear coupling between spatial modes, which have the following form:

$$f_{lmnp} = \frac{A_{\text{eff}}}{(I_l I_m I_n I_p)^{1/2}}\int\int F_l F_m F_n F_p dx dy. \tag{15}$$

In the case of strong-coupling regime, the equation of signal propagation is described by the following Manakov equation (Mecozzi et al. 2012):

$$\frac{\partial \bar{A}}{\partial z} + \beta'\frac{\partial \bar{A}}{\partial t} + i\frac{\beta''}{2}\frac{\partial^2 \bar{A}}{\partial t^2} = i\gamma\kappa|\bar{A}|^2\bar{A}, \tag{16}$$

where

$$\kappa = \sum_{k \leq l}^{M} \frac{32}{2^{\delta_{kl}}}\frac{f_{kkll}}{6M(2M+1)}, \tag{17}$$

$\bar{A}$ is the vector of slowly varying mode amplitudes, β' is the average inverse group velocity, and β'' is the average group velocity dispersion.

It should be noted that in the weak-coupling regime, the signals in different spatial modes propagate with different velocities, that is, the coefficients β_{1p} will be different for different modes. In the strong-coupling regime, the parameter β' is a scalar, so the modes will propagate at the same velocity. In addition, in the case of weak coupling, the nonlinear interaction will increase with an increase in the number of modes, since the sum on the right-hand side of Eq. (14) grows. In the case of strong mode coupling, the nonlinear parameter κ (17) in Eq. (16) will decrease, since the denominator has a term of the order of M^2.

The Influence of Nonlinear Effects on the Propagation of Optical Signals

The simulated transmission link is shown in Fig. 5 and consists of a transmitter, 10 spans of MMF, and a receiver.

Each transmitter generated 16-QAM modulated raised cosine pulses at a symbol rate of 32 Gbaud, with a roll-off factor of 0.2 and an oversampling factor of 16. The central wavelength of the emitted signal band was at $\lambda = 1550$ nm. The generated signals were subsequently launched into a transmission link that consisted of 10 spans of 100 km step-index multimode fiber each. The MMF parameters were refractive index of the core $n_{co} = 1.454$, refractive index of the cladding $n_{cl} = 1.444$, core radius $a = 7\,\mu$m, and fiber loss $\alpha = 0.2$ dB. Such fiber supports the propagation of four modes without taking into account the degenerate ones (LP01, LP11, LP02, LP21). Differential modal delays (DMD), dispersion parameters (D), and effective mode area for four modes are shown in Table 1.

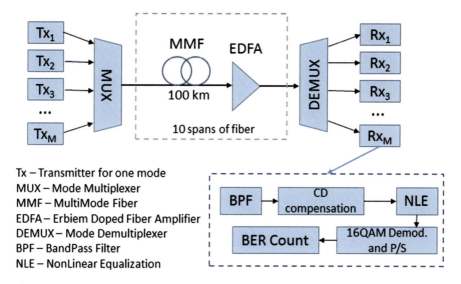

Fig. 5 Scheme of simulated transmission link

Table 1 DMD, D, and A_{eff} of four guided modes

	DMD [ps/km]	D [ps/(km–nm)]	A_{eff} [μm²]
LP01	0	22.5	102
LP11	4.2	23.2	103
LP02	6.1	3.3	121.4
LP21	7.9	17.2	122

Table 2 Coefficients of nonlinear coupling between spatial modes

–	LP01	LP11	LP02	LP21
LP01	1	0.6294	0.6871	0.4068
LP11		0.9932	0.3283	0.5585
LP02			0.8474	0.2934
LP21				0.8425

The coefficients of nonlinear coupling between all spatial modes with a step-index profile of the refractive index are shown in Table 2, the lower part of which is not filled on the grounds of $f_{mmpp} = f_{ppmm}$.

An EDFA of 4.5 dB noise figure was used to fully compensate the signal attenuation. The equations of propagation were solved numerically with a typical symmetrized split-step Fourier method. After transmission through the channel, the signals were coherently detected. Then a linear equalization was used for ideal compensation of chromatic dispersion effects. After down-conversion to one sample per symbol, the nonlinear equalization was performed based on least mean square (LMS) algorithm. To analyze the system performance, we estimated the Q-factor, associated with BER by

$$Q = 20 \log_{10}\left[\sqrt{2}\,\text{erfc}^{-1}(2\text{BER})\right], \qquad (18)$$

where BER represents the average bit error rate for all modes, obtained by direct error counting.

In the study of the influence of nonlinear effects, first we compared the mode-coupling regimes. Figure 6 shows the dependence of the quality parameter Q-factor on the initial signal power for the strong- and weak-coupling regimes for data transmission using four spatial modes. As can be seen from the figure, the weak-coupling regime shows better performance in comparison with the strong-coupling regime. This can be explained by the fact that at a strong mode coupling, the signals propagate with the same velocities, and in the course of propagation, the nonlinear interaction between them remains high (Fig. 7 left). In the case of weak coupling, the signals in different modes propagate with different velocities. This leads to the fact that the peaks of the pulses along the propagation move relative to each other (Fig. 7 right), and the nonlinear interaction between them decreases.

If we consider the weak-coupling regime in which the signals in different modes propagate with the same velocities (blue line in the Fig. 6), then the performance will be worse than in the strong-coupling regime. This is due to the fact that at a strong

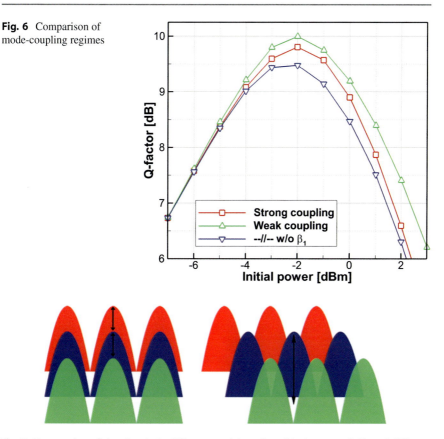

Fig. 6 Comparison of mode-coupling regimes

Fig. 7 Propagation of the signals in different spatial modes with the same (left) and different (right) velocities

mode coupling, the nonlinear parameter κ (17) will be smaller than the nonlinear part in the weak-coupling regime (14).

To investigate the effect of a differential mode delay on the transmission performance, the dependence of the bit error rate on DMD has been found. Figure 8 shows the dependence of BER on DMD for signal propagation in the case of weak coupling using two modes. As can be seen from the figure, the number of transmitted errors decreases with the increase of differential mode delay. However, BER decreases to a certain limit, corresponding to the case when in the course of propagation the optical pulse in one spatial mode reaches the neighboring pulse in another mode.

We also investigate the influence of nonlinear effects with an increase in the number of modes in the strong- and weak-coupling regimes. Figure 9a shows the dependence of Q-factor, averaged over all used modes, on the initial signal power for the propagation in one, two, three, or four modes at strong coupling. In this case, an increase in the number of modes leads to a worse transmission performance, but

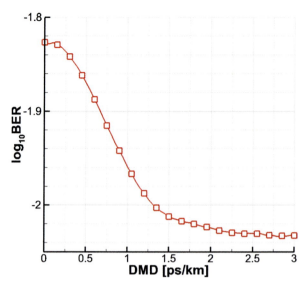

Fig. 8 BER as a function of DMD for signal propagation in the case of weak coupling using two modes

the decrease in the Q-factor slows down for a large number of modes. This is due to the fact that in the strong-coupling regime, the nonlinear parameter κ (17) will decrease as the number of used modes increases.

Figure 9b shows the dependence of the Q-factor on the initial signal power for optical signal propagation in one, two, three, or four modes in the weak-coupling regime. As can be seen from the figure, the cases of one and two modes show approximately the same Q-factor; however, when adding the signal, propagating in the third mode LP02, the transmission performance declines significantly. This is due to the fact that for LP02, the value of the dispersion parameter is small (3.34 ps/(km–nm)), and the signal propagating in this mode undergoes a smaller dispersion broadening and, therefore, is more susceptible to nonlinear distortions. Thus, the input of the LP02 mode to the Q-factor averaged over all used modes decreases the system performance. A slight improvement may be reached by adding the fourth mode LP21 with rather large dispersion parameter. In this case, the value of the Q-factor is greater than in the case of three modes.

Raman Cleanup Effect and Raman Lasing in Multimode Graded-Index Fibers

The Raman beam cleanup effect is known as the effect of stimulated Raman scattering (SRS) of a low-quality pump beam, which produces a redshifted beam with much better beam quality. Laser generation is also possible in passive fibers owing to the SRS-induced amplification of the scattered light. A value of the so-called Stokes shift is defined by the vibration quanta amounting to ∼13 THz in standard telecommunication silica fibers (Stolen et al. 1972). Similar effect was also

Fig. 9 Q-factor as a function of initial power for signal propagation in the case of strong (**a**) and weak (**b**) coupling using one, two, three, and four modes

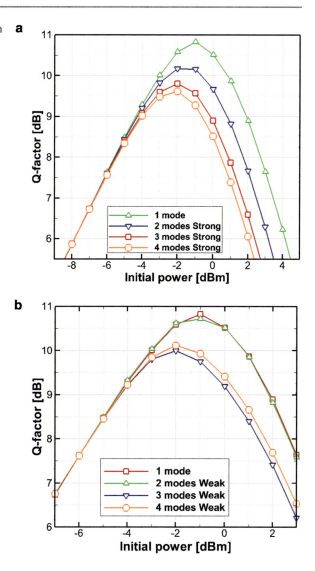

observed in another dissipative process such as stimulated Brillouin scattering (SBS) (Rodgers et al. 1999; Russell et al. 2001; Lombard et al. 2006). There are several important distinctions between the SBS and SRS. The first and most important difference is the gain bandwidth (100 MHz for SBS and 3–10 THz for SRS). This makes SBS usable only for narrowband lasers, whereas SRS is important for broadband lasers. Second, SRS amplifies both the backward and forward scattered light, whereas SBS supports backward scattering only. Third, the Stokes shift for SRS is by three orders of magnitude larger than that for SBS. Another consequence of the large Stokes shift is that, the SRS-induced Stokes beam intensity profile will deviate from that for the pump beam more quickly. Thus the fiber length required for the beam cleanup is considerably shorter than that required for SBS cleanup.

Experimental Observations and Theoretical Models of Raman Cleanup Effect

In the first experimental demonstrations, the field pattern of the Stokes waves generated in the SRS process had a ring shape of much smaller diameter than the pump one, as it was observed independently in bulk media (Komine et al. 1986; Goldhar et al. 1984; Hanna et al. 1985; Reintjes et al. 1986) and multimode graded-index (GRIN) fibers (Baldeck et al. 1987; Grudinin et al. 1988; Nesterova et al. 1981) in the 1980 s. Van den Heuvel has developed a numerical model of SRS beam cleanup in bulk media which explains the process in terms of the balance between amplification and diffraction (van den Heuvel 1995). In the case of multimode GRIN fibers, the researchers used picosecond pump radiation at wavelengths of 532 or 1064 nm with kW-level peak power. As a consequence, most of them explained the observed phenomenon as a result of self-focusing.

Chiang (1992) supposed that the significant improvement of spatial beam quality of the Stokes component generated via SRS can be explained in terms of mode competition between the various Stokes transverse modes of a graded-index fiber. He questioned the self-focusing mechanism because the nonlinear part of refractive index calculated by him was too small to modify the field confinement in the fiber (Chiang 1992, 1993). In order to check the assumption of mode competition, Chiang launched nanosecond pulses from a dye laser operating at 585 nm into a 30-m multimode graded-index (GRIN) fiber and observed that, under optimum pump launching conditions, the generated four Stokes waves propagate in the fundamental LP_{01} mode. When the angle and the position of the input fiber end relative to the pump beam were slightly changed, several low-order modes and their combinations were preferentially excited at the Stokes wavelengths (Fig. 10a). For a 1-km fiber, higher-order Stokes waves were generated in the fundamental mode even if the first Stokes wave contains a mixture of a large number of modes (Fig. 10b, c). So, the process of cascaded SRS in the fiber is accompanied by stripping off the high-order modes. Chiang explained this effect by increasing mode coupling in the long fiber which results in a steady-state mode distribution favoring low-order modes. This effect was also noticeable in Pourbeyram et al. (2013), where nanosecond radiation at 523 nm generated multiple cascaded Raman peaks extending up to 1300 nm at propagation in 1 km 50/125 multimode GRIN fiber. Although the non-diffraction-limited pump beam excites multiple modes, all Stokes peaks comprise only low-order modes. For example, the sixth Stokes component at 610 nm consists of the fundamental mode only. However, the higher-order Stokes waves start to propagate in the lower transverse modes most likely due to the combined effect of SRS and non-degenerate collinear four-wave mixing (FWM).

Chiang also proposed a first simple model of SRS beam cleanup in multimode fibers based on overlap integral of the pump field modes and the Stokes field

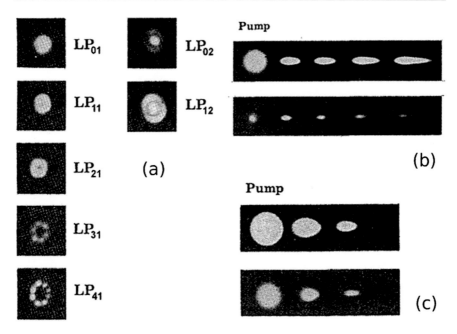

Fig. 10 (a) Images of the first Stokes mode patterns obtained in a 30-m fiber with the pump filtered out; images of the dispersed output from a 1-km fiber showing two situations: (b) the LP_{01} mode evolves at all the Stokes shifts, (c) the LP_{01} mode evolves in the second Stokes shift, while the first Stokes shift contains a number of modes

modes (Chiang 1992). He considered a short multimode GRIN fiber where the mode coupling was negligible. The main drawback of the model is that the intensity profiles of step-index fiber modes were used instead of graded-index fiber modes, in spite of that the SRS beam cleanup is absent for step-index fibers. Nevertheless, the normalized overlap integral in the calculation has a maximum value when the excited mode and the evolving mode were the same. Therefore, Chiang concluded that the preferentially excited mode usually has the highest Raman gain. This model explains the results obtained in the experiment with the 30-m MM GRIN fiber (Fig. 10).

Following Chiang, Polley et al. (Polley and Ralph 2007) calculated the effective Raman gain for the lowest-order pump and signal mode pair, considering elements of the Raman gain matrix G_{ij} for the j-th signal mode gain due to the i-th pump mode with power P_i over the effective length L_{eff}:

$$G_{ij} = \exp\left(\left(\frac{g_R}{A_{\text{eff}}}\right)_{ij} P_i L_{\text{eff}}\right). \tag{19}$$

The effective Raman gain coefficient, i.e., the ratio of Raman gain coefficient g_R to the effective area, was found for each pump and signal mode pairs (i, j) as

$$\left(\frac{g_R}{A_{\text{eff}}}\right)_{ij} = \frac{\int\int g_R(r)\psi_i^2(r,\theta)\psi_j^2(r,\theta)drd\theta}{\int\int \psi_i^2(r,\theta)drd\theta \int\int \psi_j^2(r,\theta)drd\theta}, \tag{20}$$

where ψ_i and ψ_j are the transverse electric field profiles for the j-th signal and the i-th pump mode, respectively. The field of confined LP$_{mp}$ modes in a weakly guiding graded-index fiber is given by:

$$\psi_{mp} = \cos(m\phi) R^m L_{p-1}^{(m)}(VR^2) \exp(-VR^2/2), \tag{21}$$

where $L_{p-1}^{(m)}$ represents the associated Laguerre polynomial; $R = r/a$, where a is the radius of the fiber core, V is the normalized frequency of the fiber. Therefore, the overlap integral differs from that calculated by Chiang. Nevertheless, it has the highest value for the pair of LP$_{01}$ pump and signal modes. The effective Raman gain computed for the lowest-order mode (LP$_{01}$) in 62.5 and 50 μm MM GRIN fibers appeared to be comparable to the effective Raman gain for standard SMF, amounting to 0.5 W^{-1} km^{-1}. In a graded-index fiber, the doping density of GeO$_2$ varies with radial coordinate and can be estimated from the refractive index profile. Therefore, the Raman gain coefficient, which increases with the doping, is also a radial function. Compared to SMF, the higher Raman gain coefficient due to the higher GeO$_2$ doping completely compensates for the larger effective area of the lowest-order MMF modes.

Finally, Terry et al. explained why beam cleanup does not occur in step-index fibers (Terry et al. 2007a). Their model of SRS beam cleanup describes beam cleanup in graded-index and step-index fibers in terms of their overlap integrals for a range of launching conditions. The initial power in the Stokes beam was considered as uniformly distributed over the transverse modes at single longitudinal mode, like the pump one. It was also assumed that each Stokes mode interacts with one pump mode only. The values of the calculated overlap integrals showed that the overlap of the LP$_{01}$ pump mode with the LP$_{01}$ Stokes mode is greater than the overlap of any other pair of intensity patterns. Moreover, the largest values of the overlap integral are reached when a pump mode overlaps with its corresponding Stokes mode. It becomes progressively smaller for higher-order transverse modes. According to this simple description, mode competition in a graded-index fiber favors the LP$_{01}$ Stokes mode that corresponds to Polley's table (Polley and Ralph 2007). In a step-index fiber, the overlap of the higher-order transverse modes with their respective pump modes was greater than the overlap of the LP$_{01}$ Stokes mode with the LP$_{01}$ pump mode, which corresponds to Chiang's table (Chiang 1992). Therefore, mode competition in a step-index fiber does not favor LP$_{01}$ mode.

Alternative theory of SRS beam cleanup was proposed by Murray et al. (1999). It is based on the central limit theorem in statistics and the theory of linear systems. The theorem states that repeated convolutions (or correlations) of arbitrary functions asymptotically produce a Gaussian function. By decomposing the input pump and Stokes beams in terms of their plane-wave components, Murray et al. expressed the driving nonlinear polarizations for SRS as a series of autocorrelation and cross-correlation operations between the plane-wave components of the beams. The resulting polarization is a function having a prominent central peak in the transverse plane, which favors amplification of the Stokes beam in the fundamental GRIN fiber mode. The theory is also applicable to the case of beam cleanup via stimulated Brillouin scattering in fibers.

Summarizing the above, the beam cleanup property of SRS is commonly attributed to the preferential amplification of the fundamental mode of the Stokes beam relative to the higher-order modes through its better overlap with the multimode pump beam in GRIN fibers. Since all modes in the Stokes waves grow exponentially with their own gains, the difference in power between any two Stokes modes with different gains also grows exponentially with distance. Now consider several practical applications of the SRS cleanup effect.

In 2002, Russell et al. performed a SRS beam cleanup experiment using a 300-m-long 50 μm graded-index fiber (Russell et al. 2002). In this experiment, the multimode fiber was pumped with a frequency-doubled Q-switched Nd:YAG laser at 532 nm. While the transmitted pump beam has poor beam quality ($M^2 = 20.7$), the Stokes beam has a greatly improved beam quality ($M^2 = 2.4$). The Stokes beam contains multiple Stokes shifts spanning over 100 nm. In order to calculate M^2, Russell et al. assumed that all Stokes waves have the first Stokes wavelength. Studying the dependence of the Stokes beam spot size on the GRIN fiber length, they found a transition point between long and short fibers, in which the beam cleanup properties were lost and higher-order modes were excited. Fibers longer than 100 m generate Stokes beams with Gaussian intensity profile (Fig. 11a). Under 100 m, the Stokes beam was found to fluctuate between a Gaussian-like profile and a multi-lobe profile (Fig. 11b), in agreement with Chiang's observations (Chiang 1992).

Russell et al. (2002) also proposed to apply beam cleanup properties of SRS in multimode fibers for brightness enhancement and scaling of common laser beams through the beam cleanup and combining, respectively. First successful combination and cleanup of four laser beams via stimulated Raman scattering in GRIN fibers was demonstrated by Flusche et al. (2006). Four coherent pulsed laser beams split off from a Q-switched Nd:YAG laser were combined by multi-port fiber combiner and then launched into a 2.5-km-long GRIN fiber with 100 or 200 μm core diameter. Due to the long fiber and high peak power, cascaded Stokes shifts up to 1.5 μm were generated through combined effects of SRS and non-degenerate collinear FWM. Despite of different beam quality of the transmitted pump beam in 100 μm ($M^2 = 26$) and 200 μm ($M^2 = 42$) GRIN fibers, the Stokes beam has a good quality

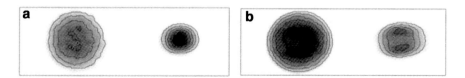

Fig. 11 Pump (left) and Stokes (right) intensity profiles for 300 m (**a**) and 75 m (**b**) fiber

with nearly the same M^2 value of 2.7. The SRS beam cleanup was also studied depending on the fiber length: M^2 of the Stokes beam slightly increases from 1.5 to 2 with shortening the fiber from 1.3 to 0.4 km.

In addition to the beam combination, the SRS beam cleanup effect was also applied to Raman fiber lasers (Baek and Roh 2004; Terry et al. 2007b), amplifiers (Polley and Ralph 2007; Rice 2002; Terry et al. 2008), and even to multimode fiber communication links (Polley and Ralph 2007). In the last case, it improves inter-symbol interference for 10-Gb/s data transmission in multi-kilometer multimode fiber link.

Raman Cleanup Effect in Raman Fiber Amplifiers and Lasers

In Raman fiber amplifiers (RFAs), as predicted by Rice (2002), the lowest-order Stokes mode launched into the core of graded-index multimode fiber grows at the expense of higher-order Stokes modes. The RFA could therefore use a multimode pump beam to amplify a single-mode Stokes signal. Because the multimode pump beam can be efficiently coupled into a multimode fiber, an RFA based on a multimode graded-index fiber offers an intriguing route to creating a highly efficient RFA with single-mode output.

Terry et al. (2008) refuted this assumption. They modeled the beam quality of the Stokes output by considering the relative gain of the Stokes modes of the fiber. The model was based on a set of coupled differential equations that describe the interaction of the pump and Stokes components. It is evident that in the absence of an external Stokes input, spontaneous Raman scattering in a long fiber uniformly seeds all the transverse Stokes modes of the fiber. Since all Stokes modes are seeded with equal amounts of power, the output power of a given Stokes mode depends upon its gain. In the case of RFA, the output power of a given Stokes mode depends on both its gain and its input power. So, it is possible that the Stokes modes with low gain and high initial power dominate over modes with high gain and low power.

The most promising application of SRS beam cleanup relates to the power scaling in high-power Raman fiber lasers (RFLs). Relatively low Raman gain, as compared to the gain in active fibers, requires a much larger length of passive fiber exceeding hundreds of meters in conventional RFLs. The power of single-mode LDs is limited by 1 W level that is comparable with the RFL threshold. That is why RFLs are usually pumped by high-power Yb- or Er-doped fiber lasers with single-mode output coupled directly to the core of a passive Raman fiber. To ensure the single-mode character of the output beam, RFLs usually use a passive single-mode fiber as the Raman gain medium. The use of single-mode fiber, however, limits the efficiency of pump coupling to the fiber, especially when a multimode pump beam is used; hence the overall conversion efficiency suffers.

A first approach to solve this problem is to use a double-clad passive fiber to allow multimode cladding pumping in a Raman fiber laser (Codemard et al. 2006). It takes advantage of low threshold and a diffraction-limited output in a single-mode core. However, the use of a single-mode core may not be desirable if power scaling is a goal due to optical damage limits. In addition, double-clad passive fibers have high attenuation losses and requires special fabrication technologies.

An alternative approach is to use multimode GRIN fibers, which permit a higher coupling efficiency and better overlap between the pump and Stokes beams in the fiber. Both factors should lead to higher conversion efficiency of the Raman laser. Since multimode GRIN fibers tend to support a fundamental fiber mode with a larger mode area, it is reasonable to expect higher-output powers. An additional advantage of using standard commercial GRIN fibers is that they are well engineered and have undergone technological maturation due to their telecom use, so that the background loss is quite low. Since the Raman laser is based on the pump-induced Raman gain in a passive fiber, it has a fundamental difference in lasing properties as compared with Er- or Yb-doped fiber lasers, namely, it features a small quantum defect, fast response of the gain on pump variations, low background spontaneous emission, as well as the absence of photo-darkening effect that is a problem in doped active fibers, especially at short wavelengths.

Baek et al. first demonstrated a nearly single-mode operation of a RFL that uses a multimode fiber as the Raman gain medium (Baek and Roh 2004). The all-fiber laser cavity includes a couple of multimode fiber Bragg gratings and commercial GRIN multimode fiber with 50-μm core diameter (Fig. 12). The fiber length of 40 m was chosen as a compromise between two requirements: ensuring a sufficient gain length for beam cleanup and minimizing the attenuation in the fiber. RFL was pumped by multimode Nd:YAG laser at 1.064 μm with beam quality parameter $M^2 \sim 7$. The beam quality of the Stokes wave was determined by measuring of the spot size variation of the beam in the vicinity of the focal point of a lens with a frame grabber system. It is $M^2 = 1.66$ at 0.8 W output power.

The slope efficiency of the first RFL based on multimode GRIN fiber was only 7%, but further optimization of the cavity resulted in enhancement of the laser efficiency and the output power (Terry et al. 2007b). In the experiment, two

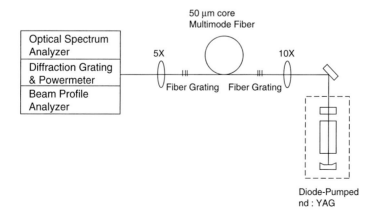

Fig. 12 Schematic diagram of the multimode RFL (Baek and Roh 2004)

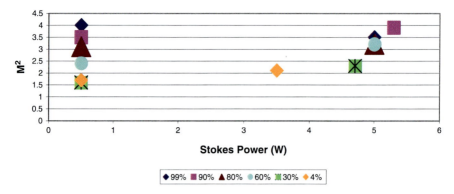

Fig. 13 Beam quality of the RFL with various output FBGs for two different Stokes output powers (Terry et al. 2007b)

temporally incoherent CW Nd:YAG lasers were combined using polarizing beam splitters and then launched into the RFL based on 2.5-km multimode GRIN fiber with 50 μm core. The cavity of the RFL was almost the same, as shown in Fig. 12. The input FBG possessed a single-mode reflectivity of 99% at the first Stokes wavelength. The peak single-mode reflectivity of the output FBG varied from 30% to 99% or it was replaced by 4% Fresnel reflectivity of the normally cleaved fiber end. It was shown that the Fresnel reflection provided higher slope efficiency of the RFL, and the M^2 value appeared to increase with the Stokes power and reflectivity of the output FBG (Fig. 13).

It is also promising to pump the multimode RFL directly with multimode diodes. This approach eliminates intermediate conversion stage and benefits from the wide wavelength coverage available from high-power laser diodes (800–980 nm). In this way it is possible to generate Stokes wavelengths which are not covered by alternative practical fiber laser sources, e.g., in 830–1000 nm range. Note that power of LDs operating at ∼915 and ∼980 nm has already exceeded the 100 W level for a single unit.

GRIN Fiber Raman Lasers Directly Pumped by Multimode Laser Diodes

This concept was developed in two directions. The first one includes RFL cavity based on bulk optical elements (Yao et al. 2015; Glick et al. 2016a, 2017). The second one follows Baek and Roh (2004) and Terry et al. (2007b) in developing an all-fiber RFL cavity by means of FBGs (Kablukov et al. 2013; Zlobina et al. 2016; Zlobina et al. 2017a) and also focuses on the all-fiber coupling of pump radiation into the laser cavity (Zlobina et al. 2017b). The last approach offers a simple and reliable all-fiber configuration of high-power RFL with good beam quality.

The typical RFL configuration with bulk-optic cavity is shown in Fig. 14 (Yao et al. 2015). Beams of two high-power multimode LDs operating at 975 nm are

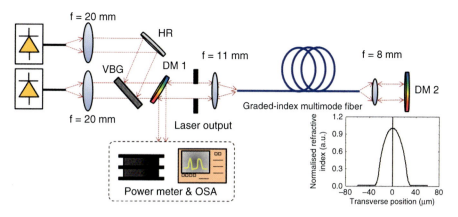

Fig. 14 Experimental setup of the CW RFL with GRIN Raman gain fiber (Yao et al. 2015)

combined into a single beam in an angled narrowband volume Bragg grating. The wavelength-combined pump light is launched into a 62.5 μm multimode GRIN fiber via a lens. Dichroic mirror placed in the far end of the fiber was used to launch pump and Stokes light back into the fiber. It has reduced reflectivity at the second Stokes wavelength to prevent cascaded Raman generation. The left side of the fiber was normally cleaved and acted as a 4% output coupler of the cavity.

The slope efficiency of such RFL reached 81% in a 1.5-km-long GRIN fiber and output power amounted to 19 W. The measured beam quality of the generated beam at 1019 nm equals to $M^2 = 5$, while it was 22 for the input pump beam. Yao et al. (2015) achieved a good agreement of the experimental and calculated power characteristics when assumed that the pump was uniformly distributed over the 1530 μm^2 effective area which is two times less than the core area of 3070 μm^2. Therefore, the pump intensity was more likely concentrated near the center of the core. When the GRIN fiber was replaced by 650 m special double-clad fiber, the output beam quality was improved to $M^2 = 1.9$, but the output power was reduced to 6 W because of comparatively high propagation loss of the specialty double-clad fiber.

Glick et al. (2016a) used a basic configuration which was similar to that shown in Fig. 14. Output Stokes power as high as 85 W was generated at 1020 nm when they used a 0.5 km GRIN fiber and 978-nm pump LD which was operating in a quasi-CW regime of 20 ms duration and 15 Hz repetition rate. The slope efficiency of the RFL equals to 70%. M^2 values for the Raman output beam were measured to be from 2.9 to 5.6, while it amounted to 12–14 for the input pump beam. In general, the Stokes beam quality deteriorates at higher-power levels, repeating the behavior of the incident pump beam (Fig. 15a). The reduction of the laser beam quality was explained as follows. When the pump power increases, the Raman lasing threshold is reached in more peripheral regions of the fiber core, so those areas also begin to participate in Raman conversion, and the effective Stokes beam diameter becomes larger, thus resulting in poorer beam quality. The brightness enhancement is determined as the ratio of the Raman output brightness to the pump brightness,

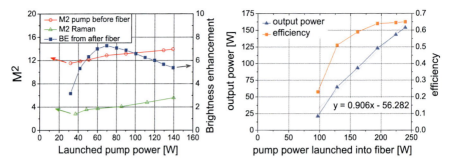

Fig. 15 (**a**) M^2 of pump and Raman beams and brightness enhancement (BE) versus launched pump power. Open circles are M^2 of pump before entering fiber, and open squares are M^2 of pump after propagating through fiber. Triangles are M^2 of Raman output. Solid circles are BE of Raman compared to pump before fiber and solid squares are BE of Raman compared to pump after fiber (Glick et al. 2016a); (**b**) Raman output power and efficiency as a function of launched pump power in 0.2-km-long RFL (Glick et al. 2017)

defined as $B = P / (M^2 \lambda)^2$, is also shown in Fig. 15a. Comparing the Raman output to the pump after its double propagation through the fiber, a maximum brightness enhancement of 7.3 was obtained which is comparable to factor of 5.2 reported in Yao et al. (2015).

Further power scaling was performed by adding the second 978-nm LD in the scheme and shortening the GRIN fiber to 0.2 km (Glick et al. 2017). One advantage of the fiber shortening was the higher threshold for the second Stokes wave. The other advantage was the lower passive losses. Reduction of fiber length resulted in an increase of the laser threshold by a factor of 2.5 compared to Glick et al. (2016a). However, the output power and slope efficiency in CW regime were also increased to 154 W and 90% correspondingly (Fig. 15b). The M^2 value of the Raman output was measured to be between 4 and 8 at increasing the output power from 20 to 154 W, while the pump beam before entering the fiber showed $M^2 = 16$–18. The maximum brightness enhancement 8.4 almost did not change with the fiber length shortening.

Despite the outstanding results in output power and slope efficiency obtained in Yao et al. (2015) and Glick et al. (2016a, 2017), the generated wavelength of ~1020 nm was not as interesting because a much higher power at this wavelength is available from YDFLs (Glick et al. 2016b). Kablukov et al. (2013) earlier reported on the first CW RFL operating below 1 μm. Pump radiation from 938 nm LD was launched via collimating lens into 62.5 μm GRIN fiber with an in-fiber cavity formed by a highly reflective FBG and 4% Fresnel reflection from the cleaved fiber end (Fig. 16a). The RFL threshold power was estimated as:

$$P_{th} = \frac{2\alpha_s L - \ln(R_1 R_2)}{2 g_R L_{eff}} \quad (22)$$

where $L_{eff} = [1 - \exp(-L\alpha_p)] / \alpha_p$ is the effective absorption length.

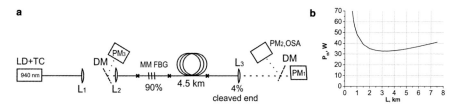

Fig. 16 (a) Experimental setup of the CW RFL with GRIN Raman gain fiber, FBG reflector (R1) and Fresnel output (R2). (b) Calculated threshold power versus fiber length in the cavity formed by mirrors with $R_1 = 90\%$ and $R_2 = 4\%$ (Kablukov et al. 2013)

Following the results of calculations, the threshold power P_{th} takes a minimum value in the range of cavity lengths between 2.5 and 4.5 km (Fig. 16b).

According to preliminary calculations, the generation threshold at the wavelength of 980 nm in the experiment with 4.5-km-long GRIN fiber was around 35 W. The forward Stokes power grows to 2.3 W when threshold of the second Stokes wave at 1025 nm is achieved. The RFL slope efficiency of 25% was not high. The low value of the second threshold was explained by the low-index transverse modes generation. The generated spectrum included three individual peaks (Fig. 17a), while the main part of the power was concentrated in the peak at 980 nm corresponding to the fundamental mode. The far-field profile of the output beam demonstrated a reduction of the beam divergence by three times compared to the pump beam (Fig. 17b). Use of an additional mirror at the fiber end which reflected residual pump power at 938 nm back to the cavity led to the increase of the output power and slope efficiency up to 3.3 W and 35%, respectively.

Babin et al. (2013) reported on endeavors to get Raman lasing in a LD-pumped passive fiber without a conventional resonator, using a positive feedback provided by a randomly distributed Rayleigh backscattering. Though the Rayleigh backscattering integral reflection coefficient is low ($\sim$0.1%), the scattered radiation is amplified after random reflections so that the integral Raman amplification may compensate the losses thus reaching laser threshold even in the absence of conventional cavity mirrors, similar to random lasing in single-mode fiber Raman lasers (Turitsyn et al. 2010). The experimental study of random lasing is performed on the basis of the RFL scheme with direct LD pumping (Fig. 16a), in which the narrowband FBG has been replaced by a broadband mirror. Moreover, instead of the 4% Fresnel end reflection (removed by cleaving the fiber with angle $>15°$ against a normal one), a distributed Rayleigh backscattering was employed for feedback. The random distributed feedback laser threshold increased to 42 W. The generated Stokes radiation at 980 nm reached 0.3 or 0.5 W in the RFL configuration with or without an additional mirror reflecting back the residual pump power. The relative reduction of the generated beam divergence was by 4.5 times as compared with that for the pump beam, i.e., higher than in the case of SRS beam cleanup only (Kablukov et al. 2013). Therefore, an additional influence of the SBS (Rodgers et al. 1999) present near threshold and the Rayleigh backscattering distributed feedback make the cleanup effect stronger. Due to the high beam quality of the generated Stokes wave, such a laser has a very low threshold of the second-order Stokes wave generation.

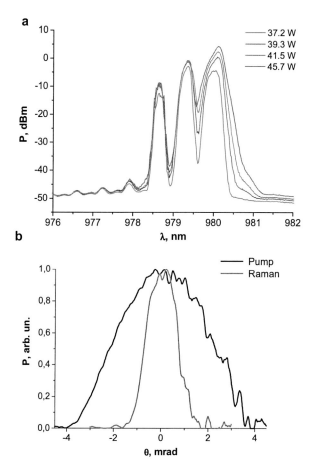

Fig. 17 (a) Output spectrum of the generated Stokes wave. (b) Output beam profile for the pump (black) and Raman (gray) lasers (Kablukov et al. 2013)

Combined Action of Raman Beam Cleanup and Mode-Selecting FBGs in GRIN Fiber Raman Lasers

A further increase of the generated power in GRIN fiber-based RFLs with FBGs was achieved by optimization of the cavity length, which was obviously larger than optimal, as the transmitted pump power was relatively low in Kablukov et al. (2013) (10% of the coupled pump power). Optimization of the GRIN fiber length and utilization of the 915 nm pumping allowed Zlobina et al. (2016) to increase the output power and slope efficiency while decreasing the laser wavelength down to 954 nm. The experimental scheme is similar to that in Fig. 16a. The linear cavity of the laser was formed by high-reflection fiber Bragg grating FBG1 and normally cleaved fiber end providing weak (~4%) Fresnel feedback, or output fiber Bragg grating FBG2 with low reflection. The multimode FBGs were written by a high-power UV argon laser in the same multimode GRIN fiber. The experimental power threshold of ∼30 W is nearly the same for two fiber lengths ($L = 3.7$ and 2.5 km), in correspondence with the calculated threshold power curve (see Fig. 16b). At the same time, the differential efficiency is higher for the shorter length amounting to 37% and 26% at $L = 2.5$ and 3.7 km, respectively. Optimization of highly reflective grating FBG1 resulted in increasing the generated power to 4.7 W and the slope efficiency to 41%. Since the cascaded generation of higher Stokes orders limits the power for the first Stokes, it is possible to increase the second Stokes threshold by substituting the Fresnel reflection by low reflective (∼3%) output FBG2 at 954 nm. The second Stokes wave was not observed in this case, but the laser threshold and generation power worsen in comparison with the normally cleaved output end.

Figure 18a shows the output Stokes spectrum measured for the scheme with (solid line) and without (dash-dot line) output FBG2. The spectrum is smooth without FBG2. Its 3 dB linewidth of 0.7 nm is defined by the FBG1 bandwidth.

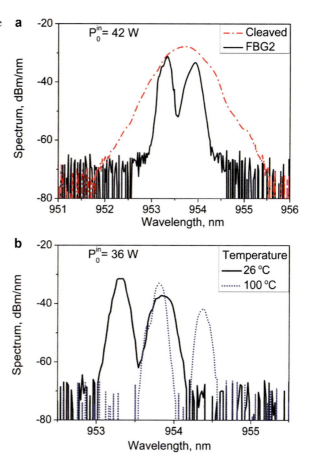

Fig. 18 Output spectra of the Stokes wave for $L = 2.5$ km RFL: (**a**) with normally cleaved fiber end (dash-dot line) or with the FBG2 (solid line); (**b**) at different temperatures of the FBG2 (Zlobina et al. 2016)

With the FBG2 the spectrum has multi-peak structure which is similar to one obtained with another multimode FBGs (Terry et al. 2007b; Kablukov et al. 2013) and is mainly defined by the FBG2. The wavelength difference $\Delta\lambda$ between two neighboring groups of the transverse modes in GRIN fiber (Gloge et al. 1973; Kawakami and Tanji 1983) can be estimated using Eq. (23) for propagation constant β from Mizunami et al. (2000) and condition $\beta = \pi/\Lambda$ for reflection at FBG with period Λ:

$$\Delta\lambda = \lambda^2 \text{NA}/\left(\pi d n_{cl}^2\right), \tag{23}$$

where λ is wavelength of the fundamental mode, NA is numerical aperture, d is the core diameter, and n_{cl} is the cladding refractive index. Estimated value of the wavelength difference $\Delta\lambda = 0.61$ nm is in a good agreement with the distance between two lines $\sim$0.62 nm experimentally observed in the case of FBG2 (see Fig. 18a). Therefore those lines correspond to two neighboring groups of low-index transverse modes which are reflected by FBG2. Heating of the FBG2 to 74 degrees resulted in synchronous shift of the two-peak structure by 0.6 nm (see Fig. 18b).

As a next step, the possibility of mode selection in graded-index fiber Raman laser by means of FBG mode-selection properties was studied in Zlobina et al. (2017a). To improve the selection of the fundamental mode of graded-index fiber, instead of conventional UV FBG2 (recorded by CW UV radiation), a FS FBG2 was spliced at the laser output. It was point-by-point recorded in the central region of the fiber core by a femtosecond (FS) laser beam (Dostovalov et al. 2016). Reflection spectra of UV FBG1 and FS FBG2 are shown in Fig. 19a. The FS FBG2 and UV FBG2 were first compared in a 2.5-km-long Raman laser (Kablukov et al. 2016) (see Fig. 19b). The measurements revealed the correspondence between the FBG reflection spectra and the generation spectra (Fig. 19). When using the highly reflective UV FBG1 with nonselective Fresnel reflection ($\sim$4%), one can observe the generation of a relatively homogeneous spectrum of $\sim$1 nm width (dashed curve), because of the relatively wide spectrum of UV FBG1 where individual resonances cannot be resolved. In the case of output UV FBG2 (which was also used in Zlobina et al. 2016), one can observe a three-peak structure with peaks spaced by $\sim$0.6 nm (see Eq. (23)), which corresponds to the reflection of three individual groups of graded-index fiber modes with small transverse indices (dot-dashed curve). The reflection spectrum is narrow for each group of modes; therefore, the generation spectra of different mode groups are well resolved. In the case of output FS FBG2, one observes a two-peak generation spectrum with a distance of $\sim$1.2 nm between the peaks (solid curve). The doubling of the distance between the peaks can be explained by the absence of the reflection peak for the second-group mode (see Fig. 19a) (because of the small overlap integral for the field of this mode with the grating recorded in the central part of the fiber core). Therefore the second group of modes (at a wavelength of 953.6 nm) is not involved in lasing. The peak of the third group at 953 nm is unstable in time and manifest itself only in the 2.5-km-long laser. Stable single-mode lasing was achieved in shorter GRIN fiber.

Fig. 19 (a) Reflection spectra of highly reflective UV FBG1 (dashes) and output FS FBG2 (Zlobina et al. 2017a). (b) Generation spectra at 1.9 W output power for 2.5-km GRIN RFL with different output coupler: 4% cleaved end (dash), UV FBG2 (dash-dot), and FS FBG2 (solid line) (Kablukov et al. 2016)

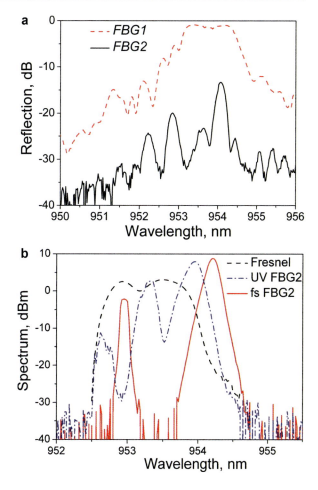

It is known (see, e.g., Mafi 2012) that the mode-group number in a graded-index fiber can be characterized by the quantity $g = 2p + |m| - 1$, where p and m are, respectively, the radial and azimuthal mode numbers. The radial (r) and azimuthal (ϕ) field distributions for modes LP_{mp} are described by the Laguerre polynomials (21) L_p^m. Hence, the first two mode groups are not degenerate and contain modes LP_{01} and LP_{11}, while the third group includes two modes: LP_{02} and LP_{21}.

The fundamental-mode diameter in the graded-index Corning 62.5/125 fiber can be estimated as ∼9.8 μm (Mafi 2012), and the photomodification region in FS FBG2 has the form of an ellipse in the fiber cross section, with characteristic sizes of principal axes of about 1 and 10 μm (see in Fig. 20a). Despite the comparable values of the mode diameter and the photomodification region size in one of the directions, the overlap integrals are small for modes with nonzero azimuthal index m. The modes with index $m = 0$ are maximally overlapped with the FBG recorded near

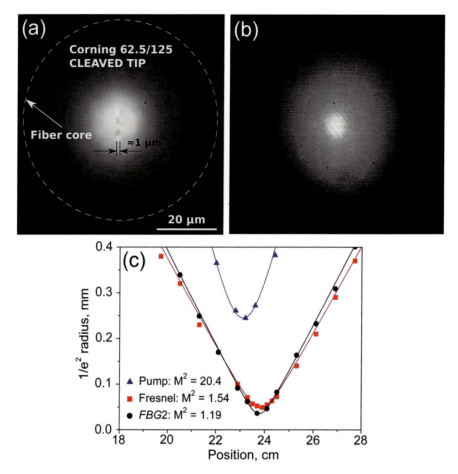

Fig. 20 (a) Microscope image of FS modification area in the core of GRIN fiber for FS FBG2. (b) Far-field image of the output beam near the Raman threshold in the RFL configuration with FS FBG2. Pump beam is gray, Stokes beam is bright. (c) M^2 measurement for residual pump radiation (triangles), Stokes generation in RFL with Fresnel reflection (squares) and FS FBG2 (circles) (Zlobina et al. 2017a)

the fiber center. Therefore, only the reflection peaks corresponding to these modes (LP_{01} in the first group and LP_{02} in the third group) arise in the reflection spectrum of the FS FBG2 and, correspondingly, in the generation spectrum of the Raman laser when this grating is installed at the output.

Shortening the GRIN fiber to 1.1 km results in the selection of a single transverse mode by means of FS FBG 2; see Fig. 20b, c (Zlobina et al. 2017a). Independent measurements of the output beam in the laser with a 1.1-km-long cavity revealed its quality factor M^2 to be better than 1.2 at output powers in the range from 5 to 10 W. This fact indicates that the lasing is close to single-mode. Power characteristics of 1.1-km single-mode RFL based on 62.5 μm GRIN fiber with bulk optics (Zlobina et al. 2017a) were compared with that for all-fiber coupling of the pump radiation (see Fig. 21 Zlobina et al. 2017b), in which the output radiation of three high-power multimode LDs at wavelength of 915 nm is combined by a 3 × 1 multimode fiber pump combiner with an output port made of 100-μm MM GRIN fiber. It is fusion spliced to the RFL cavity based on GRIN fiber with 85-μm or 62.5-μm core. The laser cavity was formed by highly reflective (UV FBG1) and output (FS FBG2) gratings inscribed in the core of the same graded-index fiber.

The obtained power and spectra in two configurations with 62.5-μm MM GRIN fiber are compared in Fig. 22. The laser generation threshold increases from ~40 to 50 W when changed from the bulk-optic pumping to the all-fiber one. The slope efficiency increases from 38% to 47% (Fig. 22a). Increase of the second threshold power from 10 to 16 W of generated power results from the elimination of Fresnel reflection at the input fiber facet after the modification from free-space to all-fiber

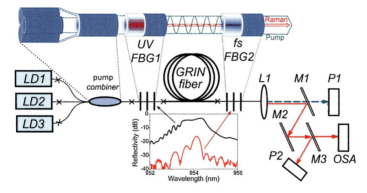

Fig. 21 Experimental setup of the all-fiber CW RFL with GRIN Raman gain fiber, in-fiber FBG reflectors and fiber pump combiner (Zlobina et al. 2017b)

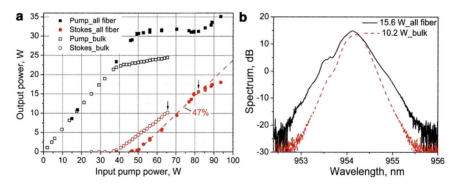

Fig. 22 (a) Power of residual pump (squares) and generated Stokes (circles) waves as a function of launched pump power for the bulk-optic (empty symbols) (Zlobina et al. 2017a) and all-fiber (filled symbols) pump coupling to the same 62.5-μm GRIN fiber with FBGs. The second Stokes thresholds are shown by arrows. (b) Corresponding generation spectra at the second Stokes threshold for the all-fiber and bulk-optic pump coupling configurations (Zlobina et al. 2017b)

pump coupling. As a result, the maximum power at 954 nm increases to about 16 W at 82 W launched pump power, whereas the generation linewidth (at −3 dB level) is left at the level of 0.4 nm (Fig. 22b). At the same time, after the transition to all-fiber configuration, the RFL operation becomes much more stable.

The combined pump power of three 915-nm LDs is sufficient to observe Raman lasing at 954 nm in 85-μm GRIN fiber, for which the threshold is ∼85 W (Fig. 23a). The second Stokes threshold grows from ∼135 to ∼180 W of coupled pump power at reducing fiber length from 1.95 to 1.5 km, whereas the first Stokes threshold remains nearly the same. As a result, maximum power at 954 nm grows from 21 to 49 W, while the residual pump power varies from 20 to 25 W only. The threshold grows by almost two times in proportion with the GRIN fiber core area increase. However, the slope efficiency of pump-to-Stokes conversion is higher for the 85-μm GRIN fiber (67%) than that for 62.5-μm one (47%) in spite of sufficiently stronger integral pump attenuation. It may be caused by a lower quality of the generated beam in this case. To check this, the output beam profile was measured (Fig. 23b) as well as beam quality factor $M^2 = 2.6$ at 32 W output power. It is almost two times of the quality factor for the 62.5-μm fiber laser ($M^2 \sim 1.3$), while the beam profile appears to be close to Gaussian. The measured M^2 value is only slightly (<10%) varying with power in 10–40 W range. The measured parameter M^2 for the transmitted pump radiation amounts to $M^2 \approx 26$, which is higher than the value measured in 62.5-μm GRIN fiber ($M^2 \approx 21$) without pump combiner (Zlobina et al. 2017a). The obtained brightness enhancement factor of 20 is higher than in bulk-optic multimode RFLs (Yao et al. 2015; Glick et al. 2016a, 2017).

The evolution of the generated spectra at ≈954 nm with increasing power in 1.5-km fiber is shown in Fig. 23c. The shape of the generated spectrum is close

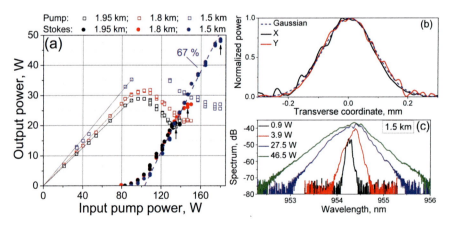

Fig. 23 (a) Output power (filled symbols) and residual pump power (empty symbols) as a function of coupled pump power (linear fit for transmitted pump in lasing-free case is shown by solid lines). The 2nd Stokes thresholds are shown by arrows. (b) Corresponding beam profile (with Gaussian fit) at 32 W output power and (c) generated spectrum as a function of output power in the 85-μm GRIN fiber of 1.5 km length (Zlobina et al. 2017b)

to a hyperbolic secant with exponential wings. Its width grows as square root of the power (both for -3 and -10 dB levels), just like the spectrum of conventional Raman lasers based on long single-mode fibers (Babin et al. 2007), owing to an interplay of self-phase modulation and dispersion. The only sign of higher-order transverse mode impurities consists in a visible asymmetry of the line that appears already at 4 W, and it does not change significantly with increasing power. This fact is in agreement with the weak dependence of the quality factor M^2 on the laser power and indicates the formation of stable group of coupled low-index modes.

To summarize, the all-fiber configuration based on conventional multimode GRIN fiber directly pumped by multimode laser diodes at 915 nm demonstrates the possibility of generating high-power radiation at new wavelengths (defined by available pump diodes). The beam quality parameter is not far from single-mode one.

Note that this is only the first step demonstrating the basic principles of the new type of CW high-power high-beam-quality LD-pumped fiber laser. The next steps toward a better pump coupling at higher powers and a better fundamental mode selection will result in further improvements in optical efficiency, output power, and brightness of such source. This makes it very attractive for applications, such as efficient/bright sources for pumping solid-state/fiber lasers, second-harmonic generation, laser displays, and biomedical imaging. Interesting fundamental challenges also appear, e.g., investigation of mode instability in a graded-index fiber. Such instability may behave differently from that in step-index fibers, thus breaking the existing limits of high power fiber lasers.

Kerr Beam Self-Cleaning

Whenever propagation involves a relatively large (e.g., greater than 10) number of guided modes, the output intensity in the transverse beam profile from a MMF exhibits an irregular speckled structure resulting from mode interference (see Fig. 24, left panel). Moreover, typical fiber perturbations such as stress, bending, or temperature variations lead to sudden variations of this speckled interference pattern. This has largely prevented so far the possibility of transporting spatially encoded information with a multimode optical fiber in a variety of applications, from communication links to image transmission and high beam quality transport in medical and industrial applications.

In fact, MMFs are often employed to scramble the spatial beam profile of a bell-shaped coherent laser beam. In March of 2015, researchers from Limoges University in France were experimentally studying spatiotemporal beam reshaping in the process of second-harmonic generation from a quadratic nonlinear crystal, pumped by a highly multimode fundamental beam (Krupa et al. 2015). In order to produce a spatially multimode speckled beam, a short piece of MMF was used. As a manifestation of scientific serendipity, when increasing the length of a GRIN MMF to above a few meters, Krupa et al. obtained the unexpected and striking result that, at sufficiently high input powers, the transverse profile of the pulses

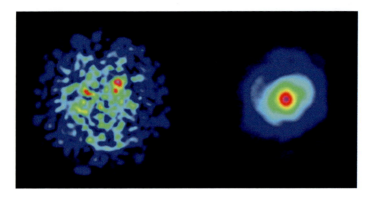

Fig. 24 Output beam from GRIN MMF: at low powers the beam is scrambled (left); at high powers the beam cleans up (right)

self-cleans up (see Fig. 24, right panel). In other words, the light beam emerging from the GRIN MMF reshapes into a bell-shaped profile, whose cross section is very close to the fundamental mode of the fiber. The beam remains self-cleaned at all distances beyond the threshold length. Most importantly, self-cleaning is very robust in the presence of mechanical perturbations such as squeezing and bending of the fiber at intermediate points. Moreover, self-cleaning vanishes when bringing down the input intensity below its threshold value. In these first experiments, 30 ps pulses propagating at 1064 nm were used, so that both chromatic and modal dispersion effects can be neglected over the involved propagation distances.

When using optical pulses (from sub-nanosecond down to 100 femtosecond durations), and GRIN MMF lengths up to a few tens of meters, the power threshold for observing Kerr self-cleaning is lower than that for SRS. Hence Kerr beam cleanup occurs before the SRS-induced beam cleanup described in the previous section. This means that one can observe Kerr-induced cleaning for a pump beam itself, before any spectral broadening or frequency conversion occurs.

Moreover, when using pulses with durations in the picosecond regime, Kerr beam cleanup occurs at powers which are about two orders of magnitude (i.e., in the 1–10 kW regime) lower than the power threshold for catastrophic self-focusing. Note also that Kerr beam cleaning is essentially a spatial effect, since it is observed with relatively long pulses propagating in the normal dispersion regime of a few meters long GRIN MMFs, so that chromatic and modal dispersion do not play a significant role in the pulse dynamics. Kerr beam cleaning has the important property to significantly increase the beam brightness at the output of a MMF.

When using an essentially lossless propagation environment (i.e., a few meters of MMF), Kerr beam cleanup has been reported so far only in GRIN fiber. That is, similarly to the case of SRS beam cleanup, the Kerr beam cleanup effect has not been observed in step-index fibers. Very interestingly, in the case of a strongly dissipative propagation environment, such as it occurs in a lossy or active Yb-doped MMF, Kerr beam self-cleaning has also been observed in a nearly step-index

MMF. Although such recent finding is not yet fully theoretically understood, it may open the way to the use of Kerr beam self-cleaning using standard MMF laser technologies.

In this section, we will first describe the model that permits the numerical simulation of Kerr beam cleanup. Next, we will discuss experiments of Kerr beam self-cleaning in lossless GRIN MMFs. Subsequently, we will present experimental studies of Kerr self-cleaning in dissipative, nearly step-index Yb-doped fibers in the presence of either loss or gain and discuss their different performances. Finally, we will outline the use of self-cleaning in a composite laser cavity including an Yb-doped fiber in the external cavity.

Theoretical Models of Spatiotemporal Dynamics

In this subsection, we present a theoretical model that allows for the efficient numerical simulation of pulse propagation in a GRIN MMF. This model permits to reproduce the effect of Kerr beam self-cleaning. The evolution in three space and one temporal coordinates of the complex envelope $A(x, y, z, t)$ of a light pulse obeys the generalized 3D nonlinear Schrödinger equation (GNLSE3D)

$$\frac{\partial A}{\partial z} - i\frac{1}{2k_0}\nabla_\perp^2 A + i\frac{\kappa''}{2}\frac{\partial^2 A}{\partial t^2} + i\frac{k_0 \Delta}{R^2}r^2 A$$
$$= i\gamma(1 - f_R)|A|^2 A + i\gamma f_R A \int h(\tau)|A(t-\tau)|^2 d\tau. \quad (24)$$

Here $k_0 = \omega_0 n_{co}/c$, $\gamma = \omega_0 n_2/c$ is the nonlinear coefficient, and $h(t)$ is the Raman temporal response of silica glass, with $f_R = 0.18$ the SRS fraction of the total cubic nonlinearity. According to our normalization units, $|A|^2$ has the physical dimension of optical intensity.

Equation 24 contains all main physical mechanisms at play: diffraction, dispersion, waveguide profile, Kerr effect, and SRS. As the pulse power is increased, it is affected by the interplay of transverse beam diffraction, waveguide potential, and the Kerr effect. Moreover, optical pulses will break up in time, which eventually leads to the generation of new frequency sidebands. However we will not describe such effects here in this section. Besides parametric wave-mixing effects, in the normal dispersion regime, the pump light experiences the competing process of SRS-induced frequency conversion to the Stokes wavelength.

The results of numerical simulations of spatial beam evolution along the GRIN MMF according to Eq. 24 are displayed in Fig. 25. Figure 25a–d shows that, in the nonlinear regime, the initially highly multimodal distribution of the laser beam evolves into a bell-shaped transverse profile, in agreement with experiments. The fraction of power carried by the fundamental mode increases by a factor ∼2.3 after 1 m of propagation distance, whereas Fig. 25a'–d' shows that at low powers, an irregular speckled intensity pattern is always observed. In simulations, we used the

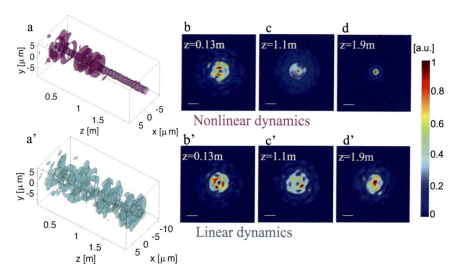

Fig. 25 Simulations showing beam self-cleaning in lossless GRIN MMF. (**a, a'**), Spatial reshaping along propagation coordinate z, showing Kerr beam cleanup in the nonlinear regime (**a**), whereas no self-cleaning occurs in the linear regime (**a'**); (**b–d, b'–d'**), 2D output distributions in nonlinear (**b–d**) and linear (**b'–d'**) regime. Spatial frames are obtained by averaging the time-integrated intensity along distance z over three consecutive samples (spaced by 5 mm) (Reproduced from Krupa et al. 2017a)

integration step of 0.05 mm, and a transverse 128×128 grid, for a spatial window of $150 \times 150\,\mu m$. We considered a standard GRIN MMF with core radius of $26\,\mu m$ and a peak core refractive index 1.47 and for a cladding index of 1.457. We used an input beam diameter of $40\,\mu m$. To reduce computational time, we rescaled the problem by using (in the nonlinear regime) a pump power of 63 kW, which is higher than the typical value in experiments, to reduce the propagation length to maximum $L = 2$ m; moreover, we used a pulse duration of 7.5 ps. To mimic the speckled near-field image observed at low pump powers, we used two different approaches, which gave similar numerical results. The results presented in Fig. 25 were obtained by blurring an input flat spatial phase front with a random uniformly distributed 2D function. In this case the multiplicative input phase noise is intended to mimic the phase shifts in the first steps of propagation in the fiber. In the second approach, we applied a coarse-step distributed coupling method starting from a coherent input condition represented by a flat phase and Gaussian intensity distribution, and we perturbed the propagation by randomly changing selected key parameters, such as the fiber core diameter or the relative refractive index difference, and/or by introducing random artificial tilts to the beam in the fiber. We introduced these numerical perturbations every 1 mm, which is much longer than the integration step.

In simulations, we stored along the propagation coordinate z the complex field $A(x, y, z, t = 0)$ at a given point in time $t = 0$, (in a reference frame co-moving with the pulse and corresponding to the pulse peak). This permits to calculate

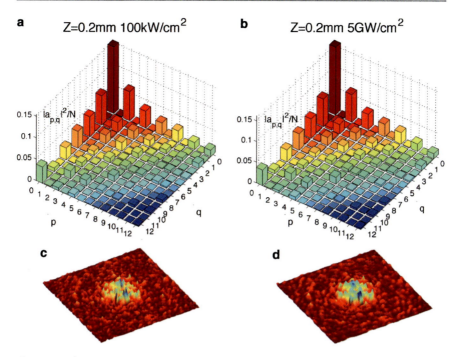

Fig. 26 Numerical beam propagation simulations in GRIN MMF at z=0.2 mm. (**a, b**) Modal distribution and (**c, d**) Instantaneous intensity profile (normalized to the local maximum) for input intensity of $100\,\text{kW/cm}^2$ (**a, c**) and $5\,\text{GW/cm}^2$ (**b, d**) (Reproduced from Krupa et al. 2017b)

the orthogonal projections of the transverse field envelope $A(x, y, z, t = 0)$ on selected elements of the basis of the fiber-guided modes. In the weak-guidance approximation, these modes can be represented by Hermite-Gauss functions, which we used for simplicity in our analysis (see, for instance, Mafi 2012). The results of a modal decomposition analysis of the output beam are illustrated in Figs. 26 and 27. The Hermite-Gauss modes are identified here by two indices $H_{p,q}(x, y)$, which refer to x and y Cartesian coordinates, respectively.

The orthogonal projections $a_{p,q}(z) =< A(x, y, z, t = 0)|H_{p,q}(x, y)^* >$ of the output beam on the fiber modes where calculated along the propagation coordinate z at $t = 0$. In Fig. 26 we illustrate the relative mode occupation $|a_{p,q}(z)|^2/N(z)$ (where $N(z) = \sum_{p,q} |a_{p,q}(z)|^2$) after 0.2 mm of propagation only, for a low-power pulse (panel a) and a high-power pulse (panel b), respectively. Panels c and d show the corresponding instantaneous (at $t = 0$) relative intensity profiles of the beams. Figure 27 shows the same decompositions but obtained after 1 m of propagation. As can be seen, nonlinearity (panels b and d) substantially modifies the modal distribution of the beam, by increasing the mode $H_{0,0}(x, y)$ content, in agreement with the hypothesis of self-cleaning toward the fundamental mode of the GRIN MMF.

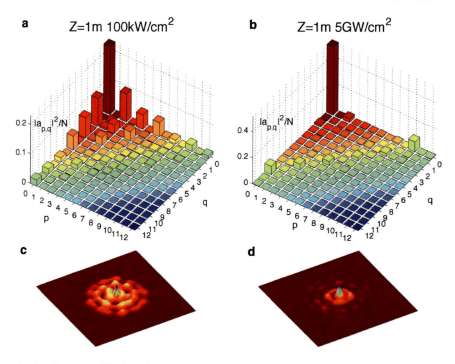

Fig. 27 Same as in Fig. 26, with z=1 m (Reproduced from Krupa et al. 2017b)

Kerr Beam Cleanup in GRIN MMF

Let us start the description of Kerr-induced spatial beam self-cleaning experiments by considering a lossless propagation environment, involving just a few meters of standard commercially available GRIN MMF. Initial experiments were carried out with 30–900 ps duration input pulses, propagating in the normal dispersion regime (Krupa et al. 2016b, 2017a, b; Wright et al. 2016). Subsequently, Kerr self-cleaning was also investigated in the femtosecond pulse regime, where the power threshold grows larger and it becomes nearly comparable with the threshold for catastrophic self-focusing (Liu et al. 2016).

Figure 28 shows details of the observed power dependence of Kerr beam cleanup in a 12m-long GRIN MMF with 52 μm of core diameter. In these experiments, 900 ps pulses with 30 kHz repetition rate from an amplified Nd:YAG microchip laser operating at 1064 nm were used (Krupa et al. 2017a, b). Panels (a–d) of Fig. 28 show the 2D near-field spatial beam pattern emerging from the fiber at 1064 nm, as a function of the output peak power (P_{pp}). Panels (a'–d') of Fig. 28 display the corresponding beam transverse profiles. At low powers ($P_{pp} = 3.7$ W), where propagation is linear, the highly multimode excitation including about 200 guided modes leads to a highly irregular speckled beam distribution (Fig. 28, panels a, a'). However, when progressively increasing the input pulse power, the output beam

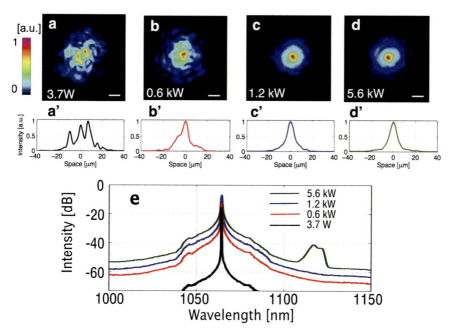

Fig. 28 Experimental spatiotemporal nonlinear dynamics in GRIN MMF. (**a–d**), Near-field images of the MMF output at the pump wavelength of 1064 nm (intensities are referred to the local maximum and are shown in linear scale; scale bar: 10 μm) showing spatial beam self-cleaning toward a well-defined bell-shaped distribution when increasing the output peak power P_{pp}. (**a'–d'**), Corresponding beam profiles versus x (y=0 section). (**e**), Corresponding output spectra. Fiber length: 12 m (Reproduced from Krupa et al. 2017a)

profile evolves into a well-defined bell-shaped spot at the beam center, whose diameter is close to the fundamental mode of the fiber, on a residual low-power background (Fig. 28, panels d, d'). Spectra in panel e of Fig. 28 reveal that the redistribution of power toward a central quasi-Gaussian profile results before any substantial spectral broadening occurs (Krupa et al. 2017a, b).

Kerr beam cleanup leads to a significant increase of the spatial quality of output beam and its brightness: a nearly threefold decrease of the output near-field beam FWHMI (full width at half maximum intensity) diameter was measured. A statistical study (involving a large number of input beams, characterized by different orientations of their linear state of polarization) confirmed that Kerr beam self-cleaning is associated with a reduction in the average beam diameter, but most importantly with a dramatic reduction in the standard deviation of the output beam size (Krupa et al. 2017a, b). It is also important to mention that, for powers above the Kerr-induced spatial beam self-cleaning threshold, one observes significant spectral and temporal reshaping of the input pulses, leading to sideband series and supercontinuum generation (Wright et al. 2016; Krupa et al. 2016a, b; Lopez-Galmiche et al. 2016).

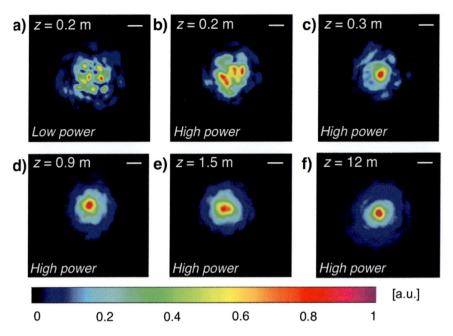

Fig. 29 Experimental evolution of near-field spatial pattern of GRIN MMF output at 1064 nm as a function of propagation distance z (**a–f**) induced by Kerr beam self-cleaning; High input power = 68 kW; Low input power = 0.007 kW; Scale bar: 10 μm (Readapted from Krupa et al. 2017a)

To gain additional insight into the beam self-cleaning process, cutback measurements as in Fig. 29 were performed. These results confirm that spatial reshaping occurs after a certain distance of propagation (here, this is only about 30 cm, owing to the relatively high input pulse powers (68 kW)). These observations prove that beam self-cleanup occurs in the presence of an initial highly multimode excitation. These observations are in quite good qualitative agreement with the numerical simulations described in the previous subsection of this chapter.

To further analyze the possible interplay between spatial and spectral degrees of freedom of the light pulses in the process of beam cleaning, one may image the fiber output face on a CCD camera through a dispersive spectroscope. Note that a standard spectrum analyzer cannot resolve the cavity mode spacing of a microchip laser. The imaging spectroscope was based on a heavily dispersive grating in a Littrow configuration. With such setup, one obtains an image that combines the transverse spatial pattern (vertical coordinate) and the frequency spectrum (horizontal coordinate). Spectral broadening leads to an image expanding along the horizontal axis. Spatial beam narrowing or broadening is instead visible in the vertical axis.

Figure 30 shows three representative sample images. Panel (a) of Fig. 30 was recorded at low powers and shows that the output spectrum is characterized by a nearly single longitudinal mode of the input laser, plus a weak satellite longitudinal

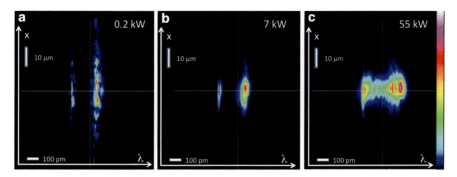

Fig. 30 Spatial (x) and spectral (λ) analysis of output wave for three pump powers: (**a**) linear propagation 0.2 kW, (**b**) high power 7 kW, (**c**) very high power 55 kW; pulse duration: 0.9 ns, fiber length: 3 m (Reproduced from Krupa et al. 2017b)

mode. In the case of linear propagation, one obtains an output beam with an extended speckled spatial pattern ($\sim$30 μm in the vertical dimension). Panel (b) of Fig. 30 shows that in the nonlinear propagation regime, beam self-cleaning leads to beam shrinking to about 10 μm in the vertical direction, with no discernible modification in the frequency spectrum. For much higher pump power values, panel (c) in Fig. 30 shows that, along with spatial compression, substantial spectral broadening appears. In these experiments we used a 3-m-long fiber, instead of the 12-m-long fiber as in Fig. 28. The difference in fiber lengths explains the higher level of power required for self-cleaning. Similar results are obtained with different powers and fiber lengths, provided that their product remains unchanged, similarly to the Kerr effect in single-mode fibers.

An important quantitative measure of the beam quality at the output of the GRIN MMF is given by its M^2 parameter. The dependence of M^2 as a function of the input beam power is illustrated in Fig. 31. Here the M^2 parameter was calculated as a second moment width function, which is more appropriate for speckled beams. Figure 31 reveals that at about 10 kW (for a fiber segment of 3 m), the beam quality reaches a stationary value close to 4: this value is then kept nearly constant even for input peak powers much above the threshold for self-cleaning. The images in the insets of Fig. 31 show corresponding transverse beam profiles.

Although numerical simulations based on the solution of Eq. 24 are indeed capable of predicting Kerr beam self-cleaning, a simple physical explanation of this effect is still missing. In fact, a few years ago Picozzi et al. have predicted, with the help of a statistical approach, the condensation into the fundamental mode (surrounded by a background equipartition of energy into the higher-order modes) of an initial highly multimode beam in the frame of the GRIN fiber model of Eq. 24 (Aschieri et al. 2011). This condensation of classical waves is associated with the generation of a Rayleigh-Jeans equilibrium distribution. This interpretation is fascinating; however it is associated with some predictions that are not yet fully corroborated by the experiments. For example, the wave condensation model

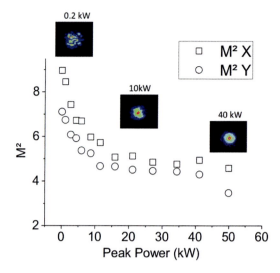

Fig. 31 M^2 measurements at the output of a 3-m-long fiber, and for the X- and Y-axis (Reproduced from Krupa et al. 2017b)

predicts that a faster convergence toward the condensed state occurs whenever the input beam is equally distributed among all modes, whereas experiments seem to indicate that Kerr beam cleanup into the fundamental mode requires an initial unbalance toward this mode in the input beam decomposition which is coupled to the MMF.

Another interpretation of Kerr beam self-cleaning has been put forward by L. Wright et al., who have conjectured that irreversible evolution toward the fundamental mode is achieved since this mode is the most unstable with respect to nonlinearity induced perturbations (Wright et al. 2016). Although this interpretation may look as a paradox, it finds its justification when considered in the broader framework of the theory of self-organized criticality, which is a universal mechanism for pattern formation and self-organization in out-of-equilibrium physical systems.

Finally, a perhaps more intuitive explanation for the nonlinear asymmetry of power flow toward the fundamental mode of the GRIN MMF may be based on the nonlinear nonreciprocity of the mode-coupling process in a GRIN MMF. The periodic oscillations of the multimode beam intensity, which occur along the fiber through mode beating, lead to Kerr effect induced longitudinal modulation of the fiber refractive index or dynamic long-period fiber grating. This grating may provide quasi-phase-matching for a variety of nonlinear four wave-mixing processes, leading to energy exchange between the fundamental and the HOMs. A simplified explanation of the physical mechanism behind the Kerr-induced spatial beam self-cleaning effect has been put forward, based on a two-mode mean-field model; see for details Krupa et al. (2017a, b). This mean-field model predicts that, for powers above a certain threshold power value, the reciprocity of mode coupling is broken, thus leading to a preferential flow of energy toward the fundamental mode.

Kerr Beam Cleanup in Step-Index Active MMF with Loss or Gain

When a few meters of MMF are used, experiments described in the previous subsection demonstrate Kerr beam cleanup in a GRIN fiber. However, similar experiments performed so far with a step-index MMF could not demonstrate self-cleaning, at least until power levels such that the fiber damaging threshold was approached. Now, an interesting question is if some modification to the linear fiber parameters could be introduced that could allow for the observation of beam self-cleaning with a step-index MMF. Again quite surprisingly and very interestingly, experiments have shown that this feature is possible, whenever the linear propagation environment is strongly dissipative, that is, in the presence of strong linear loss of gain. This is quite counterintuitive, as one would normally imagine that the delicate balance between the Kerr-induced long-period fiber grating and resulting power conversion into the fundamental mode could be disrupted in the presence of strong dissipation.

To verify such a possibility, Kerr beam self-cleaning was experimentally studied in a MMF doped by rare-earth atoms, by using the setup shown in Fig. 32. Two different configurations were employed. In the first configuration, 3-m-long, highly lossy (2.44 dB/m at 1064 nm), unpumped double-clad ytterbium-doped MMF with a nearly step-index profile was used (see Fig. 33). In the second configuration, the same Yb-doped fiber was pumped by a CW laser diode, which provided up to G=20 of gain. The Yb-doped MMF had a 55 µm core diameter, and it was surrounded by a D-shaped inner cladding of 340×400 µm size, for guiding the pump radiation (Guenard et al. 2017a).

In spite of the large overall attenuation of about 7.3 dB and the non-parabolic refractive index profile, experiments with the unpumped Yb-doped MMF still led to the observation of the Kerr beam cleanup. Figure 34 shows that, in the lossy case, the M^2 parameter of the output beam dropped from 16 (in the low-power regime) down to a value of only 2 after beam cleanup. Let us note however that the nonlinear beam reshaping process was qualitatively different from that in the GRIN MMF. In fact, the input pump power was not only coupled into the Yb-doped MMF core but also into HOMs leaking into the D-shaped cladding. At high pump powers, beam

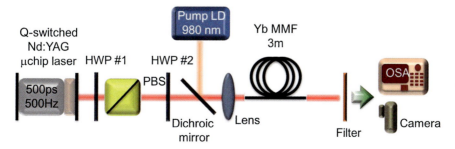

Fig. 32 Experimental setup to observe Kerr self-cleaning in Yb-doped fiber

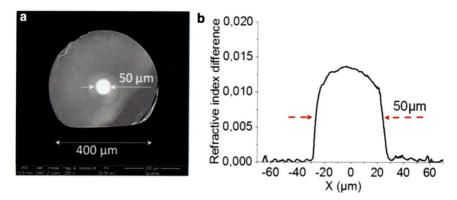

Fig. 33 Image (SEM) (**a**) of the double-clad ytterbium-doped multimode optical fiber and (**b**) its core refractive index profile (Reproduced from Guenard et al. 2017a)

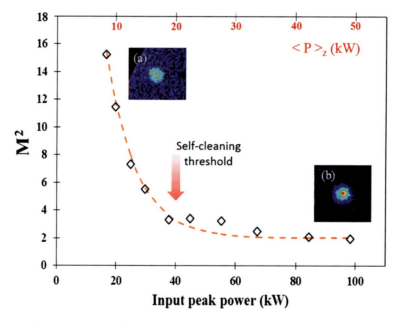

Fig. 34 M^2 measurement ($1/e^2$ diameter) of the output beam pattern versus the input peak power with the passive (unpumped) Yb-doped MMF; Insets: output beam patterns for (**a**) low power and (**b**) high power. The upper scale in red gives the path-averaged power $<P>_z$ to include the impact of the fiber losses. The dashed line is an exponential fit as a guide line for the eye (Reproduced from Guenard et al. 2017a)

cleaning into the fundamental mode of the core was accompanied by a significant reduction (by about 50%) of the energy carried by the leaky HOMs.

More interesting for the potential application to high-power MMF-based amplifiers and lasers, it is the possibility to obtain beam cleanup (with a much reduced power threshold) in the active configuration involving the diode pumped Yb-doped

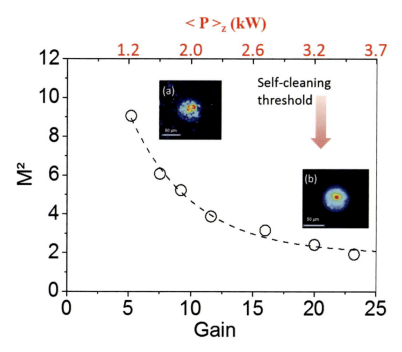

Fig. 35 M^2 measurement ($1/e^2$ diameter) of the output beam pattern versus gain of the amplified Yb-doped MMF; Input signal peak power: 0.5 kW; Insets: output beam patterns for low amplification (**a**) and high amplification (**b**) (Reproduced from Guenard et al. 2017a)

MMF. To prove this issue, in experiments, the power of the microchip Nd:YAG laser was kept fixed at 0.5 kW, and the D-shaped inner cladding of the Yb-doped MMF was pumped with a CW laser diode at 940 nm with increasing values of its power level. The insets in Fig. 35 depict the evolution of the near-field pattern at the fiber output, as a function of optical gain (G) of the Yb-doped MMF. Once again, Kerr beam cleanup was observed.

The decrease of the M^2 parameter presented in Fig. 35 further confirms the substantial improvement of the output beam quality, from a value of 9 with G=5 down to only 2 in the self-cleaning regime (for a gain G=20 or larger). Let us recall that an input peak power as high as 40 kW is necessary with the same lossy Yb-doped MMF, whereas a similar length of GRIN MMF exhibits a threshold peak power of about 7 kW.

Figure 35 shows the evolution of the M^2 upon the gain of the amplified Yb-doped MMF, and Fig. 36 provides an interesting comparison of the longitudinal power evolution in the lossy (unpumped) and in the active (pumped) Yb-doped MMF, respectively. In particular Fig. 36 reveals that, if we consider the values of the output peak power instead of the input values, both cases lead to nearly the same value of about 11 kW, which, incidentally, is also the typical output power value that is obtained for a lossless GRIN MMF of the same length. Moreover, Fig. 36 also

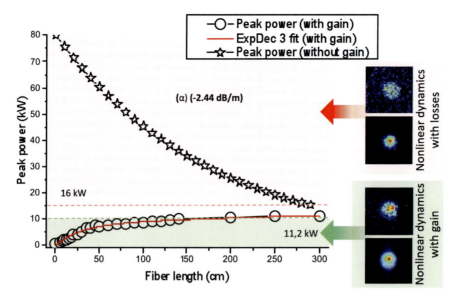

Fig. 36 Comparison of self-cleaning in the lossy and the active Yb-doped MMF. Here the power evolution along the fiber is compared in two situations leading to Kerr beam cleanup: remarkably, the output pulse power is nearly equal in both cases

reveals that the path-averaged power ($< P >_z$) is significantly reduced (by about six times) by the presence of gain with respect to the case with loss. Therefore, in the nonlinear regime the presence of loss and gain cannot be interchanged, by simply considering the value of the path-average power.

Self-Cleaning in a MMF Laser Cavity

Similar to the case of Raman beam cleanup in GRIN MMF that was described in the previous section, Kerr beam cleanup can also be usefully exploited to operate a MMF laser with a high-power, quasi-single-mode transverse output beam profile. The key difference here is that the Kerr beam cleanup occurs at the pump oscillation frequency and not at the Stokes-shifted frequency. To compensate for cavity loss, in the case of a Kerr beam self-cleaning-based MMF laser, it will be necessary to provide gain via the use of an active MMF such as the Yb-doped MMF that was described in the previous subsection. As a matter of fact, lasers based on MMFs could provide an useful alternative to large mode area (LMA) single-mode fibers for the generation of high peak power pulses and for fabricating high-power fiber amplifiers and lasers. This is mainly because of cost and modal instability issues of lasers based on LMA fibers. Kerr beam cleanup could then remove the main obstacle which has limited the use of MMFs for fiber lasers so far that is their inherent degradation of the spatial beam quality.

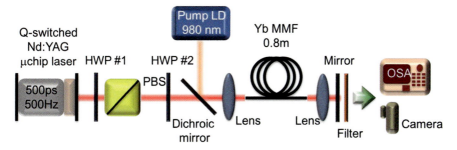

Fig. 37 Experimental setup to observe Kerr self-cleaning in composite laser cavity including Yb-doped fiber

In this section, we briefly describe the recent demonstration of Kerr beam cleanup in a coupled cavity laser (Guenard et al. 2017b), composed by a passively Q-switched Nd:YAG microchip laser plus an external cavity including the same Yb-doped MMF amplifier as presented in the previous subsection (see Fig. 37), and a polarization control (including two wave plates and a polarizer) that permits to adjust the level of coupling among the two cavities and the power injected into the MMF. Such type of coupled cavity lasers, but with single-mode fibers in the external cavity, was studied before for timing jitter reduction in Q-switched laser and supercontinuum generation (Bassri et al. 2012). Using a MMF in the external cavity, one may simultaneously shape the output beam in the spatial, spectral, and temporal domains. We obtained both Q-switched and Q-switched mode-locked operations, in combination with Kerr beam cleanup in the MMF Yb-doped fiber.

In order to form the external cavity, a mirror was inserted after the Yb-doped MMF, as shown in Fig. 37. Owing to the large length difference between the microchip cavity (∼7 mm) and the MMF cavity (∼1 m), it is always possible to find common resonance frequencies for the two cavities, without an electrooptic servo control.

Let us consider first the case of an unpumped, lossy Yb-doped MMF in the external cavity. Whenever relatively low-power pulses are launched in the MMF from the microchip laser, the two cavities are weakly coupled, and a highly multimode beam recirculates in the MMF. When cavity coupling is sufficiently increased by means of the polarization control, a sudden increase of power in the MMF section occurs. This leads to the excitation of multiple longitudinal modes in the laser spectrum (see the upper right panel in Fig. 38) and to the switching by 90° of the polarization state of the microchip laser. Simultaneously, the transverse profile of the laser beam at the MMF output switches from a speckled pattern to a clean bell-shaped profile (see the upper mid panel in Fig. 38). The abrupt change in the beam profile from the MMF is associated with the improvement of the beam quality parameter M^2 dropping down to 1.7.

Note that even with strong cavity coupling, Q-switched laser operation is preserved with unchanged repetition rate. However, the laser pulse is shortened from 525 to 225 ps (see the upper left panel in Fig. 38).

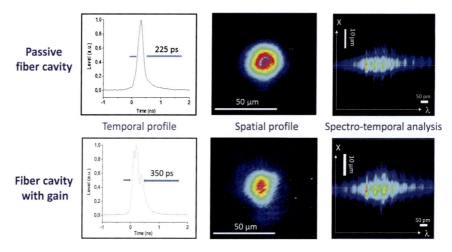

Fig. 38 Temporal, spatial, and spectral characterization of output from a MMF-based Q-switched double-cavity laser, showing pulse compression, spatial beam cleanup, and spectral broadening, respectively. Top (bottom) plots: case of a lossy (pumped) Yb-doped MMF (Readapted from Guenard et al. 2017b)

Next, consider MMF laser operation when switching on the pump laser at 940 nm, in order to change the Yb-doped MMF into an amplifier. The Yb-doped MMF gain could be varied with the pump power, up to a maximum of G=23 (for an input signal of 0.5 kW peak power). Note that the available amplifier gain is limited by the damage threshold of the fiber end face. The main impact of MMF gain is the increase of the output laser power and the addition of an ASE background to the output beam transverse profile. The bell-shaped output beam obtained for maximum gain is shown in the bottom mid panel in Fig. 38. However, the beam profile is not as clean as in the passive MMF situation. The corresponding laser spectrum exhibits a discrete structure including the longitudinal modes fixed by the coupled cavities (see the bottom right panel in Fig. 38). Still, the beam quality remains high ($M^2 = 1.8$). The temporal profile of the output pulse acquires an asymmetric shape, with a 350 ps duration (FWHMI) (see the bottom left panel in Fig. 38).

Acknowledgements The authors acknowledge financial support by the Russian Science Foundation (grant 14-22-00118) and by the Ministry of Education and Science of the Russian Federation (14.Y26.31.0017); K.K. has received funding from the European Union's Horizon 2020 research and innovation programme under the Marie Sklodowska-Curie grant agreement No. GA-2015-713694 ("BECLEAN" project).

References

A.B. Aceves, G.G. Luther, C. De Angelis, A.M. Rubenchik, S.K. Turitsyn, Phys. Rev. Lett. **75**, 73 (1995)

A.B. Aceves, O.V. Shtyrina, A.M. Rubenchik, M.P. Fedoruk, S.K. Turitsyn, Phys. Rev. A **91**, 033810 (2015)

E.R. Andresen, J.M. Dudley, D. Oron, C. Finot, H. Rigneault, J. Opt. Soc. Am. B **28**, 1716 (2011)
P. Aschieri, J. Garnier, C. Michel, V. Doya, A. Picozzi, Phys. Rev. A **83**, 033838 (2011)
S.A. Babin, D.V. Churkin, A.E. Ismagulov, S.I. Kablukov, E.V. Podivilov, J. Opt. Soc. Am. B **24**, 1729 (2007)
S.A. Babin, E.I. Dontsova, S.I. Kablukov, Opt. Lett. **38**, 3301 (2013)
S.H. Baek, W.B. Roh, Opt. Lett. **29**, 153 (2004)
P.L. Baldeck, F. Raccah, R.R. Alfano, Opt. Lett. **12**, 588 (1987)
S. Boscolo, A.I. Latkin, S.K. Turitsyn, IEEE J. Quantum Electron. **44**, 1196 (2008)
I.S. Chekhovskoy, A.M. Rubenchik, O.V. Shtyrina, M.P. Fedoruk, S.K. Turitsyn, Phys. Rev. A **94**, 043848 (2016)
I. Chekhovskoy, V. Paasonen, O. Shtyrina, M. Fedoruk, J. Comput. Phys. **334**, 31 (2017)
P.K. Cheo, A. Liu, G.G. King, IEEE Photon. Technol. Lett. **13**, 439 (2001)
K.S. Chiang, Opt. Lett. **17**, 352 (1992)
K.S. Chiang, Opt. Commun. **95**, 235 (1993)
C.A. Codemard, P. Dupriez, Y. Jeong, J.K. Sahu, M. Ibsen, J. Nilsson, Opt. Lett. **31**, 2290 (2006)
V.L. da Silva, J.P. Heritage, Y. Silberberg, Opt. Lett. **18**, 580 (1993)
A.V. Dostovalov, A.A. Wolf, A.V. Parygin, V.E. Zyubin, S.A. Babin, Opt. Express **24**, 16232 (2016)
F. Eilenberger, K. Prater, S. Minardi, R. Geiss, U. Röpke, J. Kobelke, K. Schuster, H. Bartelt, S. Nolte, A. Tünnermann, T. Pertsch, Phys. Rev. X **3**, 041031 (2013)
F. El Bassri, F. Doutre, N. Mothe, L. Jaffres, D. Pagnoux, V. Couderc, A. Jalocha, Opt. Express **20**, 1202 (2012)
A.D. Ellis, Z. Jian, D. Cotter, Approaching the non-linear Shannon limit. J. Lightwave Technol. **28**, 423–433 (2010)
R.J. Essiambre, G. Kramer, P.J. Winzer, G.J. Foschini, B. Goebel, Capacity limits of optical fiber networks. J. Lightwave Technol. **28**, 662–701 (2010)
R.J. Essiambre, R. Ryf, N.K. Fontaine, S. Randel, Breakthroughs in photonics 2012: space-division multiplexing in multimode and multicore fibers for high-capacity optical communication. IEEE Photon. J. **5**, 0701307 (2013)
T. Fan, IEEE J. Sel. Top. Quantum Electron. **11**, 567 (2005)
B.M. Flusche, T.G. Alley, T.H. Russell, W.B. Roh, Opt. Express **14**, 11748 (2006)
I. Gasulla, J. Capmany, IEEE Photon. J. **4**, 877 (2012)
Y. Glick, V. Fromzel, J. Zhang, A. Dahan, N. Ter-Gabrielyan, R.K. Pattnaik, M. Dubinskii, Laser Phys. Lett. **13**, 065101 (2016a)
Y. Glick, Y. Sintov, R. Zuitlin, S. Pearl, Y. Shamir, R. Feldman, Z. Horvitz, N. Shafir, J. Opt. Soc. Am. B **33**, 1392 (2016b)
Y. Glick, V. Fromzel, J. Zhang, N. Ter-Gabrielyan, M. Dubinskii, Appl. Opt. **56**, B97 (2017)
D. Gloge, E.A.J. Marcatili, Bell Syst. Tech. J. **52**, 1563 (1973)
J. Goldhar, M. Taylor, J. Murray, IEEE J. Quantum Electron. **20**, 772 (1984)
A.B. Grudinin, E.M. Dianov, D.V. Korbkin, A.M. Prokhorov, D.V. Khaidarov, JETP Lett. **47**, 356 (1988)
R. Guenard, K. Krupa, R. Dupiol, M. Fabert, A. Bendahmane, V. Kermene, A. Desfarges-Berthelemot, J.L. Auguste, A. Tonello, A. Barthélémy, G. Millot, S. Wabnitz, V. Couderc, Opt. Express **25**, 4783 (2017a)
R. Guenard, K. Krupa, R. Dupiol, M. Fabert, A. Bendahmane, V. Kermene, A. Desfarges-Berthelemot, J.L. Auguste, A. Tonello, A. Barthélémy, G. Millot, S. Wabnitz, V. Couderc, Opt. Express **25**, 22219 (2017b)
L. Hadzievski, A. Maluckov, A.M. Rubenchik, S. Turitsyn, Light Sci. Appl. **4**, e314 (2015)
D. Hanna, M. Pacheco, K.H. Wong, Opt. Commun. **55**, 188 (1985)
K.P. Ho, J.M. Kahn, Statistics of group delays in multimode fiber With strong mode coupling. J. Lightwave Technol. **29**, 3119–3128 (2011)
K. Igarashi, T. Tsuntani, I. Morita, 1-exabit/s×km super-nyquist-WDM multi-core-fiber transmission, in *Optical Communication (ECOC)*. Systematic Paris Region Systems and ICT Cluster, 2014, pp. 1–3

S.I. Kablukov, E.I. Dontsova, E.A. Zlobina, I.N. Nemov, A.A. Vlasov, S.A. Babin, Laser Phys. Lett. **10**, 085103 (2013)

S.I. Kablukov, E.A. Zlobina, M.I. Skvortsov, I.N. Nemov, A.A. Wolf, A.V. Dostovalov, S.A. Babin, Quantum Electron. **46**, 1106 (2016)

J.M. Kahn, K.P. Ho, M.B. Shemirani, Mode coupling effects in multi-mode fibers, in *Proceedings of the OFC/NFOEC 2012*, **OW3D.3**, 2012

S. Kawakami, H. Tanji, Electron. Lett. **19**, 100 (1983)

M. Kolesik, J.V. Moloney, Nonlinear optical pulse propagation: from Maxwells to unidirectional equations. Phys. Rev. E **70**, 036604 (2004)

H. Komine, W.H. Long, E.A. Stappaerts, S.J. Brosnan, J. Opt. Soc. Am. B **3**, 1428 (1986)

K. Krupa, A. Labruyère, A. Tonello, B.M. Shalaby, V. Couderc, F. Baronio, A.B. Aceves, Optica **2**, 1058 (2015)

K. Krupa, C. Louot, V. Couderc, M. Fabert, R. Guenard, B.M. Shalaby, A. Tonello, D. Pagnoux, P. Leproux, A. Bendahmane, R. Dupiol, G. Millot, S. Wabnitz, Opt. Lett. **41**, 5785 (2016a)

K. Krupa, A. Tonello, A. Barthélémy, V. Couderc, B.M. Shalaby, A. Bendahmane, G. Millot, S. Wabnitz, Phys. Rev. Lett. **116**, 183901 (2016b)

K. Krupa, A. Tonello, B.M. Shalaby, M. Fabert, A. Barthélémy, G. Millot, S. Wabnitz, V. Couderc, Nat. Photon. **11**, 234 (2017a)

K. Krupa, A. Tonello, B.M. Shalaby, M. Fabert, A. Barthélémy, G. Millot, S. Wabnitz, V. Couderc, Suppl. Inf. Nat. Photon. **11**, 234 (2017b)

E.A. Kuznetsov, J.J. Rasmussen, K. Rypdal, S.K. Turitsyn, Phys. D: Nonlinear Phenom. **87**, 273 (1995). *Proceedings of the Conference on the Nonlinear Schrodinger Equation*

J. Lhermite, E. Suran, V. Kermene, F. Louradour, A. Desfarges-Berthelemot, A. Barthélémy, Opt. Express **18**, 4783 (2010)

Z. Liu, L.G. Wright, D.N. Christodoulides, F.W. Wise, Opt. Lett. **41**, 3675 (2016)

L. Lombard, A. Brignon, J.P. Huignard, E. Lallier, P. Georges, Opt. Lett. **31**, 158 (2006)

G. Lopez-Galmiche, Z.S. Eznaveh, M.A. Eftekhar, J.A. Lopez, L.G. Wright, F. Wise, D. Christodoulides, R.A. Correa, Opt. Lett. **41**, 2553 (2016)

A. Mafi, J. Lightwave Tech. **30**, 2803 (2012)

A. Mecozzi, C. Antonelli, M. Shtaif, Nonlinear propagation in multi-mode fibers in the strong coupling regime. Opt. Express. **20**, 11673–11678 (2012)

S. Minardi, F. Eilenberger, Y.V. Kartashov, A. Szameit, U. Röpke, J. Kobelke, K. Schuster, H. Bartelt, S. Nolte, L. Torner, F. Lederer, A. Tünnermann, T. Pertsch, Phys. Rev. Lett. **105**, 263901 (2010)

T. Mizunami, T.V. Djambova, T. Niiho, S. Gupta, J. Lightwave Technol. **18**, 230 (2000)

S. Mumtaz, R.J. Essiambre, G.P. Agrawal, Nonlinear propagation in multimode and multicore fibers: generalization of the Manakov equations. J. Lightwave Technol. **31**, 398–406 (2013)

J.T. Murray, W.L. Austin, R.C. Powell, Opt. Mater. **11**, 353 (1999)

Z.V. Nesterova, I.V. Aleksandrov, A.A. Polnitskii, D.K. Sattarov, JETP Lett. **34**, 371 (1981)

F. Poletti, P. Horak, Description of ultrashort pulse propagation in multimode optical fibers. J. Opt. Soc. Am. B **25**, 1645–1654 (2008)

A. Polley, S. Ralph, IEEE Photon. Technol. Lett. **19**, 218 (2007)

H. Pourbeyram, G.P. Agrawal, A. Mafi, Appl. Phys. Lett. **102**, 201107 (2013)

D. Qian, M.F. Huang, E. Ip, Y.K. Huang et al., 101.7-Tb/s (370×294-Gb/s) PDM-128QAM-OFDM transmission over 3×55-km SSMF using pilot-based phase noise mitigation, in *Proceedings of the OFC/NFOEC 2011*, PDPB5, 2011

G. Rademacher, S. Warm, K. Petermann, Analytical description of cross-modal nonlinear interaction in mode multiplexed multimode fibers. IEEE Photon. Technol. Lett. **24**, 1929–1932 (2012)

L.P. Ramirez, M. Hanna, G. Bouwmans, H.E. Hamzaoui, M.Bouazaoui, D. Labat, K. Delplace, J. Pouysegur, F. Guichard, P. Rigaud, V. Kermène, A. Desfarges-Berthelemot, A. Barthélémy, F. Prévost, L. Lombard, Y. Zaouter, F. Druon, P. Georges, Opt. Express **23**, 5406 (2015)

J. Reintjes, R.H. Lehmberg, R.S.F. Chang, M.T. Duignan, G. Calame, J. Opt. Soc. Am. B **3**, 1408 (1986)

R. Rice, in *Multimode Raman Fiber Amplifier and Method*, US Patent 6363087 (2002)

D.J. Richardson, J.M. Fini, L.E. Nelson, Nat. Photon. **7**, 354 (2013)
B.C. Rodgers, T.H. Russell, W.B. Roh, Opt. Lett. **24**, 1124 (1999)
T.H. Russell, W.B. Roh, J.R. Marciante, Opt. Express **8**, 246 (2001)
T.H. Russell, S.M. Willis, M.B. Crookston, W.B. Roh, J. Nonlinear Opt. Phys. Mater. **11**, 303 (2002)
K. Saitoh, S. Matsuo, J. Lightwave Technol. **34**, 55 (2016)
R.H. Stolen, E.P. Ippen, A.R. Tynes, Appl. Phys. Lett. **20**, 62 (1972)
N.B. Terry, T.G. Alley, T.H. Russell, Opt. Express **15**, 17509 (2007a).
N.B. Terry, K.T. Engel, T.G. Alley, T.H. Russell, Opt. Express **15**, 602 (2007b)
N.B. Terry, K. Engel, T.G. Alley, T.H. Russell, W.B. Roh, J. Opt. Soc. Am. B **25**, 1430 (2008)
H. Tünnermann, A. Shirakawa, Opt. Express **23**, 2436 (2015)
S.K. Turitsyn, S.A. Babin, A.E. El-Taher, P. Harper, D.V. Churkin, S.I. Kablukov, J.D. Ania-Castañón, V. Karalekas, E.V. Podivilov, Nat. Photon. **4**, 231 (2010)
J.C. van den Heuvel, J. Opt. Soc. Am. B **12**, 650 (1995)
R.G.H. van Uden, R. Amezcua Correa, E. Antonio Lopez, F.M. Huijskens, C. Xia, G. Li, A. Schülzgen, H. de Waardt, A.M.J. Koonen, C.M. Okonkwo, Nat. Photon. **8**, 865 (2014)
A.M. Weiner, Rev. Sci. Instrum. **71**, 1929 (2000)
P. Winzer, G.J. Foschini, Outage calculations for spatially multiplexed fiber links, in *Proceedings of the OFC/NFOEC 2011*, **OThO5**, 2011
L.G. Wright, Z. Liu, D.A. Nolan, M-J. Li, D.N. Christodoulides, F.W. Wise, Nat. Photon. **10**, 771 (2016)
T. Yao, A.V. Harish, J.K. Sahu, J. Nilsson, Appl. Sci. **5**, 1323 (2015)
V.E. Zakharov, E.A. Kuznetsov, Physics-Uspekhi **55**, 535 (2012)
E.A. Zlobina, S.I. Kablukov, M.I. Skvortsov, I.N. Nemov, S.A. Babin, Laser Phys. Lett. **13**, 035102 (2016)
E.A. Zlobina, S.I. Kablukov, A.A. Wolf, A.V. Dostovalov, S.A. Babin, Opt. Lett. **42**, 9 (2017a)
E.A. Zlobina, S.I. Kablukov, A.A. Wolf, I.N. Nemov, A.V. Dostovalov, V.A. Tyrtyshnyy, D.V. Myasnikov, S.A. Babin, Opt. Express **25**, 12581 (2017b)

Shock Waves 9

Stefano Trillo and Matteo Conforti

Contents

Introduction	374
Gradient Catastrophe and Classical Shock Waves	377
Regularization Mechanisms	379
Shock Formation in Optical Fibers	381
Mechanisms of Wave-Breaking in the Normal GVD Regime	386
Shock in Multiple Four-Wave Mixing	390
The Focusing Singularity	392
Control of DSW and Hopf Dynamics	395
Riemann Problem and Dam Breaking	398
Competing Wave-Breaking Mechanisms	401
Resonant Radiation Emitted by Dispersive Shocks	403
Phase-Matching Condition	404
Steplike Pulses	406
Bright Pulses	406
Periodic Input	408
Shock Waves in Passive Cavities	409
Conclusions	411
Appendix A	411
References	416

S. Trillo (✉)
Department of Engineering, University of Ferrara, Ferrara, Italy
e-mail: stefano.trillo@unife.it

M. Conforti
CNRS, UMR 8523, PhLAM – Physique des Lasers Atomes et Molécules, University of Lille, Lille, France
e-mail: matteo.conforti@univ-lille1.fr

© Springer Nature Singapore Pte Ltd. 2019
G.-D. Peng (ed.), *Handbook of Optical Fibers*,
https://doi.org/10.1007/978-981-10-7087-7_16

Abstract

We discuss the physics of shock waves with special emphasis on the phenomena related to the field of nonlinear fiber optics. We first introduce the general mechanism commonly known as gradient catastrophe and the related concept of classical shock waves. Then we proceed to discuss the possible regularization mechanisms of the shock, and in particular the dispersive regularization, which is behind the formation of dispersive shock waves in fibers. We then discuss different possible scenarios that lead to observe the formation of dispersive shock waves in fibers, such as pulse propagation, four-wave mixing, and passive resonators, also showing that fibers allow for investigating the dispersive regime of classical problems related to the physics of shock such as the dam-break problem and the propagation of Riemann waves. We also discuss the phase-matching mechanism that induces the shock to efficiently radiate resonant radiation in the normal dispersion regime. Throughout the text we refer to the mathematical models and the approaches that are employed to describe such phenomena.

Introduction

Classical shock waves are disturbances which exhibit a steep jump of the associated physical quantities such as density, velocity, temperature, etc., which can move with their own characteristic velocity (Lax 1973; Whitham 1974; LeVeque 2004). They have been deeply investigated in connection with systems in the form of so-called conservation laws, especially in different branches of studies of fluid flows such as standard gas dynamics, hydrodynamics, or blast waves to name a few wide subareas of interest. The physical phenomenon which is behind the formation of a shock wave is the gradient catastrophe, namely, the divergence of the gradient in one point along the disturbance, which develops at a finite time (or distance, whenever the latter is used as the evolution variable instead of time). At this stage the wave is said to "break." The essential mechanism that explains why a physical system can be driven toward the gradient catastrophe is the dependence of the velocity on the local wave elevation. This effect can drive a smooth initial disturbance to develop an infinite gradient, after which the wave disturbance tends to become multivalued. Mathematically, classical shock waves are introduced with reference to the weak formulation of the underlying conservation law to remove the multivalued stage in favor of an ideal propagating jump. In the physical reality, however, shock waves exhibit a finite extension rather than being strict local jumps, across which the physical quantities change rapidly. This is usually due to the dissipative or viscous mechanisms (neglected in the conservation laws) which, once properly accounted for, are found to regularize the abrupt changes by smoothing out the variation of the physical quantities involved in the shock.

When light is considered, either a beam propagating in free space in transparent media or a pulse propagating along an optical fiber, the same type of phenomena can occur, namely, the tendency to exhibit wave-breaking, when the propagation occurs in the strong nonlinear regime. In this respect the light effectively behaves as a "photon fluid" showing remarkable similarities with the dynamics of gas or with the behavior of waves in shallow or deep water. However, since the viscous effects are usually negligible in this case, the shock waves turn out to be regularized by dispersive (or diffractive in space) effects. This leads to a completely different scenario, where instead of smoothing due to viscous effects, the shock features the onset of fast oscillations that spontaneously grow and expand around the steep gradient, thus regularizing the tendency of the gradient to diverge. Such type of nonstationary structures is known, nowadays, as dispersive shock waves (DSWs) (El and Hoefer 2016) or undular bores, borrowing a term which is more often used in the field of hydrodynamics (Peregrine 1966). The role of such structures and, more generally, the studies of the type of hydrodynamic behaviors where dispersive effects become prominent (dispersive hydrodynamics) have only recently become an area of active investigation also outside optics, for instance, in the areas of Bose-Einstein condensation of ultracold atoms (Hoefer et al. 2006), fluid dynamics (Maiden et al. 2016; Trillo et al. 2016), or spin waves in magnetic films (Janantha et al. 2017).

Historically, in optics, such behavior has been first reported in the spatial domain with reference to the propagation of laser beams in thermal defocusing media (Akhmanov et al. 1968; Whinnery et al. 1967) and analyzed in terms of geometric optics approximation. Undulatory deformations of Gaussian modes have been interpreted as aberrations of the effective lens self-induced by the beam via heating of the medium caused by weak absorption. In this area, a much deeper connection with wave-breaking and DSW formation has been established recently (Wan et al. 2007; Ghofraniha et al. 2007; Conti et al. 2009), also remarkably extending the notion of the shock, which is intrinsically a local and ordered entity, to media with nonlocal response (Ghofraniha et al. 2007) and to disturbances constituted by superposition of incoherent waves (Garnier et al. 2013; Xu et al. 2015; Randoux et al. 2017). Since we will not discuss further these achievements here, the interested reader is referred to the original literature on these advanced topics.

In time domain, shock formation and DSWs have been first realized to play a significant role in optical fibers in the 1980s (Nakatsuka et al. 1981; Tomlinson et al. 1985; Hamaide and Emplit 1988; Rothenberg and Grischkowsky 1989; Rothenberg 1989), right after optical fiber solitons started to be actively investigated (Agrawal 2013). Indeed pulses in the form of solitons rely on the perfect balance between the effect due to the Kerr-induced self-phase modulation (SPM) and the anomalous (negative) group-velocity dispersion (GVD). However, in the opposite regime of normal (positive) GVD, the SPM enforces the dispersive broadening leading to steepening the pulse fronts. Under most common situations, such steepening was found to occur symmetrically on both the leading and trailing pulse fronts (Nakatsuka et al. 1981; Rothenberg and Grischkowsky 1989), at variance with regimes where it is only one of the two fronts that steepens due to mechanism of intensity-dependent velocity (Demartini et al. 1967; Grischkowsky et al. 1973; Anderson and Lisak 1983). The subsequent stage of steepening in fibers also features the onset of fast oscillations (Tomlinson et al. 1985; Hamaide and Emplit

1988; Rothenberg and Grischkowsky 1989). At that time, however, the analogy with the breaking phenomena occurring in the spatial domain went unnoticed, nor it was properly realized that the wave-breaking regularized by GVD as observed in fibers was the manifestation of a general phenomenon predicted and investigated in other areas of physics. Indeed the study of DSWs was actually pioneered in the context of plasmas and fluids. In the 1960s indeed Sagdeev and coworkers pointed out the oscillatory nature of shock waves occurring in extremely rarefied plasmas (Moiseev and Sagdeev 1963) (this type of plasma is called collisionless, and hence at that time, DSWs have been mainly termed collisionless shocks). Few years later, in a celebrated seminal contribution, Zabusky and Kruskal (Zabusky and Kruskal 1965) numerically investigated (see also Trillo et al. 2016 for an experimental realization) the dispersive breaking of a sinusoidal wave in the framework of the Korteweg-de Vries (KdV) equation in an attempt to give an explanation to the problem of recurrences in the Fermi-Pasta-Ulam problem (Fermi et al. 1965). In plasma the dispersive type of breaking was observed in ion acoustic waves few years later (Taylor et al. 1970). Dispersive breaking was also observed in water waves under the denomination of undular bores (Hammack and Segur 1974). On the theoretical side, the first construction of a DSW solution of a nonlinear dispersive model (again the KdV) was proposed by Gurevich and Pitaevskii (1974), who exploited the so-called Whitham averaging (1965) to predict the asymptotic evolution of a steplike initial condition (shock). Such an approach can be considerably extended to more general models (Dubrovin and Novikov 1983; Kamchatnov 2000). Among these, the defocusing nonlinear Schrödinger equation (NLSE) has received a considerable theoretical interest (Gurevich and Krylov 1987; Anderson et al. 1992; El et al. 1995; Kodama and Wabnitz 1995; Quiroga-Teixeiro et al. 1995; Forest and McLaughlin 1998; Kodama 1999), clearly establishing a link with the nonlinear propagation in optical fibers in the regime of normal GVD, which is described by such a model (Agrawal 2013). However it is only more recently that the understanding of the dispersive wave-breaking phenomena in fiber optics has been substantially advanced and placed in a more general context (Biondini and Kodama 2006; Finot et al. 2008; Fratalocchi et al. 2008; Trillo and Valiani 2010; Malaguti et al. 2010; Conti et al. 2010; Wabnitz 2013; Moro and Trillo 2014), at the same time achieving remarkable experimental results (Varlot et al. 2013; Fatome et al. 2014; Xu et al. 2016, 2017; Wetzel et al. 2016; Parriaux et al. 2017) and establishing a link with applications such as supercontinuum generation and frequency comb spectroscopy (Liu et al. 2012; Millot et al. 2016).

Here, starting from the first principles, we will review the main theoretical and experimental results connected with dispersive breaking in fibers. These encompass the studies of DSWs in multiple four-wave mixing phenomena, the role of background in wave-breaking of pulses, the control of shock dynamics via simple Riemann waves, the photonic reproduction of the breaking of a dam, the competition of shock formation with other mechanisms of breaking, the prediction of radiation emitted by shocks, and the role of DSWs in nonconservative fiber environments (passive fiber resonators).

An important point to underline is that shock waves in fiber optics literature are often associated with the steepening term that arises as a higher-order correction to SPM (Anderson and Lisak 1983; Agrawal 2013). This shock-driving term, however, becomes effective only for ultrashort pulses with sub-ps durations. Conversely Kerr-induced SPM is effective for longer time scales (ps or longer pulses or modulations in the range of tens of GHz). Indeed SPM, once acting alone, does not affect the temporal waveform of pulses, while it becomes responsible for strong steepening when acting in conjunction with weak GVD. The effect of initially weak GVD then becomes quite pronounced close to the steepened fronts, inducing dispersive wave-breaking characterized by smooth evolutions toward strongly oscillating envelopes. This mechanism is the leading-order effect that is at the basis of the phenomena discussed in this chapter.

The chapter is organized as follows. In the following section, we review briefly the basic general concepts of classical shock waves and their regularization. Then, in the section entitled "Shock Formation in Optical Fibers," we discuss the details of wave-breaking phenomena in fibers, with emphasis on recent advances. The photonic analogue of the process of the rupture of a dam is studied in the section entitled "Riemann Problem and Dam Breaking." The section entitled "Competing Wave-Breaking Mechanisms" is devoted to show that certain regimes of wave-breaking can compete with modulational instability, while the problem of radiation is discussed in the section "Resonant Radiation Emitted by Dispersive Shocks." Finally, the section "Shock Waves in Passive Cavities" briefly considers the case of passive fiber resonators and is followed by the general conclusions. Details on the mathematical construction of a DSW, according to the modulation theory, are reported in Appendix A.

Gradient Catastrophe and Classical Shock Waves

The fundamental concepts related to shock waves can be introduced by considering the simplest extension to the nonlinear regime of the equation $u_z + c u_t = 0$, i.e., the transport equation or linear unidirectional wave equation for the generic wave disturbance $u = u(t, z)$, where c is the linear velocity of the waves. Note that, as commonly used in fiber optics (as well as in other areas in the field of nonlinear waves Trillo et al. 2016), we assigned the role of independent variable to the distance z instead of time t, while we continue to refer to c as a velocity in the spirit of the theory of shock waves, though, in this notation, c has the dimension of an inverse velocity. By admitting that this velocity becomes proportional (or equal, for simplicity) to the local wave elevation u, i.e., $c = c(u) = u$, we obtain the following equation known as the Hopf or inviscid Burgers equation:

$$u_z + u u_t = 0 \quad \Leftrightarrow \quad u_z + f_t(u) = 0; \; f(u) = \frac{u^2}{2}, \tag{1}$$

which takes, as shown, the form of a scalar *conservation law* characterized by a particular form of the so-called flux $f(u)$, which in this case turns out to be quadratic in u. Here, for the time being, the Hopf equation (1) is introduced as a basic toy model, though we will go back to discuss it in the following, in order to show also its practical relevance for fiber optics. Equation (1) can be solved with the method of characteristics (Whitham 1974; LeVeque 2004), i.e., given a generic initial waveform $u_0(t) = u(t, z = 0)$, the generic value at time t_0, say $\hat{u}_0 = u_0(t_0)$, is transported along the linear characteristic $t(z) = t_0 + c(u_0(t_0))z = t_0 + \hat{u}_0 z$. Along a negative slope front, these characteristics are oblique convergent lines that determine a gradient catastrophe at the finite breaking distance z_b where they first cross. For the Hopf equation, one easily finds $z_b = 1/(-m)$, where m is the maximal negative slope of $u_0(t)$ (Whitham 1974). Beyond this distance the field becomes multivalued, as shown, as an example, in Fig. 1a for a Gaussian input. It is important to note that this mechanism is driven by the nonlinearity since, in the linear case such that c is a constant, all the characteristic lines are parallel and never cross each other (this is indeed the simple case in which any input wave is transported without deformation and moves with velocity c).

A jump is introduced to overcome the problem of the multivalued u, as shown by the dashed vertical line in Fig. 1a. A classical shock wave is a piecewise smooth solution of the more general conservation law $u_z + f_t = 0$ which contains such a jump. The jump moves along a path $t_s(z)$ according to the so-called Rankine-Hugoniot (RH) condition, which gives the shock velocity:

$$V_s = \frac{dt_s(z)}{dz} = \frac{[f]}{[u]},\qquad(2)$$

where [...] is the contracted notation for the difference of the quantity inside parenthesis across the jump. Equation (2) is derived by considering that across the

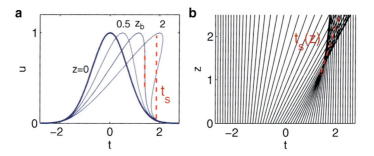

Fig. 1 (a) Gaussian input developing a gradient catastrophe at distance $z = z_b$ according to the Hopf equation. The dashed red line is the classical shock wave which has the (temporal) location $t_s = t_s(z)$. (b) Corresponding shock dynamics in the plane (t, z). The first point where characteristic lines intersect stands for the gradient catastrophe point occurring at the breaking distance $z = z_b$. The red dashed curve $t_s(z)$ emanating from this point corresponds to the classical shock wave path

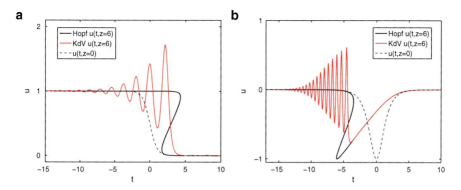

Fig. 2 Snapshots comparing the post-breaking evolutions ruled by the dispersionless (Hopf) and dispersive (KdV) model, respectively: (**a**) DSW formed by input $u(t,0) = [1-\tanh(t)]/2$ (dashed), $\varepsilon^2 = 0.05$. (**b**) Input $u(t,0) = -\text{sech}(t)$ (dashed), $\varepsilon^2 = 0.005$ giving rise to a DSW and a rarefaction wave on the negative and positive slope front, respectively

jump the differential form (1) of the conservation law is undefined, whereas the associated integral form,

$$\frac{d}{dz}\int_{t_1}^{t_2} u(t,z)dt = f(t_1,z) - f(t_2,z), \qquad (3)$$

continues to hold for any t_1, t_2 and reduces to Eq. (2), assuming $t_1 \leq t_s \leq t_2$ (Whitham 1974). In general the RH condition (2) is equivalent to select one curve in the wedge of intersecting characteristics, as shown in Fig. 1b.

On this basis, the initial jump from $u(t \leq 0) = u_L$ to $u(t > 0) = u_R$ is a classical shock wave that moves, according to Eq. (2), with velocity $V_s = (u_L^2/2 - u_R^2/2)/(u_L - u_R) = \frac{u_L + u_R}{2}$. For instance, by specializing the previous result to a unit amplitude step ($u_L = 1, u_R = 0$), the characteristic velocity turns out to be $V_s = 1/2$. We briefly recall that the additional condition $u_L > u_R$, usually known as entropy condition (originating from the language of gas dynamics), must hold for a shock solution to be valid (Whitham 1974). In the opposite case $u_L < u_R$, the solution that turns out to be compatible with the conservation law is a rarefaction wave. In the latter case, the step is smoothed out as it propagates (see Fig. 2b below for a visual example).

Regularization Mechanisms

Classical shock waves constitute an important tool in many problems of fluid dynamics. However, in several applications the impact of dissipative or dispersive phenomena cannot be neglected. One can include dissipation in Eq. (1) by adding the lowest-order even derivative, which leads to the famous Burgers equation:

$$u_z + uu_t = \alpha u_{tt}, \qquad (4)$$

which possesses, assuming boundary conditions $u(t=-\infty)=1$ and $u(t=\infty)=0$, the shock solution $u=\frac{1}{2}\{1+\tanh[\frac{1}{4\alpha}(t-\frac{1}{2}z)]\}$. This represents a smooth shock moving with velocity $v=dt/dz=1/2$. Therefore the viscous effect introduces a finite width 4α of the shock, while the velocity remains equal to the velocity obtained in the inviscid case. Indeed, in the limit $\alpha \to 0$, such solution reduces to the unit jump (classical shock wave) solution of Eq. (1) with the same velocity $v_s = 1/2$ predicted by Eq. (2). As a result, one can say that the classical shock wave is the zero dissipation limit of the shock wave propagating in the presence of losses.

Vice versa dispersive regularization turns out to be more complex and somehow more intriguing. It can be understood by considering the lowest-order dispersive correction to Eq. (1), namely, the odd derivative u_{ttt} weighted by the small coefficient ε^2 (the first odd derivative u_t is trivially removable by introducing a shifted coordinate). In this case one obtains the famous integrable KdV equation (in the weak dispersion limit), which reads

$$u_z + uu_t + \varepsilon^2 u_{ttt} = 0. \tag{5}$$

Even if $\varepsilon^2 \ll 1$, the dispersion drastically alters the dynamics. In Fig. 2, we compare the evolution ruled by the dispersionless (Eq. (1)) and the weakly dispersive (Eq. (5)) models, respectively. As shown in Fig. 2a, for a negative slope smooth front, the three-valued field ruled by Eq. (5) is regularized by dispersion through the onset of spontaneous oscillations, i.e., a nonstationary DSW that links the lower and upper constant states. In Fig. 2b we show the case of $u(t,0) = -\text{sech}(t)$, which for the KdV stands for an initial value which contains no solitons. In this case, the DSW is accompanied by the formation of a rarefaction wave on the positive slope front. Importantly, (i) the period of the oscillations scales with dispersion being $O(\varepsilon)$, as evident by comparing Fig. 2a, b; this means that the limit $\varepsilon \to 0$ the oscillation become infinitely dense, and hence one can recover the dispersionless limit only when averaging over the oscillating structure; (ii) the oscillations expand in a region bounded by the characteristic velocities of the leading and trailing edges of the DSW, which replace the single velocity V_s of the classical shock wave. These velocities can be obtained in the framework of Whitham modulation theory (Gurevich and Pitaevskii 1974; Whitham 1965; Hoefer et al. 2006; El and Hoefer 2016). The latter assumes the DSW to be constituted by a modulation of an invariant periodic solution of the underlying nonlinear dispersive model (the so-called cn−oidal or dn−oidal wave, from the name of the corresponding Jacobian elliptic functions), characterized by slowly varying parameters (compared with the oscillation average period). Averaging over the oscillations gives the so-called Whitham equations which rule the evolution of such parameters. Self-similar smooth solutions of such equations allow to calculate the edge velocities and other parameters of interest for the DSW (details on the application of such theory to the NLSE are reported in Appendix A).

This type of dispersive regularization of the shock and the general features discussed above are quite general for a wide class of models characterized by a small dispersion operator that perturbs the core dynamics ruled by a hyperbolic conservation law (or hydrodynamics model), which is responsible for driving the system toward the formation of a shock wave. In the proper regime of dispersion, the typical model that describes the nonlinear propagation in optical fibers also belongs to this class, as we discuss below. In this case the underlying conservation law turns out to be a vectorial one due to the fact that the wave envelope amplitude is complex.

Before discussing the fiber case in detail, we emphasize that there are important physical situations such that the dissipative and the dispersive regularization mechanisms act jointly. Perhaps the simplest (yet retaining great physical importance) model that describes such situation is the so-called KdV-Burgers equation:

$$u_z + uu_t - \alpha u_{tt} + \beta u_{ttt} = 0, \qquad (6)$$

which accounts for both the dissipative effect in Eq. (4) and the dispersive effect in Eq. (5) at once. Such model describes, for instance, ion acoustic waves in a dusty plasma. As demonstrated experimentally in Nakamura et al. (1999), in this case, the shock wave can have a prevalent oscillatory or smooth (monotonic) structure, depending on the relative weight of the coefficient α and β. As a matter of fact, one can calculate a precise threshold for the dissipation α, depending on the dispersion coefficient β and the shock velocity, above which the regularized shock wave becomes monotonic (Karpman 1975). In fiber optics, an experimentally relevant situation which involves the interplay of dissipative and dispersive effects is that of an optical field recirculating in a passive cavity with synchronous reinjection and undergoing shock formation, a topic that will be briefly covered at the end of the chapter.

Furthermore, we point out that, here (and in the remainder of this chapter), we have considered for sake of simplicity only models constituted by nonlinear and dispersive partial differential equations. However, shock waves and their regularized versions can develop also for *discrete* models, which describe the propagation of waves in lattices or oscillator chains. Among a large body of literature on this case, important results have been produced for the Hamiltonian models originally considered by Fermi-Pasta-Ulam (Fermi et al. 1965) or integrable versions such as the well-known Toda lattice. Dissipation can be additionally considered in such type of models, and shock waves have been explicitly discussed, for instance, in Hietarinta et al. (1995) for the Toda model and in Salerno et al. (2000) for the discrete NLSE (see also Cai et al. 1997 for a continuous version of such model), the latter being potentially of interest for coupled waveguide arrays, where coupling plays the role of effective dispersion in the system.

Shock Formation in Optical Fibers

In general, a sufficiently accurate model for describing the propagation along an optical fiber of a pulse that modulates the a carrier at frequency ω_0 is the following envelope equation, namely, a generalized NLSE (a more general model can be employed to properly model the phenomenon of fiber supercontinuum (Agrawal 2013), but the following is sufficient for our goals):

$$iE_Z + ik'E_T - \frac{k''}{2}E_{TT} + \gamma|E|^2 E + \frac{i}{\omega_0}(|E|^2 E)_T - T_R(|E|^2)_T E = 0, \quad (7)$$

where the nonlinear terms are due, in order from left to right, to the SPM owing to the nearly instantaneous Kerr effect, the so-called self-steepening due to the dispersion of nonlinearity, and the intrapulse Raman scattering with coefficient T_R, which is nothing but the characteristic time of the Raman response. It is important to emphasize that, in the literature, the self-steepening term is also termed the *shock term*, and indeed its coefficient is also indicated using the notation $T_{shock} = 1/\omega_0$, which we find misleading since DSWs can originate, as we show below, from SPM which should be regarded as the main shock-inducing term in the model, at least for common pulses with durations in the range of tens of picoseconds. Usually, the self-steepening term becomes important, in fiber optics, for much shorter (sub-psec) pulses. Conversely the effect of this term can be especially important in photonic crystal planar waveguides for longer pulses. As far as the dispersive terms are concerned, for the time being, we restrict ourselves to the effect of leading-order term or second-order dispersion or GVD with coefficient k'' (higher-order effects will be further discussed with reference to radiation driven by shock waves). When the GVD is weak, the correct way to derive a hydrodynamic limit of Eq. (7) is to make use of the Wentzel-Kramers-Brillouin (WKB) method (or, equivalently, the Madelung or geometric optics transformation) applied to the equation cast in the following semiclassical form:

$$i\varepsilon\psi_z - \beta_2\frac{\varepsilon^2}{2}\psi_{tt} + |\psi|^2\psi + i\varepsilon\gamma_S(|\psi|^2\psi)_t - \gamma_R\psi(|\psi|^2)_t = 0, \quad (8)$$

where we have conveniently introduced the normalized variables $t = (T - k'Z)/T_0$ and $z = Z/\sqrt{L_d L_{nl}}$, $\beta_2 = \text{sign}(k'') = k''/|k''|$, as well as the following smallness dispersion parameter and the normalized nonlinear coefficients:

$$\varepsilon = \sqrt{\frac{L_{nl}}{L_d}} = \frac{1}{\sqrt{N}}, \quad \gamma_S = \sqrt{\frac{\gamma P}{|k''|\omega_0}}, \quad \gamma_R = \frac{T_R}{T_0}, \quad (9)$$

where $L_{nl} = (\gamma P)^{-1}$ and $L_d = T_0^2/|k''|$ are the nonlinear and dispersion length, respectively, $N = L_d/L_{nl}$ is the soliton order, and $P = max(|E(Z = 0, T)|^2)$ and T_0 are the peak power and the duration of the input envelope $E(Z = 0, T)$. By inserting in Eq. (8) the WKB ansatz $u(t, z) = \sqrt{\rho(t, z)}\exp[iS(t, z)/\varepsilon]$ and introducing the chirp $u = -S_t$, we obtain

$$\rho_z + \left[\beta_2\rho u + \frac{3}{2}\gamma_S\rho^2\right]_t = 0, \quad (10)$$

$$u_z + \beta_2 u u_t + [\rho + \gamma_S(\rho u) - \gamma_R \rho_t]_t = \varepsilon^2 \frac{1}{4}\left[\frac{\rho_{tt}}{\rho} - \frac{(\rho_t)^2}{2\rho^2}\right]_t. \tag{11}$$

Equations (10) and (11) without approximations are fully equivalent to the NLSE (8). The so-called dispersionless limit of the generalized NLSE (analogous of the Hopf equation for the KdV) is obtained by neglecting the RHS of Eq. (11), which is of higher-order $[O(\varepsilon^2)]$ with respect to the LHS of the equations that correspond to order $O(\epsilon^1)$ and $O(\epsilon^0)$, respectively. Incidentally, note that the term in the RHS $V_{QP} = \frac{1}{4}\left[\rho_{tt}/\rho - (\rho_t)^2/2\rho^2\right]$ is equivalent to the quantum potential in the interpretation of quantum mechanics based indeed on such potential (Bohm and Hiley 1984). In particular, in quantum mechanics, V_{QP} acts along with the standard external potential, though being determined by the probability density ρ, in the framework of the standard Schrödinger equation. In this "dispersionless" approximation, Eqs. (10) and (11) have the form of the evolution equations of a Eulerian fluid, with ρ and u playing the role of the fluid density (or water elevation) and velocity, respectively. Potentially, all the nonlinear terms can contribute to forming shock singularities. In particular, the steepening term is clearly a shock-driving term (see below and Agrawal 2013; Anderson and Lisak 1983), while the Raman term was shown to support smooth shock waves (or kink solitons) similarly to the KdV-Burgers model discussed above, both in the anomalous (Agrawal and Headley III 1992) and normal (Kivshar and Turitsyn 1993; Kivshar and Malomed 1993) dispersion regimes. Furthermore the effect of Raman response on the wave-breaking dynamics of ultrashort pulses has been also addressed in Quiroga-Teixeiro et al. (1995); Conti et al. (2010). However, one must realize that the coefficients of steepening and Raman terms are normally such that $\gamma_S, \gamma_R \ll 1$, i.e., they are much smaller than the Kerr coefficient (which is scaled to one in the adopted units), unless the regime of ultrashort pulses is considered (i.e., sub-psec down to few fsec). Therefore we will mainly address the DSWs developing through the Kerr effect which is of leading-order under usual experimental conditions, whereas we leave the concomitant effects of steepening and Raman as an advanced topic, which needs additional consideration and investigation, and will not be discussed here to a further extent.

Nonetheless, the general form of Eqs. (10) and (11) is extremely useful to understand the impact of GVD over the formation of shock waves. In particular, if we consider the formal limit of strictly vanishing GVD ($\beta_2 = 0$), they reduces to the following system:

$$\rho_z + 3\gamma_S \rho \rho_t = 0; \quad u_z + [\rho + \gamma_S(\rho u) - \gamma_R \rho_t]_t = 0, \tag{12}$$

which shows that the power ρ turns out to be *decoupled* from the chirp (phase) dynamics, with the power evolution being ruled by a shock-bearing equation of the Hopf type. In the strict limit of zero GVD, therefore, one can conclude that the shock formation in the intensity profile is solely driven by the steepening term, thus occurring along the negative slope front of the pulse, as in the generic example of

Fig. 1a (Anderson and Lisak 1983; Agrawal 2013). This is consistent with the fact that, in this limit, the Kerr SPM, which is responsible for the "pressure"-like term ρ_t in the second of Eqs. (12), only induces a phase modulation or chirp, thereby not affecting the temporal profile of the intensity. However, such conclusion becomes incorrect in the presence of an arbitrarily small dispersion, i.e., a GVD of order ε^2 (with ε arbitrarily small) compared with the Kerr effect of order $O(1)$, as in Eq. (8). In fact, even if the RHS of Eq. (11) is dropped as in the dispersionless limit of the NLSE (8), the GVD affects the dynamics through terms that still appear in Eqs. (10) and (11), since they retain the same leading order of the nonlinear term. These terms turn out to be extremely important because they couple the SPM-induced chirp to the power ρ. Therefore, when GVD is arbitrarily small but nonvanishing, the steepening of the pulse fronts can occur through the Kerr effect or SPM, which becomes the dominant one whenever $\gamma_S, \gamma_R \ll 1$. In this regime, the formation of the shock can be described in the framework of the reduced quasi-linear system of two equations (Gurevich and Shvartsburg 1970; Gurevich and Krylov 1987), commonly (though improperly) denoted also as the dispersionless NLSE:

$$\rho_z + \beta_2(\rho u)_t = 0; \qquad u_z + \beta_2 u u_t + \rho_t = 0. \tag{13}$$

Shock formation occurs when Eqs. (13) are hyperbolic, i.e., in the regime of normal GVD, where $\beta_2 = 1$. In this case, the tendency to overtake is caused by the formation of a non-monotonic chirp which drives also the steepening of the power profile (Anderson et al. 1992), until the gradients become so large that the dispersive effects associated with the RHS in Eq. (11) set in.

In the normal GVD regime ($\beta_2 = 1$), Eqs. (13) are identical to the shallow water equations (SWEs), which rule in 1D the propagation of the water elevation ρ and the (vertically averaged) horizontal velocity u, with interchanged role of space and time (in hydrodynamics the evolution variable is considered to be the time t, while the longitudinal distance plays the role of the time variable in fiber optics). In hydraulics these equations are also known as Saint-Venant equations. Interestingly enough, they also have one-to-one correspondence to the so-called p-system which rule the gas dynamics for an isentropic gas with pressure law $p = \rho^2/2$. Importantly, the SWEs can also be cast in the diagonal form in terms of new variables $r^{\pm}(t, z) = u(t, z) \pm 2\sqrt{\rho(t, z)}$, which, in the language of the hyperbolic equation theory, are called Riemann invariants:

$$r_z^{\pm} + V^{\pm} r_t^{\pm} = 0, \tag{14}$$

where $V^{\pm} \equiv V^{\pm}(r^{\pm}) = (3r^{\pm} + r^{\mp})/4 = u \pm \sqrt{\rho}$ are the real eigenvelocities of the problem.

An additional formulation of Eqs. (13) refers to the form of a differential conservation law, which allows for introducing the classical shock waves with their characteristic-associated velocities by means of extending the RH condition (Eq. (2)) based on the integral formulation of the conservation law. In this case the conservation law, which can be easily derived from Eqs. (13), is no longer a scalar one but rather the following 2×2 vectorial form:

$$\mathbf{q}_z + [\mathbf{f}(\mathbf{q})]_t = 0, \qquad (15)$$

where $\mathbf{q} = (\rho, \rho u)^T$ and the flux turns out to be $\mathbf{f}(\mathbf{q}) = (\beta_2 \rho u, \beta_2 \rho u^2 + \rho^2/2)^T$. Physically, Eqs. (15) express the conservation of mass and momentum in differential form. When a classical shock (a step) is introduced, the vectorial equivalent of the RH condition (2) can be introduced. However, since this becomes a vectorial condition which involves the same shock velocity V_s in the two components, it fixes not only the shock velocity but also the admissible values of the jump (i.e., the left (ρ_L, u_L) and right (ρ_R, u_R) values are not arbitrary but satisfy a constraint, a fact that bears no similarity with the scalar case) (LeVeque 2004; Hoefer et al. 2006; Conforti et al. 2014; Xu et al. 2017). The classical shock wave introduced in this way is crucial to develop a description of its dispersive regularization, i.e., of the DSW commonly observed in optical fibers. Furthermore, it is the key ingredient for properly describing the general evolution of steplike initial conditions (see below).

In general, the fact that there are two eigenvelocities or two families of characteristics in the dispersionless NLSE (Eqs. (13), (14), and (15)) expresses the fact that the NLSE describes indeed the bidirectional type of dispersive hydrodynamics. The latter is characterized by two families of wave trains that have different characteristic speeds (which, in the laboratory frame, are naturally referred to the group velocity of light, which is removed in the passage from Eq. (7) to (8)) (El and Hoefer 2016). This is at variance with the KdV equation and its dispersionless limit that describes unidirectional hydrodynamics, as discussed in the previous section. Although Eqs. (13) rule the formation of shock wave via the mechanism of gradient catastrophe, there are important differences with the similar phenomenon ruled by a scalar conservation law (such as the Hopf equation). First, symmetric pulse envelopes used in several fiber optics applications typically exhibit two points of breaking instead of a single one. Only in some specific case these points degenerate in a single one (see discussion in the following subsection). Second, although there is a method, namely, the so-called hodograph transform (Whitham 1974; Moro and Trillo 2014), which permits to invert the role of dependent and independent variables in Eqs. (13), thus reducing the model to a linear model (LeVeque 2004; Whitham 1974), in practice, solutions are indeed difficult to be written. In general, for a generic smooth input waveform, it is challenging to predict the finite distance after which undergoes breaking. Apart from special cases for which this becomes possible (Moro and Trillo 2014), one must resort to (i) numerical simulations, (ii) approximate estimates (Anderson et al. 1992), and (iii) bind the breaking distance between a lower and upper value employing a general criterium due to Lax (Forest and McLaughlin 1998; Lax 1973).

Below, we will discuss in more details the mechanisms of wave-breaking under different excitations in the regime described by Eqs. (13), which is the one widely observed in nonlinear fiber optics experiments in the psec regime.

Mechanisms of Wave-Breaking in the Normal GVD Regime

The shallow water equations (13), or their equivalent diagonal (Eqs. (14)) or conservation law (Eqs. (15)) formulations, entail two different main breaking mechanisms depending on the input pulse shape.

For input bright and symmetric bell-shaped pulses $\psi_0 = \psi(t, z = 0)$ with a generic finite background, steepening occurs symmetrically (in contrast with the steepening ruled by the Hopf equation) until two points of gradient catastrophe occur at finite distance on both the leading and trailing edges of the input pulse. The strong gradients are regularized by the quantum potential term (RHS in Eqs. (10) and (11)), which, in order to describe the post-breaking dynamics, is equivalent to reconsider the full dimensionless NLSE

$$i\varepsilon\psi_z - \frac{\varepsilon^2}{2}\psi_{tt} + |\psi|^2\psi = 0. \tag{16}$$

The dynamics ruled by Eq. (16) is smooth at any distance. The initial stage, ruled by Eqs. (13), is dominated by the nonlinearity which causes a steepening of the tails. The physical mechanism behind such steepening is the fact that the dominant Kerr nonlinearity induces a strong self-phase modulation. The acquired phase turns out to be proportional to the intensity $|\psi_0(t)|^2$, in turn implying an instantaneous frequency deviation or chirp $\delta\omega(t) = \partial_t\phi(t) = \partial_t|\psi_0(t)|^2$. Such a chirp is turned into an instantaneous change of velocity $\delta u(t) \propto -\delta\omega(t)$, according to the fact that in normally dispersive media, the velocity decreases for increasing frequency. As a consequence the top parts of the pulse experience a larger absolute velocity toward the pulse tails (the velocity is negative on the leading edge of the pulse and positive on the trailing edge, respectively), which is at the origin of the front steepening. Such process proceeds as the self-phase modulation grows along the fiber. However the point of infinite gradient (breaking) and successive overtaking is never reached. Indeed, as the gradient along the steepened front grows sufficiently large, the weak GVD becomes effective, and spontaneous formation of oscillations starts to become visible around the strongly steepened fronts. This usually occurs slightly before the breaking distance defined in the dispersionless limit ruled by Eqs. (13). The expanding oscillations form two symmetric DSWs, as shown in Fig. 3a, b for a Gaussian input. They propagate and expand in opposite directions compared with the natural group velocity of the wave, which corresponds the line $t = 0$ in Fig. 3.

A mathematical description of each of the two DSWs is possible on the basis of Whitham modulation theory, which can be successfully formulated for the defocusing NLSE (16). According to such an approach, the DSW is described as a slowly varying modulation of the invariant nonlinear traveling-wave periodic solution (so-called dn-oidal or cn-oidal wave). The slowly varying parameters of such wave obey modulation equations which are obtained by performing a proper averaging due to Whitham (1965) over the fast oscillation. For the defocusing NLSE, the modulation equations are a set of four equations, which turns out to be hyperbolic and diagonalizable. A simple and smooth (rarefaction) solution of

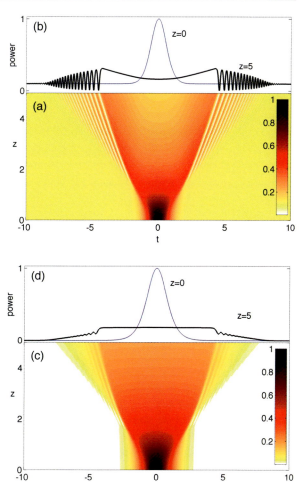

Fig. 3 False color level plot (**a, c**) and snapshots of power $\rho = |\psi|^2$ (**b, d**) of DSWs ruled by NLSE (16), $\varepsilon = 0.05$: (**a–b**) Gaussian input $\psi_0 = [0.1 + 0.9\exp(-t^2)]^{1/2}$ with 10% power pedestal; (**c–d**) input $\psi_0 = \mathrm{sech}(t)$, with no pedestal

such equations allows to characterize the modulation and hence the DSW. We report in Appendix A the details of such calculations for the interested reader. An important point is that a single DSW is always characterized by two edges, as also clear from Fig. 3a. One of the edge is such that the amplitude of the oscillations vanishes: this is the linear edge of the DSW (it corresponds to the trailing edge of the right-going DSW or the leading edge of the left-going DSW; see Fig. 3a). Over the opposite edge of the same DSW, which exhibits the deepest oscillation, the periodic solution locally reduces to a soliton (i.e., the modulus m of the Jacobian function locally tends to one): hence, this is called the soliton edge of the DSW (it corresponds to the leading edge of the right-going DSW or the trailing edge of the left-going DSW; see Fig. 3a). The modulation theory allows to calculate the two velocities of the linear and soliton edges, respectively. However, it is worth pointing out that the modulation theory is an asymptotic method which provides a good description only for large enough propagation distances and/or small enough dispersion parameter ε. In any event the dynamics of DSWs developing from smooth initial data cannot have strict correspondence with the DSW construction based on the modulation equations, since the latter implies an initial condition for the modulation equation in the form of a step (see Appendix A). Conversely, a quantitative test of the outcome of the modulation theory becomes possible by injecting steplike pulses, a case that we treat separately in the following.

It is important to emphasize that the pulse background has a strong influence on the formation of the DSW, particularly in terms of the contrast of the oscillations. This is clear from Fig. 3, where Fig. 3c, d contrasts the evolution relative to the input $\psi_0 = \mathrm{sech}(t)$ with the case of the Gaussian with background in Fig. 3a, b (note that, for a more rapidly decaying pulse than the hyperbolic secant, such as the Gaussian considered in Fig. 3a, b, the oscillations would be barely visible in the limit of vanishing background). This is the reason for which early measurements in fibers (Rothenberg and Grischkowsky 1989) reported a much lower contrast compared with later spatial experiments performed with non-zero background (Wan et al. 2007). A detailed experimental study of the effects of the background has been recently performed in Xu et al. (2016). Figure 4a summarizes the contrast of the oscillations (defined conventionally for the first fringe, as indicated in Fig. 4b), which are measured as a function of the background (in percent of the peak power of the pulse), while explicit examples for background levels corresponding to $0.1, 1, 6\%$ are shown in Fig. 4b–g, comparing experimental results and NLSE simulations. As shown, the contrast of the oscillation is rapidly enhanced by increasing the background, reaching a saturated contrast of 100% where the minimum power of the leading edge touches zero. At larger background levels, the contrast smoothly decreases since such null point shifts from the leading edge of the DSW, a phenomenon that can be accurately described in the framework of the so-called dam-breaking experiment concerning steplike initial data (see section "Riemann Problem and Dam Breaking"). A physical insight into the role of the background can be gained as follows. The oscillating wave train in the DSW can be understood as the interference phenomenon between original components with zero chirp (the background) and new frequencies generated via the nonlinearity and GVD, which coexist on the temporal regions which correspond to the steepened tails of the pulse. The resulting pattern is characterized by the difference frequency and a visibility which depends on the amplitude ratio between such frequency components, thus requiring the background to be sufficiently large to give rise to large contrasts.

When a dark input of the type $\psi_0 = \tanh(t)$ is considered, the breaking occurs in $t = 0$. In contrast with the previous breaking mechanism (where only one Riemann invariant breaks at each catastrophe point), the latter is a nongeneric mechanism which implies breaking of both Riemann invariants in the null intensity point $t = 0$ (this can be shown analytically for the similar case for which the phase profile is suppressed, i.e., $\psi_0 = |\tanh(t)|$ Moro and Trillo 2014). The emerging DSW exhibits a single fan with a narrower central black (zero-velocity) soliton and symmetric gray pairs around it (see Fig. 5a, b). The intermediate stage displayed in Fig. 5a, b shows the typical features of a DSW. However, in this case, one can

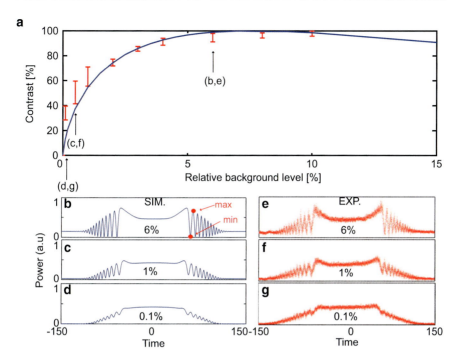

Fig. 4 Effect of pulse background on shock formation. (**a**) Contrast of the DSW fringe ($P_{max} - P_{min})/(P_{max} + P_{min})$, defined relative to the DSW leading edge, as shown in (**b**), versus percentage of the background (relative to the input peak power). (**b**)–(**g**): examples of different calculated and measured intensity profiles for 0.1, 1, and 6% background level: (**b**)–(**d**) numerics from NLSE and (**e**)–(**g**) experimental results (From Xu et al. 2016)

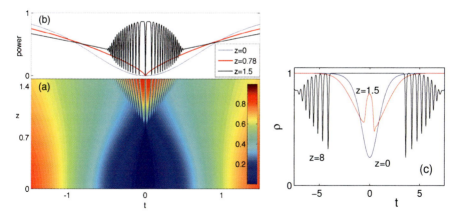

Fig. 5 (**a**,**b**) Wave-breaking occurring from $\psi_0 = \tanh(t)$, $\varepsilon = 0.005$, from numerical integration of the NLSE: (**a**) false color level plot; (**b**) snapshots at $z = 0$ (input), $z = 0.78$ (breaking), $z = 1.5$ (DSW). (**c**) Snapshots of DSW formation from a gray input $\psi_0 = w \tanh(wt) + iv$, $w^2 + v^2 = 1$. Here $v = 0.5$, $\varepsilon = 0.05$

further exploit the integrability of the NLSE and apply the inverse scattering method. This allows to show that the initial condition $\psi_0 = \tanh(t)$ contains $2N_s - 1$ solitons, where $N_s \equiv \lfloor 1/\varepsilon \rfloor$ (more precisely, one can show that $\psi_0 = \tanh(t)$ turns out to be a reflectionless potential of the scattering problem associated with the NLSE whenever $1/\varepsilon$ is integer). These solitons turn out to be a central black solitons and symmetric gray pairs with opposite velocities, which asymptotically (i.e., at normalized distance $z \sim 1/\varepsilon$) separate on the same background (Fratalocchi et al. 2008). In this case the DSW is in fact a multisoliton solution, where the solitons embedded in the initial condition start to emerge after the catastrophe, which development is ruled by the SWEs or dispersionless NLSE. It worth pointing out that for the phase suppressed input ($\psi_0 = |\tanh(t)|$), which shows the same type of breaking mechanism, only symmetric pairs of gray solitons emerge beyond the catastrophe point, while the central black soliton no longer appears in the DSW fan. Furthermore, similarly to the hyperbolic tangent case, a gray input still generates $2/\varepsilon - 1$ (narrower) gray solitons. However, in this case, the field breaks in two distinct points (see snapshot at $z = 1.5$ in Fig. 5c), from which two asymmetric DSWs emerge. These wave trains or solitonic DSWs can be shown to be asymptotically composed of $1/\varepsilon - 1$ and $1/\varepsilon$ gray solitons, respectively.

As discussed above the process of breaking depends on the shape of the input pulse, as shown in Figs. 3, 4, and 5. In particular, one can also find pulse shapes that do not lead to breaking. For instance, parabolic pulses are wave-breaking-free and led to the concept of similaritons, pulses which can be amplified while retaining their parabolic shape. On the other hand, other situations where wave-breaking phenomena lead to interesting developments involve a continuous wave with constant power and chirp modulation (Kodama 1999; Biondini and Kodama 2006; Wabnitz 2013). For instance, this leads to the generation of flaticon pulses, recently demonstrated in Varlot et al. (2013).

Finally it is worth emphasizing that, in the spectral (Fourier) domain, the wave-breaking process is characterized by a strong spectral broadening. Indeed the steepening process corresponds to the generation of high frequencies in the spectrum. Beyond the breaking distance, the spectrum does not substantially reshape, while the DSW spreads (see Fig. 15b below, for an example of this behavior). In this regime, roughly speaking, the spectrum extends up to the highest frequencies of the generally non-monochromatic oscillations (for a different estimate of the broadening, see Parriaux et al. (2017). Such a dramatic spectral broadening has been shown to be highly beneficial for several applications such as supercontinuum generation and comb spectroscopy (Finot et al. 2008; Liu et al. 2012; Millot et al. 2016).

Shock in Multiple Four-Wave Mixing

While in the previous subsection we have considered wave-breaking generated from bright or dark pulses, it was recently shown that DSWs can be generated also from periodic waves, i.e., amplitude-modulated waves which give rise, via the Kerr effect, to the phenomenon of multiple four-wave mixing (mFWM). This indicates the generation of multiple sideband orders at $\omega_0 \pm n\Omega/2$, n odd integer, which is produced via the Kerr term from an input beat signal or dual-frequency input $\omega_0 \pm \Omega/2$ (Thompson and Roy 1991). Similarly an amplitude-modulated carrier frequency ω_0, whose spectrum contains frequencies $\omega_0, \omega_0 \pm \Omega$, generates mFWM products $\omega_0 \pm n\Omega$, n integer. When such processes occur in the strong nonlinear regime which involves the generation of several mFWM orders, the field undergoes breaking. The wave-breaking mechanism turns out to be similar to the one illustrated in Fig. 5 for dark pulses. In order to understand the regime of breaking in mFWM, let us consider, for instance, the NLSE Eq. (16) subject to the initial condition $\psi_0 = \sqrt{2}\cos(\pi t/2)$. This corresponds to the celebrated numerical experiment performed by Zabusky and Kruskal for the KdV equation (Zabusky and Kruskal 1965), though performed instead for the NLSE in the present case. In this case, without loss of generality, the normalized frequency detuning is fixed to $\Omega T_0 = \pi$ by choosing $T_0 = 1/2\Delta f = \pi/\Omega$. The dynamics is ruled, in this case, by the single smallness parameter $\varepsilon = \Delta f \sqrt{4k''/(\gamma P)}$ (Trillo and Valiani 2010).

When $\varepsilon \ll 1$ the cosine input exhibits multiple points of breaking at the nulls of the power profile $|\psi_0|^2 = 2\cos^2(\pi t/2)$, i.e., at $t = (2k+1)$, $k = 0, 1, 2, \ldots$, as shown in Fig. 5a, b. The field at the wave-breaking points, displayed in Fig. 6b, is strongly reminiscent of the one generated in the hyperbolic tangent case shown in Fig. 5a. A remarkable difference is that, in the periodic case, the DSW emerging from each breaking point collides with the adjacent ones forming multiphase structures (see Fig. 6a, c). The elastic nature of the collision as well as the relationship between the darkness and the velocity of the single filaments that compose the DSW suggests that they behave as dark solitons. While solitons cannot exists in the strict sense due to the periodic nature of the problem, numerical results

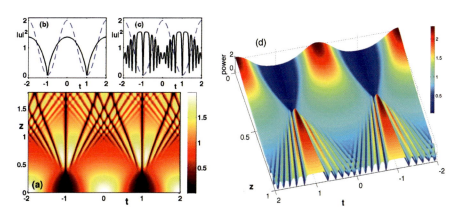

Fig. 6 (**a, b, c**) Breaking of an input cosine according to NLS Eq. (16): (**a**) level colorplot of power $|\psi|^2$; (**b–c**) snapshots of power $\rho = |\psi|^2$ at breaking distance $z = 0.34$ (**b**) and $z = 0.95$ (**c**) compared with the input (dashed blue). (**d**) Breaking for an input modulated field $\psi_0 = \sqrt{\eta} + \sqrt{2(1-\eta)}\cos(\pi t)$, $\eta = 0.8$. Here $\varepsilon = 0.04$

show that the finite-band solutions that arise in the scattering problem shrink in the limit of small ε, resembling true solitons of the infinite line problem in the limit $\varepsilon \to 0$. The number such soliton-like excitations grow as ε decreases.

Noteworthy, when the modulation has no zeros (i.e., for an imbalanced dual-frequency or triple-frequency input), the temporal locations of the breaking become non-degenerate, and breaking occurs at two distinct instants around all the minima of the input modulation, as shown in Fig. 5d. In this case, two symmetric DSWs emerge from each double breaking point, still giving rise to multiphase regions.

The phenomenon of DSW in mFWM was observed in a recent fiber experiment, by employing the Picasso platform at Laboratoire Interdisciplinaire Carnot de Bourgogne in Dijon (Fatome et al. 2014). The setup allows to reach the necessary power of the input modulation at 28 GHz, without resorting to pulses, which would hamper the visibility of the DSW of the periodic case. Furthermore, since one cannot easily measure the field along fibers which are several km long, in the experiment, the evolution of the DSW is reconstructed at finite physical propagation length $L = 6$ km, by increasing the input power of the modulated field. Changing the power P at fixed length L (and fiber parameters k'' and γ, as well as fixed modulation frequency) amounts indeed to change the normalized length $z = L/\sqrt{L_d L_{nl}} = L\sqrt{k''\gamma P/T_0}$.

The results are shown in Fig. 7. In particular the left column figures show the measured temporal traces versus the input power for three different configurations, corresponding to (a) sinusoidal input (suppressed carrier, balanced dual-frequency input), (b) dual-frequency (suppressed carrier) input with imbalanced power fractions, and (c) modulated carrier corresponding to triple-frequency input. In particular Fig. 7a confirms the scenario illustrated in Fig. 6a–c. In the other two cases, two breaking points arise around the minima of the modulated input, with preserved symmetry (in time) in the triple-frequency case and broken symmetry for the imbalanced dual-frequency (carrier suppressed) case. The simulations based on the dimensional NLS equation show a remarkable agreement in all cases (see Fig. 7d–f), without using any fitting parameter. Noteworthy, the NLSE does not need any higher-order corrections (steepening, Raman, higher-order dispersion) to correctly describe the experiment.

The Focusing Singularity

We have extensively discussed the normal GVD regime case. However it is worth mentioning that also the anomalous GVD regime ($k'' < 0$), which is described by a NLSE of the *focusing* type, can exhibit singularity formation. In this case the WKB reduction with $\beta_2 = -1$ takes the following form, obtained with the transformation $u \to -u$, (Gurevich and Shvartsburg 1970):

$$\rho_z + (\rho u)_t = 0; \quad u_z + u u_t - \rho_t = 0, \qquad (17)$$

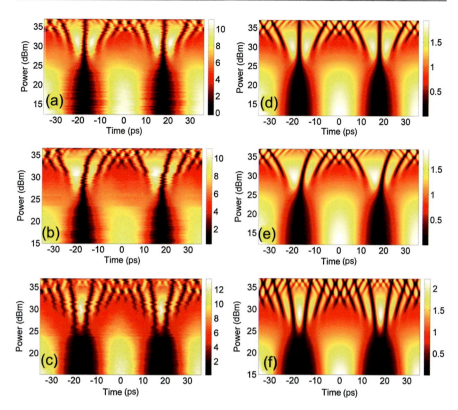

Fig. 7 DSWs arising from mFWM, showing the color level plot of the temporal profile of the modulated field against the input power P: experiment (**a, b, c**) vs. simulations (**d, e, f**) based on NLSE (7), neglecting steepening and Raman terms. (**a–d**) Sinusoidal (suppressed carrier balanced dual-frequency) input; (**b–e**) imbalanced dual-frequency (suppressed carrier) input; (**c, f**) modulated carrier (triple-frequency) input. The power is given in log units, $P(dBm) = 10\log_{10} P(mW)$ (From Fatome et al. 2014)

which shows the presence of the equivalent negative pressure term $-\rho_t$. As a consequence, this dispersionless limit turns out to be elliptic nature, which reflects the fact that the full NLSE exhibits modulational instability (MI). In this case the catastrophe implies breaking around a focus point and is termed elliptic umbilic (see Dubrovin et al. 2015 and references therein). An example, which corresponds to the implicit solution of Eqs. (17) for the initial datum $\psi_0 = \text{sech}(t)$ discussed in Gurevich and Shvartsburg (1970), Kamchatnov (2000), is illustrated in Fig. 8a–c.

In this case, the SPM induces a chirp which is initially linear around the origin and then increases its slope until eventually determines an abrupt change of sign around $t = 0$ (see Fig. 8a), somehow in a way similar to the null point for the $\psi_0 = \tanh(t)$ initial datum in the defocusing regime of the NLSE (Conti et al. 2009). This, in turn, represents a compressional wave which induces a sharp peak with steepened

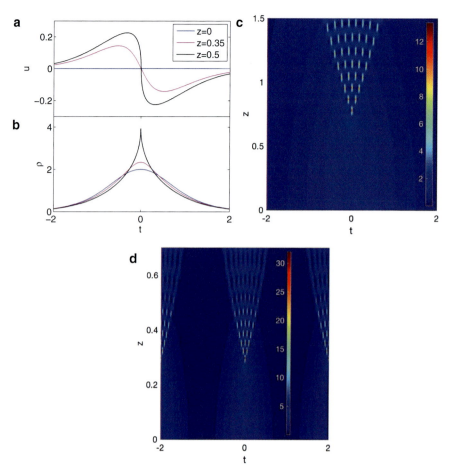

Fig. 8 Focusing catastrophe: (**a**, **b**) snapshots of chirp (hydrodynamical velocity) u (**a**) and power ρ (**b**) obtained from the dispersionless equations (Eq. (17)) with initial data $\rho_0 = \mathrm{sech}^2(t)$, $u_0 = 0$, corresponding to NLSE with initial datum $\psi_0 = \mathrm{sech}(t)$; (**c**) corresponding NLSE dynamics, $\varepsilon = 0.05$; (**d**) similar dynamics for initial datum $\psi_0 = \sqrt{2}\cos(\pi t/2)$ that leads to mFWM, $\varepsilon = 0.05$

fronts at a finite critical distance (see cusp-like structure in Fig. 8b). It is worth pointing out that this dynamics is caused by the strongly dominant nonlinearity in the NLSE, thus being of different nature from the collapse phenomenon in the critical NLSE (i.e., the NLSE with cubic nonlinearity in 1 + 2 dimensions), which occurs right above the threshold (critical norm or power in the case of beams) where nonlinearity and diffraction mutually balance (soliton or so-called Townes profile). In the present case, dealing with 1 + 1D NLSE, the dynamics at the catastrophe and beyond it is regularized by the onset of dispersion and turns out to be usually very complicated (see Fig. 8c for an example) and can exhibit zones of multiphase oscillations separated by caustics ruled by a large ensemble of bright solitons (cases not shown) (Kamvissis et al. 2003). Only under particular conditions, which we will not deepen here, the observed oscillating structure represents the focusing analogue of the one-phase DSWs featured by the defocusing case.

Incidentally, also input periodic waves, such as the cosine considered above for mFWM in the normal GVD regime (defocusing NLSE), exhibit the same type of breaking and similar post-breaking dynamics, except for the fact that now periodic (in time) points of breaking occur, as displayed in Fig. 8d. For long enough distances, adjacent structures are expected to collide and form even more complex patterns, which need further theoretical and experimental characterization.

In the presence of noise, due to both high gain and large bandwidth which characterize the MI in the semiclassical (i.e., strongly nonlinear or weakly dispersive) limit, the impact of MI-amplified fluctuations could dramatically affect the breaking process (Ghofraniha et al. 2007). In this regime the breaking dynamics would strongly benefit from MI-suppressing mechanisms such as a nonlocal type of nonlinear response. The latter, however, is not effective in fibers, at least for long pulses (in the order of few tens or hundreds of psec; in fact, the retarded response due to the Raman scattering effect is the temporal analogue of a spatial nonlocality but becomes effective for shorter pulse durations). Overall, wave-breaking in the anomalous dispersion regime essentially remains an open area of investigation especially from the experimental point of view.

Control of DSW and Hopf Dynamics

We finish this section by illustrating a regime of fiber optics propagation where the shock wave formation, though still being induced via the leading-order (Kerr) effect and in principle described by means of the NLSE, turns out to be effectively governed by the simpler Hopf or inviscid Burgers equation (1), introduced in the beginning of this chapter (Malaguti et al. 2010; Wabnitz 2013; Wetzel et al. 2016). If we consider pulses where the chirp (equivalent hydrodynamical velocity u) is not arbitrary, but rather linked to the power $\rho = |\psi|^2$ as $u = \pm 2\sqrt{\rho}$, one of the two Riemann invariants in Eqs. (14) identically vanishes. Under this constraint, the solution of Eqs. (14) is called a *simple Riemann wave* (more generally a simple wave solution of a hyperbolic systems is such that one of the Riemann invariants is constant in the region of interest). In particular, in this case, it is easy to show that either the SWEs for power and chirp ρ, u or their diagonal form (14) for the Riemann invariant reduce to an effective equation of the Hopf type. In particular, for instance, the effective equation for the power ρ or the amplitude reads as (Malaguti et al. 2010)

$$\rho_z \pm 3\sqrt{\rho}\rho_t = 0 \quad \Leftrightarrow \quad a_z \pm a a_t = 0, \quad a = 3\sqrt{\rho}, \tag{18}$$

i.e., a canonical Hopf equation for the normalized envelope amplitude $a = a(t, z)$. It is worth noting that, in this case, the Hopf equation holds for the pulse envelope amplitude (modulating the carrier ω_0). This is in contrast, for instance, with the dispersionless limit of the KdV equations introduced in the beginning of the chapter, where the variable is not an envelope variable but rather the wave amplitude itself in shallow water waves. Moreover, at variance with Eq. (12) where the steepening term dominates the dynamics, in Eq. (18), the Kerr effect prevails, and the validity of the model extends to regimes of relatively long pulses (i.e., psec to nsec). Noteworthy, the choice of the sign in Eq. (18), in turn fixed by the chirp sign, means that breaking can occur either on the negative (upper sign in Eq. (18)) or the positive (lower sign in Eq. (18)) slope front of the input waveform.

According to Eq. (18), the dynamics of a pre-chirped pulse with phase initially locked to $\phi(t, z = 0) = \pm 2\varepsilon^{-1} \int_{-\infty}^{t} \sqrt{\rho(t', z = 0)} dt'$ is ruled by the implicit solutions $\rho = \rho_0(t \mp 3\sqrt{\rho}z)$, $u = \pm 2\sqrt{\rho_0(t \mp 3\sqrt{\rho}z)}$ up to the point where the catastrophe occurs and dispersive effects appear. In this case the temporal symmetry implicit in the cases shown in Figs. 3 and 4 breaks down, since steepening occurs over one of the two fronts controlled by the choice of the upper or lower sign in the above solutions.

Figure 9 displays the dynamics of an initially (positively) chirped bright pulse, contrasting the Hopf and the full NLSE dynamics. The pulse develops a gradient catastrophe over the trailing edge (negative slope front) at a distance of 500 m, where the characteristic lines of the Hopf equation shown in Fig. 9a first intersect. This dynamics has been recently confirmed in a fiber optics experiment (Wetzel et al. 2016). A further proposed generalization of this concept refers to additional tailoring of these Riemann waves by introducing a quadratic spectral phase by means of pulse shaping techniques. This allowed to experimentally prove the effectiveness of the method in order to tailor the distance of breaking, thus opening a promising route to shock wave control in optical fibers (Wetzel et al. 2016). At the bottom line of such result, it stands the noteworthy fact that a complicated model such as the nonlinear Maxwell equations which governs the propagation of the optical field is effectively reduced to a remarkably simple model such as the Hopf equation, which in the end rules the evolution of the electric field envelope.

The validity of the Hopf model for appropriately pre-chirped waveform is obviously not limited to bright pulses. Figure 10a, b shows the case of the breaking of a dark profile (input power profile $\rho_0 = \tan h^2(t)$) with negative chirp, which still breaks on the trailing front, which, however, in this case represents the positive slope front. Such phenomenon is not restricted to any specific form of the pulse. Moreover, by combining different signs of chirp on the negative and positive temporal semiaxis, one can also enhance the rapidity of breaking compared with the unchirped case (case not shown, see Malaguti et al. 2010) or suppress the shock formation (see Fig. 10c). Therefore, not only in this regime, the nonlinear SPM induces a pure Hopf dynamics, but also it gives extra degrees of freedom to control the occurrence of breaking. Furthermore, in the presence of a tapering of the GVD (or equivalently third-order dispersion and frequency shifting), extreme high-intensity compressed pulse can develop, which has been termed optical tsunamis in analogy to a shoaling tsunami in water waves (Wabnitz 2013).

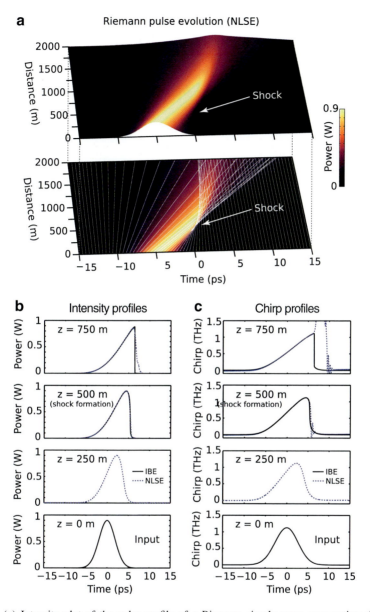

Fig. 9 (a) Intensity plot of the pulse profile of a Riemann simple wave propagating along a fiber obtained numerically from the NLSE model (top). The projected pulse intensity (bottom) is compared with the characteristic lines obtained analytically from the Hopf equation (white), showing shock formation at $z = 500$ m, where the characteristics start to intersect. Temporal profiles of the (**b**) intensity and (**c**) chirp are shown at selected distances, comparing predictions from the Hopf equation (solid black curve) with NLSE simulations (dashed blue curve) (From Wetzel et al. 2016, reporting the experimental realization of the Hopf dynamics. Here fiber parameters are $k'' = 0.8397 \, \text{ps}^2/\text{km}$, $\gamma = 11.7 \, (\text{W km})^{-1}$, and Gaussian pulse with 2.7 ps duration). (From Wetzel et al. 2016)

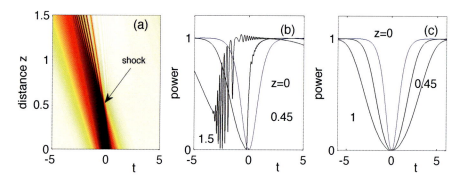

Fig. 10 (**a, b**) Evolution and snapshot of a chirped input with hyperbolic tangent amplitude profile, undergoing a gradient catastrophe on the trailing edge at the normalized distance $z \sim 0.45$ and the subsequent development into a DSW (note that a rarefaction wave develops on the leading edge); (**c**) shock suppression by a suitable choice of input chirp

Riemann Problem and Dam Breaking

In the theory of quasi-linear hyperbolic PDEs such as the SWEs, a fundamental problem is the evolution of a steplike initial condition, which is known as the Riemann problem (LeVeque 2004). The solution to such problem is indeed a building block for understanding the scenario of possible evolutions as well as for developing numerical schemes of integration (Riemann solvers). According to the general theory, a steplike initial condition of a 2×2 problem, such as the SWEs, decays into a pair of fundamental waves, which can be of the shock wave or rarefaction wave type, respectively. In general a step initial condition involves a jump in both the variables ρ and u, which vary from the left (ρ_L, u_L) to the right (ρ_R, u_R) constant state, where, without loss of generality, the step is assumed to be located in $t = 0$, so that subscripts L and R refer to $t < 0$ and $t > 0$, respectively. In terms of Riemann invariants, one has also step initial conditions from the left values $r_L^\pm = u_L \pm \sqrt{\rho_L}$ to the right values $r_R^\pm = u_R \pm \sqrt{\rho_R}$. According to the specific value of the four boundaries $r_{L,R}^\pm$, the SWEs give rise to different evolutions which involve the decay into wave pairs of the type: (i) rarefaction-rarefaction, (ii) shock-shock, and (iii) rarefaction-shock (LeVeque 2004; El et al. 1995). In particular, the latter case can be accessed by implementing only a step in the power variable ρ with the initial chirp being identically vanishing, which can be more easily accessed experimentally. In this case, the Riemann problem for the SWEs is known, in the context of hydrodynamics, as the dam-break problem, namely, the 1D evolution that follows the instantaneous removal (rupture) of a dam separating different downstream (ρ_L) and upstream $(\rho_R > \rho_L)$ levels of still water. In this case the solution to the SWEs is composed by a rarefaction wave and a classical shock wave pointing in opposite directions (upstream and downstream, respectively) separated by a constant expanding plateau. While we refer the reader to the hydrodynamic literature for its derivation, Fig. 11 illustrates the profile of ρ

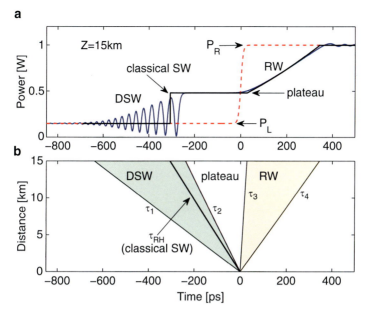

Fig. 11 (a) Snapshots comparing in real units the DSW-RW pair (solid blue curve) obtained from the NLSE with steplike input $E(T, 0) = [P_L + (P_R - P_L)(1 + \tanh(T/T_r))/2]^{1/2}$ (dashed red) with the ideal dispersionless solution of the SWEs (solid black). The weak oscillations over the top of the RW should not be confused with those of the DSW, as they appear in the numerics for a fast jump (T_r small) due to the Gibbs phenomenon. (b) Wedges in real-world time-distance plane (T, Z), corresponding to the RW (fable orange), the DSW (green), and the plateau in between (white). The boundaries correspond to slopes $\sqrt{k''\gamma P_R \tau_j}$, $j = 1, 2, 3, 4$. Here $k'' = 176\,\text{ps}^2/\text{km}$, $\gamma = 3\,(\text{W km})^{-1}$, $P_R = 1\,\text{W}$, $P_L = 150\,\text{mW}$, and $T_r = 10\,\text{ps}$, as in the experiment in Xu et al. (2017)

for such solution (see black solid line in Fig. 11a), contrasting it with the dispersive counterpart obtained from the numerical solution of the full NLSE (blue solid line in Fig. 11a) with steplike initial power profile (dashed red line in Fig. 11a). As shown, the solution arising from the SWEs or dispersionless limit gives a quantitatively good description of the NLSE dynamics for what concerns the smooth part that includes the rarefaction wave and the plateau. Conversely, the classical shock is replaced by a DSW as expected on the basis of the general principles illustrated in the introduction. In this situation, the edge velocities of the DSW can be predicted by applying the Whitham averaging theory (or modulation theory) illustrated in detail in the Appendix A. Following the general theory (see Appendix A), such velocities are conveniently expressed in terms of the self-similar variable $\tau = t/z$. In the present case, the boundary conditions for the shock are given by the intermediate constant values of the plateau $\rho_i = (\sqrt{\rho_L} + \sqrt{\rho_R})^2/4$, $u_i = \sqrt{\rho_L} - \sqrt{\rho_R}$ and the quiescent left state $\rho_L, u_L = 0$. This allows for expressing the linear (say τ_1) and the soliton (say τ_2) edge velocities as a function of ρ_L and ρ_R only, obtaining

$$\tau_2 = -\frac{\sqrt{\rho_L} + \sqrt{\rho_R}}{2}; \quad \tau_1 = \frac{\rho_L - 2\rho_R}{\sqrt{\rho_R}}. \tag{19}$$

Conversely, the Whitham averaging gives for the edge velocities of the rarefaction wave, say τ_3 and τ_4, the same expression which would be obtained in the dispersionless limit from the SWEs Eq. (13). They read as

$$\tau_4 = \sqrt{\rho_R}; \quad \tau_3 = \frac{3\sqrt{\rho_L} - \sqrt{\rho_R}}{2}. \tag{20}$$

All of such velocities define, in the plane (t, z), the wedges where the three components (rarefaction, plateau, and DSW) expand, as displayed in Fig. 11b.

Importantly, unlike experiments performed with smooth pulse waveform, the dam-break problem gives the unique opportunity to quantitatively test the Whitham modulation theory against experimental results. In this respect, it is important to emphasize that, as illustrated in the Appendix A, the modulation theory predicts for the DSW a critical transition where the DSW envelope is no longer monotone (as it is in Fig. 11a) but rather exhibits a cavitating state associated with the appearance of a vacuum point where the optical intensity vanishes (see also Appendix A, Fig. 23). Such a transition occurs when the crucial parameter $r = \rho_L/\rho_R$, namely, the ratio between the two quiescent states, decreases below the threshold value $r = r_{th} = 1/9 \simeq 0.11$. In the limit $r \to 0$ ($\rho_L \to 0$), the vacuum point shifts toward the linear edge of the DSW, but at the same time, the amplitude of the DSW oscillations vanishes. In this limit the shock disappears, and the dynamics involves a single rarefaction wave, recovering the solution of the SWEs in the so-called dry-bed case (a vanishing level of water downstream) (Kodama and Wabnitz 1995).

The dam-breaking dynamics has been recently observed by means of a full fiber setup at the University of Lille (Xu et al. 2017). The input shape, obtained by means of an electrooptic generator driven by a generator of arbitrary waveform, is shown in Fig. 12a. This shape allows for observing the formation of the rarefaction-DSW pair for the increasing step between the two states with nonvanishing powers P_L and P_R (leading edge of input pulse) while observing, at the same time, the dry-bed dynamics on the decreasing step from P_R to zero (trailing edge of input pulse). The parameter of the fiber employed in the experiment is reported in the caption of Fig. 11. The relatively large GVD ($k'' = 176\,\text{ps}^2/\text{km}$) permits to have the oscillation period of the DSW, which scales as $\sqrt{k''/\gamma P_i}$, in the tens of psec range, allowing for a good temporal resolution of the fast oscillating DSW. Another crucial feature of the experiment is the compensation of the fiber losses, which would cause a strong deviation from the expected dynamics, by means of a counterpropagating Raman pump. Figure 12b shows the output profile after the propagation through the 15 km fiber. The measured profile clearly shows the rarefaction-DSW structure, which turns out to in good agreement with the NLSE simulation and with the delays calculated from modulation theory (dashed

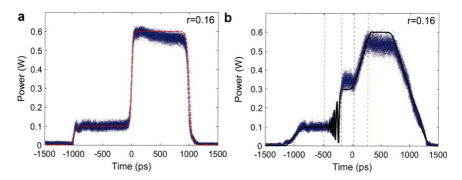

Fig. 12 Temporal traces of the whole waveform, experiment (blue dots) vs. numerics based on the NLSE (black curve; the input is in dashed red): (**a**) input power profile; (**b**) output power profile after propagation along a 15 km long fiber. The vertical dashed lines give the predicted delays of the edges of the DSW (magenta and green lines, from Eqs. (19)) and the RW (gray and orange lines, from Eqs. (20)), respectively

vertical lines). Conversely, only a rarefaction is observed on the trailing edge of the pulse.

Figure 13 shows a detail of the DSW soliton edge, obtained for a fixed $P_R = 1$ W and variable P_L (variable ratio r). The measured profiles clearly show the critical transition to cavitation at the threshold value $r = 0.11$ (in full quantitative agreement with modulation theory), where the soliton at the edge of the DSW becomes black. Further decreasing r makes the vacuum point shifting toward the linear edge of the DSW, again in good quantitative agreement with modulation theory and NLSE simulations.

Competing Wave-Breaking Mechanisms

Dispersive nonlinear wave propagation gives rise to different universal mechanisms of breaking. In addition to the formation of DSWs discussed so far, another well-known breaking mechanism of a carrier wave due to growth of low-frequency modulations is the universal modulational instability phenomenon. For the scalar NLSE, these two mechanisms are mutually exclusive. In fact, the gradient catastrophe occurs in the defocusing regime characterized indeed by a hyperbolic dispersionless limit. Conversely, MI takes place in the focusing regime where the gradient catastrophe is precluded, reflecting the elliptic dispersionless limit of Eqs. (17) discussed above. However, there are several different situations where the two mechanisms can indeed coexist and compete. For example, they can coexist in the presence of higher-order dispersion (Conforti et al. 2014) or when nonlinearly coupled modes are considered as, for example, in the case of second harmonic generation for quadratic nonlinearities, or polarization modes for the Kerr nonlinearities.

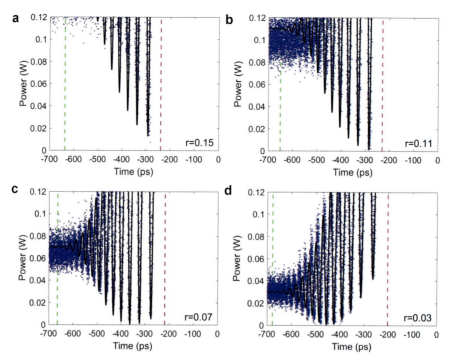

Fig. 13 (**a**–**d**) Zoom around the soliton edge of the DSW, showing the transition to cavitation, for fixed $P_R = 1$ W and different fractions $r = P_L/P_R$: (**a**) 0.15; (**b**) 0.11 (cavitation threshold); (**c**) 0.07; (**d**) 0.03. Experiment vs. theory and vertical lines as in Fig. 12b

In nonlinear fiber optics, the vectorial model of interest that permits to observe the interplay between MI and DSW formation is the vector NLSE (VNLSE), which, adopting the same normalization that leads to the NLSE (8) can be cast in the following dimensionless form:

$$i\varepsilon(\psi_j)_z - \frac{\varepsilon^2}{2}(\psi_j)_{tt} + \left(|\psi_j|^2 + X|\psi_{3-j}|^2\right)\psi_j = 0, \quad (21)$$

where $j = 1, 2$, and we assume to operate in the normal GVD regime. In contrast with the scalar case ($X = 0$) which is modulationally stable, the plane-wave solutions $u_j = \sqrt{P_j}\exp[i(P_j + XP_{3-j})z/\varepsilon]$ of Eqs. (21) are modulationally unstable provided $X > 1$. The MI gain reads as

$$g = \varepsilon^{-1}\sqrt{K^2[\sqrt{(P_1 - P_2)^2 + 4X^2 P_1 P_2} - (P_1 + P_2) - K^2]}; \quad K^2 \equiv (\varepsilon q)^2/2. \quad (22)$$

On the other hand, Eqs. (21) admit a hyperbolic dispersionless limit, as it is immediately clear in the symmetric case $\psi_1 = \psi_2$ for which Eqs. (21) reduces to a

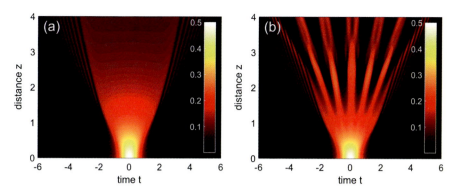

Fig. 14 Color level plot of $|\psi_1|^2$ ($|\psi_2|^2$ is similar) evolving according to Eqs. (21), comparing stable DSW formation for $X = 1$ (**a**), with the MI unstable case for $X = 2$ (**b**). Here $\varepsilon = 0.04$ and $\psi_1(0,t) = \psi_2(0,t) = [0.1 + 0.9\exp(-t^2)]/\sqrt{2}$

scalar NLSE with nonlinear coefficient $(1 + X)$. Therefore a competition between DSW and MI should be expected in this case. Restricting for definiteness to the symmetric excitation of Gaussians on pedestal, Fig. 14 shows that this is indeed the case. Here, a crossover behavior is induced by changing the cross-phase modulation coefficient X at fixed $\varepsilon = 0.04$. While for $X = 1$ (so-called Manakov case), the system exhibits the formation of stable DSW (Fig. 14a) in both modes, when $X > 1$ the onset of MI (Fig. 14b) leads to additional oscillations appearing on the flat top of the beams which dramatically affect the coherence of the DSW at large propagation distances.

Resonant Radiation Emitted by Dispersive Shocks

Bright solitons propagating in standard or photonic crystal fibers close to the zero-dispersion wavelength (ZDW) are known to emit resonant radiation (RR) in the region of normal GVD. The underlying mechanism is the resonant coupling with linear dispersive waves (DW) induced by higher-order dispersion (Akhmediev and Karlsson 1995). The emission of RR is usually thought to be a prerogative of solitons, but quite recently experimental observations (Webb et al. 2013; Conforti et al. 2015) and theoretical investigations (Conforti and Trillo 2013; Conforti et al. 2014) proved that this is not necessarily the case. In particular, the DSWs, which develop in the regime of weak dispersion, resonantly amplify DW at frequencies given by a specific phase-matching selection rule. In fact, the strong spectral broadening that accompanies wave-breaking seeds linear waves, which may be resonantly amplified, thanks to the well-defined velocity of the shock front.

Consider, for example, a standard telecom fiber (Corning MetroCor) with nonlinear and dispersion parameters as follows: $\gamma = 2.5\,\text{W}^{-1}\text{km}^{-1}$, $k_2 = 6.4\,\text{ps}^2/\text{Km}$,

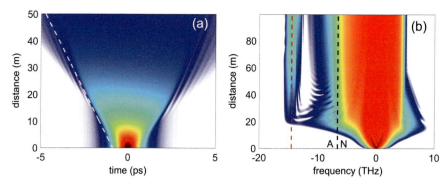

Fig. 15 Temporal (**a**) and spectral (**b**) evolution of an input sech pulse $P_0 = 600$ W, $T_0 = 850$ fs, at $\lambda_p = 1568.5$ nm (normal GVD). A/N labels anomalous/normal GVD regions, and the dashed red lines stand for the DW detuning predicted by Eq. (26) with velocity given by oblique dashed line in (**a**)

$k_3 = 0.134$ ps^3/Km, and $k_4 = -9 \times 10^{-4}$ ps^4/Km (higher-order terms are negligible), which gives a ZDW $\lambda_{ZDW} = 1625$ nm (Webb et al. 2013). Figure 15 shows the temporal and spectral propagation of a hyperbolic secant pulse in the normal dispersion regime of the fiber, obtained, for the sake of completeness, from the numerical solution of generalized NLSE (gNLSE), which accounts for higher-order nonlinear terms in full integral form (Agrawal 2013). The pulse undergoes a steepening of the leading and trailing edge, which leads to wave-breaking at a propagation distance of 20 m. After the breaking, two DSWs develop with broken symmetry (in time) due to the presence of third-order dispersion. The spectrum is broadest at the breaking point and clearly shows a narrowband peak in the anomalous dispersion region. This peak can be interpreted as a resonant radiation emitted by the leading edge of the pulse. In the next section, we will show how to predict the spectral position of this resonant radiation.

Phase-Matching Condition

We develop our analysis starting from the NLSE suitably extended to account for the effects of higher-order dispersion (HOD). In particular we extend the semiclassical form of NLSE (Eq. (8)) to include HOD terms, while we safely neglect Raman and self-steepening due to the pulse duration and power range that we consider. By defining the dispersion coefficients as $\beta_n = \partial_\omega^n k / \sqrt{(L_{nl})^{n-2}(\partial_\omega^2 k)^n}$, we recover the defocusing NLSE in the weakly dispersing form (with $\beta_2 = 1$)

$$i\varepsilon\partial_z\psi + d(i\varepsilon\partial_t)\psi + |\psi|^2\psi = 0,$$

$$d(i\varepsilon\partial_t) = \sum_{n\geq 2} \frac{\beta_n}{n!}(i\varepsilon\partial_t)^n \qquad (23)$$

$$= -\frac{\varepsilon^2}{2}\partial_t^2 - i\frac{\beta_3\varepsilon^3}{6}\partial_t^3 + \frac{\beta_4\varepsilon^4}{24}\partial_t^4 + \ldots$$

where we considered normal GVD. Note that the normalized dispersive operator $d(i\varepsilon\partial_t)$ has progressively smaller terms, weighted by powers of the parameter $\varepsilon \ll 1$ and coefficients β_n.

The process of wave-breaking ruled by Eq. (23) can be described by applying again the Madelung transformation $\psi = \sqrt{\rho}\exp(iS/\varepsilon)$. At leading-order in ε, we obtain a quasi-linear hydrodynamic reduction, with $\rho = |\psi|^2$ and $u = -S_t$ equivalent density and velocity of the flow, which can be further cast in the form (Conforti et al. 2014)

$$\rho_z + \left[\beta_2 \rho u + \frac{\beta_3}{2}\rho u^2 + \frac{\beta_4}{6}\rho u^3 + \ldots\right]_t = 0, \tag{24}$$

$$(\rho u)_z + \left[\beta_2 \rho u^2 + \frac{\beta_3}{2}\rho u^3 + \frac{\beta_4}{6}\rho u^4 + \ldots + \frac{1}{2}\rho^2\right]_t = 0. \tag{25}$$

of a conservation law $\mathbf{q}_z + [\mathbf{f}(\mathbf{q})]_t = 0$ for mass and momentum, with $\mathbf{q} = (\rho, \rho u)$, which suitably extends Eqs. (15). For small $\beta_{3,4}$, since Eqs. (24) and 25) continue to be hyperbolic, thus admitting weak solutions in the form of classical shock waves, i.e., traveling jumps from left (ρ_L, u_L) to right (ρ_R, u_R) values, whose velocity V_c can be found from the generalized RH condition (i.e., the natural extension of Eq. (2) discussed previously in the scalar case: $V_c(\mathbf{q}_L - \mathbf{q}_R) = [\mathbf{f}(\mathbf{q}_L) - \mathbf{f}(\mathbf{q}_R)]$ (LeVeque 2004). However, the jump is regularized by GVD in the form of a DSW. In this regime, the shock velocity can be identified with the velocity V_s of the steep front near the deepest oscillation (DSW leading edge), which differs from V_c and can be determined numerically (or analytically via modulation theory for steplike initial data, see Appendix A). The strong spectral broadening that accompanies steep front formation can act as an efficient seed for DW which are phase-matched to the shock in its moving frame at velocity V_s.

In order to calculate the frequency of the DW, we assume an input pump $\psi_0 = \psi(t, z = 0)$ with central frequency $\omega_p = 0$ (i.e., in real-world units ω_p coincides with ω_0, around which $d(i\varepsilon\partial_t)$ in Eq. (23) is expanded). Let us denote as $V_s = dt/dz$ the "velocity" of the SW near a wave-breaking point and as $\tilde{d}(\varepsilon\omega) = \sum_n \frac{\beta_n}{n!}(\varepsilon\omega)^n$ the Fourier transform of $d(i\varepsilon\partial_t)$. Linear waves $\exp(ik(\omega)z - i\omega t)$ are resonantly amplified when their wavenumber in the shock moving frame, which reads as $k(\omega) = \frac{1}{\varepsilon}[\tilde{d}(\omega) - V_s(\varepsilon\omega)]$ equals the pump wavenumber $k_p = k(\omega_p = 0) = 0$. Denoting also as k_{nl} the difference between the nonlinear contributions to the pump and RR wavenumber (the nonlinear contribution to the wavenumber of the resonant radiation is induced by cross-phase modulation with a non-zero background, on top of which RR propagates), respectively, the radiation is resonantly amplified at frequency detuning $\omega = \omega_{RR}$ that solves the explicit phase-matching equation (Conforti et al. 2014)

$$\sum_n \frac{\beta_n}{n!}(\varepsilon\omega)^n - V_s(\varepsilon\omega) = \varepsilon k_{nl}. \tag{26}$$

We show below that Eq. (26) correctly describes the RR emitted by a DSW. At variance with solitons of the focusing NLSE where $V_s(\omega_p = 0) = 0$ (Akhmediev and Karlsson 1995), DSWs possess non-zero velocity V_s, which must be carefully evaluated, having great impact on the determination of ω_{RR}.

Steplike Pulses

We consider first a step initial value that allows us to calculate analytically the velocity. Without loss of generality, we take $\beta_3 < 0$. Specifically, we consider the evolution of an initial jump from the left state $\rho_L, u_L = 0$ for $t < 0$ to the right state $\rho_R(< \rho_L), u_r = 2(\sqrt{\rho_R} - \sqrt{\rho_L})$ for $t > 0$ (Hoefer et al. 2006; Conforti et al. 2014). The leading edge of the resulting DSW can be approximated by a gray soliton, whose velocity can be calculated as $V_l = \sqrt{\rho_L} + u_R = 2\sqrt{\rho_R} - \sqrt{\rho_L}$ (Hoefer et al. 2006; Conforti et al. 2014).

If we account for $k_{nl} = k_{nl}^{sol} - k_{nl}^{RR} = -\frac{1}{\varepsilon}\rho_L$ arising from the soliton $k_{nl}^{sol} = \rho_l/\varepsilon$ and the cross-induced contribution $k_{nl}^{RR} = 2\rho_L/\varepsilon$ to the RR, Eq. (26) explicitly reads as

$$\frac{\beta_3}{6}(\varepsilon\omega)^3 + \frac{\beta_2}{2}(\varepsilon\omega)^2 - V_s(\varepsilon\omega) + \rho_L = 0. \tag{27}$$

Real solutions $\omega = \omega_{RR}$ of Eq. (27) correctly predicts the RR as long as $|\beta_3| < 0.5$, as shown by the NLSE simulation in Fig. 16. The DSW displayed in Fig. 16a clearly exhibits a spectral RR peak besides spectral shoulders due to the oscillating front, as shown by the spectral evolution in Fig. 16b. Perfect agreement is found between the RR peak obtained in the numerics and the prediction (dashed vertical line in Fig. 16b) from Eq. (27) with velocity $V_s = V_L$, where V_L is the leading or soliton edge velocity of the DSW that can be calculated by means of Whitham theory (see Appendix A). We also point out that k_{nl} represents a small correction, so ω_{RR} can be safely approximated by dropping the last term in Eq. (27) to yield $\varepsilon\omega_{RR} = \frac{3}{2\beta_3}(-\beta_2 \pm \sqrt{\beta_2^2 + 8V_s\beta_3/3})$, that can be reduced to the simple formula $\varepsilon\omega_{RR} = -\frac{3\beta_2}{\beta_3}$ (Webb et al. 2013) only when $\beta_3 V_s \to 0$.

Bright Pulses

The behavior of steplike initial data are basically recovered for pulse waveforms that are more manageable in experiments. As shown in Fig. 17, RR occurs also in the limit of vanishing background, allowing us to conclude that a bright pulse does not need to be a soliton (as in the focusing NLSE, $\beta_2 = -1$) to radiate. In fact, resonant amplification of linear waves occurs via SWs also in the opposite regime where the nonlinearity strongly enforces the effect of leading-order dispersion, the

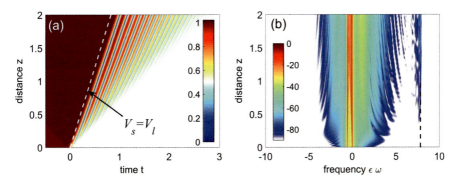

Fig. 16 Radiating DSW ruled by NLSE (23) with $\varepsilon = 0.03$, steplike input $\rho_L, \rho_R = 1, 0.5$, and 3-HOD $\beta_3 = -0.35$: (**a**) Color level plot of density $\rho(t,z)$ (the dashed line gives the DSW leading edge velocity V_l); (**b**) corresponding spectral evolution. The vertical dashed line stands for the RR predicted by Eq. (27)

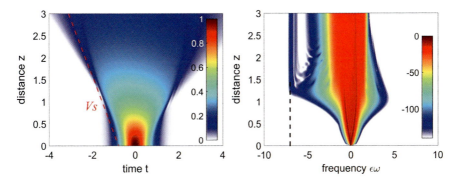

Fig. 17 (**a, b**) Temporal and spectral evolution of a Gaussian pulse without background emitting RR, for $\beta_3 = 0.35$, and $\varepsilon = 0.03$

only key ingredients being a well-defined velocity of the front and the spectral broadening that seeds the RR at phase-matching. Experimental evidence for such RR scenario was reported quite recently (Webb et al. 2013), corresponding to numerical simulations reported in Fig. 15. The physical parameters used to obtain Fig. 15, gives normalized parameters $\varepsilon \simeq 0.07$ and $\beta_3 \simeq 0.37$, typical of the wave-breaking regime ($\varepsilon \ll 1$) with perturbative 3-HOD. Since $\beta_3 > 0$, the radiating shock turns out to be the one on the leading edge ($t < 0$), and its velocity $V_s = -0.75$, inserted in Eq. (26), gives a negative frequency detuning $\Delta f_{RR} = \omega_{RR} T_0^{-1}/2\pi \simeq 13$ THz, in excellent agreement with the value reported in Webb et al. (2013). A detailed numerical study of this particular case, including Raman effects is reported in Conforti and Trillo (2013).

Periodic Input

We are interested in the evolution ruled by Eq. (23) subject to the dual-frequency initial condition

$$\psi_0 = \sqrt{\eta}\exp(i\omega_p t/2) + \sqrt{1-\eta}\exp(-i\omega_p t/2), \tag{28}$$

where we fix the normalized frequency $\omega_p \equiv \Omega T_0 = \pi$ consistently with the notation introduced in the discussion of DSWs in mFWM. Here η accounts for the possible imbalance of the input spectral lines. In this case a frequency comb is generated thanks to mFWM. Formation of DSWs, occurring in the regime of weak normal dispersion ($\beta_2 = 1$), enhances the broadening of the comb toward high-order sideband pairs. When such DSWs are excited sufficiently close to a ZDW, they are expected to generate RR, owing to phase-matching with linear waves induced by higher-order dispersion. An example of the RR ruled by TOD (we set $\beta_3 = 0.3$) is shown in Fig. 18 for an imbalanced input ($\eta = 0.3$ in Eq. (28)). The colormap evolution in Fig. 18a clearly shows that the initial waveform undergoes wave-breaking around $z \sim 0.4$. The mechanism of breaking has been discussed before in section "Shock Formation in Optical Fibers," and also analyzed in details in Fatome et al. (2014), Trillo and Valiani (2010). It involves two gradient catastrophes occurring across each minimum of the injected modulation envelope. The GVD regularizes the catastrophes leading to the formation of two DSWs, where the individual oscillations in the trains exhibits dark soliton features, moving with nearly constant darkness and velocity inversely proportional to it. Importantly, the breaking scenario is weakly affected by TOD; However, one can notice that the darkest soliton-like oscillation emits RR. This radiation has much higher frequency than the comb spacing and turns out to be generated over the CW plateau of the leading edge labeled ψ_0. This is clear from Fig. 18b, which shows the enhancement

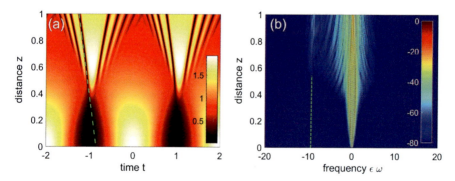

Fig. 18 RR emitted by shock with asymmetric pumping $\eta = 0.3$, and $\varepsilon = 0.04$: (a) temporal and (b) spectral colormap evolution for $\beta_3 = 0.3$; Dashed line in (a) highlights the DSW edge velocity $V = -0.4$. Vertical dashed line in (b) indicates ω_{RR} from Eq. (26) with $k_{nl} = -|\psi_0|^2/\varepsilon$

of such frequency at the distance of breaking where the strong spectral broadening associated with the shock acts as a seed for the phase-matched (resonant) frequency.

Shock Waves in Passive Cavities

Passive nonlinear cavities which are externally driven have been recently become popular, and have been implemented both in fiber rings and monolithic microresonators. They are exploited for fundamental studies as well as impactful applications such as the formation of wide-span frequency combs. Such resonators exhibit extremely rich dynamics characterized by a host of phenomena such as bistability, modulational instability and soliton formation. It has been recently shown that passive cavities admits a novel dynamical behavior featuring the formation of dispersive-dissipative shock waves (Malaguti et al. 2014). These phenomena are well captured by a mean field approach which yields a driven damped NLSE, often referred to as the Lugiato-Lefever equation (LLE) Lugiato and Lefever (1987). By accounting for HOD, a generalized LLE can expressed in dimensionless units as

$$i\varepsilon\psi_z + d(i\varepsilon\partial_t)\psi + |\psi|^2\psi = [\delta - i\alpha]\psi + iS, \quad (29)$$

where we adopt the normalization introduced in Malaguti et al. (2014). We just recall that the parameter $\varepsilon = \sqrt{L/L_d} \ll 1$ (L and $L_d = T_0^2/k''$ are the fiber (cavity) length and the dispersion length associated with time scale T_0 and GVD) quantifies the smallness of the GVD and the HOD introduced through the operator $d(i\varepsilon\partial_t) = \sum_{n\geq 2} \beta_n (i\varepsilon\partial_t)^n/n! = -\beta_2\varepsilon^2\partial_t^2/2 - i\beta_3\varepsilon^3\partial_t^3/6 + \ldots$, where the coefficients $\beta_n = \partial_\omega^n k/\sqrt{(L)^{n-2}(\partial_\omega^2 k)^n}$ [note that $\beta_2 = \text{sign}(\partial_\omega^2 k)$] are related to real-world HOD $\partial_\omega^n k$.

Let us first neglect the HOD terms ($\beta_n = 0, n > 2$). Equation (29) can exhibit a bistable response with two coexisting stable branches of CW solutions. A DSW can be seen as a fast oscillating modulated wave train that connects two sufficiently different quasi-stationary states. Starting from a cavity biased on the lower state, one can easily reach a different state on the upper branch by using an addressing external pulse with moderate power. In this regime the intracavity pulse edges undergo initial steepening, which is mainly driven by the Kerr effect, tending to form shock waves. The strong gradient associated with the steepened fronts enhances the impact of GVD, which ends up inducing the formation of wave trains that connect the two states of the front. An example is reported in Fig. 19, by using the injected field $S(t) = \sqrt{P} + \sqrt{P_p}\text{sech}(t)$ with $P = 0.0041$ and $P_p = 0.16P$.

As shown for the conservative case in the previous Section, the presence of HOD may lead the shock to radiate. We concentrate on the first relevant dispersive perturbation, i.e., third-order dispersion (TOD $\beta_3 \neq 0$), but the scenario is qualitatively similar for others order of HOD. An example is shown in Fig. 20, for the same injected field used for the example reported Fig. 19. In the temporal domain, the effect of TOD is to induce an asymmetry between the leading and

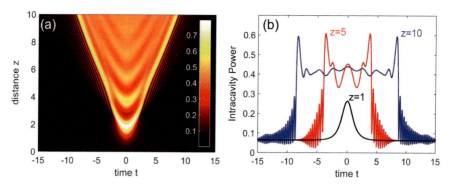

Fig. 19 DSW generation ruled by Eq. (29). (**a**) Color level plot of intracavity power $|\psi(z,t)|^2$. (**b**) Snapshots of intracavity power at difference distances. Here $\varepsilon = 0.1$, $\delta = \pi/10$, $\alpha = 0.03$

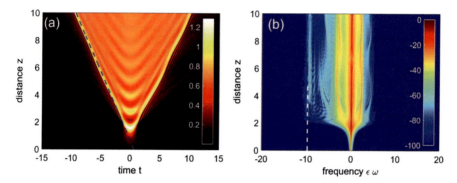

Fig. 20 Temporal (**a**) and spectral (**b**) evolution of a DSW in the cavity. Parameters: $\beta_2 = 1$, $\varepsilon = 0.1$, $\delta = \pi/6$, $\beta_3 = 0.25$, and $\alpha = 0.03$. Dashed blue line in (**a**) represents the front velocity $V_s = -1$; dashed line in (**b**) stands for the value of $\varepsilon\omega_{RR} = -9.6$ calculated from the phase-matching equation (30)

trailing fronts. The most striking feature is visible in the spectral propagation shown Fig. 19b, where an additional frequency component, well detached from the shock spectrum, is generated starting at a distance $z \approx 2$. The frequency of this radiation can be found by means of a perturbation approach, in close analogy to the one developed for the conservative case. In the limit of small losses α, we find that the frequency of the RR must satisfy the following phase-matching equation:

$$\left[\beta_3 \frac{(\varepsilon\omega)^3}{6} + \beta_2 \frac{(\varepsilon\omega)^2}{2} - \frac{(\varepsilon\omega)}{V_s} - \delta\right] + 2P_{uH} = 0, \qquad (30)$$

where P_{uH} is the power of the higher state of the front, propagating with velocity V_s where RR is shed. This equation is very similar to Eq. (27), but it contains the cavity detuning δ as an additional parameter.

Conclusions

In summary, we have shown that nonlinear fiber optics represents an ideal playground for observing dispersive hydrodynamics phenomena and specifically for making accurate experiments on dispersive shock waves. Recent theoretical studies have permitted to substantially progress the full understanding of the phenomenon. Moreover, challenging experiments have been carried out that have demonstrated the accessibility of wave-breaking phenomena in different new contexts, ranging from multiple four-wave mixing to pulse shaping experiment, supercontinuum and comb generation, observation of radiative effects, and fundamental problems of fluid flow such as dam breaking. Yet, a lot of new experiments can be envisaged that could aim at unveiling new fundamental scenarios discussed in this chapter as well as to exploit wave-breaking in fiber optics applications. Furthermore optical experiments are important also for the understanding similar phenomena of hydrodynamic origin in other contexts ranging from tidal bores in fluid dynamics, gravity waves in the atmosphere, spin waves in magnetic films, and atom collective behavior in Bose-Einstein condensates, for which the experimental implementation in the lab turns out to be more challenging.

Appendix A

In this Appendix we outline the calculation of the power (density) and chirp (velocity) profile of the DSW, according to Whitham modulation equation and the construction originally proposed by Gurevich-Pitaevskii-Krylov (Gurevich and Pitaevskii 1974; Gurevich and Krylov 1987). Our approach closely follows that of Hoefer et al. (2006), to which the reader is also referred to for further details. Below, we specifically focus on the case of a right-going DSW, which is generated by a suitable choice of the initial step. We recall, however, that also a left-going DSW is a suitable solution of the NLSE, which can be easily obtained with symmetry arguments from the case discussed below.

Let us start by considering an invariant traveling-wave periodic solution of the NLSE (Eq. (16)) of the form:

$$\psi(t,z) = \sqrt{\rho(\theta)} \exp\left[i\phi(\theta)\right], \tag{31}$$

where $\theta \equiv \frac{t-Vz}{\varepsilon}$ is a fast variable since $\varepsilon \ll 1$. By means of direct substitution into the NLSE, one can easily obtain the invariant dn-oidal solution:

$$\rho(t,z) = \lambda_3 - (\lambda_3 - \lambda_1)\mathrm{dn}^2\left(\sqrt{\lambda_3 - \lambda_1}\,\theta\,|m\right), \tag{32}$$

$$u(t,z) = \phi' = V - \sigma\frac{\sqrt{\lambda_1\lambda_2\lambda_3}}{\rho}, \tag{33}$$

which depends on the four parameters $\lambda_1, \lambda_2, \lambda_3, V$, with the additional constraint $V = \sqrt{\lambda_1 + \lambda_2 + \lambda_3}$, and $\sigma = \pm 1$. Here the wave period $L = 2K(m)/\sqrt{\lambda_3 - \lambda_1}$ is given in terms of the elliptic integral of first kind $K(m)$, with $m = \frac{\lambda_2 - \lambda_1}{\lambda_3 - \lambda_1}$. A modulation of such dn-oidal solution describes the DSW. The slow (compared to L) evolution of the parameters of such modulation is ruled by the Whitham equations, obtained by averaging the conservation laws of the NLSE over the period L (Whitham 1965). For integrable systems such as the NLSE, these equations are known to be expressible in diagonal form. For the NLSE this can be done by introducing four Riemann invariants $r_i = r_i(t, z)$, $i = 1, 2, 3, 4$, $r_1 < r_2 < r_3 < r_4$, which are a suitable combination of the original four parameters V, λ_i (Pavlov 1987):

$$\frac{\partial r_i}{\partial z} + v_i(r_1, r_2, r_3, r_4)\frac{\partial r_i}{\partial t} = 0; \quad i = 1, 2, 3, 4. \tag{34}$$

Here the velocities $v_i = v_i(r_1, r_2, r_3, r_4)$ constitute a deformation of the velocity V that depends on combinations of $\{r_i\}$ and elliptic integrals of first (i.e., $K(m)$) and second (i.e., $E(m)$) kind. For instance, the velocity $v_3 = v_3(r_1, r_2, r_3, r_4)$ that will be relevant in the following reads as

$$v_3 = V - \frac{1}{2}(r_4 - r_3)\left[1 - \frac{(r_4 - r_2)E(m)}{(r_3 - r_2)K(m)}\right]^{-1}, \tag{35}$$

and all quantities are recast as functions of r_i, viz.:

$$V = \frac{1}{4}(r_1 + r_2 + r_3 + r_4),$$

$$\lambda_3 = \frac{1}{16}(-r_1 - r_2 + r_3 + r_4)^2,$$

$$\lambda_2 = \frac{1}{16}(-r_1 + r_2 - r_3 + r_4)^2, \tag{36}$$

$$\lambda_1 = \frac{1}{16}(r_1 - r_2 - r_3 + r_4)^2,$$

$$m = \frac{(r_4 - r_3)(r_2 - r_1)}{(r_4 - r_2)(r_3 - r_1)}.$$

We are interested to describe the DSW ruled by the NLSE with initial step data corresponding to

$$\rho(t, z = 0) = \begin{cases} \rho_L, & t < 0 \\ \rho_R, & t > 0 \end{cases}, \quad u(t, z = 0) = \begin{cases} u_L, & t < 0 \\ u_R, & t > 0 \end{cases}. \tag{37}$$

A general step in amplitude decays into a combination of rarefaction and dispersive shock wave (El et al. 1995). In order to generate a pure right-going DSW, we consider that the initial value in terms of Riemann invariants $r^{\pm} = u \pm 2\sqrt{\rho}$ is characterized by $r^- = const.$, i.e., it is a right-going simple wave. The condition of a decreasing step ($\rho_L > \rho_R$) implies that the simple wave is of shock type (and not a rarefaction wave). In order to respect the simple wave assumption, we have to calculate one of the four parameters (ρ_L, ρ_R, u_L, u_R) as a functions of the others. By imposing, for example, $u_R = u_L + 2(\sqrt{\rho_R} - \sqrt{\rho_L})$, the initial value in terms of Riemann invariants (see Fig. 21) is characterized by a constant value for r^- and a decreasing steplike variation for r^+:

$$r^-(t, z=0) = u_L - 2\sqrt{\rho_L}, \tag{38}$$

$$r^+(t, z=0) = \begin{cases} r_L^+ = u_L + 2\sqrt{\rho_L}, & t < 0 \\ r_R^+ = u_L + 4\sqrt{\rho_R} - 2\sqrt{\rho_L}, & t > 0 \end{cases}, \tag{39}$$

which lead to a traveling (right-going) SW that connects the two constant states (Hoefer et al. 2006; LeVeque 2004). The corresponding DSW can be described in terms of a self-similar simple rarefaction wave of Whitham equations (34) generated by the following four-dimensional initial value that arises from initial data regularization (Hoefer et al. 2006; Kodama 1999) (see Fig. 21):

$$r_1(t, z=0) = r^- = u_L - 2\sqrt{\rho_L};$$

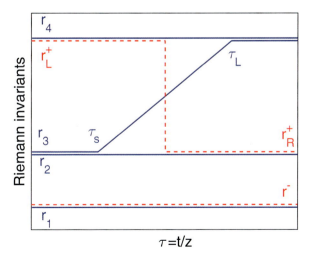

Fig. 21 Sketch of the Riemann invariants of Eqs. (34), where r_1, r_3, r_4 stay constant and $r_3 = r_3(\tau = t/z)$ varies smoothly in the range $\tau_S \leq \tau \leq \tau_L$ between the values r_R^+ and r_L^+. τ_L and τ_S correspond to the linear and soliton edges of the DSW, respectively. The dashed red lines are the initial Riemann invariants in the dispersionless limit

$$r_2(t,z=0) = r_R^+ = u_L + 4\sqrt{\rho_R} - 2\sqrt{\rho_L};$$

$$r_4(t,z=0) = r_L^+ = u_L + 2\sqrt{\rho_L}; \tag{40}$$

$$r_3(t,z=0) = \begin{cases} r_R^+ = u_L + 4\sqrt{\rho_R} - 2\sqrt{\rho_L}, & t<0 \\ r_L^+ = u_L + 2\sqrt{\rho_L}, & t>0. \end{cases}$$

In particular, the initial value (40) evolves in such a way that the Riemann variables r_1, r_2, r_4 remain constant and only $r_3 = r_3(\tau)$ varies, forming a pure rarefaction wave (owing to the fact that $r_3(z=0)$ is non-decreasing) that depends on the self-similar variable $\tau = t/z$.

Indeed all Whitham equations are formally satisfied when $r_{1,2,4}(t,z) = const.$ and $r_3(t,z) = r_3(\tau)$, provided the equation $(\tau - v_3)\, r_3' = 0$ is fulfilled. For $r_3(\tau) \neq constant$, this implies

$$\tau = v_3(r^-, r_R^+, r_3(\tau), r_L^+). \tag{41}$$

Equation (41) is a nonlinear equation in the only unknown $r_3(\tau)$, which can be solved with any root-finding method to compute the profile of the rarefaction wave. Once the value of $r_3(\tau)$ is known, the values of the parameters λ_i and V are calculated from Eqs. (36), which can be plugged into Eqs. (32) and 33) to get the DSW profile. It can be easily proved that the power and the chirp of the DSW are bounded by the envelopes $\rho^\pm$ and $u^\pm$ defined as

$$\rho^+(t,z) = \lambda_2,$$
$$\rho^-(t,z) = \lambda_1,$$
$$u^+(t,z) = V - \sigma\sqrt{\lambda_1\lambda_3/\lambda_2},$$
$$u^-(t,z) = V - \sigma\sqrt{\lambda_2\lambda_3/\lambda_1}. \tag{42}$$

An example of DSW profile is shown in Fig. 22a–c and contrasted with numerical solution of NLSE in Fig. 22b–d, with a smoothed input step in order to avoid numerical artifacts.

The velocities τ_L and τ_S of the leading (soliton) and trailing (linear) edges of the DSW correspond to the edges of this rarefaction wave and can be calculated as the limits of $v_3(r^-, r_R^+, r_3(\tau), r_L^+)$ for $r_3 \to r_R^+$ ($m \to 1$, soliton edge) and $r_3 \to r_L^+$ ($m \to 0$, linear edge), respectively:

$$\tau_L = \lim_{r_3 \to r_L^+} V_3(r^-, r_R^+, r_3, r_L^+) = \frac{2r_L^+ + r^- + r_R^+}{4} + \frac{(r^- - r_L^+)(r_R^+ - r_L^+)}{2r_L^+ - r^- - r_R^+}$$

$$\tau_S = \lim_{r_3 \to r_R^+} V_3(r^-, r_R^+, r_3, r_L^+) = \frac{2r_R^+ + r^- + r_L^+}{4} \tag{43}$$

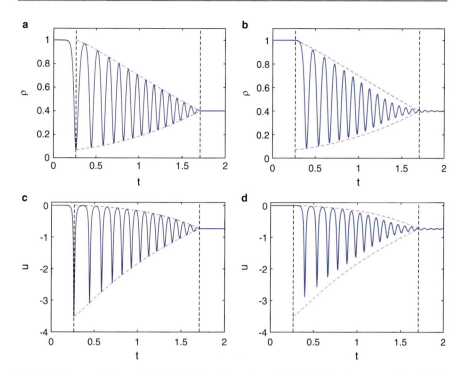

Fig. 22 Power and chirp of a DSW calculated from Eqs. (32–33) (**a**, **c**) and corresponding NLSE simulation (**b**, **d**). Magenta dashed lines depict the envelopes Eqs. (42). Parameters: $\rho_L = 1$, $\rho_R = 0.4$, $\epsilon = 0.03$, $z = 1$. Vertical dashed lines indicate the position of the soliton and linear edge of the DSW obtained from Eqs. (43)

These values are reported as vertical dashed lines in Fig. 22.

The modulation theory entails a crossover between two different regimes separated by a critical condition. In the first regime, the DSW envelopes are monotone and the DSW power never vanishes. However, below a critical value of ρ_L/ρ_R, the DSW exhibits a self-cavitating point, i.e., zero power, corresponding to a vacuum point in gas dynamics. Since the minimum power of the cnoidal wave turns out to be $\rho_{\min} = (r_1 - r_2 - r_3 + r_4)^2/16$ (Hoefer et al. 2006), the existence in the DSW of a cavitating or vacuum state $\rho_{\min} = 0$ requires $r_3 = r_1 - r_2 + r_4$. Therefore we obtain the self-similar location of the vacuum by performing the limit $r_3 \to r^- - r_R^+ + r_L^+$ in Eq. (35):

$$\tau_v = V_3(r^-, r_R^+, r^- - r_R^+ + r_L^+, r_L^+) =$$

$$= \frac{r^- + r_L^+}{2} - \frac{r_R^+ - r^-}{2}\left[1 - \frac{r_L^+ - r_R^+}{r^- - 2r_R^+ + r_L^+}\frac{E(m_v)}{K(m_v)}\right]^{-1} \quad (44)$$

$$m_v = \left(\frac{r_R^+ - r^-}{r_L^+ - r_R^+}\right)^2$$

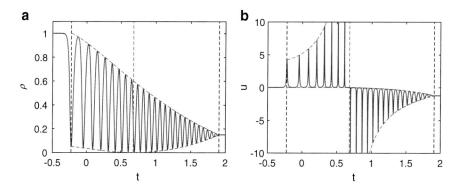

Fig. 23 Power (**a**) and chirp (**b**) of a DSW calculated from Eqs. (32) and (33) in presence of a vacuum point. Magenta dashed lines depict the envelopes Eqs. (42). Parameters: $\rho_L = 1$, $\rho_R = 0.15$, $\epsilon = 0.03$, $z = 1$. Vertical black dashed lines indicate the position of the soliton and linear edge of the DSW obtained from Eqs. (43), whereas the red dashed line highlights the position of the vacuum point from Eq. (44)

It can be shown that in order to get a solution that conserves momentum ρu, the velocity should change sign at the vacuum point (Hoefer et al. 2006). This fixes σ in Eq. (33) as:

$$\sigma(t,z) = \text{sign}(t/z - \tau_v).$$

The threshold for cavitation can be found by imposing that the vacuum point, coincides with the soliton edge, or $m_v = 1$, that gives

$$\left(\frac{\rho_R}{\rho_L}\right)_{th} = \frac{1}{4}. \tag{45}$$

At such threshold the vacuum point coincides with the soliton edge of the DSW, while decreasing the ratio ρ_R/ρ_L below the threshold makes the vacuum point progressively move along the envelope of the DSW toward its linear edge, where the oscillations vanish (this behavior has been experimentally confirmed for the Riemann dam-break problem, where the DSW develops between the lower constant state and the intermediate level generated in the dynamics; see section "Riemann Problem and Dam Breaking"). An example of non-monotonic DSW which features a vacuum point along its envelope is shown in Fig. 23a–b for $\rho_R/\rho_L = 0.15$.

References

G.P. Agrawal, *Nonlinear Fiber Optics*, 5th edn. (Academic, New York, 2013)
G.P. Agrawal, C. Headley III, Kink solitons and optical shocks in dispersive nonlinear media. Phys. Rev. A **46**, 1573 (1992)

S.A. Akhmanov, D.P. Krindach, A.V. Migulin, A.P. Sukhorukov, R.V. Khokhlov, Thermal self-action of laser beams. IEEE J. Quantum Electron. **QE-4**, 568 (1968)

N. Akhmediev, M. Karlsson, Cherenkov radiation emitted by solitons in optical fibers. Phys. Rev. A **51**, 2602–2607 (1995)

D. Anderson, S. Lisak, Nonlinear asymmetric self-phase modulation and self-steepening of pulses in long optical waveguides. Phys. Rev. A **27**, 1393 (1983)

D. Anderson, M. Desaix, M. Lisak, M.L. Quiroga-Teixeiro, Wave breaking in nonlinear-optical fibers. J. Opt. Soc. Am. B **9**, 1358 (1992)

G. Biondini, Y. Kodama, On the Whitham equations for the defocusing nonlinear Schrödinger equation with step initial data. J. Nonlinear Sci. **16**, 435–481 (2006)

D. Bohm, D.J. Hiley, Measurement understood through the quantum potential approach. Found. Phys. **14**, 255 (1984)

D. Cai, A.R. Bishop, N. Gronbech-Jensen, B. Malomed, Dark shock waves in the nonlinear Schrödinger system with internal losses. Phys. Rev. Lett. **78**, 223 (1997)

M. Conforti, S. Trillo, Dispersive wave emission from wave breaking. Opt. Lett. **38**, 3815–3818 (2013)

M. Conforti, F. Baronio, S. Trillo, Resonant radiation shed by dispersive shock waves. Phys. Rev. A **89**, 013807 (2014)

M. Conforti, A. Mussot, A. Kudlinski, S. Trillo, Parametric excitation of multiple resonant radiations from localized wavepackets. Sci. Rep. **5**, 9433 (2015)

C. Conti, A. Fratalocchi, M. Peccianti, G. Ruocco, S. Trillo, Observation of a gradient catastrophe generating solitons. Phys. Rev. Lett. **102**, 083902 (2009)

C. Conti, S. Stark, P.S.J. Russell, F. Biancalana, Multiple hydrodynamical shocks induced by the Raman effect in photonic crystal fibres. Phys. Rev. A **82**, 013838 (2010)

F. Demartini, C.H. Townes, T.K. Gustafson, P.L. Kelley, Self-steepening of light pulses. Phys. Rev. **164**, 312 (1967)

B. Dubrovin, S. Novikov, The Hamiltonian formalism of one-dimensional systems of hydrodynamic type and the Bogoliubov-Whitham averaging method. Akademiia Nauk SSSR, Doklady **270**, 781–785 (1983)

B. Dubrovin, T. Grava, C. Klein, A. Moro, On critical behaviour in systems of Hamiltonian partial differential equations. J. Nonlinear Sci. **25**, 631–707 (2015)

G.A. El, M.A. Hoefer, Dispersive shock waves and modulation theory. Physica D **333**, 11 (2016)

G.A. El, V.V. Geogjaev, A.V. Gurevich, A.L. Krylov, Decay of an initial discontinuity in the defocusing NLS hydrodynamics. Physica D **87**, 186–192 (1995)

J. Fatome, C. Finot, G. Millot, A. Armaroli, S. Trillo, Observation of optical undular bores in multiple four-wave mixing. Phys. Rev. X **4**, 021022 (2014)

E. Fermi, J. Pasta, S. Ulam, in *Collected Papers of Enrico Fermi*, vol. 2, ed. by E. Segré (The University of Chicago, Chicago, 1965), pp. 977–988

C. Finot, B. Kibler, L. Provost, S. Wabnitz, Beneficial impact of wave-breaking for coherent continuum formation in normally dispersive nonlinear fibers. J. Opt. Soc. Am. B **25**, 1938–1948 (2008)

M.G. Forest, K.T.R. McLaughlin, Onset of oscillations in nonsoliton pulses in nonlinear dispersive fibers. J. Nonlinear Sci. **8**, 43 (1998)

A. Fratalocchi, C. Conti, G. Ruocco, S. Trillo, Free-energy transition in a gas of noninteracting nonlinear wave particles. Phys. Rev. Lett. **101**, 044101 (2008)

J. Garnier, G. Xu, S. Trillo, A. Picozzi, Incoherent dispersive shocks in the spectral evolution of random waves. Phys. Rev. Lett. **111**, 113902 (2013)

N. Ghofraniha, C. Conti, G. Ruocco, S. Trillo, Shocks in nonlocal media. Phys. Rev. Lett. **99**, 043903 (2007)

D. Grischkowsky, E. Courtens, J.A. Armstrong, Observation of self-steepening of optical pulses with possible shock formation (Rb vapour). Phys. Rev. Lett. **31**, 422 (1973)

A. Gurevich, A.L. Krylov, Dissipationless shock waves in media with positive dispersion. Sov. Phys. JETP **65**, 944–953 (1987)

A. Gurevich, L. Pitaevskii, Nonstationary structure of a collisionless shock wave. Sov. Phys. JETP **38**, 291 (1974)

A. Gurevich, A. Shvartsburg, Nonstationary structure of a collisionless shock wave. Sov. Phys. JETP **31**, 1084–1089 (1970)

J.-P. Hamaide, P. Emplit, Direct observation of optical wave breaking of picosecond pulses in nonlinear single-mode optical fibres. Electron. Lett. **24**, 819 (1988)

J.L. Hammack, H. Segur, The Korteweg-de Vries equation and water waves. Part 2. Comparison with experiments. J. Fluid Mech. **65**, 289–314 (1974)

J. Hietarinta, T. Kuusela, B.A. Malomed, Shock waves in the dissipative Toda lattice. J. Phys. A. **28** 3015–3024 (1995)

M. Hoefer, M. Ablowitz, I. Coddington, E. Cornell, P. Engels, V. Schweikhard, Dispersive and classical shock waves in Bose-Einstein condensates and gas dynamics. Phys. Rev. A **74**, 023623 (2006)

P.A.P. Janantha, P. Sprenger, M.A. Hoefer, M. Wu, Observation of self-cavitating envelope dispersive shock waves in yttrium iron garnet thin films. Phys. Rev. Lett. **119**, 024101 (2017)

A. Kamchatnov, *Nonlinear Periodic Waves and Their Modulations: An Introductory Course* (World Scientific, Singapore, 2000)

S. Kamvissis, K.D.T.-R. McLaughlin, P. Miller, *Semiclassical Soliton Ensembles for the Focusing Nonlinear Schrödinger Equation* (Princeton University Press, Princeton, 2003)

V.I. Karpman, *Nonlinear Waves in Dispersive Media* (Pergamon, Oxford, 1975), p. 101

Y.S. Kivshar, B.A. Malomed, Raman-induced optical shocks in nonlinear fibers. Opt. Lett. **18**, 485 (1993)

Y.S. Kivshar, S.K. Turitsyn, Optical double layers. Phys. Rev. A **47**, R3502 (1993)

Y. Kodama, The Whitham equations for optical communications: mathematical theory of NRZ. SIAM J. Appl. Math. **59**, 2162 (1999)

Y. Kodama, S. Wabnitz, Analytical theory of guiding-center nonreturn-to-zero and return-to-zero signal transmission in normally dispersive nonlinear optical fibers. Opt. Lett. **20**, 2291 (1995)

P.D. Lax, *Hyperbolic Systems of Conservation Laws and the Mathematical Theory of Shock Waves* (SIAM, Philadelphia, 1973)

R.J. LeVeque, *Finite-Volume Methods for Hyperbolic Problems* (Cambridge University Press, Cambridge, 2004)

Y. Liu, H. Tu, S.A. Boppart, Wave-breaking-extended fiber supercontinuum generation for high compression ratio transform-limited pulse compression. Opt. Lett. **37**, 2172 (2012)

L.A. Lugiato, R. Lefever, Spatial dissipative structures in passive optical systems. Phys. Rev. Lett. **58**, 2209 (1987)

M.D. Maiden, N.K. Lowman, D.V. Anderson, M.E. Schubert, M.A. Hoefer, Observation of dispersive shock waves, solitons, and their interactions in viscous fluid conduits. Phys. Rev. Lett. **116**, 174501 (2016)

S. Malaguti, A. Corli, S. Trillo, Control of gradient catastrophes developing from dark beams. Opt. Lett. **35**, 4217–4219 (2010)

S. Malaguti, G. Bellanca, S. Trillo, Dispersive wave-breaking in coherently driven passive cavities. Opt. Lett. **39**, 2475–2478 (2014)

G. Millot, S. Pitois, M. Yan, T. Hovhannisyan, A. Bendahmane, T. Hänsch, N. Picquét, Frequency-agile dual-comb spectroscopy. Nat. Photon. **10**, 27–30 (2016)

S. Moiseev, R. Sagdeev, Collisionless shock waves in a plasma in a weak magnetic field. J. Nucl. Energy **5**, 43 (1963)

A. Moro, S. Trillo, Mechanism of wave breaking from a vacuum point in the defocusing nonlinear Schrödinger equation. Phys. Rev. E **89**, 023202 (2014)

Y. Nakamura, H. Bailung, P.K. Shukla, Observation of ion-acoustic shocks in a dusty plasma. Phys. Rev. Lett. **83**, 1602 (1999)

H. Nakatsuka, D. Grischkowsky, A.C. Balant, Nonlinear picosecond-pulse propagation through optical fibers with positive group velocity dispersion. Phys. Rev. Lett. **47**, 910 (1981)

A. Parriaux, M. Conforti, A. Bendahmane, J. Fatome, C. Finot, S. Trillo, N. Pique, G. Millot, Spectral broadening of picosecond pulses forming dispersive shock waves in optical fibers. Opt. Lett. **42**, 3044 (2017)

M.V. Pavlov, Nonlinear Schrödinger equation and the Bogolyubov-Whitham method of averaging. Theor. Math. Phys. **71**, 584 (1987)

D. Peregrine, Calculations of the development of an undular bore. J. Fluid Mech. **25**, 321–330 (1966)

M.L. Quiroga-Teixeiro, Raman-induced asymmetry of wave breaking in optical fibers. Phys. Scr. **51**, 373 (1995)

S. Randoux, F. Gustave, P. Suret, G. El, Optical random Riemann waves in integrable turbulence. Phys. Rev. Lett. **118**, 233901 (2017)

J.E. Rothenberg, Femtosecond optical shocks and wave breaking in fiber propagation. J. Opt. Soc. Am. B **6**, 2392 (1989)

J.E. Rothenberg, D. Grischkowsky, Observation of the formation of an optical intensity shock and wave breaking in the nonlinear propagation of pulses in optical fibers. Phys. Rev. Lett. **62**, 531 (1989)

M. Salerno, B.A. Malomed, V.V. Konotop, Shock wave dynamics in a discrete nonlinear Schrödinger equation with internal losses. Phys. Rev. E **62**, 8651 (2000)

R. Taylor, D. Baker, H. Ikezi, Observation of collisionless electrostatic shocks. Phys. Rev. Lett. **24**, 206 (1970)

J.R. Thompson, R. Roy, Nonlinear dynamics of multiple four-wave mixing processes in a single-mode fiber. Phys. Rev. A **43**, 4987–4996 (1991)

W.J. Tomlinson, R.H. Stolen, A.M. Johnson, Optical wave breaking of pulses in nonlinear optical fibers. Opt. Lett. **10**, 467 (1985)

S. Trillo, A. Valiani, Hydrodynamic instability of multiple four-wave mixing. Opt. Lett. **35**, 3967–3969 (2010)

S. Trillo, G. Deng, G. Biondini, M. Klein, G. Clauss, A. Chabchoub, M. Onorato, Experimental observation and theoretical description of multisoliton fission in shallow water. Phys. Rev. Lett. **117**, 144102 (2016)

B. Varlot, S. Wabnitz, J. Fatome, G. Millot, C. Finot, Experimental generation of optical flaticon pulses. Opt. Lett. **38**, 3899–3902 (2013)

S. Wabnitz, Optical tsunamis: shoaling of shallow water rogue waves in nonlinear fibers with normal dispersion. J. Opt. **15**, 064002 (2013)

W. Wan, S. Jia, J.W. Fleischer, Dispersive superfluid-like shock waves in nonlinear optics. Nat. Phys. **3**, 46–51 (2007)

K.E. Webb, Y.Q. Xu, M. Erkintalo, S.G. Murdoch, Generalized dispersive wave emission in nonlinear fiber optics. Opt. Lett. **38**, 151–153 (2013)

B. Wetzel, D. Bongiovanni, M. Kues, Y. Hu, Z. Chen, J.M. Dudley, S. Trillo, S. Wabnitz, R. Morandotti, Experimental generation of Riemann waves in optics: a route to shock wave control. Phys. Rev. Lett. **117**, 073902 (2016)

J.R. Whinnery, D.T. Miller, F. Dabby, Thermal convection and spherical aberration distortion of laser beams in low-loss liquids. IEEE J. Quantum Electron. **QE-3**, 382 (1967)

G. Whitham, Non-linear dispersive waves. Proc. R. Soc. Lond. A **283**, 238–261 (1965)

G.B. Whitham, *Linear and Nonlinear Waves* (Wiley, New York, 1974)

G. Xu, D. Vocke, D. Faccio, J. Garnier, T. Roger, S. Trillo, A. Picozzi, From coherent shocklets to giant collective incoherent shock waves in nonlocal turbulent flows. Nat. Commun. **6**, 8131 (2015)

G. Xu, A. Mussot, A. Kudlinski, S. Trillo, F. Copie, M. Conforti, Shock wave generation triggered by a weak background in optical fibers. Opt. Lett. **41**, 2656 (2016)

G. Xu, M. Conforti, A. Kudlinski, A. Mussot, S. Trillo, Dispersive dam-break flow of a photon fluid. Phys. Rev. Lett. **118**, 254101 (2017)

N.J. Zabusky, M.D. Kruskal, Interaction of "solitons" in a collisionless plasma and the recurrence of initial states. Phys. Rev. Lett. **15**, 240 (1965)

A Variety of Dynamical Settings in Dual-Core Nonlinear Fibers

10

Boris A. Malomed

Contents

The List of Acronyms	422
Introduction	422
Solitons in Dual-Core Fibers	426
The Symmetry-Breaking Bifurcation (SBB) of Solitons	426
Gap Solitons in Asymmetric Dual-Core Fibers	433
The Coupler with Separated Nonlinearity and Dispersion	436
Two Polarizations of Light in the Dual-Core Fiber	438
Solitons in Linearly Coupled Fiber Bragg Gratings (BGs)	440
Bifurcation Loops for Solitons in Couplers with the Cubic-Quintic (CQ) Nonlinearity	445
Dissipative Solitons in Dual-Core Fiber Lasers	450
Introduction	450
The Exact SP (Solitary-Pulse) Solution	452
Special Cases of Stable SPs (Solitary Pulses)	455
Stability of the Solitary Pulses and Dynamical Effects	456
Interactions Between Solitary Pulses	459
CW (Continuous-Wave) States and Dark Solitons ("Holes")	460
Evolution of Solitary Pulses Beyond the Onset of Instability	461
Soliton Stability in $\mathcal{PT}$ (Parity-Time)-Symmetric Nonlinear Dual-Core Fibers	463
Conclusion	466
References	468

Abstract

The chapter provides a survey of (chiefly, theoretical) results obtained for self-trapped modes (solitons) in various models of one-dimensional optical

---

B. A. Malomed (✉)
Faculty of Engineering, Department of Physical Electronics, School of Electrical Engineering, Tel Aviv University, Tel Aviv, Israel

ITMO University, St. Petersburg, Russia
e-mail: malomed@post.tau.ac.il

© Springer Nature Singapore Pte Ltd. 2019
G.-D. Peng (ed.), *Handbook of Optical Fibers*,
https://doi.org/10.1007/978-981-10-7087-7_70

waveguides based on a pair of parallel guiding cores, which combine the linear inter-core coupling with the intrinsic cubic (Kerr) nonlinearity, anomalous group-velocity dispersion, and, possibly, intrinsic loss and gain in each core. The survey is focused on three main topics: spontaneous breaking of the inter-core symmetry and the formation of asymmetric temporal solitons in dual-core fibers; stabilization of dissipative temporal solitons (essentially, in the model of a fiber laser) by a lossy core parallel-coupled to the main one, which carries the linear gain; and stability conditions for $\mathcal{PT}$ (parity-time)-symmetric solitons in the dual-core nonlinear dispersive coupler with mutually balanced linear gain and loss applied to the two cores.

The List of Acronyms

1D: one-dimensional
2D: two-dimensional
BG: Bragg grating
CGLE: complex Ginzburg-Landau equation
CQ: cubic-quintic (nonlinearity)
CW: continuous-wave (solution)
GPE: Gross-Pitaevskii equation
GS: gap soliton
GVD: group-velocity dispersion
MI: modulational instability
NLSE: nonlinear Schrödinger equation
$\mathcal{PT}$: parity-time (symmetry)
SBB: symmetry-breaking bifurcation
SP: solitary pulse
SPM: self-phase modulation
VA: variational approximation
WDM: wavelength-division multiplexing
XPM: cross-phase modulation

Introduction

One of basic types of optical waveguides is represented by dual-core couplers, in which parallel guiding cores exchange the propagating electromagnetic fields via evanescent fields tunneling across the dielectric barrier separating the cores (Huang 1994). In most cases, the couplers are realized as twin-core optical fibers (Digonnet and Shaw 1982; Trillo et al. 1988) or, in a more sophisticated form, as twin-core structures embedded in photonic crystal fibers (Saitoh et al. 2003). Such double fibers can be drawn by means of an appropriately shaped preform from melt, or fabricated by pressing together two single-mode fibers, with the claddings removed in the contact area. Alternatively, a microstructured fiber with a dual guiding core can be fabricated and used too (MacPherson et al. 2003).

If the intrinsic nonlinearity in the cores is strong enough, the power exchange between them is affected by the intensity of the guided signals (Jensen 1982). This effect may be used as a basis for the design of diverse all-optical switching devices (Friberg et al. 1987, 1988; Heatley et al. 1988; Królikowski and Kivshar 1996; Tsang et al. 2004; Uzunov et al. 1995; Lederer et al. 2008) and other applications, such as nonlinear amplifiers (Malomed et al. 1996a; Chu et al. 1997), stabilization of wavelength-division-multiplexed (WDM) transmission schemes (Nistazakis et al. 2002), logic gates (Wu 2004), and bistable transmission (Chevriaux et al. 2006). Nonlinear couplers also offer a setup for efficient compression of solitons by passing them into a fiber with a smaller value of the group-velocity dispersion (GVD) coefficient: as demonstrated in work (Hatami-Hanza et al. 1997), the highest quality of the soliton's compression is achieved when two fibers with different dispersion coefficients are not directly spliced one into the other, but are connected so as to form a coupler (a necessarily asymmetric one, in this case).

In addition to the simplest dual-core system, realizations of nonlinear couplers have been proposed in many other settings, including the use of the bimodal structure (orthogonal polarizations) of guided light (Trillo and Wabnitz 1988), semiconductor waveguides (Villeneuve et al. 1992), plasmonic media (Hochberg et al. 2004; Petráček 2013; Smirnova et al. 2013), and twin-core Bragg gratings (Mak et al. 1998; Tsofe and Malomed 2007; Sun et al. 2013), to mention just a few. In addition to the ubiquitous Kerr (local cubic) nonlinearity of the core material, the analysis has been developed for systems with nonlinearities of other types, including saturable (Peng et al. 1994), quadratic (alias second-harmonic-generating) (Mak et al. 1997; Shapira et al. 2011), cubic-quintic (CQ) (Albuch and Malomed 2007), and nonlocal cubic interactions (Shi et al. 2012). Unlike the Kerr nonlinearity, more general types of the self-interaction of light can be realized not in fibers (i.e., not in the *temporal domain*), but rather in planar waveguides (i.e., in the *spatial domain*). Theoretical modeling of these settings is facilitated by the fact that the respective nonlinear Schrödinger equations (NLSEs) for the evolution of the local amplitude of the electromagnetic waves takes identical forms in the temporal and spatial domains, with the temporal variable replaced by the transverse spatial coordinate, in the latter case. Generally, dual-core systems are adequately modeled by systems of two linearly coupled NLSEs, in which the linear coupling represents the tunneling of electromagnetic fields between the cores (Jensen 1982; Wright et al. 1989; Snyder et al. 1991).

Further, effective *discretization* of continuous nonlinear couplers may be provided, in the spatial domain too, by the consideration of parallel arrays of discrete waveguides (Herring et al. 2007; Hadžievski et al. 2010; Shi et al. 2013). The coupler concept was also extended for the *spatiotemporal* propagation of light in dual-core planar waveguides, with the one-dimensional (1D) NLSE replaced by its two-dimensional (2D) version, which includes both the temporal and spatial transverse coordinates (Dror and Malomed 2011).

Similar to the double-fiber waveguides for optical waves are dual-core cigar-shaped (strongly elongated) traps for matter waves in atomic Bose-Einstein condensates (BECs) (Strecker et al. 2003). Transmission of matter waves in these settings have been studied theoretically (Gubeskys and Malomed 2007; Matuszewski et al. 2007; Salasnich et al. 2010), making use of the fact that the Gross-Pitaevskii equation (GPE) for the mean-field wave function of the matter waves in BEC (Pethick and Smith 2008) is actually identical to the NLSE for electromagnetic waves in similar optical waveguides.

The abovementioned systems in optics and BEC imply lossless propagation of optical and atomic waves; hence the respective models are based on the NLSEs and GPEs which do not include dissipative terms. On the other hand, loss and gain play an important role in many optical systems, such as fiber lasers. The fundamental model of these systems is based on complex Ginzburg-Landau equations (CGLEs), i.e., an extension of the NLSE with real coefficients replaced by their complex counterparts (van Hecke 2003; Grelu and Akhmediev 2012). Accordingly, dual-core fiber lasers are described by systems of linearly coupled CGLEs (Sigler and Malomed 2005). Dissipative linearly coupled systems with the gain and loss applied to different cores are relevant too, as models admitting stable transmission (Malomed and Winful 1996; Atai and Malomed 1996, 1998b), filtering (Chu et al. 1995a), and nonlinear amplification (Chu et al. 1997) of optical pulses in fiber lasers with the cubic nonlinearity; see a brief review of the topic in Malomed (2007).

A special system is one with exactly equal gain and loss acting in the parallel-coupled cores, which are identical as concerns other coefficients (Driben and Malomed 2011a). Such settings feature the $\mathcal{PT}$ (parity-time) symmetry between the cores (for the general concept of the $\mathcal{PT}$ symmetry; see original works (Bender and Boettcher 1998; Ruschhaupt et al. 2005; El-Ganainy et al. 2007; Berry 2008; Musslimani et al. 2008; Makris et al. 2008; Klaiman et al. 2008; Longhi 2009) and reviews (Bender 2007; Makris et al. 2011; Suchkov et al. 2016; Konotop et al. 2016)). The linear spectrum of the $\mathcal{PT}$-symmetric coupler may remain purely real (i.e., it does not produce decay due to imbalanced loss or blowup due to imbalanced gain), provided that the gain-loss coefficient does not exceed a critical value (in fact, it is exactly equal to the coefficient of the linear coupling between the cores (Driben and Malomed 2011a; Alexeeva et al. 2012); see section "Dissipative Solitons in Dual-Core Fiber Lasers" below). Overall, $\mathcal{PT}$-symmetric settings may be considered as dissipative systems which are able to emulate conservative ones, as they support not only real spectra but also stable soliton families if appropriate nonlinearity is included (Musslimani et al. 2008; Suchkov et al. 2016; Konotop et al. 2016), as is shown below in detail in section "Dissipative Solitons in Dual-Core Fiber Lasers" (generic dissipative systems create isolated nonlinear states (in particular, dissipative solitons Malomed 1987b), which play the role of attractors, rather than continuous families of stable solutions).

Thus, nonlinear dual-core systems represent a vast class of settings relevant to optics, BEC, and other areas, which offer a possibility to model and predict many physically significant effects. As examples of similar systems which are well known in completely different areas of physics, it is relevant to mention tunnel-coupled pairs of long Josephson junction, which are described by systems of linearly coupled sine-Gordon equations (see original works (Mineev et al. 1981; Kivshar and Malomed 1988; Ustinov et al. 1993) and reviews (Makhlin et al. 2001; Savel'ev et al. 2010)), and the propagation of internal waves in stratified liquids with two well-separated interfaces, which are described by pairs of linearly coupled Korteweg de Vries equations, which were derived in various forms (Gear and Grimshaw 1984; Malomed 1987a; Lou et al. 2006; El et al. 2006; Espinosa-Ceron et al. 2012). The purpose of this chapter is to present a reasonably compact review of the corresponding models and results. Because the general topic is very broad, the review is limited to optical waveguides based on dual-core optical fibers. Related settings, such as those based on double planar waveguides and double traps for matter waves in BEC, are briefly mentioned in passing.

A fundamental property of nonlinear couplers with symmetric cores is the *symmetry-breaking bifurcation* (SBB), which destabilizes obvious symmetric modes (sometimes called *supermodes*, as they extend to both individual cores, which support individual modes), and gives rise to asymmetric states. The SBB was theoretically analyzed in detail for temporally uniform states (alias *continuous waves*, CWs) in dual-core nonlinear optical fibers (Snyder et al. 1991) and, in parallel, for self-trapped solitary waves, i.e., temporal solitons in the same system (Wright et al. 1989; Trillo et al. 1989; Paré and Florjańczyk 1990; Maimistov 1991; Chu et al. 1993; Akhmediev and Ankiewicz 1993a; Soto-Crespo and Akhmediev 1993; Tasgal and Malomed 1999; Chiang 1995; Malomed et al. 1996b), as well as for dual-core nonlinear fibers with Bragg gratings (BGs) written on each core (Mak et al. 1998). Some results obtained in this direction were summarized in early review by Romagnoli et al. (1992) and later in Malomed (2002). The SBB analysis was then extended to solitons in couplers with the quadratic (Mak et al. 1997) and CQ (Albuch and Malomed 2007) nonlinearities.

The Kerr nonlinearity in the dual-core system gives rise to the *subcritical* SBB for solitons, with originally unstable branches of emerging asymmetric modes going backward (in the direction of weaker nonlinearity) and then turning forward (for the classification of bifurcations, see Iooss and Joseph 1980). The asymmetric modes retrieve their stability at the turning points. On the other hand, the *supercritical* SBB gives rise to stable branches of asymmetric solitons going in the forward direction. For solitons, the SBB of the latter type occurs in twin-core Bragg gratings (Mak et al. 1998) (see section "Two Polarizations of Light in the Dual-Core Fiber" below) and in the system with the quadratic nonlinearity (Mak et al. 1997). The coupler with the intra-core CQ nonlinearity gives rise to a closed *bifurcation loop*, whose shape may be concave or convex (Albuch and Malomed 2007) (see details below in section "Solitons in Linearly Coupled Fiber Bragg Gratings (BGs)").

In models of nonlinear dual-core couplers, the SBB point can be found in an exact analytical form for the system with the cubic nonlinearity (Wright et al. 1989), and the emerging asymmetric modes were studied by means of the variational approximation (VA) (Paré and Florjańczyk 1990; Uzunov et al. 1995; Peng et al. 1994; Mak et al. 1997, 1998; Dror and Malomed 2011) and numerical calculations

(Akhmediev and Ankiewicz 1993a; Soto-Crespo and Akhmediev 1993); see also reviews Romagnoli et al. (1992) and Malomed (2002).

In addition to the studies of solitons in uniform dual-core systems, the analysis was developed for *fused couplers*, in which the two cores are joined in a narrow segment (Sabini et al. 1989). In the simplest approximation, the corresponding dependence of the coupling strength on coordinate z may be represented by the delta function, $\delta(z)$. Originally, interactions of solitons with a locally fused segment were studied in the temporal domain, *viz.*, for bright (Sabini et al. 1989; Chu et al. 1995b; Mandal and Chowdhury 2005) and dark (Afanasjev et al. 1997) solitons in dual-core optical fibers and fiber lasers (Boskovic et al. 1995). In that case, the coupling affects the solitons only in the course of a short interval of their evolution. The *spatial-domain* optical counterpart of the fused coupler is provided by a dual-core planar waveguide with a narrow coupling segment created along the coordinate (x) perpendicular to the propagation direction (z) (Akhmediev and Ankiewicz 1993b; Li et al. 2012). Dynamics of spatial optical solitons in such settings, including stationary solitons trapped by the fused segment of the coupler and scattering of incident solitons on one or several segments, was analyzed in Harel and Malomed (2014).

The objective of this chapter is to present a review of basic findings produced by studies of models developed for dual-core nonlinear optical fibers and fiber lasers, along the abovementioned directions. The review is chiefly focused on theoretical results, as experimental ones are still missing for solitons, in most cases. In section "Solitons in Dual-Core Fibers," the most fundamental results are summarized for the SBBs of solitons in dual-core fibers with identical cores. Some essential findings for asymmetric waveguides, with different cores, are included too. The results are produced by a combination of numerical and methods and analytical approximations (primarily, the variational approximation, VA). In section "Dissipative Solitons in Dual-Core Fiber Lasers," the results are presented for the creation of stable dissipative solitons in models of fiber lasers, the stabilization being provided by coupling the main core, which carries the linear gain, to a parallel lossy one. Section "Soliton Stability in $\mathcal{PT}$ (Parity-Time)-Symmetric Nonlinear Dual-Core Fibers" is focused on nonlinear dual-core couplers featuring the abovementioned parity-time ($\mathcal{PT}$) symmetry, which is provided by creating mutually balanced gain and loss in two otherwise identical cores, the main issue being stability conditions for the corresponding $\mathcal{PT}$-symmetric solitons (which can be done in an exact analytical form, in this model). The chapter is concluded by section "Conclusion."

Solitons in Dual-Core Fibers

The Symmetry-Breaking Bifurcation (SBB) of Solitons

The Formulation of the Model

The basic model of the symmetric coupler, i.e., a dual-core fiber with equal dispersion and nonlinearity coefficients in the parallel-coupled cores, is represented by a system of linearly coupled NLSEs, which are written here in the scaled form (Trillo et al. 1988), with subscripts standing for partial derivatives:

$$i u_z + \frac{1}{2} u_{\tau\tau} + |u|^2 u + K v = 0, \quad (1)$$

$$i v_z + \frac{1}{2} v_{\tau\tau} + |v|^2 v + K u = 0, \quad (2)$$

where z is the propagation distance, $\tau \equiv t - z/V_{gr}$ is the reduced time (t is the physical time, V_{gr} is the group velocity of the carrier wave (Agrawal 2007)), and u and v are amplitudes of the electromagnetic waves in the two cores; sign $+$ in front of the group-velocity-dispersion (GVD) terms, represented by the second derivatives, implies the anomalous character of the GVD in the fiber (Agrawal 2007), the cubic terms represent the intra-core Kerr effect, and K, which is defined to be positive (actually, it may be scaled to $K \equiv 1$), is the coupling constant accounting for the light exchange between the cores.

An additional effect which can be included in the model represents the temporal dispersion of the inter-core coupling, represented by its own real coefficient, K'. The accordingly modified Eqs. (1) and (2) take the form (Chiang 1995)

$$i u_z + \frac{1}{2} u_{\tau\tau} + |u|^2 u + K v + i K' v_\tau = 0, \quad (3)$$

$$i v_z + \frac{1}{2} v_{\tau\tau} + |v|^2 v + i K u + K' u_\tau = 0. \quad (4)$$

Below, this generalization of the nonlinear-coupler model is not considered in detail, as the analysis has demonstrated that the dispersion of the inter-core coupling does not produce drastic changes in properties of solitons (Rastogi et al. 2002).

Equations (1) and (2) can be derived, by means of the standard variational procedure, from the respective Lagrangian (L), which, in turns, includes the Hamiltonian of the system (H) (Malomed 2002):

$$L = \int_{-\infty}^{+\infty} \left[\frac{i}{2} \left(u^* u_z + v^* v_z \right) d\tau + \text{c.c.} \right] - H, \quad (5)$$

$$H = \int_{-\infty}^{+\infty} \left[\frac{1}{2} \left(|u_\tau|^2 + |v_\tau|^2 \right) - \frac{1}{2} \left(|u|^4 + |v|^4 \right) - K \left(u^* v + u v^* \right) \right], \quad (6)$$

where both $*$ and c.c. stand for the complex-conjugate expressions. The Hamiltonian, along with the integral energy (alias total norm) of the solution,

$$E = \frac{1}{2} \int_{-\infty}^{+\infty} \left(|u|^2 + |v|^2 \right) d\tau, \quad (7)$$

and the total momentum,

$$P = \frac{i}{2} \int_{-\infty}^{+\infty} \left[(uu_\tau^* + vv_\tau^*) + \text{c.c.} \right] d\tau, \tag{8}$$

are dynamical invariants (conserved quantities) of the system.

If the dispersion of the inter-core dispersion is included, see Eqs. (3) and (4), the additional term in the Hamiltonian density in Eq. (6) is $-(iK/2)(u^*v_\tau + v^*u_\tau - uv_\tau^* - vu_\tau^*)$, while the expression for the total energy and momentum keep the same form as defined in Eqs. (7) and (8).

Equations (1) and (2) admit obvious symmetric and antisymmetric soliton solutions (*supermodes*),

$$u = \pm v = a^{-1} \text{sech}\left(\frac{\tau}{a}\right) \exp\left(\frac{iz}{2a^2} \pm iKz\right), \tag{9}$$

where a is an arbitrary width, which determines the total energy (7), and Hamiltonian (6) of the symmetric and antisymmetric states

$$E_{\text{symm-sol}} = 2a^{-1}, \quad H_{\text{symm-sol}} = -\frac{2}{3}a^{-3} \mp 4Ka^{-1}. \tag{10}$$

In the case of $K > 0$ (that may always be fixed by definition), the antisymmetric solitons are unstable (Soto-Crespo and Akhmediev 1993), as they correspond to a maximum, rather than minimum, of the coupling term ($\sim K$) in Hamiltonian (10), therefore they are not considered below. For the symmetric solitons, the SBB, which destabilizes them and replaces them by stable asymmetric solitons, with different energies in the two cores, is an issue of major interest. The onset of the SBB of the symmetric soliton, i.e., the value of the soliton's energy at the SBB point, can be found in an exact analytical form (Wright et al. 1989). To this end, one looks for a general (possibly asymmetric) stationary solution of Eqs. (1) and (2) for solitons with propagation constant k as

$$\{u(z, \tau), v(z, \tau)\} = e^{ikz} \{U(\tau), V(\tau)\}, \tag{11}$$

where real functions U and V satisfy the following ordinary differential equations:

$$\frac{1}{2}\frac{d^2U}{d\tau^2} + U^3 + KV = kU, \tag{12}$$

$$\frac{1}{2}\frac{d^2V}{d\tau^2} + V^3 + KU = kV. \tag{13}$$

The SBB corresponds to the emergence of an *antisymmetric* eigenmode of infinitesimal perturbations,

$$\{\delta U(\tau), \delta V(\tau)\} = \varepsilon \{U_1(\tau), -U_1(\tau)\} \tag{14}$$

(ε is a vanishingly small perturbation amplitude) around the unperturbed symmetric solution of Eqs. (12) and (13), which is taken as per Eq. (9), i.e.,

$$U_0(\tau) = V_0(\tau) \equiv a^{-1}\text{sech}\left(\frac{\tau}{a}\right), \quad k = \frac{1}{2a^2} + K. \tag{15}$$

The linearization of Eqs. (12) and (13) around the exact symmetric state leads to the equation

$$\frac{1}{2}\frac{d^2 U_1}{d\tau^2} + 3U_0^2 U_1 - (K+k)U_1 = 0, \tag{16}$$

which is tantamount to the stationary version of the solvable 1D linear Schrödinger equation with the Pöschl-Teller potential. Then, with the help of well-known results from quantum mechanics (Landau and Lifshitz 1989), it is easy to find that, with the growth of the soliton's energy E, i.e., with the increase of k (see Eqs. (10) and (15)), a nontrivial eigenstate, produced by Eq. (16), appears at Wright et al. (1989)

$$E = E_{\text{bif}} \equiv 4\sqrt{K/3} \approx 2.31\sqrt{K}. \tag{17}$$

Thus, the SBB and destabilization of the symmetric solitons (15) take place precisely at point (17).

Continuous-Wave (CW) States and Their Modulational Instability (MI)

To complete the formulation of the model of the symmetric nonlinear coupler, it is relevant to mention that, besides the solitons, it admits simple continuous-wave (CW) states, with constant U and V in Eq. (11). Indeed, Eqs. (12) and (13) easily produce the full set of CW solutions: symmetric and antisymmetric ones,

$$U_{\text{symm}}^{(\text{CW})} = V_{\text{symm}}^{(\text{CW})} = \sqrt{k-K}, \quad U_{\text{anti}}^{(\text{CW})} = -V_{\text{anti}}^{(\text{CW})} = \sqrt{k+K}, \tag{18}$$

which exist, respectively, at $k > K$ and $k > -K$. With the growth of k, i.e., increase of the CW amplitude, the symmetric state undergoes the SBB at $k = 2K$, giving rise to asymmetric CW states, which exist at $k > 2K$:

$$U_{\text{asymm}}^{(\text{CW})} = \sqrt{\frac{k}{2} + \sqrt{\frac{k^2}{4} - K^2}}, \quad V_{\text{asymm}}^{(\text{CW})} = \sqrt{\frac{k}{2} - \sqrt{\frac{k^2}{4} - K^2}} \tag{19}$$

(and its mirror image, with $U \rightleftarrows V$). The SBB for CW states in models of couplers with more general nonlinearities was studied in detail in Snyder et al. (1991).

However, all the CW states are subject to the *modulational instability* (MI) (Trillo et al. 1989), which, roughly speaking, tends to split the CW into a chain of solitons. While this conclusion is not surprising in the case of the anomalous GVD in Eqs. (1) and (2), as it gives rise to the commonly known MI in the framework of the single

NLSE (Agrawal 2007), all symmetric and asymmetric CW states are modulationally unstable too in the system of linearly coupled NLSEs with the normal sign of the GVD in each equation (Tasgal and Malomed 1999), this MI being produced by the linear coupling. Because of the instability of the CW background, nonlinear couplers, even with normal GVD, cannot support stable dark solitons or domain walls, i.e., delocalized states in the form of two semi-infinite asymmetric CWs, which are transformed into each other by substitution $U \rightleftarrows V$, linked by a transient layer (Malomed 1994). Because bright solitons cannot exist in the case of the normal GVD, the development of the MI in the latter case leads to a state in the form of "optical turbulent" (Tasgal and Malomed 1999).

As concerns the temporal dispersion of the inter-core coupling (see Eqs. (3) and (4)), its effect on the MI of the CW states was studied in Li et al. (2011).

The Variational Approximation (VA) for Solitons

Asymmetric solitons, which emerge at the SBB point, cannot be found in an exact form, but they can be studied by means of the VA. This approach for solitons in nonlinear couplers was developed in works (Paré and Florjańczyk 1990; Maimistov 1991; Chu et al. 1993); see also work (Ankiewicz et al. 1993) which discussed limitations of the VA in this setting. Here, the main findings produced by the VA and their comparison with results obtained by means of numerical methods are presented as per work (Malomed et al. 1996b).

The VA is based on the following trial analytical form (*ansatz*) for the two-soliton soliton:

$$u = A \cos(\theta)\text{sech}\left(\frac{\tau}{a}\right) \exp\left(i(\phi + \psi) + ib\tau^2\right), \quad (20)$$

$$v = A \sin(\theta)\text{sech}\left(\frac{\tau}{a}\right) \exp\left(i(\phi - \psi) + ib\tau^2\right), \quad (21)$$

where real variational parameters, A, θ, a, ϕ, ψ, and b, may be functions of the propagation distance, z. In particular, $A(z)$ and $a(z)$ are common amplitude and width of the two components, chirp $b(z)$ must be introduced, as it is well known (Anderson 1983; Anderson et al. 1988), in the dynamical ansatz which allows evolution of the soliton's width, $\phi(z)$ is an overall phase of the two-component soliton, angle $\theta(z)$ accounts for the distribution of the energy between the components, and $\psi(z)$ is a relative phase between them. The shape and phase parameters form conjugate pairs, viz., (A, ϕ), (θ, ψ), and (a, b).

Note that the ansatz based on Eqs. (20) and (21) assumes that centers of the two components of the soliton are stuck together. This implies that the linear coupling between the two cores is strong, which corresponds to the real physical situation. Nevertheless, it is also possible to consider a case when the linear coupling plays the role of a small perturbation, making a two-component soliton a weakly bound state of two individual NLSE solitons belonging to the two cores (Abdullaev et al. 1989; Kivshar and Malomed 1989b; Cohen 1995).

Switching of a soliton between the two cores of the coupler was considered, on the basis of a full system of variational equations for ansatz (20), (21) in work (Uzunov et al. 1995). It was also demonstrated by Smyth and Worthy (1997) that the approximation for the switching dynamics in the nonlinear coupler can be further improved if the radiation component of the wave field is incorporated into the ansatz.

Here, the consideration is focused on the basic case of static solitons, for which ansatz (20), (21) gives rise to the following variational equations, in which all parameters of ansatz (20) and (21), but overall phase ϕ, are assumed constant:

$$\sin(2\theta)\sin(2\psi) = 0, \tag{22}$$

$$\frac{E}{3a}\cos(2\theta) - K\cot(2\theta)\cos(2\psi) = 0, \tag{23}$$

$$a^{-1} = E\left[1 - \frac{1}{2}\sin^2(2\theta)\right], \tag{24}$$

$$\frac{d\phi}{dz} = -\frac{1}{6a^2} + \frac{2E}{3a}\left(1 - \frac{1}{2}\sin^2(2\theta)\right) + \kappa\sin(2\theta)\cos(2\psi),$$

where E is the soliton's energy, which, according to its definition (7), takes value $E = A^2 a$ for ansatz (20), (21).

As it follows from Eq. (22), the static soliton may have either $\sin(2\theta) = 0$ or $\sin(2\psi) = 0$. According to the underlying ansatz, the former solution implies that all the energy resides in a single core, which contradicts Eqs. (1) and (2), hence this solution is spurious. The latter solution, $\sin(2\psi) = 0$, implies that $\cos(2\psi) = \pm 1$. As mentioned above, the solutions corresponding to $\cos(2\psi) = -1$, i.e., antisymmetric ones, with respect to the two components, are unstable. Therefore, only the case of $\cos(2\psi) = +1$, corresponding to solitons with in-phase components, is considered here. Then, width a can be eliminated by means of Eq. (24), and the remaining equation (24) for the energy-distribution angle θ takes the form of

$$\cos(2\theta)\left[\frac{E^2}{3K}\sin(2\theta)\left(1 - \frac{1}{2}\sin^2(2\theta)\right) - 1\right] = 0. \tag{25}$$

Further analysis reveals that, in the interval $0 < E^2 < E_1^2$, where

$$E_1^2 = (9/4)\sqrt{6}K \approx 5.511\,K, \tag{26}$$

the only relevant solution to Eq. (25) is the symmetric one, with $\theta = \pi/4$ (corresponding to $\cos(2\theta) = 0$) and equal energies in both components, according to Eqs. (20) and (21). When the soliton's energy attains value E_1, predicted by Eq. (26), there emerge *asymmetric* solutions with $\cos(2\theta) = \pm 1/\sqrt{3}$. When E^2 attains a slightly larger value,

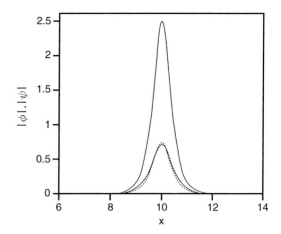

Fig. 1 A typical example of two components of a stable asymmetric soliton, with $|\phi(x)| \equiv U(\tau)$, $|\psi(x)| \equiv V(\tau)$, as per Sakaguchi and Malomed (2011). Continuous and dashed lines designate the numerically found solution and its VA-produced counterpart, respectively

$$E_2^2 = 6K, \qquad (27)$$

a *backward (subcritical) bifurcation* (Iooss and Joseph 1980) occurs, which makes the symmetric solution with $\theta = \pi/4$ unstable. The comparison with full numerical results corroborates the weakly subcritical shape of the SBB for solitons in the nonlinear coupler.

A typical asymmetric soliton is displayed in Fig. 1, and the entire bifurcation diagram is presented in Fig. 2. Note that quantity $\cos(2\theta)$, which is used as the vertical coordinate in the diagram, measures the asymmetry of the soliton, because, as it follows from Eqs. (20) and (21),

$$\cos(2\theta) \equiv \frac{E^{(1)} - E^{(2)}}{E^{(1)} + E^{(2)}}, \qquad (28)$$

where $E^{(j)}$ is the energy in the j-th core. Even without detailed stability analysis, one can easily distinguish between stable and unstable branches in the diagram, using elementary theorems of the bifurcation theory (Iooss and Joseph 1980).

Thus, the VA predicts the backward bifurcation at the soliton's energy $E_2 = \sqrt{6K} \approx 2.45\sqrt{K}$. The accuracy of the VA is characterized by comparison of this prediction with the abovementioned exactly found bifurcation value (17), the relative error being 0.057 (the analytical solutions for E_{bif} does not predict the subcritical character of the SBB).

The above consideration addressed a single soliton in the nonlinear coupler. A cruder version of the VA was used in work (Doty et al. 1995) to analyze two-soliton interactions in the same system. Accurate numerical results for the interaction were reported in Peng et al. (1998). Furthermore, it was recently demonstrated that chains of stable solitons with opposite signs between adjacent ones, in the dual-core fiber, support the propagation of *supersolitons*, i.e., self-trapped collective excitations in the chain of solitons (Li et al. 2014). The underlying chain may be built of symmetric

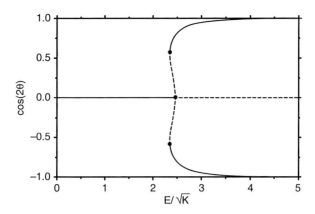

Fig. 2 The dependence of the asymmetry parameter of two-component solitons in the nonlinear coupler with identical cores, $\cos(2\theta)$, on the scaled total energy, $E/\sqrt{K}$, as predicted by the VA, see Eq. (28). The figure demonstrates a weakly subcritical SBB, solid and dashed lines designating stable and unstable states, respectively. The results are presented as per Malomed et al. (1996b)

solitons, as well as of asymmetric ones, with alternating polarities, i.e., placements of larger and smaller components in the two cores.

Gap Solitons in Asymmetric Dual-Core Fibers

Asymmetric dual-core fibers, consisting of two different cores, can be easily fabricated, and properties of solitons in them may be markedly different from those in the symmetric couplers. A general model of the asymmetric coupler is (cf. Eqs. (1), (2))

$$i u_z + q u + \frac{1}{2} u_{\tau\tau} + |u|^2 u + v = 0, \qquad (29)$$

$$i v_z - \delta \cdot \left(q v + \frac{1}{2} v_{\tau\tau} \right) + |v|^2 v + u = 0, \qquad (30)$$

where real parameter $-\delta$ accounts for the difference between GVD coefficients in the cores, and another real coefficient, $(1+\delta)q$, defines the phase-velocity mismatch between them, while a possible group-velocity mismatch can be eliminated in the equations by a simple transformation.

The effect of the asymmetry between the cores on the SBB for solitons was addressed in Malomed et al. (1996b) and Kaup (1997) (strictly speaking, in this case the subject of the analysis is spontaneous breaking of *quasi-symmetry*, which remains after lifting the exact symmetry by the mismatch between the cores). In Kaup (1997), a VA-based analytical approach was elaborated, which showed good agreement with numerical results. A noteworthy feature of the SBB in the asymmetric model is a possibility of hysteresis in a broad region, while in the symmetric system, the hysteresis occurs in the narrow bistability region between the two bifurcation points, as seen in Fig. 2. A systematic analysis of the MI of CW states in the model of asymmetric nonlinear couplers was reported in Arjunan et al. (2017).

The most interesting version of the asymmetric model is one with $\delta > 0$ in Eq. (30), i.e., with *opposite* signs of the GVD (Kaup and Malomed 1998). In this case, the substitution of $u, v \sim \exp(ikz - i\omega\tau)$ in the linearized version of Eqs. (29) and (30) yields the respective dispersion relation,

$$k = \frac{1}{4}(\delta - 1)(\omega^2 - 2q) \pm \sqrt{\frac{1}{16}(\delta + 1)^2(\omega^2 - 2q)^2 + 1}. \tag{31}$$

Self-trapped states may exist, as *gap solitons* (GSs), at values of the propagation constant, k, that belong to the *gap* in spectrum (31), i.e., such that values of ω corresponding to given k, as per Eq. (31), are unphysical (imaginary of complex). The gap always exists in the case of $\delta > 0$, as seen from typical examples of the spectra for negative and positive mismatch q, which are displayed in Fig. 3. If the formal values of ω in the gap are complex, GS's tails decay with oscillations, while for pure imaginary ω they decay monotonously. In particular, it follows from Eq. (31) that, in subgap $0 \leq k^2 < 4\delta/(1+\delta)^2$, the tails always decay with oscillations.

Solutions to Eqs. (29) and (30) for stationary GSs are sought for as $u(z, \tau) = U(\tau)\exp(ikz)$, $v(z, \tau) = V(\tau)\exp(ikz)$, with real U and V determined by equations:

$$(q-k)U + \frac{1}{2}\frac{d^2U}{d\tau^2} + U^3 + V = 0,$$

$$-(\delta q + k)V - \frac{1}{2}\delta\frac{d^2V}{d\tau^2} + V^3 + U = 0. \tag{32}$$

Approximate solutions to Eqs. (32) can be constructed by means of the VA, using the Gaussian ansatz

$$U = A \exp(-\tau^2/2a^2), \quad V = B \exp(-\tau^2/2b^2). \tag{33}$$

Energies of the two components of the soliton corresponding to this ansatz are

$$E_u \equiv \int_{-\infty}^{+\infty} |U(\tau)|^2 \, dt = \sqrt{\pi}A^2 a, \quad E_v \equiv \int_{-\infty}^{+\infty} |V(\tau)|^2 \, dt = \sqrt{\pi}B^2 b, \tag{34}$$

with the net energy $E \equiv E_u + E_v$.

Elimination of amplitudes A and B from the resulting system of variational equations leads to coupled equations for the widths a and b,

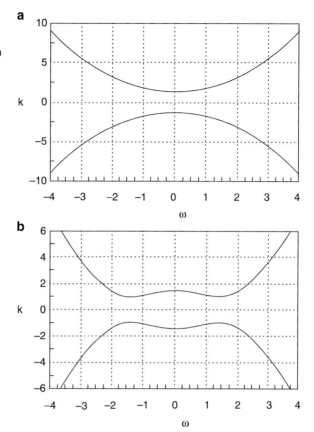

Fig. 3 Typical dispersion curves produced by Eq. (31) for the dual-core fiber with opposite signs of the GVD in the cores, corresponding to Eqs. (29) and (30) with $\delta = 1$: (**a**) $q = -1$; (**b**) $q = +1$ (as per Kaup and Malomed 1998)

$$\left[3 - 4(k-q)a^2\right]\left[3\delta + 4(k+\delta q)b^2\right] = 32(ab)^3\left(b^2 - 3a^2\right)$$
$$\left(3b^2 - a^2\right)\left(a^2 + b^2\right)^{-3}, \tag{35}$$

$$\frac{\left[3 - 4(k-q)a^2\right]\left(3a^2 - b^2\right)^2}{\left[3\delta + 4(k+\delta q)b^2\right]\left(3b^2 - a^2\right)^2} = \frac{a^3\left[\delta\left(3a^2 + b^2\right) + 4(k+\delta q)b^2\left(b^2 - a^2\right)\right]}{b^3\left[(3b^2 + a^2) + 4(k-q)a^2\left(b^2 - a^2\right)\right]}. \tag{36}$$

These equations can be solved numerically, to find a and b as functions of propagation constant k and parameters δ and q.

The results reported in Kaup and Malomed (1998) demonstrate that the GSs indeed exist in a part of the available gap, and, in most cases, they are stable. However, another part of the gap remains *empty* (there are intervals of k in the gap where no soliton can be found). A noteworthy feature of the GSs is that more than half of their net energy *always* resides in the normal-GVD component v, in spite of the obvious fact that the normal-GVD core cannot, by itself, support any

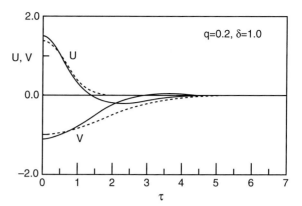

Fig. 4 A numerically found (solid lines) gap soliton solution of Eqs. (29) and (30) with oscillating decaying tails, and its VA-predicted counterpart (dashed lines), in the case of $\delta = 1$ (opposite GVD in the two cores) and $q = 0.2$. The total energy of the numerically found gap soliton is $E = 2.734$ (as per Kaup and Malomed 1998)

bright soliton. Further, a typical GS predicted by the VA (see Fig. 4) has a narrower component with a larger amplitude in the anomalous-GVD core, and a broader component with a smaller amplitude in the normal-GVD one, see Fig. 4.

As is seen from Fig. 4, the VA generally correctly approximates the soliton's core, but the simplest ansatz (33) does not take into regard the fact that, as mentioned above, the soliton's tails decay with oscillations. The contribution of the tails also accounts for a conspicuous difference of the energy share E_v/E in the normal-GVD core from the value predicted by the VA for the same net energy E: for example, in the case shown in Fig. 4, the VA-predicted value is $E_v/E = 0.585$, while its numerically found counterpart is $E_v/E = 0.516$ (but it exceeds $1/2$, as stressed above).

The Coupler with Separated Nonlinearity and Dispersion

For better understanding of the light dynamics in strongly asymmetric nonlinear couplers, it is relevant to consider the model of an extremely asymmetric dual-core waveguide, in which the Kerr nonlinearity is carried by one core, and the GVD is concentrated in the other (Zafrany et al. 2005). Such a system, although looking "exotic," can be created by means of available technologies, adjusting the zero-dispersion point of the first core to the carrier wavelength of the optical signal and using a large effective cross-section area in the first core to suppress its nonlinearity. The respective system of coupled equations is

$$iu_z + |u|^2 u + v = 0, \tag{37}$$

$$iv_z + qv + (D/2)v_{\tau\tau} + u = 0, \tag{38}$$

cf. Eqs. (1) and (2). Here the inter-core coupling coefficient is normalized to be 1, real parameter q, which may be positive or negative, is the phase-velocity mismatch between the cores, and D is the GVD coefficient, that we may be scaled to be $+1$ or -1, which corresponds to the anomalous or normal GVD, respectively. Group-velocity terms, such as $ic_1 u_\tau$ in Eq. (37) and $ic_2 v_\tau$ in Eq. (38), with some real coefficients c_1 and c_2, can be removed: the former one by the shift of the velocity of the references frame, $\tau \to \tau - c_1 z$, and the latter one by the phase transformation, $v \to v \exp(i c_2 \tau / D)$. Therefore, these terms are not included.

Looking for a solution to the linearized version of Eqs. (37) and (38) in the usual form, $\{u, v\} \sim \exp(ikx - i\omega\tau)$ with real ω, one arrives at the dispersion relation,

$$k = -\frac{1}{2}\left(\frac{1}{2}D\omega^2 - q\right) \pm \sqrt{\frac{1}{4}\left(\frac{1}{2}D\omega^2 - q\right)^2 + 1}. \qquad (39)$$

Straightforward consideration of the spectrum defined by this expression demonstrates that, in the case of the anomalous GVD ($D = +1$), it gives rise to *finite* and *semi-infinite gaps*,

$$-\frac{1}{2}\left(\sqrt{4+q^2} - q\right) < k < 0; \quad \frac{1}{2}\left(\sqrt{4+q^2} + q\right) < k < \infty, \qquad (40)$$

and in the case of the normal GVD ($D = -1$), the *semi-infinite* and *finite gaps* are

$$-\infty < k < \frac{1}{2}\left(\sqrt{4+q^2} - q\right); \; 0 < k < \frac{1}{2}\left(\sqrt{4+q^2} + q\right). \qquad (41)$$

Note that both gaps (40) are broader than their counterparts (41) at $q > 0$ and vice versa at $q < 0$.

Equation (39) can be inverted, to yield $\omega^2 = 2(Dk)^{-1}(1 + qk - k^2)$. This relation implies that, inside both the finite and semi-infinite gaps, ω^2 takes real negative values, suggesting a possibility to find exponentially localized solitons in both gaps. To realize this possibility, soliton solutions of Eqs. (37) and (38) were looked for as $\{u, v\} = \exp(ikz)\{U(\tau), V(\tau)\}$. In the case of anomalous GVD, $D = +1$, it was thus found that the semi-infinite gap is *completely filled* by *stable* solitons, while the finite bandgap remains completely empty. This result does not depend on the magnitude and sign of the mismatch parameter, q, in Eq. (38). A typical example of the stable solitons found in the semi-infinite gap is shown in Fig. 5. Naturally, the shape of the soliton in the dispersive mode (V) is much smoother than in the nonlinear one (U). Nevertheless, the shapes of both components are strictly smooth; in particular, there is no true cusp at the tip of the dispersive one.

The fact that the anomalous GVD supports stable solitons in the semi-infinite gap of the present system is not surprising, as the situation seems qualitatively similar to what is commonly known for the usual NLSE, even if the shape of the solitons is very different from that in the NLSE; see Fig. 5. More unexpected is the situation in the case of the normal GVD, $D = -1$ in Eq. (38). As found in Zafrany et al. (2005), in this case the semi-infinite gap remains empty, but the *finite* one is *completely filled* by *stable GSs*; see a typical example in Fig. 6.

It is relevant to mention that Eqs. (37) and (38) are not Galilean invariant. In accordance with this, it was not possible to create moving solitons in the framework of this system.

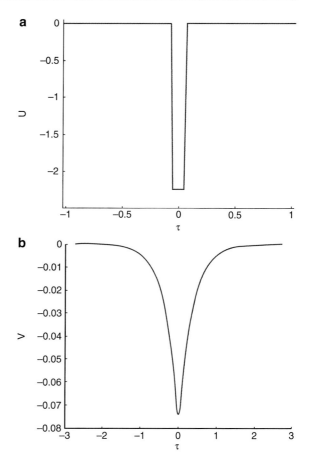

Fig. 5 An example of a stable soliton generated by the two-core system with separated nonlinearity and dispersion, based on Eqs. (37) and (38), with parameters $D = 1$ (the anomalous sign of the GVD) and mismatch $q = 0.8$. The propagation constant corresponding to this soliton is $k = 5$, which places it in the semi-infinite gap, see Eq. (40). Panels (**a**) and (**b**) display, respectively, the nonlinear- and dispersive-mode components of the soliton

The analysis was also extended for the case when the nonlinear mode has weak residual GVD, of either sign (Zafrany et al. 2005). Still earlier, a similar model was considered in Atai and Malomed (2000), which introduced a two-core system with the nonlinearity in one core and a linear BG in the other. That system creates a rather complex spectral structure, featuring three bandgaps and a complex family of soliton solutions, including the so-called *embedded* solitons, which, under special conditions, may exist in (be *embedded into*) spectral bands filled by linear waves, where, generically, solitons cannot exist (Champneys et al. 2001).

Two Polarizations of Light in the Dual-Core Fiber

A relevant extension of the model of the nonlinear coupler takes into regard two linear polarizations of light in each core. In this case, Eqs. (29) and (30) are replaced by a system of four equations Lakoba et al. (1997),

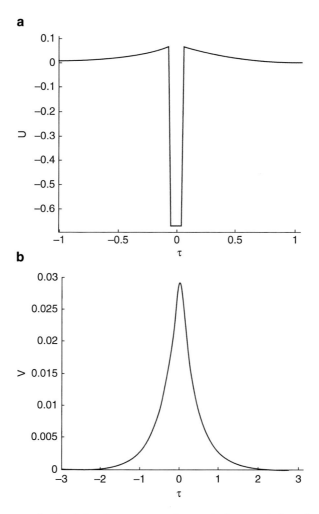

Fig. 6 The same as in Fig. 5, but for parameters $D = -1$ (the normal sign of the GVD) and mismatch $q = 0.8$. The propagation constant of this stable soliton is $k = 0.4$, placing it in the finite bandgap; see Eq. (41)

$$\begin{aligned}
i\,(u_1)_z + \tfrac{1}{2}(u_1)_{\tau\tau} + (|u_1|^2 + \tfrac{2}{3}|v_1|^2)u_1 + u_2 &= 0,\\
i\,(v_1)_z + \tfrac{1}{2}(v_1)_{\tau\tau} + (|v_1|^2 + \tfrac{2}{3}|u_1|^2)v_1 + v_2 &= 0,\\
i\,(u_2)_z + \tfrac{1}{2}(u_2)_{\tau\tau} + (|u_2|^2 + \tfrac{2}{3}|v_2|^2)u_2 + u_1 &= 0,\\
i\,(v_2)_z + \tfrac{1}{2}(v_2)_{\tau\tau} + (|v_2|^2 + \tfrac{2}{3}|u_2|^2)v_2 + v_1 &= 0,
\end{aligned} \qquad (42)$$

where fields u and v represent the two linear polarizations, the subscripts 1 and 2 label the cores, and the coupling coefficient is scaled to be $K \equiv 1$. In the case of circular polarizations (rather than linear ones), the XPM (cross-phase-modulation) coefficient 2/3 in Eq. (42) is replaced by 2.

Four-component soliton solutions to Eqs. (42) can be looked for by means of the VA based on the Gaussian ansatz:

$$u_{1,2}(z,\tau) = A_{1,2}\exp\left(ipz - a^2\tau^2/2\right), \ v_{1,2}(z,\tau) = B_{1,2}\exp\left(iqz - b^2\tau^2/2\right), \tag{43}$$

with mutually independent real propagation constants p and q. Existence regions for all the solutions in the (p,q) plane, produced by the VA for symmetric and asymmetric solitons (the asymmetry is again realized with respect to the two mutually symmetric cores), are displayed in Fig. 7, in the most essential case when the signs of amplitudes $A_{1,2}$ and $B_{1,2}$ in each polarization coincide (otherwise, all the solitons are unstable). Outside the shaded area in Fig. 7, there exist only solutions with a single polarization (i.e., with either $v_{1,2} = 0$ or $u_{1,2} = 0$), which were considered above. In particular, at the dashed-dotted borders of the shaded area, asymmetric four-component solitons (denoted by symbol AS1 in Fig. 7) carry over into the two-component asymmetric solitons of the single-polarization system. The symmetric solitons exist inside the sector bounded by straight continuous lines. The SBB, which gives rise to the asymmetric solitons AS1 and destabilizes the symmetric ones, takes place along the short-dashed curve in the left lower corner of the shaded area.

There is an additional asymmetric soliton (AS2 in Fig. 7) in the inner area confined by the dashed curve. Thus, the total number of soliton solutions changes, as one crosses the bifurcation curves in Fig. 7 from left to right, from 1 to 3 to 5. However, soliton AS2 is generated from the symmetric one by an additional SBB, which takes place *after* the symmetric soliton has already been destabilized by the bifurcation that gives rise to asymmetric soliton AS1. For this reason, soliton AS2 is always unstable, while the primary asymmetric one AS1 is stable. Further details concerning the stability of different solitons in this model can be found in Lakoba and Kaup (1997).

Solitons in Linearly Coupled Fiber Bragg Gratings (BGs)

In the systems described by the single or coupled NLSEs, the second-derivative terms account for the intrinsic GVD of the fiber or waveguide. On the contrary to this, strong *artificial dispersion* can be induced by a BG, i.e., a permanent periodic modulation of the refractive index written along the fiber (usually, the modulation is created in the fiber's cladding), the modulation period being equal to half the wavelength of the propagating light. The nonlinear optical fiber carrying the BG is adequately described by the system of coupled-mode equations for amplitudes $u(x,t)$ and $v(x,t)$ of the right- and left-traveling waves (de Sterke and Sipe 1994; Aceves 2000):

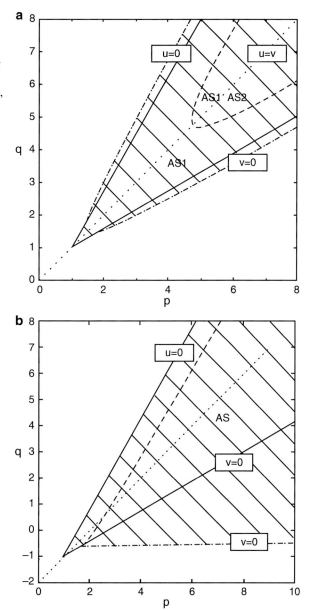

Fig. 7 Regions of existence of the symmetric and two types of asymmetric (stable, AS1, and unstable, AS2) solitons in the plane (p, q) of two propagation constants of four-component solitons (43), in the model (42) of the dual-core-fiber system carrying two linear polarizations of light. Symbols $u = 0$, $v = 0$, and $u = v$ refer to particular solutions with a single polarization and equal amplitudes of the two polarizations, respectively

$$iu_t + iu_x + \left[(1/2)|u|^2 + |v|^2\right]u + v = 0, \tag{44}$$

$$iv_t - iv_x + \left[|u|^2 + (1/2)|v|^2\right]v + u = 0. \tag{45}$$

Here, the speed of light in the fiber's material is scaled to be 1, as well as the linear-coupling constant, that accounts for mutual conversion of the right- and left-traveling waves due to the resonant reflection of light on the BG. The ratio of the XPM and SPM (self-phase-modulation) coefficients in Eqs. (44) and (45), 2 : 1, is the usual feature of the Kerr nonlinearity.

The dispersion relation of the linearized version of Eqs. (44) and (45) is $\omega^2 = 1 + k^2$, hence the existence of GSs (alias BG solitons) with frequencies belonging to the corresponding spectral bandgap, $-1 < \omega < +1$, may be expected. Indeed, although the system of Eqs. (44) and (45) is not integrable, it has a family of exact soliton solutions (Voloshchenko et al. 1981; Aceves and Wabnitz 1989; Christodoulides and Joseph 1989), which contains two nontrivial parameters, *viz.*, amplitude Q, which takes values $0 < Q < \pi$, and velocity c, which belongs to interval $-1 < c < +1$. In particular, the solution for the quiescent solitons ($c = 0$) is

$$\begin{aligned} u &= \sqrt{2/3}\,(\sin Q)\,\text{sech}\left(x \sin Q - \tfrac{1}{2}iQ\right) \cdot \exp\left(-it \cos Q\right), \\ v &= -\sqrt{2/3}\,(\sin Q)\,\text{sech}\left(x \sin Q + \tfrac{1}{2}iQ\right) \cdot \exp\left(-it \cos Q\right), \end{aligned} \tag{46}$$

where frequencies $\omega_{\text{sol}} \equiv \cos Q$ precisely fill the entire gap, while Q varies between 0 and π. Stability of the BG solitons was investigated too, the result being that they are stable, roughly, in a half of the bandgap, namely, at $0 < Q < Q_{\text{cr}} \approx 1.01 \cdot (\pi/2)$ (Malomed and Tasgal 1994; Barashenkov et al. 1998; De Rossi et al. 1998).

A natural generalization of the fiber BG is a system of two parallel-coupled cores with identical gratings written on both of them (Mak et al. 1998). The respective system of four coupled equations can be cast in the following normalized form, cf. Eqs. (44) and (45) for the single-core BG fiber and Eqs. (1) and (2) for the dual-core fiber without the BG:

$$iu_{1t} + iu_{1x} + \left(\frac{1}{2}|u_1|^2 + |v_1|^2\right)u_1 + v_1 + \lambda u_2 = 0, \tag{47}$$

$$iv_{1t} - iv_{1x} + \left(\frac{1}{2}|v_1|^2 + |u_1|^2\right)v_1 + u_1 + \lambda v_2 = 0, \tag{48}$$

$$iu_{2t} + iu_{2x} + \left(\frac{1}{2}|u_2|^2 + |v_2|^2\right)u_2 + v_2 + \lambda u_1 = 0, \tag{49}$$

$$iv_{2t} - iv_{2x} + \left(\frac{1}{2}|v_2|^2 + |u_2|^2\right)v_2 + u_2 + \lambda v_1 = 0, \tag{50}$$

where λ is the coefficient of the linear coupling between the two cores, which may be defined to be positive (unlike the models considered above, it is not possible to fix $\lambda = 1$ by means of rescaling, because the scaling freedom has been already used to fix the Bragg reflection coefficient equal to 1). The same model applies to the spatial-domain propagation in two parallel-coupled planar waveguides which carry BGs in the form of a system of parallel cores, in which case t and x play the roles of the propagation distance and transverse coordinate, respectively, while the paraxial diffraction in the waveguides is neglected.

The dispersion relation for system (47), (48), (49), and (50) contains four branches (taking into regard that ω may have two opposite signs):

$$\omega^2 = \lambda^2 + 1 + k^2 \pm 2\lambda\sqrt{1 + k^2}. \tag{51}$$

This spectrum has no gap in the case of strong inter-core coupling, $\lambda > 1$. For the weaker coupling, with $\lambda < 1$, the bandgap exists:

$$-(1 - \lambda) < \omega < +(1 - \lambda). \tag{52}$$

To populate the bandgap, solutions for zero-velocity GSs are looked for as

$$u_{1,2} = \exp(-i\omega t)\, U_{1,2}(x), \quad v_{1,2} = \exp(-i\omega t)\, V_{1,2}(x), \tag{53}$$

where relation $V_{1,2} = -U_{1,2}^*$ may be imposed (in fact, the exact GS solutions (46) in the single-core BG are subject to the same constraint). Substituting this in Eqs. (47), (48), (49), and (50) leads to two coupled equations (instead of four):

$$\omega U_1 + i\frac{dU_1}{dx} + \frac{3}{2}|U_1|^2 U_1 - U_1^* + \lambda U_2 = 0, \tag{54}$$

$$\omega U_2 + i\frac{dU_2}{dx} + \frac{3}{2}|U_2|^2 U_2 - U_2^* + \lambda U_1 = 0. \tag{55}$$

Stationary equations (54) and (55) can be derived from their own Lagrangian, with density

$$\mathcal{L} = \omega(U_1 U_1^* + U_2 U_2^*) + \frac{i}{2}\left(\frac{dU_1}{dx}U_1^* - \frac{dU_1^*}{dx}U_1 + \frac{dU_2}{dx}U_2^* - \frac{dU_2^*}{dx}U_2\right)$$
$$+ \frac{3}{4}(|U_1|^4 + |U_2|^4) - \frac{1}{2}(U_1^2 + U_1^{*2} + U_2^2 + U_2^{*2}) + \lambda(U_1 U_2^* + U_1^* U_2). \tag{56}$$

Then, in the framework of the VA the following ansatz may be adopted for the *complex* soliton solution sought for:

$$U_{1,2} = A_{1,2}\,\mathrm{sech}(\mu x) + iB_{1,2}\sinh(\mu x)\,\mathrm{sech}^2(\mu x), \tag{57}$$

with real $A_{1,2}$, $B_{1,2}$, and μ. The integration of Lagrangian density (56) with this ansatz and subsequent application of the variational procedure give rise to the following system of equations:

$$3\lambda A_{2,1} - 3(1-\omega)A_{1,2} + 3A_{1,2}^3 + \frac{3}{5}A_{1,2}B_{1,2}^2 - \mu B_{1,2} = 0, \tag{58}$$

$$\lambda B_{2,1} + \frac{3}{2}B_{1,2} - 3.857 B_{1,2}^3 + \frac{3}{5}A_{1,2}^2 B_{1,2} - \mu A_{1,2} = 0, \tag{59}$$

$$2\omega\left(A_1^2 + A_2^2\right) + \frac{2\omega}{3}\left(B_1^2 + B_2^2\right) + \left(A_1^4 + A_2^4\right) - 1.2857\left(B_1^4 + B_2^4\right)$$
$$+ \frac{2}{5}\left(A_1^2 B_1^2 + A_2^2 B_2^2\right) - 2\left(A_1^2 + A_2^2\right) + \frac{2}{3}\left(B_1^2 + B_2^2\right) + 4\lambda A_1 A_2 + \frac{4\lambda}{3}B_1 B_2 = 0, \tag{60}$$

where numerical coefficients 3.857 and 1.2857 are defined by some integrals.

A general result, following from both a numerical solution of variational equations (58), (59), and (60) and direct numerical solution of Eqs. (54) and (55), is that a symmetric mode, with $A_1^2 = A_2^2$ and $B_1^2 = B_2^2$, exists at all values of ω in the bandgap (52), and it is the single soliton solution if the coupling constant λ is close enough to 1, i.e., the bandgap (52) is narrow. However, below a critical value of λ (which depends on given ω), the symmetric solution undergoes a bifurcation, giving rise to three branches, one remaining symmetric, while two new ones, which are mirror images to each other, represent nontrivial *asymmetric* solutions.

The bifurcation can be conveniently displayed in terms of an effective asymmetry parameter,

$$\Theta \equiv \left(U_{1m}^2 - U_{2m}^2\right) / \left(U_{1m}^2 + U_{2m}^2\right), \tag{61}$$

where U_{1m}^2 and U_{2m}^2 are peak powers (maxima of the squared absolute values) of complex fields $U_{1,2}$ in the two cores (note its difference from the asymmetry parameter (28), which was defined in terms of integral energies, rather than peak powers). A complete plot of the SBB for the GSs in the present system, i.e., Θ vs. ω and λ, is displayed in Fig. 8. At $\lambda = 0$, when Eqs. (54) and (55) decouple, the numerical solution matches the exact solution (46) in one core, while the other core is empty. Note the difference of this *supercritical* (alias *forward*) SBB from its weakly *subcritical* (*backward*) counterpart for the solitons in the nonlinear coupler without the BG, which is shown in Fig. 2.

The bifurcation diagram in Fig. 8 was drawn using numerical results obtained from the solution of Eqs. (54) and (55), but its variational counterpart is very close to it, a relative discrepancy between the VA-predicted and numerically exact values of λ, at which the SBB takes place for fixed ω, being $\lesssim 5\%$. To directly illustrate the accuracy of the VA in the present case, comparison between typical shapes of a stable asymmetric soliton, as obtained from the full numerical solution and as predicted by the VA, is presented in Fig. 9.

Direct numerical test of the stability of the symmetric and asymmetric solitons in the present model has yielded results exactly corroborating what may be expected:

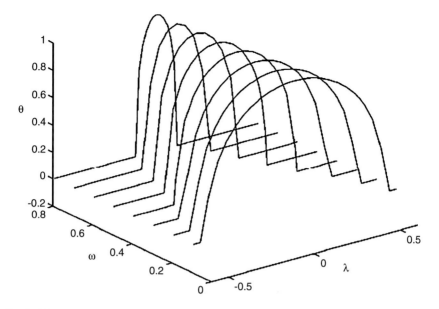

Fig. 8 The symmetry-breaking bifurcation diagram for zero-velocity gap solitons in the model of the dual-core nonlinear optical fiber with identical Bragg gratings written on both cores (as per Mak et al. 1998)

all the asymmetric solitons are stable whenever they exist, while all the symmetric solitons, whenever they coexist with the asymmetric ones, are unstable. However, all the symmetric solitons are stable prior to the bifurcation, where their asymmetric counterparts do not exist.

Lastly, it is relevant to mention that influence of a possible phase shift between the BGs, written in the parallel-coupled cores, on four-component GSs in this system was studied too (Tsofe and Malomed 2007; Sun et al. 2013). In that case, the spontaneous (intrinsic) symmetry breaking is combined with the external symmetry breaking imposed by the mismatch between the BGs.

Bifurcation Loops for Solitons in Couplers with the Cubic-Quintic (CQ) Nonlinearity

To conclude this section, it is relevant to briefly consider results obtained for the coupler with the CQ nonlinearity, i.e., a combination of competing self-focusing cubic and defocusing quintic terms in the respective system of coupled NLSEs:

$$iu_z + u_{\tau\tau} + 2|u|^2 u - |u|^4 u + \lambda v = 0, \tag{62}$$

$$iv_z + v_{\tau\tau} + 2|v|^2 v - |v|^4 v + \lambda u = 0, \tag{63}$$

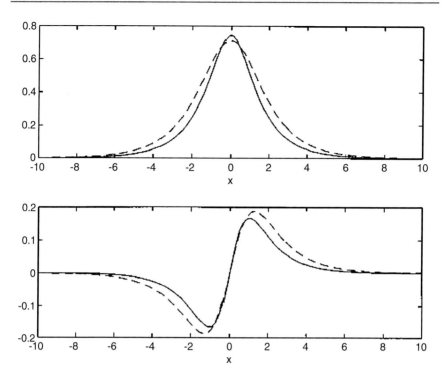

Fig. 9 Shapes of the larger component of the quiescent soliton, U_1, in the dual-core Bragg grating (as per Mak et al. 1998). The upper and lower plots show the real and imaginary parts of U_1. Here, $\omega = 0.5$ and $\lambda = 0.2$. The soliton and dashed curves show numerical and variational results, respectively

cf. the usual system of Eqs. (1) and (2) with the cubic self-focusing (in both systems, the anomalous sign of the GVD is assumed). By means of straightforward rescaling, the coefficients in front of the nonlinear and dispersive terms may be fixed, without the loss of generality, as written in Eqs. (62) and (63), while coefficient $\lambda > 0$ of the linear inter-core coupling remains a free irreducible parameter. Note that the $\mathcal{PT}$-symmetric version of the coupler with the CQ nonlinearity and solitons in it were considered too (Burlak et al. 2016).

The CQ combination of the competing nonlinearities, which is assumed in the present system, occurs in various optical media. The realization which is directly relevant to the fabrication of dual-core fibers is provided by chalcogenide glasses (Smektala et al. 2000; Ogusu et al. 2004).

The starting point of the analysis is a well-known exact soliton solution of the single CQ NLS equation (Pushkarov et al. 1979; Cowan et al. 1986), to which Eqs. (73) and (74) reduce in the symmetric case:

$$u = v = e^{ikz} U_{\text{symm}}(\tau),$$

$$U_{\text{symm}}(\tau) = \sqrt{\frac{2(k-\lambda)}{1+\sqrt{1-4(k-\lambda)/3}\cosh\left(2\sqrt{k-\lambda}\tau\right)}}, \quad (64)$$

where the propagation constant k takes values in the interval of $\lambda < k < \frac{3}{4} + \lambda$. In the limit cases of $k = \lambda$ and

$$k = \frac{3}{4} + \lambda \quad (65)$$

this solution goes over, respectively, into the trivial zero solution and into the delocalized (continuous-wave, CW) state with a constant amplitude, $u = v = \sqrt{\frac{3}{2}}\exp\left(i\left(\frac{3}{4}+\lambda\right)z\right)$. The energy of soliton (64), which is defined by the same expression (7) as above, is

$$E_{\text{symm}} = \frac{\sqrt{3}}{2}\ln\left(\frac{\sqrt{3}+2\sqrt{k-\lambda}}{\sqrt{3}-2\sqrt{k-\lambda}}\right). \quad (66)$$

Naturally, it diverges in the limit corresponding to Eq. (65).

Following the pattern of the above analysis, asymmetric stationary soliton solutions to Eqs. (62) and (63) are looked for as

$$\{u(z,\tau), v(z,\tau)\} = e^{ikz}\{U(\tau), V(\tau)\}. \quad (67)$$

It can be proved (Albuch and Malomed 2007) that only solutions with real functions $U(\tau)$ and $V(\tau)$ can be generated by the SBB from the symmetric soliton (64), hence the substitution of expressions (67) with real U and V in Eqs. (73) and (74) leads to a system

$$\frac{d^2 U}{d\tau^2} - kU + \lambda V + 2U^3 - U^5 = 0, \quad (68)$$

$$\frac{d^2 V}{d\tau^2} - kV + \lambda U + 2V^3 - V^5 = 0, \quad (69)$$

which was solved numerically, and the stability of the so found solitons was then identified by dint of direct simulations (Albuch and Malomed 2007).

The numerical solution of Eqs. (68) and (69) produces a sequence of bifurcation diagrams displayed in Figs. 10 and 11. In these diagrams, the soliton's asymmetry parameter,

$$\epsilon \equiv \frac{U_{\text{max}}^2 - V_{\text{max}}^2}{U_{\text{max}}^2 + V_{\text{max}}^2}, \quad (70)$$

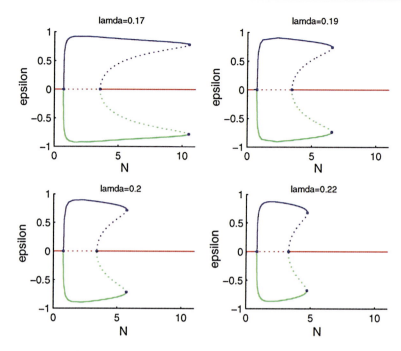

Fig. 10 A set of bifurcation diagrams for symmetric and asymmetric solitons in the plane of the total energy, defined as in Eq. (7), but denoted N here (instead of E), and the asymmetry parameter (70). The diagrams are produced by numerical solution of Eqs. (12) and (13) with the cubic-quintic nonlinearity, at different values of the linear-coupling constant, λ. Stable and unstable branches of the solutions are shown by solid and dashed curves, respectively, and bold dots indicate bifurcation points (as per Albuch and Malomed 2007)

where $U_{\max}^2$ and $V_{\max}^2$ are the peak powers of the two components of the soliton, is shown versus its total energy. Note the similarity of this definition of the asymmetry to that adopted above in the form of Eq. (61) for solitons in the dual-core BG.

A remarkable peculiarity of the present system is the existence of the *bifurcation loop*: as Figs. 10 and 11 demonstrate, the *direct* SBB, which occurs with the increase of the energy, being driven, as above, by the cubic self-focusing, is followed, at larger energies, by the *reverse bifurcation*, which takes place when the dominant nonlinearity becomes self-defocusing, represented by the quintic terms in (68) and (69). The loop exists at $0 < \lambda \leq \lambda_{\max} \approx 0.44$. The direct bifurcation is seen to be always supercritical, while the reverse one, which closes the loop, is subcritical (giving rise to the bistability and concave shape of the loop, on its right-hand side) up to $\lambda \approx 0.40$. In the interval of $0.40 < \lambda < 0.44$, the reverse bifurcation is supercritical, and the (small) loop has a convex form. The picture of the bifurcations is additionally illustrated by Fig. 12, which displays the energy of the symmetric soliton at points of the direct and reverse bifurcations.

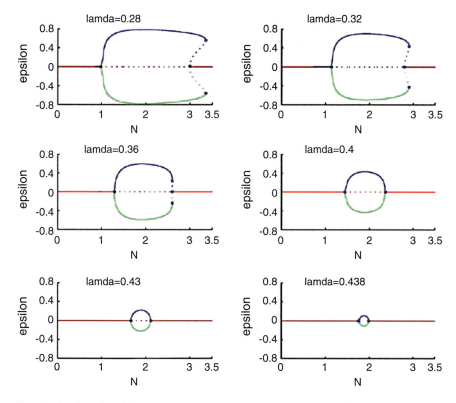

Fig. 11 Continuation of Fig. 10 to larger values of the coupling constant, λ

Stability and instability of different branches of the soliton solutions can be anticipated on the basis of general principles of the bifurcation theory (Iooss and Joseph 1980): the symmetric solution becomes unstable after the direct supercritical bifurcation, and asymmetric solutions emerge as stable ones at this point; eventually, the reverse bifurcation restores the stability of the symmetric solution. In the case when the reverse bifurcation is subcritical and, accordingly, the bifurcation loop is concave on its right side, two branches of asymmetric solutions meet at the turning points, the branches which originate from the reverse-bifurcation point being unstable. These expectations are fully borne out by direct numerical simulations (Albuch and Malomed 2007). In particular, in the case when the bifurcation loop has the concave shape, an unstable asymmetric soliton has a choice to evolve into either a still more asymmetric one or the symmetric soliton (also stable). Numerical results clearly demonstrate that unstable asymmetric solitons choose the former option, evolving into the *more asymmetric* counterparts.

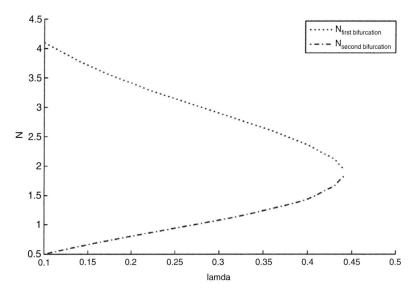

Fig. 12 Values of the energy of the symmetric soliton at which the direct and reverse bifurcations occur in Figs. 10 and 11. The two curves merge and terminate at $\lambda = \lambda_{\max} \approx 0.44$

Dissipative Solitons in Dual-Core Fiber Lasers

Introduction

Experimental and theoretical studies of fiber lasers are, arguably, the fastest developing area of the modern laser science (Richardson et al. 2010). A commonly adopted model for the evolution of optical pulses in fiber lasers is based on complex Ginzburg-Landau equations (CGLEs), which readily predict formation of dissipative solitons, alias solitary pulses (SPs), in the lasers (Grelu and Akhmediev 2012), due to the stable self-sustained balance of loss and gain, the latter provided by the lasing mechanism (typically, stimulated emission of photons by externally pumped ions of rear-earth metals which are embedded as dopants into the fiber's silica (Richardson et al. 2010)). Important applications of the CGLEs are known in many other fields, including hydrodynamics, plasmas, reaction-diffusion systems, etc., as well as other areas of nonlinear optics (Aranson and Kramer 2002; Malomed 2005).

The CGLE of the simplest type is one with the linear dispersive gain and cubic loss (which represents two-photon absorption), combined with the GVD and Kerr nonlinearity. This equation readily produces an exact analytical solution for SPs, in the form of the chirped hyperbolic secant (Hocking and Stewartson 1972; Pereira and Stenflo 1977), but they are unstable, for an obvious reason: the linear gain destabilizes the zero background around the SP. The most straightforward modification which makes the existence of stable SPs possible is the introduction of the cubic-quintic (CQ) nonlinearity, which includes linear loss (hence the zero background is stable), cubic gain (provided by a combination of the usual linear gain and saturable absorption), and additional quintic loss that provides for the overall stabilization of the model. The CGLE with the CQ nonlinearity was first proposed (in a 2D form) by Sergeev and Petviashvili (Petviashvili and Sergeev 1984). A stable SP solution in the 1D version of this equation, which is relevant to modeling fiber lasers, was first reported, in an approximate analytical form, in Malomed (1987b). These solutions were found by treating the dissipation and gain terms in the CGLE as small perturbations added to the usual cubic NLSE with the anomalous sign of the GVD. Accordingly, the SP was obtained as a perturbation of the standard NLSE soliton. In later works, SPs and their stability in the CQ CGLE were investigated in a broad region of parameters (van Saarloos and Hohenberg 1990; Malomed and Nepomnyashchy 1990; Hakim et al. 1990; Marcq et al. 1994; Soto-Crespo et al. 1996).

Another possibility to produce stable SPs, which is directly relevant to the general topic of the present chapter, is to linearly couple the usual cubic CGLE to an additional equation which is dominated by the linear loss. A coupled system of this type was first introduced in Malomed and Winful (1996), as a model of a dual-core nonlinear dispersive optical fiber, with linear gain, γ_0, in one (*active*) core, and linear loss, Γ_0, in the other (*passive*) one:

$$iu_z + \frac{1}{2}u_{\tau\tau} + |u|^2 u - i\gamma_0 u - i\gamma_1 u_{\tau\tau} + \kappa v = 0, \quad (71)$$

$$iv_z + (1/2)v_{\tau\tau} + |v|^2 v + i\Gamma_0 v + \kappa u = 0. \quad (72)$$

Here, u and v are, respectively, envelopes of the electromagnetic waves in the active and passive cores, z and τ are, as above, the propagation distance and reduced time, cf. Eqs. (1) and (2), κ is the constant of the inter-core coupling, and γ_1 accounts for dispersive loss in the active core (in other words, γ_1 determines the *bandwidth-limited* character of the linear gain). The model assumes the usual self-focusing Kerr nonlinearity and anomalous GVD in the fiber, with the respective coefficients scaled to be 1. The gain in the dual-core fiber can be experimentally realized, similar to the usual fiber lasers, by means of externally pumped resonant dopants (Li et al. 2005). Actually, the dual-core fiber may be fabricated as a symmetric one, with both cores doped, while only one core is pumped by an external light source, which gives rise to the gain in that core.

A more general system of linearly coupled CGLEs applies to a system of parallel-coupled plasmonic waveguides (Marini et al. 2011). The system was further extended for coupled 2D CGLEs, representing stabilized laser cavities (Firth and Paulau 2010; Paulau et al. 2010, 2011).

It was first theoretically predicted in Winful and Walton (1992) and Walton and Winful (1993) that, adding the parallel-coupled passive core (with loose ends) to the soliton-generating fiber laser, one can improve the stability of the output: while the soliton, being a self-trapped nonlinear mode, remains essentially confined to the active core, small-amplitude noise easily couples to the passive one, where it is radiated away through the loose ends. Independently, a similar dual-core system, with loss in the additional core but without gain in the main one, was proposed as an optical filter cleaning solitons from noise (Chu et al. 1995a). The basic idea is that, due to the action of the self-focusing nonlinearity, the soliton keeps itself in the core in which it is propagating, while the linear noise tunnels into the parallel core, where it is suppressed by the loss. It was found that the best efficiency of the filtering is attained not with very strong loss (Γ_0) in the extra core, but rather at $\Gamma_0 \sim \kappa$; see Eq. (72) (Chu et al. 1995a). Another possible application of the dual-core system with the gain in the *straight core* (the one into which the input signal is coupled) and losses in the *cross core* (the one linearly coupled to the straight core) was proposed for the design of a nonlinear amplifier of optical signals: a very weak (linear) input would pass into the cross core and would be lost there, while the input whose power exceeds a certain threshold, making it a sufficiently nonlinear mode, stays in the straight core, being amplified there (Malomed et al. 1996a).

In Malomed and Winful (1996), the possibility of the existence of stable SPs in the system of Eqs. (71) and (72) was predicted in the framework of an analytical approximation that treated both the coupling and gain/loss terms as small perturbations. The stability of the so predicted pulses was then verified by direct simulations (Atai and Malomed 1996). Further, it was demonstrated that the system may be simplified, by dropping the nonlinear and GVD terms in Eq. (72), where they are insignificant (the nonlinearity may be omitted as the amplitude of the component in the passive core is small, and the GVD is negligible, as the linear properties of the passive core are dominated by the loss term). On the other hand, an extra linear term, namely, a phase-velocity mismatch between the cores, should be added to Eq. (72), as the respective effect may be essential (see below). As a result, a pair of SP solutions for the simplified system was found in an *exact analytical form* (which is displayed below), the pulse with a larger amplitude being stable in a vast parameter region, while its counterpart with the smaller amplitude is always unstable (Atai and Malomed 1998b; Efremidis et al. 2000a). The basic results are presented, in some detail, in subsections following below. A detailed review of these and related results can be found in Malomed (2007).

The Exact SP (Solitary-Pulse) Solution

The most fundamental coupled system, which has the cubic nonlinearity in the active core only, is based on the following system, which is simplified in comparison with original equations (71) and (72) (Atai and Malomed (1998b); see also Marti-Panameno et al. (2001)).

$$iu_z + \left(\frac{1}{2} - i\gamma_1\right) u_{\tau\tau} + (\sigma + i\gamma_2) |u|^2 u - i\gamma_0 u + v = 0, \tag{73}$$

$$iv_z + k_0 v + i\Gamma_0 v + u = 0, \tag{74}$$

where $\kappa = 1$ is fixed by means of scaling, $\sigma = +1$ and -1 correspond to the anomalous and normal GVD in the active core (assuming that the actual nonlinearity is self-focusing, which is the case in optical fibers, this sign parameter may be placed in front of the cubic term, as written in Eq. (73), although σ originally appears in front of the second derivative), $\gamma_2 \geq 0$ accounts for cubic loss (two-photon absorption), and k_0 is the abovementioned phase-velocity mismatch between the cores.

The *exact* SP solution to Eqs. (73) and (74) can be found in the analytical form suggested by the well-known solution (Hocking and Stewartson 1972; Pereira and Stenflo 1977) of the cubic CGLE:

$$\{u, v\} = \{A, B\} e^{ikz} \left[\text{sech}\left(\chi \tau\right)\right]^{1+i\mu}, \tag{75}$$

where all the constants but B are real. Coefficient μ, which determines the *chirp* of the pulse, is

$$\mu = \frac{\sigma \sqrt{9(1 - 2\sigma \gamma_1 \gamma_2)^2 + 8(2\gamma_1 + \sigma \gamma_2)^2} - 3(1 - 2\sigma \gamma_1 \gamma_2)}{2(2\gamma_1 + \sigma \gamma_2)}. \tag{76}$$

The complex and real amplitudes, B and A, are given by expressions

$$B = (k - k_0 - i\Gamma_0)^{-1} A, \tag{77}$$

$$A^2 = \sigma \frac{\left[(1 - 2\sigma \gamma_1 \gamma_2)(2 - \mu^2) + 3\mu(2\gamma_1 + \sigma \gamma_2)\right] \chi^2}{2(1 + 2\gamma_2^2)}, \tag{78}$$

with the two remaining real parameters χ and k determined by one complex equation,

$$k + i\gamma_0 - (k - k_0 - i\Gamma_0)^{-1} = \left(\frac{1}{2} - i\gamma_1\right)(1 + i\mu)^2 \chi^2. \tag{79}$$

In the further analysis of the SP solutions, one may set $\gamma_2 = 0$, as the two-photon absorption is insignificant in silica fibers. Then, χ^2 can be eliminated from Eq. (79),

$$\chi^2 = \frac{8\gamma_0 \gamma_1}{8\gamma_1^2 + 3 - \sigma \sqrt{9 + 32\gamma_1^2}} \left(1 - \frac{\Gamma_0}{\gamma_0 \left[(k - k_0)^2 + \Gamma_0^2\right]}\right), \tag{80}$$

and one arrives at a final cubic equation for k:

$$k \left[(k - k_0)^2 + \Gamma_0^2\right] - (k - k_0) = \frac{\sigma \sqrt{9 + 32\gamma_1^2}}{2\gamma_1} \left[\gamma_0 (k - k_0)^2 - \Gamma_0 (1 - \gamma_0 \Gamma_0)\right], \tag{81}$$

which may give rise to one or three real solutions. Physical solutions are those which make expression (80) positive. In particular, the number of the physical solutions changes when expression (80) vanishes, which happens at $k_0^2 = (\gamma_0 \Gamma_0)^{-1} (1 - \gamma_0 \Gamma_0) (\Gamma_0 - \gamma_0)^2$.

The above results may be cast in a more explicit form in the case of no wavenumber mismatch between the cores, $k_0 = 0$. Note that the SP solution is of interest if it is stable, a necessary condition for which is the stability of the zero background, i.e., the trivial solution, $u = v = 0$. If $k_0 = 0$, necessary conditions for the stability of the zero solution are $\gamma_0 < \Gamma_0$ and $\gamma_0 \Gamma_0 < 1$. It is natural to focus on the case when gain γ_0 is close to the maximum value, $(\gamma_0)_{\max} \equiv 1/\Gamma_0$, admitted by the latter condition, i.e.,

$$0 < 1 - \gamma_0 \Gamma_0 \ll 1 \tag{82}$$

(then, condition $\gamma_0 < \Gamma_0$ reduces to $\Gamma_0 > 1$). In this case, Eq. (81) may be easily solved. The first root has small k, which yields unphysical solutions, with $\chi^2 < 0$. Two other roots for k are physically relevant ones, in which γ_0 may be replaced by $1/\Gamma_0$, due to relation (82):

$$4k = \frac{\sigma \sqrt{9 + 32\gamma_1^2}}{\Gamma_0 \gamma_1} \pm \sqrt{\frac{9 + 32\gamma_1^2}{\Gamma_0^2 \gamma_1^2} - 16 \left(\Gamma_0^2 - 1\right)}, \tag{83}$$

$$\chi^2 = \frac{8\gamma_1 k^2 \left(k^2 + \Gamma_0^2\right)^{-1}}{\Gamma_0 \left(3 + 8\gamma_1^2 - \sigma \sqrt{9 + 32\gamma_1^2}\right)}. \tag{84}$$

Expression (84) is always positive, while a nontrivial existence condition for these two solutions follows from Eq. (83): k must be real, which means that

$$9\gamma_1^{-2} + 32 > 16\Gamma_0^2 \left(\Gamma_0^2 - 1\right). \tag{85}$$

Thus, Eqs. (83) and (84), along with Eqs. (75), (76), (77), (78), (80), and (81), furnish the SP solutions in the region of the major interest, and Eq. (85) is a fundamental condition which secures the existence of these solutions.

If condition (85) holds, one has the following set of solutions: (i) the stable zero state, (ii) the broader SP with a smaller amplitude, corresponding to smaller k^2, i.e., with sign $\pm$ in expression (83) chosen opposite to σ, and (iii) the narrower pulse with a larger amplitude, corresponding to larger k^2, i.e., with $\pm$ in (83) chosen to coincide with σ. Basic principles of the bifurcation theory (Iooss and Joseph 1980) suggest that stable and unstable solutions alternate, hence, because the trivial solution is stable, the larger-amplitude narrower pulse ought to be stable too, while the intermediate broader pulse with the smaller amplitude is always unstable, playing the role of a *separatrix* between the two *attractors*. This expectation is, generally, corroborated by numerical results (Atai and Malomed 1996, 1998b), as shown in some detail below. Note that the abovementioned SP waveform (Hocking and Stewartson 1972; Pereira and Stenflo 1977), which suggested ansatz (75) for the exact solutions under the consideration, is, by itself, *always unstable* as the solution of the single cubic CGL equation.

Special Cases of Stable SPs (Solitary Pulses)

There are two particular cases of physical interest that should be considered separately. The first corresponds to the model with $\gamma_1 = 0$ (negligible dispersive loss). In this case, the above SP solution may only exist in the system with anomalous GVD, $\sigma = +1$. Special consideration of this case is necessary because the above formulas are singular for $\gamma_1 = 0$. An explicit result for this situation can be obtained *without* adopting condition (82): the solutions take the form of Eq. (75) with $\mu = 0$ (no chirp), Eq. (78) being replaced by $A^2 = \chi^2$, while Eqs. (79) and (78) become

$$k - k_0 = \pm\sqrt{\Gamma_0 \gamma_0^{-1}(1 - \gamma_0 \Gamma_0)},$$

$$\chi^2 = 2k_0 \pm 2(\gamma_0 + \Gamma_0)\sqrt{(\gamma_0 \Gamma_0)^{-1}(1 - \gamma_0 \Gamma_0)}. \qquad (86)$$

Two solutions corresponding to both signs in Eqs. (86) exist simultaneously, i.e., $\chi^2 > 0$ holds for both of them, provided that $k_0^2 > (\gamma_0 \Gamma_0)^{-1}(1 - \gamma_0 \Gamma_0)(\gamma_0 + \Gamma_0)^2$. However, one can check that this inequality contradicts the stability conditions of the zero state, therefore only *one* solution given by Eqs. (86) may exist in the case of interest. General principles of the bifurcation theory (Iooss and Joseph 1980) suggest that this single nontrivial solution is automatically unstable in the case of $\gamma_1 = 0$, once the trivial one is stable.

The other specially interesting case is that of zero GVD, corresponding to the physically important situation when the carrier wavelength is close to the zero-dispersion point (Agrawal 2007) of the optical fiber. In this case, Eq. (74) does not change its form, while Eq. (73) is replaced by

$$iu_z - iu_{\tau\tau} + |u|^2 u - i\gamma_0 u + v = 0 \qquad (87)$$

(in the absence of the GVD, both signs of σ are equivalent, hence $\sigma = +1$ is fixed here, and normalization $\gamma_1 \equiv 1$ may be adopted). Explicit solutions can be obtained, as well as in the general case considered above, by assuming $k_0 = 0$ and taking $\gamma_0 \Gamma_0$ close to 1. Then, expressions (78) and (76) are replaced by $\mu = \sqrt{2}$, $A^2 = 3\sqrt{2}\chi^2$, while solutions (83) and (84) take the form of

$$k = \sqrt{2}\Gamma_0^{-1} \pm \sqrt{2\Gamma_0^{-2} - \Gamma_0^2 + 1}, \quad \chi^2 = \Gamma_0^{-1}\left(\Gamma_0^2 + k^2\right)^{-1} k^2, \qquad (88)$$

and the existence condition (85) becomes very simple, $\Gamma_0^2 < 2$. Thus, in the zero-GVD case, both SP solutions (88) exist simultaneously, suggesting that the one with the larger value of k^2 may be *stable*.

Lastly, it may be interesting to consider another particular case, with normal GVD, $\sigma = -1$, and small dispersive-loss coefficient, $\gamma_1 \ll 1$. In this case, there are two nontrivial SP solutions, the one with the larger amplitude that has the chance to be stable, being

$$\mu = -\frac{3}{2}\gamma_1^{-1}, A^2 = \frac{3}{2}(\gamma_1 \Gamma_0)^{-1}, k = -\frac{3}{2}\gamma_1^{-1}, \chi^2 = \frac{4}{3}\left(\Gamma_0^{-1}\gamma_1\right). \tag{89}$$

The large value of μ in this solution implies that the pulse is strongly chirped. Obviously, the latter solution disappears in the limit of $\gamma_1 \to 0$, in accordance with the abovementioned negative result (no stable SP) for the case of $\gamma_1 = 0$.

Stability of the Solitary Pulses and Dynamical Effects

As mentioned above, the SP cannot be stable unless its background, $u = v = 0$, is stable. To explore the stability of the zero solution, infinitesimal perturbations are substituted in the linearized version of Eqs. (73) and (74), $\{u, v\} = \{u_1, v_1\} e^{i(qz-\omega t)}$, where ω and q are an arbitrary real frequency and the corresponding propagation constant (generally speaking, a complex one). The stability condition is $\text{Im}(q) = 0$, which must hold at all real ω. This condition leads an inequality that should be valid at all $\omega^2 \geq 0$,

$$\Gamma_0(\gamma_0 - \gamma_1 \omega^2)\left[1 + \frac{(2k_0 + \omega^2)^2}{4(\Gamma_0 - \gamma_0 + \gamma_1 \omega^2)^2}\right] \leq 1. \tag{90}$$

If the GVD coefficient is zero, i.e., Eq. (73) is replaced by Eq. (87), expression $(2k_0 + \omega^2)^2$ in Eq. (90) is replaced by $4k_0^2$. An obvious corollary of Eq. (90) is

$$\gamma_0 < \Gamma_0, \tag{91}$$

i.e., the trivial solution may be stable only if the loss in the passive core is stronger than the gain in the active one. Further, in the case of zero wavenumber mismatch between the cores, $k_0 = 0$, which was considered above (see Eqs. (83) and (84)), a simple necessary stability condition is obtained from Eq. (90) at $\omega = 0$ (as was already mentioned above, see Eq. (82)): $\gamma_0 \Gamma_0 > 1$.

Stability regions of the SP solutions in the parameter space of the system were identified in Atai and Malomed (1996) and Efremidis et al. (2000a) in a numerical form, combining the analysis of the necessary stability condition of the zero background, given by Eq. (90), and direct simulations of Eqs. (73) and (74) for perturbed SPs. As said above, the case of major interest is the one with $k_0 = \gamma_2 = 0$

Fig. 13 The stability region (shaded) in the parameter plane of the exact solitary-pulse solution (75) of Eqs. (73) and (74), for the case of $k_0 = \gamma_2 = 0$ and $\gamma_0 \Gamma_0 = 0.9$, as per Atai and Malomed (1998b) and Malomed (2007). Panels (**a**) and (**b**) display results for the anomalous and normal GVD, respectively, i.e., $\sigma = +1$ and $\sigma = -1$ in Eq. (73). The separate curve shows the existence boundary given by Eq. (85)

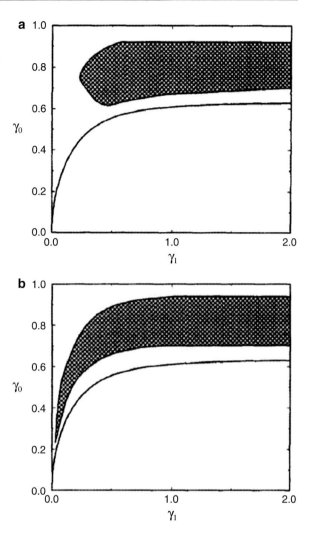

and $\gamma_0 \Gamma_0$ close to 1. For this case, the stability regions are displayed in Fig. 13 and separately in Fig. 14 for the zero-GVD system. In each case, there is a single stable SP (but the system is a *bistable* one, as the stable SP always coexists with the stable zero solution).

Another cross section of the stability region in the full three-dimensional parameter space of the model is represented by region II in Fig. 15, for the normal-GVD case ($\sigma = -1$), with $\gamma_1 = 1/18$ (this value is a typical one corresponding to physically relevant parameters of optical fibers). Note that this plot reveals a very narrow region (area III), in which the zero solution is stable, while the SP is not.

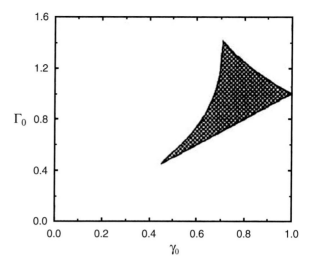

Fig. 14 The same as in Fig. 13, but for the exact solitary-pulse solution (88), in the system with zero dispersion (as per Atai and Malomed 1998b and Malomed 2007)

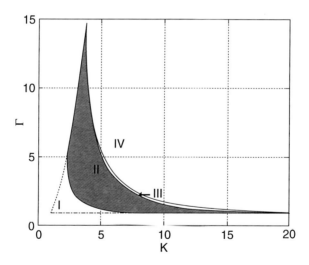

Fig. 15 Stability regions of the zero solution and existence and stability regions for the exact solitary-pulse solution (75) of Eqs. (73) and (74), as per Atai and Malomed (1998b) and Malomed (2007), in the plane of parameters $K \equiv 1/\gamma_0$ and $\Gamma \equiv \Gamma_0/\gamma_0$, in the model with the normal GVD ($\sigma = -1$), $k_0 = \gamma_2 = 0$, and $\gamma_1 = 1/18$. Region I: the zero background is unstable. Region II: the solitary pulse is stable. Region III: the zero background is stable, while the solitary pulse is not, decaying to zero in direct simulations. Region IV: the solitary-pulse solution does not exist. In the region located outside of region I, which is bordered by curves $\gamma_0 \Gamma_0 = 1$ (the dotted curve) and $\Gamma_0 = \gamma_0$ (the dotted-dashed curve), the zero solution is certainly unstable, as condition (82) does not hold in that region. In region I, zero background is unstable even though condition (82) holds in this region

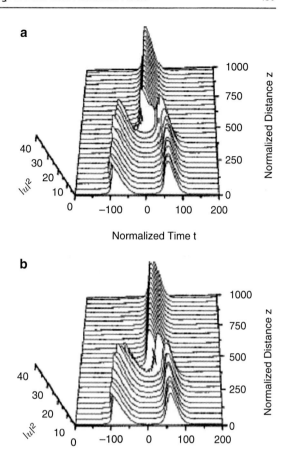

Fig. 16 Merger of chirped solitary pulses in the system of Eqs. (73) and (74) with normal GVD ($\sigma = -1$). Other parameters are $k_0 = \gamma_2 = 0$, $\gamma_0 = 0.2$, $\Gamma_0 = 0.8$, and $\gamma_1 = 1/18$. The initial phase shift between the pulses is $\Delta\phi = 0$ (**a**) and $\Delta\phi = \pi$ (**b**). The results are displayed as per Atai and Malomed (1998a) and Malomed (2007)

Interactions Between Solitary Pulses

It is well known that the sign of the interaction between ordinary solitons is determined by the phase shift between them, $\Delta\phi$: the interaction is attractive for in-phase soliton pairs, with $\Delta\phi = 0$, and repulsive for $\Delta\phi = \pi$ (Kivshar and Malomed 1989a). However, this rule does not apply to the SPs in the present model with the normal sign of the GVD ($\sigma = -1$ in Eq. (73)), which feature strong chirp in their phase structure (see, e.g., the expression for chirp μ in solution (89), with $\gamma_1 \ll 1$). It was found (Efremidis et al. 2000a) that, irrespective of the value of $\Delta\phi$, the SPs in the normal-GVD model *attract* each other and eventually merge into a single pulse, as shown in Fig. 16.

On the other hand, in-phase pairs of the SPs in the anomalous-GVD system, with $\sigma = +1$, readily form robust bound states (Atai and Malomed 1998a). Three-pulse bound states were found too, but they are unstable against symmetry-breaking perturbations, which split them according to the scheme $3 \to 2 + 1$.

CW (Continuous-Wave) States and Dark Solitons ("Holes")

In addition to the SPs, Eqs. (73) and (74) also admit CW states with constant amplitudes,

$$u = a \exp(ikz - i\omega\tau), \quad v = b \exp(ikz - i\omega\tau). \tag{92}$$

The propagation constant and amplitudes of this solution can be easily found in the case of $\gamma_2 = 0$:

$$b = (k - k_0 - i\Gamma_0)^{-1} a,$$

$$k - k_0 = \pm\sqrt{\Gamma_0 \tilde{\gamma}_0^{-1} (1 - \Gamma_0 \tilde{\gamma}_0)},$$

$$\sigma a^2 = k_0 \pm \sqrt{(\Gamma_0 \tilde{\gamma}_0)^{-1} (1 - \Gamma_0 \tilde{\gamma}_0)} (\Gamma_0 - \tilde{\gamma}_0), \tag{93}$$

where $\tilde{\gamma}_0 \equiv \gamma_0 - \gamma_1 \omega^2$. According to this, at given ω there may exist two CW states with different amplitudes, provided that $k_0^2 \geq (\Gamma_0 \tilde{\gamma}_0)^{-1} (1 - \Gamma_0 \tilde{\gamma}_0) (\Gamma_0 - \tilde{\gamma}_0)^2$, and a single one in the opposite case. The CWs may be subject to the MI (modulational instability), which was investigated in detail (Afanasjev et al. 1997; Ganapathy et al. 2006). In particular, all CWs are unstable at $k_0 = 0$, although the character of the MI is different for the normal and anomalous signs of the GVD (Ganapathy et al. 2006). Direct simulations demonstrate that the development of the MI splits the CW state into an array of SPs.

At $k_0 \neq 0$, there is a parameter region in the normal-GVD regime (with $\sigma = -1$) where the CW solutions are stable, which suggests to explore solutions in the form of dark solitons (which are frequently called "holes," in the context of the CGLEs (Nozaki and Bekki 1984; Sakaguchi 1991)). Such solutions of Eqs. (73) and (74) can be found in an exact analytical form based on the following ansatz (Efremidis et al. 2000b) (cf. the form of exact solution (75) for the bright SP):

$$u = \frac{(2 - e^{2\chi\tau}) A}{(1 + e^{2\chi\tau})^{1-i\mu}} e^{ikz - i\mu\chi\tau}, \quad v = \frac{u}{k_0 - k + i\Gamma_0}, \tag{94}$$

with $\mu = (3/4\gamma_1) + \sqrt{(9/4\gamma_1)^2 + 2}$ (in the case of $\gamma_2 = 0$), the other parameters, A, χ, and k, being determined by cumbersome algebraic equations. Numerical analysis demonstrates that dark solitons (94) are stable in a small parameter region, as shown in Fig. 17.

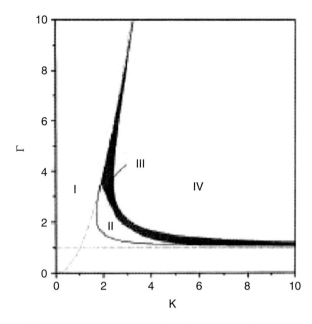

Fig. 17 Regions of the existence and stability of the CW state and dark soliton (94), produced by Eqs. (73) and (74) in the same parameter plane as in Fig. 15, in the case of $\sigma = -1$, $\gamma_2 = 0$, $\gamma_1 = 1/7$, and $k_0 = 2\gamma_2$. The existence region of the CW solution is confined by the dashed lines, $\Gamma_0 = 1$ and $\gamma_0 \Gamma_0 = 1$. In regions I and IV, the CW is unstable. In region II, it is stable, while the dark soliton is not. In region III, both the CW and dark-soliton solutions are stable. The results are displayed as per Efremidis et al. (2000a) and Malomed (2007)

Evolution of Solitary Pulses Beyond the Onset of Instability

Direct simulations of the system with normal GVD, $\sigma = -1$ in Eq. (74), demonstrate that unstable SPs (in the case when stable solutions do not exist) either decay to zero (if the zero background is stable) or blow up, initiating a transition to a "turbulent" state, if the background is unstable. A different behavior of unstable SPs was found in Sakaguchi and Malomed (2000) in the model with the anomalous GVD ($\sigma = +1$). If the instability of the zero background is weak, it does not necessarily lead to the blowup. Instead, it may generate a small-amplitude background field featuring regular oscillations. In that case, the SP sitting on top of such a small-amplitude background remains completely stable, as shown in Fig. 18. The proximity of this state to the stability border is characterized by the *overcriticality parameter*,

$$\epsilon \equiv (\gamma_0 - (\gamma_0)_{\text{cr}}) / (\gamma_0)_{\text{cr}}, \tag{95}$$

where $(\gamma_0)_{\text{cr}}$ is the critical size of the linear gain at which the instability of the zero solution sets in, at given values of γ_1, k_0, and Γ_0. An example of the stable pulse

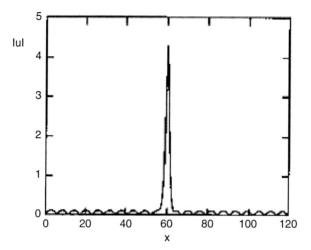

Fig. 18 An example of a stable solitary pulse in the system of Eqs. (73) and (74), with the anomalous GVD ($\sigma = +1$), which exists on top of the small-amplitude background featuring regular oscillations, in the case when the zero background is weakly unstable (the respective overcriticality is $\epsilon = 0.025$; see Eq. (95)). Parameters are $k_0 = 0$, $\gamma_0 = 0.54$, $\Gamma_0 = 1.35$, and $\gamma_1 = 0.18$. The results are displayed as per Sakaguchi and Malomed (2000) and Malomed (2007)

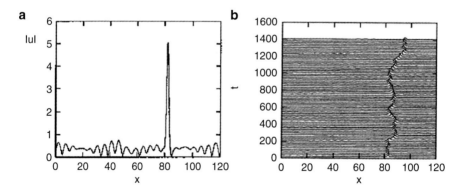

Fig. 19 (a) An example of a solitary pulse which remains stable, as a whole, on top of the background featuring chaotic oscillations, at overcriticality $\epsilon = 0.157$. Parameters are the same as in Fig. 18, except for $\gamma_0 = 0.61$. (b) The random walk of the stable pulse from (a), driven by its interaction with the chaotic background. The results are displayed as per Sakaguchi and Malomed (2000) and Malomed (2007)

found on top of the finite background, which is displayed in Fig. 18, pertains to $\epsilon = 0.025$.

At larger but still moderate values of the overcriticality, such as $\epsilon = 0.157$ in Fig. 19, the background oscillations become chaotic, while keeping a relatively small amplitude. As a result of the interaction with this chaotic background, the

SP remains stable as a whole, featuring a random walk. The walk shows a typically diffusive behavior, with the average squared shift in the τ-direction growing linearly with z (Sakaguchi and Malomed 2000). The randomly walking pulses may easily form bound states, which then feature synchronized random motion. Finally, at essentially larger values of the overcriticality, $\epsilon \gtrsim 1$, the system goes over into a turbulent state, which may be interpreted as a chaotic gas of solitary pulses (Sakaguchi and Malomed 2000).

Soliton Stability in $\mathcal{PT}$ (Parity-Time)-Symmetric Nonlinear Dual-Core Fibers

The dual-core fibers with equal (mutually balanced) gain and loss in the cores offer a natural setting for the realization of the $\mathcal{PT}$ symmetry, in addition to other optical media, where this symmetry was proposed theoretically (Ruschhaupt et al. 2005; El-Ganainy et al. 2007; Musslimani et al. 2008; Makris et al. 2008, 2011; Klaiman et al. 2008; Longhi 2009; Suchkov et al. 2016; Konotop et al. 2016; Burlak et al. 2016) and implemented experimentally (Guo et al. 2009; Ruter et al. 2010).

The basic model of the $\mathcal{PT}$-symmetric nonlinear coupler is based on the equations similar to Eqs. (1) and (2), in which $\gamma > 0$ is the gain-loss coefficient, and the coefficient of the inter-core coupling (K in Eqs. (1) and (2)) is scaled to be 1 (Driben and Malomed 2011a; Alexeeva et al. 2012):

$$i u_z + (1/2)u_{tt} + |u|^2 u - i\gamma u + v = 0, \tag{96}$$

$$i v_z + (1/2)v_{tt} + |v|^2 v + i\gamma v + u = 0. \tag{97}$$

Note that the $\mathcal{PT}$-balanced gain and loss in this system correspond to the border between stable and unstable settings: if the loss coefficient in Eq. (97) is replaced by an independent one, $\Gamma > 0$ (different from the gain factor γ in Eq. (97)), the zero solution, $u = v = 0$, is unstable at $\gamma > \Gamma$ and may be stable at $\gamma < \Gamma$; see Eq. (91).

Obviously, *any* solution to the NLSE (with a frequency shift), $iU_z + (1/2)U_{tt} + |U|^2 U \pm \sqrt{1-\gamma^2} U = 0$, gives rise to two *exact* solutions of the $\mathcal{PT}$-symmetric system, provided that condition $\gamma \leq 1$ holds:

$$v = \left(i\gamma \pm \sqrt{1-\gamma^2}\right) u = U(z,t). \tag{98}$$

For $\gamma = 0$, solutions (98) with + and − amount, respectively, to symmetric and antisymmetric modes in the dual-core coupler, therefore the respective solutions (98) may be called $\mathcal{PT}$-symmetric and $\mathcal{PT}$-antisymmetric ones. In the limit of $\gamma = 1$, two solutions (98) reduce to a single one, $v = iu = U(z,t)$. In particular, $\mathcal{PT}$-symmetric and antisymmetric solitons, with arbitrary amplitude η, are generated by the NLSE solitons,

$$U(z,t) = \eta \exp\left(i\left(\eta^2/2 \pm \sqrt{1-\gamma^2}\right)z\right) \operatorname{sech}(\eta t). \tag{99}$$

As concerns stability of the solitons in this system, it is relevant to compare it to the stability in the usual coupler model, with $\gamma = 0$. As explained above (see Eq. (17)), the symmetric solitons in the nonlinear coupler are unstable against the spontaneous symmetry breaking at

$$\eta^2 > \eta_{\max}^2 (\gamma = 0) \equiv 4/3 \tag{100}$$

(Wright et al. 1989), while antisymmetric solitons are always unstable (Soto-Crespo and Akhmediev 1993) (although their instability may be weak).

The analysis of the instability of the usual two-component symmetric solitons against antisymmetric perturbations, $\delta u = -\delta v$, which leads to the exact result (100) (see Eqs. (14) and (16), can be extended for the $\mathcal{PT}$-symmetric system. The respective perturbation δu at the critical point, $\eta^2 = \eta_{\max}^2$, obeys the linearized equation,

$$\left\{4\sqrt{1-\gamma^2} - d^2/dt^2 + \eta_{\max}^2 \left[1 - 6\operatorname{sech}^2(\eta_{\max}t)\right]\right\}\delta u = 0, \tag{101}$$

which is the respective generalization of Eq. (16). This equation with the Pöschl-Teller potential is solvable, yielding

$$\eta_{\max}^2(\gamma) = (4/3)\sqrt{1-\gamma^2}. \tag{102}$$

This analytical prediction was verified by direct simulations of the perturbed evolution of the $\mathcal{PT}$-symmetric solitons. The simulations were run by adding finite initial antisymmetric perturbations, at the amplitude level of $\pm 3\%$, to the symmetric solitons. For $\mathcal{PT}$-antisymmetric solitons, the instability boundary was identified solely in the numerical form, by running the simulations with initial symmetric perturbations. The results are summarized in Fig. 20, as per Driben and Malomed (2011a). The numerically identified stability border for the symmetric solitons goes somewhat below the analytical one (102) because the finite perturbations used in the simulations are actually not quite small. Taking smaller perturbations, one can obtain the numerical stability border approaching the analytical limit. For instance, at $\gamma = 0.5$, the perturbations with relative amplitudes $\pm 5\%$, $\pm 3\%$, and $\pm 1\%$ give rise to the stability border at $\eta_{\max}^2 = 1.02$, 1.055, and 1.08, respectively, while Eq. (102) yields $\eta_{\max}^2 \approx 1.15$ in the same case. As concerns the $\mathcal{PT}$-antisymmetric solitons, a detailed analysis demonstrated that they are completely unstable (Alexeeva et al. 2012), while the numerically found boundary delineates an area in which the instability is very weak.

It is relevant to stress that, being the stability boundary of the $\mathcal{PT}$-symmetric solitons, the present system, unlike the usual dual-core coupler (see above), cannot support asymmetric solitons, as the balance between the gain and loss is obviously

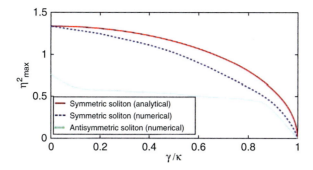

Fig. 20 The analytically predicted stability border (102) for the $\mathcal{PT}$-symmetric solitons and its counterpart produced by systematic simulations of the perturbed evolution of the solitons, as per Driben and Malomed (2011a). An effective numerical stability border for the $\mathcal{PT}$-antisymmetric solitons is shown too, although all the antisymmetric solitons are, strictly speaking, unstable (Alexeeva et al. 2012) (the numerically identified stability area for them implies very weak instability). The solitons with amplitude η are stable at $\eta < \eta_{\max}$. In this figure, γ/κ is identical to γ, as the inter-core coupling coefficient is fixed by scaling to be $\kappa = 1$; see the text

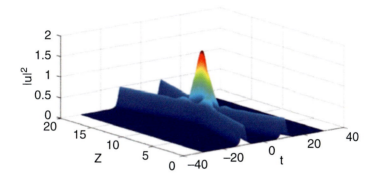

Fig. 21 The elastic collision between stable $\mathcal{PT}$-symmetric solitons with $\eta = 0.7$, boosted by frequency shift $\chi = \pm 5$ at $\gamma = 0.7$; see Eq. (103). The figure is shown as per Driben and Malomed (2011a)

impossible for them. Accordingly, the instability of solitons with $\eta > \eta_{\max}$ leads to a blowup of the pumped field, u, and decay of the attenuated one, v, in direct simulations (not shown here).

The presence of the gain and loss terms in Eqs. (96) and (97) does not break their Galilean invariance, which suggests to consider collisions between moving stable solitons, setting them in motion by means of *boosting*, i.e., replacing

$$\{u, v\} \to \{u, v\} \exp(\pm i \chi t) \qquad (103)$$

in the initial state ($z = 0$), with arbitrary frequency shift χ. Simulations demonstrate that the collisions are always elastic; see a typical example in Fig. 21.

The limit case of $\gamma = 1$ may be considered as one of "supersymmetry," because the inter-core coupling constant and gain-loss coefficients are equal in Eqs. (96) and (97) in this case. According to Eq. (102), the stability region of the solitons in the supersymmetric systems shrinks to nil. The linearization of Eqs. (96) and (97) with $\gamma = 1$ around the NLSE solution (98) leads to the following equations for perturbations δu and δv:

$$\hat{L}(\delta u + i \delta v) = 0, \quad (104)$$

$$\hat{L}(\delta u - i \delta v) = -2i\kappa (\delta u + i \delta v), \quad (105)$$

where $\hat{L}\delta u \equiv \left[i\partial_z + (1/2)\partial_{tt} + 2|U|^2\right]\delta u + U^2 \delta u^*$ is the NLSE linearization operator. If the underlying NLSE solution is stable by itself, Eq. (104) produces no instability, while Eq. (105) gives rise to a *resonance*, as $(\delta u + i\delta v)$ is a zero mode of operator $\hat{L}$. According to the linear-resonance theory (Landau and Lifshitz 1988), the respective perturbation $(\delta u - i\delta v)$ is unstable, growing $\sim z$ (rather than exponentially). Direct simulations of Eqs. (96) and (97) with $\gamma = 1$ confirm that the solitons are unstable, the character of the instability being consistent with its subexponential character (Driben and Malomed 2011a).

The "supersymmetric" solitons may be stabilized by means of the *management* technique (Malomed 2006), which, in the present case, periodically reverses the common sign of the gain-loss and inter-core coupling coefficients, between $\gamma = K = +1$ and -1 (recall K is the coefficient of the inter-core coupling, which was scaled to be 1 above, and may now jump between $+1$ and -1). Flipping γ between $+1$ and -1 implies switch of the gain between the two cores, which is possible in the experiment. The coupling coefficient, κ, cannot flip by itself, but the signal in one core may pass π-shifting plates, which is tantamount to the periodic sign reversal of K.

Ansatz (98) still yields an exact solutions of Eqs. (96) and (97) with coefficients $\gamma = K$ subjected to the periodic management. On the other hand, the replacement of K by the periodically flipping coefficient destroys the resonance in Eq. (105). In simulations, this management mode indeed maintains robust supersymmetric solitons (Driben and Malomed 2011b).

Conclusion

Dual-core optical fibers is a research area which gives rise to a great variety of topics for fundamental theoretical and experimental studies, as well as to a plenty of really existing and potentially possible applications to photonics, including both traditional optics and plasmonics. While currently employed devices based on dual-core waveguides operate in the linear regime (couplers, splitters, etc.), the use of the intrinsic nonlinearity offers many more options, chiefly related to the use of self-trapped robust modes in the form of solitons. In terms of fundamental studies, solitons in dual-core fibers are the subject of dominant interest.

Theoretical studies of solitons in these systems had begun about three decades ago (Friberg et al. 1987, 1988; Heatley et al. 1988; Trillo et al. 1988; Trillo and Wabnitz 1988; Wright et al. 1989; Paré and Florjańczyk 1990; Snyder et al. 1991; Maimistov 1991; Winful and Walton 1992; Walton and Winful 1993; Romagnoli et al. 1992; Akhmediev and Ankiewicz 1993a; Soto-Crespo and Akhmediev 1993; Chu et al. 1993; Ankiewicz et al. 1993; Akhmediev and Ankiewicz 1993b). The earlier works and more recent ones have produced a great advancement in this field, with the help of analytical and numerical methods alike. The present chapter is focused on reviewing the results obtained, chiefly, in three most essential directions: the spontaneous symmetry breaking of solitons in couplers with identical cores and the formation of asymmetric solitons; the creation of stable dissipative solitons in gain-carrying nonlinear fibers (actually, fiber lasers), stabilized by coupling the active (pumped) core to a parallel lossy one; and the stability of solitons in $\mathcal{PT}$-symmetric couplers, with equal strengths of gain and loss carried by the parallel cores.

While the analysis of these areas has been essentially completed, taking into account both early and recent theoretical results, there remain many directions for the extension of the studies. In particular, a natural generalization of dual-core fibers is provided by multi-core arrays, which allow the creation of self-trapped modes which are discrete and continuous along the directions across the fiber array and along the fibers, respectively. These modes include semi-discrete solitons, which may be expected and used in many settings (Aceves et al. 1994a, b, 1995; Matsumoto et al. 1995; Blit and Malomed 2012). Another generalization implies the transition from 1D to 2D couplers, represented by dual-core planar optical waveguides. The consideration of spatiotemporal propagation in such a system makes it possible to predict the existence of novel species of 2D stable "light bullets" (spatiotemporal solitons (Silberberg 1990)), such as spatiotemporal vortices (Dror and Malomed 2011), solitons realizing the optical emulation of spin-orbit coupling (Kartashov et al. 2015; Sakaguchi and Malomed 2016), and 2D $\mathcal{PT}$-symmetric solitons (Burlak and Malomed 2013; Sakaguchi and Malomed 2016).

The most challenging problem is that, as yet, there are very few experimental results reported for solitons in dual-core and multi-core systems. One of experimental findings is the creation of semi-discrete "light bullets" (Minardi et al. 2010), including ones with embedded vorticity (Eilenberger et al. 2013) (actually, in a transient form), in three-dimensional arrays of fiber-like waveguides permanently written in bulk samples of silica. Further development of experimental studies in this vast area is a highly relevant objective.

Acknowledgements I thank Professor Gang-Ding Peng for his invitation to join the production of this volume and to write the present chapter.

References

F.K. Abdullaev, R.M. Abrarov, S.A. Darmanyan, Dynamics of solitons in coupled optical fibers. Opt. Lett. **14**, 131–133 (1989)

A.B. Aceves, Optical gap solitons: past, present, and future; theory and experiments. Chaos **10**, 584–589 (2000)

A.B. Aceves, S. Wabnitz, Self-induced transparency solitons in nonlinear refractive periodic media. Phys. Lett. A **141**, 37–42 (1989)

A.B. Aceves, C. De Angelis, A.M. Rubenchik, S.K. Turitsyn, Multidimensional solitons in fiber arrays. Opt. Lett. **19**, 329–331 (1994a)

A.B. Aceves, C. De Angelis, G.G. Luther, A.M. Rubenchik, Multidimensional solitons in fiber arrays. Opt. Lett. **19**, 1186–1188 (1994b)

A.B. Aceves, G.G. Luther, C. De Angelis, A.M. Rubenchik, S.K. Turitsyn, Energy localization in nonlinear fiber arrays: collapse-effect compressor. Phys. Rev. Lett. **75**, 73–76 (1995)

V.V. Afanasjev, B.A. Malomed, P.L. Chu, Dark soliton generation in a fused coupler. Opt. Commun. **137**, 229–232 (1997)

G.P. Agrawal, *Nonlinear Fiber Optics*, 4th edn. (Academic, San Diego, 2007)

N. Akhmediev, A. Ankiewicz, Novel soliton states and bifurcation phenomena in nonlinear fiber couplers. Phys. Rev. Lett. **70**, 2395–2398 (1993a)

N. Akhmediev, A. Ankiewicz, Spatial soliton X-junctions and couplers. Opt. Commun. **100**, 186–192 (1993b)

L. Albuch, B.A. Malomed, Transitions between symmetric and asymmetric solitons in dual-core systems with cubic-quintic nonlinearity. Math. Comput. Simul. **74**, 312–322 (2007)

N.V. Alexeeva, I.V. Barashenkov, A.A. Sukhorukov, Y.S. Kivshar, Optical solitons in $\mathcal{PT}$-symmetric nonlinear couplers with gain and loss. Phys. Rev. A **85**, 063837 (2012)

D. Anderson, Variational approach to nonlinear pulse propagation in optical fibers. Phys. Rev. A **27**, 3135–3145 (1983)

D. Anderson, M. Lisak, T. Reichel, Asymptotic propagation properties of pulses in a soliton-based optical-fiber communication system. J. Opt. Soc. Am. B **5**, 207–210 (1988)

A. Ankiewicz, N. Akhmediev, G.D. Peng, P.L. Chu, Limitations of the variational approach in soliton propagation in nonlinear couplers. Opt. Commun. **103**, 410 (1993)

I.S. Aranson, L. Kramer, The world of the complex Ginzburg-Landau equation. Rev. Mod. Phys. **74**, 99–143 (2002)

G. Arjunan, B.A. Malomed, M. Arumugam, U. Ambikapathy, Modulational instability in linearly coupled asymmetric dual-core fibers. Appl. Sci. **7**, 645 (2017)

J. Atai, B.A. Malomed, Stability and interactions of solitons in two-component systems. Phys. Rev. E **54**, 4371–4374 (1996)

J. Atai, B.A. Malomed, Bound states of solitary pulses in linearly coupled Ginzburg-Landau equations. Phys. Lett. A **244**, 551–556 (1998a)

J. Atai, B.A. Malomed, Exact stable pulses in asymmetric linearly coupled Ginzburg–Landau equations. Phys. Lett. A **246**, 412–422 (1998b)

J. Atai, B.A. Malomed, Bragg-grating solitons in a semilinear dual-core system. Phys. Rev. E **62**, 8713–8718 (2000)

I.V. Barashenkov, D.E. Pelinovsky, E.V. Zemlyanaya, Vibrations and oscillatory instabilities of gap solitons. Phys. Rev. Lett. **80**, 5117 (1998)

C.M. Bender, Making sense of non-Hermitian Hamiltonians. Rep. Prog. Phys. **70**, 947 (2007)

C.M. Bender, S. Boettcher, Real spectra in non-Hermitian Hamiltonians having $\mathcal{PT}$ symmetry. Phys. Rev. Lett. **80**, 5243–5246 (1998)

M.V. Berry, Optical lattices with $\mathcal{PT}$ symmetry are not transparent. J. Phys. A **41**, 244007 (2008)

R. Blit, B.A. Malomed, Propagation and collisions of semidiscrete solitons in arrayed and stacked waveguides. Phys. Rev. A **86**, 043841 (2012)

A. Boskovic, S.V. Chernikov, J.R. Taylor, Spectral filtering effect of fused fiber couplers in femtosecond fiber soliton lasers. J. Mod. Opt. **42**, 1959–1963 (1995)

G. Burlak, B.A. Malomed, Stability boundary and collisions of two-dimensional solitons in $\mathcal{PT}$-symmetric couplers with the cubic-quintic nonlinearity. Phys. Rev. E **88**, 062904 (2013)

G. Burlak, S. Garcia-Paredes, B.A. Malomed, $\mathcal{PT}$-symmetric couplers with competing cubic-quintic nonlinearities. Chaos **26**, 113103 (2016)

A.R. Champneys, B.A. Malomed, J. Yang, D.J. Kaup, "Embedded solitons": solitary waves in resonance with the linear spectrum. Phys. D **152–153**, 340–354 (2001)

D. Chevriaux, R. Khomeriki, J. Leon, Bistable transmitting nonlinear directional couplers. Mod. Phys. Lett. B **20**, 515–532 (2006)

K.S. Chiang, Intermodal dispersion in two-core optical fibers. Opt. Lett. **20**, 997–999 (1995)

D.N. Christodoulides, R.I. Joseph, Slow Bragg solitons in nonlinear periodic structures. Phys. Rev. Lett. **62**, 1746–1749 (1989)

P.L. Chu, B.A. Malomed, G.D. Peng, Soliton switching and propagation in nonlinear fiber couplers: analytical results. J. Opt. Soc. Am. B **10**, 1379–1385 (1993)

P.L. Chu, G.D. Peng, B.A. Malomed, H. Hatami-Hansa, I.M. Skinner, Time domain soliton filter based on a semidissipative dual-core coupler. Opt. Lett. **20**, 1092–1094 (1995a)

P.L. Chu, Y.S. Kivshar, B.A. Malomed, G.D. Peng, M.L. Quiroga-Teixeiro, Soliton controlling, switching, and splitting in fused nonlinear couplers. J. Opt. Soc. Am. B **12**, 898–903 (1995b)

P.L. Chu, B.A. Malomed, G.D. Peng, Passage of a pulse through a nonlinear amplifier. Opt. Commun. **140**, 289–295 (1997)

G. Cohen, Soliton interaction and stability in nonlinear directional fiber couplers. Phys. Rev. E **52**, 5565–5573 (1995)

S. Cowan, R.H. Enns, S.S. Rangnekar, S.S. Sanghera, Quasi-soliton and other behaviour of the nonlinear cubic-quintic Schrödinger equation. Can. J. Phys. **64**, 311–315 (1986)

A. De Rossi, C. Conti, S. Trillo, Stability, multistability, and wobbling of optical gap solitons. Phys. Rev. Lett. **81**, 85–88 (1998)

C.M. de Sterke, J.E. Sipe, Gap solitons. Prog. Opt. **33**, 203–260 (1994)

M.J.F. Digonnet, H.J. Shaw, Analysis of a tunable single-mode optical fiber coupler. IEEE J. Quantum Electron **18**, 746–754 (1982)

S.L. Doty, J.W. Haus, Y. Oh, R.L. Fork, Soliton interactions on dual-core fibers. Phys. Rev. E **51**, 709–717 (1995)

R. Driben, B.A. Malomed, Stability of solitons in parity–time-symmetric couplers. Opt. Lett. **36**, 4323–4325 (2011a)

R. Driben, B.A. Malomed, Stabilization of solitons in $\mathcal{PT}$ models with supersymmetry by periodic management. EPL **96**, 51001 (2011b)

N. Dror, B.A. Malomed, Symmetric and asymmetric solitons and vortices in linearly coupled two-dimensional waveguides with the cubic-quintic nonlinearity. Phys. D **240**, 526–541 (2011)

N. Efremidis, K. Hizanidis, B.A. Malomed, H.E. Nistazakis, D.J. Frantzeskakis, Stable transmission of solitons in the region of normal dispersion. J. Opt. Soc. Am. B **17**, 952–958 (2000a)

N. Efremidis, K. Hizanidis, H.E. Nistazakis, D.J. Frantzeskakis, B.A. Malomed, Stabilization of dark solitons in the cubic Ginzburg-Landau equation. Phys. Rev. E **62**, 7410–7414 (2000b)

F. Eilenberger, K. Prater, S. Minardi, R. Geiss, U. Röpke, J. Kobelke, K. Schuster, H. Bartelt, S. Nolte, A. Tünnermann, T. Pertsch, Observation of discrete, vortex light bullets. Phys. Rev. X **3**, 041031 (2013)

G.A. El, R.H.G. Grimshaw, N.F. Smyth, Unsteady undular bores in fully nonlinear shallow-water theory. Phys. Fluids **18**, 027104 (2006)

R. El-Ganainy, K.G. Makris, D.N. Christodoulides, Z.H. Musslimani, Theory of coupled optical $\mathcal{PT}$-symmetric structures. Opt. Lett. **32**, 2632–2634 (2007)

A. Espinosa-Ceron, B.A. Malomed, J. Fujioka, R.F. Rodriguez, Symmetry breaking in linearly coupled KdV systems. Chaos **22**, 033145 (2012)

W.J. Firth, P.V. Paulau, Soliton lasers stabilized by coupling to a resonant linear system. Eur. Phys. J. D **59**, 13–21 (2010)

S.R. Friberg, Y. Silberberg, M.K. Oliver, M.J. Andrejco, M.A. Saifi, P.W. Smith, Ultrafast all-optical switching in dual-core fiber nonlinear coupler. Appl. Phys. Lett. **51**, 1135–1137 (1987)

S.R. Friberg, A.M. Weiner, Y. Silberberg, B.G. Sfez, P.S. Smith, Femtosecond switching in dual-core-fiber nonlinear coupler. Opt. Lett. **13**, 904–906 (1988)

R. Ganapathy, B.A. Malomed, K. Porsezian, Modulational instability and generation of pulse trains in asymmetric dual-core nonlinear optical fibers. Phys. Lett. A **354**, 366–372 (2006)

J.A. Gear, R. Grimshaw, Weak and strong-interactions between internal solitary waves. Stud. Appl. Math. **70**, 235–258 (1984)

P. Grelu, N. Akhmediev, Dissipative solitons for mode-locked lasers. Nat. Photonics **6**, 84–92 (2012)

A. Gubeskys, B.A. Malomed, Symmetric and asymmetric solitons in linearly coupled Bose-Einstein condensates trapped in optical lattices. Phys. Rev. A **75**, 063602 (2007)

A. Guo, G.J. Salamo, D. Duchesne, R. Morandotti, M. Volatier-Ravat, V. Aimez, G.A. Siviloglou, D.N. Christodoulides, Observation of $\mathcal{PT}$-symmetry breaking in complex optical potentials. Phys. Rev. Lett. **103**, 093902 (2009)

Lj. Hadžievski, G. Gligorić, A. Maluckov, B.A. Malomed, Interface solitons in one-dimensional locally coupled lattice systems. Phys. Rev. A **82**, 033806 (2010)

V. Hakim, P. Jakobsen, Y. Pomeau, Fronts vs. solitary waves in nonequilibrium systems. Europhys. Lett. **11**, 19–24 (1990)

A. Harel, B.A. Malomed, Interactions of spatial solitons with fused couplers. Phys. Rev. A **89**, 043809 (2014)

H. Hatami-Hanza, P.L. Chu, B.A. Malomed, G.D. Peng, Soliton compression and splitting in double-core nonlinear optical fibers. Opt. Commun. **134**, 59–65 (1997)

D.R. Heatley, E.M. Wright, G.I. Stegeman, Soliton coupler. Appl. Phys. Lett. **53**, 172–174 (1988)

G. Herring, P.G. Kevrekidis, B.A. Malomed, R. Carretero-González, D.J. Frantzeskakis, Symmetry breaking in linearly coupled dynamical lattices. Phys. Rev. E **76**, 066606 (2007)

M. Hochberg, T. Baehr-Jones, C. Walker, A. Scherer, Integrated plasmon and dielectric waveguides. Opt. Exp. **12**, 5481–5486 (2004)

L.M. Hocking, K. Stewartson, On the nonlinear response of a marginally unstable plane parallel flow to a two-dimensional disturbance. Proc. R. Soc. Lond. A **326**, 289–313 (1972)

W.P. Huang, Coupled-mode theory for optical waveguides: an overview. J. Opt. Soc. Am. A **11**, 963–983 (1994)

G. Iooss, D.D. Joseph, *Elementary Stability and Bifurcation Theory* (Springer, Berlin, 1980)

S.M. Jensen, The nonlinear coherent coupler. IEEE J. Quantum Electron **18**, 1580–1583 (1982); A.A. Maier, Optical transistors and bistable devices utilizing nonlinear transmission of light in systems with unidirectional coupled waves. Sov. J. Quantum Electron **12**, 1490–1494 (1982)

Y.V. Kartashov, B.A. Malomed, V.V. Konotop, V.E. Lobanov, L. Torner, Stabilization of solitons in bulk Kerr media by dispersive coupling. Opt. Lett. **40**, 1045–1048 (2015)

D.J. Kaup, B.A. Malomed, Gap solitons in asymmetric dual-core nonlinear optical fibers. J. Opt. Soc. Am. B **15**, 2838–2846 (1998)

D.J. Kaup, T.I. Lakoba, B.A. Malomed, Asymmetric solitons in mismatched dual-core optical fibers. J. Opt. Soc. Am. B **14**, 1199–1206 (1997)

Y.S. Kivshar, B.A. Malomed, Dynamics of fluxons in a system of coupled Josephson junctions. Phys. Rev. B **37**, 9325–9330 (1988)

Y.S. Kivshar, B.A. Malomed, Dynamics of solitons in nearly integrable systems. Rev. Mod. Phys. **61**, 763–915 (1989a)

Y.S. Kivshar, B.A. Malomed, Interaction of solitons in tunnel-coupled optical fibers. Opt. Lett. **14**, 1365–1367 (1989b)

S. Klaiman, U. Günther, N. Moiseyev, Visualization of branch points in $\mathcal{PT}$-symmetric waveguides. Phys. Rev. Lett. **101**, 080402 (2008)

V.V. Konotop, J. Yang, D.A. Zezyulin, Nonlinear waves in $\mathcal{PT}$-symmetric systems. Rev. Mod. Phys. **88**, 035002 (2016)

W. Królikowski, Y.S. Kivshar, Soliton-based optical switching in waveguide arrays. J. Opt. Soc. Am. B **13**, 876–887 (1996)

T.I. Lakoba, D.J. Kaup, Stability of solitons in nonlinear fiber couplers with two orthogonal polarizations. Phys. Rev. E **56**, 4791–4802 (1997)

T.I. Lakoba, D.J. Kaup, B.A. Malomed, Solitons in nonlinear fiber couplers with two orthogonal polarizations. Phys. Rev. E **55**, 6107–6120 (1997)

L.D. Landau, E.M. Lifshitz, *Mechanics* (Nauka Publishers, Moscow, 1988)

L.D. Landau, E.M. Lifshitz, *Quantum Mechanics* (Nauka Publishers, Moscow, 1989)

F. Lederer, G.I. Stegeman, D.N. Christodoulides, G. Assanto, M. Segev, Y. Silberberg, Discrete solitons in optics. Phys. Rep. **463**, 1–126 (2008)

C. Li, G. Xu, L. Ma, N. Dou, H. Gu, An erbium-doped fibre nonlinear coupler with coupling ratios controlled by pump power. J. Opt. A Pure Appl. Opt. **7**, 540–543 (2005)

J.H. Li, K.S. Chiang, K.W. Chow, Modulation instabilities in two-core optical fibers. J. Opt. Soc. Am. B **28**, 1693–1701 (2011)

Y. Li, W. Pang, S. Fu, B.A. Malomed, Two-component solitons under a spatially modulated linear coupling: inverted photonic crystals and fused couplers. Phys. Rev. A **85**, 053821 (2012)

P. Li, L. Li, B.A. Malomed, Multisoliton Newton's cradles and supersolitons in regular and parity-time-symmetric nonlinear couplers. Phys. Rev. E **89**, 062926 (2014)

S. Longhi, Bloch oscillations in complex crystals with $\mathcal{PT}$ symmetry. Phys. Rev. Lett. **103**, 123601 (2009)

S.Y. Lou, B. Tong, H.C. Hu, X.Y. Tang, Coupled KdV equations derived from two-layer fluids. J. Phys. A Math. Gen. **39**, 513–527 (2006)

W.N. MacPherson, J.D.C. Jones, B.J. Mangan, J.C. Knight, P.S.J. Russell, Two-core photonic crystal fibre for Doppler difference velocimetry. Opt. Commun. **223**, 375–380 (2003)

A.I. Maimistov, Propagation of a light pulse in nonlinear tunnel-coupled optical waveguides. Kvantovaya Elektron (Moscow) **18**, 758–761 (1991) [Sov. J. Quantum Electron **21**, 687–690 (1991)]

W.C.K. Mak, B.A. Malomed, P.L. Chu, Soliton coupling in waveguide with quadratic nonlinearity. Phys. Rev. E **55**, 6134–6140 (1997)

W.C.K. Mak, B.A. Malomed, P.L. Chu, Solitary waves in coupled nonlinear waveguides with Bragg gratings. J. Opt. Soc. Am. B **15**, 1685–1692 (1998)

Y. Makhlin, G. Schön, A. Shnirman, Quantum-state engineering with Josephson-junction devices. Rev. Mod. Phys. **73**, 357–400 (2001)

K.G. Makris, R. El-Ganainy, D.N. Christodoulides, Z.H. Musslimani, Beam dynamics in $\mathcal{PT}$ symmetric optical lattices. Phys. Rev. Lett. **100**, 103904 (2008)

K.G. Makris, R. El-Ganainy, D.N. Christodoulides, Z.H. Musslimani, $\mathcal{PT}$-symmetric periodic optical potentials. Int. J. Theor. Phys. **50**, 1019–1041 (2011)

B.A. Malomed, Leapfrogging solitons in a system of coupled Korteweg – de Vries equations. Wave Motion **9**, 401 (1987a)

B.A. Malomed, Evolution of nonsoliton and "quasiclassical" wavetrains in nonlinear Schrödinger and Korteweg – de Vries equations with dissipative perturbations. Phys. D **29**, 155–172 (1987b)

B.A. Malomed, Optical domain walls. Phys. Rev. E **50**, 1565–1571 (1994)

B.A. Malomed, Variational methods in fiber optics and related fields, in *Progress in Optics*, vol. 43, ed. by E. Wolf (North Holland, Amsterdam, 2002), pp. 71–193

B.A. Malomed, Complex Ginzburg-Landau equation, in *Encyclopedia of Nonlinear Science*, ed. by A. Scott (Routledge, New York, 2005), pp. 157–160

B.A. Malomed, *Soliton Management in Periodic Systems* (Springer, New York, 2006)

B.A. Malomed, Solitary pulses in linearly coupled Ginzburg-Landau equations. Chaos **17**, 037117 (2007)

B.A. Malomed, A.A. Nepomnyashchy, Kinks and solitons in the generalized Ginzburg-Landau equation. Phys. Rev. A **42**, 6009–6014 (1990)

B.A. Malomed, R.S. Tasgal, Vibration modes of a gap soliton in a nonlinear optical medium. Phys. Rev. E **49**, 5787–5796 (1994)

B.A. Malomed, H.G. Winful, Stable solitons in two-component active systems. Phys. Rev. E **53**, 5365–5368 (1996)

B.A. Malomed, G.D. Peng, P.L. Chu, A nonlinear optical amplifier based on a dual-core fiber. Opt. Lett. **21**, 330–332 (1996a)

B.A. Malomed, I.M. Skinner, P.L. Chu, G.D. Peng, Symmetric and asymmetric solitons in twin-core nonlinear optical fibers. Phys. Rev. E **53**, 4084–4091 (1996b)

B. Mandal, A.R. Chowdhury, Solitary optical pulse propagation in fused fibre coupler – effect of Raman scattering and switching. Chaos, Solitons Fractals **24**, 557–565 (2005)

P. Marcq, H. Chaté, R. Conte, Exact solutions of the one-dimensional quintic complex Ginzburg-Landau equation. Phys. D **73**, 305–317 (1994)

A. Marini, D.V. Skryabin, B.A. Malomed, Stable spatial plasmon solitons in a dielectric-metal-dielectric geometry with gain and loss. Opt. Exp. **19**, 6616–6622 (2011)

E. Marti-Panameno, L.C. Gomez-Pavon, A. Luis-Ramos, M.M. Mendez-Otero, M.D.I. Castillo, Self-mode-locking action in a dual-core ring fiber laser. Opt. Commun. **194**, 409–414 (2001)

M. Matsumoto, S. Katayama, A. Hasegawa, Optical switching in nonlinear waveguide arrays with a longitudinally decreasing coupling coefficient. Opt. Lett. **20**, 1758–1760 (1995)

M. Matuszewski, B.A. Malomed, M. Trippenbach, Spontaneous symmetry breaking of solitons trapped in a double-channel potential. Phys. Rev. A **75**, 063621 (2007)

S. Minardi, F. Eilenberger, Y.V. Kartashov, A. Szameit, U. Röpke, J. Kobelke, K. Schuster, H. Bartelt, S. Nolte, L. Torner, F. Lederer, A. Tünnermann, T. Pertsch, Three-dimensional light bullets in arrays of waveguides. Phys. Rev. Lett. **105**, 263901 (2010)

M.B. Mineev, G.S. Mkrtchyan, V.V. Shmidt, On some effects in a system of 2 interacting Josephson junctions. J. Low Temp. Phys. **45**, 497–505 (1981)

Z.H. Musslimani, K.G. Makris, R. El-Ganainy, D.N. Christodoulides, Optical solitons in $\mathcal{PT}$ periodic potentials. Phys. Rev. Lett. **100**, 030402 (2008)

H.E. Nistazakis, D.J. Frantzeskakis, J. Atai, B.A. Malomed, N. Efremidis, K. Hizanidis, Multi-channel pulse dynamics in a stabilized Ginzburg-Landau system. Phys. Rev. E **65**, 036605 (2002)

K. Nozaki, N. Bekki, Exact solutions of the generalized Ginzburg-Landau equation. J. Phys. Soc. Jpn. **53**, 1581–1582 (1984)

K. Ogusu, J. Yamasaki, S. Maeda, M. Kitao, M. Minakata, Linear and nonlinear optical properties of Ag-As-Se chalcogenide glasses for all-optical switching. Opt. Lett. **29**, 265–267 (2004)

C. Paré, M. Florjańczyk, Approximate model of soliton dynamics in all-optical fibers. Phys. Rev. A **41**, 6287–6295 (1990)

P.V. Paulau, D. Gomila, P. Colet, N.A. Loiko, N.N. Rosanov, T. Ackemann, W.J. Firth, Vortex solitons in lasers with feedback. Opt. Exp. **18**, 8859–8866 (2010)

P.V. Paulau, D. Gomila, P. Colet, B.A. Malomed, W.J. Firth, From one- to two-dimensional solitons in the Ginzburg-Landau model of lasers with frequency-selective feedback. Phys. Rev. E **84**, 036213 (2011)

G.D. Peng, P.L. Chu, A. Ankiewicz, Soliton propagation in saturable nonlinear fiber couplers – variational and numerical results. Int. J. Nonlin. Opt. Phys. **3**, 69–87 (1994)

G.D. Peng, B.A. Malomed, P.L. Chu, Soliton collisions in a model of a dual-core nonlinear optical fiber. Phys. Scr. **58**, 149–158 (1998)

N.R. Pereira, L. Stenflo, Nonlinear Schrödinger equation including growth and damping. Phys. Fluids **20**, 1733–1734 (1977)

C.J. Pethick, H. Smith, *Bose-Einstein Condensation in Dilute Gases*, 2nd edn. (Cambridge University Press, Cambridge, 2008)

J. Petráček, Nonlinear directional coupling between plasmonic slot waveguides. Appl. Phys. B **112**, 593–598 (2013)

V.I. Petviashvili, A.M. Sergeev, Spiral solitons in active media with excitation thresholds. Dokl. AN SSSR **276**, 1380–1384 (1984) [Sov. Phys. Doklady **29**, 493 (1984)]

K.I. Pushkarov, D.I. Pushkarov, I.V. Tomov, Self-action of light beans in nonlinear media: soliton solutions. Opt. Quant. Electr. **11**, 471–478 (1979)

V. Rastogi, K.S. Chiang, N.N. Akhmediev, Soliton states in a nonlinear directional coupler with intermodal dispersion. Phys. Lett. A **301**, 27–34 (2002)

D.J. Richardson, J. Nilsson, W.A. Clarkson, High power fiber lasers: current status and future perspectives. J. Opt. Soc. Am. B **27**, B63–B92 (2010)

M. Romagnoli, S. Trillo, S. Wabnitz, Soliton switching in nonlinear couplers. Opt. Quantum Electron **24**, S1237–S1267 (1992)

A. Ruschhaupt, F. Delgado, J.G. Muga, Physical realization of $\mathcal{PT}$-symmetric potential scattering in a planar slab waveguide. J. Phys. A Math. Gen. **38**, L171 (2005)

C.E. Ruter, K.G. Makris, R. El-Ganainy, D.N. Christodoulides, M. Segev, D. Kip, Observation of parity-time symmetry in optics. Nat. Phys. **6**, 192–195 (2010)

J.P. Sabini, N. Finalyson, G.I. Stegeman, All-optical switching in nonlinear X junctions. Appl. Phys. Lett. **55**, 1176–1178 (1989)

K. Saitoh, Y. Sato, M. Koshiba, Coupling characteristics of dual-core photonic crystal fiber couplers. Opt. Exp. **11**, 3188–3195 (2003)

H. Sakaguchi, Hole solutions in the complex Ginzburg-Landau equation near a subcritical bifurcation. Progr. Theor. Phys. **86**, 7–12 (1991)

H. Sakaguchi, B.A. Malomed, Breathing and randomly walking pulses in a semilinear Ginzburg-Landau system. Phys. D **147**, 273–282 (2000)

H. Sakaguchi, B.A. Malomed, Symmetry breaking of solitons in two-component Gross-Pitaevskii equations. Phys. Rev. E **83**, 036608 (2011)

H. Sakaguchi, B.A. Malomed, One- and two-dimensional solitons in $\mathcal{PT}$-symmetric systems emulating spin–orbit coupling. New J. Phys. **18**, 105005 (2016)

L. Salasnich, B.A. Malomed, F. Toigo, Competition between the symmetry breaking and onset of collapse in weakly coupled atomic condensates. Phys. Rev. A **81**, 045603 (2010)

S. Savel'ev, V.A. Yampol'skii, A.L. Rakhmanov, F. Nori, Terahertz Josephson plasma waves in layered superconductors: spectrum, generation, nonlinear and quantum phenomena. Rep. Prog. Phys. **73**, 026501 (2010)

A. Shapira, N. Voloch-Bloch, B.A. Malomed, A. Arie, Spatial quadratic solitons guided by narrow layers of a nonlinear material. J. Opt. Soc. Am. B **28**, 1481–1489 (2011)

X. Shi, B.A. Malomed, F. Ye, X. Chen, Symmetric and asymmetric solitons in a nonlocal nonlinear coupler. Phys. Rev. A **85**, 053839 (2012)

X. Shi, F. Ye, B. Malomed, X. Chen, Nonlinear surface lattice coupler. Opt. Lett. **38**, 1064–1066 (2013)

A. Sigler, B.A. Malomed, Solitary pulses in linearly coupled cubic-quintic Ginzburg-Landau equations. Phys. D **212**, 305–316 (2005)

Y. Silberberg, Collapse of optical pulses. Opt. Lett. **22**, 1282–1284 (1990)

F. Smektala, C. Quemard, V. Couderc, A. Barthélémy, Non-linear optical properties of chalcogenide glasses measured by Z-scan. J. Non-Cryst. Solids **274**, 232–237 (2000)

D.A. Smirnova, A.V. Gorbach, I.V. Iorsh, I.V. Shadrivov, Y.S. Kivshar, Nonlinear switching with a graphene coupler. Phys. Rev. B **88**, 045443 (2013)

N.F. Smyth, A.L. Worthy, Dispersive radiation and nonlinear twin-core fibers. J. Opt. Soc. Am. B **14**, 2610–2617 (1997)

A.W. Snyder, D.J. Mitchell, L. Poladian, D.R. Rowland, Y. Chen, Physics of nonlinear fiber couplers. J. Opt. Soc. Am. B **8**, 2102–2112 (1991)

J.M. Soto-Crespo, N. Akhmediev, Stability of the soliton states in a nonlinear fiber coupler. Phys. Rev. E **48**, 4710–4715 (1993)

J.M. Soto-Crespo, N.N. Akhmediev, V.V. Afanasjev, Stability of the pulselike solutions of the quintic complex Ginzburg–Landau equation. J. Opt. Soc. Am. B **13**, 1439–1449 (1996)

K.E. Strecker, G.B. Partridge, A.G. Truscott, R.G. Hulet, Bright matter wave solitons in Bose-Einstein condensates. New J. Phys. **5**, 73.1 (2003)

S.V. Suchkov, A.A. Sukhorukov, J. Huang, S.V. Dmitriev, C. Lee, Y.S. Kivshar, Nonlinear switching and solitons in $\mathcal{PT}$-symmetric photonic systems. Laser Photon. Rev. **10**, 177–213 (2016)

Y. Sun, T.P. White, A.A. Sukhorukov, Coupled-mode theory analysis of optical forces between longitudinally shifted periodic waveguides. J. Opt. Soc. Am. B **30**, 736–742 (2013)

R.S. Tasgal, B.A. Malomed, Modulational instabilities in the dual-core nonlinear optical fiber. Phys. Scr. **60**, 418–422 (1999)

S. Trillo, S. Wabnitz, Coupling instability and power-induced switching with 2-core dual-polarization fiber nonlinear coupler. J. Opt. Soc. Am. B **5**, 483–491 (1988)

S. Trillo, S. Wabnitz, E.M. Wright, G.I. Stegeman, Soliton switching in fiber nonlinear directional couplers. Opt. Lett. **13**, 672–674 (1988)

S. Trillo, G. Stegeman, E. Wright, S. Wabnitz, Parametric amplification and modulational instabilities in dispersive nonlinear directional couplers with relaxing nonlinearity. J. Opt. Soc. Am. B **6**, 889–900 (1989)

S.C. Tsang, K.S. Chiang, K.W. Chow, Soliton interaction in a two-core optical fiber. Opt. Commun. **229**, 431–439 (2004)

Y.J. Tsofe, B.A. Malomed, Quasisymmetric and asymmetric gap solitons in linearly coupled Bragg gratings with a phase shift. Phys. Rev. E **75**, 056603 (2007)

A.V. Ustinov, H. Kohlstedt, M. Cirillo, N.F. Pedersen, G. Hallmanns, G. Heiden, Coupled fluxon modes in stacked Nb/AlO$_x$/Nb long Josephson junctions. Phys. Rev. B **48**, 10614–10617 (1993)

I.M. Uzunov, R. Muschall, M. Gölles, Y.S. Kivshar, B.A. Malomed, F. Lederer, Pulse switching in nonlinear fiber directional couplers. Phys. Rev. E **51**, 2527–2537 (1995)

M. van Hecke, Coherent and incoherent structures in systems described by the 1D CGLE: experiments and identification. Phys. D **174**, 134–151 (2003)

W. van Saarloos, P.C. Hohenberg, Pulses and fronts in the complex Ginzburg-Landau equation near a subcritical bifurcation. Phys. Rev. Lett. **64**, 749–752 (1990)

A. Villeneuve, C.C. Yang, P.C.J. Wigley, G.I. Stegeman, J.S. Aitchison, C.N. Ironside, Ultrafast all-optical switching in semiconductor nonlinear directional couplers at half the band-gap. Appl. Phys. Lett. **61**, 147–149 (1992)

Y.I. Voloshchenko, Y.N. Ryzhov, V.E. Sotin, Stationary waves in nonlinear, periodically modulated media with large group retardation. Zh. Tekh. Fiz. **51**, 902–907 (1981) [Sov. Phys. Tech. Phys. **26**, 541–544 (1982)]

D.T. Walton, H.G. Winful, Passive mode locking with an active nonlinear directional coupler: positive group-velocity dispersion. Opt. Lett. **18**, 720–722 (1993)

H.G. Winful, D.T. Walton, Passive mode locking through nonlinear coupling in a dual-core fiber laser. Opt. Lett. **17**, 1688–1690 (1992)

E.M. Wright, G.I. Stegeman, S. Wabnitz, Solitary-wave decay and symmetry-breaking instabilities in two-mode fibers. Phys. Rev. A **40**, 4455 (1989)

Y.D. Wu, Coupled-soliton all-optical logic device with two parallel tapered waveguides. Fiber Integr. Opt. **23**, 405–414 (2004)

A. Zafrany, B.A. Malomed, I.M. Merhasin, Solitons in a linearly coupled system with separated dispersion and nonlinearity. Chaos **15**, 037108 (2005)

Part III
Optical Fiber Fabrication

Advanced Nano-engineered Glass-Based Optical Fibers for Photonics Applications

11

M. C. Paul, S. Das, A. Dhar, D. Dutta, P. H. Reddy, M. Pal, and A. V. Kir'yanov

Contents

Introduction	479
Importance of the Nano-engineered Glass-Based Optical Fiber	483
The Basic Material of Nano-engineered Glass-Based Optical Fiber	484
Importance of Ceramic Oxides in Nano-engineered Glass-Based Optical Fiber	485
Mechanism to Develop Nano-engineered Glass-Based Optical Fiber	485
Fiber Drawing Process	486
Fabrication of Erbium-Doped Nano-engineered Zirconia-Yttria-Alumina-Phospho-Silica (ZYAPS) Glass-Based Optical Fiber	486
Material Characterization of Erbium-Doped Nano-engineered ZYAPS Glass-Based Optical Preform and Fiber	488
The Optical Performance of Erbium-Doped Nano-engineered ZYAPS Glass-Based Optical Fiber	491
Fabrication of Erbium-Doped Nano-engineered Scandium-Phospho-Yttria-Alumina-Silica (SPYAS) Glass-Based Optical Fiber	494
Material Characterization of Erbium-Doped Nano-engineered SPYAS Glass-Based Optical Preform and Fiber	495
The Optical Performance of Erbium-Doped Nano-engineered SPYAS Glass-Based Optical Fiber	498

M. C. Paul (✉) · S. Das · A. Dhar · D. Dutta · M. Pal
Fiber Optics and Photonics Division, CSIR-Central Glass and Ceramic Research Institute, Kolkata, India
e-mail: mcpal@cgcri.res.in; dshyamal@cgcri.res.in; anirband@cgcri.res.in; dielsalder6023@gmail.com

P. H. Reddy
Academy of Scientific and Innovative Research (AcSIR), IR-CGCRI Campus, Kolkata, India
e-mail: phvr159@gmail.com

A. V. Kir'yanov
Centro de Investigaciones en Optica, Guanajuato, Mexico
e-mail: alejandrokir@gmail.com

Fabrication of Multielement (P-Yb-Zr-Ce-Al-Ca) Fiber for Moderate-Power Laser Application... 503
    Material Characterization of Multielement (P-Yb-Zr-Ce-Al-Ca) Optical Preform and Fiber... 507
    The Optical Performance of Multielement (P-Yb-Zr-Ce-Al-Ca) Optical Fiber......... 509
Fabrication of Chromium-Doped Nano-phase Separated Yttria-Alumina-Silica (YAS) Glass-Based Optical Fiber.. 513
    Material Characterization of Chromium-Doped Nano-phase Separated YAS Glass-Based Optical Preform and Fiber...................................... 514
    The Optical Performance of Chromium-Doped Nano-phase Separated YAS Glass-Based Optical Fiber.. 518
Conclusions.. 523
Future Work.. 525
References... 525

Abstract

Nano-engineered glass-based silica optical fibers doped with different rare earth ions and transition metal ions is a new research area with potential application towards fiber-based amplifier, laser, and sensors. This chapter describes the basic material, fabrication techniques, and related material as well as the optical properties of nano-engineered glass-based optical fibers doped with erbium/ytterbium, erbium/scandium, ytterbium, and chromium for photonics applications. Accordingly, we present the development of erbium (Er)-doped zirconia-yttria–alumina-phospho-silica glass-based optical fibers and their application towards multichannel amplification in the C-band region under core as well as cladding pumped configuration. A new class of Er-doped nano-engineered scandium-yttria-alumina-silica glass-based optical fibers with proper annealing is also discussed which exhibited a flat-gain spectrum suitable for broad-band optical amplification in the C+ L band region having a maximum gain of 39.25 dB at 1560 nm. Application of materials towards a burning issue in case of high power Yb-doped laser called photodarkening has been reported where developed multielement (P-Yb-Zr-Ce-Al-Ca) nano-phase separated silica-glass-based optical fiber reveals that multielement-Yb-doped fiber is a promising candidate for laser applications with enhanced photodarkening resistivity. Lastly focused fabrication and properties of transition metal chromium-doped nano-phase separated yttria-alumina-silica glass-based optical fibers which may be suitable as saturable absorber with potential application towards Q-switching for 1–1.1-µm all-fiber ytterbium lasers.

Keywords

Nano-engineered glass · Rare-earth doped nano-particles · Optical fiber · Saturable absorber · Photodarkening effect · Fiber laser · Optical amplifier · Nonlinear refractive index · Broadband light source

Introduction

Nano structure of optical material plays an important role at current nano-science for manipulating and enhancing light-matter interactions for improving fundamental device properties. Now-a-days fiber laser, optical amplifier, and broadband light source are the most essential opto-electronic devices in modern communication systems. Recently an interesting research area being attempted across the globe to develop optical devices comprised of rare earth (RE) and different transition metal-doped nano-structured materials-based optical fibers for their use in telecommunications as broadband light source, optical amplifier, and high power fiber lasers with low photo-darkening phenomena. Accordingly, research on nanostructured optical fibers elaborated by incorporating dielectric metallic nanoparticles, silicon nanoparticles, semiconductor nanoparticles, phase-separated dielectric nonmetallic nanoparticles, or quantum dots in an amorphous matrix attracted much attention. These nanoparticles dispersed in a silica matrix-based optical fibers exhibit large nonlinear optical properties and offer a great potential for optical amplification and lasing application as the RE ions doping concentration can be higher than conventional amorphous medium. Moreover, the energy transfer process in these rare-earths doped nanoparticles containing optical fiber leads to exotic luminescence properties.

The well-known silica-based erbium-doped fiber amplifiers (EDFA) can produce gain in the C (1530–1565 nm) (Mears et al. 1987) and L bands (1570–1605 nm) (Massicott et al. 1990). The EDF based on fluoride and tellurite fibers are also reported to be applicable in the C and L bands, while the thulium-doped fiber amplifiers (TDFA) exhibit gain in the S band (Kasamatsu et al. 1999). Furthermore, the praseodymium-doped fiber amplifiers can operate in the O band (1260–1360 nm) (Ohishi et al. 1991) and holmium-doped ones can access the U band (1625–1675 nm) window and beyond. On the other hand, fiber Raman amplifiers can cover a broad spectral region, provided a suitable fiber and pump source to be used. However, neither of the above mentioned amplifiers can fully cover the 1100–1500 nm range with a single "active" fiber to be used. Bismuth- and chromium-doped fibers having pronounceable absorption and fluorescence bands within the 1100–1500 nm range are currently being investigated.

Erbium (Er)-doped fiber amplifiers (EDFA) have become established, since the invention in 1987 (Mears et al. 1987), as a standard component in telecommunication networks, facilitating information exchange worldwide due to their high performance and cost effectiveness. Broad gain at important low-dispersion wavelengths covering the telecom C-band (1525–1565 nm) and L-band (1565–1610 nm) (Atkins et al. 1989; Sun et al. 1997), low noise (Wysocki et al. 1997), low-loss and compatibility with fiber lightwave systems (Giles and Desurvire 1991; Bergano and Davidson 1996) make EDFA an excellent fit for signal amplification with high optical efficiency at various points of such networks. Furthermore, EDFA, especially in the eye-safe spectral region 1500–1700 nm, continue to be useful in a wide range of other applications, including optical communication

(Zimmerman and Spiekmann 2004), range finding, remote sensing (Wysocki et al. 1994), ultra-high bit-rate telecom transmission systems (Zhou et al. 2010), and free-space communications (Ma et al. 2009). The importance of the development work becomes evident from expanding global optical amplifier market which reached $900 million in 2012 and is anticipated to reach $2.8 billion by 2019 (NEW YORK 2013). To achieve high power Er-doped amplifier, it is essential to dope Er-ions at high concentration level which requires use of different glass host other than pure silica to overcome clustering problem. Accordingly, over the past several years extensive researches have been carried out on different materials for incorporation of Er-ions, such as silica glass modified with incorporation of GeO_2, Al_2O_3 and P_2O_5, telluride glass, phosphate glass chalcogenide, bismuthate, fluorozirconate, lithium niobate, lanthanum (Wang et al. 2003; Yamada et al. 1996; Jha et al. 2000; Naftaly et al. 2000; Jiang et al. 2000; Harun et al. 2010). Literature survey shows that different materials have different impacts on the performance of the developed waveguide or amplifier. Some materials allow higher Er concentrations to be realized without detrimental effects of concentration quenching (Snoeks et al. 1996) and cluster formation (Gill et al. 1996), thereby providing higher gain for a compact device. Although many nonsilica-based fibers doped with Er offer high quantum efficiencies for 1.5-μm transition, their material properties may not be suitable for practical devices. Moreover, owing to the difference in melting temperatures, such fibers cannot be spliced with standard single-mode fibers using a standard splicing machine. On the contrary, silica-based fiber doped with erbium continues to be the most preferable choice due to its proven reliability and compatibility with conventional fiber-optic components (Cheng et al. 2009; Harun et al. 2009).

Due to unavailability of high power single mode pump lasers, development of erbium-doped fiber (EDF) is a challenging task and researchers are employing a double-clad erbium co-doped with ytterbium fiber amplifier (EYDF) in which the signal light propagates in the core and pump light propagates in the first cladding around the core. Such a double-clad EYDF amplifier can use watt-class multimode laser diodes, with promise of realizing amplifier of high output power. Compared with EDF, the co-doping with Yb ions considerably improves the pump absorption. In EYDF, the ytterbium can absorb a pump photon within its spectral band of 800–1100 nm and be excited to the $^2F_{5/2}$ state, from which it can transfer energy to $^4I_{11/2}$ state of an erbium ion. The ions in $^4I_{11/2}$ then make a nonradiative transition to $^4I_{13/2}$, forming a population inversion between $^4I_{13/2}$ and $^4I_{15/2}$, thereby producing amplification of incident optical signal around 1550 nm.

An all-fiber format is of great interest outside the laboratory environment. Among single mode fiber sources at these wavelengths, the highest output power of 150 W with 33% efficiency was achieved using Er/Yb co-doped fibers cladding pumped at 915 or 976 nm (Jeong et al. 2005). However, for the maximum output power, amplified spontaneous emission (ASE) from Yb ions at 1 μm reaches 50% of the laser power at 1560 nm, leading to a rollover of the slope efficiency and raising the difficult issue of getting rid of the parasitic (non-eye-safe) radiation. In the recent past, significant success has been obtained in power scaling of in-band pumped Er-doped fiber (EDF) lasers and amplifiers (Zhang et al. 2011; Jebali et al. 2012;

Supradeepa et al. 2012; Lim et al. 2012). In such approach, the reduced quantum defect leads to a very attractive lasing slope efficiency of 75% (Jebali et al. 2012). However, the electrical to optical efficiency of the pump diode sources or Er–Yb fiber lasers at 1532 nm (Zhang et al. 2011; Jebali et al. 2012) and Raman fiber lasers at 1480 nm (Supradeepa et al. 2012) is at least two times lower than that of usual pump diodes at 976 nm, making the overall electrical to optical efficiency in the demonstrated in band pumped lasers lower than 15%. It appears that the double-clad (DC) Yb-free EDF pumped at 976 nm is the simplest architecture. Slope efficiencies of 24% (Kuhn et al. 2011a) in a single-mode fiber and 30% in a multimode fiber have been reported (Kuhn et al. 2011b). By decreasing the pump-cladding diameter and using the suitable core glass matrix, a slope efficiency of 32% (>7 W at 1575 nm with potential power scaling) in a single-mode EDF has been demonstrated (Kotov et al. 2012). Very recently L. V. Kotov et al. demonstrated 75 W 40% efficiency single-mode all-fiber erbium-doped laser cladding pumped at 976 nm (Kotov et al. 2013). Thus, it is essential to carry out basic research in order to identify a suitable glass host for fabrication of EDF for higher optical amplification efficiency.

Ytterbium (Yb)-doped fiber lasers due to their many folds advantages, namely, simple energy level diagram, small quantum defect, long lifetime at the excited state (~1 ms), broad absorption (~800–1080 nm), and emission (~970–1200 nm) bands (Zervas and Codemard 2014) over the conventional bulk state solid state lasers, find variety of applications covering materials processing (cutting, grinding, and engraving) medical and military applications (Jauregui et al. 2013) using continuous wave (CW) power level as high as 100 kW have been reported (Fomin et al. (2014); Gapontsev et al. 2014). Nevertheless, research is continuing to achieve higher power for specific industrial and defense applications. This can be achieved by enhancement of Yb concentration with improved waveguide design and optimization of laser cavity design to minimize different nonlinear effects, e.g., stimulated brillouin scattering (SBS), stimulated Raman scattering (SRS) besides photodarkening (PD) phenomenon. Photodarkening has been a hot topic of research interest since 2005 after Koponen et al. (2005) first reported about photodarkening in details, although prior to that Paschota et al. reported about unsaturable loss in Yb-doped fiber (Paschotta et al. 1997). The photodarkening effect can be defined as a time dependent increase of background absorption in a RE-doped fiber's core which supposedly originated from a multistep multiphoton absorption process, with the main threats to the efficiency of the Yb-doped gain media leading to degradation of laser performance and long-term stability. Photodarkened Yb-doped fibers demonstrate absorption centered in UV and extends to the VIS region with its tail stretching towards NIR region, the latter leads to degradation of the fibers lasing property (Mattsson 2011). However, PD is not a solely Yb-specific problem, but is as well observed in other RE ions like Tm, Tb, Ce, Pr doped into silica fiber (Broer et al. 1993; Atkins and Carter 1994; Behrens and Powell 1990). Accordingly, it is important to control and diminish PD in order to achieve enhanced power level and to improve the performance and reliability of Yb-doped high-power fiber lasers.

Although the actual mechanism behind the origin of PD in Yb-doped silica is still an open subject of research, in general formation of defect centers, also known as "color center," seems to be the prime reasons for PD. Different theories like formation of oxygen deficiency centers (ODC) (Yoo et al. 2007), Yb^{2+}-related charge-transfer-states (CTS) were reported by different groups (Engholm and Norin 2008; Guzman Chavez et al. 2007). On the other hand, different processes related to PD mitigation (completely or partially) were also reported, namely, oxygen or hydrogen (Jaspara et al. 2006) loading, temperature annealing (Soderlund et al. 2009), exposure to UV or VIS light (Guzman Chavez et al. 2007; Manek-Honninger et al. 2007) using pump conditions (Jetschke et al. 2007), etc. However, none of the abovementioned methods can be considered as a practical solution as all of them directly impact on fiber coating materials; so research is still pursuing to find out an acceptable route to mitigate PD. It has been observed that higher Al-doping level (Kitabayashi et al. 2006), lower Yb-ion concentration (Morasse et al. 2007), co-doping with phosphorous (P) (Jetschke et al. 2008; Sahu et al. 2008) and cerium (Ce) (Engholm et al. 2009; Malmstrom et al. 2010) can increase PD resistivity. Higher Al doping level increases numerical aperture (NA) and degrades beam quality and thus is not a worth solution. Use of lower Yb concentration unavoidably leads to longer fiber to be used for lasing, which strengthens unwanted nonlinear effects and ultimately prevents high-power operation. In this context, co-doping of Yb-doped fiber with P, though enabling superior PD resistivity, is associated with increased base loss, lower emission, and absorption cross sections besides a central dip formation during fabrication, leading to degradation of lasing performance. Furthermore, Ce-doping could be really an attractive option to reduce PD, but it increases NA and has constraint for practical applications. Very recently, a glass composition with equal Al and P ratio was shown (Unger et al. 2009) to be a promising alternative to enhance PD resistivity, but this composition generally produces triangular refractive index (RI) profile and inhomogeneity along the length. Considering all the above referred information, searching for a relevant solution is going on to alter core glass composition for minimizing formation of PD-related color center even at a higher Yb ion concentration.

Fiber-based amplifiers which cover the 1100–1500 nm range with a single "active" fiber are still under development as both Bi- and Cr-doped fiber fabrication which can operate in this region is challenging. Nevertheless, between Bi and Cr, Cr-doped fiber fabrication is the most challenging task. Among the transition metal (TM) doped fibers, Cr^{4+}:YAG doped fiber having broadband fluorescence that covers 1100–1600 nm range (Lo et al. 2005; Huang et al. 2006, 2013; Lai et al. 2011) is of great importance for implementing broadly tunable solid-state lasers for 1400–1500 nm spectral range as well as a reliable saturable absorber for ytterbium- and neodymium-doped lasers at 1000–1100 nm range (Angert et al. 1988; Eilers et al. 1993; Borodin et al. 1990; Il'ichev et al. 1998a, b; Mellish et al. 1998; Dussardier et al. 2011). On the other hand, Cr^{4+}:YAG doped fibers can fit a variety of applications (Alcock 2013), including injection-seeded optical parametric oscillators (OPOs), spectroscopy, optical coherence tomography (OCT), "eye-safe" ranging, optical time domain reflectometry, tissue welding as well as pumping EDFA and erbium doped fiber lasers. Accordingly, Cr^{4+} as active ion and yttrium-aluminum garnet (YAG) as a host is a potential candidate which needs further research. Embedding chromium into a glassy (silicate) matrix directly was not so successful to meet the practical applications highlighted above due to the fact that chromium exhibits different oxidation states in silicates while there are definitive technological difficulties in attempt to stabilize a wanted state (especially Cr^{4+} ions). Each oxidation state has its own potential applications: e.g., Cr^{4+} ions can be used as an "active" medium for lasing and amplifying around 1400 nm (Zhavoronkov et al. 1995; Dong et al. 2001) while Cr^{3+}/Cr^{4+} co-doping – for realizing tunable solid-state lasers (Nikolov et al. 2004; Sennaroglu 2002) usefulness of the Cr^{4+} species as saturable absorbers (Dussardier et al. 2011; Kück 2001). However, the different oxidation states arise due to change in glass compositions, processing temperature, and upon an employed preparation technology. Several attempts have been made to fabricate chromium-doped silica fibers by different authors worldwide. For instance, earlier workers (Lo et al. 2005; Huang et al. 2006, 2013; Lai et al. 2011) reported the laser-heated pedestal growth (LHPG) method to fabricate a single-crystal fiber, but the process implementation was intricate due to the technological complexity associated with single-crystal fiber drawing. Another method known as the rod-in-tube (RIT) technique was exploited which results in better glass uniformity with small core dimension (Huang et al. 2007, 2010). Nevertheless, increasing chromium concentration in Cr:YAG rod in the RIT method was difficult, and thus powder-in-tube (PIT) method was introduced to increase chromium ions concentration further (Huang et al. 2011). Many researchers (Felice et al. 2000; Dvoyrin et al. 2003; Glazkov et al. 2002; Abramov et al. 2014) fabricated silica-glass-based Cr^{4+}-doped fibers based on alumino-silica, germano-silica, and gallium-silica glasses and optical fibers, obtained using the well-known Modified Chemical Vapor Deposition (MCVD) coupled with solution doping (SD) technique, though they were not able to demonstrate broadband emission of Cr^{4+} ions. The only successful approaches to get near-IR (1 μm) fluorescence stemming from Cr^{4+} ions presence were proposed in Zhuang et al. (2009), and Wang and Shen (2012), but in that case the network permitting stabilization of Cr^{4+} species was glass ceramics. Thus, fabrication of Cr-doped fiber with a suitable glass host to stabilize Cr^{4+} ion through conventional modified chemical vapor deposition (MCVD) coupled with solution doping process is a challenging task.

In this book chapter, we have thus focused to develop suitable glass host materials comprised of nano-engineered phase separated silica-based optical fiber, fabricated through conventional MCVD-SD technique to tackle the abovementioned problems. The principles associated with fabrication techniques along with material and optical characterization as well as fiber performance will be discussed in details.

Importance of the Nano-engineered Glass-Based Optical Fiber

Since pure silica is not a suitable matrix for rare earth (RE) ions due to low solubility of RE ($\sim$ few hundreds of ppm) higher RE concentration leads to phase separation and reduction in fiber performance. If the phase of micron-size aggregates is separated from a homogeneous glass matrix, it causes a fatal increase of spectral attenuation of the fibers and must be prevented to develop low loss RE-doped fibers. Modification of glass matrices mainly with Al_2O_3, P_2O_5 or GeO_2 has been practiced in silica fibers for years with the aim to increase the solubility of REs in the fiber core as well as to enhance the emission property. However, such co-dopants are not sufficient to achieve high concentration RE-doping without clustering phenomena for making of specialty fibers under high power applications. Successful insulation of RE-ions from matrix vibrations is required by appropriate ion-site engineering through the formation of phase-separated nano-particles.

This approach is proposed via encapsulation of dopants inside glassy or crystalline nanoparticles embedded in the fiber glass under suitable doping host composition using ceramic oxides, such as Al_2O_3, Sc_2O_3, ZrO_2, and Y_2O_3 followed by post thermal annealing process. In such nano-engineering glass-based optical fibers, the basic material silica serves as a support for providing optical and mechanical properties to the fiber, whereas the spectroscopic properties would be controlled by the composition along with nature of nano-particles. The thermal shock resistance of the nano-engineering host of REs will be increased under high power applications. The sizes of nano-particles should be kept within 5–10 nm ranges to reduce the scattering loss as well as increases the optical transparency of the glass.

The Basic Material of Nano-engineered Glass-Based Optical Fiber

The nano-engineering glass is based on multielements doped into silica glass matrix containing ceramic oxides, such as Al_2O_3, ZrO_2, Y_2O_3 along with RE such as Er, Yb, Tm, Ce and transition metals such as Cr, Sc. In some cases, minor amount of P_2O_5 is used for making of fiber laser and optical amplifier where P_2O_5 acts as nucleating agent to accelerate the growth of formation of phase separated particles upon heating through thermal perturbation owing to the higher field strength difference (>0.31) between Si^{4+} and P^{5+}. Accordingly, in this chapter, the fabrication of the following nano-engineering glass-based optical fibers have been discussed.

- Er-doped nano-engineered scandium-yttria-alumina-silica (SYAS) glass-based optical fiber for brand optical amplification with flat gain
- Erbium-doped zirconia-yttria-alumina-phospho-silica glass-based optical fiber for high power optical amplification
- Er-doped nano-engineered scandium-yttria-alumina-silica (SYAS) glass-based optical fiber for broadband optical communication system
- Fabrication of multielement (ME) (P-Yb-Zr-Ce-Al-Ca) nano-phase separated silica glass-based core region of optical fiber using conventional modified chemical

vapor deposition (MCVD) process coupled with solution doping technique and study of their photodarkening resistant behavior
- Fabrication of yttria-alumina-silica nano-engineering glass host-based chromium-doped phase-separated particles containing optical fibers using the MCVD process in conjunction with the SD technique along with study of their spectroscopic properties

Importance of Ceramic Oxides in Nano-engineered Glass-Based Optical Fiber

The use of different ceramic oxides has multiple advantages as mentioned below.

- It helps to reduce the clustering phenomena of RE ions significantly by enhancing solubility of RE compared to pure silica glass matrix.
- Increases the refractive index of silica glass due to the larger sizes of Al, Zr, and Y cations as well as their enhanced ionic polarizability.
- Increases the optical nonlinearity of the doping host of REs because of the formation of large number of nonbridging oxygen (NBO) and formation of phase-separated nano-particles. Nonbridging oxygens have high ionicity and are easily distorted by applied optical-electric field as the defects generated from the phase-separated nano-particles.
- Zirconia (ZrO_2) doping into silica glass increases its physical and chemical properties, including hardness, wear resistance, low coefficient of friction, elastic modulus, chemical inertness, ionic conductivity, and electrical properties.
- Y_2O_3 serves as an attractive host material for laser application since it is a refractory oxide with a melting point of 2380 °C a very high thermal conductivity, $k_{Y2O3} = 27$ W/mK, two times YAG's one, $k_{YAG} = 13$ W/mK. Another interesting property allowing radiative transitions between electronic levels is the dominant phonon energy of 380 cm^{-1} which is one of the smallest phonon cut-offs among oxides.
- Scandium oxide is a cubic crystal belonging to the Ia3 space group (space group number 206) having the lattice constant of 9.8459 °A while $Er^{3+}:Sc_2O_3$ sesquioxide is a promising candidate for high energy, eye-safe solid-state lasers as it can operate with a very low quantum defect and high thermal conductivity Peters et al. (2002; Ter-Gabrielyan et al. (2008)).

Mechanism to Develop Nano-engineered Glass-Based Optical Fiber

The following diagram shows the general steps involved towards development of nano- engineering glass-based optical fibers:

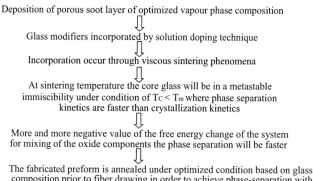

Deposition of porous soot layer of optimized vapour phase composition
⇩
Glass modifiers incorporated by solution doping technique
⇩
Incorporation occur through viscous sintering phenomena
⇩
At sintering temperature the core glass will be in a metastable immiscibility under condition of $T_c < T_m$ where phase separation kinetics are faster than crystallization kinetics
⇩
More and more negative value of the free energy change of the system for mixing of the oxide components the phase separation will be faster
⇩
The fabricated preform is annealed under optimized condition based on glass composition prior to fiber drawing in order to achieve phase-separation with smaller particle size to reduce loss characteristics of final fiber

Fiber Drawing Process

Fiber drawing from annealed preform was done using fiber drawing tower as shown in Fig. 1, with on-line dual resin coating. Depending upon the diameter and fiber drawing speed, the furnace temperature is set. At the same time, fiber is drawn down by a capstan puller at certain speed. After cooling down, the bare fiber is coated with two layers of UV–curable acrylate resin. High-quality coating with good coating concentricity and accurate dimensions is essential for getting high quality fibers.

The purpose of the coating on the optical fiber is to protect the fiber from contact and foreign particles as these reduce the life of the fiber in terms of strength and static fatigue. Besides strength preservation, the coating must protect from microbending by being concentric and bubble free around the fiber and by having a stable performance in different environments. Out of two layers of UV-curable acrylate, a soft inner layer protects the fiber and a hard outer layer ensures good mechanical properties. The movement of the coating liquid occurs surrounding the surface of the fiber where fine bubbles can be trapped at the interface of fiber surface and hard coating. This can be avoided by controlling pressure and keeping the viscosity of the liquid constant during the drawing process before it enters the UV curing oven. Finally the fibers are wound on a precision spooling machine.

Fabrication of Erbium-Doped Nano-engineered Zirconia-Yttria-Alumina-Phospho-Silica (ZYAPS) Glass-Based Optical Fiber

Erbium-doped optical fibers made based on ZYAPS glass employing modified chemical vapor deposition (MCVD) process in combination with the solution doping technique. Figure 2a shows the deposition of porous silica soot layer within the pure silica substrate tube of dimension 20 mm with thickness of 1.5 mm.

Fig. 1 Photograph showing the view of the optical fiber drawing tower with picture of the neck down of preform coming out from the heated furnce before drawing the fiber

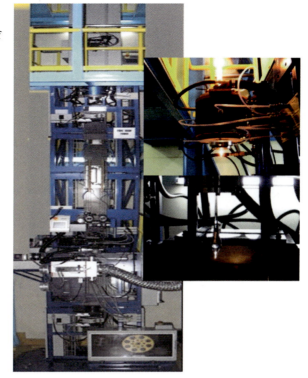

Figure 2b shows the actual collapsing process in MCVD set-up. The glass formers incorporated by the MCVD process are SiO_2 and P_2O_5 along with the glass modifiers Al_2O_3, ZrO_2, Er_2O_3, and Y_2O_3, which are incorporated by the solution doping technique using an alcohol-water mixture of suitable strength (1:5) to form the complex molecules $ErCl_3.6H_2O$, $AlCl_3.6H_2O$, $YCl_3.6H_2O$, and $ZrOCl_2$ $8H_2O$. Small amounts of Y_2O_3 and P_2O_5 are added during the fabrication stage, which serve as a nucleating agent to increase the phase separation for generating Er_2O_3-doped microcrystallites in the core matrix of the optical preform (Paul et al. 2010, 2011; Pal et al. 2011). The inclusion of the Y_2O_3 particulates into the host matrix serves the additional purpose of slowing down or eliminating changes in the ZrO_2 crystal structure which exists in three distinct crystalline structures in a bulk glass matrix depending on the fabrication temperature. Another critical stage is annealing of the fabricated preform which was performed at 1100 °C under optimized conditions in a closed furnace under heating and cooling rates of 20 °C/min to generate Er_2O_3-doped ZrO_2 rich nano-crystalline particles and subsequently fibers were drawn with on-line resin coating using fiber drawing tower. Details of representative fibers are presented in Table 1.

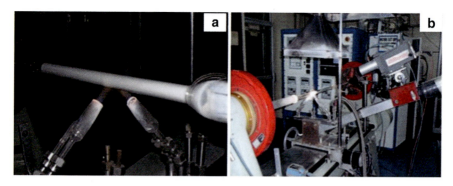

Fig. 2 (a) Deposition of porous phospho-silica layer inside the silica substrate tube; (b) Collapsing is in progress for making Zr-EDF. (Reprinted with permission from reference 53, M. C. Paul, M. Pal, S. Das, A. Dhar and S. K. Bhadra; Journal of Optics 45, 260 (2016), Copyright @ The Optical Society of India, 2016, publisher Springer India)

Table 1 Nano-crystalline zirconia yttria alumina silica particles doped optical fibers

Fiber No	Composition of doping host	NA	Core diameter (μm)	Er ion concentration (ppm wt)
MEr-1	SiO_2-Al_2O_3-Y_2O_3-ZrO_2-P_2O_5-Er_2O_3	0.17	10.5	2800
MEr-2	SiO_2-Al_2O_3-Y_2O_3-ZrO_2-P_2O_5-Er_2O_3	0.19	10.0	3888
MEr-3	SiO_2-Al_2O_3-Y_2O_3-ZrO_2-P_2O_5-Er_2O_3	0.21	10.3	4320

Material Characterization of Erbium-Doped Nano-engineered ZYAPS Glass-Based Optical Preform and Fiber

During fiber drawing at around 2000 °C, the nano-crystalline host of ZrO_2 was preserved in the silica glass matrix as confirmed by the TEM analyses with EDX spectra and electron diffraction patterns as shown in Fig. 3. Therefore, it could be ascertained that during fiber drawing at the high temperature further vitrification could be avoided. This clearly confirms the existence of ZrO_2 crystallites within the host matrix. The average particle sizes are in the range 5–8 nm. The core and cladding geometry of the fiber was inspected by high resolution optical microscope (Olympus BX51). The core was homogeneous and had no observable defects at the interface between the core and the silica cladding. The average dopant percentages of the fiber samples were measured by electron probe microanalyses (EPMA) as shown in Table 2.

Three different types of nano-engineered glass-based Zr-EDFs were fabricated with a variation of the doping levels of different co-dopants. The cross-sectional view of one of the fabricated Zr-EDF (MEr-2) is shown in Fig. 4a. It can be seen that it does not contain any core-clad imperfection which may affect bending loss

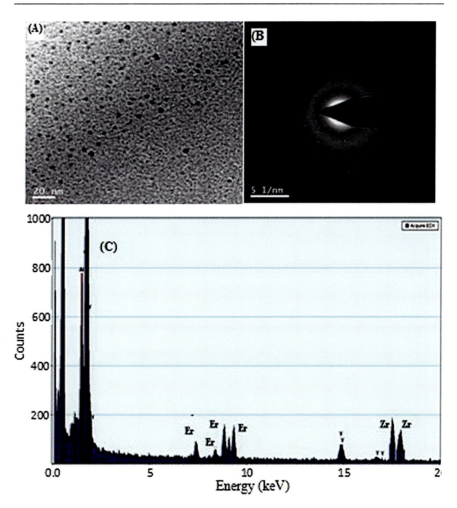

Fig. 3 (a) TEM image of fiber core glass shows nano-crystallites (black spots) (b) electron diffraction pattern of nanoparticle of Zr-EDF (MEr-2) and (c) EDX analyses showing presence of Er Y, Al and Zr. (Reprinted with permission from reference 53, M. C. Paul, M. Pal, S. Das, A. Dhar and S. K. Bhadra; Journal of Optics **45,** 260 (2016), Copyright @ The Optical Society of India, 2016, publisher Springer India)

Table 2 Doping levels within core region of the fibers

ID	Al_2O_3 (mol%)	ZrO_2 (mol%)	Er_2O_3 (mol%)
MEr-1	0.25	0.65	0.155
MEr-2	0.26	1.47	0.195
MEr-3	0.24	2.10	0.225

of the fiber. The refractive-index profile of fiber MEr-1 is shown in Fig. 4b. The spectral attenuation curve of fiber MEr-2 is shown in Fig. 4c. The Zr-EDFs show small variation of the fluorescence lifetime at different pump powers varying from 50 to 300 mW.

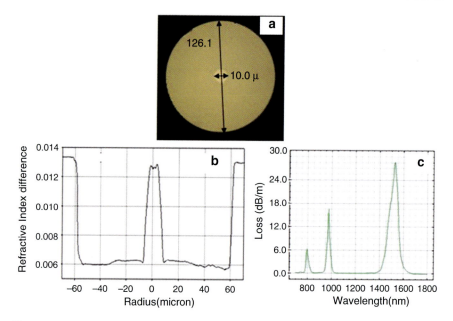

Fig. 4 (**a**) Cross-sectional view (MEr-2), (**b**) refractive index profile (MEr-1) and (**c**) Loss curve of EDF (MEr-2). (Reprinted with permission from reference 53, M. C. Paul, M. Pal, S. Das, A. Dhar and S. K. Bhadra; Journal of Optics 45, 260 (2016), Copyright @ The Optical Society of India, 2016, publisher Springer India)

The fluorescence lifetimes of drawn fibers are determined and found to be around 8 and 10 ms under 100 mw pump power at 976 nm wavelength. MEr-1 shows a slightly lower fluorescence lifetime of around 8.0 ms as the fiber contains lowest doping level of ZrO_2 among the three fibers under study. With a combination of both Zr and Al along with other co-dopants, we achieved high Er-concentration of 4320 ppm in the fiber MEr-3 without any phase separations and concentration quenching effect of REs. One of the conventional methods to test for quenching is the measurement of fluorescence lifetime of Er-ions at different pump powers. There must be concentration quenching of Er-ions if the fluorescence lifetime differs too much with different pump power (Keiichi Aiso et al. 2001).

In order to check this property, we have measured the fluorescence lifetime of two different doped EDF hosts at different pump powers as shown in Fig. 5. The curves show that there was not much noticeable change of measured lifetime of Zr-EDF (MEr-3) that only changes from 10.65 to 10.70 ms with increasing pump power from 25 to 250 mW as in Fig. 5. This confirms that the lifetime of alumina-silica glass-based EDF normally varies from 10.25 to 10.58 ms. This phenomenon indicates that nano-crystalline ZrO_2 prevents the formation of Er–Er bonds which reduces the quenching process of Er ions even at high doping levels of around 4320 ppm. The success is related to co-doping of ZrO_2 with Al_2O_3. Both aluminum and zirconium

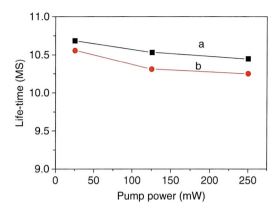

Fig. 5 Fluorescence lifetime of three EDFs. (**a**) MEr-3 and (**b**) alumina-silica glass-based EDF at different pump powers. (Reprinted with permission from reference 53, M. C. Paul, M. Pal, S. Das, A. Dhar and S. K. Bhadra; Journal of Optics **45,** 260 (2016), Copyright @ The Optical Society of India, 2016, publisher Springer India)

ions surround the Er-ions and form a solvation shell thereby adjusting the charge balance and improving the solubility of Er ions into the host. This inhibits clustering of the Er-ions.

The Optical Performance of Erbium-Doped Nano-engineered ZYAPS Glass-Based Optical Fiber

Four-channel amplification at different input signal levels 20, 25, and 30 dBm/ch at different pump powers (40, 50, and 60 mW) using one of the Zr-EDFs (MEr-3) published by our group (Pal et al. 2011) has been also evaluated. The amplified signals and output signal to noise ratio (SNR) were measured for three different pump powers of 40, 50, and 60 mW at 20 dBm/ch input signal level.

Signal amplifications were increased in each channel and observed more rapid in the shorter wavelength side of C-band as pump power was increased, and the output SNRs are always >31 dB for each channel. The nature of signal amplifications and SNRs were measured for three different input signal levels (−20, −25, and −30 dBm/ch) for constant pump power of 50 mW (Pal et al. 2011). It is observed that there is appreciable optical gain of at least 20 dB for −20 dBm/ch and a minimum output SNR of 21.8 dB for 30 dBm/ch input signal level. Both the gain and output SNRs would increase if pump power was further increased. The measured noise figures were within 4.2 to 4.7 dB. The critical passive losses of erbium-doped optical fiber in EDFA system arise from the OH − absorption that is present in doped glass which gives rise to strong absorption at 1380 nm wavelength. The presence of OH degrades the gain performance of the fibers due to the energy transfer from $^4I_{13/2}$ level to OH ions. The OH absorption losses of all the three fibers at 1380 nm wavelength are kept within 200–250 dB/km. On the other hand, splicing losses with standard SMF-28 fiber induces further loss in EDFA system which also reduces the pump efficiency of the fiber. All the splicing losses of all the three fibers are observed to be 0.04–0.05 dB.

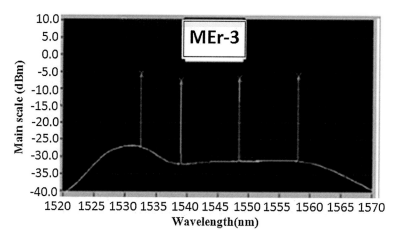

Fig. 6 Four-channel amplification in MEr-3 fiber with input signal level: −30 dBm/ch. (Reprinted with permission from reference 53, M. C. Paul, M. Pal, S. Das, A. Dhar and S. K. Bhadra; Journal of Optics 45, 260 (2016), Copyright @ The Optical Society of India, 2016, publisher Springer India)

The experimental results revealed that the Zr-EDF (MEr-3) is quite suitable for multichannel small-signal amplification. In our experiment, we achieved the minimum gain of 22.5 ± 0.5 dB for the input signal level of −30 dBm/ch where about 1.5 m length of the fiber was used in the experiment. The measured results of output power and SNR are shown in Fig. 6. We obtained gain difference (max-to-min gain) of less than 2 dB in MEr-3 sample. The maximum noise level was −28.5 dBm/nm for MEr-3. Moreover, MEr-3 showed an output SNR value of >22 dB. The motivation for choosing nanocrystalline zirconia dispersed in the silica glass matrix is to get low noise as well as better flatness of the gain spectrum at high doping level of Er ions. A major factor that influences the quenching process in RE-doped materials is multiphonon relaxation. Zirconium oxide possesses a stretching vibration at about 470 cm^{-1}, which is very low compared with that of Al_2O_3 (870 cm^{-1}) and SiO_2 (1100 cm^{-1}) (Patra et al. 2002). Hence, much attention is given to select such host materials with low phonon energy in order to improve the erbium ion solubility with reduced concentration quenching process and enhance the gain flatness and output SNRs for multichannel small-signal amplification.

As stated earlier the composition of the doping host should be engineered properly for reducing the clustering effect of Er ions. We have tried to develop new compositional glass host for doping of Er ions with high ZrO_2 content with the large cladding pump region. The microscopic image of hexagonal shaped low RI coated Er-doped fiber HPE-2 core and cladding spectral absorption curves are given in Fig. 7a, b, c.

The core absorption is around 80 dB/m at 980 nm wavelength, whereas cladding absorption is around 2.0 dB/m at 980 nm. The different dopants are distributed uniformly along the diameter of the fiber, which is confirmed by electron probe microanalysis (EPMA) shown in Fig. 8a. A preliminary result of optical amplification

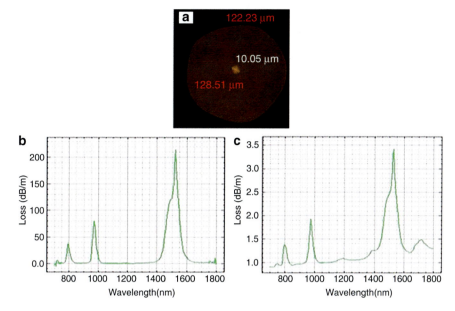

Fig. 7 (**a**) Cross-sectional view, (**b**) core absorption loss, and (**c**) cladding absorption loss of high concentration of erbium-doped fiber (HPE-2). (Reprinted with permission from reference 53, M. C. Paul, M. Pal, S. Das, A. Dhar and S. K. Bhadra; Journal of Optics **45,** 260 (2016), Copyright @ The Optical Society of India, 2016, publisher Springer India)

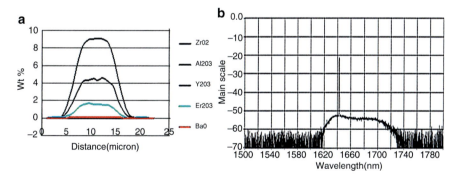

Fig. 8 (**a**) Distribution of different dopants across the core diameter and (**b**) amplification curve of EDF (HPE-2) under 4 W pump power at 980 nm having I/P Signal +6 to 9 dBm with O/P signal of 500 mW. (Reprinted with permission from reference 53, M. C. Paul, M. Pal, S. Das, A. Dhar and S. K. Bhadra; Journal of Optics **45,** 260 (2016), Copyright @ The Optical Society of India, 2016, publisher Springer India)

using a 4 m length of fiber under 4 W pump power having I/P signal +6 to 9 dBm is shown in Fig. 8b. The wideband ASE was observed for the yttria stabilized zirconia-alumina-phospho-silica glass host with hexagonal shaped cladding structure coated with low RI resin under different cladding pump power is given in Fig. 9.

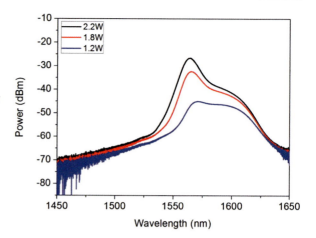

Fig. 9 ASE output for 6 m HPEr-5 for different pump powers for the wavelength range 1450–1650 nm. (Reprinted with permission from reference 53, M. C. Paul, M. Pal, S. Das, A. Dhar and S. K. Bhadra; Journal of Optics **45**, 260 (2016), Copyright @ The Optical Society of India, 2016, publisher Springer India)

A wide band emission from 1530 to 1620 nm is observed. As the zirconia-rich Er_2O_3-doped phase-separated particles with partially crystalline character possesses lower phonon energy which gives rise to increase of larger emission cross section (Pal et al. 2011). The introduction of Zr^{4+} allows one to avoid formation of Er^{3+} clusters in silica host and consequently to make more intense luminescence (Paul et al. 2011). Replacing the intermediate Al_2O_3 by a modifier ZrO_2, the number of nonbridging oxygen is expected to increase which makes the silica network structure more open. As a result, Zr-EDF has wider emission spectra as compared to silica-EDF, especially at the longer wavelengths at 1620 nm because of its larger emission cross-section (Paul et al. 2010, 2011). The widening of the ASE spectra towards a longer wavelength is believed to be a result of the Stark level of the Er^{3+} ions in the Zr-EDF, which is separated to a larger degree due to intense ligand field. This is due to inhomogeneous energy level degeneracy that the ligand field of the zirconia host glass induced as a result of site-to-site variations, also known as the Stark effect, causing the widened optical transitions.

Fabrication of Erbium-Doped Nano-engineered Scandium-Phospho-Yttria-Alumina-Silica (SPYAS) Glass-Based Optical Fiber

Development of solid-state $Er:Sc_2O_3$ laser started recently although relevant spectroscopic investigations are still relatively rare due to limited availability of high optical quality material Fechner et al. (2008). Er-doped scandium-phospho-yttria-alumina-silica (SPYAS) glass-based fiber was fabricated using Modified Chemical Vapor Deposition (MCVD) process (Nagel et al. 1982) coupled with solution doping technique, by passing $SiCl_4$ and $POCl_3$ vapors through a slowly rotating high purity silica glass tube of outer diameter of 20 mm and inner diameter of 17 mm. An external flame source moved along the length of the tube as it rotated to heat it up under an optimized temperature of around 1550 ± 10 °C, monitored by an IR pyrometer, synchronously moved with the oxy-hydrogen burner. This high temperature caused chloride vapors to oxidize, resulting in deposition of porous phospho-silica layer along the inner diameter of the tube. The glass formers, SiO_2 and P_2O_5, were incorporated in the core matrix of the optical fiber through the MCVD process, with small amount of P_2O_5 serving as a nucleating agent, to increase the phase separation along with the generation of Er-enriched microcrystallites. Then, the glass modifiers, Al_2O_3, Er_2O_3, Sc_2O_3, and Y_2O_3, were added through the solution doping technique using an alcohol/water mixture of suitable strength (1:5) to form the complex molecules $ErCl_3.6H_2O$, $AlCl_3.6H_2O$, $YCl_3.6H_2O$, and $ScCl_3.6H_2O$, diffusing into the silica glass matrix of the tube. Note that incorporating a small amount of Y_2O_3 into the host matrix served to slow down or stop the changes in the crystalline structure formed and thereby preserving the mechanical strength and integrity of the final fiber.

The fabricated preform was then cut into two portions. First portion of the preform was drawn into a fiber strand with diameter 125.0 ± 0.5 μm through a conventional fiber drawing tower. The second portion of the preform was thermally annealed at 1200 °C for 3 h with heating and cooling rate of 15 °C/min in a closed furnace that maintained an optimized heating cycle and then was drawn to fiber. Note that the thermal annealing was a terminating step in the formation of the nano-engineered scandium-phospho-yttrium-alumina-silica (SPYAS) glass-based optical fiber. Because of the phase separation phenomenon, slight brownish coloration was observed in the core glass of the fiber. Once the fiber has been drawn, the primary and secondary coatings were added to the fiber; their uniformity was assured by controlling the flow pressure of the inlet gases into the coating resin vessels during the drawing process, as well as the proper alignment of the coating cup units. To ensure that a high-quality fiber was produced, its diameter was controlled throughout the drawing process (Bhadra and Ghatak 2013).

Material Characterization of Erbium-Doped Nano-engineered SPYAS Glass-Based Optical Preform and Fiber

The optical microscopic views, TEM bright-field images, and electron diffraction (ED) patterns of Er-doped nano-engineered SPYAS glass-based fibers drawn from both the pristine and the annealed performs along with Sc-free EDF are demonstrated in Figs. 10 and 11, respectively.

The microscopic views of the fibers cross-sections were measured by a high-resolution optical microscope (Olympus BX51), connected to a high-resolution digital color camera. As seen from Figs. 10 and 11a, both the fibers have core of 12.5 μm and are homogeneous without any core-clad imperfections, which otherwise might lead to a certain loss at bending. As also seen from Fig. 11a, the color of core glass is slightly brownish yellow in the fiber drawn from the annealed preform, being a result of the formation of glass-ceramic nature of the doping host. The morphology of the fiber core glass was studied by means of

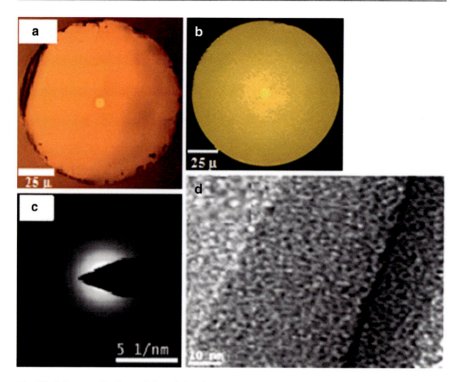

Fig. 10 Microscopic views of the pristine SPAYS (**a**) and Sc-free standard (**b**) Er-doped fibers; ED pattern (**c**) and TEM picture (**d**) of the SPAYS fiber. (Reprinted with permission from reference 75, P. H. Reddy, S. Das, D. Dutta, A. Dhar, A. V. Kir'yanov, M. Pal, S. K. Bhadra and M. C. Paul, Physica Status Solidi A 1700615 (2018), copyright @ 2018 WILEY-VCH Verlag GmbH & Co. KGaA, Weinheim)

high-resolution transmission electron microscopy (TEM) (Tecnai G2 30ST, FEI Company, USA), producing the bright-field images shown in Figs. 10 and 11b. The nature of nanoparticles was evaluated from the ED patterns. For the fiber drawn from the pristine preform, the formation of nano-phase separated amorphous nanoparticles is presented in Figs. 10b and 11c. In turn, for the fiber drawn from the annealed preform, crystalline nature in core-glass is evidenced by Fig. 11b, c. Furthermore, the TEM along with ED pattern suggest the formation of crystalline nanoparticles of size 8–12 nm.

Average doping distributions in the core region of the Er-doped SPYAS glass-based optical fiber were measured by Electron Probe Micro-Analyzer (EPMA), as shown in Fig. 12. From the distribution profiles shown, the following estimates of doping levels have been made: 1.256 wt% of Sc_2O_3, 5.85 wt% of Al_2O_3, 0.567 wt% of Y_2O_3, and 0.285 wt% of Er_2O_3.

The Refractive Index (RI) profile of the Er-doped SPYAS glass-based optical fiber along with Sc free EDF, shown in Fig. 13, was measured using a fiber analyzer

11 Advanced Nano-engineered Glass-Based Optical Fibers for... 497

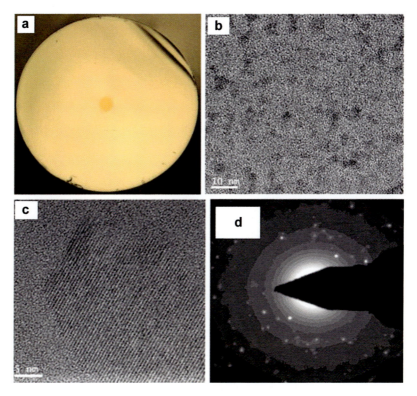

Fig. 11 (**a**) Microscopic view, (**b**) TEM picture, (**c**) high-resolution TEM, and (**d**) ED pattern of Er-doped nano-engineered SPYAS glass-based optical fibers drawn from the annealed perform. (Reprinted with permission from reference 75, P. H. Reddy, S. Das, D. Dutta, A. Dhar, A. V. Kir'yanov, M. Pal, S. K. Bhadra and M. C. Paul, Physica Status Solidi A 1700615 (2018), copyright @ 2018 WILEY-VCH Verlag GmbH & Co. KGaA, Weinheim)

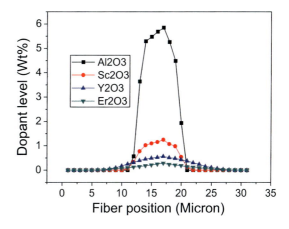

Fig. 12 Elemental analysis of the Er-doped SPYAS glass-based optical fiber. (Reprinted with permission from reference 75, P. H. Reddy, S. Das, D. Dutta, A. Dhar, A. V. Kir'yanov, M. Pal, S. K. Bhadra and M. C. Paul, Physica Status Solidi A 1700615 (2018), copyright @ 2018 WILEY-VCH Verlag GmbH & Co. KGaA, Weinheim)

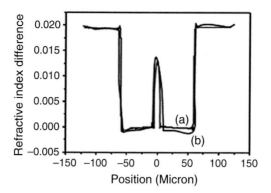

Fig. 13 RI Profile of the Er-doped nano-engineered SPYAS glass-based (**a**) and Sc-free standard Er-doped fiber (**b**). (Reprinted with permission from reference 75, P. H. Reddy, S. Das, D. Dutta, A. Dhar, A. V. Kir'yanov, M. Pal, S. K. Bhadra and M. C. Paul, Physica Status Solidi A 1700615 (2018), copyright @ 2018 WILEY-VCH Verlag GmbH & Co. KGaA, Weinheim)

(NR-9200, EXFO, Canada). The average Numerical Aperture (NA) of the Er-doped nano-engineered fiber and Sc free standard Er-doped fiber (core diameter: 9.25 μm) was estimated to be 0.18 and 0.19, respectively.

The Optical Performance of Erbium-Doped Nano-engineered SPYAS Glass-Based Optical Fiber

Absorption spectra as shown in Fig. 14 were obtained using a white-light source with fiber output and an optical spectrum analyzer (OSA) with a 0.5 nm resolution. 3.0 cm length of fiber was used in actual experiment, and Fig. 14a shows the measured absorption (loss) spectra of Er-doped SPYAS glass-based optical fiber and Sc-free EDF where the bands peaking at ∼980 and ∼1530 nm adhere to the well-known Er^{3+} absorption transitions. On the other hand the absorption loss spectra of fiber drawn from the annealed SPYAS preform indicate multiple sharp peaks within the ∼980 and ∼1530 nm bands (Fig. 14b), which may signify that Er ions are present in environment of crystalline nanoparticles.

Fluorescence spectra of the fiber samples were obtained by pumping at 980 nm wavelength under 250-mW pump power, in a lateral geometry. A fiber coupled mode pump laser diode at 976 nm is used as pump source. Such fiber pigtailed pump source was connected to the 980 nm arm of 980/1550 nm WDM coupler. A short piece of fiber around 4.5 cm having a peak small absorption less than 0.2 dB was fusion spliced to the 980/1550 nm arm of the WDM, its exit end was broken and immersed in an index matching fluid to avoid reflections. The fluorescence generated from the surface side of fiber collected at the WDM's 1550 nm arm using the optical spectrum analyzer AQ-6315 (Ando) and was recorded using a computer attached to the OSA. An InGaAs photodetector was connected to record

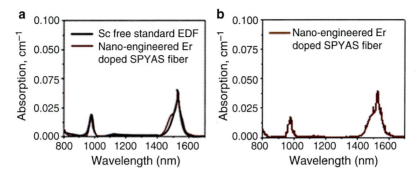

Fig. 14 Absorption spectra of the Sc-free standard Er-doped fiber/Er-doped pristine SPYAS glass-based fiber (**a**) and nano-engineered Er-doped SPYAS-based fiber (**b**). (Reprinted with permission from reference 75, P. H. Reddy, S. Das, D. Dutta, A. Dhar, A. V. Kir'yanov, M. Pal, S. K. Bhadra and M. C. Paul, Physica Status Solidi A 1700615 (2018), copyright @ 2018 WILEY-VCH Verlag GmbH & Co. KGaA, Weinheim)

fluorescence decay time. The 976 nm laser diode was modulated externally by an acousto-optic modulator in the course of the lifetime measurement. We used less than 5 cm long fibers to avoid amplified spontaneous emission and reabsorption.

The nano-engineered Er-doped SPYAS glass-based fiber shows a very broad fluorescence spectrum as well as strong emission signal at 1550 nm with increasing pump power at 980 nm compared to the Er-doped pristine SPYAS glass-based optical fiber and Sc-free standard Er-doped optical fiber shown in Figs. 15 and 16, respectively.

It was observed that, with increasing pump power up to 350 mW at 980 nm, the intensity of 1.5-μm emission increased significantly in the case of the nano-engineered SPYAS fiber, as evident from Fig. 16. The enhancement in fluorescence intensity in the annealed fiber is due to the presence of crystalline nanoparticles, where scandium (Sc) enhances the crystal field strength of Er ions as the ionic radius of Sc is smaller than that of Er ions (Fornasiero et al. 1998) and also, because of diminished up-conversion phenomena as yttrium (Y) cations substitute Er ions, preventing up-conversion involved neighboring Er ions due to similar ionic radius of Er and Y ions (Lo Savio et al. 2013). The fluorescence life time measurement curves of erbium-doped fibers drawn before and after thermal annealing were shown in Fig. 17.

The fluorescence lifetimes of the $^4I_{13/2} \rightarrow {}^4I_{15/2}$ emission of the pristine and annealed fibers are found to be 10.159 ms and 11.3 ms, respectively. The longer lifetimes of the $^4I_{13/2}$ levels together with the enhanced intensity of the $^4I_{13/2} \rightarrow {}^4I_{15/2}$ emission in the annealed fiber sample compared to the base pristine EDF suggest a scandium environment for Er^{3+} ions according to their incorporation in the nanocrystalline phase. Moreover, the maximum phonon energy of Sc_2O_3 is reported around 625 cm^{-1} (Palambo and Pratesi 2004). The low phonon energy along with the surrounding crystalline environment of erbium ions favors to enhance the fluorescence lifetime.

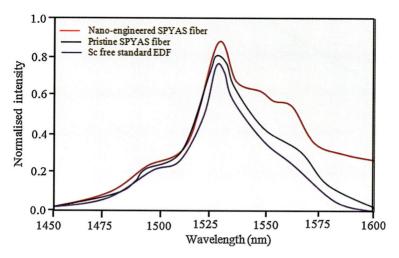

Fig. 15 Fluorescence spectra of the Er-doped pristine, nano-engineered SPYAS glass-based optical fibers and Sc-free standard Er-doped optical fiber. (Reprinted with permission from reference 75, P. H. Reddy, S. Das, D. Dutta, A. Dhar, A. V. Kir'yanov, M. Pal, S. K. Bhadra and M. C. Paul, Physica Status Solidi A 1700615 (2018), copyright @ 2018 WILEY-VCH Verlag GmbH & Co. KGaA, Weinheim)

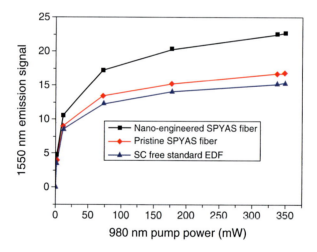

Fig. 16 The effect of 980 nm pump power on 1550 nm emission signal intensity of the pristine SPYAS fiber, nano-engineered SPYAS fiber and Sc-free standard Er-doped optical fiber (**c**). (Reprinted with permission from reference 75, P. H. Reddy, S. Das, D. Dutta, A. Dhar, A. V. Kir'yanov, M. Pal, S. K. Bhadra and M. C. Paul, Physica Status Solidi A 1700615 (2018), copyright @ 2018 WILEY-VCH Verlag GmbH & Co. KGaA, Weinheim)

The experimental setup is shown in Fig. 18 to investigate the gain and noise figure characteristics of the fabricated SPYAS-based EDF. In the experiment, tunable laser source (TLS) as an input signal was varied from 1520 to 1620 nm wavelength. The variable optical attenuator (VOA) is used to obtain the specific and accurate input

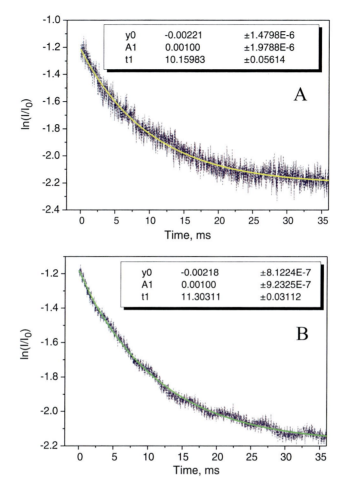

Fig. 17 Fluorescence decay curve of Er-doped nano-engineered SPYAS glass-based optical fibers drawn from the (**a**) pristine and (**b**) annealed preforms. (Reprinted with permission from reference 75, P. H. Reddy, S. Das, D. Dutta, A. Dhar, A. V. Kir'yanov, M. Pal, S. K. Bhadra and M. C. Paul, Physica Status Solidi A 1700615 (2018), copyright @ 2018 WILEY-VCH Verlag GmbH & Co. KGaA, Weinheim)

power to the cavity. An isolator is placed after input power to prevent a backward ASE noise from entering the first stage and parasitic lasing which cause degradation of the optical gain of amplifiers. Ten meter lengths of the newly developed SYAS-based EDFs drawn from both the pristine and annealed performs are used as a gain medium. The EDF is forward pumped by a 980 nm laser diode via a 980/1550 nm wavelength division multiplexing (WDM) coupler. The results were analyzed and measured by optical spectrum analyzer (OSA), which was located at the end of the configuration setup of the amplifier.

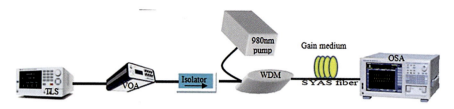

Fig. 18 Experimental setup for measuring the gain and noise figure of SPYAS glass-based fiber. (Reprinted with permission from reference 75, P. H. Reddy, S. Das, D. Dutta, A. Dhar, A. V. Kir'yanov, M. Pal, S. K. Bhadra and M. C. Paul, Physica Status Solidi A 1700615 (2018), copyright @ 2018 WILEY-VCH Verlag GmbH & Co. KGaA, Weinheim)

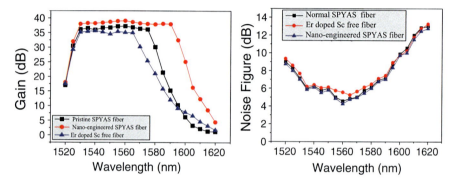

Fig. 19 Comparison of gain and noise figure spectra between Er-doped nano-engineered SYAS glass-based optical fiber drawn from pristine and annealed performs. (Reprinted with permission from reference 75, P. H. Reddy, S. Das, D. Dutta, A. Dhar, A. V. Kir'yanov, M. Pal, S. K. Bhadra and M. C. Paul, Physica Status Solidi A 1700615 (2018), copyright @ 2018 WILEY-VCH Verlag GmbH & Co. KGaA, Weinheim)

In Fig. 19, we show the gain and noise-figure (NF) spectra of 10-m pieces of the Er-doped nano-engineered SPYAS glass-based fibers, drawn from the pristine (red curve) and the annealed (black curve) preforms, measured under 175-mW laser-diode pumping at 980 nm using a − 15-dBm signal as launch. As seen from the Fig. 19, the maximum 39.45 dB gain has been obtained at 1560 nm with the fiber drawn from the annealed preform. Furthermore, the nano-engineered optical fiber shows an average gain of 38.675 dB within the spectral gain flatness region. Such kind of nano-engineered optical fiber shows a better gain variation of ±0.70 dB within 1530-1590 nm wavelength compared to the standard Sc-free Er-doped fiber, which is important for a broad band optical amplifier. On the other hand, the variation of the NF of the pristine Er-doped fiber, nano-engineered SPYAS glass-based fiber and standard Sc-free Er-doped fiber within the spectral gain flatness region is found to be (4.61–7.15), (4.35–6.85), and (5.5–7.67) dB, respectively, shown in Table 3. Figure 19 shows that the values of NF decrease from the nano-engineered SPYAG glass-based fiber to the standard Sc-free EDF.

Table 3 Amplification parameters of Er-doped nano-engineered and printine SPYAS glass-based fibers along with Er-doped Sc-free standard fiber

Amplification parameters	Er-doped nano-engineered SPYAS fiber	Er-doped normal SPYAS fiber	Er-doped Sc-free fiber
Maximum gain at 1560 nm	39.45 dB	37.35 dB	35.89 dB
Flat-gain region	1530–1590 nm	1530–1575 nm	1530–1565 nm
Flat-gain(average value)	38.675 dB	36.765 dB	36.5 dB
Gain variation (dB)	±0.70	±0.67	±0.82
Noise fig. (NF) (dB)	4.35–6.85	4.61–7.15	5.5–7.67

The improvement of gain performance in the first fiber is most probably explained due to the presence of annealing. Furthermore, the ligand field of the Sc subsystem induces the Stark effect because of site-to-site variations of the surrounding environment of Er ions due to inhomogeneous energy level degeneracy which may cause the widened optical transitions for getting broad band flat gain spectra.

On the other hand, the passive loss of such kind of fiber arises from OH content giving rise to strong absorption at 1380 nm. Due to the presence of hydroxyl groups, energy transfer from $^4I_{13/2}$ level of Er^{3+} to OH ions degrades the gain performance of the fibers. Here the energy levels of $^4I_{15/2}$ and $^4I_{13/2}$ are generally split more widely in SPYAS fiber drawn from annealed preform suggesting that the crystal field around Er ions sites is stronger in Sc_2O_3 than in Y_2O_3 due to the smaller ionic size of Sc^{3+} than that of Y^{3+} which bring the neighboring oxygens closer to Er^{3+} surrounded by Sc_2O_3. The results suggest that such kind of Er-doped nano-engineered fiber-based amplifiers are expected to be useful in broadband optical communication systems.

Fabrication of Multielement (P-Yb-Zr-Ce-Al-Ca) Fiber for Moderate-Power Laser Application

Fabrication of high power laser is an important research topic due to its wide range of applications considering materials processing, defense application, biomedical uses, etc. Although power level in the range of 1 kW had already achieved, researchers from different laboratories around the world are looking for higher power level for some specific application in the strategic field. Nevertheless, the bottleneck to achieve higher power is associated with nonlinear effects like SBS, SRS, four wave mixing, and photodarkening which must be controlled to get long term stability at high power level. Accordingly, it is inevitable to search for an improved glass host which could provide better thermal stability and thus accordingly we have presented a multielement (ME) fiber comprising Yb to test its performance under moderate power level.

As mentioned earlier, the preform fabrication was carried out employing conventional MCVD-solution doping technique starting with Suprasil F-300 grade silica tube of dimension 20 mm (OD) with tube thickness of 1.5 mm. Different process parameters such as the porous core layer deposition temperature, solution composition, sintering environment, and temperature were precisely optimized in order to achieve the desired core glass composition. The process was initiated by deposition of numbers of matched cladding layers composed of pure silica at around 1850 ± 10 °C. Subsequently, phosphosilicate porous layer was deposited at 1300 ± 10 °C using the back-pass deposition technique to avoid complete sintering of the deposited soot layer. The presintering of the soot layer was one of the important process steps to achieve the desired dopant level in the final fiber, with temperature being precisely controlled and monitored using an IR pyrometer, moved synchronously with an oxy-hydrogen burner along the tube in forward direction. The porous soot layer was then soaked with an alcoholic solution comprising different dopants, namely $AlCl_3.6H_2O$ (Alfa Aesar), $YbCl_3.6H_2O$ (Alfa Aesar), $CeCl_3.6H_2O$ (Alfa Aesar), $ZrOCl_2.6H_2O$ (Alfa Aesar), and $CaCl_2$ (Alfa Aesar), for fixed time span in different preform runs. The soaked tube was then dried under inert gas (nitrogen) flow and rejoined for subsequent processing. The soaked layer was then oxidized and dehydrated, followed by sintering employing the optimized conditions. The final preform was obtained through collapsing of the processed tube, followed by annealing of the preform at around 800–900 °C for fixed time span prior to fiber drawing. Drawing of low RI coated fiber of bare diameter 125 ± 0.5 μm was made from the fabricated preform after grinding to double D-shaped/hexagonal structure, employing a drawing tower (Heathway, UK). Hexagonal cladding has been achieved by grinding the initial fabricated preform equally from six phases followed by polishing prior to drawing of fiber. The inner cladding from flat to flat surface and vertex to vertex of hexagonal-shaped fiber is found to be 120 μm and 130 μm, respectively, while that of the outer polymer cladding is about 250 μm. The core/clad boundary of the fabricated fibers was inspected using a high-resolution optical microscope.

The use of different dopants in ME (P-Yb-Zr-Ce-Al-Ca) core-glass composition is to achieve superior properties due to change of local environment which can, in turn, tune spectroscopic properties. Al-doping is known to enhance RE solubility by providing a solvation shell (Aria et al. 1986), whereas P_2O_5 doping helps in nucleation to growth of nano-crystals during annealing step (Varshneya (1994); Paul et al. 2012; d'Acapito et al. 2008; Rawson (1967)). CaO co-doping in multielement glass is known to stabilize Yb^{3+} state (Singh et al. 2013; Sugiyama et al. 2013) and thereby reduces the formation of detrimental Yb^{2+} ions which is one of the main reasons for the PD effect. Ca^{2+} ion along with Al^{3+} stabilizes Yb^{3+} valence state with charge compensation in the silica glass. Ce_2O_3 is known to prevent formation of Yb^{2+} through charge transfer mechanism (Jetschke et al. 2016; Engholm et al. 2009) with the temporary oxidation of Ce^{3+} to Ce^{4+} which helps to trap the holes or electrons. The addition of ZrO_2 plays significant role by providing thermal stability, thermo-chemical resistance, as well as optical transparency (Patra et al. 2002).

Table 4 Properties of fibers under investigation

ID	Core composition	NA (±0.01)	Core diameter (±0.1) μm	Cladding absorption @ 976 nm, dB/m
A	SiO_2-Yb_2O_3-P_2O_5-Al_2O_3	0.15	10.4	7.9
B	SiO_2-Yb_2O_3-P_2O_5-Al_2O_3-Ce_2O_3-CaO-ZrO_2	0.15	10.3	8.3

Table 5 Comparison of elemental analysis of fibers under investigation

ID	Yb (mol%)	Al (mol%)	P (mol%)	Ca (mol%)	Ce (mol%)	Zr (mol%)
A	05	19	56	–	–	–
B	06	18	55	11	07	03

Amounts of the dopants incorporated in various preforms and their distributions across their cores were measured using a polished slice of thickness 1.5 mm using an electron probe microanalysis (EPMA). Formation of phase separated regions within the fiber core was detected using a high resolution transmission electron microscope (Technai G^2, Japan) along with measuring electron diffraction profiles in a selected area. The fiber parameters of standard phospho-alumino-silicate (PAS) glass-based Yb-doped fiber and multielements (ME) doped PAS glass-based fibers are given in Table 4. Table 5 presents the elemental composition of fabricated fibers. The main target was to investigate the role of multielements in standard phospho-alumino-silicate glass on lasing and photodarkening properties of the Yb-doped fiber laser where both the fibers maintained almost fixed Yb-content, as evident from their pretty close cladding absorption around 976 nm. It is also worth to note that both fibers also have approximately same P_2O_5 and Al_2O_3 content and thus any variation in the final performance of fiber B is due to addition of Ca, Ce, and Zr in glass structure.

The measurements of absorption and fluorescence spectra, using an optical spectrum analyzer (OSA) of the Ando type, as well as the measurements of fluorescence lifetimes and analysis of the resultant fibers in the sense of high-power applications, were made. The absorption spectra were measured applying a standard cut-back procedure with the use of a white-light source; Yb fluorescence spectra and kinetics were obtained in-cladding pumping of the fibers by a pig-tailed diode laser at 976-nm wavelength decay in the lateral detecting arrangement. The Yb-doped fibers were tested under the laser configuration, schematically presented in Fig. 20. A double D-shaped fiber in double clad structure was pulled in 125 μm inner cladding diameter. The large inner cladding diameter was chosen to enable an efficient pump launch from the high-power pump laser diodes. The fiber was end-pumped by a 976 nm laser diode. Pump launch end of the fiber was cleaved perpendicular to the fiber axis to provide 4% Fresnel reflection to form the laser cavity. At the other end, a high reflective mirror (100%) at signal band was used to close the laser cavity.

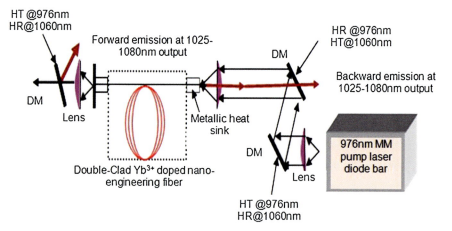

Fig. 20 Schematic of laser setup. (Reprinted with permission from reference 2, A. Dhar, M. C. Paul, S. Das, H. P. Reddy, S. Siddiki, D. Dutta, M. Pal and A. V. Kir'yanov; Physica Status Solidi A, 214, 4–7 (2017), copyright @ 2017 WILEY-VCH Verlag GmbH & Co. KGaA, Weinheim)

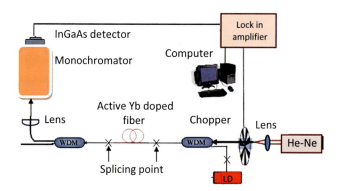

Fig. 21 Schematic of experimental PD setup. (Reprinted with permission from reference 2, A. Dhar, M. C. Paul, S. Das, H. P. Reddy, S. Siddiki, D. Dutta, M. Pal and A. V. Kir'yanov; Physica Status Solidi A, 214, 4–7 (2017), copyright @ 2017 WILEY-VCH Verlag GmbH & Co. KGaA, Weinheim)

The PD of the fibers was evaluated by monitoring the transmitted probe power at 633 nm through the test fiber under 976 nm irradiation. The PD measurement setup is presented in Fig. 21. A fiber-coupled single-mode 976 nm laser diode was used as a pump source. The output end of the pump fiber was spliced to one port of a wavelength-division multiplexing (WDM) coupler, and the pump beam was delivered to the test fiber by splicing the output end of the WDM coupler and the test fiber.

A He-Ne laser at 633 nm was used as a probe beam once coupled to the test fiber through the WDM. The probe beam propagated in the same direction as the pump beams. The output end of the test fiber was spliced to another WDM coupler to separate the pump and probe beam. The probe beam was chopped by a mechanical chopper, and the output power was detected by a photo-detector and lock-in amplifier after passing through a monochromator. We used very short piece (1–2 cm) of the test fiber to avoid any effect of amplified spontaneous emission on the data and to diminish possible re-absorption increasing at PD. The pump power was maintained to provide ~40% of population inversion of Yb^{3+} ions along the test fibers.

Material Characterization of Multielement (P-Yb-Zr-Ce-Al-Ca) Optical Preform and Fiber

Refractive index profile of different preform/fiber samples was found to have a central dip, especially featured in GeO_2 and P_2O_5 doped performs fabricated using the MCVD process. An example of RI profile (RIP) of fiber sample B is presented in Fig. 22 which reveals no central dip which is achieved due to use of multielement glass host composition.

The analysis of core sample under high resolution transmission electron microscope (HRTEM) revealed that phase-separation occurs in preform sample B but not in A. Presence of liquid miscibility gap in the phase diagram of our selected glass composition is the main reason behind achieving the phase-separation within the core area. Interestingly, nano-particles within the phase-separated core region

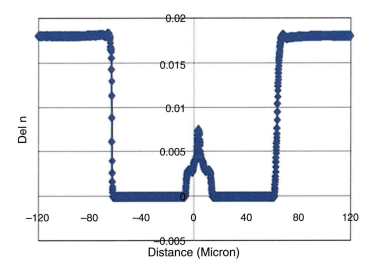

Fig. 22 Refractive index profile of fiber B. (Reprinted with permission from reference 2, A. Dhar, M. C. Paul, S. Das, H. P. Reddy, S. Siddiki, D. Dutta, M. Pal and A. V. Kir'yanov; Physica Status Solidi A, 214, 4–7 (2017), copyright @ 2017 WILEY-VCH Verlag GmbH & Co. KGaA, Weinheim)

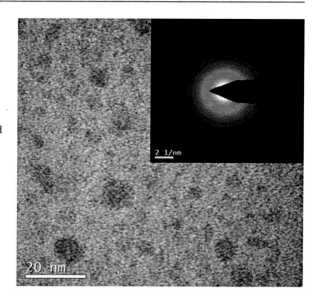

Fig. 23 HRTEM analysis data with ED pattern of ME-Yb fiber (sample B) showing particles of size range 7–10 nm. (Reprinted with permission from reference 2, A. Dhar, M. C. Paul, S. Das, H. P. Reddy, S. Siddiki, D. Dutta, M. Pal and A. V. Kir'yanov; Physica Status Solidi A, 214, 4–7 (2017), copyright @ 2017 WILEY-VCH Verlag GmbH & Co. KGaA, Weinheim)

retained only in sample B in its fiber stage. This clearly indicates better thermal stability of core glass composition in preform sample B and the effect of addition of ZrO_2. The HRTEM data of annealed fiber sample B are presented in Fig. 23 (main frame) together with the selected area electron diffraction (SAED) pattern (inset) which indicates the phase-separated nano-particles ranging from 7 to 10 nm with partially crystalline nature. The electron diffraction X-ray (EDX) analysis associated with HRTEM confirms the presence of Zr ion in the fiber core.

The elemental distribution of different dopants in the fiber core of fiber sample B measured using EPMA is presented in Fig. 24 while for fiber sample A is presented in Fig. 25. The result obtained reveals that there is no central dip formation in case of P_2O_5 which is generally observed in P_2O_5-doped fiber sample prepared through the MCVD process. This type of dopant distribution curve was observed in our fabricated ME optical fibers without any depression of P_2O_5 level and probably occurred due to suppression of evaporation of P_2O_5 in the presence of Zr, Ca and Ce.

The spectra of absorption loss were obtained using a white-light source through Bentham spectral attenuation setup with a fiber output and optical spectrum analyzer (OSA) turned to a 0.5 nm resolution. The spectrum for fiber sample B was compared with standard PAS fiber (sample A) to understand the effect of different dopants (Ca, Zr, and Ce) comprising of similar Yb_2O_3 concentration, ~0.15 mol%. The combined data presented in Fig. 26 clearly reveal that although the 975-nm peak height is virtually the same, there is a variation in the shoulder around 915-nm and this variation is the characteristic of the core glass composition.

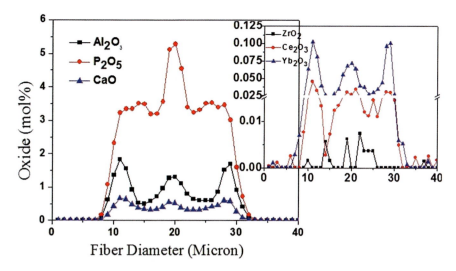

Fig. 24 Dopants distributions in ME-Yb fiber (sample B) measured using EPMA. (Reprinted with permission from reference 2, A. Dhar, M. C. Paul, S. Das, H. P. Reddy, S. Siddiki, D. Dutta, M. Pal and A. V. Kir'yanov; Physica Status Solidi A, 214, 4–7 (2017), copyright @ 2017 WILEY-VCH Verlag GmbH & Co. KGaA, Weinheim)

Fig. 25 Dopants distributions in standard Yb fiber (sample A) measured using EPMA. (Reprinted with permission from reference 2, A. Dhar, M. C. Paul, S. Das, H. P. Reddy, S. Siddiki, D. Dutta, M. Pal and A. V. Kir'yanov; Physica Status Solidi A, 214, 4–7 (2017), copyright @ 2017 WILEY-VCH Verlag GmbH & Co. KGaA, Weinheim)

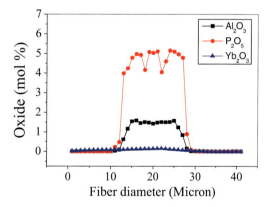

The Optical Performance of Multielement (P-Yb-Zr-Ce-Al-Ca) Optical Fiber

To study the lasing performance, we have measured the output power using the experimental setup described earlier in Fig. 20. It is clearly observed that the output power linearly increased with the launched pump power. The output power reached 37.5 W at launching 50 W of pump power, representing good slope efficiency of 76% with respect to the launched pump power. The comparison of output powers obtained from standard fiber A and ME fiber B is presented in Fig. 27. The output

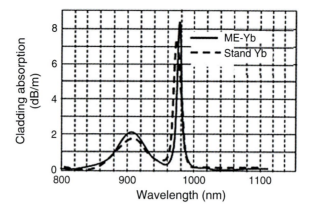

Fig. 26 Comparison of cladding absorption of standard phospho-alumino-silicate Yb fiber (sample A) with ME Yb fiber (sample B). (Reprinted with permission from reference 2, A. Dhar, M. C. Paul, S. Das, H. P. Reddy, S. Siddiki, D. Dutta, M. Pal and A. V. Kir'yanov; Physica Status Solidi A, 214, 4–7 (2017), copyright @ 2017 WILEY-VCH Verlag GmbH & Co. KGaA, Weinheim)

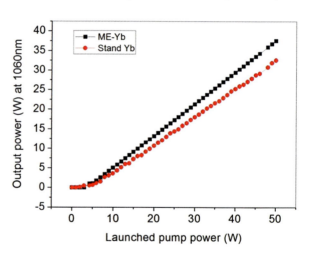

Fig. 27 Laser output power of standard phospho-aluminosilicate Yb fiber (sample A) and ME Yb fiber (sample B). (Reprinted with permission from reference 2, A. Dhar, M. C. Paul, S. Das, H. P. Reddy, S. Siddiki, D. Dutta, M. Pal and A. V. Kir'yanov; Physica Status Solidi A, 214, 4–7 (2017), copyright @ 2017 WILEY-VCH Verlag GmbH & Co. KGaA, Weinheim)

power in both A and B reported here is not the maximum but can be further enhanced by increasing launched pump power.

In general, the induced photodarkening (PD) loss is proportional to the inversion level of the Yb^{3+} ions. However, it is well known fact that the selected host material can influence PD significantly; e.g.; phosphosilicate glass is known to lead to suppression of PD loss in a significant amount, as compared to the aluminosilicate counterpart. The PD behavior of both fibers was thus carried out to evaluate the effect of host-glass composition with special emphasis on effect of dopant ions like Ca, Ce and Zr ions present in fiber B. The small signal absorption in different fibers was around 7–8 dB/m. The temporal characteristics of the transmitted probe power @633 nm are represented in Fig. 28 for standard Yb-doped fiber with our fabricated

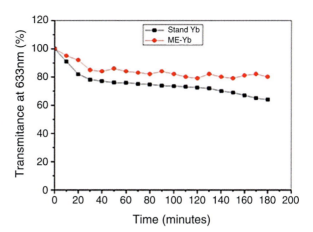

Fig. 28 PD induced loss of standard phospho-aluminosilicate Yb fiber (sample A) and ME Yb fiber (sample B). (Reprinted with permission from reference 2, A. Dhar, M. C. Paul, S. Das, H. P. Reddy, S. Siddiki, D. Dutta, M. Pal and A. V. Kir'yanov; Physica Status Solidi A, 214, 4–7 (2017), copyright @ 2017 WILEY-VCH Verlag GmbH & Co. KGaA, Weinheim)

ME-Yb-doped fiber. The observed result clearly indicates that the PD induced loss is significantly reduced in the Yb-doped ME (P_2O_5-ZrO_2-Al_2O_3-CeO_2-CaO) fiber.

A relation has been established between the absorption coefficient change of probe power @633 nm and the time under pumping at 976 nm. The experimental curves are fitted by exponential decay curve (red color) shown in Figs. 29 and 30 where the $\Delta\alpha(t)$ is the absorption coefficient change in units of dB/m, the constant $\Delta\alpha_{sat}$ is the saturation parameter and τ is the time constant. The following relation is given below with time.

$$-\Delta\alpha(t) = \Delta\alpha_{sat} + \Delta\alpha_{sat} \cdot e^{-(t/\tau)}.$$

The photodarkening-induced absorption coefficient change at 633 nm for standard Yb-doped fiber (A) is found to be very fast compared to multielements glass-based Yb-doped fiber (B) as presented in Figs. 29 and 30. The photodarkening phenomenon of Yb-doped fibers show spectrally broad transmission loss centering at the visible wavelengths and extending up to the pump and signal wavelength region due to formation or creation of color centers which causes the main absorption band at visible region tailoring to the NIR region. Therefore, monitoring of transmitted power of 633 nm visible light with time under pumping at 976 nm will give an indication about the photodarkening behavior of Yb-doped fibers. So, we have monitored the transmitted probe power of 633 nm visible light with time (Koponen et al. 2006).

In case of high power operation of fiber laser, there is a probability of reduction of Yb^{3+} to Yb^{2+} ions with formation of free holes. These holes then get trapped at various defects and imperfections in the glass structure with the result of an induced absorption. At the same time, Yb^{2+} ions have possibility to oxidize and transform back to Yb^{3+} ion with the release of an electron. To overcome this problem, Ce, Ca, and Zr are added into the doping host of Yb whose already explained in the fabrication part.

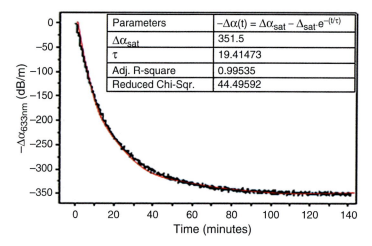

Fig. 29 Photodarkening induced absorption coefficient change $\Delta\alpha$ of standard fiber A at 633 nm with time at 10.0 W pump power. (Reprinted with permission from reference 2, A. Dhar, M. C. Paul, S. Das, H. P. Reddy, S. Siddiki, D. Dutta, M. Pal and A. V. Kir'yanov; Physica Status Solidi A, 214, 4–7 (2017), copyright @ 2017 WILEY-VCH Verlag GmbH & Co. KGaA, Weinheim)

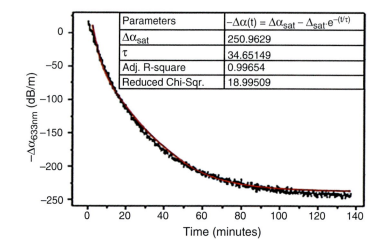

Fig. 30 Photodarkening induced absorption coefficient change $\Delta\alpha$ of multielement glass-based fiber B at 633 nm with time at 10.0 W pump power. (Reprinted with permission from reference 2, A. Dhar, M. C. Paul, S. Das, H. P. Reddy, S. Siddiki, D. Dutta, M. Pal and A. V. Kir'yanov; Physica Status Solidi A, 214, 4–7 (2017), copyright @ 2017 WILEY-VCH Verlag GmbH & Co. KGaA, Weinheim)

In support of our observation regarding PD experiment, we compared the peak originated around 220–230 nm associated with the formation of Ytterbium Oxygen Defect Center (YbODC) (Yoo et al. 2007; Engholm and Norin 2008). The peak intensity related to YbODC was found to reduce appreciably in our ME-Yb-doped

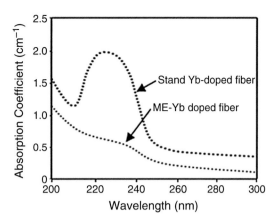

Fig. 31 Comparison of transmission loss in UV region between standard phospho-aluminosilicate Yb-fiber (sample A) with ME-Yb fiber (sample B). (Reprinted with permission from reference 2, A. Dhar, M. C. Paul, S. Das, H. P. Reddy, S. Siddiki, D. Dutta, M. Pal and A. V. Kir'yanov; Physica Status Solidi A, 214, 4–7 (2017), copyright @ 2017 WILEY-VCH Verlag GmbH & Co. KGaA, Weinheim)

optical fiber sample with respect to standard alumino-phospho silicate glass-based optical fiber (A), as evident from Fig. 31. Ytterbium-doped alumina-silica glass has a strong absorption band near 230 nm, known as a charge transfer band. This absorption band corresponds to the transfer of an electron from the nearby ligands (oxygen) to the ytterbium ion with the formation of a divalent Yb ion and a localized hole left behind on the surrounding ligands. The probability for the creation of mobile charges decreases with addition of ZrO_2 in the glass structure and hence reduces the formation of induced color centers.

This fact indicates that the presence of ME core glass reduces the rate of formation of YbODC centers. PD in the Yb-doped aluminosilicate takes place through the breaking of ODCs under two-photon absorption which gives rise to release of free electrons. The released electrons may then be trapped at Al or Yb sites and form a color center resulting in PD. To avoid that situation, we have selected ME core glass to reduce the formation of YbODC significantly. Additionally, formation of nano-phase separated region is expected to facilitate shifting of a lone electron to the nearest neighbor nonbridging oxygen (NBO) atom bonded to Yb ions into the positive charge vacancy zones under high-pump power and therefore enhances PD resistivity significantly. As the field strength of the modifier cation enhances either through decreasing ionic radius or increasing valence charge, which may give the perturbation of aluminosilicate network more strongly. Such perturbation occurs because of the energetically stabilization of Yb ions provided by closer association of negatively charged species, in particular NBO.

Fabrication of Chromium-Doped Nano-phase Separated Yttria-Alumina-Silica (YAS) Glass-Based Optical Fiber

Chromium (Cr)-doped optical fiber is very important in terms of its application as saturable absorber and in optical coherence tomography (OCT) covering a wavelength region of 1.1–1.6 μm range. Nevertheless, it is important to stabilize

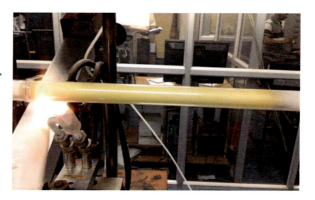

Fig. 32 Appearance of yellow coloration during oxidation stage after solution soaking. (Reprinted with permission from reference 17, D. Dutta, A. Dhar, A. V. Kir'yanov, S. Das, S. Bysakh and M. C. Paul; Phys. Status Solidi A 212 1836 (2015), copyright @ 2015 WILEY-VCH Verlag GmbH & Co. KGaA, Weinheim)

Cr^{4+} inside the suitable glass host structure to achieve the abovementioned broad band fluorescence. Although there are some literature available which deals with silica glass-based Cr^{4+}-doped fibers based on alumino-silica, germano-silica, and gallium-silica glasses, fabricated using the well-known MCVD-solution doping (SD) technique but nobody able to report broadband emission of Cr^{4+} ions unlike bulk glass systems as it is extremely difficult to stabilize Cr^{4+} in silica glass-based optical fiber. Accordingly, here we have presented fabrication of nano-phase separated yttria-alumina-silica glass-based optical fibers through the MCVD process in conjunction with a solution doping technique under suitable thermal annealing conditions.

Chromium-doped preform fabrication process starts with the deposition of porous silica layers formed due to oxidation of $SiCl_4$ vapor in presence of oxygen inside a waveguide silica tube of an outer dimension sized by 20 mm with thickness of 1.5 mm at a suitable temperature (Fig. 32). The temperature during deposition was monitored using an IR pyrometer which moves synchronously with an oxy-hydrogen burner along the tube in forward direction. Different dopant ions, namely, Al, Cr, Mg, and Y, were introduced into the porous frit employing the SD technique (Townsend et al. 1987).

Material Characterization of Chromium-Doped Nano-phase Separated YAS Glass-Based Optical Preform and Fiber

The concentrations of the dopants inside the cores were varied by means of altering the composition of soaking solution comprising $AlCl_3$, $6H_2O$ (Alfa Aesar), $CrCl_3$, $6H_2O$ (Alfa Aesar), $MgCl_2$, $6H_2O$ (Alfa Aesar) and $Y(NO_3)_3$, xH_2O (Alfa Aesar), dissolved in a mixture of water and ethanol. After soaking for a stipulated time span, the soaked tube was air dried, remounted on the glass working lathe followed by subsequent processing to obtain the final fiber.

Refractive index profiles (RIP) of the fabricated preforms and fibers were measured using preform analyzer (PK-2600, Photon Kinetics, USA) and a fiber

Table 6 Details of representative fibers

Fiber no.	Core composition	NA ±0.01	Core-dia (±0.02) μm
Cr-2	SiO_2-Al_2O_3-MgO-Y_2O_3-Cr_2O_3	0.21	10.5
HCr-2	SiO_2-Al_2O_3-MgO-Y_2O_3-Cr_2O_3	0.21	10.5

analyzer (NR-9200, EXFO, Canada). The amounts of incorporated dopants and their distributions across the cores were measured using an electron probe microanalysis (EPMA) using a polished sample of thickness 1.5 mm. The microstructure of the porous deposit was evaluated using a scanning electron microscope (SEM) while in the fiber was measured using a high resolution transmission electron microscope (HRTEM) (Technai G^2, Japan) along with selected area electron diffraction (SAED) profile. The absorption spectra of the fabricated fibers in the ranges of 300–1100 nm and 400–1600 nm were measured using a Bentham setup (UK) and an Optical Spectrum Analyzer (OSA) (AQ6370B, Yokogawa, Japan), respectively. The emission spectra of samples with thicknesses of ∼1 cm were measured by exciting the samples at 370 and 474 nm using a Xenon lamp (Edinburgh Instrument, UK). Accordingly, the fluorescence spectra of fiber samples were measured using the OSA at excitation by an Ytterbium fiber laser (wavelength 1064 nm, IPG Polus, USA), in frontal geometry. This characterization method was repeated for different fiber samples before and after annealing to investigate its effect with respect to changes in concentration of chromium. The heat-treated fibers are assigned by the symbol "HCr." The annealing conditions (temperature, time span, and heat cycles) were optimized based on our initial characterization results.

The detailed results obtained using one representative fiber sample is presented here (Table 6). The average numerical aperture (NA) of the fabricated Cr-2 was found to be around 0.21 with good uniformity along its overall length (∼20 cm) and the representative RIP profile of the core region is demonstrated in Fig. 33. The dopants distributions measured using EPMA inside the fiber core are presented in Fig. 34a and b which reveal quite uniform profiles with concentrations of the dopants being ∼8.5 wt% (Al_2O_3), ∼0.75 wt% (Y_2O_3), ∼0.013 wt% (Cr_2O_3), and ∼2.0 wt% (MgO). It is worth noting here that a divalent co-dopant (Mg^{2+} in our case) has to be incorporated for generating tetrahedral coordination of chromium as it facilitates stabilization of the Cr^{4+} species.

The nature and the size of the phase-separated particles within the core area (annealed and un-annealed) as well as in the final fiber samples were studied using the HRTEM and SAED analyses. The TEM image of the active core glass with ED pattern for Cr-2 nonannealed preform sample is shown in Fig. 35a. The HRTEM image of Cr-2 reveals absence of any phase-separated regions, whereas the HRTEM image of the fiber drawn from same preform after annealing (HCr-2) at 1200 °C exhibits presence of nano-phase separated inclusions, as evident from Fig. 35b. This nonuniformity reduces to below 5 nm in the case of the fiber drawn from the annealed preform with significant reduction in nonhomogeneity in the core glass shown in Fig. 36b. Since there is no phase separation in the nonannealed preform,

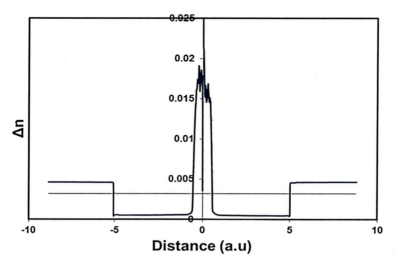

Fig. 33 RIP profile of chromium-doped fiber Cr-2. (Reprinted with permission from reference 17, D. Dutta, A. Dhar, A. V. Kir'yanov, S. Das, S. Bysakh and M. C. Paul; Phys. Status Solidi A 212 1836 (2015), copyright @ 2015 WILEY-VCH Verlag GmbH & Co. KGaA, Weinheim)

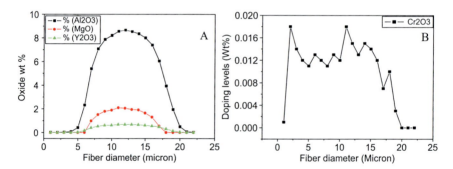

Fig. 34 (**a**) Distributions of different dopants (Al_2O_3, MgO, and Y_2O_3) across the core area of fiber Cr-2 and (**b**) distribution of Cr_2O_3 across the core area of fiber Cr-2. (Reprinted with permission from reference 17, D. Dutta, A. Dhar, A. V. Kir'yanov, S. Das, S. Bysakh and M. C. Paul; Phys. Status Solidi A 212 1836 (2015), copyright @ 2015 WILEY-VCH Verlag GmbH & Co. KGaA, Weinhcim)

the dark boundary surrounding the bright circle in the SAED pattern is not so prominent. The phase-separated inclusions appear as black spots with average size ranging from 10 to 60 nm. The corresponding size distribution for nano-phase in the 30–60 nm range is presented in Fig. 36b (average size 20–30 nm). The spot EDX on the globule and the matrix clearly indicates that the phase-separated nano-phase is rich in Al and Y, thus indicating the composition is possibly YAS glass. In case of the fiber drawn from the annealed samples as in Fig. 35b, those inclusions transform into very fine spots that tend to dissolve into the parent glassy phase with average

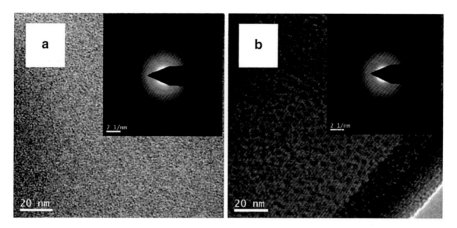

Fig. 35 TEM pictures along with ED pattern of chromium-doped glass-based (**a**) nonannealed preform (Cr-2) and (**b**) fiber drawn from the annealed (HCr-2) preform. (Reprinted with permission from reference 17, D. Dutta, A. Dhar, A. V. Kir'yanov, S. Das, S. Bysakh and M. C. Paul; Phys. Status Solidi A **212** 1836 (2015), copyright @ 2015 WILEY-VCH Verlag GmbH & Co. KGaA, Weinheim)

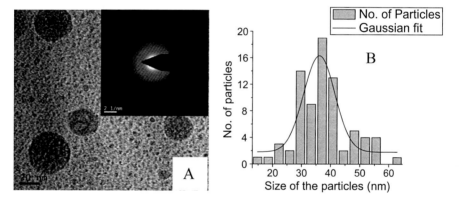

Fig. 36 TEM pictures along with ED pattern of chromium-doped glass-based preform: (**a**) annealed preform (HCr-2) and (**b**) particle size distribution in annealed preform (HCr-2). (Reprinted with permission from reference 17, D. Dutta, A. Dhar, A. V. Kir'yanov, S. Das, S. Bysakh and M. C. Paul; Phys. Status Solidi A 212 1836 (2015), copyright @ 2015 WILEY-VCH Verlag GmbH & Co. KGaA, Weinheim)

size around 5 nm shown in Fig. 36a. A possible reason of decrease in particle size is re-melting and stretching of the separated phase during the fiber's drawing.

The SEM images of porous structure before and after solution doping (SD) were examined in order to get an idea about nonuniformity in the dopants distributions in the final core. Additionally, a comparison of the EDX analysis of porous soot after making SD with that of the final one reveals that significant amount of chromium ions was lost due to evaporation during sintering/collapsing. The colorless soot particles during drying in the presence of oxy-hydrogen flame (tube's outside temperature ∼800 °C) first changes to yellow color as seen from Fig. 32 (due to the formation of Cr(VI) species) and is converted into clear glass after complete sintering. We observed faint greenish color inside the core for Cr_2O_3 concentration above 0.15 wt%.

Due to the small size of nano-phases observed at the fiber stage (Fig. 36b), a spot EDX analysis could not be performed and so the exact composition of these nano-size phases could not be determined. It can only be proposed that those granules with dark appearance may possess elements having higher atomic mass. According to our earlier work (Halder et al. 2013) on Tm-Yb co-doped nano-phase separated yttrium-alumino-silicate (YAS) core glass fibers, it was found that most of Tm-Yb ions concentrate within the black granular nano-YAS phases and that those granular nano-YAS phases are composed of Al_2O_3, Y_2O_3, and SiO_2 along with REs (Tm and Yb). Here we observed similar black granular phase separation and due to the similarity of the two glasses, we can presume that the black spots are chromium-rich zones within YAS glass which also observed in the case of the RE-doped nano-YAS matrix (Halder et al. 2013).

The Optical Performance of Chromium-Doped Nano-phase Separated YAS Glass-Based Optical Fiber

The absorption spectra of the fibers drawn from preforms Cr-2 (nonannealed) and HCr-2 (annealed) were measured using short (a few to tens cm) lengths of fiber; the resultant spectra are shown in Fig. 37. Both spectra can be equally decomposed into at least four bands in the VIS and near-IR, peaked at ∼430 (I), ∼630 (II), ∼860 (III), and ∼1030 (IV) nm and additionally an intensive band (V) in the UV within 300–400 nm is present in the spectra (inset to the figure). Based on literature data (Wang and Shen 2012; Murata et al. 1997), the absorption bands of both Cr-2 and HCr-2 fibers having Mg and Cr-doped alumino-silica glass is given below. The bands I and II are assigned to the (d-d) transitions $^4A_2$-$^4T_1$ and $^4A_2$-$^4T_2$, $^4A_2$-$^2T_1$, and $^4A_2$-^2E of Cr^{3+} ions in octahedral coordination, while band V to the charge-transfer transition of Cr^{6+} ions in octahedral coordination. The bands III and IV generate due to the presence of Cr^{4+} in tetrahedral coordination with transitions $^3A_2$-$^3T_1$ and $^3A_2$-$^3T_2$. In the meantime, the first of these transitions covers partially band II; so this spectral band (centered at ∼630 nm) seems to comprise the contributions from both Cr^{4+} (in the most degree) and Cr^{3+} (in the least degree) ions.

It is found from Fig. 37 that although the basic nature of the absorption spectra did not change after annealing, the extinctions of all the absorption bands in VIS to near-IR (I to IV) decrease. The same phenomenon happened for the band V located in the UV (compare curves 1 and 2). This clearly points out that certain structural rearrangements have occurred inside the fiber as a result of thermal treatment. Since the drop of extinction in all the bands' maxima is virtually equal and measured to be ∼3.0–3.5 times (notice that the band widths do not demonstrate changes), we can conclude that no serious transformations among Cr^{4+}, Cr^{3+}, and Cr^{6+} ions

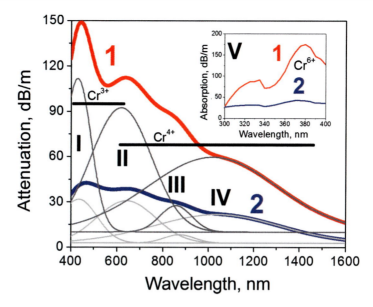

Fig. 37 Absorption spectra of chromium-doped fiber samples drawn from the preforms (**a**) before (Cr-2, curve 1) and (**b**) after (HCr-2, curve 2) annealing. Dashed lines show the bands I to IV that fit the experimental spectra 1 and 2 (a deconvolution procedure using Gaussian shape of the bands was used). (Reprinted with permission from reference 17, D. Dutta, A. Dhar, A. V. Kir'yanov, S. Das, S. Bysakh and M. C. Paul; Phys. Status Solidi A 212 1836 (2015), copyright @ 2015 WILEY-VCH Verlag GmbH & Co. KGaA, Weinheim)

happened (Dvoyrin et al. 2003), but rather a kind of interdiffusion process (Huang et al. 2013) of the dopants (including chromium ions) and Si (from the adjacent cladding region) took place at thermal annealing with the result being overall drop of extinction in all bands I to V. Another reason behind the phenomenon could be "generation" of different chromium species in lower valences (+1 and +2) at annealing, but the available literature data provide no evidence for such kind of process to go further. Furthermore, an explanation behind negligible conversion of Cr^{4+} species to Cr^{3+} at the thermal treatment that we applied seems to be a stabilizing effect, given by the presence of Mg in the core glass network, which is a species mainly involved Cr^{4+} ions formation. At the same time, the partial re-structuration of the core glass occurs from nonannealed to annealed stage of optical preform Cr-2—HCr-2 through suitable thermal annealing process. One of the consequences of the latter process is demonstrated by an experiment on excitation of fluorescence in Cr-2 and HCr-2 fibers at 474-nm pumping. The result is presented in Fig. 38.

The emission spectra at 474-nm excitation of the fibers exhibited a complex, but qualitatively similar to each other, spanning 500 to 800 nm spectral range. This fluorescence is associated with the presence of Cr^{3+} ions, as already reported in literature (Murata et al. 1997). However, its intensity was found to significantly

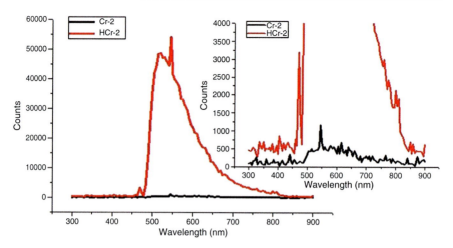

Fig. 38 Emission spectra of chromium-doped fiber samples before (Cr-2) and after (HCr-2) annealing process under pumping at 474 nm wavelength. In inset showing the enlarged view of emission spectra. (Reprinted with permission from reference 17, D. Dutta, A. Dhar, A. V. Kir'yanov, S. Das, S. Bysakh and M. C. Paul; Phys. Status Solidi A 212 1836 (2015), copyright @ 2015 WILEY-VCH Verlag GmbH & Co. KGaA, Weinheim)

enhanced under optimized annealing around 1200 °C (i.e., in sample HCr-2), as evident from Fig. 38. We guess that the main cause of the fluorescence enhancement effect is nano-phase re-structuration of the core glass (taking place as a result of thermal treatment), positively affecting the fluorescent ability of the emitter, in this case being Cr^{3+} ions.

Quite short pieces of fibers (2.5 and 5.5 cm) were used in this experiment, providing 2.5–3.0 dB absorption of pump light. The spectra of fluorescence spanning from ~900 to ~1400 nm (centered at ~1100 nm) were detected in both cases in vastly the same experimental conditions and so are worth of direct comparison. We present here the data obtained at ~50 mW pump power launched into the fibers (the spectrum of pump is shown by curve 3). Cr^{4+} fluorescence spectra recorded at higher pump powers (up to ~2.5 W) revealed almost no change in shape and negligible rise of intensity, given that saturating of fluorescence power happens at very low pump power, measured by 5–10 mW. As seen from Fig. 39, the fluorescence spectra are quite similar for Cr-2 and HCr-2 fibers, which shows that annealing does not lead to any pronounced worsening of the fluorescent ability of Cr^{4+} after thermal treatment. Emphasized that the fluorescence spectra in Fig. 39 match well the ones reported earlier for other types of Cr, Mg co-doped alumina-silica glass and fiber, also ascribed to Cr^{4+} ions (Huang et al. 2010, 2013; Wang and Shen 2012).

Let's consider the results of an experiment demonstrating the bleaching effect in Cr-2 and HCr-2 fibers under the action of 1.064-μm pump light as shown in Figs. 40a, b, respectively. In the experiment, we employed the same setup as was used in the fluorescence spectra measurements (Fig. 39), but now we measured

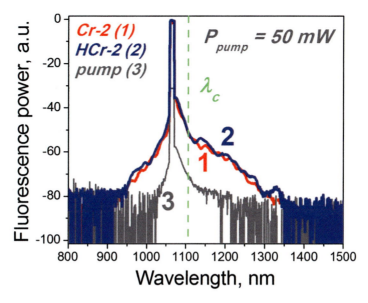

Fig. 39 Emission spectra of chromium-doped fiber samples drawn from the preform before (Cr-2, curve 1) and after (HCr-2, curve 2) annealing. The spectrum of pump light is given in (a) by curve 3 (pump power is 50 mW in all cases). λ_c marks the central wavelength of the emission spectra. (Reprinted with permission from reference 17, D. Dutta, A. Dhar, A. V. Kir'yanov, S. Das, S. Bysakh and M. C. Paul; Phys. Status Solidi A **212** 1836 (2015), copyright @ 2015 WILEY-VCH Verlag GmbH & Co. KGaA, Weinheim)

transmission coefficients (T) of pump light with power P after propagation through pieces (a few to tens cm in length L) of Cr-2 and HCr-2 fibers; these data were then proceeded for getting the nonlinear (dependent on incidence pump power) "resonant" absorption coefficients α of the fibers, applying formula $\alpha(P) = -\ln[T(P)]/L$ (i.e., the quantities provided in Fig. 40) in function of P.

It is seen from the figures that in both fibers a pronounceable bleaching effect, viz., a decrease of absorption against pump power is observed. Moreover, the effective saturating (bleaching) pump power (P_{sat}) is measured by a few (5–10) mW only, which is quite promising for utilizing of these or similar chromium-doped fibers as a Q-switch fiber for near-IR pulsed applications. Given that core diameter of both Cr-2 and HCr-2 fibers is $\sim$10 μm (Table 6), the saturating intensity at 1.06 μm is estimated to be around 10 kW/cm^2. Furthermore, if one assumes that the decay of the near-IR fluorescence of Cr^{4+} ions (Fig. 39) is around 10 μs, the ground-state absorption cross-section of Cr^{4+} ions in the fibers at $\sim$1.0–1.1 μm (i.e., via transition $^3A_2$-$^3T_2$ band IV; Figure 40) can be estimated to be $\sim$10^{-18} cm^2. Such value is comparable with the ones reported elsewhere (V. Felice et al. 2000) for chromium-doped alumina-silica glasses.

Finally, a brief discussion on the backgrounds we employed to make the fibers ought to be highlighted. As chromium is doped into YAS phase-separated particles through the solution doping technique, it substitutes aluminum and normally

Fig. 40 Dependences (points) of resonant absorption coefficient a of Cr-2 (**a**) and HCr-2 (**b**) fibers upon 1064 nm pump power P. Squares (at P = 0) mark the small-signal absorption values (a_0) at 1064 nm, also denoted in insets where near-IR part of the fibers absorption spectra are reproduced (Fig. 5). P_{sat} are the saturation pump power at which a is decreased twice relatively to a_0 values. (Reprinted with permission from reference 17, D. Dutta, A. Dhar, A. V. Kir'yanov, S. Das, S. Bysakh and M. C. Paul; Phys. Status Solidi A 212 1836 (2015), copyright @ 2015 WILEY-VCH Verlag GmbH & Co. KGaA, Weinheim)

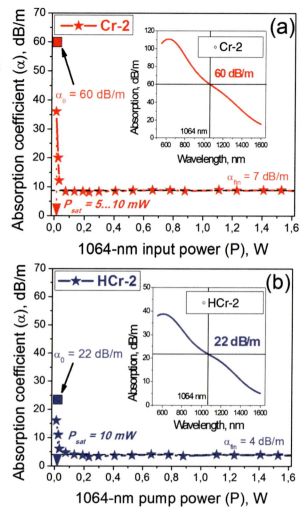

occupies the octahedral site, which is seen as the characteristic band in Fig. 37, ascribed to Cr^{3+} species. On the other hand, when the divalent co-dopant Mg^{2+} is introduced into glass, it substitutes Y^{3+} in dodecahedral site and leaves a negatively charged vacancy in the lattice. During the sintering process of soaked layer under slightly reducing environment, the oxygen-excess defect centers (such as super oxide ion O_2^- and peroxy bonding –O–O–) serve as an oxidizing agent for conversion $Cr^{3+} \rightarrow Cr^{4+}$. The formation of these oxygen-excess defect centers can be generated in the presence of a large content of glass modifier cations such as Mg. Cr^{4+} has a larger ionic size (0.041 nm) than Si^{4+} (0.026 nm) in coordination 4 and may be associated with bridging and/or nonbridging oxygen of the silica network (Shannon 1976).

However, in solid state oxides of aluminum, yttrium, and chromium cations are incorporated more preferably in the +3 oxidation state in an octahedral environment of the nearest oxygen atoms. Similarly, in the silica glass network that contains aluminum, chromium atoms can replace silicon atoms and be stabilized in the tetrahedral environment. During this transformation aluminum atoms form negatively charged complexes $[AlO_{4/2}]^{2-}$ (Murata et al. 1997). In silica glass, chromium ions presented in the second coordination sphere of aluminum and yttrium ions can serve as charge compensators which contribute to stabilization of aluminum in the tetrahedral site. At the same time, such charge compensation also favors stabilization of Cr^{4+} in the metastable tetrahedral site (Murata et al. 1997). Such a situation arises during the fiber drawing process with fast glass quenching. Thus, the effects of co-doping core glass with Al and Mg upon formation and stabilization of Cr^{4+} ions (and also upon hardening against decay to Cr^{3+} ions at thermal annealing) seem to be equally positive.

Presently, we are targeting at optimizing the chromium-doped fabrication conditions and also the core glass composition in order to achieve stable Cr^{4+} species in higher concentrations in the fiber stage and, thereafter, to get broader near-IR fluorescence and the bleaching effect (Fig. 40) of a higher contrast at pumping at ~ 1.06 μm or at near wavelength.

Conclusions

In conclusions, we have focused the basic material and optical properties of advanced nano-engineered glass-based optical fibers doped with erbium/ytterbium, erbium/scandium, ytterbium, and chromium for photonics applications such as multichannel optical amplification in C-band, broadband optical amplification in the C+ L band region, fiber laser with enhanced PD resistivity, and broad-band source at NIR.

A new class of optical fiber based on Er-doped crystalline zirconia yttria alumina phospho silica nano-particles made through a solution doping technique followed by a MCVD process with optimization of the process parameters at different stages of fabrication. Glass modifiers and nucleating agents are added into the host glass followed by a proper annealing of the preform to generate uniform distribution of crystalline zirconia yttria alumina-phospho silica nano-particles in the core region of optical fiber. Such new type of Zr-EDF was analyzed as multichannel amplification in the C-band region where this particular fiber exhibits maximum gain difference <2 dB with output SNR of >22 dB at the input signal level of -30 dBm/channel amplification under pumping at 980 nm wavelength.

Another novel material design-based optical fiber developed through incorporation of Er ions into nano-engineered scandium-yttria-alumina-silica (SYAS) glass through the MCVD process coupled with the solution doping technique and followed by thermal annealing of the fabricated perform in order to enhance the fluorescent ability along with optical gain in the C- and L-bands. Such kind of phenomenon have been established owing to the suitable thermal annealing of the preform, as compared to the one drawn from pristine (not-annealed) preform. The absorption spectra along with the SAED patterns from the TEM analyses suggest that the revealed enhancements in the fiber drawn from the annealed preform occurs through the modification of surrounding environment of Er^{3+} ions within or in proximity to the crystalline nanoparticles which generated during thermal annealing step. The enhancement of the fluorescence lifetime of $^4I_{13/2}$ level also signifies about the crystalline environment of the surrounding Er ions induced by proper thermal annealing of pristine preform. The formation of scandium-ultra rich nano-crystalline environment, possessing low photon energy around the erbium ions, enhanced the fluorescence intensity. Such kind of nano-engineered glass reduces the noise figure around 4.35 dB and provides broadband optical flat gain with an average value of 38.675 dB, varied by less than ±0.7 dB spanning over a broad wavelength region of 1530–1590 nm compared to the pristine and Sc free Er-doped fibers. Such kind of nano-engineered glass-based Er-doped fiber will be useful for making highly efficient optical amplifiers, suitable for present broadband optical communication systems.

Yb-doped new glass composition-based nano-engineered multielements (P-Zr-Ce-Al-Ca)-doped optical fibers have been developed and compared the lasing as well as 633 nm absorption loss performance with standard alumino-silicate Yb-doped fiber under moderate power level. Such kind of fiber indicates that the ME-Yb-doped fiber is a promising candidate for high-power laser applications with enhanced PD resistivity.

Chromium-doped phase-separated yttria-alumina-silica glass-based fiber containing Cr(IV) was fabricated by thermal annealing of the preform before fiber drawing process. The extinction coefficients in all the absorption bands corresponding to Cr^{3+}, Cr^{6+}, and Cr^{4+} ions are reduced 3 times in case of fibers drawn from the annealed preform with a possible reason for this being inter-diffusion of co-dopants in core and Si of cladding. The annealed preform's fiber exhibits more intense broadband emission (500–800 nm) than the nonannealed one, which is mainly due to the presence of Cr^{3+} ions inside the nano-structured YAS core glass. In this glass, it is possible to increase the content of Cr^{4+} species in order to generate more wideband near-IR fluorescence at room temperature and to get a more prominent bleaching effect with a higher contrast at 1.0–1.1 μm pumping.

Such nano-structuration of the doping host of doped fiber through post-thermal annealing of the fabricated preforms followed by fiber drawing will serve as a new route to "engineer" the local dopant environment. All the results of such kind of advanced nano-engineered glass-based optical fibers will be very much useful for making of optical fiber-based devices such as lasers, amplifiers, and sensors, which can now be realized with silica glass. In general, the composition of doped region of nano-engineered glass is crucial, and based on selected composition, it has a great potential for development of fiber laser, optical amplifier, as well as broad-band source for modern optical communication system.

Future Work

Although nano-material doping technology has opened a new way in developing novel specialty optical fibers, as far as we know, there are few work about optical fiber amplification based on silica fiber doped with nano-semiconductor materials. In future work, we will demonstrate a novel special silica fiber doped with InP and ZnS semiconductor nano-particles into the core. Due to the nano size, semiconductor nano particles will show remarkable quantum confinement effect and size tunable effect, which may provide excellent amplification features. Since this particular field is relatively in its initial stage, the future work will also target the fabrication of optical fiber comprised of suitable host of advanced nano-engineered Er/Yb and Tm/Yb co-doped modified silica glass containing 90% SiO_2 through MCVD technique followed by solution doping process. The main aspect of the targeted work will reveal the enhancement of energy transfer efficiency from Yb to Er and Tm to achieve high lasing efficiency based fiber laser at NIR region through modification of the surrounding environments of RE ions embedding into different nano-engineered glass hosts.

References

A.N. Abramov, M.V. Yashkov, A.N. Guryanov, M.A. Melkumov, D.A. Dvoretskii, I.A. Bufetov, L.D. Iskhakova, V.V. Koltashev, M.N. Kachenyuk, M.F. Tursunov, Inorg. Mater. **50**, 1283 (2014)
K. Aiso, Y. Tashiro, T. Suzuki, T. Yagi, Furukawa Rev. **20**, 41 (2001)
J. Alcock, IEEE Photon Soc. News Lett. June, vol. **27**, (2013)
N.B. Angert, N.I. Borodin, V.M. Garmash, V.A. Zhitnyuk, A.G. Okhrimchuk, O.G. Siyuchenko, A.V. Shestakov, Sov. J. Quantum Electron. **18**, 73 (1988)
K. Aria, H. Namikawa, K. Kumata, T. Honda, Y. Ishii, T. Handa, J. Appl. Phys. **59**, 3430 (1986)
G.R. Atkins, A.L.G. Carter, Opt. Lett. **19**, 874 (1994)
C.G. Atkins, J.F. Massicott, J.R. Armitage, R. Wyatt, B.J. Ainslie, S.P. Craig-Ryan, Electron. Lett. **25**, 910 (1989)
E.G. Behrens, R.C. Powell, JOSA B **7**, 1437 (1990)
N.S. Bergano, C.R. Davidson, J. Lightwave Technol. **14**, 1299 (1996)
S. K. Bhadra, A. K. Ghatak (eds.), *Guided Wave Optics and Photonic Devices* (CRC Press, New York, 2013)
N.I. Borodin, V.A. Zhitnyuk, A.G. Okhrimchuk, A.V. Shestakov, Izv. Akad. Nauk SSSR, Ser. Fiz. **54**, 1500 (1990)
M.M. Broer, D.M. Krol, D.J. Digiovanni, Opt. Lett. **18**, 799 (1993)
X.S. Cheng, R. Parvizi, H. Ahmad, S.W. Harun, IEEE Photon. J. **1**, 259 (2009)
Y.C. Huang, Y.K. Lu, J.C. Chen, Y.C. Hsu, Y.M. Huang, S.L. Huang, W.H. Cheng, *Broadband emission from Cr-doped fibers fabricated by drawing tower,* Opt. Exp. **14**, 8492-8497 (2006)
F. d'Acapito, C. Maurizio, M.C. Paul, T.S. Lee, W. Blanc, B. Dussardier, Mat. Sci. Eng. B **146**, 167 (2008)
A. Dhar, M.C. Paul, S. Das, H.P. Reddy, S. Siddiki, D. Dutta, M. Pal, A.V. Kir'yanov, Phys. Status Solidi A **214**, 1 (2017)
J. Dong, P. Deng, Y. Liu, Y. Zhang, J. Xu, W. Chen, X. Xie, Appl. Opt. **40**, 4303 (2001)
B. Dussardier, J. Maria, P. Peterka, Appl. Opt. **50**, E20 (2011)
D. Dutta, A. Dhar, A.V. Kir'yanov, S. Das, S. Bysakh, M.C. Paul, Phys. Status Solidi A **212**, 1836 (2015)

V.V. Dvoyrin, V.M. Mashinsky, V.B. Neustruev, E.M. Dianov, A.N. Guryanov, A.A. Umnikov, J. Opt. Soc. Am. B **20**, 280 (2003)

H. Eilers, W.M. Dennis, W.M. Yen, S. Kück, K. Peterman, G. Huber, W. Jia, IEEE J. Quantum Electron. **29**, 2508 (1993)

M. Engholm, L. Norin, Opt. Express **16**, 1260 (2008)

M. Engholm, P. Jelger, F. Laurell and L. Norin, *Improved photodarkening resistivity in ytterbium-doped fiber lasers by cerium codoping*, Optics Letters. **34**(8), 1285–87 (2009)

M. Fechner, R. Peters, A. Kahn, K. Petermann, E. Heumann, G. Huber, in *Conference on Lasers and Electro-Optics/Quantum Electronics and Laser Science Conference and Photonic Applications Systems Technologies* (OSA Technical Digest (CD), Optical Society of America, 2008), paper CtuAA3

V. Felice, B. Dussardier, J.K. Jones, G. Monnom, D.B. Ostrowsky, et al., Eur. Phys. J. Appl. Phys. **11**, 107 (2000)

V. Fomin, V. Gapontsev, E. Shcherbarkov, A. Abramov, A. Ferin, D. Mochalov, Int. Conf. Laser Optics (2014), St. Petersburg, Russia, 2014 (IEEE, St. Petersburg), p. 1.

L. Fornasiero, K. Petermann, E. Heumann, G. Huber, Opt. Mater. **10**, 9 (1998)

V. Gapontsev, A. Avdokhin, P. Kadwani, HYPERLINK I. Samartsev, N. Platonov, R. Yagodkin *SM green fiber laser operating in CW and QCW regimes and producing over 550W of average output power*, Proc. SPIE 8964, Nonlinear Frequency Generation and Conversion: Materials, Devices, and Applications XIII, 896407 (20 February 2014); https://doi.org/10.1117/12.2058733; San Francisco, California, United States

C.R. Giles, E. Desurvire, J. Lightwave Technol. **9**, 147 (1991)

D.M. Gill, L. McCaughan, J.C. Wright, Spectroscopic site determinations in erbium-doped lithium niobate. Phys. Rev. B **53**, 2334 (1996)

V.I. Glazkov, K.M. Golant, Y.S. Zavorotny, V.F. Lebedev, A.O. Rybaltovskii, Glass Phys. Chem. **28**, 201 (2002)

A.D. Guzman Chavez, A.V. Kir'yanov, Y.O. Barmenkov, N.N. Il'ichev, Laser Phys. Lett. **4**, 734 (2007)

A. Halder, M.C. Paul, S.W. Harun, S.K. Bhadra, S. Bysakh, J. Lumin. **143**, 393 (2013)

S.W. Harun, N. Tamchek, S. Shahi, H. Ahmad, Progr. Electromagn. Res. C **6**, 1 (2009)

S.W. Harun, R. Parvizi, X.S. Cheng, A. Parvizi, S.D. Emami, H. Arof, H. Ahmad, Opt. Laser Technol. **42**, 790 (2010)

Y.-C. Huang, J.-S. Wang, Y.-K. Lu, W.-K. Liu, K.-Y. Huang, S.-L. Huan, W.-H. Cheng, Opt. Express **15**, 14382 (2007)

Y.-C. Huang, J.-S. Wang, Y.-S. Lin, T.-C. Lin, W.-L. Wang, Y.-K. Lu, S.-M. Yeh, H.-H. Kuo, S.-L. Huang, W.-H. Cheng, IEEE Photon. Techn. Lett. **22**, 914 (2010)

Y.-C. Huang, J.-S. Wang, K.-M. Chu, T.-C. Lin, W.-L. Wang, T.-L. Chou, S.-M. Yeh, S.-L. Huang, W.-H. Cheng, OSA/OFC/NFOEC, paper OWS1 (2011)

Y.-C. Huang, C.-N. Liu, Y.-S. Lin, J.-S. Wang, W.-L. Wang, F.-Y. Lo, T.-L. Chou, S.-L. Huang, W.H. Cheng, Opt. Express **21**, 4790 (2013)

N.N. Il'ichev, A.V. Kir'yanov, P.P. Pashinin, Quantum Electron. **28**, 147 (1998a)

N.N. Il'ichev, A.V. Kir'yanov, E.S. Gulyamova, P.P. Pashinin, Quantum Electron. **28**, 17 (1998b)

J. Jaspara, M. Andrejco, D. DiGiovanni, in *Conference of Laser and Electro Optics,* Long Beach, California United States, 21–26 May 2006, ISBN: 1-55752-813-6, OSA Technical Digest (CD) (Optical Society of America, 2006), paper CTuQ5.

C. Jauregui, J. Limpart, A. Tunnermann, Nat. Photonics **7**, 861 (2013)

M. A. Jebali, J. Maran, S. LaRochelle, S. Chatigny, M. Lapointe, E. Gagnon, in *Conference on Lasers and Electro-Optics 2012: A&T* (Optical Society of America, 2012), paper JTh1I

Y. Jeong, J.K. Sahu, D.B.S. Soh, C.A. Codemard, J. Nilsson, Opt. Lett. **30**, 2997 (2005)

S. Jetschke, S. Unger, U. Ropke, J. Kirchhof, Opt. Express **15**, 14838 (2007)

S. Jetschke, S. Unger, A. Schwuchow, M. Leich, J. Kirchhof, Opt. Express **16**, 15540 (2008)

S. Jetschke, S. Unger, A. Schwuchow, M. Leich, M. Jager, Opt. Express **24**, 13009 (2016)

A. Jha, S. Shen, S. Naftaly, Phys. Rev. B **62**, 6215 (2000)

S. Jiang, B.-C. Hwang, T. Luo, K. Seneschal, F. Smektala, S. Honkanen, J. Lucas 4, N. Peyghambarian, Proc. OFC, vol. PD5-1 (2000)

T. Kasamatsu, Y. Yano, H. Sekita, Opt. Lett. **24**, 1684 (1999)

T. Kitabayashi, M. Ikeda, M. Nakai, T. Sakai, K. Himeno, K. Ohashi, in *Conference of Laser and Electro Optics, OThC5* (2006)

J. Koponen, M. Soderlund, S. Tammela, H. Po, in *Conference of Laser and Electro Optics Europe*, (Optical Society of America, 2005), paper CP2-2-THU

J. Koponen, M. Söderlund, H. Hoffman, S. Tammela, Opt. Express **14**, 11539 (2006)

L.V. Kotov, M.E. Likhachev, M.M. Bubnov, O.I. Medvedkov, D.S. Lipatov, N.N. Vechkanov, A.N. Guryanov, Quantum Electron. **42**, 432 (2012)

L.V. Kotov, M.E. Likhachev, M.M. Bubnov, O.I. Medvedkov, M.V. Yashkov, A.N. Guryanov, J. Lhermite, S. Février, E. Cormier, Opt. Lett. **38**, 2230 (2013)

S. Kück, Appl. Phys. B Lasers Opt. **72**, 515 (2001)

V. Kuhn, D. Kracht, J. Neumann, P. Wessels, IEEE Photon. Technol. Lett. **23**, 432 (2011a)

V. Kuhn, D. Kracht, J. Neumann, P. Wessels, *CLEO/Europe and EQEC 2011 Conference Digest* (OSA Technical Digest (CD), Optical Society of America, 2011b), paper CJ7_5

C.-C. Lai, C.-P. Ke, S.-K. Liu, D.-Y. Jheng, D.-J. Wang, M.-Y. Chen, Y.-S. Li, P.S. Yeh, S.-L. Huang, Opt. Lett. **36**, 784 (2011)

E.-L. Lim, S.-u. Alam, D.J. Richardson, Opt. Express **20**, 13886 (2012)

R. Lo Savio, M. Miritello, A. Shakoor, P. Cardile, K. Welna, L.C. Andreani, D. Gerace, T.F. Krauss, L. O'Faolain, F. Priolo, M. Galli, Opt. Express **21**, 10278 (2013)

C.Y. Lo, K.Y. Huang, J.C. Chen, C.Y. Chuang, C.C. Lai, S.L. Huang, Y.S. Lin, P.S. Yeh, Opt. Lett. **30**, 129 (2005)

J. Ma, M. Li, L. Tan, Y. Zhou, S. Yu, Q. Ran, Opt. Express **17**, 15571 (2009)

M. Malmstrom, P. Jelger, M. Engholm, F. Laurell, in *Conference of Laser and Electro Optics Europe*, San Jose, California, USA, 2010, OSA Technical Digest (CD) (Optical Society of America, California, 2010) p. JTuD68

I. Manek-Honninger, J. Boullet, T. Cardinal, F. Guillen, M. Podgorski, R.B. Doua, F. Salin, Opt. Express **15**, 1606 (2007)

J.F. Massicott, J.R. Armitage, R. Wyatt, B.J. Ainslie, S.P. Craig-Ryan, Electron. Lett. **26**, 1645 (1990)

K.E. Mattsson, Opt. Express **19**, 19797 (2011)

R.J. Mears, L. Reekie, I.M. Jauncey, D.N. Payne, Electron. Lett. **23**, 1026 (1987)

R. Mellish, S.V. Chernikov, P.M.W. French, J.R. Taylor, Electron. Lett. **34**, 552 (1998)

B. Morasse, S. Chatigny, E. Gagnon, C. Hovington, J.-P. Martin, J.-P. de Sandro, Proc. SPIE **6453**, Proc. SPIE 6453, 64530H-64 530H, 6453-17, (2007)

T. Murata, M. Torisaka, H. Takebe, K.J. Morinaga, J. Non-Cryst. Solids **220**, 139 (1997)

M. Naftaly, S. Shen, A. Jha, Appl. Opt. **39**, 4979 (2000)

S.R. Nagel, J.B. MacChesney, K.L. Walker, IEEE Trans. Microw. Theory Techn **30**, 305 (1982)

NEW YORK, Jan. 30, 2013 /PRNewswire/ – www.Reportlinker.com

I. Nikolov, X. Mateos, F. Güell, J. Massons, V. Nikolov, P. Peshev, F. Diaz, Opt. Matter. **25**, 53 (2004)

Y. Ohishi, T. Kanamori, T. Kitagawa, S. Takahashi, E. Snitzer, G.H. Sigel Jr., Opt. Lett. **16**, 1747 (1991)

M. Pal, M.C. Paul, S.K. Bhadra, S. Das, S. Yoo, M.P. Kalita, A.J. Boyland, J.K. Sahu, J. Lightwave Technol. **29**, 2110 (2011)

G. Palambo, R. Pratesi, Comprehensive series in photochemistry and photobiology, in *Laser and Current Optical Techniques in Biology*, vol. 4, (Royal Society of Chemistry, Cambridge, UK, 2004)

R. Paschotta, J. Nilsson, P.R. Barber, J.E. Caplen, A.C. Tropper, D.C. Hanna, Opt. Comm. **136**, 375 (1997)

A. Patra, C.S. Friend, R. Kapoor, R.N. Prasad, J. Phys, Chem. B **106**, 1909 (2002)

M.C. Paul, S.W. Harun, N.A.D. Huri, A. Hamzah, S. Das, M. Pal, S.K. Bhadra, H. Ahmad, S. Yoo, M.P. Kalita, A.J. Boyland, J.K. Sahu, Performance comparison of Zr-based and bi-based erbium-doped fiber amplifiers. Opt. Lett. **35**, 2882 (2010)

M.C. Paul, S.W. Harun, N.A.D. Huri, A. Hamzah, S. Das, M. Pal, S.K. Bhadra, H. Ahmad, S. Yoo, M.P. Kalita, A.J. Boyland, J.K. Sahu, J. Lightave Technol. **28**, 2919 (2011)

M. C. Paul, A. V. Kir'yanov, S. Bysakh, S. Das, M. Pal, S. K. Bhadra, M. S. Yoo, A. J. Boyland, J. K. Sahu, in *Selected Topics on-Optical-Fiber-Technology*, ed. by Dr. Moh. Yasin; InTech, Janeza Trdine 9, 51000 Rijeka, Croatia, (2012). https://doi.org/10.5722/26394. ISBN 978-953-51-0091-1

M.C. Paul, M. Pal, S. Das, A. Dhar, S.K. Bhadra, J. Opt. **45**, 260 (2016)

V. Peters, A. Boltz, K. Petermann, G. Huber, J. Cryst. Growth **237–239**, 879 (2002)

H. Rawson (ed.), *Inorganic Glass-Forming Systems* (Academic Press, London, 1967)

P. H. Reddy, S. Das, D. Dutta, A. Dhar, A. V. Kir'yanov, M. Pal, S. K. Bhadra, M. C. Paul, Phys. Status Solidi A, 1700615, vol. **251** (2018)

J. K. Sahu, S. Yoo, A. J. Boyland, M. P. Kalita, C. Basu, A. S. Webb, C. Sones, J. Nilsson, D. Payne, in *Conference of Laser and Electro Optics JTuA27* San Jose, California, USA 2008, OSA Technical Digest (CD) (Optical Society of America, California, 2008), p. JTuA27 (2008)

A. Sennaroglu, Quantum Electron. **26**, 287–352 (2002)

R.D. Shannon, Acta Crystallogr. Sect. A: Cryst. Phys., Diffr.,Theor. Gen. Cryst. **32**, 75 (1976)

V. Singh, V.K. Rai, K. Al-Shammery, M. Haase, S.H. Kim, Appl. Phys. A Mater. Sci. Process. **113**, 747 (2013)

E. Snoeks, P.G. Kik, A. Polman, Opt. Mater. **5**, 159 (1996)

M.J. Soderlund, J.J.M. Ponsoda, J.P. Koplow, S. Honkanen, Opt. Express **17**, 9940 (2009)

S. Sugiyama, Y. Fujimoto, M. Murakami, H. Nakano, T. Sato, H. Shiraga, Electron. Lett. **49**, 148 (2013)

Y. Sun, J.W. Sulhoff, A.K. Srivastava, J.L. Zyskind, T.A. Strasser, J.R. Pedrazzani, C. Wolf, J. Zhou, J.B. Judkins, R.P. Espindola, A. Vengsarkar, Electron. Lett. **33**(23), 1965 (1997)

V. R. Supradeepa, J. W. Nicholson, K. Feder, in *Conference on Lasers and Electro-Optics 2012: S&I* (OSA Technical Digest (CD), IEEE, Optical Society of America, 2012), paper CM2N

N. Ter-Gabrielyan, L.D. Merkle, G.A. Newburgh, M. Dubinskii, A. Ikesue, *Advanced Solid-State Photonics* (Optical Society of America, Washington, DC, 2008)., paper TuB4

J.E. Townsend, S.B. Poole, D.N. Payne, Electron. Lett. **23**, 329 (1987)

S. Unger, A. Schwuchow, S. Jetschke, J. Kirchhof, Proc. SPIE **7212**, 72121B (2009)

A. K. Varshneya (ed.), *Fundamentals of Inorganic Glasses* (Academic Press, Boston, 1994)

J.-S. Wang, F.-H. Shen, J. Non-Crystal. Sol. **358**, 246 (2012)

B. S. Wang, G. Puc, R. Osnato, B. Palsdottir, Proc. SPIE Asia Pacific Opt. Commun. pp. 161–166 (2003)

P.F. Wysocki, M.J.F. Digonnet, B.Y. Kim, H.J. Shaw, J. Lightwave Technol. **12**, 550 (1994)

P.F. Wysocki, J.B. Judkins, R.P. Espindola, M. Andrejco, A.M. Vengsarkar, IEEE Photon. Technol. Lett. **9**, 1343 (1997)

M. Yamada, T. Kanamori, Y. Terunuma, K. Oikawa, M. Shimuzu, S. Sudo, K. Sagawa, IEEE Photon. Technol. Lett. **8**, 882 (1996)

S. Yoo, C. Basu, A.J. Boyland, C. Sones, J. Nilsson, J.K. Sahu, D. Payne, Opt. Lett. **32**, 1626 (2007)

M.N. Zervas, C.A. Codemard, IEEE J. Quantum Electron. **20**, 0904123 (2014)

J. Zhang, V. Fromzel, M. Dubinskii, Opt. Express **19**, 5574 (2011)

N.I. Zhavoronkov, V.P. Mikhailov, N.V. Kuleshov, B.I. Minkov, A.S. Avtukh, Quantum Electron. **25**, 31 (1995)

X. Zhou, J. Yu, M.F. Huang, Y. Shao, T. Wang, P. Magill, M. Cvijetic, L. Nelson, M. Birk, G. Zhang, S. Ten, H.B. Matthew, S.K. Mishra, J. Lightwave Technol. **28**, 456 (2010)

Y. Zhuang, Y. Teng, J. Luo, B. Zhu, Y. Chi, E. Wu, H. Zeng, J. Qiu, Appl. Phys. Lett. **95**, 111913 (2009)

D.R. Zimmerman, H.S. Spiekmann, J. Lightwave Technol. **22**, 63 (2004)

Fabrication of Negative Curvature Hollow Core Fiber

12

Muhammad Rosdi Abu Hassan

Contents

From Conventional Fibers to Photonics Crystal Fibers	530
Photonic Crystal Fiber	530
Background of Photonics Crystal Fiber	531
Development of Hollow Core Fiber	531
Development of Negative Curvature Hollow Core Fibers	533
The Importance Negative Curvature	534
Guiding Mechanism	535
Antiresonant Reflecting Optical Waveguide (ARROW)	535
Marcatili and Schmeltzer's Model	537
Coupled-Mode Model	538
Fabrication of Fiber	538
Fabrication of Negative Curvature Hollow Core Fiber	539
Stack and Draw	539
Design and Properties of Fiber	542
Conclusion	545
References	545

Abstract

In this chapter, we describe a review covering the development of the negative curvature hollow core fiber for the mid-IR region. The topics cover various types of hollow core fiber and their improvement made in term of attenuation of fiber, followed by a description of the guiding mechanism of the negative curvature hollow core fiber (NC-HCF) using antiresonant reflecting optical waveguide (ARROW) mechanism. Then, we present the general fabrication steps and the

---

M. R. Abu Hassan (✉)
Centre for Optical Fibre Technology (COFT), School of Electrical, Electronic Engineering, Nanyang Technological University, Singapore, Singapore, Singapore
e-mail: muhdrosdi22@gmail.com

© Springer Nature Singapore Pte Ltd. 2019
G.-D. Peng (ed.), *Handbook of Optical Fibers*,
https://doi.org/10.1007/978-981-10-7087-7_75

fabrication process for negative curvature fiber. In the second part of the chapter, the design and properties of the hollow core applied in the other research work are presented.

Keywords

Fiber optics · Microstructured optical fiber · Hollow core fiber · Photonics crystal fiber

From Conventional Fibers to Photonics Crystal Fibers

Optical fiber is a key component of modern telecommunication technologies. This technology has been attracting much attention since the first single-mode fiber with loss <20 dB/km were made in 1970 (Kaiser and Astle 1974). Currently, conventional optical fiber forms the backbone of our global telecommunication network. Optical fibers perform very well in telecom applications but also are becoming more prominent for non-telecom technologies such as a laser, sensing, biomedical applications, and a host of other fields.

Even though conventional optical fibers have achieved a number of successes in the existing telecommunication technologies, there is a fundamental limitation which is related to the two materials structure used in the conventional fibers. There is a need to make a new class of fiber that can be as competitive as conventional fiber and offer significantly improved performance in some respects.

Photonic Crystal Fiber

Over the past 25 years, conventional optical fibers have revolutionized communications, transmitting more information over greater distances than could ever be achieved in copper wires, and are also vital in many technologies such as imaging endoscopes and high power laser transport for cutting and drilling applications.

Optical fibers have evolved into many types of structure to overcome the limitation of existing fiber. In 1996, Knight et al. fabricate new class of fiber using a single material which operates by total internal reflection as a guiding mechanism (Knight et al. 1996a). These new classes of fibers have the potential to outperform conventional fiber in certain aspects due to high flexibility to modify the structure and parameters. Some parameters of the fiber such as dispersion can be modified and tailored to suit the desire applications by designing different fiber structure.

An innovation was made by J.C. Knight and his group in 1996 where they managed to produce the photonic crystal fibers (PCF) which consist of a pure silica core surrounded by a silica-air photonic crystal material with a hexagonal symmetry (Knight et al. 1996b). Since then PCFs have captured research interests very rapidly due to the increasing demand for high-performance optical fibers. Two major limitations of conventional optical fibers that have led to the fabrication of photonic crystal fibers are the smallness of refractive-index difference, Δn which causes bend loss (0.5 dB at 1550 nm in Corning SMF-28 for one turn around a mandrel 32 mm in diameter), and also the reliance on total internal reflection so that guidance in a hollow-core is impossible.

These new fibers, known collectively as microstructured fibers, can be made entirely from one type of glass as they do not rely on dopants for guidance. Instead, the cladding region is peppered with many small air holes that run the entire fiber length. These fibers are typically separated into two classes defined by the way in which they guide light:

1. Photonic band-gap fibers, in which guidance in a hollow core can be achieved via photonic band-gap effects.
2. Holey fibers, in which the core is solid and light is guided by a modified form of total internal reflection as the air holes lower the effective refractive index of the cladding relative to that of the solid core.

Background of Photonics Crystal Fiber

Fiber optics changed radically with the first demonstration (Knight et al. 1996a, 1997) of a new class of optical fibers that can have a variety of structures. Photonic Crystal Fibers (PCFs) as known as Microstructured Optical Fibers (MOFs), which have modified cladding structure in order to provide a high index contrast between core and cladding.

In 1974, it was first observed by P. Kaiser and H.W. Astle that low-loss single material fibers can be fabricated entirely from silica and the lowest steady-state loss of about 3 dB/km at a wavelength of 1.1 μm was obtained (Kaiser and Astle 1974). The basic idea explains that light could be trapped inside a hollow fiber core by creating a periodic wavelength-scale lattice of microscopic holes in the cladding glass – a "photonic crystal." In 1996, Knight et al. reported the first photonic crystal fibers which consist of a pure silica core surrounded by a silica-air photonic crystal material with a hexagonal symmetry (Knight et al. 1997).

Then, followed the Endlessly Single Mode (ESM) PCF, which the fiber will guide only the fundamental mode (Birks et al. 1997). Since then PCFs have captured research interests very rapidly due to the increasing demand on high-performance optical fibers. By now, we have seen optical fibers that can have an endless variety of structures of PCF – large mode area (Knight et al. 1998a), dispersion controlled (Mogilevtsev et al. 1998), hollow core (Cregan 1999) birefringent (Ortigosa-Blanch et al. 2000), and multicore (Il 2000) (Fig. 1).

Development of Hollow Core Fiber

The hollow core optical waveguides refer to long cylindrical tubes and rectangular stripes made of dielectric or even metal. The potential of hollow core fiber made of a metal material for long-haul electromagnetic wave transmission theoretically has

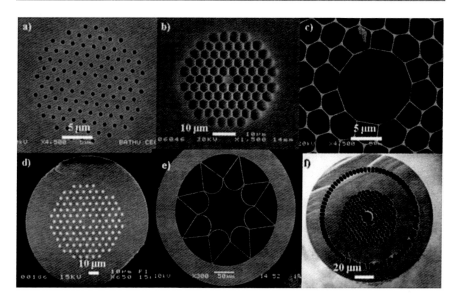

Fig. 1 Scanning electron microscope images of photonic crystal fibers: (**a**) solid core fiber, (**b**) high air-filling fraction solid core fiber, (**c**) hollow core fiber, (**d**) solid band-gap fiber consisting of a mixture of doped silica glasses, (**e**) a negative curvature fiber, and (**f**) double-clad hollow core fiber. This figure is reproduced from Russell (2003)

been proposed by Carson et al. (1936). Later, the feasibility of using the dielectric HCF in telecommunications has been studied by Marcatili and Schmeltzer (1964). They demonstrated analytically that the trade-off between modal leaky loss and a bending loss would fundamentally limit the application of dielectric hollow core fiber for long-haul optical signal transmission.

In the late twentieth century, the development of hollow-core photonic bandgap fiber (HC-PBG) speeded up the development of HCFs in both theory and applications immensely (Knight et al. 1998b; Russell 2006).

The state-of-the-art HC-PBGs was demonstrating extremely low transmission loss, comparable to that of commercial optical fibers (Roberts et al. 2005). They also have a very high damage threshold (Jaworski et al. 2013), very low group-velocity dispersion, and an optical nonlinearity that is many orders of magnitude less than that of competing fibers (Ouzounov et al. 2003).

Kagome fiber also is a part of HCF family. "Kagome" fiber was firstly reported in 2002 (Benabid et al. 2002). The Kagome fiber usually exhibits multiple transmission bands and overall covers a broader spectral range than HC-PBGs. Numerical simulations show that the "Kagome" lattice supports no photonic bandgap which makes the Kagome fiber distinct (Russell 2006; Benabid and Roberts 2011).

Another type of HC fiber is the negative curvature hollow core fiber (NC-HCF). With the simple single ring cladding design and negative curvature core wall, NC-HCF combine the strength and durability of silica with record low attenuation in the 3–4 μm spectral band, where their attenuation can be 10,000 times less than that of bulk silica (Yu et al. 2012; Wheeler et al. 2014; Yu and Knight 2013; Bei et al. 2013; Kim et al. 2016). In this chapter, we describe a high performance of the negative curvature hollow core fiber (NC-HCF). Then, follow it by the characterization of the performance of NC-HCF.

Development of Negative Curvature Hollow Core Fibers

Negative curvature hollow-core fiber (NC-HCF) is a novel type of hollow core fiber, which is characterized by the negative curvature of core wall shape (Yu and Knight 2016). Such fiber exhibits multiple spectral transmission bands and can guide light in the hollow core with low attenuation over lengths of hundreds of meters. The exceptional qualities of NC-HCF, such as low transmission loss, extended transmission wavelength, and low latency and nonlinearity lend these fibers to a wide range of applications. The low-loss transmission windows in NC-HCF can span from visible to mid-infrared wavelengths (Yu and Knight 2016) and have been successfully applied in the transmission of high power lasers (Shephard et al. 2015), the design of mode-locked fiber lasers (Harvey et al. 2015), and study of nonlinear optics in gas (Wang et al. 2014a). Moreover, the simple structure of fiber cladding design allowing tuning flexibility of NC-HCF's design and dimension to meet specific wavelength and application requirement (Setti et al. 2013; Kolyadin et al. 2013; Belardi and Knight 2014b; Shiryaev 2015) (Fig. 2).

In 2010, Kagome hollow-core fiber with negative curvature core boundary results in lower attenuation than conventional kagome fiber (Wang et al. 2010). Results from that, a series of subsequent experiments have been done and confirmed the importance of core wall shape in the reduction of attenuation in such fibers (Wang et al. 2011). In 2011, a group led by Pryamikov attempt to fabricate the simple structure of hollow core which used only eight capillaries as a cladding. This fiber is the first NC-HCF made and operated in 3 μm wavelength (Pryamikov et al. 2011).

In 2013, Fei Yu reported transmission in 3.1-microns region with minimum attenuation of 34 dB/km. He called "core ice-cream shape of cladding" that has curvature towards the core that affects the attenuation of the fiber. Later, the fiber was successfully applied to invasive surgical laser procedure in delivering high energy microsecond pulses at 2. 4 μm (Jaworski et al. 2013). Kolyadin et al. purposed simple design of the cladding structure of the NC-HCF to improve the attenuation in the mid-IR region. He reported the open boundary of core wall in NC-HCF, which is the contactless capillary in the cladding. He demonstrated low loss transmission of light in the mid-infrared spectrum range from 2.5 to 7.9 μm (Kolyadin et al. 2013). In 2014, Belardi et al. demonstrated significantly reduced the bending loss of the negative curvature fibers in 2014 (Belardi and Knight 2014a).

NC-HCF can be designed to guide light in an empty or gas-filled hollow core. Since as much as 99% of the optical power in these fibers can travel in air and not in the glass, they do not suffer from the same limitations to loss as conventional fibers and can exhibit drastically reduced optical nonlinearity, making them promising candidates for future ultra-low loss transmission fibers.

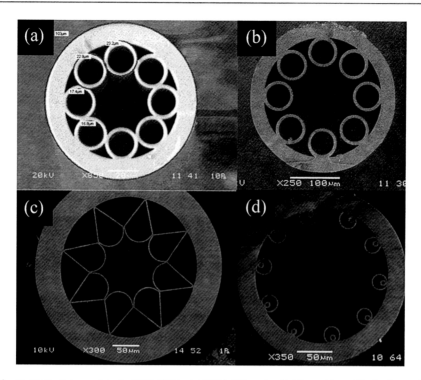

Fig. 2 SEM pictures of typical NC-HCFs: (**a**) NC-HCF made of silica, with capillaries in touch in the cladding [1]; (**b**) NC-HCF made of silica, with nontouching capillaries in the cladding [5]; (**c**) NC-HCF made of silica with ice cream cone-shape capillary in the cladding [3]; (**d**) NC-HCF made of silica, with non-touching capillaries. Extra capillaries are added to reduce the coupling between the core and cladding modes [33]

The NC-HCF has applied in many applications. Setti et al. were realized terahertz guidance in polymethyl-methacrylate in 2013 (Setti et al. 2013). Besides that, NC-HCF also has implemented in high power laser (Urich et al. 2013a, b) delivery and have been applied in the study of gas-light interaction.

We will present the steps to fabricate NC-HCF. We start with describing the guiding mechanism of the NC-HCF using Antiresonant Reflecting Optical Waveguide (ARROW) mechanism. Then, we will explain the fiber draw technique and the NC-HCF technique fabrication process. The characteristics of the NC-HCF will be described later.

The Importance Negative Curvature

NC-HCFs are characterized by the negative curvature of the core wall, which has been numerically and experimentally demonstrated to effectively reduce the attenuation. Numerical studies showed that in NC-HCF, an increase of negative curvature influences the mode attenuation and even bending loss in a complex way (Belardi and Knight 2013). It has been confirmed that a large curvature of the core wall (small radius of curvature) can help decrease the overlap of core mode field with fiber materials to 10^{-4}. This has been experimentally and numerically demonstrated in silica NC-HCFs (Yu and Knight 2013; Belardi and Knight 2013). Despite recent efforts in the analytical analysis (Pryamikov 2014), the function of the core wall shape is not yet clear. It appears that both the properties of the high-index and of the low-index cladding modes are affected by the curvature.

Guiding Mechanism

NC-HCF has a simple structure as shown in Fig. 3. It has no periodic cladding and does not possess a photonic bandgap, which explains the leaky nature of such fiber. The leaky nature and antiresonant guiding mechanism make it natural to compare NC-HCF with Kagome fiber. Both have multiple transmission bands and similar attenuation figures at comparable wavelengths (Debord et al. 2013; Alharbi et al. 2013).

Antiresonant Reflecting Optical Waveguide (ARROW)

The guiding mechanism of NC-HCF can be simply explained using Antiresonant Reflecting Optical Waveguide (ARROW) model. The ARROW model was firstly proposed in 1986 to explain the enhanced confinement in a planar waveguide with a series of high-low index regions forming the cladding (Duguay et al. 1986; White et al. 2002; Litchinitser et al. 2003). In 2002, it was used to provide a simple way to understand the light transmission in hollow-core photonic bandgap HC-PBG fiber

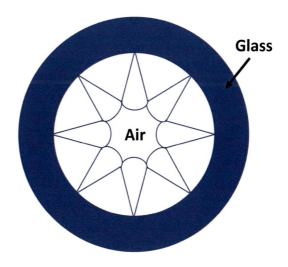

Fig. 3 Schematic representation of the cladding of a negative curvature hollow core fiber. The core is surrounded by the negative curvature cladding that guides the light in the core

regarding waveguide theory more rather than the tight binding theory (Litchinitser et al. 2002). Soon it was being applied to a range of microstructured HCF's.

The ARROW model is based on consideration of the individual modes of the high index regions. It approximates the cladding of HC-PBG as an array of high and low refractive index layers. Each higher refractive index layer can be considered as a Fabry–Perot resonator as shown in Fig. 4.

For one-dimensional case, when the wavelength of light in the core matches a resonant wavelength of the Fabry–Perot, the light will leak out of the core and confine in the high-index region, n_1, but if they are in anti-resonance wavelength with high-index region, n_1, the light will be confined back to the low-index core, n_2, of the waveguide as shown in Fig. 4.

In the resonance in the 2-D waveguide, when the wavelength of light in the core matches a resonant wavelength of the Fabry–Perot cavity, the light will leak out of the core through the high index layer, n_1. As the wavelength is far away from the resonance of the cladding, the light will be reflected back and more strongly confined in the core, n_2 of the waveguide. NC-HCF possesses multiple transmission bands, and the band edges are determined by the resonance wavelengths of cladding as described by the ARROW model (Fig. 5).

Fig. 4 Schematic diagrams of a one-dimensional anti-resonant planar waveguide (the core is running vertically). In these cases, high index regions, n_1, are darker grey and the low index background material, n_2, is white

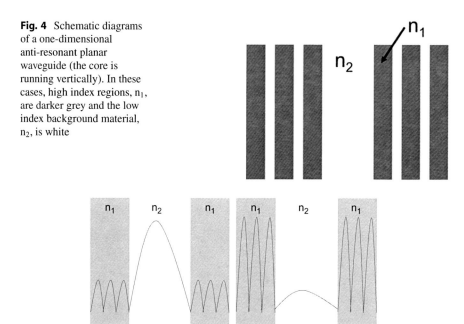

Fig. 5 Schematic representations of the intensity profile in the core region and first high index regions in the one-dimensional planar waveguide. In both cases, high index regions, n_1, are darker grey and the low index background material n_2 is white; (**a**) is the anti-resonant case where light is confined to the core region by anti-resonant reflections and (**b**) is when the wavelength is close to a resonance of the high index regions corresponding to the peak transmission of the Fabry–Perot cavity

The ARROW model described the waveguide characteristics based on ray optics principle. ARROW cannot predict the loss magnitude or any other band details either. There are other models that can be used to explain the guidance mechanism of NC-HCFs and other leaky HCF, which is Marcatili and Schmeltzer model (Marcatili and Schmeltzer 1964), and the coupled mode model (Snyder and Love 2012). Although the ARROW model is far from perfection, it is still widely acknowledged as one of the most effective and frequently used models to understand leaky HCFs.

Marcatili and Schmeltzer's Model

In 1964, Marcatili and Schmeltzer first analytically studied the mode properties of the dielectric HCF (Marcatili and Schmeltzer 1964). In their pioneering work, they derived formulas of mode attenuation and bending loss, which revealed the basic properties of modes in HCFs. An HCF consists of a circular core surround by an infinite homogenous non-absorptive dielectric medium of the higher refractive index than the core material (usually gas/vacuum). Due to the inverted refractive index difference, total internal reflection cannot occur when the incident light from the core at the interface between the core and cladding. The partial reflectivity at the core boundary indicates an inevitable loss of light propagating in the core. The mode attenuation and the corresponding propagation constant are written as (Marcatili and Schmeltzer 1964);

$$A_{vm} = \left(\frac{u_{vm}}{2\pi}\right)^2 \frac{\lambda^2}{r^3} \operatorname{Re}(V_v) \tag{1}$$

$$\beta_{vm} = \frac{2\pi}{\lambda} \left\{ n_{\text{core}} - \frac{1}{2}\left(\frac{u_{vm}}{2\pi}\right)^2 \left[1 + \operatorname{Im}\left(\frac{(V_v \lambda)}{\pi r}\right)\right] \right\} \tag{2}$$

Here u_{vm} is the m_{th} zero of the Bessel function J_{v-1}, and v and m are azimuthal and radial number of modes; λ is the wavelength; r is the core radius; n_{core} is the refractive index of core medium; and V_v is the constant determined by the cladding refractive index and mode order (Marcatili and Schmeltzer 1964). Equations (1) and (2) are the most basic formulas to understand modes in all leaky HCFs. As r/λ grows bigger, the attenuation of modes is quickly reduced. For a mode of a specific order, a higher propagation constant is required to satisfy the transverse resonance condition in a bigger core. Such larger core would give an increased glancing angle of light incident at the core boundary, which results in a higher Fresnel reflection. As a result, the mode in a larger core has a smaller attenuation. Marcatili and Schmeltzer's model fails when complex structures are introduced to the cladding of HCF. The multiple reflections from the structured cladding can reduce or increase the attenuation of leaky modes by constructive/destructive interference, and it also affects the dispersion of modes.

Coupled-Mode Model

The coupled-mode model applies coupled mode theory (Snyder and Love 2012) to analyze the properties of HCFs based on Marcatili and Schmeltzer's model. In real HCFs, the cladding is no longer an infinite and homogeneous medium but has a complex configuration of refractive index distribution. The cladding modes are a set of modes including both dielectric modes localized in the higher index regions and leaky air modes inside lower index regions (Couny et al. 2007; Vincetti and Setti 2010). In the coupled-mode model, the properties of core mode are interpreted as results of the longitudinal coupling with those cladding modes. This method was successfully applied in analyzing the formation of bandgaps in HC-PBG (Birks et al. 2006). By this method, inhibited coupling was proposed to explain the guidance in Kagome fiber (Benabid and Roberts 2011). In 2007, Argyros applied the coupled mode theory to the square lattice polymer HCF and quantitatively analyzed the formation of band edges of leaky HCF (Argyros and Pla 2007). They pointed out that the absolute phase matching is not necessary to achieve an effective coupling between the cladding and core modes. $\Delta \beta \sim 10^{-4}$ m was found to be the threshold to estimate the transmission band edge, which matched well with the experimental measurement (Argyros and Pla 2007). In 2012, Vincetti and Setti applied this method to NC-HCF and presented details of cladding mode features in NC-HCF (Vincetti and Setti 2010). Later, they used this method to demonstrate that the geometry of cladding elements was important to determine the confinement loss of HCFs. The polygonal shaped tube in the cladding adds extra loss due to the Fano-like coupling between the core and cladding modes (Vincetti and Setti 2012). This can be used to explain the different spectral features between NC-HCFs and Kagome fibers.

Fabrication of Fiber

Vapor-phase fabrication techniques are commonly used to fabricate conventional single mode fibers. In these processes, glass is created from high purity component gases using modified chemical vapor deposition (MCVD) process. The fabrication procedure for Photonic Crystal Fiber (PCF) differs radically from that used for conventional step index fiber as the microstructure usually comprises glass and air.

Photonic Crystal Fiber is unlike conventional fiber in that it is usually manufactured from only one material, the most commonly pure silica. As the guiding mechanism differs between the conventional optical fibers and photonic crystal fibers, the fabrication process also differs as the cladding requires a proper structural alignment in order to obtain PCFs with different advantages. Here, we are using the stack-and-draw method to fabricate novel PCF structures. We will discuss much detail about this and other techniques in the next section.

Fabrication of Negative Curvature Hollow Core Fiber

Negative Curvature Hollow Core Fiber (NC-HCF) is unlike conventional fiber is typically fabricated from only one material, the most commonly pure silica. The fabrication procedure for negative curvature hollow core fiber differs completely from that used for conventional step index fiber as the microstructure usually comprises glass and air. The simple one layer cladding of NC-HCF makes it easy to stack but require more attention to details when we draw to fiber. As the guiding mechanism differs between the conventional optical fibers, the fabrication process also differs as the cladding requires a proper structural arrangement to obtain high-performance NC-HCFs.

We are mainly using the stack-and-draw technique to fabricate our NC-HCF structures. This method is inherited from photonics crystal fiber (PCF) fabrication technique (Birks et al. 1995). There are other techniques to form the preform such as extrusion (Kiang et al. 2002; Kumar et al. 2002) and sol-gel casting (Bise and Trevor 2005; van Eijkelenborg et al. 2001) that are not discussed in this chapter.

Stack and Draw

The fibers are fabricated using an adapted stack and draw procedure starting from commercially available synthetic fused silica tubing. The main reason to use the stack and draw technique because, it offers high flexibility in designing the fiber structure, low cost, speed and repeatability. Moreover, this flexibility allows us to design a negative curvature fiber. Figure 6 shows a scanning electron microscope (SEM) images of a few types of NC-HCFs with different cladding structures.

The stack and draw technique and fiber drawing of NC-HCF that used in this work are shown in Fig. 7. In the stack and draw technique, the general step is to stack silica capillary tubes or solid silica rods into the desired pattern to produce a preform. The precision and uniformity while stacking capillaries and rods are essential because these will affect the properties and quality of fiber structure after the preform is drawn to fiber.

All materials used in fabrications presented in this thesis was F300, synthesized fused silica material from Heraeus which has a low concentration of OH group (Humbach et al. 1996). We started with two tubes different outer and inner diameter was used. Tube 1 has an outer diameter (OD) 16.03 mm and 12.27 mm of inner diameter. Tube 2 has an outer diameter (OD) 25.1 mm and 23.39 mm of inner diameter. Tube 1 will be a jacket tube, and tube two will draw to a smaller diameter to stack inside the Tube 2.

Stack

The fiber was fabricated using the stack and draw technique, by placing eight identical capillaries drawn from thin wall silica tube (Suprasil F300, Heraeus) inside a larger jacketing tube. To fabricate NC-HCF, two short capillaries (blue) or rods

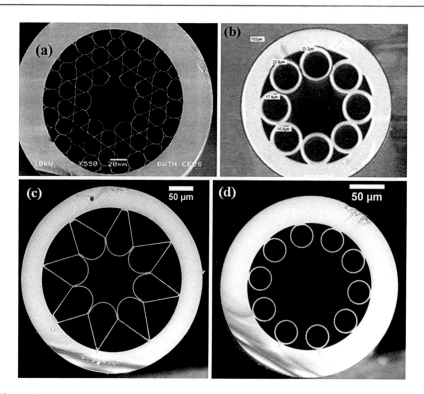

Fig. 6 Scanning electron microscope pictures of few types of negative curvature fiber: (**a**) hypocycloid-core Kagome hollow-core photonic crystal fiber (Wang et al. 2010), (**b**) is negative curvature fiber from Pryamikov group (Pryamikov et al. 2011), (**c**) ice cream shape NC-HCF from Yu et al. (2012), and (**d**) NC-HCF made of silica, with nontouching capillaries fabricated by Belardi and Knight (2014b)

need to be inserted in the core region at the ends of the stack to support the cladding and prevent the stack from collapsing and to make sure the stack tight and does not slip out during the drawing process as shown in the Fig. 7a.

Drawing of Cane

The cane is an intermediate preform which is drawn from the stack (Fig. 8b). The cane is essentially a complete optical waveguide but on the millimeter scale. The drawing of the cane is a necessary and important step in the fabrication of NC-HCF and other HCFs. One of the reason is to avoid a drastic scale change from the stack of the preform (centimeter scale) to final end-product (fiber) in micrometer scale. In the reference, Chen and Birks (2013) mention that preventing the structure from deformation becomes more problematic in the large-scale change. Conclusions, drawing fiber from the smaller cane more rather than a larger (in diameter) stack is a best practice to obtain a high-quality NC-HCFs. In our fabrication of NC-HCF, the cane size was usually about 3–4 cm in diameter.

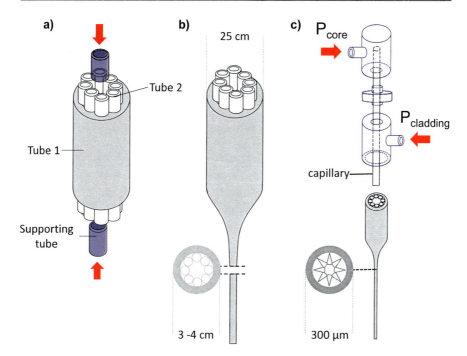

Fig. 7 Three stages of the stack and draw technique of NC-HCF. (**a**) The first stage is a stack to the desire cane. To maintain the cladding structure, two short capillaries (blue tube) are to insert on both ends as support. (**b**) Drawing of canes. The middle part of the stack without supporting capillaries is drawn into canes of smaller diameter (typically cane diameter is 3–4 cm). (**c**) Fiber fabrication from canes. Different pressures are applied to different regions of core and cladding

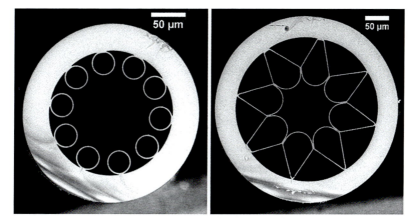

Fig. 8 Scanning electron micrographs of the two different forms of hollow fiber used in the laser system (Abu Hassan et al. 2016). Left: gain fiber with transmission at 1.53 and 3.1 μm wavelengths. Right: feedback fiber with low loss at 3.1 μm

Table 1 Fiber drawing parameters to obtain NC-HCF for 3.16 μm wavelength

Feed rate (mm/min)	Draw speed (m/min)	Temperature (°C)	P core (kPa)	P cladding (kPa)
50	6.6	1910	0.1	6

Drawing of Fiber

Next, the cane is drawn to the fiber size using the fiber drawing tower as shown in Fig. 7c. In this process, the end of the cane is fitted to the brass metal holder. This custom made holder has a small hole to apply pressure into the cane to maintain air holes from collapse in the drawing process. The pure nitrogen used for pressurization. The pressurization is crucial not only to avoid the air holes from collapse during the drawing process but also the key to the formation of the negative curvature core wall. The negative curvature core wall can affect the attenuation and the mode profile of the hollow core fiber.

While the fiber has been drawn, the feed speed of the cane, the draw speed of the fiber, the furnace temperature, and the gas pressure must be carefully adjusted, or the fine structure of fiber can be easily deformed otherwise. In the drawing stage, polymer material will be coated around the final fiber for protection and enhancement of the mechanical strength of the optical fiber. The fabrication parameter to obtain the NC-HCF are shown in this book are presented in Table 1.

Design and Properties of Fiber

Fiber Design

The silica-based hollow fibers used in the experiments in the Abu Hassan et al. (2016) offer a complementary approach to fiber lasers in the mid-IR and are shown in Fig. 8. The work in Abu Hassan et al. (2016), we used two different hollow core fibers with slightly different designs. First, is the hollow core with un-touching capillaries in the cladding and second is the "ice cream shape" in the cladding. This fiber will be our gain fiber and feedback fiber respectively. The similarity features of both fibers are negative curvature core boundary wall. The negative curvature core boundary provides a lower attenuation in interested wavelength for both fibers (Yu and Knight 2013).

First, the anti-resonant hollow core fiber used as a gain fiber was fabricated and reported in Belardi and Knight (2014a). This fiber is chosen because it has low attenuation at the pump band around 1.53 μm and low attenuation at the lasing wavelength around 3.1 μm as shown in Fig. 7. The gain fiber has a core diameter of ∼109 μm; an outer diameter of ∼233 μm, an average "cladding capillaries" diameter of 27.9 μm, and a silica wall thickness of ∼2.4 μm (Fig. 9).

Secondly, the negative curvature fiber (NCF) was specially designed to provide the lowest possible attenuation at the lasing wavelength. The NCF was used as a feedback fiber. It has a core diameter of ∼95 μm, an outer diameter of ∼357 μm, as shown in Fig. 10b. We will discuss more technique used to obtain NCF attenuation in the next section.

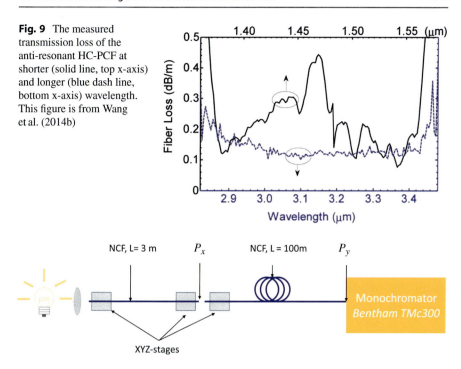

Fig. 9 The measured transmission loss of the anti-resonant HC-PCF at shorter (solid line, top x-axis) and longer (blue dash line, bottom x-axis) wavelength. This figure is from Wang et al. (2014b)

Fig. 10 Schematic of the modified version of the conventional cut-back technique to determine attenuation for feedback fiber

Attenuation Measurement: Cutback Method

Attenuation is among of the main characteristics of the fiber. Attenuation of an optical fiber is the loss of optical power as a result of leakage, scattering, bending, and other loss mechanisms as the light travel through the fiber. The cutback method is often used for measuring the total attenuation of an optical fiber. The cutback method involves comparing the optical power transmitted through a longer piece of fiber to the power transmitted through a shorter piece of the fiber.

The cutback method requires that a test fiber of known length (L) be cut back to a shorter length (L_c). The cutback method begins by measuring the output power, P_y, of the test fiber of known length (L). Without disturbing the input conditions, the test fiber is cut back to a shorter length (L_c). The output power, P_x, of the short test fiber (L_c) is then measured, and the fiber attenuation is calculated. The attenuation of the cut piece can be determined by:

$$\alpha \, (\text{dB/m}) = \frac{10 \log_{10}(P_x/P_y)}{L - L_c} \quad (3)$$

The standard cutback technique is a destructive technique which means the fiber under test to determining certain optical fiber transmission characteristics, such as attenuation and bandwidth. To avoid cutting the feedback fiber, we have measured

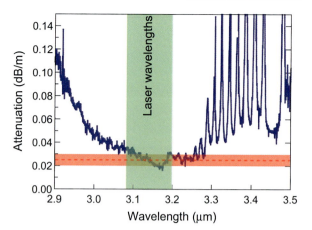

Fig. 11 Attenuation curve for the feedback fiber showing attenuation of 0.025 ± 0.005 dB/m over the laser wavelength band. Sharp features at 3.3 μm and beyond arise from small quantities of HCl gas present in the fiber core. The pink band indicates the uncertainty in the measurement of minimum attenuation

the fiber attenuation accurately by using a modified version of the conventional cut-back technique in which we use low loss butt-coupling (see Fig. 11), which is repeatable, instead of cutting the fiber, which is not. Butt-coupling has a very low loss (<0.1 dB) in these fibers because of the low numerical aperture, large core size and absence of Fresnel reflections. The uncertainty in the loss measurement is ±5 dB/km. The fiber was loosely coiled on the bench with a bend radius of 40 cm to avoid additional loss caused by bending.

In this section, all the cut-back measurements were performed by using a tungsten lamp; it has a broad spectrum from near-UV to mid-IR. In our experiments, all the spectra were measured and recorded by the optical spectrum analyzer (OSA) *Ando AQ 6315A* (wavelength range from 350 to 1750 nm) and a scanning monochromator *Bentham TMc300* (wavelength range from 250 nm to 5.4 μm).

Figure 11 shows the attenuation curve for the feedback fiber, showing attenuation of 0.025 ± 0.005 dB/m over the laser wavelength band. Sharp features at 3.3 μm and beyond arise from small quantities of HCl gas present in the fiber core. The presence of trace amounts of HCl gas in our fiber would appear to be reasonable given that our starting material is F300 synthetic fused silica, which contains 1450 ppm of chlorine (Haken et al. 2000), and our measurement is over 100 m in length. We can remove those absorption lines by purging the fiber with nitrogen.

The red color band indicates the uncertainty in the measurement of attenuation. Then, we remeasure the attenuation using single wavelength laser to validate the accuracy of our measurement. Our feedback fiber (ice-cream cone shape) has one of the lowest attenuations reported for an optical fiber in this spectral band (Abu Hassan et al. 2016).

The disadvantage of the tungsten lamp is its limited luminance especially at wavelengths above 3 μm in the mid-infrared spectral region. To validate our attenuation measurement for feedback fiber using a tungsten lamp, we used our own CW 3 μm laser as a source to re-measure the fiber attenuation. The cut-back measurements were performed by replacing the tungsten lamp with our stable CW

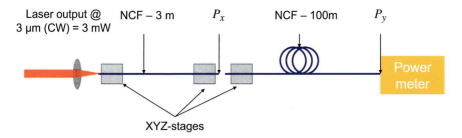

Fig. 12 Schematic of the modified version of the conventional cut-back technique used to determine attenuation for feedback fiber

Table 2 Readings of the attenuation of the NC-HCF using single wavelength laser

P_x (mW)	P_y (mW)	Loss (dB/m)
0.600	0.303	0.0297
0.605	0.309	0.0292
0.615	0.304	0.0306

3 μm laser and the output measured using a power meter as shown in Fig. 12. Measurements of optical powers were performed using a thermal power meter (*Ophir 3A-SH*).

Table 2 lists the readings of the attenuation at 3.12 μm using stable CW laser. We recorded three readings with the average attenuation 0.0298 dB/m, slightly higher compared to the attenuation using tungsten lamp due to a smaller fiber coil radius causing the bending loss increase. However, the attenuation is still in the acceptable range and shows an agreement with the previous measurement.

Conclusion

This chapter shows the study of design and properties of negative curvature fiber. The step by step fabrication method of NC-HCF using stack and draw was shown. The cut-back measurement is the main method to determine the attenuation of NC-HCF experimentally.

References

M.R. Abu Hassan et al., Cavity-based mid-IR fiber gas laser pumped by a diode laser. Optica **3**(3), 218 (2016). https://www.osapublishing.org/abstract.cfm?URI=optica-3-3-218. Accessed 7 July 2016

M. Alharbi et al., Hypocycloid-shaped hollow-core photonic crystal fiber. Part II: cladding effect on confinement and bend loss. Opt. Express **21**(23), 28609–28616 (2013). http://www.osapublishing.org/viewmedia.cfm?uri=oe-21-23-28609&seq=0&html=true. Accessed 26 Jan 2016

A. Argyros, J. Pla, Hollow-core polymer fibres with a Kagome lattice: potential for transmission in the infrared. Opt. Express **15**(12), 7713 (2007). https://www.osapublishing.org/oe/abstract.cfm?uri=oe-15-12-7713. Accessed 7 Dec 2016

J. Bei et al., Reduction of scattering loss in fluoroindate glass fibers. Opt. Mater. Express **3**(9), 1285 (2013). http://www.opticsinfobase.org/abstract.cfm?URI=ome-3-9-1285

W. Belardi, J.C. Knight, Effect of core boundary curvature on the confinement losses of hollow antiresonant fibers. Opt. Express **21**(19), 21912 (2013). https://www.osapublishing.org/oe/abstract.cfm?uri=oe-21-19-21912. Accessed 7 Dec 2016

W. Belardi, J.C. Knight, Hollow antiresonant fibers with low bending loss. Opt. Express **22**(8), 10091–10096 (2014a). http://www.osapublishing.org/viewmedia.cfm?uri=oe-22-8-10091&seq=0&html=true. Accessed 15 Oct 2015

W. Belardi, J.C. Knight, Hollow antiresonant fibers with reduced attenuation. Opt. Lett. **39**(7), 1853–1856 (2014b). http://www.osapublishing.org/viewmedia.cfm?uri=ol-39-7-1853&seq=0&html=true. Accessed 26 Jan 2016

F. Benabid, P.J. Roberts, Linear and nonlinear optical properties of hollow core photonic crystal fiber. J. Mod. Opt. **58**(2), 87–124 (2011). http://www.tandfonline.com/doi/abs/10.1080/09500340.2010.543706. Accessed 7 Dec 2016

F. Benabid et al., Stimulated Raman scattering in hydrogen-filled hollow-core photonic crystal fiber. Science **298**(5592), 399–402 (2002)

T.A. Birks, P.J. Roberts, P.S.J. Russell, D.M. Atkin, T.J. Shepherd, et al., Full 2-D photonic bandgaps in silica/air structures. Electron. Lett. **31**(22), 1941–1943 (1995)

T.A. Birks, J.C. Knight, P.S. Russell, Endlessly single-mode photonic crystal fiber. Opt. Lett. **22**(13), 961–963 (1997). http://www.ncbi.nlm.nih.gov/pubmed/18185719

T.A. Birks, G.J. Pearce, D.M. Bird, Approximate band structure calculation for photonic bandgap fibres. Opt. Express **14**(20), 9483–9490 (2006). http://www.ncbi.nlm.nih.gov/pubmed/19529335

R.T. Bise, D.J. Trevor, Sol-gel derived microstructured fiber: fabrication and characterization, in *OFC/NFOEC Technical Digest. Optical Fiber Communication Conference, 2005*, vol. 3 (2005), p. 3. Available at: http://ieeexplore.ieee.org/lpdocs/epic03/wrapper.htm?arnumber=1501298

J.R. Carson, S.P. Mead, S.A. Schelkunoff, Hyper-frequency wave guides – mathematical theory. Bell Syst. Tech. J. **15**(2), 310–333 (1936). http://ieeexplore.ieee.org/lpdocs/epic03/wrapper.htm?arnumber=6772987. Accessed 7 Dec 2016

Y. Chen, T.A. Birks, Predicting hole sizes after fibre drawing without knowing the viscosity. Opt. Mater. Express **3**(3), 346 (2013). http://www.osapublishing.org/viewmedia.cfm?uri=ome-3-3-346&seq=0&html=true. Accessed 7 Mar 2016

F. Couny et al., Identification of Bloch-modes in hollow-core photonic crystal fiber cladding. Opt. Express **15**(2), 325 (2007). https://www.osapublishing.org/abstract.cfm?URI=oe-15-2-325. Accessed 7 Dec 2016

R.F. Cregan, Single-mode photonic band gap guidance of light in air. Science **285**(5433), 1537–1539 (1999). http://www.sciencemag.org/cgi/doi/10.1126/science.285.5433.1537. Accessed 7 Nov 2013

B. Debord et al., Hypocycloid-shaped hollow-core photonic crystal fiber. Part I: arc curvature effect on confinement loss. Opt. Express **21**(23), 28597–28608 (2013). http://www.osapublishing.org/viewmedia.cfm?uri=oe-21-23-28597&seq=0&html=true. Accessed 26 Jan 2016

M.A. Duguay et al., Antiresonant reflecting optical waveguides in SiO_2-Si multilayer structures. Appl. Phys. Lett. **49**(1), 13 (1986). http://scitation.aip.org/content/aip/journal/apl/49/1/10.1063/1.97085. Accessed 26 Jan 2016

U. Haken et al., Refractive index of silica glass: influence of fictive temperature. J. Non-Cryst. Solids **265**(1–2), 9–18 (2000). http://www.sciencedirect.com/science/article/pii/S0022309399006973. Accessed 8 Mar 2016

C. Harvey et al., in *Reducing Nonlinear Limitations of Ytterbium Mode-Locked Fibre Lasers with Hollow-Core Negative Curvature Fibre*. CLEO: 2015 (OSA, Washington, DC, 2015),

p. STh1L.5. Available at: https://www.osapublishing.org/abstract.cfm?uri=CLEO_SI-2015-STh1L.5. Accessed 23 Mar 2016

O. Humbach et al., Analysis of OH absorption bands in synthetic silica. J. Non-Cryst. Solids **203**, 19–26 (1996). http://www.sciencedirect.com/science/article/pii/0022309396003298. Accessed 7 Mar 2016

D.O. Il, Experimental study of dual-core photonic crystal fibre. Electron. Lett. **36**(16), 1358–1359 (2000)

P. Jaworski et al., Picosecond and nanosecond pulse delivery through a hollow-core negative curvature fiber for micro-machining applications. Opt. Express **21**(19), 22742–22753 (2013). http://www.ncbi.nlm.nih.gov/pubmed/24104161

P. Kaiser, H.W. Astle, Low-loss single-material fibers made from pure fused silica. Bell Syst. Tech. J. **53**(6), 1021–1039 (1974). http://ieeexplore.ieee.org/lpdocs/epic03/wrapper.htm?arnumber=6774079. Accessed 16 Jan 2019

K.M. Kiang et al., Extruded singlemode non-silica glass holey optical fibres. Electron. Lett. **38**(12), 546 (2002). http://digital-library.theiet.org/content/journals/10.1049/el_20020421

W.H. Kim et al., Recent progress in chalcogenide fiber technology at NRL. J. Non-Cryst. Solids **431**, 8–15 (2016). https://doi.org/10.1016/j.jnoncrysol.2015.03.028

J.C. Knight et al., All-silica single-mode optical fiber with photonic crystal cladding. Opt. Lett. **21**(19), 1547 (1996a). https://www.osapublishing.org/abstract.cfm?URI=ol-21-19-1547. Accessed 15 Mar 2018

J.C. Knight et al., Pure silica single-mode fiber with hexagonal photonic crystal cladding. Proc. Opt. Fiber Commun. Conference (1996b), pp. 339–342

J.C. Knight et al., All-silica single-mode optical fiber with photonic crystal cladding: errata. Opt. Lett. **22**(7), 484–485 (1997). http://www.ncbi.nlm.nih.gov/pubmed/18183242

J.C. Knight et al., Large mode area photonic crystal fibre. Electron. Lett. **34**(13), 1347 (1998a). http://link.aip.org/link/ELLEAK/v34/i13/p1347/s1&Agg=doi

J.C. Knight et al., Photonic band gap guidance in optical fibers. Science **282**(5393), 1476–1478 (1998b)

A.N. Kolyadin et al., Light transmission in negative curvature hollow core fiber in extremely high material loss region. Opt. Express **21**(8), 9514–9519 (2013). http://www.osapublishing.org/viewmedia.cfm?uri=oe-21-8-9514&seq=0&html=true. Accessed 21 Jan 2016

V.V.R. Kumar et al., Extruded soft glass photonic crystal fiber for ultrabroad supercontinuum generation. Opt. Express **10**(25), 1520–1525 (2002). http://www.ncbi.nlm.nih.gov/pubmed/19461687

N.M. Litchinitser et al., Antiresonant reflecting photonic crystal optical waveguides. Opt. Lett. **27**(18), 1592 (2002). http://www.osapublishing.org/viewmedia.cfm?uri=ol-27-18-1592&seq=0&html=true. Accessed 26 Jan 2016

N.M. Litchinitser et al., Resonances in microstructured optical waveguides. Opt. Express **11**(10), 1243–1251 (2003). http://www.ncbi.nlm.nih.gov/pubmed/19465990

E.A.J. Marcatili, R.A. Schmeltzer, Hollow metallic and dielectric waveguides for long distance optical transmission and lasers. Bell Syst. Tech. J. **43**(4), 1783–1809 (1964). http://ieeexplore.ieee.org/lpdocs/epic03/wrapper.htm?arnumber=6773550. Accessed 28 Dec 2015

D. Mogilevtsev, T.a. Birks, P.S. Russell, Group-velocity dispersion in photonic crystal fibers. Opt. Lett. **23**(21), 1662–1664 (1998). http://www.ncbi.nlm.nih.gov/pubmed/18091876

A. Ortigosa-Blanch et al., Highly birefringent photonic crystal fibers. Opt. Lett. **25**(18), 1325–1327 (2000). http://www.ncbi.nlm.nih.gov/pubmed/18066205

D.G. Ouzounov et al., Generation of megawatt optical solitons in hollow-core photonic band-gap fibers. Science **301**(5640), 1702–1704 (2003). http://www.ncbi.nlm.nih.gov/pubmed/14500976. Accessed 5 Mar 2013

A.D. Pryamikov, in *Negative Curvature Hollow Core Fibers: Design, Fabrication, and Applications*, ed. by S. Ramachandran (2014), p. 89610I. Available at: http://proceedings.spiedigitallibrary.org/proceeding.aspx?doi=10.1117/12.2041653. Accessed 7 Dec 2016

A.D. Pryamikov et al., Demonstration of a waveguide regime for a silica hollow – core microstructured optical fiber with a negative curvature of the core boundary in the spectral region

>3.5 μm. Opt. Express **19**(2), 1441–1448 (2011). http://www.ncbi.nlm.nih.gov/pubmed/21263685

P.J. Roberts et al., Ultimate low loss of hollow-core photonic crystal fibres. Opt. Express **13**(1), 236 (2005). https://www.osapublishing.org/oe/abstract.cfm?uri=oe-13-1-236. Accessed 7 Dec 2016

P. Russell, Photonic crystal fibers. Science **299**(5605), 358–362 (2003). http://www.ncbi.nlm.nih.gov/pubmed/12532007. Accessed 16 Jan 2019

P.S.J. Russell, Photonic-crystal fibers. J. Lightwave Technol. **24**(12), 4729–4749 (2006). http://ieeexplore.ieee.org/document/4063429/. Accessed 7 Dec 2016

V. Setti, L. Vincetti, A. Argyros, Flexible tube lattice fibers for terahertz applications. Opt. Express **21**(3), 3388–3399 (2013). http://www.ncbi.nlm.nih.gov/pubmed/23481799; http://www.osapublishing.org/viewmedia.cfm?uri=oe-21-3-3388&seq=0&html=true. Accessed 26 Jan 2016

J.D. Shephard et al., Silica hollow core microstructured fibers for beam delivery in industrial and medical applications. Front. Phys. **3**, 24 (2015). http://journal.frontiersin.org/article/10.3389/fphy.2015.00024/abstract. Accessed 23 Mar 2016

V.S. Shiryaev, Chalcogenide glass hollow-core microstructured optical fibers. Front. Mater. **2**, 24 (2015). http://journal.frontiersin.org/article/10.3389/fmats.2015.00024/abstract. Accessed 13 Nov 2015

A.W. Snyder, J. Love, *Optical Waveguide Theory* (Springer, New York, 2012). https://books.google.com/books?hl=en&lr=&id=DCXVBwAAQBAJ&pgis=1. Accessed 20 Apr 2016

A. Urich et al., Flexible delivery of Er:YAG radiation at 2.94 μm with negative curvature silica glass fibers: a new solution for minimally invasive surgical procedures. Biomed. Opt. Express **4**(2), 193–205 (2013a). http://www.osapublishing.org/viewmedia.cfm?uri=boe-4-2-193&seq=0&html=true. Accessed 15 Apr 2016

A. Urich et al., Silica hollow core microstructured fibres for mid-infrared surgical applications. J. Non-Cryst. Solids **377**, 236–239 (2013b). http://www.sciencedirect.com/science/article/pii/S0022309313001166. Accessed 20 Apr 2016

M. van Eijkelenborg et al., Microstructured polymer optical fibre. Opt. Express **9**(7), 319–327 (2001). http://www.ncbi.nlm.nih.gov/pubmed/19516722

L. Vincetti, V. Setti, Waveguiding mechanism in tube lattice fibers. Opt. Express **18**(22), 23133–23146 (2010). http://www.ncbi.nlm.nih.gov/pubmed/21164654. Accessed 7 Dec 2016

L. Vincetti, V. Setti, Extra loss due to Fano resonances in inhibited coupling fibers based on a lattice of tubes. Opt. Express **20**(13), 14350 (2012). https://www.osapublishing.org/oe/abstract.cfm?uri=oe-20-13-14350. Accessed 7 Dec 2016

Y.Y. Wang et al., in *Low loss broadband transmission in optimized core-shape Kagome hollow-core PCF*. Conference on Lasers and Electro-Optics (CLEO) and Quantum Electronics and Laser Science Conference (QELS), 2010 (2010), pp. 4–5

Y.Y. Wang et al., Low loss broadband transmission in hypocycloid-core Kagome hollow-core photonic crystal fiber. Opt. Lett. **36**(5), 669–671 (2011)

Z. Wang, F. Yu, et al., Efficient 1.9 μm emission in H2-filled hollow core fiber by pure stimulated vibrational Raman scattering. Laser Phys. Lett. **11**(10), 105807 (2014a). http://iopscience.iop.org/article/10.1088/1612-2011/11/10/105807. Accessed 3 Feb 2016

Z. Wang, W. Belardi, et al., Efficient diode-pumped mid-infrared emission from acetylene-filled hollow-core fiber. Opt. Express **22**(18), 21872 (2014b). https://www.osapublishing.org/oe/abstract.cfm?uri=oe-22-18-21872

N.V. Wheeler et al., Low-loss and low-bend-sensitivity mid-infrared guidance in a hollow-core-photonic-bandgap fiber. Opt. Lett. **39**(2), 295–298 (2014). http://www.ncbi.nlm.nih.gov/pubmed/24562130

T.P. White et al., Resonance and scattering in microstructured optical fibers. Opt. Lett. **27**(22), 1977–1979 (2002). http://www.ncbi.nlm.nih.gov/pubmed/18033417

F. Yu, J.C. Knight, Spectral attenuation limits of silica hollow core negative curvature fiber. Opt. Express **21**(18), 21466–21471 (2013). http://www.ncbi.nlm.nih.gov/pubmed/24104021; http://www.osapublishing.org/viewmedia.cfm?uri=oe-21-18-21466&seq=0&html=true. Accessed 28 Sept 2015

F. Yu, J. Knight, Negative curvature hollow core optical fiber. IEEE J. Sel. Top. Quantum Electron. **22**(2), 1–11 (2016). http://opus.bath.ac.uk/47694/3/07225120.pdf. Accessed 25 Jan 2016

F. Yu, W.J. Wadsworth, J.C. Knight, Low loss silica hollow core fibers for 3–4 μm spectral region. Opt. Express **20**(10), 11153–11158 (2012). http://www.ncbi.nlm.nih.gov/pubmed/22565738

Optimized Fabrication of Thulium Doped Silica Optical Fiber Using MCVD

13

S. Z. Muhamad Yassin, Nasr Y. M. Omar, and Hairul Azhar Bin Abdul Rashid

Contents

Introduction	552
Thulium Doped Fibers	553
Fabrication Methods of Silica Fibers	555
MCVD-Solution Doping Technique	556
Fabrication and Characterization of Optical Fiber Preforms	559
Soot Deposition Temperature	561
Mechanism of Soot Deposition	562
Soot Characteristics: Physisorption and Scanning Electron Microscope (SEM) Measurements	564
Effect of Soot Condition on the Final Preform Characteristics	569
Alumina, Gallia, and Baria Solution Doped Silica Preforms	572
Spectroscopic Characteristics of Thulium Doped Fibers (TDF)	578
Absorption	579
Lifetime	580
Conclusions	582
References	583

Abstract

This chapter describes the fabrication of thulium (Tm) doped silica fibers using the modified chemical vapor deposition (MCVD) technique coupled with solution doping. Section "Introduction" provides an introduction to rare earth (RE)-doped fiber amplifiers. Section "Thulium Doped Fibers" reviews

---

S. Z. Muhamad Yassin
Photonics Laboratory, Telekom Research and Development, Cyberjaya, Malaysia
e-mail: shahrinzen@tmrnd.com.my

N. Y. M. Omar (✉) · H. A. B. Abdul Rashid
Faculty of Engineering, Multimedia University, Cyberjaya, Malaysia
e-mail: nasr-omar@hotmail.com; hairul@mmu.edu.my

© Springer Nature Singapore Pte Ltd. 2019
G.-D. Peng (ed.), *Handbook of Optical Fibers*,
https://doi.org/10.1007/978-981-10-7087-7_76

recent developments for thulium doped fibers. Sections "Fabrication Methods of Silica Fibers" and "MCVD-Solution Doping Technique" outline the common fabrication techniques of silica fibers and the MCVD-solution doping method, respectively. Section "Fabrication and Characterization of Optical Fiber Preforms" describes in details the experimental procedures used in this work for the fabrication and characterization of thulium-doped optical fiber preforms. The preforms characterization results are also provided in section "Fabrication and Characterization of Optical Fiber Preforms." The spectroscopic characteristics of thulium ions in the fabricated silica fibers are given in section "Spectroscopic Characteristics of Thulium Doped Fibers (TDF)." The conclusions drawn from this work are provided in section "Conclusions."

Keywords

Thulium · Barium · Gallium · Optical fiber · Optical preform · Modified chemical vapor deposition · Solution doping · Fluorescence decay lifetime

Introduction

The first demonstration of light amplifications in fibers by Snitzer (1961) has inspired many subsequent works to realize fiber lasers and amplifiers as practical optical devices (Koester and Snitzer 1964; Stone and Burrus 1973). The growth in this field of technology is encouraged by its potential applications in optical communication. The commercial interest combined with the available semiconductor pump laser and some optical passive devices such as wave division multiplexer and optical isolator have assisted the development of such fiber-based active devices. The major breakthrough came in 1987 when a group from Southampton University demonstrated a method for fabricating the high quality rare earth doped silica fibers (Payne 1987). Since then, much work has been focused on the development and the exploitation of the erbium doped fiber amplifier (EDFA). This development has revolutionized the telecommunication industry as EDFA has replaced electronic repeaters in fiber-based networks.

As the demand for bandwidth has increased dramatically over the years, the current conventional communication window (i.e., 1530–1560 nm) will not be sufficient to cater for the exponential traffic growth. This motivates the search for possible new communication windows or the improvement of the operations of the old ones. Suggestions for reusing the O-band (i.e., 1310 nm) and the S-band (i.e., 1450–1510 nm) or even introducing a new band (1800–2100 nm) have been proposed as viable solutions. To realize this, other types of rare earth fiber amplifiers are needed to provide gain within the respective wavelength range analogous to EDFA in the conventional C-band. Among the potential options are thulium and praseodymium doped fiber amplifiers (TDFA and PDFA). However, unlike erbium, the near infra-red emission of thulium and praseodymium ions inside silica fibers suffers from high multi-phonon transition loss. The decrease of the emission efficiency to below 10% in silica glass makes them impractical as optical active devices. This drawback has motivated the use of other types of glass hosts that exhibit a low multiphonon transition loss (i.e., low phonon energy glass). It has been demonstrated that the use of halide or chalcogenide glass as the host material yields above 90% of emission efficiency out of Tm and Pr ions (Hewak et al. 1993; Walsh and Barnes 2004).

As for the current optical communications, the use of low loss silica fibers has been widely accepted due to the abundant availability of silica and the simplicity of its fabrication process which make silica-based fibers highly cost effective. Due to bias towards the fiber communications technology, the incentive is towards developing active fibers that are compatible with the standard silica fibers. The fiber composition, design, and its compatibility with standard telecommunication silica fibers are among the important criteria in developing active fibers. Soft glasses (e.g., halide, telluride, and chalcogenide) exhibit low thermal stabilities which make them incompatible to be spliced to communication grade silica fibers (Sudo 1997). In addition, the chemical instability of the soft glass system contributes to the complexity of its fabrication process which in turn prevents a practical mass production. Besides, most soft glasses have low mechanical durability. It is for these reasons that researchers have to consider silica or any composition that has a thermal processing temperature comparable to silica as the host of the active ions.

Thulium Doped Fibers

In principle, any type of glass can be transformed into fibers. However, with regard to fibers' characteristics and fabrication, there are some practical prerequisites to them being a viable solution. These include the doping of the ions being homogeneous, the host phonon energy being suitable for the support of emission, the refractive index profile having a high degree of controllability, the optical loss being low, the shapes and sizes of the fibers cross sections being able to be precisely controlled, and the chemical and mechanical durability of the materials being high.

To date, numerous efforts have been made to produce active fibers that have thermal properties compatible with silica and exhibit amplifications in near infrared region. Apart from rare-earth elements, metals such as bismuth and chromium in silica have been known to provide broad emission in the range of 1000 to 1700 nm (Abramov et al. 2014; Dianov 2013). However, the realization of this type of fiber amplifier as a feasible optical active device has yet to be discovered. Thulium in silica on the other hand has captured recent interest due to its potential of emitting light from the decay of two of its energy manifolds (i.e., $^3H_4$ and $^3F_4$). These transitions are responsible for emissions at 800 nm (i.e., $^3H_4$ to $^3H_6$), 1480 nm (i.e., $^3H_4$ to $^3F_4$), and 1800 nm (i.e., $^3F_4$ to $^3H_6$) wavelengths. Table 1 shows several initiatives done to date to improve the emission of Tm ions in silica as well as recent efforts to develop different glass systems as Tm ions hosts that have thermal properties compatible with silica.

It can be noticed from Table 1 that numerous initiatives utilize Al as one of the modifier elements to improve the emission of Tm ions in silica glass systems.

Table 1 Recent developments of thulium doped fibers (TDF) compatible with silica

Ref.	Modifiers/glass type	Fabrication method	Lifetime $^3H_4$ (μs)	Other remarks
Walsh and Barnes (2004)	n.a	n.a	20	n.a
Cole and Dennis (2003)	Ga	Vapor phase	32	n.a
Dennis and Cole (2001)	n.a	n.a	55	8dB S-BandGain
Faure et al. (2004)	Al	Solution doping	50	n.a
	Pure silica	n.a	14	n.a
Simpson (2008)	Al	Solution doping	32	n.a
	Al/Sb	Solution doping	21	n.a
	Al/Sn	Solution doping	20	n.a
	Ge	Solution doping	21	n.a
Faure et al. (2007)	Al	Solution doping	50	n.a
	Ge	Solution doping	28	n.a
	P	Solution doping	9	n.a
Halder et al. (2013)	YASNanoparticle	Solution doping	n.a	Up conversion emission
Vermillac et al. (2017)	LaF$_3$Nanoparticle	Solution doping	58	n.a
Halder et al. (2015)	Y/Al/Ge/PNanoparticle	Solution doping	n.a	Up conversion emission
Ohara et al. (2002)	Bismuth glass	n.a	n.a	12dB gain
Aitken et al. (2006)	Aluminate glass	Crucible melt quench	300	n.a
Jander and Brocklesby (2004)	YAS Glass	Crucible melt quench	100	n.a

The presence of alumina in silica glass is also known to assist in the incorporation of rare earth ions inside silica glass (Arai et al. 1987). In addition, the incorporation of Al into silica improves the lifetime of Tm ions by at least 3 folds as compared to pure silica (Faure et al. 2004, 2007; cf. Table 1). Several other glass modifiers have been doped in silica TDF for the same reasons although not all of them produce a significant change to Tm spectroscopic properties. For instance, the doping of Sb and Sn has no effect on increasing the Tm ions lifetime (Simpson 2008; cf. Table 1), while high doping of Ge (i.e., 20 mol%) increases the $^3H_4$ lifetime by 14 μs as compared to pure silica (Faure et al. 2004, 2007; cf. Table 1). In contrast, the doping of P shortens the lifetime of Tm ions to 9 μs (Faure et al. 2007; cf. Table 1). The above observations suggest that the material selection plays an important role in improving the Tm ions lifetime. As can be observed from Table 1, in addition to Al, modifiers that show significant improvements in Tm ions spectroscopic properties are Y, Ga, and Bi.

In order to realize silica as a feasible host to Tm active ions, a modification to the silica glass network is essential. The phonon energy of the environment surrounding the Tm ions in silica has to be reduced. This is done by incorporating metal oxides together with rare earth ions inside the silica glass matrix. The presence of metal oxides is known to create clusters of low phonon energy environments inside the silica glass system (Cole and Dennis 2003; Greaves 1985). By occupying these low phonon energy sites, the multiphonon transition loss of Tm ions due to silica vibrations can be reduced which consequently increases the efficiency of the emission. Here, three dopant modifiers are chosen to realize this concept (i.e., Al, Ga, and Ba). Aluminum is the most common modifier used for this purpose and has shown promising results (cf. Table 1). Numerous reports have demonstrated a concentration limit of 8 to 9 mol% of Al_2O_3 inside silica preforms using both solution doping and vapor phase technique (Schuster et al. 2014; Tang et al. 2008). The purpose for Al inclusion here is to act as a benchmark for the in-house fabrication method and to compare it with other proposed dopant modifiers that will be used (i.e., Ga and Ba). Initially, all of the abovementioned proposed modifiers will be doped at the highest concentration limit in order to recognize the effect of each modifier on the emission characteristics of Tm ions in silica.

Fabrication Methods of Silica Fibers

The fabrication of rare-earth doped fibers differs with different types of glass host. Although soft glasses such as halide and chalcogenides are attractive, the complexity of their fabrication process limits their mass production. On the other hand, multicomponent silicate glass has a fairly simpler production process. For silica fibers, the simplicity of the fabrication process and the availability of materials have assisted the successful mass production of transmission fibers. This has encouraged researchers to use silica fibers as hosts of amplification ions.

The fabrication of high-purity silica mainly implements chemical vapor delivery technology as the basis of its process. It is generally classified into two categories: inside vapor phase oxidation and outside vapor phase oxidation. Methods that apply the outside vapor phase oxidation are the outside vapor deposition (OVD) and vapor axial deposition (VAD). Techniques that apply the inside vapor phase oxidation include modified chemical vapor deposition (MCVD) and plasma chemical vapor deposition PCVD. Among these technologies, MCVD allows a high degree of alteration to its process. This is why it is a common choice for rare-earth fiber fabrication. The doping of rare earth and silica glass modifier materials using the vapor phase has been a challenge since these materials have low vapor pressure and tend to solidify before reaching the reaction zone. The vapor flows are also subjected to inconsistency, which reduces the preform repeatability. However, to date, many initiatives have been conducted to resolve this setback since it offers simplicity and uniformity as compared to the solution based doping technique. Nevertheless, the solution-doping technique is still the most reliable technique to dope fiber preforms with rare earth and glass modifier oxide materials.

MCVD-Solution Doping Technique

A schematic of an MCVD system is illustrated in Fig. 1. An MCVD system is comprised of two major components: the gas and material delivery system and the lathe. The gas deposition system is a delivery system used to transport glass-forming reagent to the reaction zone. Typical glass-forming reagents used in the MCVD system are $SiCl_4$, $GeCl_4$, $POCl_3$, and BBr_3. These materials, which are liquid at room temperature, have a high saturated vapor pressure. This makes them suitable for mass transportation. These reagents are placed in a container called a "bubbler," which allows the carrier gas (i.e., O_2) to enter the container and carry the vapor form of the reagent to the deposition zone. Additionally, several other inert gases are included to assist either deposition (e.g., He) or dehydration (e.g., Cl_2) processes. On the lathe, a high-purity silica substrate tube is placed and heated to the reaction temperature using an oxy-hydrogen burner. The lathe rotates the substrate tube to ensure a radial uniformity of the deposited soot layer to be obtained at the inner wall of the tube. The burner moves transversely along the tube's longitudinal axis to allow the deposition to take place along its length. The deposited layer is then sintered to a transparent glassy layer and the substrate tube is collapsed at elevated temperatures into a solid rod called preform.

The solution-doping technique is widely known as a reliable technique for doping rare earth elements inside high-purity silica. Unlike glass-forming materials, most glass modifiers including rare earths do not have precursor materials that exist in a liquid phase at room temperature. In order to dope these materials inside silica, the solution-doping technique was first introduced by Townsend et al. (1987). The method uses a porous layer of deposited silica soot as the absorbent for the rare earth or modifier salt solutions. These solutions impregnate the soot and upon drying, the rare earth or modifier salt remains on the surface or inside the voids of the soot.

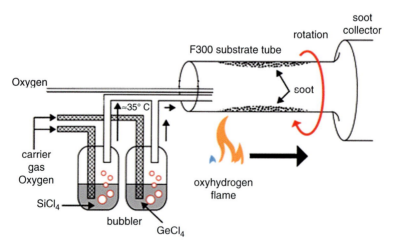

Fig. 1 A schematic diagram of the MCVD process

With proper heating and O_2 gas flow, the salt is then converted to the corresponding oxide. However, the amount of dopant introduced by the solution-doping technique can be inconsistent from one preform to another due to the unclear relationship between the nature of the porous soot and the amount of salt that remains after soaking. Hence, the optimization of the solution-doping process has been the subject of numerous studies (e.g., Dhar et al. 2006, 2008; Khopin et al. 2005; Kirchhof et al. 2003; Petit et al. 2010). In particular, Kirchhof et al. (2003) have identified that the amount of dopant that is incorporated inside the silica glass, F (mol%), can be estimated from the solution strength, c (molar), and the relative density of the porous soot, d_r, as given by

$$F(mol\%) = \frac{cV_s \left(1/d_r - 1\right)}{1 + cV_s \left(1/d_r - 1\right)} \qquad (1)$$

where V_s is the molar volume of SiO_2. The value of $\left(1/d_r - 1\right)$ represents the ratio between the empty volume and the solid volume. From Eq. 1, the relationship between the dopant concentration versus the soot density and the solution strength can be plotted as shown in Figs. 2 and 3, respectively. According to Kirchhof et al. (2003), the amount of dopant intake has a minor dependency on the type of dopant. As can be seen from the figures, the change of soot density at the lower density region causes a larger change in dopant intake than the higher density region. Additionally, if the soot density is high, the solution concentration plays a minor role in governing the amount of dopant intake. In order to achieve a high dopant intake, the soot density must be low. However, a small fluctuation will induce a large discrepancy in the final dopant concentration of the preform. This effect becomes more detrimental at higher solution strengths. Soot of low density is very sensitive to temperature variation and is prone to external disruptions, which cause inconsistency in the dopant concentration of the fabricated preforms.

In addition, the solution doping technique is known to suffer from uniformity and reproducibility issues due to the inconsistent morphology and fragile nature of the solution-impregnated soot layer. The dopant uniformity has been observed to deteriorate at high dopant concentrations (Atkins and Windeler 2003; Khopin et al. 2005). As illustrated in Fig. 2, for any particular solution strength, the concentration of dopants retained in the soot layer also decreases exponentially as the soot density increases. Besides, the soot density has been observed to change linearly with the soot process temperature (i.e., soot deposition temperature) (Petit et al. 2010). The dopant concentration is therefore more sensitive to temperature variations at lower soot deposition temperatures. This behavior restricts the dopant concentration to a lower limit if a considerable uniformity is to be achieved. To increase the uniformity, Atkins and Windeler (2003) have suggested the use of an internal heat source gas, which results in the formation of an additional soot layer that decreases the sensitivity of the soot-to-temperature variations. However, in their study, the overall uniformity of the final preform was not demonstrated and other factors such as external forces that may affect the uniformity of the final preform were not included.

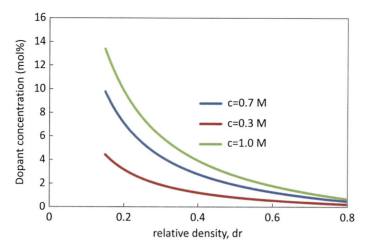

Fig. 2 Estimated dopant concentration as a function of the relative density of the soot for three solution strengths

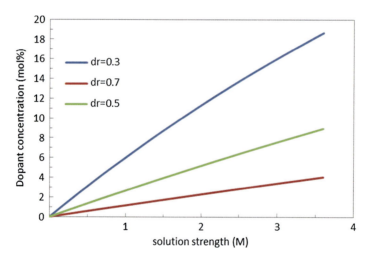

Fig. 3 Estimated dopant concentration as a function of solution strength for soot of three different relative densities

In addition, Kirchhof et al. (2003) have reported that the use of higher density soot, which is deposited at a higher temperature, enhances the longitudinal uniformity. However, increasing the soot density decreases the dopant concentration due to the reduction of the soot porosity. Hence, prior to the doping of glass modifiers (i.e., Ba and Ga), the in-house MCVD-solution doping method is optimized and improved in order to achieve the highest dopant intake and better uniformity. The optimization of the fabrication process is carried out using the common modifier (i.e., Al).

Fabrication and Characterization of Optical Fiber Preforms

The fabrication of optical fiber preforms using the modified chemical vapor deposition (MCVD) with solution-doping technique has been widely established. This is a common method used by most specialty optical fibers manufacturers to fabricate rare earth doped fibers such as erbium-doped fibers and ytterbium-doped fibers. Its flexibility to incorporate numerous types of glass modifiers and rare earth ions in the preform has made it the most preferred fabrication technique for specialty optical fibers. However, several fabrication parameters need to be optimized since each MCVD machine is unique.

This section describes in details each experimental procedure used for the fabrication of fiber preforms. The first part of the fabrication process consists of the optimization of soot condition for solution doping using MCVD. This was conducted to determine the optimum capability of the MCVD machine for doping glass modifiers using solution-doping technique. The optimum fabrication parameters for the soot are determined based on the refractive index profile of the fabricated preforms. In the first stage of experiments, the soot deposition temperature was optimized and several preforms were fabricated. The detailed procedures of soot deposition and preform fabrication are described below. Once the soot condition was optimized, several metal oxides such as the oxides of gallium, barium, and aluminum were incorporated inside the fibers preforms.

In all experiments, Heraeus F300 synthetic silica substrate tubes (25 mm OD × 19 mm ID) were used for the soot deposition. Prior to mounting the tube onto the lathe, it was rinsed with isopropanol/acetone to remove any organic contaminants. The tube was then mounted onto the lathe and surface impurities on the tube inner wall were etched using SF_6 at 2050 °C. Several layers of pure silica were then deposited and sintered acting as a barrier to avoid any impurities from diffusing into the soot. For soot deposition, $SiCl_4$ vapor from a bubbler system heated at 34 °C was carried to the substrate tube using 100sccm of O_2. In addition, 700sccm of O_2 and 75sccm of He were flowed simultaneously to promote $SiCl_4$ oxidation and improve the heat distribution in the hot zone, respectively. Concurrently, the substrate tube surface temperature was initially held at 1550 °C by an oxy-hydrogen burner traversing at the speed of 125 mm/min in the same direction as the gas flow (i.e., forward deposition). At this stage, the lathe spindle speed was increased from 30 rpm to 50 rpm to enhance the radial uniformity of the deposited soot. The deposited soot was subsequently removed from the substrate tube and subjected to morphological analysis (vide infra). Five preform tubes were prepared using the above-mentioned procedure, but with different soot deposition temperature and different number of deposited layers (Table 2). One layer of soot was deposited inside three different tubes at 1650 °C, 1750 °C, and 1800 °C, which were labeled as preform tubes P043, P046, and P023, respectively. In addition, two layers of soot were deposited inside preform tube P032 at 1800 °C by traversing the burner along the tube twice in the forward direction. The same steps were repeated for preform tube P051, but the soot deposition temperature was set to 1700 °C and 1750 °C for the first and second deposition passes, respectively. Some of the

Table 2 List of fabricated preforms for optimization of soot condition

Tube/preform #	Deposition temperature	# of layer	$SiCl_4$ flow
P023	1800	Single	100
P043	1650	Single	100
P046	1750	Single	100
P032	1800	Dual	100
P051	1700/1750	Dual	100

resultant soot was collected from preform tubes P032 and P051 and was then subjected to SEM morphology analysis. For all of the preform tubes, the deposited soot was soaked for 90 min with a 1.2 M $AlCl_3.6H_2O$/0.025 M $TmCl_3.6H_2O$ ethanol: water (9:1) solution. After the solution was drained from the tube, the soot was dried by flowing a gentle stream of nitrogen gas for 1 h. The tube was then heat treated in a tube furnace, where the temperature was raised to 800 °C at a rate of 5 °C/min. This temperature was maintained for 30 min and the tube was then allowed to cool down to room temperature naturally. The heat treatment is an essential step for removing residual solvent and converting the precursor salts to their corresponding oxides. Next, the tube was remounted onto the lathe and was further heated at 1500 °C while flowing 1000 sccm of O_2 in order to oxidize any precursor salts that might have remained in the preform tube. The doped silica soot was then gradually sintered with six burner passes starting from 1600 °C and up to 2100 °C in 100 °C increment. Subsequent to sintering, the tube was collapsed to a solid preform in a conventional manner.

The characterization of the soot morphology can be done using several techniques. One of the most important parameters of the soot is its density. The density usually determines the amount of the final dopant. Several methods have been introduced to measure the relation between the soot morphology and the applied deposition parameters. Among them are Brunauer-Emmett-Teller (BET) and Barrett-Joyner-Halenda (BJH) methods, which provide the surface area and the pore volume of the soot, respectively (Brunauer et al. 1938). These two parameters have a direct relationship with the soot density. The deposited soot (cf. Table 2) is characterized to identify the effect of different deposition parameters on the morphology of the deposited soot. This in turn helps explain the final characteristics of the fabricated preforms. 5 g of each deposited soot was scraped off of the substrate tube and was prepared for analysis using porosimetry analyzer (ASAP 2020 Micromeritics). Prior to measurement, each sample was packed inside a glass tube and was dehydrated for 450 min to remove any form of moisture. During measurement, each sample was exposed to nitrogen (N_2) gas, which is used as an adsorbate. The adsorption isotherm curves obtained from the measurements were analyzed using BET and BJH methods to obtain the surface area and porosity information.

Alternatively, scanning electron microscopy (SEM) coupled with image processing has been used by some researchers to estimate the particle and pore size distribution of the soot (e.g., Dhar et al. 2006; Tang et al. 2006). This technique measures the soot characteristics only at a single point on the preform and assumes that the soot characteristics are the same along the entire preform. A small slice of the deposited tube was cut and analyzed using SEM. The top and cross sectional views of the soot were captured as images that can be further analyzed to characterize the morphology of the deposited soot. To facilitate comparisons, all SEM images were captured using the same magnification.

Based on previous reports (Blanc et al. 2008; Faure et al. 2004; Simpson 2008), a high concentration of a metal glass modifier serves to improve the emission characteristics of thulium ions in the near infrared region. The main objective is thus to fabricate preforms with the highest possible concentration of glass modifiers using MCVD coupled with solution-doping technique. The glass modifier to be co-doped with thulium is chosen to be baria (BaO), gallia (Ga_2O_3), or alumina (Al_2O_3). The choice of materials was made following their successful incorporation inside silica glass previously (Cole and Dennis 2003; Poole 1988; Sen et al. 2010). The fabrication method used was similar to that used to fabricate P051 (cf. Table 2). As described below, the soot deposited using these fabrication parameters is optimum for a high dopant intake and uniformity. For doping with Ba or Ga, two solution-doped preforms were fabricated using two aqueous solutions of the respective dopant precursor. One of the preforms was doped with a solution nearly saturated with the dopant ions, whereas the other one was doped using a solution of moderate concentration. The moderately concentrated Ba solution had also a 0.3 M $AlCl_3.6H_2O$. For doping with Al, an aqueous solution of 3.2 M $AlCl_3.6H_2O$ (i.e., a near saturation solution) was used. Preforms fabricated using moderate Al concentration (i.e., 1.2 M) are those described above. In addition, each solution had a 0.025 M $TmCl_3.6H_2O$. The details for each solution/soaking process are listed in Table 3.

Soot Deposition Temperature

Table 4 presents the observations made during the $SiCl_4$ oxidation at temperatures ranging from 1550 to 2100 °C. With 100sccm of O_2 carrying $SiCl_4$ vapor from the

Table 3 Solution mixture, concentrations, and soaking time used to soak five solution doped preforms

Preform #	Modifier precursor	Concentration (Molar)	Active dopant precursor	Concentration (molar)	Soaking time (min)
P053	$AlCl_3.6H_2O$	3.2	$TmCl_3.6H_2O$	0.025	90
P057	$BaCl_2.H_2O$	1.4	$TmCl_3.6H_2O$	0.025	90
P085	$BaCl_2.H_2OAlCl_3.6H_2O$	0.90.3	$TmCl_3.6H_2O$	0.025	90
P058	$Ga(NO_3)_3.xH_2O$	2.3	$TmCl_3.6H_2O$	0.025	90
P060	$Ga(NO_3)_3.xH_2O$	1.5	$TmCl_3.6H_2O$	0.025	90

Table 4 Observation during $SiCl_4$ oxidation at different temperatures

Temperature (°C)	Observation
1550	No deposition
1600	White cloud
1650	Soot deposition
1700	Soot deposition
1750	Soot deposition
1800	Soot deposition
1900	Partly sintered
2100	Fully sintered

bubbler, no noticeable reaction was observed inside the tube when it was heated at 1550 °C with the burner moving at 125 mm/min. At 1600 °C, however, the formation of aerosols was visible downstream of the hot zone. This indicates that the used temperature (i.e., 1600 °C) is sufficient to promote the reaction between $SiCl_4$ and O_2 in the MCVD system. Wood et al. (1978) and Binnewies and Jug (2000) reported a starting reaction temperature of 700 °C and 900 °C, respectively. In both of these studies, the reported temperature was that of the reactant gasses, whereas in this work the reported temperature is that of the tube outer surface. Simpkins et al. (1979) and Walker et al. (1980) demonstrated the existence of a radial temperature gradient from the tube inner wall to the center of the tube at different longitudinal distances from the burner position. In our MCVD system, the temperature displayed by the pyrometer is the temperature of the glass tube outer surface that is located near the burner position. Dhar et al. (2006) reported that the reaction between $SiCl_4$ and O_2 in the MCVD was observed to start at 1200 °C by using a thinner substrate tube (i.e., 20/17 OD/ID). The lower reaction temperature is attributed to the substrate tube thinner wall. Kirchhof and Funke (1986) reported that the temperature difference between the inner and the outer surface of the glass tube was largely dependent on its thickness and the type of materials used. As the temperature was increased beyond 1600 °C until 1950 °C, the formation of silica soot, which was eventually deposited on the tube inner wall, was observed (Table 4). Above 2000 °C, the deposited soot was fully consolidated into a transparent glass layer (Table 4).

Mechanism of Soot Deposition

As schematically illustrated in Fig. 4, in MCVD, the deposition process comprises two sequential steps. In the first step, the cool $SiCl_4$ vapor approaches the hot zone of the burner and as the vapor is brought to the reaction temperature, submicrometer particles (i.e., silica soot) are formed by the oxidation of $SiCl_4$. These particles can grow in size up to few microns depending on process parameters such as precursor and carrier gas flow rates and reaction temperature (Park et al. 1999). In the second step, the formed particles are deposited inside the substrate tube downstream from the reaction hot zone where the wall is at lower temperature than the gas. In MCVD,

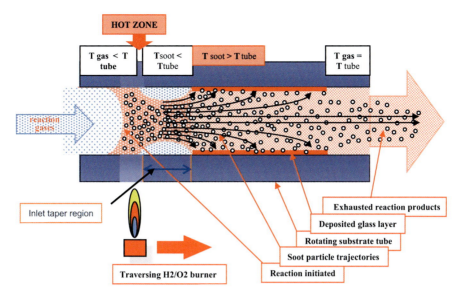

Fig. 4 The deposition mechanism in MCVD

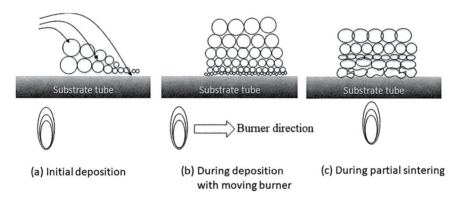

Fig. 5 Phases of soot particle evolution during deposition

thermophoresis is considered to be the main deposition mechanism by which a particle present in a temperature gradient experiences a net force in the direction of decreasing temperature (Walker et al. 1980). Further downstream, the gas and wall temperatures equilibrate causing the deposition process to stop and the nondeposited particles to flow to the exhaust tube (Park et al. 1999; Walker et al. 1980).

Park et al. (1999) demonstrated that larger particles were deposited further downstream from the burner position when a high reaction temperature was used (i.e., 1700 °C). In contrast, at low deposition temperature regime, Tang et al. (2007) demonstrated that bigger particles were deposited near the burner, while smaller ones were deposited further downstream (Fig. 5a). As the burner moves downstream, bigger particles will be deposited on top of the smaller ones (Fig. 5b). The movement of the burner downstream will also induce partial sintering to the deposited particles as the burner passing below the deposited soot heats the tube wall to the maximum temperature (Fig. 5c). This process is also expected to fuse several soot particles together to form bigger particles and consequently to increase the density of the soot. In this work, the same soot pattern as that observed by Tang et al. (2007) is expected to be obtained because the deposition process was conducted at a comparable temperature regime.

Soot Characteristics: Physisorption and Scanning Electron Microscope (SEM) Measurements

Figure 6a, b is the isotherm curves of the excess soot collected from the exhaust tube and the deposited soot scrapped off of the substrate tube, respectively. Both of the soot samples were produced at the temperature of 1700 °C. It should be mentioned that in our fabrication process, the burner was not allowed to pass below the exhaust tube and, therefore, the soot collected in the exhaust tube did not experience partial sintering as that deposited inside the substrate tube. The isotherm curve was obtained from 46 measurements at different pressures within 2 h 11 min of elapsed time. Each measurement was taken after the system had reached equilibrium for every change in pressure. As can be seen from Fig. 6, both adsorption isotherm curves exhibit a type IV isotherm (Sing et al. 1985). A similar shape of isotherm was also observed by Fan et al. (2008) and Rio et al. (2012). This isotherm indicates that the samples are mesoporous in nature. Typically, a mesoporous structured sample has a pore size of 5 to 20 nm (Gregg and Sing 1982; Sing et al. 1985), which induces capillary condensation of N_2 at high pressure. This phenomenon is observed by the existence of a hysteresis loop near the saturation pressure p_o (Fig. 6a, b). The hysteresis loop occurs when the adsorption and desorption isotherms do not coincide over a certain region of external pressures (Naumov 2009). The hysteresis loop was categorized as H_3 type hysteresis and is usually associated with the presence of nonrigid slit shaped pores (Naumov 2009).

For samples that exhibit a type IV isotherm curves, the BET and BJH analyses have been reported to provide the most appropriate solution to determine the effective surface area and the pore volume, respectively (Gregg and Sing 1982). These soot characteristics generally indicate the amount of solution that can be impregnated inside the soot and sequentially the final dopant concentration in the preform. Dhar et al. (2008) suggested two mechanisms of the solution doping: surface adsorption and pore retention that may have a correlation with the soot surface area and pore volume, respectively. One of these mechanisms may dominate the other based on the fabrication process and parameters. Table 5 shows the effective surface area and the pore volume of the soot samples produced at different temperatures. These samples consist of soot that was scrapped off of the substrate tube and a sample of excess soot that was taken from the exhaust tube.

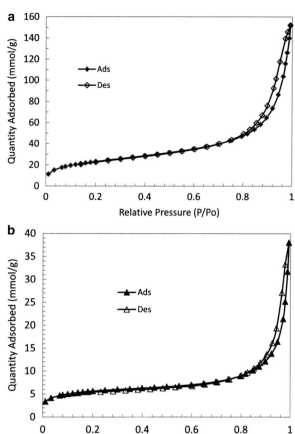

Fig. 6 N₂ adsorption isotherm curves for soot deposited at 1700 °C. (**a**) and (**b**) refer to the soot collected from the exhaust tube and the substrate tube, respectively

Table 5 BET surface area and BJH pore volume of the deposited soot

Substrate tube #	Temperature (°C)	BET surface area(m² g⁻¹)	BJH pore volume(cm³ g⁻¹)
P043	1650	35.1	0.1014
T044	1700	24.5	0.0640
P046	1750	20.9	0.0557
P023	1800	18.0	0.0480
NA	1700 (exhaust tube)	78.7	0.2070
P051	1700/1750	23.2	0.0523

The BET analysis shows that the excess soot has a very large effective surface area compared to the soot that was deposited on the substrate tube (cf. Table 5). This is due to the partial sintering that the soot on the substrate tube experienced during forward deposition. The additional heat that has been supplied by the forward passing burner onto the soot right after the deposition behaves as a heat treatment

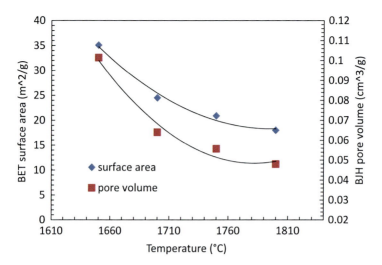

Fig. 7 Soot BJH pore volume and BET surface area evolution with deposition temperature for single layer deposition. Solid lines are a guide for the eye

that stimulates the soot particles to fuse to one another forming bigger particles thus reducing the total surface area. This phenomenon intensifies when the soot is exposed to higher temperatures (cf. Fig. 7). For example, the soot surface area was reduced significantly from 35 m^2 g^{-1} to 18 m^2 g^{-1} when the deposition temperature was increased from 1650 to 1800 °C (Table 5). In addition, the BET analysis of the dual-layer soot indicates that its average surface area (23.2 m^2 g^{-1}, Table 5) is close to that of the single layer soot deposited between 1700 and 1750 °C (24.5 and 20.9 m^2 g^{-1}, respectively (Table 5)). In a dual-layer soot structure, the additional layer is deposited on top of a porous surface instead of a solid one (i.e., the tube wall). The layer upon layer structure reduces the heat transfer that, in turn, reduces the amount of sintered particles. Thus, the temperature gradient across the dual-layer soot causes the average particle size of the deposited soot to be smaller than the single-layer soot for the same deposition temperature.

The pore volume of the soot is another characteristic that is affected by the deposition temperature. Using BJH analysis, the pore volume is observed to decrease as the deposition temperature increases (Table 5). This is because the heat from the burner supplies enough energy to cause the pore structure to collapse and the soot to density. It is believed that there is an inversely proportional relationship between the pore volume and the density of a porous sample (Zielinski and Kettle 2013). The increment of soot density with temperature that was observed by Petit et al. (2010) is in agreement with the pore volume pattern obtained from BJH measurements carried out in this work. For the dual-layer soot structure, the pore volume is found to be comparable to the single layer soot deposited between 1750 and 1800 °C (Table 5). The reason behind this difference in pore volume is unclear. It is suspected that due to the smaller size soot that is deposited at the bottom of each layer, the voids on the surface of the first layer are filled with the smaller particles that are produced by the second layer which effectively reduces the total pore volume (Tang et al. 2007). It is noteworthy to mention that a soot with a smaller pore volume is expected to be less sensitive to temperature variations and is stronger at withstanding external forces, hence improving the longitudinal uniformity.

Figure 8 shows the SEM images of the soot samples deposited at temperatures ranging from 1650 to 1800 °C using both single and dual pass deposition method. Both the top view and cross sectional side view images of the soot are displayed for each deposition temperature. As can be observed from Fig. 8a–c top view, larger soot particles were obtained at higher deposition temperatures, which is in agreement with the surface area measurements. As seen in Fig. 5b, during deposition, larger soot particles are deposited on top of smaller particles creating a gradient of increasing particle size from the tube wall to the exposed inner surface (Tang et al. 2007). However, this gradation vanishes soon after the burner passes under the deposited soot owing to the heat supplied by the burner that causes the particles to fuse to one another forming larger particles. On the other hand, at the same longitudinal position, there exists a gradient of decreasing heat distribution towards the tube axis that gives a counter effect to the aforementioned particle fusion. Tang et al. (2007) observed a greater particle fusion effect for soot nearer to the tube wall than that nearer to the exposed inner surface. In contrast, in this work, no gradation of increasing particle size is observed (Fig. 8a–c side view). This indicates that the oxy-hydrogen burner supplied enough heat to cause the particles to fuse to one another.

Figure 8e shows the soot size of the dual layer P051 (top and side views). From the top view image, the soot size is considerably similar to P023 (Fig. 8c) and P046 (Fig. 8b), which is in agreement with the BET results. However, an obvious difference between the dual and single layers is observed from the side view images. The formation of the dual layer soot is apparent and the difference between both layers is noticeable. It is obvious that the top layer has higher porosity from its "fluffy" appearance. It is believed that the radial particle size distribution of the single layer deposition that was observed by Tang et al. (2007) also applies for the case of dual layer deposition. The pattern of dual layer soot is illustrated in Fig. 9 whereby the smaller size soot is located at the bottom of each layer. As compared to the single layer, it is believed that for the dual layer the gradient of increasing particle size along the radial axis decreases by half as each layer introduces individual pattern of particle size. This will produce an innermost layer (furthermost from the tube wall) with smaller particles as compared to the single layer deposition with the same total $SiCl_4$ flow (Fig. 9). It is also believed that the average particle size will also be smaller than that of the single pass having the same total $SiCl_4$ flow and deposition temperature. This is because the second layer is deposited on top of the first layer which is not a solid surface (i.e., a porous layer). This makes the heat transfer from the burner to the second layer becomes less effective compared to the single layer. In single layer deposition, the soot is deposited on a tube wall that has a higher thermal conductivity than a porous surface (Davarzani et al. 2011).

Fig. 8 Top and side views of all deposited soot (Table 2): (**a**), (**b**), (**c**), (**d**), and (**e**) refers to 1650, 1750, 1800, 1800 (dual layer), and 1700/1750 (dual layer) °C, respectively

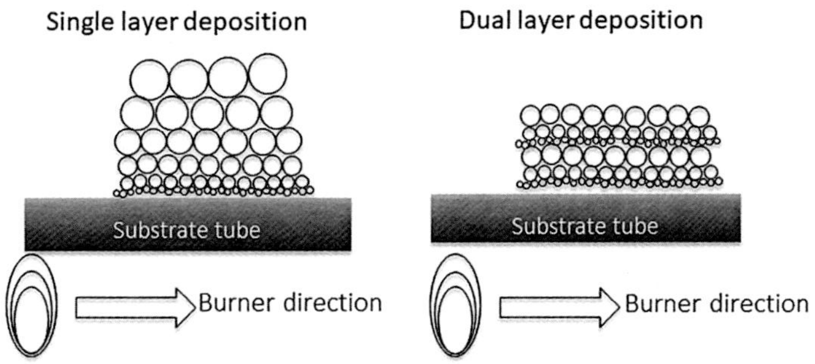

Fig. 9 Expected difference between dual- and single-layer soot patterns

Figure 8c, d compares the innermost layer (furthermost from the substrate wall) morphology of the single-layer soot to that of the dual-layer soot deposited at the same temperature (i.e., 1800 °C). It can be observed that the particles of the single soot layer are larger and more interconnected (fused) between one another. This is due to the partial fusion effect that takes place when the burner passes the soot layer. The soot nearer to the tube wall experiences greater partial fusion since it is exposed to higher temperature. It can also be observed that the innermost soot layer of the dual layer deposition, which is further form the tube wall (due to thickness), exhibits smaller particles and lesser interconnection.

Effect of Soot Condition on the Final Preform Characteristics

The characteristics of the fabricated preforms (cf. Table 2) are discussed to determine which soot deposition parameters offer the best soot performance in terms of dopant intake and longitudinal uniformity. Once the optimum soot is identified, the choice of glass modifier to be incorporated inside the preform is extended to other types of dopants besides aluminum such as gallium and barium.

Figure 10 shows the longitudinal refractive index difference of the fabricated Al_2O_3 doped preforms along a length of 23 cm. According to Kirchhof et al. (2003) and Vienne et al. (1996), the doping of Al_2O_3 inside silica increases the refractive index by 2.3×10^{-3} per mol% of Al_2O_3. The effect of refractive index change due to rare earth (RE) oxide incorporation is negligible because of the much lower concentration of RE_2O_3 in comparison to Al_2O_3. A summary of preforms characteristics is displayed in Table 6 including parameters such as average refractive index difference, Δn_{av}; maximum index difference variations, R; relative standard deviations of index difference, %RSD; and average core diameter, D_{core} for each of the fabricated preforms.

For the single layer soot preforms, the refractive index difference, which is related to the amount of dopant incorporated in the preform, is higher for lower deposition temperatures (Table 6). This result is in agreement with reports from several previous authors (e.g., Khopin et al. 2005; Tang et al. 2007). The higher

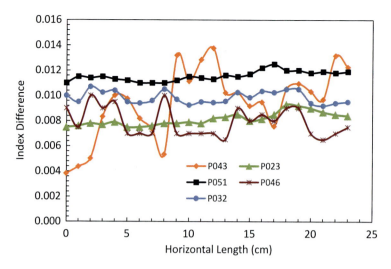

Fig. 10 Longitudinal refractive index difference of the fabricated preforms along 23 cm. Preforms P043, P046, and P023 represent the single-layer soot deposited at 1650, 1750, and 1800 °C, respectively. Preforms P032 and P051 represent the dual-layer soot deposited at 1800 and 1700/1750 °C, respectively

Table 6 Characteristics of the fabricated preforms

Preform #	T_{dep} (°C)	Δn_{av} ($\times 10^{-2}$)	Average mol%	R ($\times 10^{-3}$)	RSD (%)	D_{core} (mm)
P043	1650	0.944	4.1	0.99	29.6	1.32
P046	1750	0.791	3.4	0.35	14.3	1.31
P023	1800	0.813	3.5	0.18	6.6	1.18
P032	1800	0.985	4.3	0.15	4.8	1.42
P051	1700/1750	1.150	5.0	0.15	3.5	1.43

dopant intake for preforms with lower soot deposition temperature is due to the higher surface area and pore volume of the soot (cf. Table 5). However, the uniformity of the preform deteriorates when a lower deposition temperature is applied. An increase of R values from 0.18 to 0.99 is observed when the deposition temperature is reduced from 1800 °C to 1650 °C (Table 6). This observation is in agreement with Kirchhof et al. (2003) and Petit et al. (2010) that reported an exponential reduction of dopant intake with soot density increment and a linear increase of soot density with deposition temperature increment. This means that the variation of concentration is more severe at lower deposition temperatures as compared to the higher ones due to the features that govern the soot density. Even though the measurement of soot density is not included in this work, the features that govern the soot density are expected to be the soot surface area and the pore volume because both are reduced exponentially with increased deposition temperature (cf. Table 5). This observation is in agreement with the experimental observation and the model proposed by Kirchhof et al. (2003) which was briefly explained in section "MCVD–Solution Doping Technique." Hence, for a single layer deposition method, a higher incorporation of dopant is achievable but at the expense of preform longitudinal uniformity degradation.

In light of the above, a dual layer soot deposition method is proposed to improve the dopant intake as well as the longitudinal uniformity of the solution doped preforms. Multiple soot layer method assisted by a presintering pass has been previously proposed by Paul et al. (2010) to fabricate a large core fiber using solution doping technique. This method, used at optimum parameters, has been found to promote uniformity of the soot porosity across the soot layer thus improving the dopant radial distribution. For this work, since the fabricated preform core is small, the radial uniformity is not of a major concern compared to the longitudinal uniformity. As shown in Table 6 and Fig. 10, the additional layer is found to improve both the dopant concentration and the longitudinal uniformity of the fabricated preforms. An improvement of 22% in index difference, 21% in dopant intake and 27% in longitudinal uniformity (%RSD) is achieved by dual layer soot deposition (P032) compared to single layer deposition (P023) fabricated at the same temperature (1800 °C). To further increase the dopant intake, a lower process temperature is proposed for the dual layer soot deposition (P051). A lower deposition temperature of 50 °C for the first pass compared to the second pass is also used to promote the radial uniformity. In this case, smaller particles are deposited in the first pass compared to the second pass due to the lower deposition temperature. However, this difference in particle size is eliminated during the partial sintering of the burner second pass. This is because the soot on the bottom layer experiences higher degree of particle fusion compared to the one on the top due to temperature gradient inside the tube. As shown in Table 6, a further 17% increment in both the average refractive index and dopant concentration is achieved by lowering and varying the deposition temperatures of the dual layer soot preform. However, the average Δn fluctuation, R, is similar to that of P032 (i.e., 0.15×10^{-3}) even though the %RSD value improves by 27%. An R value of 0.15×10^{-3} corresponds to a 0.02 fluctuation in the numerical aperture, NA, of the final fiber along 23 cm of the preform. The improvement of P051 characteristics as compared to the single layer soot preform is attributed to the higher soot surface area and pore volume as aforementioned (cf., Table 5).

The average core diameter of all the fabricated preforms is also presented in Table 6. Since two layers of soot are deposited, the average D_{core} of P032 and P051 is larger as compared to the rest of the single-layer soot preforms. It is also observed that for the same number of soot layers, the core size increases with Al_2O_3 content. The evolution of D_{core} along the preform length is depicted in Fig. 11. It can be observed that the D_{core} pattern closely emulates the longitudinal Δn pattern along the length of the preform.

The concentration of Al dopant is also determined using energy-dispersive X-ray spectroscopy (EDX) point ID measurements. Figure 12 displays an example of EDX analyses carried out for P051. As can be noticed from the figure, the Al distribution profile across the preform's core closely resembles the refractive index profile (RIP). It should be mentioned that thulium is not detected by this method since its concentration falls below the EDX detection limit.

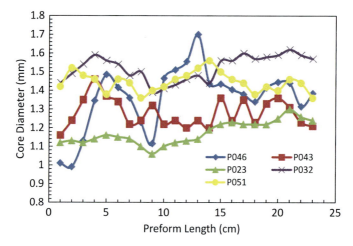

Fig. 11 Core diameter variation along the fabricated preforms

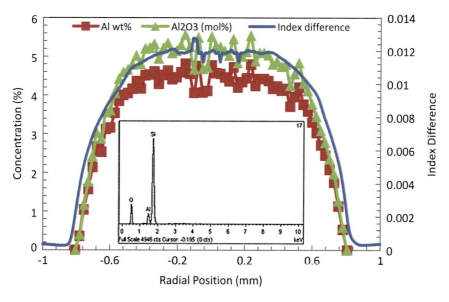

Fig. 12 EDX measurements of Al concentration across the core of P051. Inset is the EDX spectrum

Alumina, Gallia, and Baria Solution Doped Silica Preforms

Table 7 lists a summary of the preforms fabricated in this work. The characteristics of the individual preforms are given in sections "Aluminum Doped Preforms," "Gallium Doped Preforms," and "Barium Doped Preforms."

Table 7 Summary of the fabricated Ga, Ba, and Al preforms

Preform #	Modifier type	Modifier(mol%)	$\Delta n_{mean}(\times 10^{-3})$	%RSD	Soot type
P053	Al	6.1-8.3	15.4	13.5	Dual-layer
P057	Ba	3.0	10.0	7.3	Dual-layer
P085	Ba/Al	1.2/1.6	5.5	5.7	Dual-layer
P058	Ga	2.7	8.8	9.5	Dual-layer
P060	Ga	1.7	5.6	6.3	Dual-layer

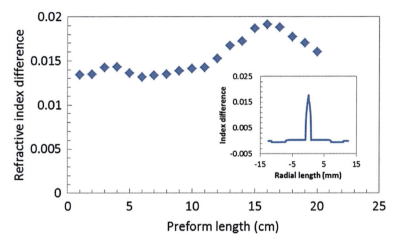

Fig. 13 Longitudinal Δn along P053 length. Inset is the RIP at 17 cm from the preform inlet

Fig. 14 Longitudinal image of P053

Aluminum Doped Preforms

The RIP of Al doped P053 is presented in Fig. 13. Using a 3.2 M Al solution, a maximum Δn value of 0.019, which corresponds to 8.3 mol%, is achieved. However, the longitudinal Δn fluctuates from 0.014 to 0.019 with %RSD of 13.4%. The lower end of Δn distribution corresponds to about 6.1 mol% of Al_2O_3. As can be observed from Fig. 14, some parts of the preform are phase separated due to the high aluminum concentration. The low uniformity of the preform may be attributed to the high solution strength that is used to impregnate the soot layer. Preforms that are soaked using high concentration solutions are more sensitive to the difference in density of the deposited soot layer as described above.

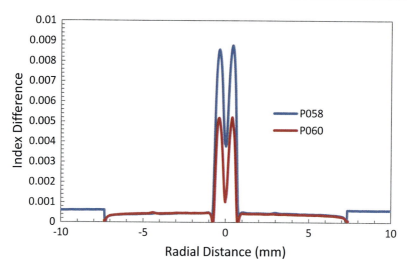

Fig. 15 Refractive index profiles of P058 and P060

Gallium Doped Preforms

Figure 15 displays the refractive index profiles of the fabricated gallium doped preforms P058 and P060. In addition, Fig. 16 illustrates the longitudinal refractive index profiles along the length of P058 and P060. The higher index difference of P058 is attributed to the higher concentration of $Ga(NO_3)_3 \cdot H_2O$ solution used in soaking the porous layer of the preform (i.e., 2.3 M, Table 3). However, the gallium doped preforms exhibit a large dip at the center of the preforms' core. The index dip is a consequence of the evaporation of the dopant material inside the core. It is known that the evaporation of the dopant becomes more rapid during the collapse stage where the tube is at its highest temperature (Cognolato 1995). From Fig. 15, the collapse process for P058 and P060 is found to result in the loss of 45% of Ga_2O_3 from the center of the preforms' core.

Figure 17a, b shows the side view of P058 and P060, respectively. As can be seen from Figure 17a, a slight a slight opalescence appears in the center of P058 between the lengths of 1 to 10 cm measured from the preform input end. The occurrence of opalescence indicates that the P058 preform core is phase separated. Phase separation only occurs for glass systems of multiple compositions (Shelby 2005). It is common for a multicomponent glass to be immiscible (i.e., a homogeneous mixture is thermodynamically unfavorable) at a certain composition concentration and temperature. The EDX measurements shown in Fig. 18 are carried out for samples taken from the opalescent region of P058. It can be deduced from the figure that the phase separation occurs when $\sim$ 2.7 mol% of Ga_2O_3 is doped inside the preform core. This shows that gallium has a lower solubility in silica as compared to aluminum. However, no sign of opalescence is observed for P060 (Fig. 17b), which indicates that phase separation is absent at the respective Ga_2O_3 concentration

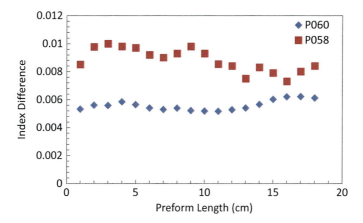

Fig. 16 Longitudinal refractive index difference along the length of P058 and P060

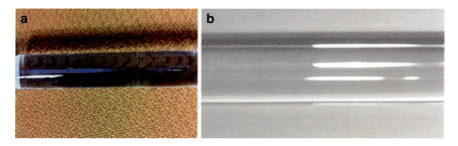

Fig. 17 a, b Longitudinal view of P058 and P060, respectively

(i.e., ~ 1.7 mol%). Schneider and Syms (1998) reported a gallium silica glass system having a 10% P_2O_5 that was fabricated using sol gel technique. They observed a clear glass (miscible glass system) as long as the Ga_2O_3 concentration was below 2.5 mol%.

The EDX spectrum of P058 is depicted in Fig. 19. There are 5 elements detected in the core (i.e., Si, Ga, O, C, and Cl). The peaks at 1.739, 9.241, and 0.525 keV are assigned to Si, Ga, and O, respectively. The small peaks at 0.277 and 2.621 keV are attributed to C and Cl, respectively. The existence of C element is due to the coating layer applied to the sample to avoid the accumulation of charges on the nonconducting surface of the measured sample. The presence of Cl may be attributed to residual chlorosiloxanes (i.e., intermediate compounds of the $SiCl_4$ oxidation reaction) that may have been trapped inside the silica matrix.

Barium Doped Preforms

The refractive index profiles of barium doped preforms P057 and P085 are shown in Fig. 20. P057 has a higher Δn than P085 due to the higher concentration of $BaCl_2 \cdot H_2O$ solution used in the solution doping process (1.4 vs. 0.9 M, respectively,

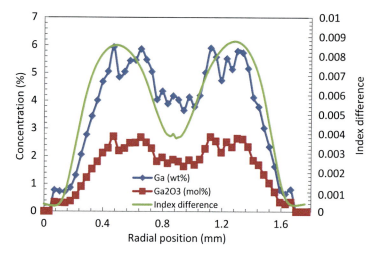

Fig. 18 EDX concentrations of gallium and gallium oxide (primary axis); and the corresponding refractive index difference (secondary axis) for P058

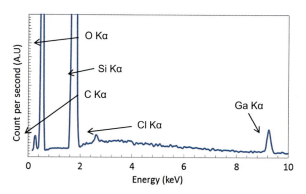

Fig. 19 Energy-dispersive X-ray (EDX) spectrum of P058

Table 3). The dip in the center of the RIP of P057 is not as large as that observed for the gallium preforms (cf. Fig. 15) indicating lesser BaO evaporation during the collapse process. However, the RIP of P085 exhibits a distortion in the center of the core, which is indicative of crystal formation during the fabrication process. It is also worth noting that the core diameter of P057 is 57% higher than that of P085. This may be attributed to the higher barium concentration and the occurrence of phase separated core in P057 (Sudo 1997).

The longitudinal images of P057 and P085 are depicted in Fig. 21a, b, respectively. As observed in this figure, P057 possesses a slight bluish-milky white opalescent core that is indicative of the presence of phase separation. This is in agreement with the observation made by Seward et al. (1968). Utilizing normal and fast quenching techniques, Seward et al. obtained the liquid-liquid immiscibility gap in the range of 4 to 28 mol% of BaO. Even though a clear glass was observed at

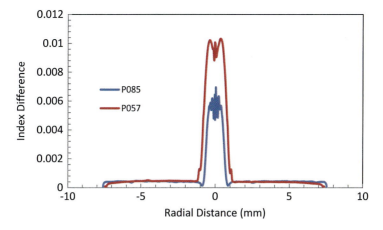

Fig. 20 Refractive index profiles of P057 and P085

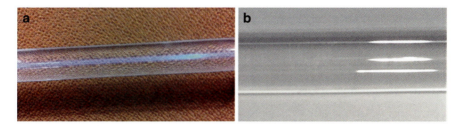

Fig. 21 Images of P057 (**a**) and P085 (**b**)

2 mol% of BaO, barium-rich particles sized between 20 to 60 nm were observable in the system indicating the occurrence of phase separation. For the MCVD coupled with solution doping technique used in this work, the opalescence is observed at ∼ 3 mol% of BaO. The opalescence at such low BaO concentration may be attributed to the slower cooling experienced by the glass due to the slow moving burner which gives ample time for nucleation and growth to take place. The phase separation phenomenon in $BaO-SiO_2$ system can be suppressed by either fast cooling from the temperature above the immiscible gap or by adding Al_2O_3 to the system. The former technique can be conducted during fiber pulling, while the latter one motivated the fabrication of P085 whereby Al_2O_3 was added to the preform by using a solution containing both Ba and Al ions (cf. Table 3). The concentration of the dopant precursors was chosen so that the final concentration (mol%) for both oxides inside the preform would be similar and that the solution did not reach its saturation point. As observed in Fig. 21b, the addition of aluminum to the $BaO-SiO_2$ system successfully suppresses the phase separation and the presence of opalescent core is not observed along the entire preform length.

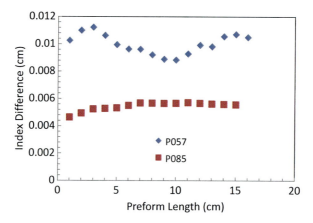

Fig. 22 Longitudinal refractive index difference along P057 and P085 length

The longitudinal refractive index difference of P057 and P085 along the preform length is shown in Fig. 22 with average Δn values of 0.010 and 0.0055, respectively. The longitudinal uniformity of P085 is found to be better than P057 by 14%. This is due to the use of higher solution concentration in P057 as compared to P085. As discussed in section "MCVD–Solution Doping Technique," high concentration solutions impose higher fluctuation of dopant concentration for a small change of soot density along the preform as compared to weaker ones. The refractive index difference and its corresponding oxide concentration for P057 and P085 are depicted in Fig. 23a, b, respectively.

The EDX spectrum for P057 is shown in Fig. 24. The inset in the figure shows the spectrum obtained for P085. Both of these preforms exhibit four common peaks at 1.739, 4.46, 4.82, and 0.525 keV corresponding to Si, Ba, Ba, and O, respectively. The bombarded electrons induce two characteristic X-ray emissions that are assigned to barium (i.e., 4.46 and 4.82 keV). For P085, there are two extra peaks at 0.277 and 1.487 keV corresponding to carbon and aluminum, respectively. The carbon peak is believed to originate from the carbon coated layer that is used to prevent charging of the glass by the electron beam.

Spectroscopic Characteristics of Thulium Doped Fibers (TDF)

After the fabrication process of the optical fiber, which consists of preform making and fiber pulling, is completed, the thulium ions spectroscopy in the modified silica fiber is analyzed. The effect of modifiers on the absorption spectrum and the fluorescence decay lifetime of thulium ions is also observed and analyzed. These two measurements provide information on how feasible the modification of the silica structure improves the overall characteristics of the TDF.

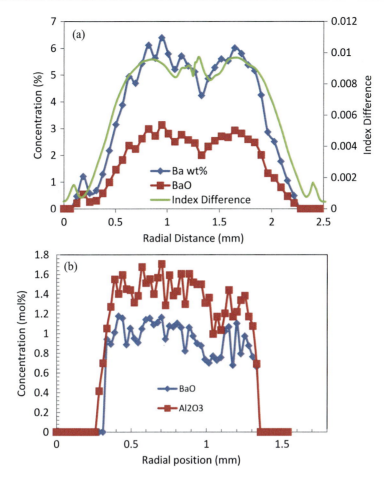

Fig. 23 (a) EDX concentration of barium and barium oxide (primary axis); and the corresponding index difference (secondary axis) of P057. (b) EDX concentration of dopant modifiers inside P085

Absorption

Figure 25 shows the normalized absorption spectra for a number of fabricated Tm doped fibers. The range of the measured spectra is from 600 to 1750 nm. It can be observed that within this range, there exist four main absorption peaks that correspond to the ground state absorption of thulium ions. These peaks represent the energy manifolds of thulium ions listed in Table 8. The thulium concentration can be determined from the absorption spectrum of the fiber (Hanna et al. 1990) and is also listed in Table 8 for the tested fibers. As observed from this table, the amount of Tm ions doped in the fibers is dependent on the type of modifier that is co-doped in the glass matrix. By using the same strength of $TmCl_3$ inside the solution (i.e., 0.025 M), the highest concentration of Tm ions is obtained by Ba/Al

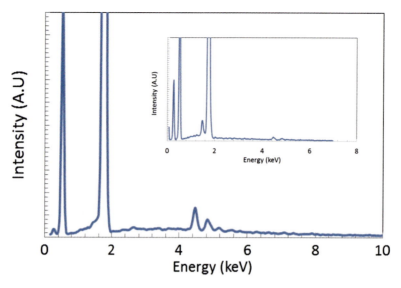

Fig. 24 Energy-dispersive X-ray (EDX) spectrum for P057. The inset is the EDX spectrum obtained for P085

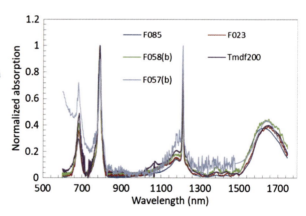

Fig. 25 Normalized absorption spectra of various Tm doped fibers. Tmdf200 is the commercially available Tm doped fiber obtained from OFS/Furukawa Company

doping (8791 ppm) followed by Al (3408 ppm), Ga (2383 ppm), and Ba (629 ppm) (Table 8). This indicates the low solubility of rare earth ions inside $BaO-SiO_2$ glass matrix.

Lifetime

The values of Tm ions lifetime for the tested fibers are listed in Table 9. The decay lifetime is the time for the intensity to drop to $1/e$ from its maximum value. This is accurate for a single exponential decay whereby the lifetime value is taken as the decay constant, τ of the decay waveform (i.e., $I(t) = I_o e^{t/\tau}$). However, for multisites

Table 8 Thulium energy manifold peak absorption and the respective estimated concentration

	$^3F_4$	$^3H_5$	$^3H_4$	$^3F_2, ^3F_3$	Concentration (ppm)
	α_p (dB/m)	α_p (dB/m)	α_p (dB/m)	α_p (dB/m)	
F023	29.9	50.9	81.8	30.7	3408
F057	7.0	19.7	15.1	7.29	629
F058	24.0	39.5	57.2	22.6	2383
F085	76.2	130.2	211.0	59.6	8791
Tmdf200	30.4	64.8	79.5	37.8	3312

Table 9 Measured lifetimes of various Tm doped fibers

Fiber#	Modifier type	Modifier concentration(mol%)	Lifetime (μs)$^3H_4$	Lifetime (μs)$^3F_4$
F023	Al	3.5	27	490
F057	Ba	3.0	28	477
F058	Ga	2.7	30	561
F085	Ba/Al	1.2/1.6	23	360
Tmdf200	n.a	n.a.	33	572
Pure silica[1]	–	–	12.4	327

[1] Simpson (2008)

luminescence of active centres, where there exist multiple decay constants, a new method is needed to better represent the fluorescence decay curve. Aronson (2006) proposed the use of the time constant of the single exponential curve fit as the lifetime value. Nevertheless, for this work, the lifetime value is determined using the conventional technique.

The longest $^3H_4$ and $^3F_4$ lifetimes of the tested fibers are obtained from Tmdf200 fiber (33 and 572 μs, respectively) followed by F058 (30 and 561 μs, respectively). Although the concentration of Ga is quite low compared to alumina (2.7 vs. 3.5 mol%, respectively), the obtained $^3H_4$ lifetime is longer (30 vs. 27 μs, respectively). This is due to Ga being heavier than Al which in turn lowers the nonradiative decay. However, the setback of Ga is that it can only be doped at low concentrations due to its lower solubility in silica as compared to alumina. Hence, the higher solubility of Al in silica (up to 9 mol%) serves it as a better candidate than Ga. Faure et al. (2007) fabricated an 8 mol% aluminosilicate fiber that exhibited a $^3H_4$ lifetime of 54 μs. In addition to Al and Ga, the $^3H_4$ lifetime of 3 mol% Ba TDF (i.e., F057) is found to be 28 μs (Table 9), which falls between that of Ga and Al with comparable dopant concentration.

The shortest lifetime for both $^3H_4$ and $^3F_4$ transitions is demonstrated by F085 (23 and 360 μs, respectively). This is due to the low dopant (Al and Ba) concentration inside the fiber. Hence, the nonradiative decay is higher in F085 as compared to fibers with higher dopant concentration and the lifetime is consequently shorter. In addition, the high concentration of Tm ions in F085 as compared to other

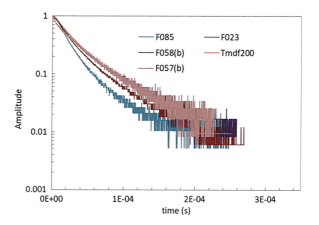

Fig. 26 Decay spectra of $^3H_4$ energy level for various Tm doped fibers

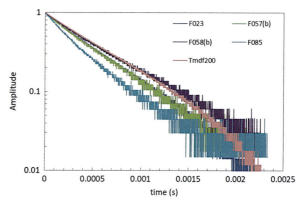

Fig. 27 Decay spectra of $^3F_4$ energy level for various Tm doped fibers

fibers may increase the cross relaxation and co-operative up-conversion rates which consequently decrease the measured fluorescence lifetime.

Figures 26 and 27 provide a better visualization for the $^3H_4$ and $^3F_4$ decay lifetimes of the tested fibers, respectively. It can be observed that the decay lifetime of $^3F_4$ closely resembles that of $^3H_4$ for the tested fibers. This can be explained by the high phonon energy of the glass that consistently influences the nonradiative decay rates for each Tm energy level.

Conclusions

The main objective of this work is to improve the Tm ions emission in silica fibers. This is done by doping the TDF with modifier metals such as Al, Ba, and Ga with the aim to reduce the high phonon energy environment surrounding Tm ions in pure silica system. The fibers were fabricated using an in-house MCVD-solution doping system. In order to maximize the concentration of the proposed modifier ions inside the glass preform, the solution doping method was optimized and improved.

A deposition temperature in the range of 1700 to 1800 °C was found to be effective in producing a fairly high dopant intake and an acceptable longitudinal uniformity. The dopant intake and the longitudinal uniformity of the preforms were then further improved by introducing an additional layer of soot (i.e., dual-layer soot) deposited at 1800 °C. This dual-layer structure is believed to have a different morphology as compared to the single layer one which in turn increased the dopant intake as well as the longitudinal uniformity of the preform by 21% and 27%, respectively. Further improvement was also obtained by using deposition temperatures of 1700 and 1750 °C for the first and second soot layers, respectively. This resulted in a further 17% increase in dopant intake. This type of dual-layer soot was then used to fabricate all the Al, Ba, and Ga doped preforms. Using this improved soot, three preforms doped individually with Ga, Ba, and Al were fabricated using the highest solution concentration. All of the fabricated preforms exhibited opalescence core indicating that the amount of the modifier oxide had reached its limit inside silica. The maximum concentrations of Ga, Ba, and Al attained inside these preforms are $\sim$ 2.7, 3.0, and 8.3 mol%, respectively.

Compared with pure silica host, lifetime measurements showed that the lifetime of the $^3H_4$ manifold increased by $\sim$ 218%, 226%, and 242% when doping with alumina, baria, and gallia, respectively. A similar improvement was also obtained for the $^3F_4$ manifold radiative transition with increments of 150%, 146%, and 172%, respectively. This showed significant improvement in obtaining radiative emission using the modifiers proposed in this work.

References

A.N. Abramov, M.V. Yashkov, A.N. Guryanov, M.A. Melkumov, D.A. Dvoretskii, I.A. Bufetov, L.D. Iskhakova, V.V. Koltashev, M.N. Kachenyuk, M.F. Torsunov, Inorg. Mater. **50**, 1283 (2014)
B.G. Aitken, M.L. Powley, R.M. Morena, B.Z. Hanson, J. Non-Cryst. Solids **352**, 488 (2006)
K. Arai, H. Namikawa, Y. Ishii, H. Imai, H. Hosono, Y. Abe, J. Non-Cryst. Solids **95**, 609 (1987)
J. E. Aronson, Ph.D. thesis, University of Southampton, 2006
R. M. Atkins, R. S. Windeler, U. S. Patent No. US20030167800A1 (11 Sept 2003)
M. Binnewies, K. Jug, Eur. J. Inorg. Chem. **2000**, 1127 (2000)
W. Blanc, T.L. Sebastian, B. Dussardier, C. Michel, B. Faure, M. Ude, G. Monnom, J. Non-Cryst. Solids **354**, 435 (2008)
S. Brunauer, P.H. Emmett, E. Teller, J. Am. Chem. Soc. **60**, 309 (1938)
L. Cognolato, J. Phys. IV **5**, 975 (1995)
B. J. Cole, M. L. Dennis, U. S. Patent No. US6667257B2 23 Dec 2003
H. Davarzani, M. Marcoux, M. Quintard, Int. J. Therm. Sci. **50**, 2328 (2011)
M. Dennis, B. Cole, S-band amplification in a thulium doped silicate fiber, in *Optical Fiber Communication Conference and International Conference on Quantum Information, 2001*. OSA Technical Digest Series (Optical Society of America, 2001), Anaheim, California, paper TuQ3. https://doi.org/10.1364/OFC.2001.TuQ3
A. Dhar, M.C. Paul, M. Pal, A.K. Mondal, S. Sen, H.S. Maiti, R. Sen, Opt. Express **14**, 9006 (2006)
A. Dhar, A. Pal, M.C. Paul, P. Ray, H.S. Maiti, R. Sen, Opt. Express **16**, 12835 (2008)
E.M. Dianov, J. Lightw. Technol. **31**, 681 (2013)
W. Fan, M.A. Snyder, S. Kumar, P.-S. Lee, W.C. Yoo, A.V. McCormick, R.L. Penn, A. Stein, M. Tsapatsis, Nat. Mater. **7**, 984 (2008)

B. Faure, W. Blanc, B. Dussardier, G. Monnom, P. Peterka, Thulium-doped silica-fiber based S-band amplifier with increased efficiency by aluminum co-doping, in *Optical Amplifiers and Their Applications/Integrated Photonics Research, Technical Digest (CD)* (Optical Society of America, 2004), Francisco, California, paper OWC2. https://doi.org/10.1364/OAA.2004.OWC2

B. Faure, W. Blanc, B. Dussardier, G. Monnom, J. Non-Cryst. Solids **353**, 2767 (2007)

G.N. Greaves, J. Non-Cryst. Solids **71**, 203 (1985)

S.J. Gregg, K.S.W. Sing, *Adsorption, Surface Area and Porosity* (Academic Press, London, 1982)

A. Halder, M.C. Paul, S.W. Harun, S.K. Bhadra, S. Bysakh, S. Das, M. Pal, J. Lumin. **143**, 393 (2013)

A. Halder, M.C. Paul, S.K. Bhadra, S. Bysakh, S. Das, M. Pal, Sci. Adv. Mater. **7**, 631 (2015)

D.C. Hanna, I.R. Perry, J.R. Lincoln, J.E. Townsend, Opt. Commun. **80**, 52 (1990)

D.W. Hewak, R.S. Deol, J. Wang, G. Wylangowski, J.A.M. Neto, B.N. Samson, R.I. Laming, W.S. Brocklesby, D.N. Payne, A. Jha, M. Poulain, S. Otero, S. Surinach, M.D. Baro, Electron. Lett. **29**, 237 (1993)

P. Jander, W.S. Brocklesby, IEEE J. Quantum Electron. **40**, 509 (2004)

V.F. Khopin, A.A. Umnikov, A.N. Gur'yanov, M.M. Bubnov, A.K. Senatorov, E.M. Dianov, Inorg. Mater. **41**, 303 (2005)

J. Kirchhof, A. Funke, Cryst. Res. Technol. **21**, 763 (1986)

J. Kirchhof, S. Unger, A. Schwuchow, Fiber lasers: materials, structures and technologies, in *Proc. SPIE4957, Optical Fibers and Sensors for Medical Applications III* (2003), pp. 1–15

C.J. Koester, E. Snitzer, Appl. Opt. **3**, 1182 (1964)

S. Naumov, *Hysteresis Phenomena in Mesoporous Materials* (Universität Leipzig, Leipzig, 2009)

S. Ohara, N. Sugimoto, Y. Kondo, K. Ochiai, Y. Kuroiwa, Y. Fukasawa, T. Hirose, H. Hayashi, S. Tanabe, Bi_2O_3-based glass for S-band amplification, in *Proceedings of SPIE4645, Rare-Earth-Doped Materials and Devices VI* (2002), pp. 8–15

K.S. Park, B.W. Lee, M. Choi, Aerosol Sci. Technol. **31**, 258 (1999)

M.C. Paul, B.N. Upadhyaya, S. Das, A. Dhar, M. Pal, S. Kher, K. Dasgupta, S.K. Bhadra, R. Sen, Opt. Commun. **283**, 1039 (2010)

D.N. Payne, Electron. Lett. **23**, 1026 (1987)

V. Petit, A. Le Rouge, F. Béclin, H. El Hamzaoui, L. Bigot, Aerosol Sci. Technol. **44**, 388 (2010)

S. B. Poole, Fabrication of Al_2O_3 co-doped optical fibres by a solution-doping technique, in *Fourteenth European Conference on Optical Communication* (1988) (ECOC 88), pp. 433–436

D. Río, A. Aguilera-Alvarado, I. Cano-Aguilera, M. Martínez-Rosales, S. Holmes, Mater. Sci. Appl. **3**, 485 (2012)

V.M. Schneider, R.R.A. Syms, Electron. Lett. **34**, 1849 (1998)

K. Schuster, S. Unger, C. Aichele, F. Lindner, S. Grimm, D. Litzkendorf, J. Kobelke, J. Bierlich, K. Wondraczek, H. Bartelt, Adv. Opt. Technol. **3**, 447 (2014)

R. Sen, A. Dhar, M. C. Paul, H. S. Maiti, Patent No. WO2010109494A2, (30 Sept 2010).

T.P. Seward, D.R. Uhlmann, D. Turnbull, J. Am. Ceram. Soc. **51**, 278 (1968)

J.E. Shelby, *Introduction to Glass Science and Technology* (Royal Society of Chemistry, Cambridge, 2005)

P.G. Simpkins, S. Greenberg-Kosinski, J.B. MacChesney, J. Appl. Phys. **50**, 5676 (1979)

D. A. Simpson, Ph.D. Thesis, Victoria University, 2008

K.S.W. Sing, D.H. Everett, R.A.W. Haul, L. Moscou, R.A. Pierotti, J. Rouquérol, T. Siemieniewska, Pure Appl. Chem. **57**, 603 (1985)

E. Snitzer, J. Appl. Phys. **32**, 36 (1961)

J. Stone, C.A. Burrus, Appl. Phys. Lett. **23**, 388 (1973)

S. Sudo (ed.), *Optical Fiber Amplifiers: Materials, Devices, and Applications* (Artech House, Boston, 1997)

F.Z. Tang, P. McNamara, G.W. Barton, S.P. Ringer, J. Non-Cryst. Solids **352**, 3799 (2006)

F.Z. Tang, P. McNamara, G.W. Barton, S.P. Ringer, J. Am. Ceram. Soc. **90**, 23 (2007)

F.Z. Tang, P. McNamara, G.W. Barton, S.P. Ringer, J. Non-Cryst. Solids **354**, 1582 (2008)

J.E. Townsend, S.B. Poole, D.N. Payne, Electron. Lett. **23**, 329 (1987)

M. Vermillac, H. Fneich, J.-F. Lupi, J.-B. Tissot, C. Kucera, P. Vennéguès, A. Mehdi, D.R. Neuville, J. Ballato, W. Blanc, Opt. Mater. **68**, 24 (2017)

G.G. Vienne, W.S. Brocklesby, R.S. Brown, Z.J. Chen, J.D. Minelly, J.E. Roman, D.N. Payne, Opt. Fiber Technol. **2**, 387 (1996)

K.L. Walker, F.T. Geyling, S.R. Nagel, J. Am. Ceram. Soc. **63**, 552 (1980)

B. Walsh, N. Barnes, Appl. Phys. B Lasers Opt. **78**, 325 (2004)

D.L. Wood, J.B. Macchesney, J.P. Luongo, J. Mater. Sci. **13**, 1761 (1978)

J. M. Zielinski, L. Kettle, Physical characterization: surface area and porosity (Intertek Chemicals and Pharmaceuticals (2013). www.intertek.com/chemicals. Accessed 11 Jan 2018

Microfiber: Physics and Fabrication

14

Horng Sheng Lin and Zulfadzli Yusoff

Contents

Introduction	588
Principle	588
Wave Equation for Microfiber	588
Adiabaticity Criterion	594
Fabrication Techniques	597
Fabrications of Meso Taper	597
Fabrications of Short Taper	604
Fabrications of Long Taper	607
Application in Structural Health Monitoring	608
Microfiber-Based IMZI Sensor Packaging	609
Microfiber-Based IMZI Sensor Deployment	613
Summary	618
References	618

Abstract

In this chapter, several essential concepts for the understanding of microfiber are provided. The effective refractive indices of core mode and cladding modes corresponding to various diameters are comprehensively explained based on wave equation. Subsequently, the effective refractive indices are related to the adiabaticity criterion of microfiber based on the upper boundary of taper angle.

H. S. Lin
Universiti Tunku Abdul Rahman, Sungai Long Campus, Kajang, Malaysia
e-mail: linhs@utar.edu.my

Z. Yusoff (✉)
Multimedia University, Persiaran Multimedia, Cyberjaya, Malaysia
e-mail: zulfadzli.yusoff@mmu.edu.my

© Springer Nature Singapore Pte Ltd. 2019
G.-D. Peng (ed.), *Handbook of Optical Fibers*,
https://doi.org/10.1007/978-981-10-7087-7_74

Following that, an overview of microfiber fabrication techniques is reviewed. Eventually, the most recent deployment of microfiber sensor for structural health monitoring application is demonstrated.

Introduction

A century ago, a fiber with an estimated diameter of $2\,\mu m$, which was known as quartz thread, was drawn by Richard Threlfall (1898). It was fabricated for mechanical applications but not for light transmissions. The application of optical fiber in the fiber-optic communication industry was proposed by the Nobel laureate, Charles Kao, and his colleague George Hockham after discovering the possibility of low-loss lightwave guiding with high-purity glasses in 1966 (Kao and Hockham 1966). Since then, standard optical fibers with a diameter of $125\,\mu m$ have been installed worldwide in the fiber-optic communication system.

The continuous research has been executed to alter the physical and optical properties of the optical fiber. It can be achieved by manipulating the geometrical and refractive index profiles of the optical fiber. The most significant work to alter the geometrical property of optical fiber is by tapering down the diameter of optical fiber to a microfiber. The microfiber was first reported in the 1980s (Burns et al. 1985, 1986; Love and Henry 1986; Love 1987). In the following decades, it was comprehensively studied on its adiabaticity and the fabrication approaches (Love et al. 1991; Black et al. 1991; Birks and Li 1992). The researchers found its possible application as optical couplers (Takeuchi and Noda 1992; Bilodeau et al. 1987; Dimmick et al. 1999), filters (Lacroix et al. 1986; Alegria et al. 1998), and sensors (Bobb et al. 1990) in the same decades.

In this chapter, the microfiber adiabaticity delineation criterion is deduced based on wave equation. Subsequently, after the fabrication, the microfiber adiabaticity is justified. The following section classifies the microfiber fabrication techniques into three categories based on its waist length, illustrating each technique through examples in between the subsections. In the final section, in conjunction to the emergence of industrial revolution 4.0, the microfiber-based inline Mach-Zehnder interferometric (IMZI) sensor is designed with the consideration to cater Internet of things (IoT) integration. The deployment of microfiber-based IMZI sensor is demonstrated starting from the work of sensor fabrication to the field test sensing result on concrete structural beam.

Principle

Wave Equation for Microfiber

A microfiber is also known as fiber taper. It has the varying fiber diameter with z-dependent refractive index profile of $n(x, y, z)$ which can be illustrated as Fig. 1a. The local-mode fields are approximately constructed by modeling the fiber as a

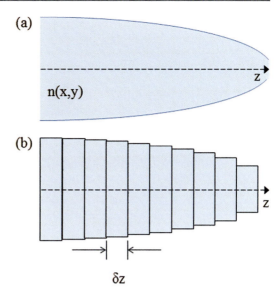

Fig. 1 (a) A nonuniform diameter varies along axial direction (z-axis) with refractive index profile of $n(x, y, z)$. (b) An approximation model with a series of cylindrical sections in the length of δz

series of cylindrical sections as shown in Fig. 1b. Hence, it is important to solve the effective refractive indices of the corresponding modes so that propagation local modes for each section are determined.

Principally, a two-layer step-index fiber model is employed to simulate light propagation in optical fiber, and the layers are core and cladding. Light field is well confined in the core with surrounding evanescent field that is bounded within the cladding. In other words, light field does not reach external surface of the cladding in two-layer step-index fiber model as shown in Fig. 2. However, this model is not valid for microfiber consisting of the untapered ends, taper, and tapered waist. The light field expands in the taper region, and it is no longer confined close to the core. A three-layer step-index fiber model is required to simulate light interaction with the third layer, which is the surrounding layer such as air, coating, etc. Furthermore, in the tapered waist region, the core is so small in diameter that it can be neglected, and the two-layer step-index fiber model can again be used for cladding-surrounding.

In a SMF, the fundamental mode, also known as LP mode, LP_{01} is the only propagation mode in the fiber. The propagation mode is known as a core mode if its effective refractive index is bounded within the condition of $n_{\text{eff}} > n_{\text{co}}$, where n_{co} is the refractive index of core. The weakly-guiding approximation is applied to identify the refractive index of the core mode. The approximation is used due to only small deviation between the core and cladding refractive indices. By solving a set of scalar wave equations with the continuity of the solution and its first derivative at the core-cladding interface, the propagation constant of each local LP modes along the fiber taper is obtained (Tsao et al. 1989). In the propagation of electromagnetic waves along z-direction, the axial axis of fiber must satisfy the scalar wave equation in cylindrical coordinate (r, ϕ, z) as

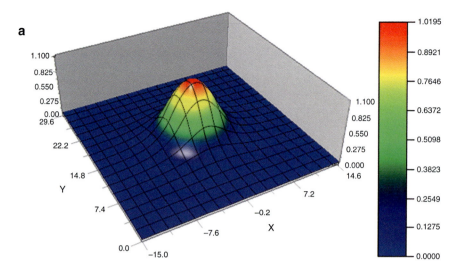

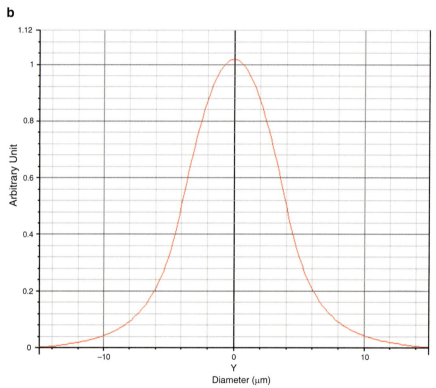

Fig. 2 (**a**) Light field is confined close to the diameter of the core with infinite-cladding geometry. (**b**) Cross section of the beam profile

$$\left\{ \frac{\partial^2}{\partial r^2} + \frac{1}{r^2} \frac{\partial}{\partial r} + \frac{1}{r^2} \frac{\partial^2}{\partial \phi^2} + k^2 n^2 - \beta^2 \right\} \psi = 0 \qquad (1)$$

where ψ is the electric or magnetic field, n is the refractive index profile, the wave number $k = 2\pi/\lambda$ in terms of free space wavelength λ, and the propagation constant $\beta = k n_{\text{eff}}$.

The core mode exists in an infinite-cladding geometry, and its n_{eff} is bounded in the condition of $n_{\text{cl}} < n_{\text{eff}} < n_{\text{co}}$, where n_{cl} is the refractive index of cladding. The thickness of the cladding is assumed to be infinite, and the cladding-air interface is negligible. The modal parameters are defined as

$$u_1 = k \sqrt{n_{\text{co}}^2 - n_{\text{eff}}^2} \qquad (2a)$$

$$w = k \sqrt{n_{\text{eff}}^2 - n_{\text{cl}}^2} \qquad (2b)$$

and the solutions of Eq. 1 are given as

$$\psi = \begin{cases} A J_\upsilon(u_1 r) e^{i\upsilon\phi}, & \text{if } 0 < r < \rho_{\text{co}}; \\ C K_\upsilon(wr) e^{i\upsilon\phi}, & \text{if } \rho_{\text{co}} < r < \rho_{\text{cl}}. \end{cases} \qquad (3)$$

where A and C are constants and ρ_{co} and ρ_{cl} are the core and cladding radii, respectively. J_υ and K_υ are the υth-order Bessel function of the first kind and modified Bessel function of the second kind, respectively. A and C are related with the continuity of electromagnetic field ψ and the first derivative $\frac{\partial \psi}{\partial r}$ at the core boundary $r = \rho_{\text{co}}$. This leads to a set of eigenvalue equations, and the mode condition is stated as

$$u_1 \frac{J_{\upsilon+1}(u_1 \rho_{\text{co}})}{J_\upsilon(u_1 \rho_{\text{co}})} = w \frac{K_{\upsilon+1}(w \rho_{\text{co}})}{K_\upsilon(w \rho_{\text{co}})} \qquad (4)$$

The eigenvalues solved by Eq. (4) are known as $\beta_{\upsilon m}$ where υ is the order of the Bessel function and m is the root number of Eq. (4). Similarly, the subscript υm is indicated in the LP mode as $\text{LP}_{\upsilon m}$.

Cladding mode with effective index of n_{eff} is guided within the cladding bounded with core and air medium which satisfies the condition of $n_{\text{air}} < n_{\text{eff}} < n_{\text{cl}}$, where n_{air} is the refractive index of the air. Hence, a three-layer step-index fiber model is applied to evaluate the light field on two boundary conditions, which are core-cladding and cladding-air interfaces (Erdogan 1997). The modal parameters are defined as

$$u_1 = k \sqrt{n_{\text{co}}^2 - n_{\text{eff}}^2} \qquad (5a)$$

$$u_2 = k \sqrt{n_{\text{cl}}^2 - n_{\text{eff}}^2} \qquad (5b)$$

$$w_3 = k\sqrt{n_{\text{eff}}^2 - n_{\text{air}}^2} \tag{5c}$$

Therefore, Eq. (1) is solved as

$$\psi = \begin{cases} AJ_\upsilon(u_1 r)e^{i\upsilon\phi}, & \text{if } 0 < r < \rho_{\text{co}}; \\ [BJ_\upsilon(u_1 r) + CY_\upsilon(u_2 r)]e^{i\upsilon\phi}, & \text{if } \rho_{\text{co}} < r < \rho_{\text{cl}}; \\ DK_\upsilon(w_3 r)e^{i\upsilon\phi}, & \text{if } r > \rho_{\text{cl}}. \end{cases} \tag{6}$$

where A, B, C, and D are just constants while Y_υ is the υth-order of Bessel function of second kind. Similarly, the A, B, C, and D constants are related to the continuity of electromagnetic field ψ and the first derivative $\frac{\partial \psi}{\partial r}$ at the core boundary $r = \rho_{\text{co}}$ and cladding boundary $r = \rho_{\text{cl}}$. This leads to two sets of eigenvalue equations, and the mode conditions are stated in the matrix form at core boundary, $r = \rho_{\text{co}}$

$$\Re_1 \upsilon_1 = \Re_2 \upsilon_2 \tag{7}$$

where

$$\Re_1 \triangleq \begin{bmatrix} J'_\upsilon(u_2\rho_{\text{co}}) & Y'_\upsilon(u_2\rho_{\text{co}}) & \frac{\sigma_2}{u_2 n_{\text{cl}}^2 \rho_{\text{co}}} J_\upsilon(u_2\rho_{\text{co}}) & \frac{\sigma_2}{u_2 n_{\text{cl}}^2 \rho_{\text{co}}} Y_\upsilon(u_2\rho_{\text{co}}) \\ 0 & 0 & J_\upsilon(u_2\rho_{\text{co}}) & Y_\upsilon(u_2\rho_{\text{co}}) \\ \frac{\sigma_1}{u_2\rho_{\text{co}}} J_\upsilon(u_2\rho_{\text{co}}) & \frac{\sigma_1}{u_2\rho_{\text{co}}} Y_\upsilon(u_2\rho_{\text{co}}) & -J'_\upsilon(u_2\rho_{\text{co}}) & -Y'_\upsilon(u_2\rho_{\text{co}}) \\ J_\upsilon(u_2\rho_{\text{co}}) & Y_\upsilon(u_2\rho_{\text{co}}) & 0 & 0 \end{bmatrix} \tag{8a}$$

$$\upsilon_1 \triangleq \begin{bmatrix} B_1 \\ C_1 \\ B_2 \\ C_2 \end{bmatrix} \tag{8b}$$

$$\Re_2 \triangleq \frac{u_1^2}{u_2} J_\upsilon(u_1\rho_{\text{co}}) \begin{bmatrix} J & \frac{\sigma_2}{u_1^2 n_{\text{co}}^2 \rho_{\text{co}}} \\ 0 & \frac{n_{\text{cl}}^2}{u_2 n_{\text{co}}^2} \\ \frac{\sigma_1}{u_1^2 \rho_{\text{co}}} & -J \\ \frac{1}{u_2} & 0 \end{bmatrix} \tag{8c}$$

$$\upsilon_2 \triangleq \begin{bmatrix} A_1 \\ A_2 \end{bmatrix} \tag{8d}$$

with

$$J \triangleq \frac{J'_\upsilon(u_1\rho_{\text{co}})}{u_1 J_\upsilon(u_1\rho_{\text{co}})} \tag{9a}$$

$$\sigma_1 \triangleq i\upsilon n_{\text{eff}} \frac{\varepsilon_o}{\mu_0} \tag{9b}$$

$$\sigma_2 \triangleq i\upsilon n_{\text{eff}} \frac{\mu_0}{\varepsilon_o} \tag{9c}$$

and at cladding boundary, $r = \rho_{\text{cl}}$,

$$\mathfrak{R}_3 \upsilon_1 = \mathfrak{R}_4 \upsilon_3 \tag{10}$$

where

$$\mathfrak{R}_3 \triangleq \begin{bmatrix} J'_\upsilon(u_2\rho_{\text{cl}}) & Y'_\upsilon(u_2\rho_{\text{cl}}) & \frac{\sigma_2}{u_2 n_{\text{cl}}^2 \rho_{\text{cl}}} J_\upsilon(u_2\rho_{\text{cl}}) & \frac{\sigma_2}{u_2 n_{\text{cl}}^2 \rho_{\text{cl}}} Y_\upsilon(u_2\rho_{\text{cl}}) \\ 0 & 0 & J_\upsilon(u_2\rho_{\text{cl}}) & Y_\upsilon(u_2\rho_{\text{cl}}) \\ \frac{\sigma_1}{u_2\rho_{\text{cl}}} J_\upsilon(u_2\rho_{\text{cl}}) & \frac{\sigma_1}{u_2\rho_{\text{cl}}} Y_\upsilon(u_2\rho_{\text{cl}}) & -J'_\upsilon(u_2\rho_{\text{cl}}) & -Y'_\upsilon(u_2\rho_{\text{cl}}) \\ J_\upsilon(u_2\rho_{\text{cl}}) & Y_\upsilon(u_2\rho_{\text{cl}}) & 0 & 0 \end{bmatrix} \tag{11a}$$

$$\mathfrak{R}_4 \triangleq \frac{w_3^2}{u_2} K_\upsilon(w_3\rho_{\text{cl}}) \begin{bmatrix} K & \frac{\sigma_2}{w_3^2 n_{\text{air}}^2 \rho_{\text{cl}}} \\ 0 & -\frac{n_{\text{cl}}^2}{u_2 n_{\text{air}}^2} \\ \frac{\sigma_1}{w_3^2 \rho_{\text{cl}}} & -K \\ -\frac{1}{u_2} & 0 \end{bmatrix} \tag{11b}$$

$$\upsilon_3 \triangleq \begin{bmatrix} D_1 \\ D_2 \end{bmatrix} \tag{11c}$$

with

$$K \triangleq \frac{K'_\upsilon(w_3\rho_{\text{cl}})}{w_3 K_\upsilon(w_3\rho_{\text{cl}})} \tag{12}$$

To consider both boundaries at once in a mode condition for the three-layer step-fiber model, it is written as

$$\mathfrak{R}_3 \mathfrak{R}_1^{-1} \mathfrak{R}_2 \upsilon_2 = \mathfrak{R}_4 \upsilon_3 \tag{13}$$

and the solution is simplified as

$$\zeta_o = \zeta'_o \tag{14}$$

where

$$\zeta_o = \frac{1}{\sigma_2} \frac{u_2\left(JK + \frac{\sigma_1\sigma_2 u_{21} u_{32}}{n_{\text{cl}}^2 \rho_{\text{co}} \rho_{\text{cl}}}\right) p_v - Kq_v + Jr_v - \frac{1}{u_2}s_v}{-u_2\left(\frac{u_{32}}{n_{\text{cl}}^2 \rho_{\text{cl}}}J - \frac{u_{21}}{n_{\text{co}}^2 \rho_{\text{co}}}K\right) p_v + \frac{u_{32}}{n_{\text{cl}}^2 \rho_{\text{cl}}}q_v + \frac{u_{21}}{n_{\text{co}}^2 \rho_{\text{co}}}r_v} \tag{15a}$$

$$\zeta'_o = \sigma_1 \frac{u_2\left(\frac{u_{32}}{n_{\text{cl}}^2 \rho_{\text{cl}}}J - \frac{u_{21} n_{\text{air}}^2}{n_{\text{cl}}^2 \rho_{\text{co}}}K\right) p_v - \frac{u_{32}}{\rho_{\text{cl}}}q_v - \frac{u_{21}}{\rho_{\text{co}}}r_v}{u_2\left(\frac{n_{\text{air}}^2}{n_{\text{cl}}^2}JK + \frac{\sigma_1\sigma_2 u_{21} u_{32}}{n_{\text{co}}^2 \rho_{\text{co}} \rho_{\text{cl}}}\right) p_v - \frac{n_{\text{air}}^2}{n_{\text{co}}^2}Kq_v + Jr_v - \frac{n_{\text{cl}}^2}{u_2 n_{\text{co}}^2}s_v} \tag{15b}$$

with

$$p_v \triangleq J_v(u_2\rho_{\text{cl}})Y_v(u_2\rho_{\text{co}}) - J_v(u_2\rho_{\text{co}})Y_v(u_2\rho_{\text{cl}}), \tag{16a}$$

$$q_v \triangleq J_v(u_2\rho_{\text{cl}})Y'_v(u_2\rho_{\text{co}}) - J'_v(u_2\rho_{\text{co}})Y_v(u_2\rho_{\text{cl}}), \tag{16b}$$

$$r_v \triangleq J'_v(u_2\rho_{\text{cl}})Y_v(u_2\rho_{\text{co}}) - J_v(u_2\rho_{\text{co}})Y'_v(u_2\rho_{\text{cl}}), \tag{16c}$$

$$s_v \triangleq J'_v(u_2\rho_{\text{cl}})Y'_v(u_2\rho_{\text{co}}) - J'_v(u_2\rho_{\text{co}})Y'_v(u_2\rho_{\text{cl}}), \tag{16d}$$

$$u_{21} \triangleq \frac{1}{u_2^2} - \frac{1}{u_1^2} \tag{16e}$$

$$u_{32} \triangleq \frac{1}{u_2^2} + \frac{1}{w_3^2} \tag{16f}$$

The cladding to core diameter ratio is constant, and the n_{eff} of the LP$_{vm}$ is solved by using Eq. (14) and plotted as a function of cladding diameter as shown in Fig. 3. According to Fig. 3, when the SMF cladding diameter is tapering down, the n_{eff} of the LP$_{01}$, which is guided in the core, is reducing accordingly and approaching the cladding refractive index. Once the n_{eff} is less than the cladding refractive index at cladding diameter of 51.1 µm, LP$_{01}$ is no longer guided in the core; instead the mode propagates in the cladding, whereas LP$_{02}$ can only be guided in the cladding and its n_{eff} reduces with the tapering down of cladding diameter yet it is still bounded below the cladding refractive index. Theoretically, at cladding diameter less than 51.1 µm, only cladding modes survive with n_{eff} tremendously decreasing with a smaller cladding diameter, while no mode survives in core.

Adiabaticity Criterion

The adiabaticity of a microfiber is determined through its local taper angle, $\Omega(z)$. A microfiber is referred to as:

- adiabatic if $\Omega(z)$ is small enough that there is no optical power transfers between different LP modes when light propagates through the microfiber.

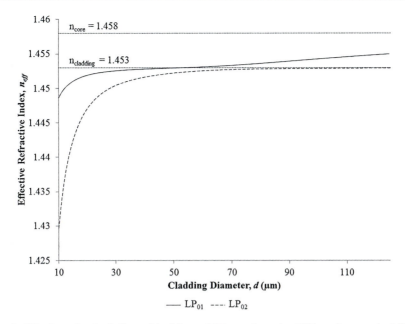

Fig. 3 Effective refractive indices of the LP$_{01}$ and LP$_{02}$ modes at λ = 1550 nm for a standard SMF fiber taper

Fig. 4 Fiber taper structure

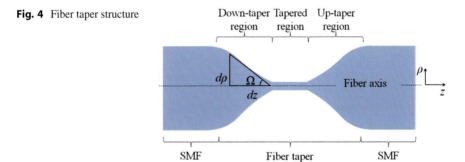

- nonadiabatic if optical power transfers exist between different modes and the optical power is no longer conserved in the microfiber. When an axisymmetric taper with a large $\Omega(z)$ is fabricated, the LP$_{01}$ mode mostly couples into a higher-order mode whose propagation constant is closest to that of the fundamental mode.

From Fig. 4, the local taper length-scale $z_t \approx dz$ is referred to as the length from the taper waist to the reference point along the taper axis, and $\Omega(z)$ is the taper angle from the reference point. The taper angle must satisfy the trigonometry of $\tan\{\Omega(z)\} = d\rho/dz$. Since $\Omega(z) \ll 1$, therefore $\Omega(z)$ is expressed as

$$\Omega(z) \approx \frac{d\rho(z)}{dz} \approx \frac{\rho(z)}{z_t(z)} \qquad (17)$$

The coupling length between the two (fundamental and higher-order) modes is known as the beat length $z_b(z)$:

$$z_b(z) = \frac{2\pi}{\beta_1(z) - \beta_2(z)} \qquad (18)$$

where $\beta_1(z)$ and $\beta_2(z)$ are the propagation constants of the LP_{01} and LP_{02}, respectively. If $z_t \gg z_b$ everywhere along the fiber taper length, the LP_{01} mode propagates adiabatically with negligible loss. On the other hand, there will be significant coupling of the optical power from the LP_{01} mode to the LP_{02} modes if $z_t \ll z_b$ and it propagates nonadiabatically. An approximate delineation between adiabatic and nonadiabatic taper is provided by the condition $z_t = z_b$, and this can be expressed mathematically as

$$\Omega(z) = \frac{\rho(z)[\beta_1(z) - \beta_2(z)]}{2\pi} \qquad (19)$$

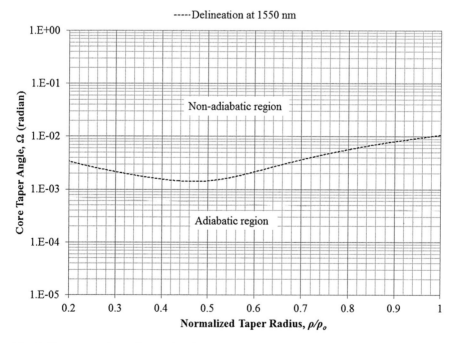

Fig. 5 The taper angle profile as a function of the normalized tapered radius $\rho(z)/\rho_o$ at 1550 nm based on Eq. (19)

Based on Eq. (19), the difference of n_{eff} between LP_{01} and higher-order modes LP_{0m} where $m > 1$ determines the local taper angle $\Omega(z)$. The n_{eff} of each mode is identified from the dispersion curve at various cladding diameters as shown and mentioned in Fig. 3. It is related to the adiabaticity delineation criterion based on Eq. (19) and plotted as Fig. 5. The figure shows the delineation of the adiabatic and nonadiabatic taper at the respective normalized radius where ρ_o is the default fiber radius. The $\Omega(z)$ is relatively larger in a larger radius, and it is interpreted that the light tends to have less mode-to-mode coupling or tends to conserve optical power in larger radius of the fiber. For a gentle biconical taper, the taper is considered adiabatic as long as the taper shape curve lies below the delineation curve. On the other hand, if any part of the taper profile crosses the delineation curve, the taper is considered nonadiabatic.

Fabrication Techniques

The microfiber fabrication techniques are classified into three categories based on the waist length, L_w, of the microfiber. Waist length is defined as the length of the tapered region referring to Fig. 4 in the previous section. The categories are fabrication techniques for short taper ($L_w < 10$ cm), meso taper (10 cm $\leqslant L_w \leqslant 10$ m), and long taper ($L_w > 10$ m).

In conjunction with the previous subsection, the cladding to core diameter ratio for a microfiber is constant based on mass conservation law. In other words, the core is proportionally tapered down relative to the cladding tapering shape. It is worth noting that the cladding to core diameter ratio for a SMF is about 16 and hence the core diameter is considered negligible while cladding diameter is tapering down at micron size (air clad condition). Thereafter, the figures illustrated in the following subsections consider cladding diameter only.

This section focuses on the fabrication of meso taper and short taper. Therefore, the meso taper and short taper fabrication techniques are prioritized to be reviewed and followed by long taper fabrication technology. It is still crucial to introduce long taper fabrication technology in the following so that it serves as the reference on how long taper fabrication technology evolved to fabricate meso taper. The overview of fabrication techniques is described in the following subsections.

Fabrications of Meso Taper

Echoed from the motivation in the development of meso taper fabrication rig, it is important to increase the waist length while tapering down the microfiber in order to maximize the effective nonlinear length and in turn maximize the microfiber nonlinearity as well. Alternatively, the smaller diameter of microfiber deduces to higher concentration of optical power which serves to increase the microfiber nonlinearity. However, generally, the later approach is not in favor of the generation of supercontinuum source. It is due to highly nonlinear fiber such as chalcogenide

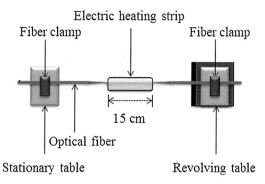

Fig. 6 Schematic of meso taper fabrication setup reported by Shi et al. (2006)

fiber having low melting point at around 260 °C which could be easily burnt down with such high optical power concentration in the microfiber. Therefore, in 2006, the fabrication of MT has emerged for the first time to address the study of nonlinearity characteristic in optical fiber (Shi et al. 2006). The fabrication setup is illustrated as Fig. 6. The pulled fiber passes above an electric strip heater with a length of 15 cm. The long heating region of the electric strip heater allows the fabricated fiber taper to achieve an equal length of uniform tapered region. In order to manipulate the length of uniform tapered region, the length of the electric heating strip must be varied accordingly.

Apart from the abovementioned fabrication setup, Vukovic et al. (2008) are known to manufacture the longest microstructured MT of a uniform waist length of 30 cm by using a feedback loop in the system. The reported setup and working principle are the same as the optical fiber drawing system which will be discussed in section "Fabrications of Long Taper." However, it is set up in horizontal axis instead of vertical axis.

The fabrication setup of both Shi et al. (2006) and Vukovic et al. (2008) is implemented in single pass mechanism. Since then, there is no meso taper fabrication with single pass mechanism reported till the recent work in 2017. A meso taper fabrication rig was developed similar to the setup of Shi et al. (2006). However, the electric strip was replaced with heating filament that traveled along the waist length in a single pass. The fiber taper was reported to have a total maximum waist length of 20 cm. Supercontinuum generation with such waist length was made possible by compensating with large-mode-area charcogenide photonics crystal fiber (Petersen et al. 2017).

It is worth to highlight that a MT fabrication setup proposed by Graf et al. (2009) uses multiple passes mechanism, which is also known as flame-brushing technique. This technique will be discussed in detail in the following subsection. In the reported fabrication rig, a stage coupled with a pair of stepper motors pulls the fiber while another traveling stage flame-brushes the fiber with a commercial mini-torch. The actual waist length is not reported, but the author claimed that 1 m long fiber taper could be potentially fabricated from the rig.

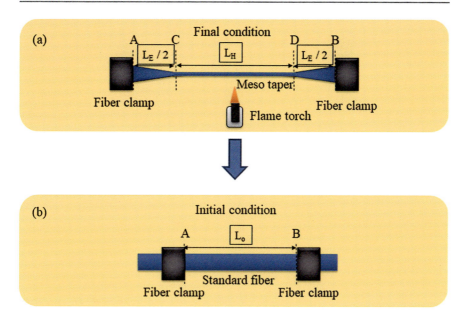

Fig. 7 Illustration of MT fabrication based on mass conservation law in (**a**) final condition and (**b**) initial condition

The MT Shape and the Design of Fabrication System

Based on the mass conservation law, the volume of fiber before and after the fabrication process is conserved which means that the volume of the fiber between A and B in the model (Fig. 7) is conserved in both final and initial conditions. AB in the initial condition is noted as the initial length, L_o. CD is the heating length, L_H, brushed by the flame torch, while AC and DB are the elongated length, L_E. In the modeling, a reverse case of modeling is considered since the targeted diameter waist, D_w, is first desired and the concern is to determine the initial condition and t that required to produce the desired shape.

The decaying radius function or shape of taper transition (Birks and Li 1992) is expressed as

$$r_w(z) = r_o \cdot e^{\frac{-L_E}{2L_o}} \qquad (20)$$

where r_o is the initial radius of fiber. Therefore, L_E is calculated by arranging Eq. 20:

$$L_E = 2L_o \cdot \ln \frac{r_o}{r_w} \qquad (21)$$

where r_w is the radius of taper waist.

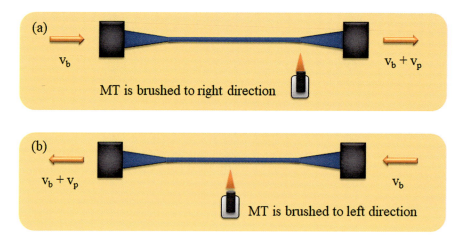

Fig. 8 Illustration of flame-brushing mechanism with stationary flame torch while MT is brushed to (**a**) right and (**b**) left directions

In the proposed mechanism of fabrication machine, the flame torch is static, while both of the stages travel to brush as well as to pull the fiber as illustrated in Fig. 8. For instance, if the MT is brushed to right direction, the right traveling stage leads the left traveling stage with the pulling velocity, v_p, while both of the traveling stages are having the same brushing velocity, v_b. Indirectly, t is estimated by considering v_p only since v_b is compensated with both of the traveling stages brushing at the same direction:

$$T = \frac{L_E}{v_p} \qquad (22)$$

In order to be able to better comprehend the relation of the taper shape and fabrication mechanism, five samples have been modeled as listed in Table 1. The respective shapes of the tapers are superimposed to have better illustration in Fig. 9. The parameters of the modeling are set as:

Initial diameter, D_o (μm)	Pulling velocity, v_p (μm/s)	Brushing velocity, v_b (mm/s)	Heating length, L_H (mm)
125	144	14.4	100

From Fig. 9, it is plotted with the diameter of the fiber taper along the length of taper at fiber axis. Obviously, all of the simulated samples have the same waist length, L_w, of 10 cm. This is due to the fact that L_H is set at 10 cm, and it is always equal to L_w of the taper. Besides that, it is also observed that the narrower the r_w, the longer the L_E of the taper at the same initial radius. Its mathematical relation is expressed in the following paragraph.

Table 1 Table of simulated sample at various t

Simulated sample	Processing time, $T(s)$	Diameter, $D_w(\mu m)$
A	750	72.896
B	920	64.498
C	1140	55.050
D	1430	44.676
E	1840	33.256

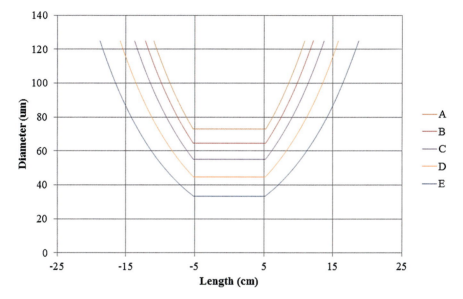

Fig. 9 The simulated profile of various D_w at L_H of 10 cm

In the design of the MT fabrication machine, the total brushing length determines the total length of the fabrication machine. Therefore, calculation has been done based on 1200 mm total brushing length to evaluate the performance of machine with respect to r_w and L_H. Based on Fig. 10, L_R is the remaining length left when the fiber moves to a side in order to heat the full length of the heat zone. The equation for the total brushing length, L_T, can be expressed as the equation given below.

$$L_T = L_H + L_E + L_R \qquad (23)$$

where L_R is actually the same as the length of heat zone, L_H. Hence, total mechanism length can also be written as

$$L_T = 2L_H + L_E \qquad (24)$$

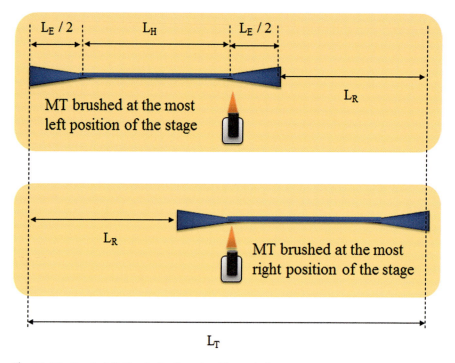

Fig. 10 The length definition in the flame-brushing technique

Based on mass conservation law, the tapered radius, r_w, is expressed as

$$r_w \propto r_o \sqrt{\frac{L_H}{L_H + L_E}} \qquad (25)$$

where r_o is the initial radius of the fiber. In order to relate to the total brushing length, the tapered radius is

$$r_w \propto r_o \sqrt{\frac{L_H}{L_T - L_H}} \qquad (26)$$

Figure 11 demonstrates the L_T required corresponding to L_H. The graph is plotted for three different sets of waist diameter, D_w, of 100, 80, and 60 μm. It can be seen from the graph that L_H is linearly proportional to L_T. Besides that, for larger D_w the plotted lines are right-shifted. This indicates an increase in the L_H for greater D_w. The potential achievable L_H corresponding to the D_w is 0.225 m for 60 μm, 0.347 m for 80 μm, and 0.468 m for 100 μm. In the simulation, for the taper of 100 μm, the maximum achievable L_H is close to 0.5 m, due to the fact that L_H is

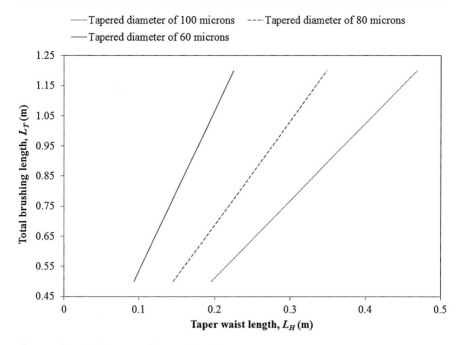

Fig. 11 Graph of L_T versus L_H at various D_w

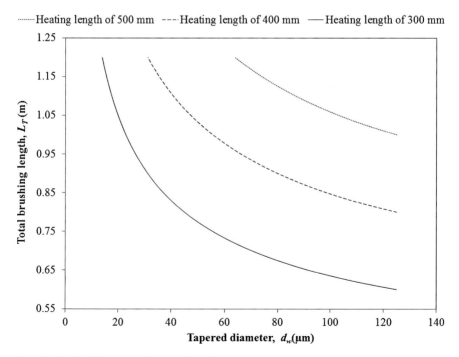

Fig. 12 Graph of L_T versus D_w at various L_H

always equal to L_w. This puts the maximum achievable L_w of MT that can be fully covered by the stage to about 0.5 m.

In Fig. 12, the graph has been plotted for L_T versus D_w. It is observed that for increasing L_H, the plotted lines shift to the right where the relationship of D_w and L_H is just the same as Fig. 11. It can be seen from the graph that L_T is having inverse square relationship with D_w. This explains that the potential achievable D_w corresponding to the L_H is 14 μm for 300 mm, 32 μm for 400 mm, and 63 μm for 500 mm.

Fabrications of Short Taper

In the 1960s, when optical waveguide theory was established, the optical waveguiding properties of taper-drawn fiber were investigated. Initially, most of the taper-drawing works were done on standard glass fiber due to the availability of high-purity glass and light launching system. The fabrication rigs were mostly set up with flame- or laser-heated tapering configurations (Kakarantzas et al. 2001; Dimmick et al. 1999). The research direction was to taper the optical fiber down to a nanometric diameter without compensating the optical loss so that its optical properties and applications were investigated. In this case, fiber taper with nanometric diameter scale was commonly bounded with $L_t < 10$ cm which is known as short taper. Therefore, short and biconical taper fabrication techniques are related to a mature technology development, namely, flame-brushing technique.

Decades later, the research interest was diverted to the fabrication of arbitrary taper profiles with nonuniform waists (Baker and Rochette 2011; Pricking and Giessen 2010). The interest was motivated by the engineering of the propagation characteristic in a taper. Moreover, with such technology, the tapers with various tapering shape could be reproduced in a taper rig (Felipe et al. 2012). The mentioned arbitrary taper fabrication techniques were nonetheless inspired by the short taper fabrication technique as described in the following subsections.

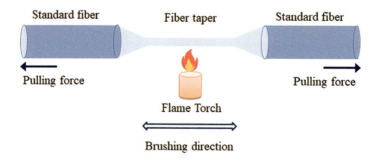

Fig. 13 Illustration of flame-brushing fiber tapering process (Birks and Li 1992)

Flame-Brushing Technique

A typical flame-brushing fabrication system is shown in Fig. 13. The hydrogen gas torch is used to provide flame-based heat to the system. The flame provides a high enough temperature for fiber melting and stretching. The technique is based on a small flame moving back and forth, also known as flame-brushing, under a fiber which is being stretched axially. The fiber is elongated with reduced diameter at the brushing region which is also known as hot zone. The control of the flame movement and the fiber pulling speed can be used to define desired fiber taper shape. Generally, it fabricates a microfiber or nanofiber with both ends attached to the standard fiber. This fiber taper structure is known as biconical fiber taper. It is worth noting that this technique is able to achieve the longest and most uniform nanowires (Brambilla et al. 2006).

Modified Flame-Brushing Technique

The configuration of the modified flame-brushing technique is similar to the conventional flame-brushing technique except that the flame is replaced by a different heat source such as CO_2 laser (Ward et al. 2006), sapphire capillary heated source by a CO_2 laser beam (Sumetsky et al. 2004), or a graphite filament, also known as graphite microheater (Brambilla et al. 2005; Wang and Mies 2009; Mägi et al. 2007). This is an elegant approach that reduces the contaminants (especially OH content) caused by the combustion of the flame source. The repeatability of this technique is further improved since the heating power is stable to provide uniform heating. Furthermore, the temperature of the heat source is controlled precisely to fabricate various optical fibers with different melting points (Mägi et al. 2007). Figure 14 shows one of the modified flame-brushing technique setups applying CO_2 laser as heat source.

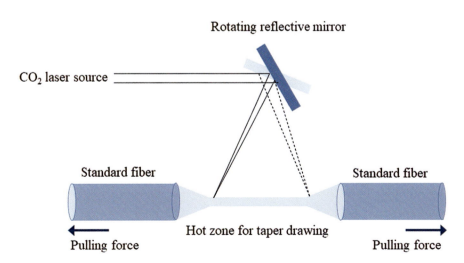

Fig. 14 Schematic of modified flame-brushing fabrication rig (Ward et al. 2006)

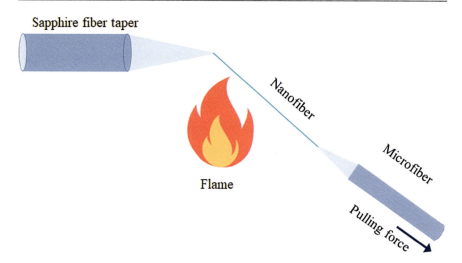

Fig. 15 Schematic of drawing nanofiber from a microfiber (Tong et al. 2003)

Self-Modulated Taper-Drawing Technique

Self-modulated taper-drawing technique is well-known for fabricating nanofibers with radii as small as 50 nm as reported in *Nature* (Tong et al. 2003). This is a two-step drawing process:

1. A standard fiber is pre-tapered to a microfiber using a conventional flame-pulling technique.
2. A sapphire fiber taper with a tip diameter of around 100 μm is used to absorb the heat from the flame and then wind the pre-tapered microfiber at the tip of the sapphire fiber taper. The heating is confined in a small volume to maintain steady temperature distribution. By applying force perpendicularly to the axis of the heated sapphire fiber taper, the microfiber is drawn to form nanofiber as shown in Fig. 15.

At the end of the process, the nanofiber is connected to a standard fiber, while the other end is left freestanding. The drawback of this process is that the fabrication procedure is rather complex and the nanofiber has a relatively high loss which is about 0.1 db/mm for the diameter of 800 nm at 1550 nm wavelength.

Direct Drawing from Bulk Technique

Direct drawing from bulk technique is not applicable to the material in fiber form, yet it is able to pull nanofibers from a bulk glass. In this technique, a sapphire rod is heated and put into contact with bulk glass, where the bulk glass experiences localized softening. The molten bulk glass is removed by the sapphire rod in small amount. Another sapphire rod is promptly brought into contact with the molten glass and then drawn away to form a fiber taper with micrometric or nanometric diameter.

Table 2 Comparison of the short taper fabrication techniques

Fabrication techniques	Features	Ref.
Flame-brushing	Small flame-brushing under an optical fiber which is being stretched Accurate control of taper shape	Brambilla et al. (2006)
Modified flame-brushing	Similar to flame-brushing technique except with a different heat source Allows tapering of wide range of low softening temperature glass	Sumetsky et al. (2004)
Self-modulated taper drawing	Two-step process Relatively high loss The radius can be drawn as small as 10 nm	Tong et al. (2003)
Direct drawing from bulk	Low cost setup The fiber form of material is not required It is difficult to control the uniformity and radius of taper	Tong et al. (2006)

The technique has successfully fabricated fiber tapers from tellurite and phosphate glasses (Tong et al. 2006) and from polymers (Xing et al. 2008).

Comparison of Short Taper Fabrication Techniques

In order to let the reader comprehensively compare the short taper fabrication techniques, the abovementioned techniques are consolidated into the comparison table, and it is listed as Table 2.

Fabrications of Long Taper

The only method to fabricate long taper is by using optical fiber drawing method. In this method, a fiber preform is heated to melt into high-viscosity liquid form, and thereafter an optical fiber is drawn according to the desired diameter optical fiber. If the drawing speed is manipulated, the tapering diameter is controlled precisely by the motorized drum spooling. According to Fig. 16, the fiber diameter detector sends instantaneous feedback signals to control circuit which further optimizes the uniformity of optical fiber via controlling the motor speed and gas flow rate.

In the 1980s, control methods were vigorously studied to solve the technical problems of optical fiber drawing. The mentioned technical problems are uniformity of optical fiber diameter, high-speed drawing and coating for cost reduction, and fabrication of high tensile optical fiber (Imoto et al. 1989; Paek 1986). Therefore, at that time, the research direction was to overcome the aforementioned problems on the drawn optical fiber (tapered region in a fiber taper as illustrated in Fig. 4), whereas there was not much effort carried out to study and characterize the tapering region in a long taper which is critical in having efficient mode coupling between multimodes and single mode based on the adiabaticity characteristic of the fiber

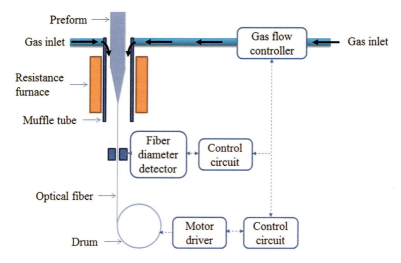

Fig. 16 Schematic of optical fiber drawing system (Imoto et al. 1977)

taper structure. Furthermore, in order to commercialize the standard optical fiber, the tapered region was easily drawn to a kilometer in length which is the long taper ($L_t > 10$ m), yet it was rare to taper down the optical fiber to a diameter of less than 100 μm.

Application in Structural Health Monitoring

The fiber-based Mach-Zehnder interferometric effect has been observed and studied in the abrupt biconical fiber taper since the late 1980s (Lacroix et al. 1988). In the following years, the inline Mach-Zehnder interferometric (IMZI) was fabricated based on long-period grating pairs (Allsop et al. 2002), two-taper structure (Tian et al. 2008a), mode-mismatch structure (Tripathi et al. 2009), and core-offset structure (Tian et al. 2008b). The fabrication process of the two-taper structure is the simplest and fastest yet has the comparable sensing performance among the mentioned structures. The IMZI has been extensively used for fiber sensor applications such as refractive index (RI) sensing (Gao et al. 2013; Tian et al. 2008b), strain sensing (Choi et al. 2007; Liu and Wei 2007; Shi et al. 2012; Wu et al. 2012; Men et al. 2011; Lu and Chen 2010), temperature sensing (Wu et al. 2012; Liu and Wei 2007; Shi et al. 2012; Jiang et al. 2011; Men et al. 2011; Lu and Chen 2010; Sun et al. 2015; Ou et al. 2013; Frazão et al. 2010; Zhou et al. 2011; Wang et al. 2013; Shen et al. 2012), and curvature sensing (Wang et al. 2013; Chen et al. 2013; Sun et al. 2015; Ou et al. 2013; Song et al. 2013; Dong et al. 2011; Monzon-Hernandez et al. 2011; Ma et al. 2014; Gong et al. 2011; Dass and Jha 2015; Shen et al. 2012; Ni et al. 2012; Frazão et al. 2010; Zhou et al. 2011). Generally, most researches characterized the curvature sensitivity of IMZI based on the wavelength shift-curvature relation (Wang et al. 2013; Ma et al. 2014; Chen et al. 2013; Sun et al. 2015; Ou et al. 2013; Song et al. 2013; Dong et al. 2011; Monzon-Hernandez et al. 2011; Gong et al. 2011; Frazão et al. 2010; Zhou et al. 2011), but only a few have been reported on intensity-curvature relation (Gong et al. 2011; Monzon-Hernandez et al. 2011; Ma et al. 2014; Shen et al. 2012; Dass and Jha 2015; Ni et al. 2012). It is worth noting that fiber-optic sensor that responds to wavelength signature has grown tremendously since with the development of fiber Bragg grating sensor. For the in-field application, the wavelength interrogation system is developed to demodulate wavelength and multiplex multiple sensors. However, the implementation cost of the system is comparably higher. Nevertheless, for the in-field application, the optical sensing system based on intensity-curvature relation is practical, and the advantages are low cost, simplicity of implementation, and possibility of being multiplexed, although intensity interrogation technique might suffer from fluctuations of optical power or other external disturbances. However, these types of errors can be prevented by calibrating the sensors before use.

Following the emergence of industrial revolution 4.0, the integration of the Internet of things (IoT), big data, and cloud computing provides a smart solution platform to monitor the structural health of the asset. Moreover, with the sufficiency of monitoring data, machine learning is able to recommend decision and hence take the prevention approach to minimize the loss of life and resources. Therefore, as aforementioned, the intensity-curvature sensing nature of microfiber-based IMZI makes Plug"N"Sense system feasible and practical to be implemented. The Plug"N"Sense system which is actually supported by wireless sensor network is vital in collecting information from a vast environment with benefits of long-distance sensing, connecting all the sensor data, reduced wiring, fast and easy installation, etc. Once the sensing data is streamed to cloud, cloud computing provides scalable processing power and huge data storage to analyze all the sensed information and make decisions from these information. The following subsections describe the deployment of microfiber-based IMZI sensor starting from the work of sensor fabrication to the field test sensing result.

Microfiber-Based IMZI Sensor Packaging

The process of fabricating the microfiber-based IMZI from the optical fiber has compromised the optical fiber's rigidity, and its fragile nature poses a limitation for its field sensing applications. Some form of packaging is required to ensure the rigidity of the fiber taper and guarantee its durability. Teflon coating has been proposed (Monzon-Hernandez et al. 2011), which promised to maintain the overall performance of the microfiber. Although this is a very rigid and flexible material, the coating cost and its nonadhesive characteristic make it difficult to be deployed for curvature sensing application. Besides that, polymer tube (Guo et al. 2009) and perspex packaging (Jasim et al. 2015) have been reported in appropriately packaging the microfiber. The encapsulating adhesive method in the polymer tube is only applicable to adiabatic microfiber, whereas the perspex sheet packaging design is too bulky and rigid that the small bending effect could not be effectively coupled to the IMZI.

A compact and cost-effective packaged IMZI sensor has been demonstrated by using polypropylene slab (Lin et al. 2016). Prior to designing the package of IMZI sensor, an experiment had been carried out to characterize the RI of the IMZI sensor to determine the encapsulating material. In the experiment, Cargille oil was applied in between the tapers without immersing the tapers with the intention of observing the effect of surrounding RI along the interferometer length only. In order to prevent tapers from immersing into the Cargille oil, the fiber is positioned along the width of the microscope slide which is 25 mm shorter than the interferometer length as shown in Fig. 17a. Cargille oil with different RI values at a room temperature of 24 °C is used for testing.

The Fourier transform of the output spectrum (inset of Fig. 17b) gives the spatial frequency spectrum. Figure 17b shows the spatial frequency's dominant peak at around $0.1\,\text{nm}^{-1}$, which corresponds to the mentioned interferometer length, $l = 5$ cm. Besides that, while the IMZI is surrounded with RI less than that of the cladding, there are multiple spatial frequency peaks that correspond to different orders of cladding modes responding to the SRI. As the stripped fiber between tapers is immersed into SRI greater than that of the cladding, the multiple spatial frequency peak is eliminated, and the spatial frequency's dominant peak is suppressed. In other words, as long as the SRI is greater than that of the cladding, the Cargille oil acts as an absorption layer that allows single cladding mode to be weakly guided in the IMZI structure. In this condition, when the fiber is slightly bent, the surviving cladding mode significantly penetrates through the absorption layer. As a result, the fringe visibility, $K = 2\sqrt{I_1 I_2}/(I_1 + I_2)$, decreases accordingly. This manifests that the absorption layer coated IMZI is highly sensitive to the curvature sensing.

Apparently, the coating of standard SMF has the required nature where the RI of inner coating and outer coating is 1.4786 and 1.5294, respectively, at wavelength 1550 nm. Therefore, in the design of IMZI packaging, the interferometer length remains unstripped during the tapering process. Besides that, cyanoacrylate is suitable to be used as an adhesive epoxy in packaging the IMZI sensor because the RI of such adhesive epoxy is generally higher than that of the cladding. The IMZI sensor is sandwiched in between two polypropylene slabs with cyanoacrylate bonded to the unstripped fiber. The packaged IMZI is shown as inset of Fig. 18. Prior to the in-field curvature test, the output spectrum of packaged IMZI sensor with and without bending is recorded. From Fig. 18, it is obvious that there is only single cladding mode which survives in the packaged IMZI sensor and such mode is suppressed due to the slight bending effect. Subsequently, as the sensor is bent, the decrease of the fringe visibility leads to the increase of the optical power at the particular trough wavelength. The sensitivity of the sensor is calculated as $85.2\,\text{dB/m}^{-1}$, and the minimum

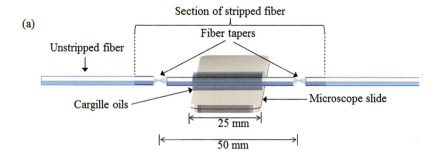

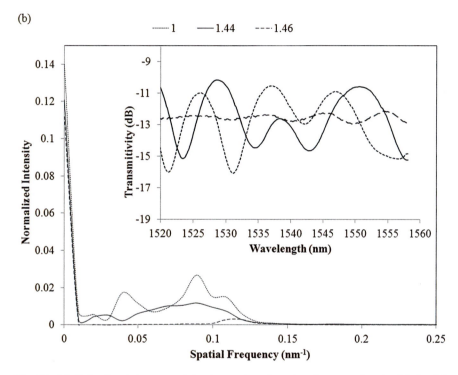

Fig. 17 (a) Refractive index test of the immersion along interferometer length. (b) Spatial frequency spectral of IMZI responding to various SRI. Inset is the corresponding spectral output (Lin et al. 2016)

detectable curvature is $35 \times 10^{-5}\,\text{m}^{-1}$ which is in the radius of 2.86 km. This implies that the packaged IMZI sensor is highly sensitive to small curvature. A comparison between the reported IMZI and packaged IMZI sensor is listed as Table 3.

The accuracy of the packaged IMZI is questionable because of the cross talk induced by the fluctuations in ambient temperature during curvature tests. It is utmost important to be able to distinguish between these two parameters if the

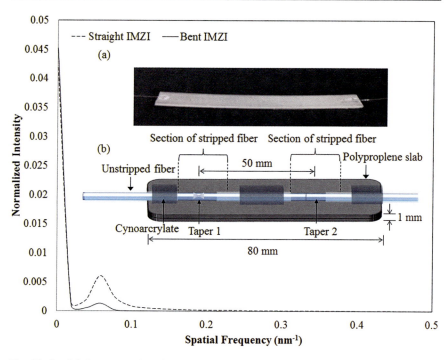

Fig. 18 Spatial frequency of packaged IMZI with and without bending. Inset consists of (**a**) photograph and (**b**) model of packaged IMZI sensor (Lin et al. 2016)

Table 3 Comparison of the performance of the reported IMZI (Lin et al. 2016)

Ref.	Fiber type	Interferometer structure	Curvature sensitivity
Shen et al. (2012)	Single-mode and polarization-maintaining fiber (PMF)	Core-offset with fusion splicing of PMF	$-0.882\,\text{dB/m}^{-1}$
Gong et al. (2011)	Single-mode and multimode fiber	Single-mode-multimode-single-mode structure	$-130.37\,\text{dB/m}^{-1}$
Dass and Jha (2015)	Single-mode fiber	Micron wire-assisted IMZI structure	$-11.92\,\text{dB/m}^{-1}$
Ni et al. (2012)	Single-mode and photonic crystal fiber (PCF)	Tapered PCF	$8.35\,\text{dB/m}^{-1}$
Lin et al. (2016)	Single-mode fiber	Packaged IMZI	$85.2\,\text{dB/m}^{-1}$

sensors were to be deployed in the field. Therefore, a multiparameter simultaneous measurement technique is demonstrated (Raji et al. 2016).

The individual parameter discrimination is achieved by manipulating the unequal sensitivities of optical power to temperature and curvature obtained at two wavelengths within the sensing spectrum. The sensor exhibits high curvature sensitivities

Fig. 19 The matching of experimental results with the prediction model (Raji et al. 2016)

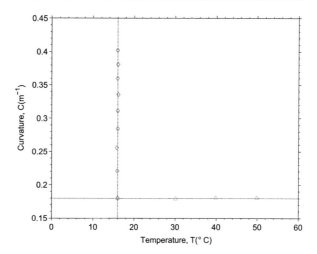

of 11.83 and 12.49 dBm/m^{-1} for curvature ranging from 0.18 to 0.4 m^{-1} and high temperature sensitivity of 0.0829 and 0.0833 dBm/°C^{-1}, for both wavelengths. The maximum error obtained from the predicting curvature and temperature is found to have a rms deviation of 0.1801 m^{-1} and 0.0826 °C^{-1}, respectively, as shown in Fig. 19. With this simultaneous sensing technique, the packaged IMZI sensor can be deployed for structural health monitoring due to the merits of high reproducibility, practical field application, and high accuracy.

Microfiber-Based IMZI Sensor Deployment

A lightweight foamed concrete structural (LWFCS) beam with 0.4% of polypropylene fiber was prepared for the field test to characterize the performance of IMZI sensors under different tensile capacities.

Figure 20a, b shows the installation of sensor onto the rebar, whereas (c) and (d) show the geometry of the LWFCS beam. The IMZI sensors were attached onto the midspan of the bottom rebar by means of a cyanoacrylate adhesive, connected to commercial optical loss test set (OLTS, brand FiberO), which consists of a laser source and optical power meter. Operating wavelength 1550 nm was used in the measurement, which is the standard communication wavelength in the optical communication industry and easily accessed compared to other optical wavelengths. Noted that, the 1550 nm wavelength has a low transmission loss 0.18 dB/km in SMF. As shown in Fig. 20a, a lead wire alloy foil strain gauge (brand TML, model FLK-6-11) was attached on the same rebar the opposite side of the IMZI sensors for measurement comparison.

Both of the top and bottom surfaces of rebars were grinded superficially to create flat surfaces for IMZI sensor and strain gauge, to ensure a strong adhesion. The strong adhesion ensures that the strain experienced by the rebars is able to be

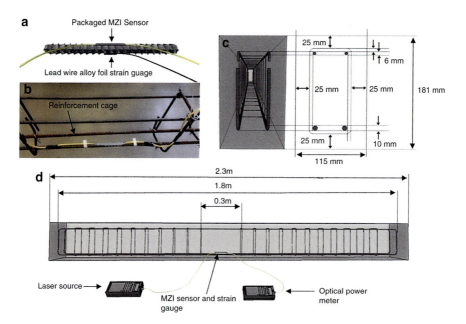

Fig. 20 (**a**) and (**b**) Installation of IMZI sensor and strain gauge on the rebar; (**c**) and (**d**) geometry of the LWFCS beam (Png et al. 2018)

coupled to the devices efficiently. The LWFCS beams were casted with concrete and left for 28 days in referring to ASTM C192 (2007) and ASTM C31 (2008) standards. This is to ensure that the beams are fully cured before proceeding to the field test.

Figure 21 depicts the installation of LWFCS beams onto the 4-point bending test machine with three linear variable displacement transducers (LVDTs) installed at the bottom of the beams to record the vertical deflection upon the continuous loading. During the field test, the loader imposed a loading force of 6 kN to the LWFCS beams upon initialization. The loading force was followed by an increment of 1 kN per step until the beams' failure. Simultaneously, the resultant optical powers and strains were measured by the IMZI sensor and strain gauges. The optical power variation was calibrated to the curvature changes experienced by the LWFCS beams for every set of experiment.

Figure 22 shows the resultant optical powers of IMZI sensor which was mapped to measurements of the strain gauge. To elaborate the optical power variation in a clear manner, the data representation was done by sectioning the whole plot into three discrete regions, namely, the linear sensing region, internal cracking region, and yielding region. Linear sensing regions include the power variations during the elastic deformation of the rebar, whereby the optical powers are responding directly to the curvatures of the rebars. As shown in the subsets of Figure 22, the resultant optical powers of IMZI sensors are linearly proportional to the respective

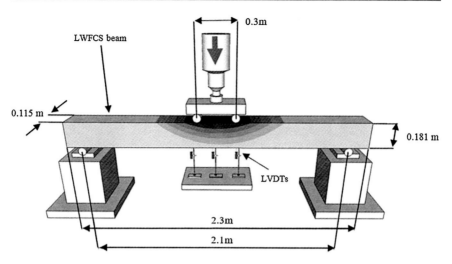

Fig. 21 Installation of reinforced LWFCS beam on the 4-points bending test machine (Png et al. 2018)

curvature changes. The IMZI sensor achieved good agreement with the respective strain gauges, with correlation coefficients of −0.956.

Internal cracking regions comprise the power fluctuations when the curvatures of LMFCS beams are high enough to cause internal cracking; thereby fluctuations of optical powers were observed. During internal cracking, shear deformation breaks the concretes into granular fragments contributes. The fragments impose nonuniform movements and forces to the upper surface of IMZI sensors; thereby the optical powers of sensors started to fluctuate accordingly. A higher percentage of polypropylene fiber enhance the tensile capacity and cohesiveness of concrete and hence reduce the free-moving granular fragments inside the concrete.

Yielding region consists of the responding optical power during the yielding of rebar. At this point, elastic deformation ended and continued with the inelastic deformation; the elongation of rebar increased tremendously and caused the failure of beam immediately. However only IMZI sensor managed to detect the yielding which is shown in Fig. 22. Optical power in IMZI sensor dropped significantly as the same time as the strain gauge.

Figure 23 depicts the surface cracks of beam respectively at each successive loading. The labels of hairline cracks and macrocracks differentiate the degree of surface crack. Hairline crack is a minor crack that can be recovered by implementing epoxy injection for curing. Macrocrack marks as the severe crack found on the beam surface which is irrevocable. The macrocrack of beams was spotted at loading of 21.5 kN.

By referring to Fig. 22, the IMZI sensors are able to detect the internal cracking before the macrocrack, where the first detection of the internal crack is marked as the crack point. The early detection of internal cracking before the macrocrack has an

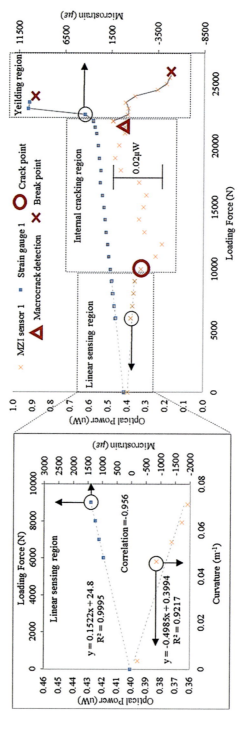

Fig. 22 Mapping of optical power of sensor to micro-strain of strain gauge in beam with 0.4% polypropylene (Png et al. 2018)

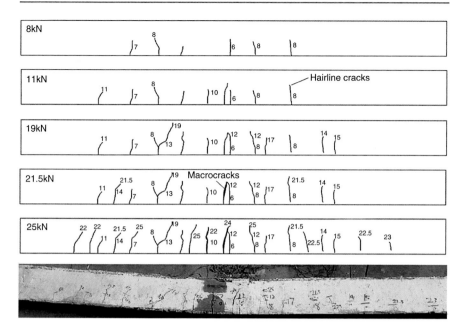

Fig. 23 Mapping of fracture pattern on beam to the successive strength. (Represented using red circles in Fig. 22) (Png et al. 2018)

Table 4 Summary table of crack point, macrocrack detection, break point, and curvature sensitivities for the comparison of IMZI sensor and strain gauge (Png et al. 2018)

	IMZI sensor	Strain gauge
Crack point (kN)	10	NA
Macrocrack detection (kN)	21.5	NA
Break point (kN)	26	24
Sensitivity $(Wm^{-1})/(N^{-1})$	-0.499	0.152

advantage over the conventional strain gauge in concrete monitoring. In the aspect of maximum sensing capacity, the IMZI sensors are varying with the strain gauges. The failures of the IMZI sensors and strain gauges were labeled as break point in Fig. 22. A summarization comprises the sensitivities and crack points tabulated in Table 4.

The implementation of IMZI sensor into the LWFCS beams was successful as the sensor managed to detect the elongation and yielding of rebar as linearly as the lead wire alloy foil strain gauges. More importantly, the double-sided sensing property of IMZI sensors managed to detect the internal cracking within the concretes before any earlier sighting of the macrocracks, which has advantage over the conventional strain gauge in concrete monitoring.

Summary

Mode condition for the three-layer step-fiber model is considered for both cladding and core boundaries to identify the effective refractive indices at various diameters. Subsequently, the effective refractive indices are related to the adiabaticity criterion of microfiber based on the upper boundary of taper angle.

Next, the microfiber fabrication techniques are reviewed to be classified into three categories based on the microfiber waist length, namely, short taper, meso taper, and long taper. The fabrication techniques for short taper and long taper are considered to be employed in fabricating MT with longer waist length.

Finally, the latest deployment of microfiber-based IMZI sensor is introduced. The packaging process of the IMZI is reviewed and justified. Besides that, the method to resolve the multiple sensing parameter-induced cross talk has been identified. Subsequently, during the IMZI deployment, the internal cracking within the concretes is detected before any earlier sighting of the macrocracks, which has advantage over the conventional strain gauge in concrete monitoring.

References

C. Alegria, R. Feced, M. Zervas, R. Laming, Acousto-optic effect in optical fibre tapered structures for the design of filters, in *IEE Colloquium on New Developments in Optical Amplifiers (Ref. No. 1998/492)*, IET (1998), pp. 11–1

T. Allsop, R. Reeves, D.J. Webb, I. Bennion, R. Neal, A high sensitivity refractometer based upon a long period grating Mach–Zehnder interferometer. Rev. Sci. Instrum. **73**(4), 1702–1705 (2002)

C. Baker, M. Rochette, A generalized heat-brush approach for precise control of the waist profile in fiber tapers. Opt. Mater. Express **1**(6), 1065–1076 (2011)

F. Bilodeau, K. Hill, D. Johnson, S. Faucher, Compact, low-loss, fused biconical taper couplers: overcoupled operation and antisymmetric supermode cutoff. Opt. Lett. **12**(8), 634–636 (1987)

T. Birks, Y. Li, The shape of fiber tapers. J. Lightwave Technol. **10**(4), 432–438 (1992)

R. Black, S. Lacroix, F. Gonthier, J. Love, Tapered single-mode fibres and devices. II. Experimental and theoretical quantification, in *Optoelectronics, IEE Proceedings J*, vol. 138 (IET, 1991), pp. 355–364

L. Bobb, P. Shankar, H. Krumboltz, Bending effects in biconically tapered single-mode fibers. J. Lightwave Technol. **8**(7), 1084–1090 (1990)

G. Brambilla, F. Koizumi, X. Feng, D. Richardson, Compound-glass optical nanowires. Electron. Lett. **41**(7), 400–402 (2005)

G. Brambilla, F. Xu, X. Feng, Fabrication of optical fibre nanowires and their optical and mechanical characterisation. Electron. Lett. **42**(9), 517–519 (2006)

W. Burns, M. Abebe, C. Villarruel, Parabolic model for shape of fiber taper. Appl. Opt. **24**(17), 2753–2755 (1985)

W. Burns, M. Abebe, C. Villarruel, R. Moeller, Loss mechanisms in single-mode fiber tapers. J. Lightwave Technol. **4**(6), 608–613 (1986)

X. Chen, Y. Yu, X. Xu, Q. Huang, Z. Ou, J. Wang, P. Yan, C. Du, Temperature insensitive bending sensor based on in-line Mach-Zehnder interferometer. Photon. Sensors. **4**(3), 193–197 (2013)

H. Choi, M. Kim, B. Lee, All-fiber Mach-Zehnder type interferometers formed in photonic crystal fiber. Opt. Express **15**(9), 5711–5720 (2007)

S. Dass, R. Jha, Micron wire assisted inline Mach-Zehnder interferometric curvature sensor. Photon. Technol. Lett. IEEE **99**, 1 (2015)

T. Dimmick, G. Kakarantzas, T. Birks, P. Russell, Carbon dioxide laser fabrication of fused-fiber couplers and tapers. Appl. Opt. **38**(33), 6845–6848 (1999)

B. Dong, J. Hao, Z. Xu, Temperature insensitive curvature measurement with a core-offset polarization maintaining photonic crystal fiber based interferometer. Opt. Fiber Technol. **17**(3), 233–235 (2011)

T. Erdogan, Cladding-mode resonances in short-and long-period fiber grating filters. JOSA A **14**(8), 1760–1773 (1997)

A. Felipe, G. Espíndola, H. Kalinowski, J. Lima, A. Paterno, Stepwise fabrication of arbitrary fiber optic tapers. Opt. Express **20**(18), 19893–19904 (2012)

O. Frazão, S. Silva, J. Viegas, J.M. Baptista, J.L. Santos, J. Kobelke, K. Schuster, All fiber Mach–Zehnder interferometer based on suspended twin-core fiber. IEEE Photon. Technol. Lett. **17**(22), 1300–1302 (2010)

S. Gao, W. Zhang, Z. Bai, H. Zhang, Ultrasensitive refractive index sensor based on microfiber-assisted U-shape cavity. Photon. Technol. Lett. IEEE **25**(18), 1815–1818 (2013)

Y. Gong, T. Zhao, Y.-J. Rao, Y. Wu, All-fiber curvature sensor based on multimode interference. Photon. Technol. Lett. IEEE **23**(11), 679–681 (2011)

J.C. Graf, S.A. Teston, P.V. de Barba, J. Dallmann, J.A. Lima, H.J. Kalinowski, A.S. Paterno, Fiber taper rig using a simplified heat source and the flame-brush technique, in *Microwave and Optoelectronics Conference (IMOC), 2009 SBMO/IEEE MTT-S International* (IEEE, 2009), pp. 621–624

T. Guo, L. Shao, H.-Y. Tam, P.A. Krug, J. Albert, Tilted fiber grating accelerometer incorporating an abrupt biconical taper for cladding to core recoupling. Opt. Express **17**(23), 20651–20660 (2009)

K. Imoto, S. Aoki, M. Sumi, Novel method of diameter control in optical-fibre drawing process. Electron. Lett. **13**(24), 726–727 (1977)

K. Imoto, M. Sumi, G. Toda, T. Suganuma, Optical fiber drawing method with gas flow controlling system. J. Lightwave Technol. **7**(1), 115–121 (1989)

A.A. Jasim, M. Dernaika, S.W. Harun, H. Ahmad, A switchable figure eight Erbium-Doped fiber laser based on inter-modal beating by means of non-adiabatic microfiber. J. Lightwave Technol. **33**(2), 528–534 (2015)

L. Jiang, J. Yang, S. Wang, B. Li, M. Wang, Fiber Mach–Zehnder interferometer based on microcavities for high-temperature sensing with high sensitivity. Opt. Lett. **36**(19), 3753–3755 (2011)

G. Kakarantzas, T. Dimmick, T. Birks, R. Le Roux, P. Russell, Miniature all-fiber devices based on CO_2 laser microstructuring of tapered fibers. Opt. Lett. **26**(15), 1137–1139 (2001)

K. Kao, G.A. Hockham, Dielectric-fibre surface waveguides for optical frequencies, in *Proceedings of the Institution of Electrical Engineers*, vol. 113 (IET, 1966), pp. 1151–1158

S. Lacroix, F. Gonthier, J. Bures, All-fiber wavelength filter from successive biconical tapers. Opt. Lett. **11**(10), 671–673 (1986)

S. Lacroix, F. Gonthier, R. Black, J. Bures, Tapered-fiber interferometric wavelength response: the achromatic fringe. Opt. Lett. **13**(5), 395–397 (1988)

H. Lin, Y. Raji, J. Lim, S. Lim, M. Mokhtar, Z. Yusoff, Packaged in-line Mach–Zehnder interferometer for highly sensitive curvature and flexural strain sensing. Sens. Actuators A Phys. **250**, 237–242 (2016)

Y. Liu, L. Wei, Low-cost high-sensitivity strain and temperature sensing using graded-index multimode fibers. Appl. Opt. **46**(13), 2516–2519 (2007)

J. Love, Spot size, adiabaticity and diffraction in tapered fibres. Electron. Lett. **23**(19), 993–994 (1987)

J. Love, W. Henry, Quantifying loss minimisation in single-mode fibre tapers. Electron. Lett. **22**(17), 912–914 (1986)

J. Love, W. Henry, W. Stewart, R. Black, S. Lacroix, F. Gonthier, Tapered single-mode fibres and devices. I. Adiabaticity criteria, in *Optoelectronics, IEE Proceedings J*, vol. 138 (IET, 1991), pp. 343–354

P. Lu, Q. Chen, Asymmetrical fiber Mach–Zehnder interferometer for simultaneous measurement of axial strain and temperature. Photon. J. IEEE **2**(6), 942–953 (2010)

L. Ma, Y. Qi, Z. Kang, S. Jian, All-fiber strain and curvature sensor based on no-core fiber. IEEE Sens. J. **14**(5), 1514–1517 (2014)

E. Mägi, L. Fu, H. Nguyen, M. Lamont, D. Yeom, B. Eggleton, Enhanced Kerr nonlinearity in sub-wavelength diameter $As_2 Se_3$ chalcogenide fiber tapers. Opt. Express **15**(16), 10324–10329 (2007)

L. Men, P. Lu, Q. Chen, Femtosecond laser trimmed fiber taper for simultaneous measurement of axial strain and temperature. Photon. Technol. Lett. IEEE **23**(5), 320–322 (2011)

D. Monzon-Hernandez, A. Martinez-Rios, I. Torres-Gomez, G. Salceda-Delgado, Compact optical fiber curvature sensor based on concatenating two tapers. Opt. Lett. **36**(22), 4380–4382 (2011)

K. Ni, T. Li, L. Hu, W. Qian, Q. Zhang, S. Jin, Temperature-independent curvature sensor based on tapered photonic crystal fiber interferometer. Opt. Commun. **285**(24), 5148–5150 (2012)

Z. Ou, Y. Yu, P. Yan, J. Wang, Q. Huang, X. Chen, C. Du, H. Wei, Ambient refractive index-independent bending vector sensor based on seven-core photonic crystal fiber using lateral offset splicing. Opt. Express **21**(20), 23812–23821 (2013)

U. Paek, High-speed high-strength fiber drawing. J. Lightwave Technol. **4**(8), 1048–1060 (1986)

C.R. Petersen, R.D. Engelsholm, C. Markos, L. Brilland, C. Caillaud, J. Trolès, O. Bang, Increased mid-infrared supercontinuum bandwidth and average power by tapering large-mode-area chalcogenide photonic crystal fibers. Opt. Express **25**(13), 15336–15348 (2017)

W. Png, H. Lin, C. Pua, J. Lim, S. Lim, Y. Lee, F. Rahman, Feasibility use of in-line Mach-Zehnder interferometer optical fibre sensor in lightweight foamed concrete structural beam on curvature sensing and crack monitoring. Struct. Health Monit. **17**, 1277–1288 (2018)

S. Pricking, H. Giessen, Tapering fibers with complex shape. Opt. Express **18**(4), 3426–3437 (2010)

Y. Raji, H. Lin, S. Ibrahim, M. Mokhtar, Z. Yusoff, Intensity-modulated abrupt tapered fiber Mach-Zehnder interferometer for the simultaneous sensing of temperature and curvature. Opt. Laser Technol. **86**, 8–13 (2016)

C. Shen, C. Zhong, Y. You, J. Chu, X. Zou, X. Dong, Y. Jin, J. Wang, H. Gong, Polarization-dependent curvature sensor based on an in-fiber Mach-Zehnder interferometer with a difference arithmetic demodulation method. Opt. Express **20**(14), 15406–15417 (2012)

L. Shi, X. Chen, H. Liu, Y. Chen, Z. Ye, W. Liao, Y. Xia, Fabrication of submicron-diameter silica fibers using electric strip heater. Opt. Express **14**(12), 5055–5060 (2006)

J. Shi, S. Xiao, M. Bi, L. Yi, P. Yang, Discrimination between strain and temperature by cascading single-mode thin-core diameter fibers. Appl. Opt. **51**(14), 2733–2738 (2012)

H. Song, H. Gong, K. Ni, X. Dong, All fiber curvature sensor based on modal interferometer with waist enlarge splicing. Sens. Actuators A Phys. **203**, 103–106 (2013)

M. Sumetsky, Y. Dulashko, A. Hale, Fabrication and study of bent and coiled free silica nanowires: self-coupling microloop optical interferometer. Opt. Express **12**(15), 3521–3531 (2004)

B. Sun, Y. Huang, S. Liu, C. Wang, J. He, C. Liao, G. Yin, J. Zhao, Y. Liu, J. Tang et al., Asymmetrical in-fiber Mach-Zehnder interferometer for curvature measurement. Opt. Express **23**(11), 14596–14602 (2015)

Y. Takeuchi, J. Noda, Novel fiber coupler tapering process using a microheater. IEEE Photon. Technol. Lett. **4**(5), 465–467 (1992)

R. Threlfall, *On Laboratory Arts* (Macmillan and Company, London/New York, 1898)

Z. Tian, S. Yam, J. Barnes, W. Bock, P. Greig, J. Fraser, H. Loock, R. Oleschuk, Refractive index sensing with Mach–Zehnder interferometer based on concatenating two single-mode fiber tapers. Photon. Technol. Lett. IEEE **20**(8), 626–628 (2008a)

Z. Tian, S.-H. Yam, H. Loock, Single-mode fiber refractive index sensor based on core-offset attenuators. Photon. Technol. Lett. IEEE **20**(16), 1387–1389 (2008b)

L. Tong, R. Gattass, J. Ashcom, S. He, J. Lou, M. Shen, I. Maxwell, E. Mazur et al., Subwavelength-diameter silica wires for low-loss optical wave guiding. Nature **426**(6968), 816–819 (2003)

L. Tong, L. Hu, J. Zhang, J. Qiu, Q. Yang, J. Lou, Y. Shen, J. He, Z. Ye, Photonic nanowires directly drawn from bulk glasses. Opt. Express **14**(1), 82–87 (2006)

S.M. Tripathi, A. Kumar, R.K. Varshney, Y. Kumar, E. Marin, J.-P. Meunier, Strain and temperature sensing characteristics of single-mode–multimode–single-mode structures. J. Lightwave Technol. **27**(13), 2348–2356 (2009)

C.Y. Tsao, D.N. Payne, W.A. Gambling, Modal characteristics of three-layered optical fiber waveguides: a modified approach. JOSA A **6**(4), 555–563 (1989)

N. Vukovic, N. Broderick, M. Petrovich, G. Brambilla, Novel method for the fabrication of long optical fiber tapers. Photon. Technol. Lett. IEEE **20**(14), 1264–1266 (2008)

B. Wang, E. Mies, Review of fabrication techniques for fused fiber components for fiber lasers, in *Proceedings of SPIE*, vol. 7159, 71950A (2009)

R. Wang, J. Zhang, Y. Weng, Q. Rong, Y. Ma, Z. Feng, M. Hu, X. Qiao, Highly sensitive curvature sensor using an in-fiber Mach-Zehnder interferometer. IEEE Sens. J. **13**(5), 1766–1770 (2013)

J. Ward, D. OShea, B. Shortt, M. Morrissey, K. Deasy, S. Nic Chormaic, Heat-and-pull rig for fiber taper fabrication. Rev. Sci. Instrum. **77**(8), 083105–083105 (2006)

D. Wu, T. Zhu, K.S. Chiang, M. Deng, All single-mode fiber Mach–Zehnder interferometer based on two peanut-shape structures. J. Lightwave Technol. **30**(5), 805–810 (2012)

X. Xing, Y. Wang, B. Li, Nanofibers drawing and nanodevices assembly in poly (trimethylene terephthalate). Opt. Express **16**(14), 10815–10822 (2008)

Y. Zhou, W. Zhou, C.C. Chan, W.C. Wong, L.-Y. Shao, J. Cheng, X. Dong, Simultaneous measurement of curvature and temperature based on PCF-based interferometer and fiber Bragg grating. Opt. Commun. **284**(24), 5669–5672 (2011)

Flat Fibers: Fabrication and Modal Characterization

15

Ghafour Amouzad Mahdiraji, Katrina D. Dambul, Soo Yong Poh, and Faisal Rafiq Mahamd Adikan

Contents

Introduction	624
Flat Fiber Fabrication	626
Flat Fiber Drawing Repeatability	628
Flat Fibers with Different Dimensions	629
Characterization of Flat Fibers: Mode Propagation	629
Multimode Propagation in Flat Fibers	629
Single-Mode Propagation in Flat Fibers	632
Conclusion	635
References	636

G. Amouzad Mahdiraji (✉)
School of Engineering, Taylor's University, Subang Jaya, Selangor, Malaysia

Flexilicate Sdn. Bhd., University of Malaya, Kuala Lumpur, Malaysia
e-mail: ghafouram@gmail.com

K. D. Dambul
Faculty of Engineering, Multimedia University, Cyberjaya, Selangor, Malaysia
e-mail: katrina@mmu.edu.my

S. Y. Poh
Integrated Lightwave Research Group, Department of Electrical Engineering, Faculty of Engineering, University of Malaya, Kuala Lumpur, Malaysia
e-mail: sysypoh@gmail.com

F. R. Mahamd Adikan
Flexilicate Sdn. Bhd., University of Malaya, Kuala Lumpur, Malaysia

Integrated Lightwave Research Group, Department of Electrical Engineering, Faculty of Engineering, University of Malaya, Kuala Lumpur, Malaysia
e-mail: rafiq@um.edu.my

© Springer Nature Singapore Pte Ltd. 2019
G.-D. Peng (ed.), *Handbook of Optical Fibers*,
https://doi.org/10.1007/978-981-10-7087-7_73

Abstract

This chapter presents the fabrication and modal characterization of flat fibers, a specialty optical fiber which has the advantages of both planar waveguides and standard optical fibers. In the introduction section, the demand for flat fibers is discussed including its applications. The main text of this chapter discusses the drawing process of flat fibers including issues such as flat fiber drawing repeatability and drawing of flat fibers with different dimensions. Next, the modal characterization of flat fibers is discussed, with a focus on the light propagation in flat fibers. Both simulation and experimental results show that the light propagation in flat fibers is inherently multimode propagation. However, it has been shown experimentally that adding two defect eyes at the sides of the core could produce flat fibers with single-mode propagation.

Keywords

Flat fibers · Optical fiber fabrication · Optical fibers characterization · Multifunctional sensing fiber · Single mode fiber

Introduction

Optical fiber sensors can be used to detect physical changes in the environment surrounding the fiber. For example, a fiber biosensor can be used to detect the concentration of specific biological or chemical substance. However, the fiber sensor detection capability is limited as it can only detect a single specific target of sensing. Therefore, a combination of fiber sensors is often required to combine the optical signals (Zhang et al. 2013) and to allow multiple substances to be detected. The multifunctional property of planar waveguides gives rise to multipurpose usage in the same chip. In planar waveguides, light is guided on defined paths on a planar substrate. Planar waveguide is the basic component in integrated optic devices such as optical splitters, optical couplers, interferometers, and gratings (Buck 1995; Saleh and Teich 2007). Optical components such as couplers, filters, modulators, amplifiers, detectors, or even breakthrough diffraction limit plasmonics can be built on the same chip (Bian and Gong 2013). These optical devices can be fabricated on the flat surface of materials such as silica (Li and Henry 1996), silicon, semiconductor crystal, or plastic (Dutton 1998). However, the optical components often require advanced fabrication technology such as photolithography, UV writing, nonlinear resonators, and super-resolution imaging, quantum information, and light harvesting (Ferretti et al. 2013; Kildishev et al. 2013). Other fabrication methods for planar waveguides include etching and epitaxial growth (Singh 1996; Hunsperger 2009), but these methods are often costly and require a high precision (Dutton 1998). The limitations of planar waveguides are high propagation and coupling loss, rigid substrate and can be fabricated for only a short length (maximum of 10–15 cm).

An alternative solution to planar waveguides with added advantage of low propagation loss and high flexibility is fulfilled by a specialty optical fiber known as flat fiber (FF). FF was introduced in 2007 and was initially developed as alternative to planar waveguides due to its low-cost fabrication technique (Webb et al. 2007). The FF structure is almost similar to a buried channel planar waveguide, with a higher refractive index core layer and a lower refractive index at the cladding. The higher core refractive index allows light to continuously propagate inside the FF's core via total internal reflection phenomena (Buck 1995; Kasap 2001). This unique fiber design fills the gap in integrated optic devices as a transport and sensing medium. FF can be described as a planar waveguide fiber with the advantages of having both optical fibers and planar waveguide's properties, such as mechanically flexible, multifunctionality, long length fabrication, low coupling and propagation loss, etc. (Adikan et al. 2012). In addition, flat fibers also have the advantages of cylindrical optical fiber such as chemically inert, ability to withstand high ambient temperatures, and at the same time having a large flat surface area for material processing at low manufacturing costs (Kalli et al. 2015).

Due to the interesting properties of FF, there is a research demand and interest to develop technology for smart structures, integrated photonic circuits, and devices for optofluidic applications (Riziotis et al. 2014). Applications have been demonstrated via femtosecond laser micromachining, such as fabrication of optical structures, including ring and disk resonators and Mach-Zehnder interferometer. FFs solve the difficulties in producing substrate with such properties, where multistage process are required and the cost of manufacturing equipment limits the research area to well-funded groups (Duncan 2002; Scifres 1996).

Inscription of Bragg gratings, ring resonator, disc resonator, Mach-Zehnder interferometer, and microfluidic channel has been demonstrated on a FF using femtosecond laser inscription (Kalli et al. 2015). The inscription is done by having a focused beam from a femtosecond laser translated into a bulk transparent material which caused the refractive index of the core material to increase due to the nonlinear absorption process (Kalli et al. 2015).

Besides femtosecond laser, a more popular approach for developing optical/photonic devices on FFs is direct UV writing (Holmes et al. 2008; Adikan 2007). Studies were done to characterize UV-written ($\lambda_{UV} = 244$ nm) chips by studying the effective index value obtained from the grating experiment. Y-splitters/combiner, Bragg gratings, ring resonator, and straight channel are examples of devices that can be fabricated on FFs by using UV-written method (Adikan et al. 2012). Single beam direct UV writing can also be used to define waveguide channels in the core of FF (Holmes et al. 2008; Adikan 2007). But the core needs to be photosensitive and this can be done by doping the core with germanium and/or boron. Using direct grating writing where Bragg gratings are written on the core of the fiber and applying the principle of shifting the Bragg wavelength, evanescent field sensors have been demonstrated on FFs (Holmes et al. 2008).

The losses in FFs are not as low as cylindrical optical fibers and are typically in the region of 0.1 dB/m. However, this depends on its fabrication methods. Over the years, improvements have been made to realize this technology (Kalli et al. 2015). Flat fibers can also be cleaved quite efficiently, unlike its predecessor of planar silica on silicon samples which needs to be polished for laser inscription.

It is often desirable for lab-on-a-chip, optofluidics, and biosensor to "access" the guiding light so that the evanescent field can interact with an external fluid sample, typically for refractive index measurements. To improve the evanescent field in conventional fibers, in most of the cases, polishing of optical fibers to the core-cladding boundary is favored, a method which is very time consuming and inefficient. On the contrary, the potential of connecting the waveguide core with the surface using a femtosecond laser waveguide offers an alternative method that is less time consuming. In this case the "connection" distance between the core and surface is short, perhaps only 50 μm, and waveguide losses will not be significant. FF offers stronger evanescent field inherently due to the thinner cladding and much wider guiding area compared to conventional fiber and has the potential for fusion splicing with standard optical fiber as used in high-power fiber applications (Adikan et al. 2012).

Flat Fiber Fabrication

The fabrication of FFs begins with the fabrication of fiber preform. Nowadays, most preforms are fabricated using vapor deposition methods such as modified chemical vapor deposition (MCVD), outer vapor deposition (OVD), plasma-activated chemical vapor deposition (PAVD), and vapor axial deposition. These methods ensure quality and pureness of the preform during the dopant deposition process. Maintaining the purity of the preform minimizes the fiber attenuation due to absorption and scattering.

For fabrication of FF, a glass tube or hollow preform is used. The next stage of fabrication is to draw the hollow preform using a standard drawing tower. The drawing of FF is similar to other conventional optical fibers where the top end of the preform is clamped to the preform holder or feeder and the tip of the other end is placed inside the furnace. The motorized preform stage will slowly feed the preform down at a specified preform feeding speed into a furnace. The furnace typically uses a graphite heating element due to its thermal and mechanical properties (Payne and Gambling 1976). The condition inside the furnace should be maintained to be free from impurities and oxides built up since graphite has a high oxidation rate at high temperature. This can be done by purging the furnace with argon gas.

Next, the furnace will be heated to the preform's softening temperature, which depends on the preform material and its dopants. When the preform surface temperature exceeds the preform's softening temperature, the neck-down process occurs. The preform's low viscosity from the high temperature combined withbreak the deformation due to the draw tension and the force of gravity gradually reduces the preform's diameter and the preform starts to drop down from the bottom of the furnace. The dropped part will be drawn by the fiber drawing tractor or capstan at a specified fiber drawing speed. The diameter of the fiber can be controlled based on the desired fiber diameter (d_{out}), fiber drawing speed (v_{out}), fiber preform outer diameter (D_{in}), and the preform feeding speed (V_{in}) as shown in the following simplified mass conservation equation:

$$D_{in}^2 V_{in} = d_{out}^2 v_{out} \qquad (1)$$

The neck-down profile is a function of the fiber drawing conditions such as furnace temperature, draw speed, feed speed and vacuum pressure, physical and material properties of the preform, and the type of heating element used in the furnace (Paek and Runk 1978; Choudhury and Jaluria 1998; Cheng and Jaluria 2002; Mawardi and Pitchumani 2010). It also determines the mechanical and optical characteristics of the optical fiber.

For the case of FF, since the preform is in a hollow form, the output fiber is initially in capillary form. Fabrication of FF begins when vacuum pressure is applied at the top of the hollow preform. As soon as vacuum pressure is applied, the preform neck-down region experiences vacuum force, causing the region to collapse or flatten. In this process, the circular capillary shape converts into flat shape, which is called flat fiber.

FF dimensions can be approximated (without considering the temperature effect) from the capillary dimensions. The thickness of the fabricated FF will be equal to the outer diameter of the capillary minus the inner diameter of the capillary. The width of the core is the inner radius of the capillary multiply with π and the width of the cladding will be equal to the width of the core and the thickness of the fiber.

FF can be drawn in two methods, either in single-stage or two-stage drawing. In single-stage drawing, a bulky hollow preform is directly drawn into a desired FF diameter. In this method, the diameter conversion rate from preform diameter to the output capillary diameter (or FF dimension) is very high, for example, drawing a 1 mm diameter capillary directly from a 25 mm diameter tube. In a two-stage method, a bulky size hollow tube preform is first drawn into a 2–3 mm diameter capillary and the same capillary is redrawn into the desired FF diameter, where the conversion ratio from the capillary to the final fiber is much smaller than that of single-stage FF fabrication.

Single-stage FF drawing method allows the fabrication of FF with a variety of dimensions by adjusting the drawing temperature or tension especially if a thick fiber core is required. Figure 1 shows an example of a thick FF with a dimension of 0.173 × 0.341 mm and a core width of 0.189 mm, which was fabricated using single-stage drawing method. The fiber was drawn directly from a hollow Ge-doped preform which has an OD/ID of 23 mm/15.4 mm at a furnace temperature of 2000 °C. However, it may be difficult to control the diameter of FF using single-stage drawing especially if a smaller diameter is required, for example, fabrication of a 125 μm fiber directly from a 25 mm preform.

The two-stage FF drawing method is more suitable for fabricating thinner cladding FF with a thinner core. Figure 2 shows an example of a FF fabricated in two-stage drawing method using the same preform used for the FF in Fig. 1. In this fabrication, the Ge-doped preform with OD/ID of 23 mm/15.4 mm was

Fig. 1 FF with a dimension of 0.173 × 0.341 mm fabricated in single-stage from a hollow Ge-doped preform with OD/ID of 23 mm/15.4 mm

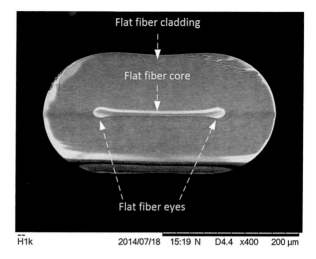

Fig. 2 FF with a dimension of 0.082 × 0.269 mm with a core width of 0.186 mm fabricated in two-stage from a Ge-doped preform with OD/ID of 23 mm/15.4 mm

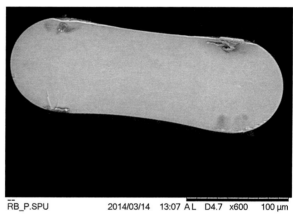

first drawn into a 3.75 mm diameter capillary, and in the second-stage drawing method, the 3.75 mm capillary is drawn into the desired FF with a dimension of 0.082 × 0.269 mm at a furnace temperature of 1960 °C with applied vacuum pressure of 5 kPa.

Flat Fiber Drawing Repeatability

In general, the fabrication of FF is repeatable if all the fabrication parameters and the fiber preform used are similar. Fabrication of a FF with dimension of 0.385 × 0.107 mm is repeated over three times using a pure silica preform with an OD/ID of 25 mm/19 mm. The drawing parameters including furnace temperature, tension, vacuum pressure, feed rate, and draw speed were kept constant in all three experimental trials. The drawing parameters were first specified based on the drawing of a capillary with a 0.3 mm diameter. When the drawing of the capillary at this diameter has stabilized, vacuum pressure is then applied to ensure a consistent FF dimension. In each of the three experimental trials, seven samples were randomly selected from the drawn FFs. From a total of 21 samples, a maximum variation of around 5.5% (or ±2.7%) is observed in the FF dimensions. It should be noted that this variation is observed in the condition that the dimension of the fiber was not on the auto-control mode; instead, the system parameters were set once at the beginning of the fabrication till the end for all the three fabrication trials.

Flat Fibers with Different Dimensions

Five different dimensions of FFs were fabricated using the same Ge-doped preform. Two-stage drawing method was employed, where the preform was first drawn into a capillary with a diameter of around 3.8 mm and then this capillary was redrawn into five different dimensions of FFs. In this fabrication, the drawing tension for all fibers was fixed within 20–30 g and the vacuum pressure was varied from 5 kPa in the smallest dimension up to 10 kPa in the largest dimension. Dimensions of the five FFs were measured as 0.177 × 0.037 mm, 0.256 × 0.052 mm, 0.357 × 0.073 mm, 0.522 × 0.103 mm, and 0.762 × 0.154 mm as shown in Fig. 3a–e, respectively. The average width-to-thickness ratio of the FFs was around 4.87 with a variation of around ±2.56%. This shows consistent dimension ratio in fabricating different dimensions of FF using the same preform. It should be noted that due to insufficient vacuum pressure in these fabrications, the FFs were not fully collapsed along the fiber width and left two small defect holes in the structure.

Characterization of Flat Fibers: Mode Propagation

Multimode Propagation in Flat Fibers

FFs are inherently multimode fibers. Mode propagation of the two FFs shown in Fig. 4 is analyzed in simulation using the software COMSOL Multiphysics. The mode profiles of both FFs obtained via simulation (Fig. 5) suggested multimode guidance mainly due to the large core area, as the modes are guided in both horizontal (n_{slow}) and vertical (n_{fast}) direction in the core.

Tables 1 and 2 show the n_{eff} and power distribution per mode in the core and eye area in the 5 μm and 1.6 μm core thickness FFs, respectively. The results are calculated in power flow, time average (W/m^2).

Referring to Table 1, for the FF core thickness of 5 μm as the power values imply, all modes from the fundamental to higher-order modes (up to 12) have high power

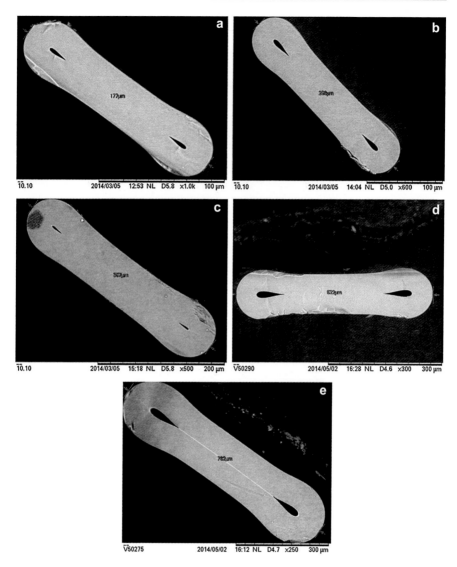

Fig. 3 Five different dimensions of FFs fabricated from a 3.8 mm diameter capillary. Dimension of the fibers are (**a**) 0.177 × 0.037 mm, (**b**) 0.256 × 0.052 mm, (**c**) 0.357 × 0.073 mm, (**d**) 0.522 × 0.103 mm, (**e**) 0.762 × 0.154 mm

in the FF core area. Referring to Table 2, for the FF core thickness of 1.6 μm, there is still a significant amount of power in the flat fiber core area from the fundamental to higher-order modes (up to 12) as well. These simulation results confirm that both FFs with thin or thick core are highly multimode fiber. This also signifies that even by reducing the core thickness, the multimode properties still exist in the slow axis, due to the wide dimensional area.

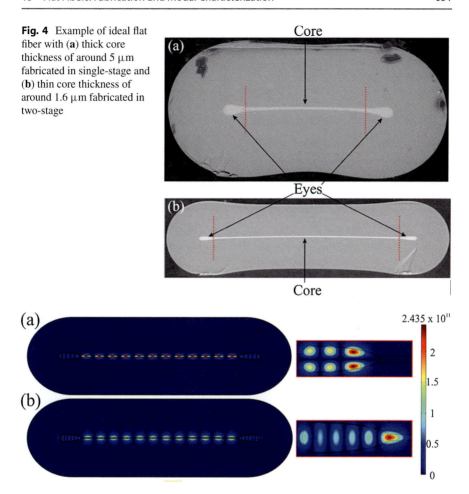

Fig. 4 Example of ideal flat fiber with (**a**) thick core thickness of around 5 μm fabricated in single-stage and (**b**) thin core thickness of around 1.6 μm fabricated in two-stage

Fig. 5 Mode profile of the two FFs in Fig. 4. (**a**) Thick core thickness of around 5 μm and (**b**) thin core thickness of around 1.6 μm. (Inset) One of the possible guiding modes in the eye region

For validation, a simple test is performed on one of the FF used for the simulation above, i.e., the flat fiber with a thick core thickness of 5 μm. Figure 6 shows the mode operation in the FF captured by charged-coupled device at three different launching conditions, i.e., (a) incident light applied into the flat fiber's left eye, (b) light applied into the center core, and (c) light applied into the right eye. It can be observed experimentally that the core and eyes were being guided in all three launching conditions. The fiber shows multimode operation in all launching conditions, as observed in the simulations shown in Fig. 5. By applying light into the FFs' eye, light travels into the central core area as well as into the other eye area and vice versa. Even though light is applied with different launching positions, the mode operation of the FF always supports multimode propagation and does not reduce into single-mode operation.

Table 1 n_{eff} and power distribution per mode in the FF core and eye area for the 5 μm core thickness

	n_{eff}	n_{core} (W/m^2)	n_{eye} (W/m^2)
Fundamental mode	1.445405	335.2701	0.9709
Second-order mode	1.445401	333.5621	3.4509
Third-order mode	1.445395	331.4029	6.5269
Twelfth-order mode	1.445237	322.3463	20.521

Table 2 n_{eff} and power distribution per mode in the FF core and eye area for the 1.6 μm core thickness

	n_{eff}	n_{core} (W/m^2)	n_{eye} (W/m^2)
Fundamental mode	1.444272	80.7868	2.6213
Second-order mode	1.444268	79.6326	7.5825
Third-order mode	1.444262	78.7959	11.5385
Twelfth-order mode	1.444097	77.6759	15.0127

Single-Mode Propagation in Flat Fibers

Single-mode propagation is observed in FF in a condition where a FF with two defect eyes, similar to the FF shown in Fig. 3e, and a high-index material (refractive index matching oil) is filled inside the defect eye hole. Now, by considering a refractive index difference between the matching oil in the defect eye holes compared to the central core, the simulated mode propagation in the central core of the FF would be as presented in Fig. 7. The mode propagation in Fig. 7 suggests that by applying a suitable (higher) refractive index in the eye holes compared to the fiber central core, the fiber tends to show a single-mode operation. The higher-order modes in this condition are effectively leaked into the eyes.

The simulation results in Fig. 7 seems to suggest that the higher-order modes after leaking into the eyes were trapped there and not returning back into the central core area. This phenomenon that exists in this FF has never been observed in the ideal FFs. This phenomenon is further clarified by experimental results as seen in Fig. 8.

Figure 8 depicts three different launching conditions of SMF coupled into the FF with defect eye holes (without filling the index matching oil inside the defect holes) (Poh et al. 2017). In Fig. 8a, c, light was launched into the fiber's left and right eyes, respectively. It was observed that the light propagating from the eye area does not propagate into the FF central core area. This was unlike what was observed in the ideal FF without defect eye holes presented in Fig. 6. Figure 8b shows the launching condition into the FF central core area. In this experiment, the incident light from the SMF is being moved slightly from one end to another end of the central core. In all swiping positions, the light propagating from the central core area never travelled into the eyes unlike as observed in the FF without defect eye holes shown in Fig. 6. However, in all conditions, the resulting mode guidance in the central core was always multimode as shown in Fig. 8f.

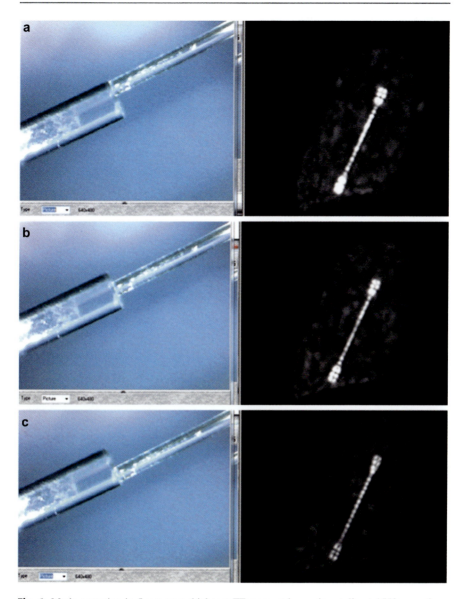

Fig. 6 Mode operation in 5 μm core thickness FF measured experimentally at 1550 nm, where the incident light applied from three different launching positions of the FF core. (**a**) Light applied into the left eye area, (**b**) light applied into the central core area, and (**c**) light applied into the right eye area

Fig. 7 Mode propagation in the proposed flat fiber with two defect holes where the holes are filled with optimum refractive index of 1.44550

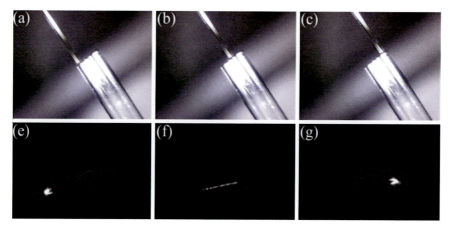

Fig. 8 Various launching conditions for the FF with defect air holes. (**a**) Light applied from left eye hole, (**b**) light applied into the fiber central core area, and (**c**) light applied from the right eye hole. Figures (**e**), (**f**) and (**g**) show the respective launching image captured by charged-coupled device (Poh et al. 2017)

By applying a higher index matching oil into the FF's eye holes, the multimode propagation in the FF reduces to single-mode operation as illustrated in Fig. 9. The dotted red curves in Fig. 9 depicts a multimode behavior measured in the FF before oil infiltration in the eye holes. In this condition, light is strongly confined in both the fast and slow axis of the FF. As soon as the FF eye holes were filled with refractive index matching oil, the higher-order modes were successfully reduced as shown by the solid green lines. It could be possible that the higher-order modes are attenuated through leaky mode mechanism and reduced to single-mode operation.

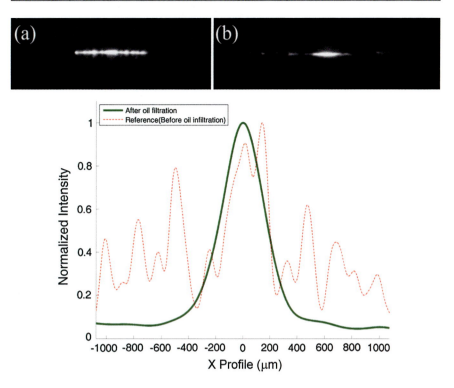

Fig. 9 Mode profiles before and after oil infiltration into the FF air holes. Insets show charged-couple device images captured (**a**) before and (**b**) after oil infiltration (Poh et al. 2017)

Conclusion

Fabrication of flat fiber is shown to be practically feasible using the conventional fiber drawing tower. Fabrication of this fiber is shown to be repeatable with a variation of around ±2.7% by employing the same drawing parameters and the preform material. FFs with different dimensions but almost similar width-to-thickness ratio (±.56% variation) can be fabricated by keeping the drawing tension constant. FFs with variety of dimensions can be fabricated when the drawing conversion from preform diameter to the fiber dimension is large and by varying the drawing tension or furnace temperature. FFs have inherently multimode propagation due to large core area or core width. However, single-mode FF is shown to be possible by applying a suitable refractive index material in the FF eye holes. This material would be either liquid form for short fiber length or would be a glass rod (with slightly higher refractive index than the FF central core) placed in the fiber during the FF fabrication. FFs would be a good alternative to planar waveguide with the advantages of flexibility, long fabrication feasibility, lower loss and ease of cleaving.

References

F.R.M. Adikan, *Direct UV-Written Waveguide Devices* (University of Southampton, Southampton, 2007)

F.R.M. Adikan, S.R. Sandoghchi, Y. Chong Wu, R.E. Simpson, M.A. Mahdi, A.S. Webb, J.C. Gates, C. Holmes, Direct UV written optical waveguides in flexible glass flat fiber chips. Sel. Top. Quant. Electron., IEEE J. **18**, 1534–1539 (2012)

Y. Bian, Q. Gong, Multilayer metal–dielectric planar waveguides for subwavelength guiding of long-range hybrid plasmon polaritons at 1550 nm. J. Opt. **16**, 015001 (2013)

J.A. Buck, *Fundamentals of Optical Fibers*, (Wiley, 1995)

X. Cheng, Y. Jaluria, Effect of draw furnace geometry on high-speed optical fiber manufacturing. Numer. Heat Transf. A: Appl. **41**, 757–781 (2002)

S.R. Choudhury, Y. Jaluria, Practical aspects in the drawing of an optical fiber. J. Mater. Res. **13**, 483–493 (1998)

H. Duncan, Fabrication of substrates for planar waveguide devices and planarwaveguide devices, Google Patents, (2002)

H.J. Dutton, *Understanding Optical Communications* (Prentice Hall PTR, New Jersey, 1998)

S. Ferretti, V. Savona, D. Gerace, Optimal antibunching in passive photonic devices based on coupled nonlinear resonators. New J. Phys. **15**, 025012 (2013)

C. Holmes, F.R.M. Adikan, A.S. Webb, J.C. Gates, C.B. Gawith, J.K. Sahu, P.G. Smith, D.N. Payne, *Evanescent Field Sensing in Novel Flat Fiber* (Optical Society of America, 2008), pp. CMJJ3

R.G. Hunsperger, *Integrated Optics Theory and Technology* (Springer, Berlin, 2009)

K. Kalli, C. Riziotis, A. Posporis, C. Markos, C. Koutsides, S. Ambran, A.S. Webb, C. Holmes, J.C. Gates, J.K. Sahu, P.G.R. Smith, Flat fibre and femtosecond laser technology as a novel photonic integration platform for optofluidic based biosensing devices and lab-on-chip applications: Current results and future perspectives. Sensors Actuators B Chem. **209**, 1030–1040 (2015)

S.O. Kasap, *Optoelectronics and Photonics: Principles and Practices* (Pearson, Boston, 2001)

A.V. Kildishev, A. Boltasseva, V.M. Shalaev, Planar photonics with metasurfaces. Science **339**, 1232009 (2013)

Y.P. Li, C.H. Henry, Silica-based optical integrated circuits. IEE Proc. Optoelectron. **143**, 263–280 (1996)

A. Mawardi, R. Pitchumani, Optical fiber drawing process model using an analytical neck-down profile. IEEE Photonics J **2**, 620–629 (2010)

U. Paek, R. Runk, Physical behavior of the neck-down region during furnace drawing of silica fibers. J. Appl. Phys. **49**, 4417–4422 (1978)

D.N. Payne, W.A. Gambling, A resistance-heated high temperature furnace for drawing silica-based fibres for optical communications. Am. Ceram. Soc. Bull. **55**, 195–197 (1976)

S.Y. Poh, G.A. Mahdiraji, Y.M. Sua, F. Amirkhan, D.C. Tee, K.S. Yeo, F.R.M. Adikan, Single-mode operation in flat fibers slab waveguide via modal leakage. IEEE Photonics J **9**, 1–9 (2017)

C. Riziotis, K. Kalli, C. Markos, A. Posporis, C. Koutsides, A.S. Webb, C. Holmes, J.C. Gates, J.K. Sahu, P.G.R. Smith, Flexible glass flat-fibre chips and femtosecond laser inscription as enabling technologies for photonic devices. At Photonics West (2014), United States, pp. 89820G–89820G-89828

B.E. Saleh, M.C. Teich, Semiconductor photon detectors. Fundam. Photonics, 644–695 (2007)

D.R. Scifres, C.A. San Jose, Multiple core fiber laser and optical amplifier, Google Patents, United States Patent 5566196, Application Number: 08/330262 (1996)

J. Singh, *Optoelectronics: An Introduction to Materials and Devices* (McGraw-Hill College, New York, 1996)

A. Webb, F.R.M. Adikan, J. Sahu, R. Standish, C. Gawith, J. Gates, P.G. Smith, D. Payne, MCVD planar substrates for UV-written waveguide devices. Electron. Lett. **43**, 517–519 (2007)

X. Zhang, A. Hosseini, X. Lin, H. Subbaraman, R.T. Chen, Polymer-based hybrid-integrated photonic devices for silicon on-chip modulation and board-level optical interconnects. Sel. Top. Quant. Electron., IEEE J. **19**, 196–210 (2013)

3D Silica Lithography for Future Optical Fiber Fabrication

16

Gang-Ding Peng, Yanhua Luo, Jianzhong Zhang, Jianxiang Wen, Yushi Chu, Kevin Cook, and John Canning

Contents

Introduction	638
Conventional Silica fiber Fabrication	640
3D Silica fiber Fabrication	641
3D Fabrication (3D Printing)	641
3D silica lithography	642
3D Silica fiber Fabrication	643
Challenges and Pathways to 3D Silica fiber Fabrication	644
Challenges for 3D Silica fibers	644
Pathways for 3D Silica fibers	645
Initial Results	649
Conclusion	650
References	650

G.-D. Peng (✉)
Photonics and Optical Communications, School of Electrical Engineering and Telecommunications, University of New South Wales, Sydney, NSW, Australia
e-mail: g.peng@unsw.edu.au

Y. Luo
Photonics and Optical Communications, School of Electrical Engineering and Telecommunications, University of New South Wales, Sydney, NSW, Australia

Key Laboratory of Optoelectronic Devices and Systems of Ministry of Education and Guangdong Province, Shenzhen University, Shenzhen, China
e-mail: yanhua.luo1@unsw.edu.au

J. Zhang
Key Lab of In-fiber Integrated Optics, Ministry of Education, Harbin Engineering University, Harbin, China
e-mail: zhangjianzhong@hrbeu.edu.cn

J. Wen
Key Laboratory of Specialty fiber Optics and Optical Access Networks, Shanghai University, Shanghai, China
e-mail: wenjx@shu.edu.cn

© Springer Nature Singapore Pte Ltd. 2019
G.-D. Peng (ed.), *Handbook of Optical Fibers*,
https://doi.org/10.1007/978-981-10-7087-7_79

Abstract

Conventional silica fiber fabrication using chemical vapor deposition (CVD) has been a successful optical fiber fabrication platform, producing the bulk of commercial optical fibers and fiber amplifiers that form the backbone of today's Internet. As the Internet evolves into a so-called ubiquitous "Internet of things," or IoT, the role of optical fibers is expanding from a mainly passive telecommunications transmission medium to host for fiber sensing, fiber devices and lasers, and beyond. This is creating a demand for increasingly sophisticated optical fibers. Unfortunately, conventional silica fiber fabrication has limited capability in both material and structure flexibility for diverse and custom-designed functionalities. This chapter discusses the research and development of 3D silica lithography using digital light processing and related technologies for future fabrication of specialty silica optical fibers with greater freedom in both structure and material formation. Showing enormous prospects and identifying challenges and pathways based on recent reported progress in 3D printing glass, the new era of additive manufacturing, or 3D printing, technologies offer great potential to produce key disruptive silica fiber technology for diverse and sophisticated needs in future fiber communication and sensing systems. The first printed optical preforms are reported here.

Keywords

Optical Fiber fabrication · Additive fiber manufacturing · 3D fiber printing · 3D silica lithography · Specialty optical fiber · Silica optical fiber · Structured optical fibers · Doped optical fibers

Introduction

Silica is an ideal platform material for many important photonic applications – optical fibers, waveguides, fiber amplifiers, and lasers. It has superior optical, mechanical, chemical, and thermal properties over all other fiber media. For optical

---

Y. Chu
Key Laboratory of In-Fiber Integrated Optics, Ministry Education of China, Harbin Engineering University, Harbin, China

Photonics and Optical Communications, School of Electrical Engineering and Telecommunications, UNSW, Sydney, NSW, Australia

interdisciplinary Photonics Laboratories (iPL), Global Big Data Technologies Centre (GBDTC), Tech Lab, School of Electrical and Data Engineering, University of Technology Sydney, Sydney, NSW, Australia
e-mail: yushi.chu@student.unsw.edu.au

K. Cook · J. Canning
interdisciplinary Photonics Laboratories (iPL), Global Big Data Technologies Centre (GBDTC), Tech Lab, School of Electrical and Data Engineering, University of Technology Sydney, Sydney, NSW, Australia
e-mail: kevin.cook@uts.edu.au; john.canning@uts.edu.au

fiber telecommunication applications, silica fibers are fabricated using traditional chemical vapor deposition (CVD) or rod and tube stacking. Conventional silica fiber fabrication based on CVD dominates the current mass production of commercial optical fibers and is primarily responsible for the huge success in telecommunication, fiber amplifier, and laser applications.

Based upon the success in fast and cost-effective optical fiber-based global telecommunication infrastructures that ensure the world being increasingly connected through the Internet, more and more functionalities and capabilities, such as control, sensing, monitoring, and diagnosing, are being integrated into the worldwide Internet, in the form of "Internet of things" (IoT). The IoT cover applications across many fields such as advanced industrial manufacturing, structural health monitoring (SHM), electricity system management and diagnosis, oil and gas exploration and storage, self-driving cars, transportation and tracking, artificial intelligence, big data, and so on. This is creating new opportunities that demands innovative research and technological advances from optical fiber and photonics. As a result, with regard to current demand of huge quantities of mainly telecommunication fibers, the future demand for optical fibers will become increasingly more diverse, focusing on applications and the sophistication demanded in their functionalities, while much less on quantity.

In recent years many researchers have focused on the R&D of specialty optical fibers with custom structure and material designs for acquiring superior and/or new functionalities. Many novel specialty optical fibers were reported (Bai et al. 2012; Luo et al. 2012, 2018; Glavind et al. 2014; Sathi et al. 2015; Gong et al. 2018; Anuszkiewicz et al. 2018; Gao et al. 2018; Firstov et al. 2018; Chen et al. 2019; Han et al. 2019a). All these works employ conventional preform fabrication techniques, by either CVD or rod and tube stacking, and followed by ultrahigh temperature drawing to make doped fibers and structured fibers. Many of these specialty optical fibers, such as aperiodic Fresnel fibers (Canning et al. 2003; Martelli and Canning 2007), doped PCFs (Wang et al. 2015; Sánchez-Martín et al. 2008), spun Hi-Bi PCFs (Michie et al. 2007), and doped spun Hi-Bi PCFs (Luo et al. 2018), have custom design structures and pose a great challenge for simple stacking. The material composition is also seriously challenged by having any off-center regions and cores and also in simply getting stuff into thin capillaries. Although these structured optical fibers are potentially superior to conventional fibers, they are highly constrained by the overwhelming manual labor involved in the preform fabrication phase. Consequently, the greatest challenge in optical fiber development is achieving desired, optimal structures and material compositions for functionalities that are not possible using currently existing means.

In summary, current fiber fabrication technology, based on CVD and/or rod and tube stacking, finds it very difficult to achieve both sophisticated structures and advanced material compositions simultaneously. This limit would hinder further specialty fiber development with superior and new functionalities. The recent demonstration of an actively doped spun photonic crystal fiber is a typical example where both material and structure need to be properly custom designed (Luo et al. 2018). Another example is a multiple-core doped fiber targeted for synthetic beam high-power fiber lasers (Zhang et al. 2011). Such multiple-core doped fibers require the combination of structure (multiple cores) and materials (erbium doping) in fiber fabrication simultaneously for new, specialty functionalities.

Additive manufacturing (3D printing) technologies (Bikas et al. 2016) allow new possibilities in fiber fabrication. The first reports in this field on 3D printing plastic fibers (Cook et al. 2015, 2016; Canning et al. 2016; Flanagan et al. 2018) immediately opened up access to novel, sophisticated fiber designs for local area networks. This work was transformative turning 3D printing into something concrete and tangible, the beginnings of the next major disruption in optical communications waveguide technologies. The recent development of 3D silica printing lithography (Kotz et al. 2016, 2017, 2018; Nguyen et al. 2017; Destino et al. 2018) for glass applications shows enormous potential to extend this work to fabricating transparent silica with greater freedom in both structure and material formation. However, although current demonstrations using polymer can be easily applied to lower-temperature soft glasses such as borosilicate or chalcogenides, silica is a very high-temperature material posing significant challenges compared to polymer. High purity silica is also extremely difficult to shape and dope, something presently inaccessible to 3D printing. As of the time of writing, there are no reports of 3D printed silica optical fiber preforms or fibers although there are recent works toward 3D printing optical glasses (Cooperstein et al. 2018). Nevertheless, we believe that this is feasible today and explore how the ground work for a future of 3D printing silica fibers can be done with an existing technology.

This chapter describes general aspects of 3D silica lithography for specialty optical fiber fabrication. In section "Conventional Silica fiber Fabrication," conventional silica fiber fabrication is briefly overviewed. In section "3D Silica fiber Fabrication," recent developments of 3D printing-related technologies, especially 3D printing silica lithography and 3D printing fiber fabrication, have been summarized. Finally, the challenges and pathways are identified and discussed from the perspective of 3D silica lithography specifically for specialty silica optical fiber fabrication. Examples of the first preform printing of glass are presented.

Conventional Silica fiber Fabrication

Silica glass: Silica glass is one of the most important materials in photonics and fibers due to its unmatched optical transparency; outstanding mechanical, chemical, and thermal resistance; as well as its thermal and electrical insulating properties (Ikushima et al. 2000). No other materials permit the transfer of light over thousands of kilometers with minimal loss using silica fiber amplifiers. While plastic optical fibers are ideal for local area networks and short distance transmission, silica remains the cornerstone of the Internet backbone and makes up many modern photonic materials and applications, including those that support the Internet: long-haul optical fibers and waveguides, fiber amplifiers, and lasers.

CVD: Currently, high-quality fused silica with special material compositions is made synthetically most commonly through a CVD process. Other processes do exist, but so far vapor-based CVD is the only viable ultrapure technology for high-quality silica optical fibers with ultralow loss for telecommunication fibers.

Doped fibers are manufactured by vapor-based CVD technology such as modified CVD (MCVD), outside vapor deposition (OVD), vapor axial deposition (VAD), plasma CVD (PCVD), and atomic layer deposition (ALD) (Wen et al. 2015). These CVD-based fabrication techniques can make fibers with certain material compositions for special material-related properties, but they cannot make fibers with sophisticated structures for special structure-related properties. They are also restricted during preform lathe fabrication to centered cores.

Solution doping: MCVD combined with solution doping is widely used for doped silica fiber fabrication (e.g., (Muhamad Yassin et al. 2019; Unger et al. 2019)). Solution doping process consists of deposition of porous (soot) core layer of appropriate thickness and composition over the cladding layer in a preform tube, followed by soaking the tube with an aqueous/alcoholic solution of dopants for a certain period. Then the tube is dried and sintered to consolidate the soot core layer to clear glass. The tube is finally collapsed into a fiber preform for fiber drawing. The exact dopant composition, concentration, and distribution achieved in fiber core are highly dependent on soot composition and porosity, solution preparation and soaking, and preform sintering and collapsing and difficult to be precisely controlled. Through the high-temperature sintering and collapsing processes, dopants experience significant diffusion and evaporation that result in dopant concentration and index distribution in the radial direction (Unger et al. 2019). In addition high-dopant concentration is often associated with problem of clustering.

Rod and tube stacking: Silica fibers with sophisticated structures such as structured and photonic crystal fibers are normally made with rod and tube stacking (Knight et al. 1996; Russell 2003). This manual stacking technique is limited to physical capillary placement that leads to targeted periodic structures with defects such as undesirable holes, interstitials, and surfaces that demand additional processing to remove them.

In summary, further fiber developments and applications are hindered by these current fabrication techniques that are difficult to achieve simultaneously. These include sophisticated structures and/or special material compositions for introducing novel and superior functionalities and features.

3D Silica fiber Fabrication

3D Fabrication (3D Printing)

3D fabrication or 3D printing is an additive manufacturing technology in which material is most commonly fused and solidified under computer control to create a 3D object. Additional materials may be added (such as liquid molecules or powder grains (Bikas et al. 2016)). Objects can be of almost any shape or geometry and typically are produced in sequential layers using digital model data from a 3D model or another electronic data source such as an additive manufacturing file (**AMF**). There are many different technical variations, including stereo-lithography (**3D lithography**) or fused deposition modeling (**FDM**) – FDM can be considered a sophisticated hot glue gun on a computer controlled xyz stage. Unlike material removed from a stock in the conventional machining process, 3D printing or additive manufacturing builds a 3D object from a computer-aided design (**CAD**) or AMF, by successively adding material layer by layer.

3D printing technology offers great opportunities for developing new fibers such as novel structured optical fibers. 3D printing has many advantages over traditional fabrication including greater structural design freedom, faster prototyping, lower processing costs for small production volumes, and more efficient material usage. 3D printing is now used in both rapid prototyping and additive manufacturing. It is revolutionizing many areas from manufacturing, science, engineering, and medicine to art. 3D printing examples include biomimetic scaffolds (Bittner et al. 2018), catalyst architectures (Zhu et al. 2018), reactionware (Symes et al. 2012), piezoelectric polymer microsystems (Han et al. 2019b), ultracompact multi-lens objectives (Gissibl et al. 2016), and microlenses (Thiele et al. 2017).

3D silica lithography

Recent development of a simple silica material technology based upon combining nanoparticle solutions and 3D light projection using Australian technology by Kotz et al. of Karlsruhe Technology Institute (KIT), in Nature in 2017 (Kotz et al. 2017) and a few related works (Kotz et al. 2016, 2018; Nguyen et al. 2017; Destino et al. 2018), forms the basis of a simple solution to 3D silica lithography for silica fiber fabrication. This particular 3D lithography relies on "LiqGlass," proposed and demonstrated recently by Kotz et al. as shown in Fig. 1 (Kotz et al. 2016). LiqGlass is simply a mixture of silica nanoparticles within a traditional monomer solution. The printing technology, through a chain of well-known material processing including material formulation, initiation, photopolymerization, polymer debinding, silica sintering, and silica vitrification, is readily converted into a 3D printed new silica material technology (Kotz et al. 2017). It exploits polymer-based 3D printing to fabricate parts and then uses annealing and subsequent sintering to burn off the polymer material and fuse the silica content. The work reported thus far has generated reasonably high transparency silica.

Besides KIT in Germany, the Lawrence Livermore National Laboratory (LLNL) has also developed similar 3D silica materials, referred as "silica inks" where silica is suspended in aqueous organic solution (Nguyen et al. 2017) and silica sol-gel (Destino et al. 2018). The Hebrew University of Jerusalem has also reported a traditional sol-gel method that produced silica ink for 3D printing lithography (Cooperstein et al. 2018). All these methods have in common the goal of enabling low-temperature additive manufacture of silica glasses through polymer writing, followed by appropriate annealing and sintering. They are all similar to the sol-gel methods previously employed for silica fiber fabrication and therefore are expected to suffer from similar limitations including the ultimate low loss possible, compared to standard CVD methods. However, they avoid costly equipment that would otherwise be required to undertake direct glass 3D printing. A major challenge with these approaches is that results so far have been relatively small volume, although the brute-force approach has allowed larger samples to be produced. Here, we focus on the higher-resolution methods of silica in polymer, followed by sintering. Other reported approaches involve brute-force application of ultrahigh temperature molten silica using FDM writing where viscosity through a nozzle restricts the resolution to the mm domain (Inamura et al. 2018) – nevertheless, this method has the advantage of immediate scalability and ultimately may be the preferred route for some fiber designs. Other methods combining the best of both worlds will likely come along – it is our goal here to report the first 3D printed glass preforms, introducing and launching a potentially disruptive technology approach to optical fiber fabrication more broadly, regardless of the final technology used.

3D Silica fiber Fabrication

3D printing could revolutionize optical fiber fabrication by realizing desirable material doping and structure design of preforms, and even optical fibers drawn directly from the printer, that have not been possible to date. Canning and colleagues first proposed and reported proof of principle demonstrations of additive manufacturing of optical fiber preforms using FDM: 3D printed preforms of modified acrylonitrile butadiene styrene (**ABS**) were made and subsequently drawn into an air-structured optical fiber in 2015 (Cook et al. 2015), coreless optical fibers made from ABS, and then polyethylene terephthalate glycol (**PETG**) using low-cost 3D printer (Cook et al. 2016) and step index polymer optical fibers (**POFs**) with ABS for the core and modified PETG for the cladding in 2016 (Canning et al. 2016). These demonstrated that it was not only possible to fabricate preforms but also to draw high-quality fiber directly from low-cost micro heaters in a 3D printer. This already offers a major disruption in the production of plastic optical fibers and signals what is eventually possible across *the entire optical fiber and waveguide fabrication ecosystem*, potentially disrupting the existing industrial base.

This 3D fiber fabrication triggered worldwide research: examples of various designs using additive fabrication include hollow-core POF drawn from 3D printed poly(methyl methacrylate) PMMA preform by Zubel et al. of DTU

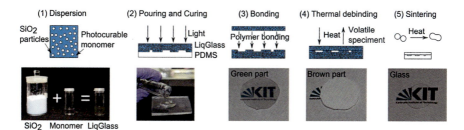

Fig. 1 Schematic work flow of the silica-based LiqGlass fabrication process by Kotz et al. (Kotz et al. 2016). This process includes five steps: (1) dispersion of amorphous silica nanopowder into the photocurable monomer mixture; (2) pouring of LiqGlass against a PDMS template and curing with UV light; (3) if multilayer glass parts are required, several cured LiqGlass sheets can be bonded using partial curing; (4) removal of organic binder by thermal debinding; (5) sintering of the powder compound to a highly transparent and dense glass body

(Zubel et al. 2016); 3D printed THz polymer waveguides (Ma et al. 2016); 3D printed microstructured ABS POF (Marques et al. 2017) and mid-IR hollow-core microstructured PETG POF (Talataisong et al. 2018a); a suspended-core polymer optical fiber that has been extruded and directly drawn from a microstructured 3D printer nozzle by using ABS (Talataisong et al. 2018b); and 3D printed microstructured POFs of ABS and polycarbonate (PC) with custom-designed cross sections (Toal et al. 2017). Combining the rod-in-tube techniques, optical fibers with special shape extruded PC cores and ABS cladding by 3D printing have been drawn, which have lower loss compared with the previous fibers made (Zhao et al. 2015). Furthermore, functional optical or magnetic dopants have been introduced into cyclic olefin polymer (COP) to fabricate functional and structured optical fiber (Kaufman et al. 2016).

So far all 3D printed optical fibers have very high loss due to high intrinsic material absorption and fabrication defects such as bubbles. However, once drawn under appropriate conditions, air is removed, and the extruded fiber was shown to have identical material loss to the best polymer optical fibers (Cook et al. 2016). Although the parameters were far from optimized, the results were tremendously exciting because a key objective of the original work was to show that this can be done at very low cost: the rapid evolution of 3D printing is ongoing, and so what are currently exorbitantly priced 3D printing systems today will be low-cost desktop systems tomorrow. With that will be a commensurate drop in attenuation and improvement in material consistency – the potential for reducing labor costs is enormous. Nonetheless, while it will be eventually suitable for LAN applications and other short distance sensor and communications, the intrinsic polymer material absorption would be prohibitive for longer-distance communication and device applications. As discussed above, silica is an intrinsically superior platform material than polymers in terms of optical transparency. Hence, it was proposed that 3D silica lithography should and could be developed to fabricate silica optical fibers (Peng 2018).

Challenges and Pathways to 3D Silica fiber Fabrication

Here, the challenges, opportunities, and pathways are identified and discussed in the development of 3D silica lithography appropriate for specialty silica optical fiber fabrication that achieve completely tailorable free-form custom design structure and material fabrication.

Challenges for 3D Silica fibers

Though recent progress in the development of 3D printing transparent silica glass is very promising (Kotz et al. 2016, 2017, 2018; Nguyen et al. 2017; Destino et al. 2018; Cooperstein et al. 2018), there remain many scientific and technological challenges for this to be suitable for fiber fabrication. An initial experimental assessment of the 3D silica material process reported in these papers by the authors has identified several key challenges toward producing either the required structure or the required material composition for specialty optical fiber fabrication, including:

- In-depth understanding and knowledge on the physical and chemical processes and relevant principles and mechanisms involved in multicomponent material systems is not there.
- Process information and data for determining reactions, interactions, and relations in silica lithography that enable accurate theoretical and predictive understanding and modeling are lacking.
- There has been no systematic study on how structures conserve and change with different material compositions and recipes and under different process conditions.
- Introducing functional guests (e.g., Er, Yb, Bi) into silica by silica lithography is not explored or resolved.
- How will optical fibers from 3D silica lithography compare with fibers from the existing CVD techniques?

In summary, at least two overarching challenges remain:

- Gaining fundamental insight and understanding of the chemical and physical interactions within mixed materials: nanosilica and nanoparticles, monomers, additive agents such as thermal and UV initiators, and chain transfer agents. All these are involved in chemical, optical, thermal, and sintering processes.
- Systematic assessment and exploration of the appropriate material compositions, process conditions, and parameters to realize 3D silica lithography fiber fabrication and to apply these in multiple applications.

By working through the systematic R&D stage, essential knowledge and technology will be gained on material formulation, preparation, and subsequent optical and thermal processing.

Pathways for 3D Silica fibers

Although there remain unanswered questions on quality of material composition and structure formation through organic sintering, producing optical fiber preforms by 3D silica lithography is feasible. Taking into considerations of the existing fiber technologies and the silica-based 3D printing ink technologies, all the key elements for developing 3D silica fiber fabrication are available:

Silica nanomaterials	MCVD (modified CVD) nanoparticles/soot technology
Material doping technology	Solution doping technique
Silica lithography technology	LiqGlass, silica inks
3D silica preform fabrication	3D lithography
fiber fabrication	Draw tower

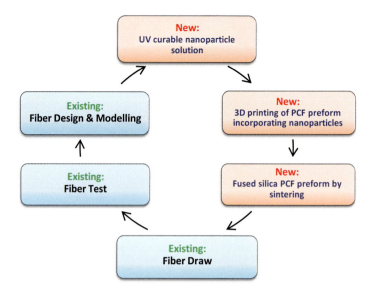

Fig. 2 Pathway to optical fiber fabrication: 3D silica lithography-based specialty fiber platform technology for great freedom in structure and material formation

3D silica fiber fabrication can follow a new and practical fiber fabrication path, depicted in Fig. 2, that integrates both structured fiber technology and doped fiber technology to realize 3D printing specialty silica fibers that require custom design structure and material compositions. Using a generic case of doped PCF fabrication for illustrative purposes, this pathway follows new and important steps:

1. Material preparation and processing: UV curable nanoparticle solution
2. 3D printing of structured and doped silica preform
3. Sintering and consolidation of silica preform

One realization of the possible and practical pathways to future 3D silica fiber fabrication, following the glass printing work initiated in Germany, is shown in Fig. 3. In this way, by replacing traditional CVD technologies such as modified CVD (MCVD) with 3D lithography in optical preform fabrication, greater freedom in structure and material design for specialty optical fiber fabrication will be possible.

As shown in Fig. 3, several key steps are to be taken to realize 3D silica fiber fabrication:

1. Material and processing formulation/development
    - Material selection – monomers, dopants, nanoparticles, etc.
    - Material processing – mixing, dispersing, initiation/sensitization
    - Material testing, characterization, and assessment

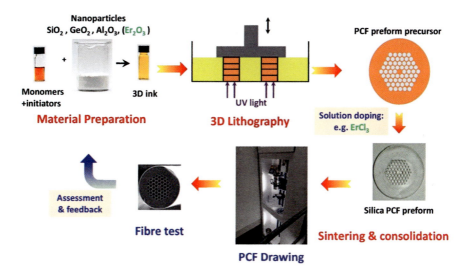

Fig. 3 Work flow of 3D silica lithography optical fiber fabrication. The process can be carried out in two different pathways: one for structured optical fibers and the other for both structured and doped optical fibers. Doping is achieved by mixing dopants with nanoparticles during material preparation or by a solution doping procedure before sintering

2. 3D silica lithography development
   - Photopolymerization process design – parameters, conditions, models, and rules
   - Photopolymerization test and evaluation
3. Preform fabrication and processing technology
   - Sintering and consolidation process design – parameters, conditions, models, and rules
     - "Carrier," the supporting polymers, removal
     - "Host," silica nanoparticle or silica matrix, sintering, and vitrification
     - "Guest," active elements or dopants, dispersion
     - Defect, water reduction techniques
   - Preform testing, characterization, and evaluation
4. fiber drawing, characterization, and evaluation

The process can be classified into two pathways, as illustrated in Fig. 3:

Path 1. Lithographic fabrication for structured optical fibers, sometimes called photonic crystal fibers, when the lattice is periodic, without active dopants, through processes involving material preparation, photopolymerization, thermal carrier removal, host sintering, and consolidation. In this pathway any doping of active ions or elements is ignored and the process follows the arrows. 3D silica lithography process can follow the carrier-host approach as reported in the development of the LiqGlass fabrication (Peng 2018). The material preparation is done by appropriately selecting and formulating monomers, initiators, chain transfer agents, and silica nanoparticles into a UV curable solution or "3D ink." The solution will be cured or solidified into a fiber polymer preform – the "carrier" by UV-induced photopolymerization in a stereo-lithography 3D printer. Here the "carrier" refers to the organic components (e.g., monomers and polymers) throughout the process, while the "host" refers to the inorganic components (e.g., nanosilica particles and silica glass). Then a silica PCF preform is produced upon thermal debinding of the "carrier" and the sintering and consolidation of silica nanoparticles – the "host" – into silica matrix and made ready for drawing into optical fibers. Any specialty need regarding structure and refractive index in both preform and fiber fabrications is to be taken into consideration.

Path 2. Lithographic fabrication for structured and doped optical fibers through similar processes to Path 1, with the addition of solution doping (1) in the initial material mixing process or (2) before the sintering process – thus replacing a passive core with an active core doped with laser-active elements: bismuth and rare earths such as Er, Yb, Tm, and so on. In fact, doping of active elements in the two ways is quite different: (1) doping is achieved by mixing with nanoparticles during material preparation, and (2) doping is through a solution doping procedure before sintering. In (1), oxides or chlorides of active elements, such as Er_2O_3 or $ErCl_3$, are mixed with host nanoparticles. In (2), however, only

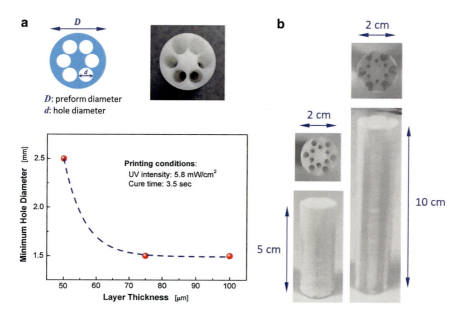

Fig. 4 The first 3D printing of silica optical fiber preforms. These are produced by using mixed monomer and silica nanoparticles and using 3D lithography printing to create the preform. (**a**) Silica optical fiber preform design and fabrication conditions. (**b**) Printed silica optical fiber preforms

non-oxide-soluble active elements, such as $ErCl_3$ and $YbCl_3$, through solution doping, can be mixed with the host matrix when it is in the form of "soot" – a partially consolidated silica.

Initial Results

We used the Australian commercial 3D lithography printer to test the KIT approach for silica glass fabrication. Figure 4a shows the minimum hole size can be achieved under the experimental fabrication conditions – it is essentially repeating the KIT approach but scaling to larger dimensions. Figure 4b shows the first results for the

Original Preform		Processed Preform	
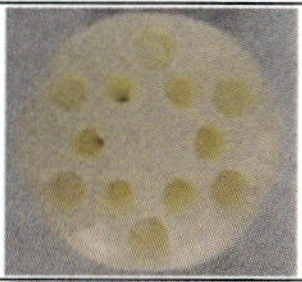	D= 20 mm L= 50 mm	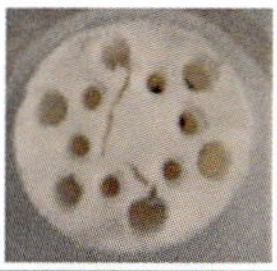	D= 19 mm L= 47 mm
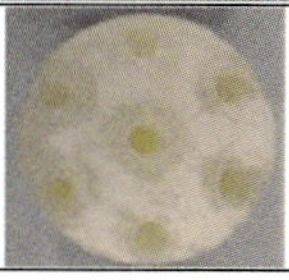	D= 20 mm L= 50 mm	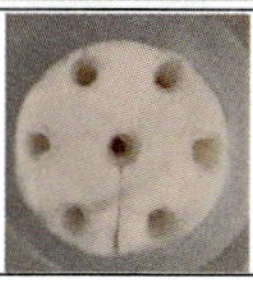	D= 18 mm L= 47 mm

(a) 3D printed silica preforms before and after sintering at 600 °C.

Original Preform		Processed Preform	
	D= 27 mm L= 20 mm		D= 18 mm L= 13 mm
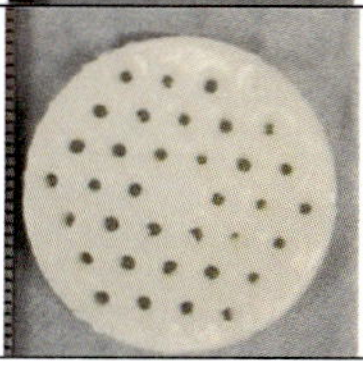	D= 27 mm L= 6 mm		D= 18 mm L= 4 mm

(a) 3D printed silica preforms before and after sintering at 1300 °C.

Fig. 5 Comparison of the 3D printed silica optical fiber preforms before and after different sintering processes. The preforms experience significant color change and shrinkage as the polymer is burnt out through sintering. Notably, the preform structures remain intact. (**a**) 3D printed silica preforms before and after sintering at 600 °C. (**b**) 3D printed silica preforms before and after sintering at 1300 °C

manufacture of structured silica glass preforms. The hybrid polymer/silica preform photopolymerized from a mixture of nanoparticles and monomer is sintered at high temperature to burn out the polymer material and allow glass consolidation. The white coloring reflects this sintering process, arising from scattering of air bubbles – further higher-temperature processing fuses the preform. Details of this work will be reported soon elsewhere.

This fabrication process is followed by sintering to remove the organic component leading to shrinkage so care must be taken to avoid cracking of the preforms. Figure 5 summarizes the preliminary results of the 3D printed silica optical fiber preforms after different sintering processes. The preforms experience significant color and size changes through the sintering processes. It is very encouraging to see that the original preform structures have been kept reasonably. The white coloration of the remaining silica reflects porous structure scattering light suggesting further sintering may be needed – alternatively, this can be achieved during the fiber drawing process (to be reported elsewhere). This indicates that further work to improve and optimize sintering is required. A multistep sintering process has been developed to better control the processes involved.

Conclusion

Recent development of 3D silica printing creates great opportunity for optical fiber fabrication. The key advantage of new 3D silica lithography, or 3D printing, is its capability in fabrication optical fibers with both custom design structure and material for superior photonic functionalities. Though there remain several significant challenges, there exist practical and feasible pathways to this. Demonstrating this feasibility are the first silica-based optical preforms produced with commercially available additive manufacturing simply by adding nanoparticles to a monomer solution prior to photopolymerization. High-temperature sintering removes the polymer, leaving silica behind. Combined with recent results in polymer and other softer materials, the start of the new revolution in silica fiber fabrication is now well underway.

References

A. Anuszkiewicz, R. Kasztelanic, A. Filipkowski, G. Stepniewski, T. Stefaniuk, B. Siwicki, D. Pysz, M. Klimczak, R. Buczynski, Fused silica optical fibers with graded index nanostructured core. Sci. Rep. **8**, 12329 (2018)

N. Bai, E. Ip, Y.-K. Huang, E. Mateo, F. Yaman, M.-J. Li, S. Bickham, S. Ten, J. Liñares, C. Montero, V. Moreno, X. Prieto, V. Tse, K. Man Chung, A.P.T. Lau, H.-Y. Tam, C. Lu, Y. Luo, G.D. Peng, G. Li, T. Wang, Mode-division multiplexed transmission with inline few-mode fiber amplifier. Opt. Express **20**, 2668–2680 (2012)

H. Bikas, P. Stavropoulos, G. Chryssolouris, Additive manufacturing methods and modelling approaches: a critical review. Int. J. Adv. Manuf. Technol. **83**(1–4), 389–405 (2016)

S.M. Bittner, J.L. Guo, A. Melchiorri, A.G. Mikos, Three-dimensional printing of multilayered tissue engineering scaffolds. Mater. Today **21**, 861–874 (2018)

J. Canning, E. Buckley, K. Lyytikainen, All-fiber phase-aperture zone plate fresnel lenses. Electron. Lett. **39**(3), 311–312 (2003)

J. Canning, M.A. Hossain, C. Han, L. Chartier, K. Cook, T. Athanaze, Drawing optical fibers from 3D printers. Opt. Lett. **41**(23), 5551–5554 (2016)

Y. Chen, N. Zhao, J. Liu, Y. Xiao, M. Zhu, K. Liu, G. Zhou, Z. Hou, C. Xia, Y. Zheng, Z. Chen, Yb3+−doped large-mode-area photonic crystal fiber for fiber lasers prepared by laser sintering technology. Opt. Mater. Express **9**, 1356–1364 (2019)

K. Cook, J. Canning, S. Leon-Saval, Z. Reid, M.A. Hossain, J.-E. Comatti, Y. Luo, G.D. Peng, Air-structured optical fiber drawn from a 3D-printed preform. Opt. Lett. **40**, 3966–3969 (2015)

K. Cook, G. Balle, J. Canning, L. Chartier, T. Athanaze, M.A. Hossain, C. Han, J.-E. Comatti, Y. Luo, G.D. Peng, Step-index optical fiber drawn from 3D printed preforms. Opt. Lett. **41**, 4554–4557 (2016)

I. Cooperstein, E. Shukrun, O. Press, A. Kamyshny, S. Magdassi, Additive manufacturing of transparent silica glass from solutions. ACS Appl. Mater. Interfaces **10**, 18879–18885 (2018)

J.F. Destino, N.A. Dudukovic, M.A. Johnson, D.T. Nguyen, T.D. Yee, G.C. Egan, A.M. Sawvel, W.A. Steele, T.F. Baumann, E.B. Duoss, T. Suratwala, R. Dylla-Spears, 3D printed optical quality silica and silica–titania glasses from sol–gel feedstocks. Adv. Mater. Technol. **3**, 1700323 (2018)

S.V. Firstov, S.V. Alyshev, K.E. Riumkin, A.M. Khegai, A.V. Kharakhordin, M.A. Melkumov, E.M. Dianov, Laser-active fibers doped with bismuth for a wavelength region of 1.6–1.8 μm. IEEE J. Select. Topics Quantum Electron. **24**, 0902415 (2018)

P. Flanagan, K. Cook, J. Canning, 3D Printed Photonic Ribs: A New Platform for Devices, Sensors and More. Asia Pacific Optical Sensors (APOS2018), Matsue City, Shimane, Japan, 2018

S. Gao, Y. Wang, W. Ding, D. Jiang, S. Gu, X. Zhang, P. Wang, Hollow-core conjoined-tube negative-curvature fiber with ultralow loss. Nat. Commun. **9**, 2828 (2018)

T. Gissibl, S. Thiele, A. Herkommer, H. Giessen, Two-photon direct laser writing of ultracompact multi-lens objectives. Nat. Photonics **10**, 554 (2016)

L. Glavind, S. Buggy, J. Canning, S. Gao, K. Cook, Y. Luo, G.D. Peng, B.F. Skipper, M. Kristensen, Long-period gratings for selective monitoring of loads on a wind turbine blade. Appl. Opt. **53**, 3993–4001 (2014)

C. Gong, Y. Gong, X. Zhao, Y. Luo, Q. Chen, X. Tan, Y. Wu, X. Fan, G.D. Peng, Y.J. Rao, Distributed fiber optofluidic laser for chip-scale arrayed biochemical sensing. Lab Chip **18**, 2741–2748 (2018)

J. Han, E. Liu, J. Liu, Circular gradient-diameter photonic crystal fiber with large mode area and low bending loss. J. Opt. Soc. Am. A **36**, 533–539 (2019a)

M. Han, H. Wang, Y. Yang, C. Liang, W. Bai, Z. Yan, H. Li, Y. Xue, X. Wang, B. Akar, H. Zhao, H. Luan, J. Lim, I. Kandela, G.A. Ameer, Y. Zhang, Y. Huang, J.A. Rogers, Three-dimensional piezoelectric polymer microsystems for vibrational energy harvesting, robotic interfaces and biomedical implants. Nat. Electron. **2**, 26–35 (2019b)

A.J. Ikushima, T. Fujiwara, K. Saito, Silica glass: a material for photonics. J. Appl. Phys. **88**, 1201–1213 (2000)

C. Inamura, M. Stern, D. Lizardo, P. Houk, N. Oxman, Additive manufacturing of transparent glass structures. 3D Print. Addit. Manuf. **5**, 269–283 (2018)

J.J. Kaufman, C. Bow, F.A. Tan, A.M. Cole, A.F. Abouraddy, 3D printing preforms for fiber drawing and structured functional particle production, in *Photonics and Fiber Technology 2016 (ACOFT, BGPP, NP)*, (Optical Society of America, Sydney, 2016), p. AW4C.1

J.C. Knight, T.A. Birks, P.S.J. Russell, D.M. Atkin, All-silica single-mode optical fiber with photonic crystal cladding. Opt. Lett. **21**, 1547–1549 (1996)

F. Kotz, K. Plewa, W. Bauer, N. Schneider, N. Keller, T. Nargang, D. Helmer, K. Sachsenheimer, M. Schäfer, M. Worgull, C. Greiner, Liquid glass: A facile soft replication method for structuring glass. Adv. Mater. **28**(23), 4646–4650 (2016)

F. Kotz, K. Arnold, W. Bauer, D. Schild, N. Keller, K. Sachsenheimer, T.M. Nargang, C. Richter, D. Helmer, B.E. Rapp, Three-dimensional printing of transparent fused silica glass. Nature **544**, 337 (2017)

F. Kotz, N. Schneider, A. Striegel, A. Wolfschläger, N. Keller, M. Worgull, W. Bauer, D. Schild, M. Milich, C. Greiner, D. Helmer, Glassomer – processing fused silica glass like a polymer. Adv. Mater. **30**(22), 1707100 (2018)

Y. Luo, J. Wen, J. Zhang, J. Canning, G.D. Peng, Bismuth and erbium codoped optical fiber with ultrabroadband luminescence across O-, E-, S-, C-, and L-bands. Opt. Lett. **37**, 3447–3449 (2012)

Y. Luo, Y. Chu, K. Cook, G. Tafti, S. Wang, W. Wang, Y. Tian, J. Canning, G.D. Peng, Spun high birefringence bismuth/erbium co-doped photonic crystal fiber with broadband polarized emission. Invited presentation, Asia Com. & Phot. Conf. ACP Hangzhou, China, 2018

T. Ma, H. Guerboukha, M. Girard, A.D. Squires, R.A. Lewis, M. Skorobogatiy, 3D printed hollow-core terahertz optical waveguides with hyperuniform disordered dielectric reflectors. Adv. Opt. Mater. **4**, 2085–2094 (2016)

T.H.R. Marques, B.M. Lima, J.H. Osório, L.E.d. Silva, C.M.B. Cordeiro, 3D Printed Microstructured Optical Fibers, in 2017 SBMO/IEEE MTT-S International Microwave and Optoelectronics Conference (IMOC), 2017

C. Martelli, J. Canning, Fresnel fibers with omni-directional zone cross-sections. Opt. Express **15**(7), 4281–4286 (2007)

A. Michie, J. Canning, I. Bassett, J. Haywood, K. Digweed, M. Åslund, B. Ashton, M. Stevenson, J. Digweed, A. Lau, D. Scandurra, Spun elliptically birefringent photonic crystal fiber. Opt. Express **15**, 1811–1816 (2007)

S.Z. Muhamad Yassin, N.Y.M. Omar, H.A. Abdul-Rashid, Optimized fabrication of thulium doped silica optical fiber using MCVD, in *Handbook of Optical Fibers*, ed. by G. D. Peng, (Springer Singapore, Singapore, 2019). https://doi.org/10.1007/978-981-10-7087-7_76-1

D.T. Nguyen, C. Meyers, T.D. Yee, N.A. Dudukovic, J.F. Destino, C. Zhu, E.B. Duoss, T.F. Baumann, T. Suratwala, J.E. Smay, R. Dylla-Spears, 3D-printed transparent glass. Adv. Mater. **29**, 1701181 (2017)

G.D. Peng, Research and development of new generation optical fibers, in International Workshop on Photonic Fibers and Applications, Shanghai, China, 2018

P.S.J. Russell, Photonic crystal fibers. Science **299**(5605), 358–362 (2003)

J.A. Sánchez-Martín, J.M. Álvarez, M.A. Rebolledo, S. Torres-Peiró, A. Díez, M.V. Andrés, Erbium-doped photonic crystal fibers: fabrication and characterization. AIP Conf. Proc. **1055**, 99–102 (2008)

Z.M. Sathi, J. Zhang, Y. Luo, J. Canning, G.D. Peng, Improving broadband emission within Bi/Er doped silicate fibers with Yb co-doping. Opt. Mater. Express **5**, 2096–2105 (2015)

M.D. Symes, P.J. Kitson, J. Yan, C.J. Richmond, G.J.T. Cooper, R.W. Bowman, T. Vilbrandt, L. Cronin, Integrated 3D-printed reactionware for chemical synthesis and analysis. Nat. Chem. **4**, 349 (2012)

W. Talataisong, R. Ismaeel, T.H.R. Marques, S. Abokhamis Mousavi, M. Beresna, M.A. Gouveia, S.R. Sandoghchi, T. Lee, C.M.B. Cordeiro, G. Brambilla, Mid-IR hollow-core microstructured fiber drawn from a 3D printed PETG preform. Sci. Rep. **8**, 8113 (2018a)

W. Talataisong, R. Ismaeel, S.R. Sandoghchi, T. Rutirawut, G. Topley, M. Beresna, G. Brambilla, Novel method for manufacturing optical fiber: extrusion and drawing of microstructured polymer optical fibers from a 3D printer. Opt. Express **26**, 32007–32013 (2018b)

S. Thiele, K. Arzenbacher, T. Gissibl, H. Giessen, A.M. Herkommer, 3D-printed eagle eye: Compound microlens system for foveated imaging. Sci. Adv. **3**, e1602655 (2017)

P.M. Toal, L.J. Holmes, R.X. Rodriguez, E.D. Wetzel, Microstructured monofilament via thermal drawing of additively manufactured preforms. Addit. Manuf. **16**, 12–23 (2017)

S. Unger, F. Lindner, C. Aichele, K. Schuster, Rare-earth-doped laser fiber fabrication using vapor deposition technique, in *Handbook of Optical Fibers*, ed. by G. D. Peng, (Springer Singapore, Singapore, 2019). https://doi.org/10.1007/978-981-10-7087-7_47-1

L. Wang, D. He, S. Feng, C. Yu, L. Hu, J. Qiu, D. Chen, Phosphate ytterbium-doped single-mode all-solid photonic crystal fiber with output power of 13.8 W. Sci. Rep. **5**, 8490 (2015)

J. Wen, J. Wang, Y. Dong, N. Chen, Y. Luo, G.D. Peng, F. Pang, Z. Chen, T. Wang, Photoluminescence properties of Bi/Al-codoped silica optical fiber based on atomic layer deposition method. Appl. Surf. Sci. **349**, 287–291 (2015)

X. Zhang, X. Zhang, Q. Wang, J. Chang, G.D. Peng, In-phase supermode selection in ring-type and concentric-type multicore fibers using large-mode-area single-mode fiber. J. Opt. Soc. Am. A **28**, 924–933 (2011)

Q. Zhao, F. Tian, X. Yang, S. Li, J. Zhang, X. Zhu, J. Yang, Z. Liu, Y. Zhang, T. Yuan, L. Yuan, Optical fibers with special shaped cores drawn from 3D printed preforms. Optik **133**, 60–65 (2015)

C. Zhu, Z. Qi, V.A. Beck, M. Luneau, J. Lattimer, W. Chen, M.A. Worsley, J. Ye, E.B. Duoss, C.M. Spadaccini, C.M. Friend, J. Biener, Toward digitally controlled catalyst architectures: hierarchical nanoporous gold via 3D printing. Sci. Adv. **4**, eaas9459 (2018)

M.G. Zubel, A. Fasano, G. Woyessa, K. Sugden, H.K. Rasmussen, O. Bang, 3D-printed PMMA preform for hollow-core POF drawing, in Proceedings of the 25th International Conference on Plastic Optical Fibers, 2016, p. 6

Part IV
Active Optical Fibers

Rare-Earth-Doped Laser Fiber Fabrication Using Vapor Deposition Technique

17

Sonja Unger, Florian Lindner, Claudia Aichele, and Kay Schuster

Contents

Introduction	658
Preform Technologies	658
MCVD Process Combined with Solution Doping for Rare-Earth and Aluminum Incorporation	659
MCVD Process Combined with Gas Phase Doping for Rare-Earth and Aluminum Incorporation	669
Conclusion	675
References	675

Abstract

Rare-earth (RE)-doped silica-based fiber lasers and amplifiers with very high output power and excellent beam quality are efficient devices for a variety of applications in industry, science, and medicine. Extreme power densities beyond the kW level and complicated fiber structures put high demands on fiber material properties and the preparation technology concerning both efficient laser operation and high-power stability. Silica is the preferred basic material. The host properties can be improved using co-dopants, which influence the optical properties (as refractive index distribution, absorption and emission behavior, optical background losses).

Over the years, alternative technologies to the common modified chemical vapor deposition (MCVD) process have been developed to incorporate REs and the most important co-dopant aluminum (Al) into silica. These fabrication

---

S. Unger · F. Lindner · C. Aichele · K. Schuster (✉)
Department of Fiber Optics, Leibniz Institute of Photonic Technology (Leibniz IPHT), Jena, Germany
e-mail: sonja.unger@leibniz-ipht.de; florian.lindner@leibniz-ipht.de; claudia.aichele@leibniz-ipht.de; kay.schuster@leibniz-ipht.de

© Springer Nature Singapore Pte Ltd. 2019
G.-D. Peng (ed.), *Handbook of Optical Fibers*,
https://doi.org/10.1007/978-981-10-7087-7_47

methods differ in process control and regarding the properties of the manufactured preforms and fibers, such as geometry, incorporated level of REs and co-dopants, homogeneity of dopant concentration, and refractive index distribution. Here, two different technologies are described in detail. The MCVD process combined with solution doping is the most widely used and successful technique due to its simplicity and versatility. However, this technology has several limitations concerning geometry, doping and refractive index homogeneity, and the incorporation of very high REs and Al concentrations. One option to overcome these limitations is the MCVD process combined with gas phase doping for REs and Al.

Advantageous and detrimental effects of the relevant technology must be carefully considered for an optimal design for a high-power laser fiber.

Introduction

Rare-earth (RE)-doped silica-based fiber lasers and amplifiers have been shown to be suitable for a variety of applications in industry, science, and medicine. They can yield very high-power output in the multi-kW range both in the pulsed and CW regimes with high efficiency, reliability, and beam quality (Richardson et al. 2010; Jeong et al. 2004; Popp et al. 2011; Unger et al. 2013; Langner et al. 2011). This progress is due to new design concepts, such as nonsymmetrical double clads and large-mode-area (LMA) cores (Tünnermann et al. 2000), and to the microstructuring of the fibers using "air clads," inner "holey" clads or solid "multifilament" cores (Tünnermann et al. 2005; Wirth et al. 2011; Canat et al. 2008). Extreme power densities beyond the kW level and complicated fiber structures put high demands on fiber material properties and the preparation technology concerning both efficient laser operation and high-power stability.

Quartz glass and high-silica glasses made by applying gas phase deposition processes are extremely important as host glasses because of their high glass stability, high purity, and low optical losses. However, the option of incorporating the RE ions (such as neodymium (Nd), ytterbium (Yb), erbium (Er), thulium (Tm), or holmium (Ho)) in silica is limited, and it is difficult to achieve a high doping level of such ions. By adding co-dopants such as aluminum (Al) and/or phosphorus (P), germanium (Ge), or boron (B), the solubility of the RE ions in silica can be improved, and thus the RE content increased without phase separation and crystallization.

Preform Technologies

Vapor phase deposition methods have been proven to provide layered structures with excellent material quality and extremely low attenuation (e.g., as low as 0.2 dB/km). For the preparation of high-silica passive fiber preforms, the following methods have been developed (Tosco 1990; Mac Chesney et al. 1990; Izawa 2000; Nagel et al. 1982):

- Modified chemical vapor deposition (MCVD)
- Plasma-activated chemical vapor deposition (PCVD)
- Outside vapor deposition (OVD)
- Vapor axial deposition (VAD)

In these vapor deposition processes, high-purity raw materials, liquids, or gaseous dopants with high vapor pressures at room temperature are needed. In contrast to silica and the common dopants Ge, P, B, and F for preparation of high-silica passive fiber preforms, volatile precursor compounds do not exist for the RE dopants and the most important co-dopant Al that can be vaporized at or slightly above room temperature.

For this reason, the following alternative processes have been developed for the incorporation of RE and Al into silica optical fibers:

- Conventional MCVD process combined with solution doping (Townsend et al. 1987; Kirchhof et al. 2003), gas phase doping (Schuster et al. 2014), or atomic layer deposition (ALD) (Montiel i Ponsoda et al. 2012)
- Flame hydrolysis deposition methods RE/Al vapor doping OVD process (Wang et al. 2008) or direct nanoparticle deposition (DND) (Tammela et al. 2006)
- Aerosol doping technique for the MCVD and OVD process (Morse et al. 1991)
- Powder sinter technology (REPUSIL) (Schuster et al. 2014)
- Crucible melting technology (Schuster et al. 2014)
- Sol-gel technique (Pedrazza et al. 2007)
- Surface-plasma chemical vapor deposition (SPCVD) (Saveley and Golant 2015)

These fabrication methods differ in process control (such as efficiency, versatility, process conditions) and regarding the properties of the manufactured active preform and fiber, such as geometry (core, cladding and preform diameter, preform length), the incorporated level of the REs and co-dopants, the homogeneity of the dopant concentration and refractive index distribution, and the optical quality.

In the following, two different methods, the MCVD process in combination with the solution doping and with the gas phase technique, are described in detail that apply to preparation of RE-doped optical fibers. Furthermore, the influence of different co-dopants and preparation conditions on the properties in active fibers is depicted.

MCVD Process Combined with Solution Doping for Rare-Earth and Aluminum Incorporation

One option of incorporating REs and Al ions in the silica matrix is to supply them via a liquid phase. This method, so-called solution doping, in combination with the common MCVD process is the most widely used and successful technique due to its simplicity and versatility (Kirchhof et al. 2003).

Active core deposition is achieved in the following way:

At first, a flocculent deposit is formed from a gaseous mixture of $SiCl_4/O_2$ (or together with $POCl_3/GeCl_4/BCl_3$, etc.) by a reverse motion of the MCVD burner relative to the gas flow at a temperature of 1600 °C on the inner surface of the quartz glass carrier tube. This porous layer has a very low density (ca. 3% compared with a fully densified layer) and is not stable if treated with a solution that peels off. Therefore, the ($P_2O_5/GeO_2/B_2O_3$, etc. doped) SiO_2 deposit is pre-sintered in a pure oxygen atmosphere in order to generate a sufficient stability with a relative density around 20%. Subsequently the porous layer is impregnated with an aqueous or alcohol solution of RE and Al salts (mostly the well-soluble chlorides or nitrates are used) and dried via preparation steps at room temperature and about 1000 °C. Again in a pure oxygen atmosphere, the solid salts remain in the porous silica. Next, the porous layer is treated with a Cl_2/O_2 atmosphere to eliminate contaminations (OH, 3d elements). For an effective purification process, the temperature must be optimized in the range of 1000 °C–1400 °C (depending on the layer composition) so that no remarkable sintering of the porous layer is carried out. During subsequent high-temperature treatment, the (doped) silica is consolidated into a fused glassy layer, and the salts are converted to RE and Al oxides and incorporated into the silica matrix. In order to complete the fiber preform, the coated tube is collapsed via several burner passes at temperatures above 2000 °C normally under a Cl_2/O_2 atmosphere to a solid rod (preform) where the doped layer has formed the preform core.

Basic investigations have shown that the doping concentration of RE and Al in the preform (after consolidation and collapsing) is dependent on the solution concentration and the relative density of the porous layer (Kirchhof et al. 2003). The content is increased with the increase in the solution concentration and reduction of the relative density. The density of the porous layer, which has already been described above, for a sufficient stability for the following preparation steps should be approximately 20%. Practically speaking, there is no advantage to increasing the density, because then the doping concentration is reduced. The conditions for the pre-sintering process must be controlled as carefully as possible in order to meet the intended density. This, in particular, is dependent on the composition on the porous layer, i.e. on the content of common dopants (such as P, Ge, B, etc.). In Fig. 1, the change in the relative density is shown in dependence on the temperature for different layer compositions.

The MCVD/solution doping process is very flexible and permits the incorporation of all RE elements. In this process, the co-dopants have manifold and complicated influences on the properties of the laser fiber. They change the refractive index, the absorption and emission properties of the RE ions, and the background loss of the fiber glass. Moreover, they influence the chemical processes of the glass preparation and glass properties as viscosity, thermal and diffusion behavior, and the atomic point defect concentration during preform and fiber fabrication.

Knowledge of these strong interactions between REs and co-dopants is very important for the successful fabrication of the defined fiber core compositions and

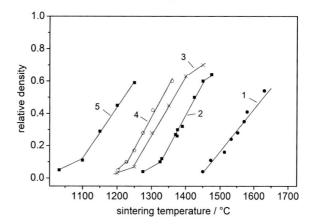

Fig. 1 Relative density in dependence on sintering temperature and layer composition
(1) SiO$_2$
(2) 0.99 SiO$_2$–0.01P$_2$O$_5$
(3) 0.80 SiO$_2$–0.20 GeO$_2$
(4) 0.90 SiO$_2$–0.10 B$_2$O$_3$
(5) 0.92 SiO$_2$–0.08 P$_2$O$_5$

structures for high-efficiency laser fibers. Over several years, these interactions have been intensely investigated. They are described in detail more in the following.

Refractive Index and Diffusion Properties of RE-Doped Fibers

The refractive index change (Δn) of the fiber core is determined by the RE content and the concentration of the co-dopants. In most cases, an additivity rule is valid with respect to the molar composition with constant increments which are given in Kirchhof et al. (2003), (2005) for GeO$_2$ (+13), P$_2$O$_5$ (+9), B$_2$O$_3$ (−5), SiF$_4$ (−50), Al$_2$O$_3$ (+22), and RE$_2$O$_3$ (+67), always in accordance with the relationship between Δn 10^{-4} and the molar concentration of the respective oxide (or fluoride).

In the case of Al/P co-doping, the additivity rule is no longer valid. Here, a relatively complicated course of Δn is observed depending on the concentration of Al and P, where, however, the refractive index is mostly lower than expected from the additivity rule. This behavior is described in detail in Unger et al. (2007). Similar results were also reported earlier by Bubnov et al. (2009) and in more detail by Di Giovanni et al. (1989), who explained the index drop by the formation of AlPO$_4$ complexes in the glass. The physical properties of such glasses strongly differ from the single components and lead to a behavior similar to the SiO$_2$ (or Si$_2$O$_4$) network group. This phenomenon seems to provide a very valuable meaning for the decrease in the core NA (numerical aperture) despite high core doping. The influence of RE$_2$O$_3$ on physical properties within this glass is still not fully clear.

Diffusion processes during preparation determine the refractive index distribution, geometry, and the NA of the preform core. The real concentration and refractive index profiles in the preform and fiber are always diffusion-influenced profiles. Under the influence of the high temperature during the sintering and collapsing process, a remarkable diffusion effect of the dopants takes place in the radial direction.

In the case of pure Al co-doping, this leads to a certain broadening of the radial profile of the refractive index and concentration distribution, whereas in the case of pure P, Ge, or B co-doping, a strong local depletion is observed, which is connected with the formation of a large "dip" in the center of the preform, caused by evaporation of volatile oxides of these co-dopants.

The REs are governed by the respective co-dopants. So, the concentration distribution of REs also shows a "dip" if either P, Ge, or B is present in excess. This is not caused by evaporation of the REs but by an abnormal diffusion of the RE ions against the concentration gradient. In Al (also together with low P content) co-doped fibers, a "dip" in REs is missing. Interestingly, the dip is also missing in Al/P co-doped fibers with a ratio of 1.

Simultaneously, in high P co-doped fibers, a certain concentration gradient of P remains along the total preform length in consequence of vaporization and redeposition of P beyond the moved heating zone (Kirchhof et al. 2004a). In the case of an Al/P ratio of 1, these effects are pronounced with respect to the refractive index, which makes this system more difficult to control, despite the advantage discussed above. In order to mitigate these problems of MCVD/solution doping, other preform preparation technologies such as, e.g. gas phase deposition of Al and REs together with P, should be considered (Bubnov et al. 2009). Figure 2 shows different radial refractive index and concentration profiles of preform samples using the MCVD/solution doping (Zimer et al. 2011).

Knowledge of the diffusion coefficients of the dopants is important to understand and optimize the process. Therefore, over a span of several years, the diffusion behavior of different dopants (such as P (Kirchhof et al. 2004b), Al and Yb (Unger et al. 2011)) in silica depending on the temperature (1600 °C–2000 °C) and dopant concentration has been investigated.

Background Losses in RE-Doped Fibers

Background losses of the fiber core material (measured between the absorption bands of the RE ions) can become a serious problem for the high-power fiber laser. On the one hand, the intensity of the laser radiation and thus the laser efficiency is reduced. On the other hand, a parasitic absorption gives rise to additional thermal load and eventually to fiber damage. By all means, such losses restrict the possibility of tailoring the laser design by a freely eligible fiber.

There are different causes of background losses. Naturally, they are influenced by impurities, especially by 3d elements. At first glance, the background losses seem to be a question of the quality of the raw materials and of a suitable process control. The MCVD/solution doping process route includes a purification step with chlorine, which is used in the first instance to "dry" the materials, to reach a low OH content (<1 mol ppm). It is widespread knowledge that this step can also improve the purity concerning metals because volatile metal chlorides are formed and removed from the silica in a porous stage, promoted by a small particle or layer size and by the relatively high diffusion rate of these metals in the silica network, compared with the usual network formers. Thus a very high purity of the fiber material is obtained, with background losses as low as 10 dB/km in the NIR region. However, besides the extrinsic loss caused by impurities, there can be several contributions to the background loss level by intrinsic effects which are described in Kirchhof et al. (2003, 2005) and summarized in the following:

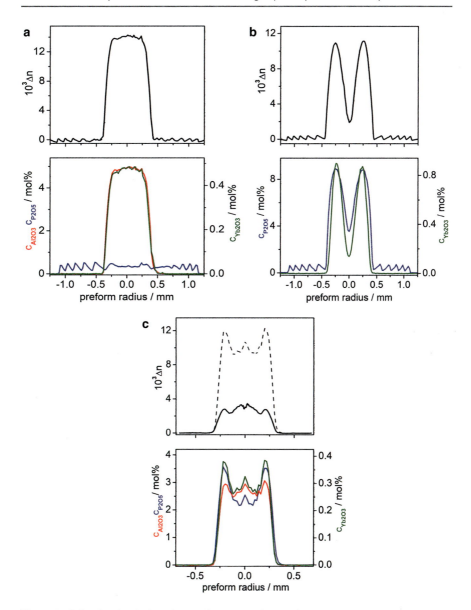

Fig. 2 Radial refractive index (Δn) and concentration profiles for (**a**) Al co-doping, (**b**) P co-doping, and (**c**) Al/P co-doping

(A) Losses due to the limited solubility of the REs:

Silica is the preferred material in fiber technology; however, in the first instance, it is not well suited as laser host, among other things, because of the small solubility for RE dopants. Fortunately, the host properties can be improved using co-dopants (Al, P, Ge, B, etc.) in excess of the RE dopant. Nevertheless, with increasing RE concentration, a loss increase is also generally observed, often abruptly starting above a certain concentration. This loss contribution depends on the kind and amount of the co-dopants and is of importance if RE concentrations in the mol% range have to be carried out. These high losses are connected with phase separation which can be visually and microscopically observed.

(B) Losses due to crystallization:

Although both pure quartz glass and high-silica glasses with the common dopants are very stable glasses, RE incorporation leads to an enhanced crystallization sensitivity. To investigate the glass stability (Kirchhof et al. 2003), preforms with different compositions were annealed at temperatures between 1400 °C and 1800 °C. Without REs, there was no tendency toward crystallization in the different host glass systems. Similar stability was found for RE_2O_3-Al_2O_3-SiO_2 compositions (investigated for Nd and Yb). However, the presence of GeO_2, P_2O_5, or B_2O_3 led to strong recrystallization (formation of cristobalite phases) connected with strong background loss increases and degradations of the fluorescence intensity and lifetime.

For the fiber drawing process, it is essential to optimize the conditions (such as drawing temperature, tension, velocity) for the corresponding fiber material in order to prevent such crystallizations. Specific annealing of the fiber material can also be necessary for special technological routes, e.g. for the preparation of large-mode-area (LMA) or birefringent laser fibers and for microstructured or photonic crystal fibers where the temperature history is more complicated than in the case of a simple solid-core single-mode preform.

(C) UV-induced losses:

The RE doped silica materials are also influenced by UV radiation, which plays a certain role if UV-cured coatings are applied or if UV-induced Bragg gratings shall be inscribed directly into the RE doped fiber core. This effect of UV radiation on the fiber loss was demonstrated in the case of Nd and Yb, where not only the high-intensity excimer laser but even the UV lamp radiation induces remarkable loss.

(D) Rare-earth-specific losses:

For many years, it has been well known that excess losses in Ge doped silica fibers depend on drawing conditions, especially in the case of high Ge content. In Nd doped fibers, a background loss was found which also shows a drawing dependence very similar to the behavior of the Ge related effects. Thus high drawing tension (or low drawing temperatures) leads to reduced additional losses. Systematic investigations concerning the RE_2O_3-Al_2O_3-SiO_2 system, doped with yttrium (Y), lanthanum (La), cerium (Ce), praseodymium (Pr), Nd, samarium (Sm), gadolinium (Gd), dysprosium (Dy), Ho, Er, Tm, and Yb, have clarified that this loss type occurs only for Pr, Nd, and Sm.

The addition of other REs (such as La and Ce) leads to a remarkable decrease in the background loss, and by co-doping with relatively low Ge amounts, the additional loss can be totally avoided. Interestingly, co-doping with P is without effect on the background loss. Moreover, it was found that the additional loss can be removed to a certain degree by reducing conditions, such as helium (He) or carbon monoxide (CO) during the collapsing process of the preform.

The reasons for this RE specific loss are still unclear. Some facts point out that the loss is related to the oxygen-deficient center (ODC) absorption at a wavelength of about 240 nm in such a manner that a high ODC absorption in the fiber core or in the inner cladding reduces the loss contribution (as is the case with Ge or with collapsing under a reducing atmosphere). Some arguments speak for fiber damage by thermal UV radiation during the drawing process as already discussed for the interpretation of additional losses in high Ge doped fibers. However, the intensity of the thermal radiation generated at 2000 °C–2200 °C in the heated preform is very low compared with UV intensities which are needed in order to increase the losses in the fiber by UV lamps or an excimer laser. The previous observations yield some evidence that this loss contribution is intrinsic and related to atomic defects.

Furthermore, in the Yb_2O_3-Al_2O_3-SiO_2 system, it was observed that a process modification can lead to detrimental effects. So in contrast to Nd, Pr, or Sm doped fibers, a sintering or collapsing in a reducing atmosphere is followed by a strong increase in the loss in consequence of induced absorptions at short wavelengths by formation of Yb^{2+} ions (Kirchhof et al. 2006). This has resulted in a degradation of the fluorescence intensity and lifetime and a reduction in the laser efficiency. In the normal MCVD/solution doping process, this effect is not essential because all preform preparation steps are carried out in an oxidizing atmosphere. For other preparation routes of RE doped materials, e.g. by melting or sintering porous bodies, this effect could become important.

Absorption and Emission Properties of RE-Doped Fibers

The kind and concentration of the co-dopant can strongly influence absorption and emission properties of a laser fiber. This was intensively investigated in Yb and Nd-doped fibers (Unger et al. 2013; Kirchhof et al. 2003, 2005; Zimer et al. 2011) and is described here in more detail for Yb doped fibers.

As shown in Fig. 3, the kind of co-dopant in Yb doped fibers influences the shape of the absorption spectrum but particularly the absolute absorption values. From pure Al co-doping to pure P co-doping, the absolute value of the absorption cross section is reduced approximately by a factor of two at the main pump wavelengths around 915 nm and 976 nm, accompanied by spectral changes, with a more or less systematic transition for intermediate ratios. In Al/P (1:1) co-doped fibers, the absorption is about 25% lower compared to pure Al co-doping. Co-doping with Ce (in an Al_2O_3-SiO_2 host) shows no influence on the absorption cross section of Yb^{3+},

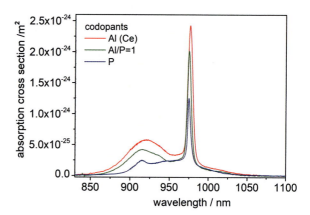

Fig. 3 Yb^{3+} absorption cross section dependent on co-dopants

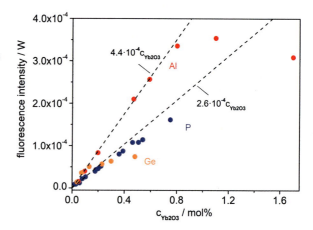

Fig. 4 Fluorescence intensity at 1075 nm depending on Yb concentration and co-dopants

and the absorption cross section of Yb^{3+} is independent of the Yb concentration in the investigated concentration region until 1.7 mol% Yb$_2$O$_3$.

The emission behavior, on the contrary, is affected at higher Yb concentrations. This can be most clearly recognized by investigations of the dependence of the fluorescence intensity (rather than the fluorescence lifetimes) on the Yb concentration. The intensities (measured at 1075 nm) show a linear increase with the Yb concentration at low values but a remarkable deviation from the linearity at higher concentrations for Al$_2$O$_3$-SiO$_2$ (around 4 mol% Al$_2$O$_3$) pointing to strong concentration quenching. In P$_2$O$_5$-SiO$_2$ (around 8 mol% P$_2$O$_5$), this effect cannot be clearly observed, but investigations at lower phosphorus concentrations (2–4 mol% P$_2$O$_5$) – not shown here – provide also evidence of intensity decrease and concentration quenching. In GeO$_2$-SiO$_2$ (around 12 mol% GeO$_2$, here not considered in detail), this phenomenon appears already at fairly low Yb concentrations, which is one reason for the low laser efficiency in the GeO$_2$-SiO$_2$ host (see Fig. 4).

The nature of concentration quenching is still unclear. Paschotta et al. (1997) have already reported about a similar effect which was found by a non-saturable absorption in different Yb doped fibers.

Photodarkening

The increased attenuation in Yb doped silica-based laser fibers induced by the pump radiation, called photodarkening (PD), turned out to be a critical factor for high-power laser action (Koponen et al. 2005). These pump-induced losses can impair the long-term power stability and efficiency of fiber lasers and amplifiers at high-power loading. Recently, the thermal load caused by the absorption of PD color centers at the pump, and laser wavelength was identified as an additional issue, which can lead to refractive index changes in the laser core, mode instabilities, and degradation of the beam quality in nominal single-mode working laser or amplifier systems during high-power operation (Jauregui et al. 2015). Low PD loss is indispensable for the long-term power stability and efficiency of fiber lasers and amplifiers.

Several investigations have dealt with the temporal evolution of the induced loss increase, which can be described in most cases by a stretched exponential function (Koponen et al. 2005). Therefore, the kinetics of PD is characterized by three parameters: the time constant, the equilibrium loss (reached at long irradiation times), and the stretching exponent (Jetschke et al. 2007). It was found that the essential factor for PD is the number of excited Yb ions rather than the power density (Kitabayashi et al. 2006). Some models have been developed to understand qualitatively and quantitatively the reasons for the spectral change, which is believed to be caused by different atomic defects in the doped fiber glass (Mattsson 2011). Even if the principal understanding is still insufficient and needs further examination, the work to date points out some ways to decrease PD. PD can be controlled by low inversion and low Yb concentration (Jetschke et al. 2007), by higher temperatures (Leich et al. 2011), or by special fiber designs (Mattsson 2009).

The most important and useful effect from a practical point of view, however, is the influence of co-dopants. In general co-dopants are needed to improve the glass properties and the optical properties when incorporating the RE ions into the silica host. These can also affect the PD kinetics on a large scale both favorably and unfavorably (Kirchhof et al. 2009). Up to now, the following glass compositions have been found to reach a practically vanishing PD: (a) pure P co-doping (Kirchhof et al. 2009), (b) Al/P co-doping up to an atomic ratio of one between Al and P (Jetschke et al. 2008), and (c) Ce co-doping of the Al-silica host (Unger et al. 2013; Jetschke et al. 2016).

In Fig. 5a, the equilibrium loss of PD (measured at a wavelength of 633 nm by accelerated tests as described in Jetschke et al. (2007)) is shown as a function of the co-doping ratio of P and Al (Zimer et al. 2011; Jetschke et al. 2008). Starting with a pure Al-silica glass host, the equilibrium loss is linearly decreased with an increasing amount of P up to an atomic ratio of 1 between P and Al, and its values keep very low in the P excess branch. This means that the superior properties of the P silica host also remain in the case of the partial admixture of Al, exactly up to a ratio of one.

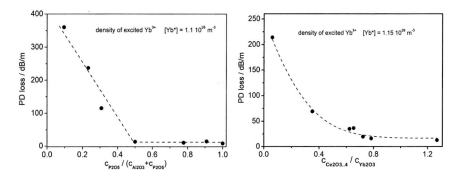

Fig. 5 PD equilibrium loss in (**a**) Al/P co-doped fibers and (**b**) Ce/Al co-doped fibers

Figure 5b demonstrates the possibility of reducing PD by Ce co-doping into the Al-silica host. Here, a continuous decrease in the equilibrium loss with an increasing amount of Ce is observed. Considering high-power laser fibers, the effect of PD should not only be mitigated but really eliminated. Therefore, the concentration of Ce has to be increased to high values, comparable with the Yb concentration, in order to scale down the equilibrium loss at 633 nm below 10 dB/m at 70% inversion. However, at high Ce concentrations, thermal issues arise from the interaction with excited Yb ions. A Ce/Yb ratio between 0.5 and 0.7 is suggested to be optimal for fibers with Yb concentrations of up to 0.4 mol% Yb_2O_3. Fibers with higher Yb and Ce concentrations seem to be less important for practical use in laser applications due to their strong thermal load (Jetschke et al. 2016).

Investigations have also shown that Tm trace impurities influence the PD behavior of Yb doped fibers. Tm co-doping with more than 10 mol ppm can strongly accelerate the process and also increase the PD loss. It is therefore necessary to use extreme purity raw materials for the preparation of Yb laser fibers with expected very low PD (Jetschke et al. 2013).

Based on these basic investigations, the deposition process and core composition have been optimized for the manufacturing of active single- or low-mode fibers with low background loss, high efficiency, reliability, and beam quality to be used in the wavelength range of 1 μm–2 μm. In Table 1, different types of high-silica RE doped fibers (with RE ions, co-dopants, emission wavelength) prepared using the MCVD/solution doping technique are summarized.

However, the MCVD/solution doping technology has several limitations concerning the geometry, doping and refractive index homogeneity, and the incorporation of very high RE and Al concentrations (in excess of 2 mol% RE_2O_3 and 7 mol% Al_2O_3). For example, this technique only permits the deposition of cores that possess an excellent optical quality of up to a diameter of about 1.5 mm–2 mm, depending on the co-dopants. The implementation of large core/cladding ratios is also limited.

Table 1 Types of high silica RE doped fibers

RE ions	Co-dopants	Emission wavelength/μm
Nd^{3+}	Al, Ge	1.0–1.1
Yb^{3+}	Al	1.0–1.1
Yb^{3+}	Al, P	1.0–1.1
Yb^{3+}	Al, Ce	1.0–1.1
Er^{3+}	Al, Ge	1.5–1.6
Er^{3+}/Yb^{3+}	P, Al	1.5–1.6
Tm^{3+}	Al	1.7–2.1
Ho^{3+}	Al	1.7–2.1
Tm^{3+}/Ho^{3+}	Al	1.7–2.1

MCVD Process Combined with Gas Phase Doping for Rare-Earth and Aluminum Incorporation

In addition to preparation routes such as the RE/Al vapor doping OVD process (Wang et al. 2008) and powder sinter technology (REPUSIL) (Schuster et al. 2014), there is another option available to overcome the geometric and homogeneity limitations of the MCVD/solution doping process. In the case of this alternative technology, the active core diameter in the preform can be significantly increased via the deposition of RE and Al in the gas phase in the MCVD process. To this end, solid precursors of metal organic complexes of RE (such as RE-(tetramethylheptanedione)) and Al (such as Al-(acetylacetonate)) or Al chloride ($AlCl_3$) are converted to the gas phase via high-temperature evaporation, as first reported in 1990 by Tumminelli et al. (1990).

There are two different concepts for the delivery of evaporated RE and Al precursors up to the quartz glass process tube where deposition takes place.

The so-called MCVD/chemical in crucible technique (CIC, see Fig. 6) was developed at the Optoelectronics Research Centre (ORC, University of Southampton) (Boyland et al. 2011). In the case of this internal delivery system, the evaporation of the precursors takes place directly within the quartz glass process tube. The precursors are provided as powder or impregnated porous silica frit, placed in a crucible or directly in the tube and heated with a burner or electrical heater. An inert carrier gas flow of helium (He) removes the evaporated precursors directly to the hot zone of the MCVD burner. Then, deposition and consolidation are carried out together with silica and common dopants (such as P, Ge, B, etc.). The advantage of this setup is the closely spaced evaporation of the precursor and the reaction zone. Precursor delivery lines are not necessary, nor do problems with condensation of the precursor inside the lines or contamination in the case of corrodible transport lines occur. High-NA, low-loss Al-silicate fibers (up to 16 mol% Al_2O_3) and highly efficient Yb doped phosphosilicate fibers were able to be prepared with this method (Boyland et al. 2011).

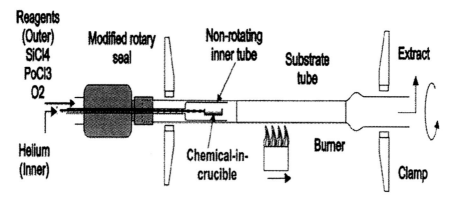

Fig. 6 Schematic of the MCVD/chemical in crucible (CIC) technique for preform fabrication (Boyland et al. 2011)

However, it is not possible to produce defined vapor mixtures of several precursors (e.g., the combination of Al and Yb). Each precursor needs its own crucible and its own evaporation temperature and carrier gas flow for a well-defined precursor transport. But the structural setup of the CIC technique only allows the evaporation of one precursor besides the common dopants.

Instead of this method, delivery of the gaseous precursors takes place from external chemical sources. This vapor system is commercially available and used worldwide in different facilities successfully for the preparation of active fiber preforms. The gas phase process for the preparation of the active core layers is shown in Fig. 7 (here for Al and Yb doping) and is characterized by the following steps (Schuster et al. 2014; Lindner et al. 2014).

The solid starting materials, which are placed at several plates in separated evaporators, are vaporized (carried out in a heated cabinet) at temperatures close to or slightly above the melting temperature and below the sublimation temperature, respectively, depending on the precursor used, to achieve an acceptable vapor pressure. Typically the evaporation temperatures are in the range of 130 °C–140 °C for the $AlCl_3$ used here and 190 °C–200 °C for Yb-(tetramethylheptanedione) [$Yb(thd)_3$], respectively. The gaseous precursors are delivered together with the carrier gas He and additional oxygen separately through heated lines to the inside of the quartz glass reactor tube and mixed here with the other common gaseous halides (such as $SiCl_4$, $POCl_3$, $GeCl_4$, etc.). To prevent condensations of the precursors inside the lines, the cabinet and all lines (from evaporation to injection into the quartz glass tube) must be heated above the evaporation temperatures of about 200 °C. The deposition, consolidation, and collapsing into a preform are carried out in the usual manner as described in Schuster et al. (2014) and Lindner et al. (2014).

In this gas phase doping process, it is possible to use two different Al precursors: the metal organic complex compound Al-(acetylacetonate) and the pure, anhydrous $AlCl_3$. Both starting materials have certain advantages and disadvantages. The metal

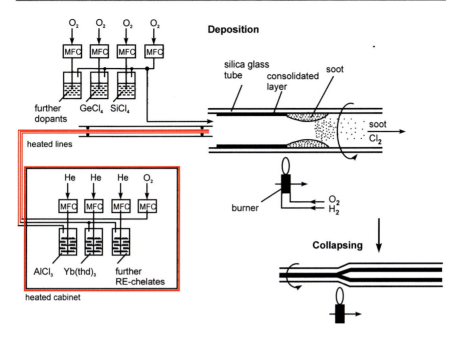

Fig. 7 Schematic of the MCVD process combined with gas phase doping by an external chemical source for RE and Al doped preforms

organic complex compounds of Al and RE are noncorrosive and easy to handle. They show similar evaporation properties, as evaporation temperatures in the range of 190 °C–200 °C. For an efficient laser fiber, an excess of Al compared to Yb is necessary (Al to Yb ≥ 5). To achieve high Al concentrations, carrier gas flows that are essentially higher than Yb are required. As a result, the total gas flow is also strongly increased, which induces a very long taper range in the preform and lower deposition rates. Effectively, only low Al contents are carried out with this Al precursor.

Alternatively, the pure, anhydrous $AlCl_3$ has a considerably higher vapor pressure at a significantly lower temperature (<150 °C). Therefore, lower carrier gas flows are needed, and the incorporation of Al into silica is more efficient with a very high doping level. However, $AlCl_3$ can react violently with humidity under formation of the aggressive compound hydrogen chloride (HCl), which corrodes the metal lines. The metal chloride that is formed (e.g., iron chloride) can be transported with the other gases into the process tube and incorporated in the silica matrix. These impurities result in strong absorptions in the preform, followed by increased background losses in the fiber (see also Bubnov et al. 2009). The handling of $AlCl_3$ is much more difficult than for the organic compounds; the material selection for the components of the equipment has to be done very carefully.

Systematic investigations concerning incorporation of Yb and Al into silica by the gas phase doping technique have been carried out in Lindner et al. (2014).

Preforms and fibers were prepared using the above-described technology in a wide range of Yb and Al concentrations depending on the process conditions (such as evaporation temperatures, gas flows, and collapsing conditions). In the following, the results of these investigations are summarized.

Refractive Index Behavior and Concentration Distribution

The prepared preform samples show very homogenous radial distribution concerning the refractive index change (Δn) and dopant concentrations of Al_2O_3 and Yb_2O_3 without layer structure in contrast to samples prepared using MCVD/solution doping. Typical refractive index and concentration profiles and a cross section of a preform core (diameter about 2 mm) are presented in Fig. 8. A preform core diameter up to 5 mm has already been achieved in Saha et al. (2014) and Sekiya et al. (2008).

The refractive index of the fiber core is determined by the Yb and Al concentration. An additivity rule also applies here with respect to the molar concentration with constant increments. It was found that the increment value for Yb (+67) is the same, but the increment for Al is slightly lower (+18) compared to the earlier estimated values in preform samples prepared using the MCVD/solution doping (Kirchhof et al. 2003, 2005) (see above). The causes for this deviant behavior of Al may be in structural differences, but it is not finally clarified at the moment.

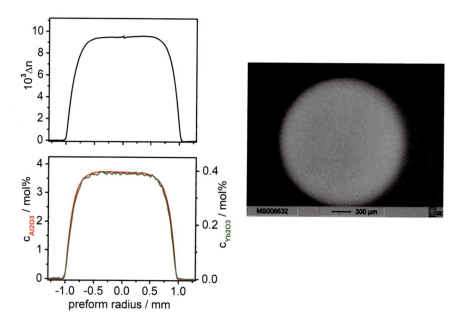

Fig. 8 Radial refractive index (Δn) and concentration profiles of Yb and Al (left) and cross section of a preform core shown by backscattered electrons (right)

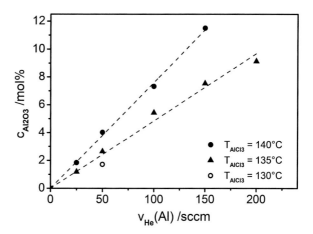

Fig. 9 Al concentration in preform samples depending on the evaporation temperature of AlCl$_3$ T$_{Al}$ and the carrier gas flow v$_{He}$ (Al)

Incorporation of Yb and Al into Silica via Gas Phase

The incorporation of Al without Yb into silica via the gas phase is shown in Fig. 9 depending on the carrier gas flow v$_{He}$ (Al) and the evaporation temperature of AlCl$_3$ T$_{Al}$ in the range of 130 °C–140 °C. It can be observed that the Al content increases with increasing T$_{Al}$ and v$_{He}$ (Al). A maximum Al content of about 12 mol% Al$_2$O$_3$ was able be achieved here. Higher doping concentrations of Al up to 18 mol% Al$_2$O$_3$ were already successfully performed with good homogeneity concerning the core NA in Bubnov et al. (2009), Boyland et al. (2011), and Saha et al. (2014).

The incorporation of Yb combined with Al into silica via the gas phase depends on a number of different parameters (such as evaporation temperatures of the Yb and Al precursor, the carrier gas flows and gas flows of both precursors, and the total gas flow). Figure 10 shows the Yb and Al concentration depending on the carrier gas flow of v$_{He}$(Yb) at two different carrier gas flows of v$_{He}$(Al) (sscm – standard cubic centimeters per minute). The Yb content increases practically linearly with the carrier gas flow of v$_{He}$(Yb) irrespective of the carrier gas flow of v$_{He}$(Al). In contrast, the Al content is not constant but rather decreases continuously with increased v$_{He}$(Yb). The reason for this reduction of the Al content could be the rising total gas flow. But this is not clear at the moment. Further basic investigations are necessary to clarify the reason for this interaction between Yb and Al.

Absorption and Emission Properties of the Preforms and Fibers

The investigations have shown that the Yb^{3+}-related absorption and emission properties (such as absorption and emission spectra, absorption cross section, fluorescence intensity, and lifetime) are comparable to Yb doped aluminosilicate preforms and fibers with a similar core composition made using the MCVD/solution doping technique (Lindner et al. 2014; Unger et al. 2014).

However, in contrast, the values of the background loss at 1200 nm of the drawn fibers are still increased slightly. The lowest values were measured to be 18 dB/km at 1200 nm. The OH content (calculated from the absorption band at 1390 nm) is also

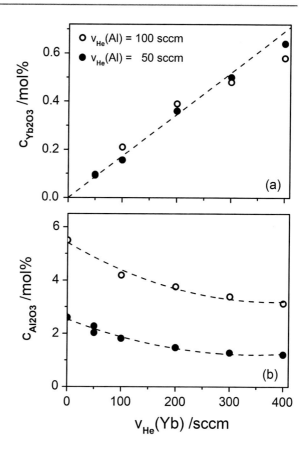

Fig. 10 Yb (**a**) and Al (**b**) concentration depending on the carrier gas flow of $v_{He}(Yb)$ and $v_{He}(Al)$

increased and varies in the range between 1.5 mol ppm and 5 mol ppm compared with samples prepared by MCVD/solution doping (<1 mol ppm OH). To remove impurities and reduce the OH content, an efficient purification step with chlorine (performed in an MCVD/solution doping process) is not possible here because of the fully densified layer. Further investigations are necessary to improve the technology to reduce the OH content and to avoid the incorporation of impurities.

Laser Behavior of the Fibers

Laser experiments have demonstrated an efficient lasing up to 200 W, showing an excellent slope efficiency of about 80% in Yb/Al (Unger et al. 2014), as well as in Yb/P (Boyland et al. 2011), doped fibers, which are comparable to fibers with similar concentrations made via the MCVD/solution doping and REPUSIL technique. Much higher laser output powers with excellent beam quality and slope efficiency of 80% in the kW range have been achieved in an Yb/Al/Ce co-doped double-clad fiber (Zheng et al. 2017).

Compared to the MCVD/solution doping method, this gas phase deposition technology makes it possible to produce efficient preforms with large core diameters (up to about 5 mm), homogenous radial distribution concerning the refractive index, and dopant concentrations by the continuous deposition of several layers. The doping level of Al can be increased significantly up to about 18 mol% Al_2O_3. Furthermore, it is possible to implement active fibers with high core to cladding ratios in high-power multimode and LMA fibers or to establish these preforms as active basic materials for the production of solid "multifilament" cores in microstructured fibers.

However, at present it is difficult to incorporate all RE elements with sufficient concentration because of the absence of suitable precursors. Furthermore, until now this gas phase doping process is less investigated compared to the MCVD/solution doping method. Therefore, it is required to continue with further systematically basic investigations concerning the incorporation of Yb and Al and their interaction, the interaction with other co-dopants (such as P, F, B, etc.), and the minimization of PD losses by suitable core compositions. Other research works should focus on the efficient incorporation of further important RE ions via the gas phase, using Ce doping to reduce PD losses, Tm for the preparation of laser fibers for a wavelength of around 2 μm.

Conclusion

Rare-earth doped silica based fibers are used to an increasing degree for high-power lasers and amplifiers in industry, science, and medicine. Silica is the preferred basis material, but the host properties can be improved using co-dopants. These co-dopants influence the optical properties such as refractive index distribution, absorption and emission behavior, and optical background losses. Advantageous and detrimental effects of the relevant preparation technology must be carefully considered for an optimal design for a high-power laser fiber.

References

A.J. Boyland, A.S. Webb, S. Yoo, F.H. Mountfort, M.P. Kalita, R.J. Standish, J.K. Sahu, D.J. Richardson, D.N. Payne, J. Lightw. Technol. **29**, 912 (2011)
M. Bubnov, V.N. Vechkanov, A.N. Guryanov, K.V. Zotov, D.S. Lipatov, M.E. Likhachev, M.V. Yashkov, Inorg. Mat. **45**, 444 (2009)
G. Canat, S. Jetschke, S. Unger, L. Lombard, P. Bourdon, J. Kirchhof, V. Jolivet, A. Dolfi, O. Vasseur, Opt. Lett. **33**, 2701 (2008)
D.J. Di Giovanni, J.B. Mac Chesney, T.Y. Kometani, J. Non-Cryst. Solids **113**, 58 (1989)
T. Izawa, IEEE J. Sel. Top. Quantum Electron. **6**, 1220 (2000)
C. Jauregui, H.-J. Otto, F. Stutzki, J. Limpert, A. Tünnermann, Opt. Exp. **23**(16), 20203 (2015)
Y. Jeong, J.K. Sahu, D.N. Payne, J. Nilsson, Opt. Exp. **12**, 6088 (2004)
S. Jetschke, S. Unger, U. Röpke, J. Kirchhof, Opt. Exp. **15**, 14838 (2007)
S. Jetschke, S. Unger, A. Schwuchow, M. Leich, J. Kirchhof, Opt. Exp. **16**, 15540 (2008)
S. Jetschke, S. Unger, A. Schwuchow, M. Leich, J. Fiebrandt, M. Jäger, J. Kirchhof, Opt. Exp. **21**, 7590 (2013)
S. Jetschke, S. Unger, A. Schwuchow, M. Leich, M. Jäger, Opt. Exp. **24**, 13009 (2016)

J. Kirchhof, S. Unger, A. Schwuchow, Proc. SPIE **4957**, 1 (2003)
J. Kirchhof, S. Unger, A. Schwuchow, S. Jetschke, B. Knappe, Proc. SPIE **5350**, 222 (2004a)
J. Kirchhof, S. Unger, J. Dellith, J. Non-Cryst. Solids **345–346**, 234 (2004b)
J. Kirchhof, S. Unger, A. Schwuchow, S. Jetschke, B. Knappe, Proc. SPIE **5723**, 261 (2005)
J. Kirchhof, S. Unger, A. Schwuchow, S. Grimm, V. Reichel, J. Non-Cryst. Solids **352**, 2399 (2006)
J. Kirchhof, S. Unger, S. Jetschke, A. Schwuchow, M. Leich, V. Reichel, Proc. SPIE, 7195 (2009)
T. Kitabayashi, M. Ikeda, M. Nakai, T. Sakai, K. Himeno, K. Ohashi, Optical Fiber Communication Conference/National Fiber Optic Engineers Conference, OThC5 (2006)
J.J. Koponen, M.J. Söderlund, S.K.T. Tammela, H. Po, Proc. SPIE **5990**, 599008 (2005)
A. Langner, M. Such, G. Schötz, S. Grimm, F. Just, M. Leich, C. Mühlig, J. Kobelke, A. Schwuchow, O. Mehl, O. Strauch, R. Niedrig, B. Wedel, G. Rehmann, V. Krause, Proc. SPIE **7914**, 79141U (2011)
M. Leich, S. Jetschke, S. Unger, J. Kirchhof, J. Opt. Soc. Am. B **28**, 65 (2011)
F. Lindner, C. Aichele, A. Schwuchow, M. Leich, A. Scheffel, S. Unger, Proc. SPIE **8982**, 89820R (2014)
J. Mac Chesney, D.J. Di Giovanni, J. Am, Ceram. Soc. **73**, 3527 (1990)
K.E. Mattsson, Opt. Exp. **17**, 17855 (2009)
K.E. Mattsson, Opt. Exp. **19**, 19797 (2011)
J.J. Montiel i Ponsoda, I. Norin, C. Ye, M.J. Söderlund, A. Tervonen, S. Honkanen, Opt. Exp. **20**, 25085 (2012)
T.F. Morse, A. Kilian, L. Reinhart, W. Risen, J.W. Cipolla, J. Non-Cryst. Solids **129**, 93 (1991)
S.R. Nagel, J.B. Mac Chesney, K.L. Walker, IEEE J. Quantum Electron. **QE-18**, 459 (1982)
R. Paschotta, J. Nilsson, P.R. Barber, J.E. Caplen, A.C. Tropper, D.C. Hanna, Opt. Commun. **136**, 375 (1997)
U. Pedrazza, V. Romano, V. Romano, Opt. Mat. **29**, 905 (2007)
A. Popp, A. Voss, T. Graf, S. Unger, J. Kirchhof, H. Bartelt, Laser Phys. Lett. **8**, 887 (2011)
D.J. Richardson, J. Nilsson, W.A. Clarkson, J. Opt. Soc. Am. B **27**, 63 (2010)
M. Saha, A. Pal, R. Sen, IEEE Photon. Techn. Lett. **26**, 58 (2014)
E.A. Saveley, K.M. Golant, Opt. Mat. Exp. **5**, 2337 (2015)
K. Schuster, S. Unger, C. Aichele, F. Lindner, S. Grimm, D. Litzkendorf, J. Kobelke, J. Bierlich, K. Wondraczek, H. Bartelt, Adv. Opt. Techn. **3**, 447 (2014)
E.H. Sekiya, P. Barua, K. Saito, A.J. Ikushima, J. Non-Cryst. Solids **354**, 4737 (2008)
S. Tammela, M. Söderlund, J. Koponen, V. Philippov, P. Stenius, Proc. SPIE **6116**, 61160G (2006)
F. Tosco, CSELT, *Fiber Optics Communication Handbook* (TAB Professional and Reference Books, Blue Ridge Summit, 1990)
E. Townsend, S.B. Poole, D.N. Payne, Electron. Lett. **23**, 329 (1987)
R.P. Tumminelli, B.C. Mc Collum, E. Snitzer, J. Lightw. Technol. **8**, 1680 (1990)
A. Tünnermann, H. Zellmer, W. Schöne, A. Giesen, K. Contag, in High-Power Diode Lasers, ed. by R. Diehl (Springer, Berlin/Heidelberg, 2000), p. 369
A. Tünnermann, S. Höfer, A. Liem, J. Limpert, M. Reich, F. Röser, T. Schreiber, H. Zellmer, Proc. SPIE **5709**, 301 (2005)
S. Unger, A. Schwuchow, J. Dellith, J. Kirchhof, Proc. SPIE **6469**, 646913 (2007)
S. Unger, J. Dellith, A. Scheffel, J. Kirchhof, Phys. Chem. Glasses Eur. J. Glass Sci. Technol. B **52**, 41 (2011)
S. Unger, A. Schwuchow, S. Jetschke, S. Grimm, A. Scheffel, J. Kirchhof, Proc. SPIE **8621**, 862116 (2013)
S. Unger, F. Lindner, C. Aichele, M. Leich, A. Schwuchow, J. Kobelke, J. Dellith, K. Schuster, H. Bartelt, Laser Phys. **035103**, 24 (2014)
J. Wang, S. Gray, D.T. Walton, M. Li, X. Chen, A. Liu, L.A. Zenteno, Proc. SPIE **6890**, 689006 (2008)

C. Wirth, O. Schmidt, A. Kliner, T. Schreiber, R. Eberhardt, A. Tünnermann, Opt. Lett. **36**, 3061 (2011)

J. Zheng, W. Zhao, B. Zhao, C. Hou, Z. Li, G. Li, Q. Gao, P. Ju, W. Gao, S. She, P. Wu, W. Li, Opt. Mat. Exp. **7**, 1259 (2017)

H. Zimer, M. Kozak, A. Liem, F. Flohrer, F. Doerfel, P. Riedel, S. Linke, R. Horley, F. Ghiringhelli, S. Demoulins, M. Zervas, J. Kirchhof, S. Unger, S. Jetschke, T. Peschel, T. Schreiber, Proc. SPIE **7914**, 791414 (2011)

Powder Process for Fabrication of Rare Earth-Doped Fibers for Lasers and Amplifiers

18

Valerio Romano, Sönke Pilz, and Hossein Najafi

Contents

Introduction	680
Optical Glass and Fibers	682
Modern Optical Fibers	683
Technological Changes in Optical Glass Fiber Production Techniques	684
Production of Active Fibers by Rare Earth Activation	686
Rediscovering Powder Techniques for Fiber Production	687
Powder Technologies for Fiber Production	688
Powder-in-Tube (PIT)	689
Improving Homogeneity	691
Refractive Index Control by Simultaneous Addition of Al_2O_3 and P_2O_5	691
Producing the Core Material Outside of the Preform	692
General Granulate Considerations	694
Granulated Silica	694
Powder-in-Tube (PIT) Technique	694
Doping Concentration	697
Granulated Silica Material Production: The Oxides Approach	700
Powder Synthesis by Mixing Oxides	700
Post-Processing: Oxides Derived Granulate	701
Fibers Based on the Oxides Approach	703
Granulated Silica Material Production: The Sol-Gel Approach	703
Powder Synthesis Using the Sol-Gel Process	703

V. Romano (✉)
Institute for Applied Laser, Photonics and Surface Technologies (ALPS), Bern University of Applied Sciences, Burgdorf, Switzerland

Institute of Applied Physics (IAP), University of Bern, Bern, Switzerland
e-mail: valerio.romano@iap.unibe.ch

S. Pilz · H. Najafi
Institute for Applied Laser, Photonics and Surface Technologies (ALPS), Bern University of Applied Sciences, Burgdorf, Switzerland
e-mail: soenke.pilz@bfh.ch; hossein.najafi@bfh.ch

© Springer Nature Singapore Pte Ltd. 2019
G.-D. Peng (ed.), *Handbook of Optical Fibers*,
https://doi.org/10.1007/978-981-10-7087-7_51

Post-Processing: Sol-Gel-Derived Granulate 707
Fibers Based on the Sol-Gel Approach 710
Solubility and Homogeneity of Rare Earth Elements 710
Thermodynamic Properties of Rare Earth Ion-Doped Silica Powder (T_g, T_x, T_c) 716
References 720

Abstract

The use of powder-based technologies for the production of rare earth (RE)-doped fibers and preforms is discussed. Although these technologies cannot compete with vapor-based technologies such as modified chemical vapor deposition (MCVD) with respect to purity of the silica material obtained, they offer a high degree of versatility with respect to the material composition and the obtainable topology of microstructured fibers.

The production of core rods starting from powder technologies and the powder-in-tube method are discussed. The challenges when using powder-based technologies lie in obtaining homogeneously doped and co-doped material as well as avoiding scattering by ion clusters. To reach a homogeneous distribution of dopants, the use of the sol-gel technology is discussed. Especially the incorporation of aluminum (Al) and phosphorus (P) to enhance the solubility of the rare earth activators as well as to control the index raise is found to be considerably eased.

Considerations from materials science point of view are made and serve as guidelines to understand the process. In this context, extremely precise characterization techniques such as wavelength dispersive x-ray fluorescence (WDXRF), scanning transmission electron microscopy with high-angle annular dark-field (STEM-HAADF), and differential thermal analysis (DTA) are discussed in order to mature the tuning of glass composition and drawing process. The thermodynamic properties of the doped glass powders discussed here could be crucial in assessing the thermal stability of the glass, required cooling rate, and its susceptibility to temperature changes during vitrification, devitrification, and fiber drawing steps.

Introduction

No other optical components depend so much on glass quality as optical fibers. This is self-evident when using optical fibers; in general long or very long propagation distances are involved, while in other optical devices and instruments, light travels only for a few centimeters through glass.

Traditionally, glass production has been based on powders that are processed at elevated temperatures. As historically documented (Rasmussen 2012), these processes in the middle ages yielded astonishingly good glass for the technology used with high transparency. Venetian "cristallo" glass reached transmission losses of merely 10 dB/m to 100 dB/m (see Fig. 1) which was already sufficient to produce

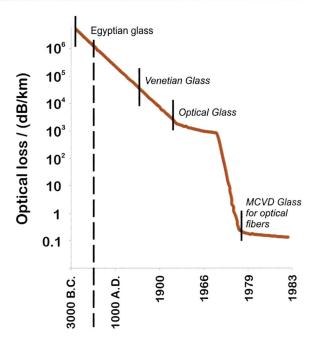

Fig. 1 Historical reduction of loss as a function of time (Nagel 1987)

optical lenses (Nagel 1987). The secret for the excellent quality of "cristallo" glass was the raw material underneath. Venetians used quartz pebbles from the Trentino and Adige rivers to produce their powders and mixed them with soda ash from plants (Rasmussen 2012). Moreover, it is documented that Galileo Galilei was in search of ways to improve the transparency of the glasses by using crystalline quartz as base for the powders and tartaric acid as a soda source (Sagredo 1618).

Although powder-based glass production was very successful in terms of quantity to quality ratio, it wouldn't have been good enough to produce optical fibers.

It was only in the second half of the twentieth century with the adoption of the modified chemical vapor deposition (MCVD) technique for the fiber preform production when the loss levels became low enough to allow for light propagation across kilometers inside the optical fiber. This achievement was ingenious but had a high price: first, at the beginning, it restricted the choice of optical materials mostly to silica and germanium, and second, it set the cylindrically symmetric shape as the de facto fiber structure. Although this was a high limitation if one considers the high variety of microstructures and materials employed today, the reward was enormous. By virtue of the intrinsic filtering effect of the vapor deposition method that results from the high difference in vapor pressure of the liquid silica and germanium precursors and the metallic impurities, it has been possible to develop material of such high purity where the limit is set only by Rayleigh scattering (Nagel et al. 1982) of the index fluctuation in the fiber core when drawing.

It was only under the pressure of modern fiber paradigms – the need for other materials, larger cores, higher dopant levels, and the ability to depart from the circular symmetry or to structure the fibers in their cross region – that one started to seriously think about other production methods, e.g., powder-based methods (e.g., powder-in-tube or granulated silica method).

In this part, we will discuss some silica powder-based fiber production technologies to produce rare earth (RE)-doped fibers. As co-dopants, we will limit our considerations to aluminum (Al) and phosphorus (P) that give us the possibility to control the solubility of the rare earths or other activation dopants and of the refractive index.

Optical Glass and Fibers

The improvement in quality of optical glass (quality being understood as low optical losses and high transmission to produce optical instruments) reached a plateau between the end of the nineteenth and the first half of the twentieth centuries with losses in the range of 1000 dB/km (Nagel 1987). Such glass was pure enough for most optical components like lenses or prisms with their dimensions of some millimeters along the optical path. However, these losses were still too high to even imagine the use of such glass for the production of long optical fibers.

Glass fibers, in their early implementation, were simple thin cylindrical rods without cladding. The necessary index step needed to achieve total internal reflection (TIR) for guiding light was given by the glass-air transition at their surface. The losses were high, and in perfect fibers with no microbubbles, they consisted of the intrinsic material losses of 1000 dB/km and other losses deriving from imperfections at the surface.

From the point of view of fiber structure and implementation of the TIR effect, one can identify the birth of the modern optical fiber around 1951. The proposed fibers consisted of a core and cladding (Hecht 1999); the cladding had a lower refractive index. From today's perspective, this might be regarded as a small development step; however, this substantial improvement to the fiber's transmission properties (Hecht 1999) was not obvious. Despite this improvement, the losses were still high. But they were low enough for applications such as fiber bundles for image transmission (Hecht 1999).

In their study on dielectric fiber surface waveguides, (Kao and Hockam 1966) concluded that impurities were the reason for the optical losses and therefore postulated that by reducing the impurities the losses would decrease.

In the sequel, a most significant milestone was reached in 1970 (Kapron et al. 1970) when fibers with losses below 10 dB/km were drawn. Thereby optical fibers, for the first time, entered the regime where one could imagine using them for long-distance transmission of light signals. Once the process had been initiated, the fabrication technology was improved until in 1979 losses as low as 0.2 dB/km (at 1.55 μm wavelength) (Miya et al. 1979) were obtained. One must highly treasure that this loss figure is very close to the fundamental limit given by Rayleigh scattering. The availability of such low loss fibers led to revolutionary developments in the field of telecommunication. The technology used to produce the preforms for such fibers was modified chemical vapor deposition (MCVD).

Modern Optical Fibers

Modern optical fibers have become complex micro-nanowaveguide devices that can consist of specially tailored core and cladding regions. The tailoring can be done:

(i) At the level of the materials used to build the fiber to fit the wavelength range to be transmitted and to control the index step between core and cladding regions.
(ii) It can be done at the fiber structure level by intelligently creating regular microstructures across the fibers.

The latter allows exploiting a variety of effects that go well beyond the simple TIR to cover a vast field of applications.

Apart from guiding light pulses for transmission of information, modern optical fibers span an application range that goes from generating continuous (CW) and pulsed laser light at high power levels (up to kilowatts average power) if their core is appropriately doped with optically active elements. They can be used for measuring and sensing and can be arranged and spliced together or structured longitudinally to obtain special optical functions such as interferometers or act as mirrors (e.g., fiber Bragg gratings (FBG)). Solid-core photonic crystal fibers allow to obtain fibers with large cores that conserve a high beam quality or allow to obtain wavelength-independent single-mode behavior over a wide range of wavelengths. Hollow core fibers in their different variants can guide light pulses with peak intensities in the range of hundreds of TW/cm^2 or offer possibilities for dispersion compensation to name just a few.

A crucial step for telecommunication was realized in the late 1980s with the introduction of fiber-optical amplification (Mears et al. 1987; Desurvire et al. 1987). Although fibers could transmit optical signals over hundreds of kilometers with low losses, after very long distances (e.g., transoceanic communications), signals had still to be amplified and regenerated. Although this could be done electrically, mainly for transoceanic connections, better ways had to be found. The solution was found with the fiber-optic amplifier. In a rough approximation, one can see this device as a piece of active fiber spliced in between two passive telecommunication fiber pieces. The pump power for the amplifier would be transported by the fiber itself, and there would be no need for additional electrical installations to regenerate the signals electronically.

Maximum power levels of generated light in single-mode fiber lasers and amplifiers were about 100 W and the pump light was coupled into the core. This required the pump light to have the same high quality (focusability) as the generated or amplified light, and, hence, the power limit was mostly given by the maximum pump power that could be coupled into the small single-mode core.

A remarkable progress step was done in 1989 when it was proposed to inject the pump light into the cladding of the active fiber that was surrounded by a lower refractive index region (double-clad fibers). In general, this low-index region could be a low-index polymer or a further cladding consisting of fluorinated glass (Snitzer et al. 1988). In this configuration the pump can be multimode and much higher power levels can be coupled into the active fiber. Later, together with the development of large mode area (LMA) fibers with core diameters up to 25 μm, fiber lasers have reached CW power levels in the multi-kilowatt range. As we will see below, this was a strong driving force for the development of alternative fiber production techniques.

Technological Changes in Optical Glass Fiber Production Techniques

Hand in hand with its evolution and the changing paradigms, the optical fiber went through different production techniques. While in the early stages, up to the first middle of the twentieth century, fibers were produced by heating up glass and pulling it into fine fibers – a very old technique – or by extrusion of a low viscosity melt through appropriate nozzles, later fiber production techniques were more elaborated. A good overview is given in Méndez and Morse (2006).

However, to enter its main application field, telecommunication, fiber production techniques that minimized the transmission losses as much as possible had to be found. This task was difficult since at that time the main source of loss was not yet known.

Developing a core-cladding structure in 1951 solved the problem of scattering at the fiber surface and probably would also solve the problem of cross talk from one fiber to its neighbors in the fiber bundles, but the intrinsic losses of 1000 dB/km could not be undercut.

With the disruptive hypothesis of Kao and coworkers in 1966 that the main responsibility for the 1 dB/m fiber loss plateau was absorption by impurities (Kao and Hockham 1966), the search for a high-purity fiber production technique started.

It was very quickly recognized that this had to be a non-powder-based technique. The limits of powder purification had been reached, and high-temperature treatment for melting the powders was prone to introduce contaminations.

Chemical Vapor Deposition (CVD)

Chemical vapor deposition (CVD) offered itself as a valid route to high-purity materials. The crucial point that makes this method so well suited to achieve high purity is that the deposition of the desired material combination is performed from gas phase (Oh and Paek 2012).

For CVD, the starting materials (in general liquid precursors) are heated up to their evaporation temperature and are then brought in contact with the heated substrate. During this process, they can chemically react until they finally coat the hot substrate.

This method yields intrinsically materials of highest purity since during evaporation the different vapor pressures of the individual materials can be used to discriminate among deposition material and impurities. In general, impurities (e.g., iron (Fe)) at given temperatures have vapor pressures that are orders of magnitude lower than the materials to be deposited and are naturally excluded from the deposition process.

CVD was adapted to the production of fiber preforms in several variants. One of the first variants to be adopted was called modified chemical vapor deposition (MCVD).

Modified Chemical Vapor Deposition (MCVD)

In MCVD the materials that will form the core (typically silica doped with germanium (Ge) or phosphorus (P) for telecommunication fibers) are evaporated from liquid precursors and deposited layer by layer onto the inner wall of a pure silica glass tube (see Fig. 2). The tube rotates on a lathe around its longitudinal axis, and the gaseous precursors are mixed inside the tube. The tube is heated by a flame from a traveling burner that moves back and forth along the tube. The chemicals react on the heated surface of the inner surface of the tube and, depending on the transformation temperatures of the deposited materials and the temperature to which the tube is heated by the traveling flame, are deposited there either as a porous soot or as a vitrified layer.

In general, for MCVD-based silica telecommunication fibers, the involved materials are SiO_2 as a base material and GeO_2 and P_2O_5 as co-dopants to control the optical properties of the core (e.g., raising the index).

The MCVD process and the involved chemical reactions are very well described in Unger et al. (2014).

For telecommunication, the choice of MCVD was a stroke of genius. With their simple cylindrically symmetric shape and a small core size between 4 μm and

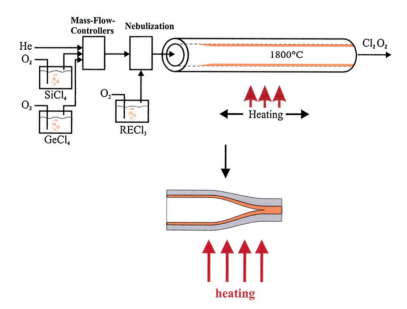

Fig. 2 MCVD apparatus for silica deposition and addition of dopants and co-dopants (Unger et al. 2014)

10 μm for single-mode operation at 1.55 μm wavelength, there were no further serious challenges to the production of telecommunication fiber preforms and fibers.

As a matter of fact, this synthetic silica- and Ge-doped core material based on MCVD is the only optical material where the limit of the ultimate theoretically possible losses has (almost) been reached (Miya et al. 1979).

Production of Active Fibers by Rare Earth Activation

The fascinating success of the passive optical fiber for telecommunication drawn from preforms produced by MCVD was so remarkable that MCVD was adopted also for the production of active fibers. In 1985/1987 when the erbium-doped fiber amplifier for optical communication (Mears et al. 1987; Desurvire et al. 1987) was conceived to ease the use of the optical fiber in long-distance telecommunication, it was quite natural to start from MCVD preforms to produce fibers with active cores.

In 1983, for the first time, rare earth (RE) doping of single-mode fibers had been demonstrated. Hegarty and coworkers studied the fundamental mechanisms of RE ions in amorphous hosts (Hegarty et al. 1983). The experiments were done on fibers fabricated by MCVD. They had a 6 μm core of pure silica (SiO_2), doped with 10 ppm of Nd, surrounded by a depressed index fluorine-doped silica cladding. The background loss of the fiber, away from any Nd absorption peak, was relatively high for MCVD (8 dB/km at 6.38 μm).

A few years later, further improvements in using the MCVD technique to fabricate (RE-doped single-mode) fibers were achieved by Poole and coworkers at the University of Southampton, UK (Poole et al. 1985), resulting in RE-doped fibers with low background loss. The fiber was 2 m in length, with the cleaved fiber ends butted directly to mirrors (highly reflective at the lasing wavelength and transmissive at the pump wavelength), as the pump light was injected through one of the ends of the fiber.

All the necessary ingredients now being in place, the development of low-loss single-mode fiber lasers was followed shortly thereafter by that of fiber amplifiers.

Erbium (Er)-doped single-mode fiber amplifiers for amplification of 1.55 μm signals were simultaneously developed in 1987 at the University of Southampton and at AT&T Bell Laboratories (Philippe et al. 1999).

After establishing the Er-doped fiber amplifier as an omnipresent and essential optical component in telecommunication systems, the way was paved for other fiber laser applications. However, the fiber production methods were all gravitating around the original paradigm of loss minimization.

Furthermore, the extremely successful MCVD fiber production technique for telecommunication encouraged to stick with this technique and just slightly modify it for the production of other fibers.

While MCVD with its intrinsically high-purity level offered a good basis for passive and active single-mode fibers, for certain other applications, it was not the best choice. There were several bottlenecks such as the concentration limit in rare earth (RE) content since rare earths are not easily soluble in pure silica (Arai et al. 1986). This was relaxed by the addition of aluminum as a co-dopant, but it was necessary to introduce a significant amount of aluminum atoms per RE atom (a typical atomic ratio is 7–10 to 1). However, practical experiments showed that with solution doping, the amount of dopants and co-dopants together was limited to only about 15 wt.%. For the solution doping method, the inner wall of a silica tube was coated with pure silica in a conventional CVD way. The resulting layer was then porous and it was impregnated with a liquid solution of the desired dopants and co-dopants. The tube was then collapsed to a preform at high temperature.

Apart from the concentration limitations given by the difficulty to impregnate a porous layer with a solution, there were other drawbacks of MCVD for the production of active fibers. The elevated temperatures during the preform collapse led to evaporation of the co-dopants (of innermost layers), e.g., in the case of GeO_2 co-doping. Typically, this is visible in the refractive index dip in the center of fibers produced by MCVD. Furthermore, the fiber geometry was restricted to cylindrically symmetric structures and thick layers needed for future large mode area (LMA) fibers could not be easily achieved.

Rediscovering Powder Techniques for Fiber Production

Especially fiber lasers and amplifiers boosted the development of active fibers and fiber-based components. In fact, the first fiber laser was built already in 1961, immediately after Elias Snitzer's publication in 1961 about a theoretical description of single-mode fibers whose core would be so small it could only carry light with one waveguide mode (Snitzer 1961). Snitzer was able to demonstrate the coupling of a laser beam into a thin glass fiber. The transmitted light was sufficient for medical applications, but for communication applications the light loss became too high (Snitzer 1961).

An active fiber is a fiber whose core contains a laser active material. As mentioned before, erbium (Er)-doped fibers are mainly interesting for telecommunication because of their amplification at 1.55 μm wavelength. At this wavelength, the superposition of fundamental losses in silica incidentally has a minimum: on one side, the losses given by Rayleigh scattering decrease with λ^{-4}, and on the other side, the losses given from water absorption start to increase. So, the 1.55 μm wavelength is very well suited for optical fiber communications.

However, for laser applications also, other wavelengths and thus dopants are of interest. In fact, the first fiber laser was based on neodymium (Nd) as active element with a wavelength at 1064 nm. The choice of the active material fell on neodymium probably because of its four-level laser system and since laser emission can be obtained at very low pump power levels.

In the early 1990s, it became quite clear that there was more than just curiosity behind the fiber laser if some difficulties could be solved (such as the ability to precisely tailor the step index between core and cladding in order to obtain bigger core sizes at low-order modes); further one wanted to be able to introduce significantly higher dopant levels or to use other fiber materials rather than just silica-based materials (see Fig. 3). Fused silica limited the wavelength range to the

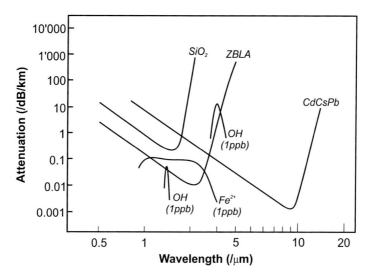

Fig. 3 Theoretical losses of lightguide material versus wavelength. BeCaKAl are fluoride glasses made with those cations; ZBLA represents a zirconium barium lanthanum aluminum fluoride glass: CdCsPb is a chloride glass made from those cations (Fanderlik 1983)

near-infrared (IR) region up to the 2 μm range (holmium (Ho) and thulium (Tm) lasers) but precluded a priori the use of wavelengths above 2.1 μm.

Another field was the doping of fibers with metals and transition metal elements (e.g., bismuth (Bi), copper (Cu), etc.).

This can be easily done with powder-based technologies, while it is more difficult to attempt it by MCVD.

Powder Technologies for Fiber Production

When rethinking the direct use of powders for producing fibers or fiber preforms, one has to make several considerations.

On the side of the "caveats" and challenges, one should take notice of the fact that powders, with their enormous surface areas, are prone to the adsorption of atmospheric contaminants, in general OH-groups. Care must be taken to remove such contamination. Depending on the light wavelength involved and the amount of OH-groups tolerated, the measures (see sections "Post-Processing: Oxides Derived Granulate" and "Thermodynamic Properties of Rare Earth Ion-Doped Silica Powder (T_g, T_x, T_c)") to be taken can range from simple thermal treatment at high temperatures (typically above 1000 °C for silica), or treatments assisted by chlorine respectively deuterium at 800 °C–1000 °C could represent valid possibilities.

Apart from the challenges with OH-groups, when melting the powders at high temperatures, (micro)bubbles can be formed which in turn scatter light and thus contribute to undesired background losses (see subsections "Powder-in-Tube (PIT) Technique" and "Powder Synthesis by Mixing Oxides").

Another caveat is an increased scattering due to microcrystalline parts in the powders that can form during powder production (e.g., sintering, see subsection "Fine Milling and Sintering" in section "Post-Processing: Sol-Gel-Derived Granulate"). Furthermore, if the powders are not homogeneous (see sections "Powder Synthesis by Mixing Oxides" and "Post-Processing: Oxides Derived Granulate"), scattering can occur from the resulting refractive index fluctuations.

On the side of the chances offered by using powder methods is the possibility to be able to use wide range of dopant materials, to achieve almost arbitrary material compositions including very high doping concentrations (on the basis of using traditional thermal equilibrium glass production methods). With traditional glass production methods, the glassy state is obtained by tailoring the composition of the glass components in such a way that even at low cooling rates, no crystallization or devitrification occurs.

One main opportunity offered by the combination of the powder-in-tube (PIT) technique with the peculiarities of the fiber drawing process is given by the high quenching rate when the fiber leaves the furnace. If one makes sure that the powders in the tube reach at least their melting temperature inside the tube, virtually any powder composition can be drawn to a glass. When drawing the preform into a fiber, it is heated up slowly, and if one takes care of designing the powder respectively granulate composition in such a way that it melts (melting temperature of the granulate $T_{f,\,gr}$) before the containing silica preform tube softens (onset melting temperature of the tube $T_{xf,\,tu}$, see subsection "Powder-in-Tube (PIT) Technique"), then the melt will have enough time to form a homogeneous viscous liquid. The fiber will then be pulled away from the preform neck region at speeds of typically 0.2 m/s to 1 m/s (for special optical fibers). Outside the drawing furnace, the fiber (with its tiny volume to surface ratio) quickly cools down within some few meters. Cooling rates of 5500°K/s have been calculated and measured during the first 300 ms after fiber formation from the neck and up to 75,000°K/s during the first 10 ms at moderate drawing speeds (Paek and Kurkjian 1975). In this way glass formation occurs via rapid quenching of a melt and not via glass composition. This allows the production of material compositions that do not form glasses with standard equilibrium glass production techniques.

It is well known that in some cases the quality of regular fibers can be improved by active cooling the fiber immediately when it leaves the hot furnace. It has also been observed that silica powder showing crystallization from heat treatment (e.g., sintering) is transformed into amorphous glass during fiber drawing with the powder-in-tube (PIT) method (see subsection "Improving Homogeneity" and subsection "Fine Milling and Sintering" in section "Post-Processing: Sol-Gel-Derived Granulate").

Powder-in-Tube (PIT)

The arguments furnished above led to the development of a technique called powder-in-tube (PIT) technique. Besides offering the possibility of mixing glass components that otherwise are immiscible, it allows to obtain high dopant concentrations. This was first successfully demonstrated by Ballato and Snitzer in 1995 (Ballato and Snitzer 1995). They were able to produce:

> fibers that contained 54wt.% Tb_2O_3, 27wt.% SiO_2, 18wt.% Al_2O_3, and 1wt.% Sb_2O_3 (as a fining agent) were successfully produced by the powder-in-tube procedure. A complete fusion between the silica cane and capillary was observed in a test section of 250m of fiber. This optical assessment agreed well with electron microscopy results, which indicated core–clad fusion without interfacial crystallization.
>
> A major concern was that unfined bubbles would be found in the core. These could lead to voids in the drawn fiber. This concern did not turn out to be a problem, as no scattering centers could be visually observed. The test section contained a continuous high rare-earth concentration core, which had cooled from a melt at the preform neckdown region. (Ballato and Snitzer 1995)

A typical implementation of the powder-in-tube method (PIT) is depicted in Fig. 4. It consists of filling a central capillary tube with doped granulated silica (doped granular silica material) and putting it in the center of a bigger second tube filled with undoped granulated silica (e.g., pure silica granulate) that has the same refractive index as the tubes.

In a similar way, Renner and coworkers were able to draw active silica fibers doped with neodymium and co-doped with aluminum (Nd-/Al-doped) for laser application. Typical background losses that can be obtained by this method are of the order of 1 dB/m (Renner-Erny et al. 2007). Renner and coworkers obtained 0.75 dB/m measured at 633 nm wavelength (Renner-Erny et al. 2007).

Unhomogeneously distributed dopants and the formation of microbubbles can lead to refractive index fluctuations and thus to scattering. This effect is observed mostly in the case when the drawing temperature is not clearly higher than the

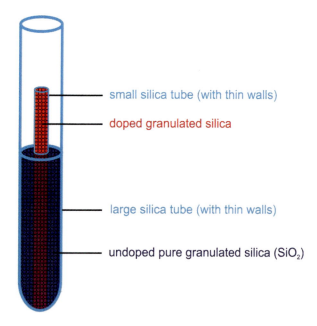

Fig. 4 Preform for powder-in-tube (PIT) technique as described by Renner and coworkers (Renner-Erny et al. 2007)

melting temperature of the powder respectively granulate mixture in the preform tube (see section "Improving Homogeneity").

Improving Homogeneity

There are several ways to improve the homogeneity of the doped material used for the powder-in-tube (PIT) technique. One of them is to solidify the powder mixture outside the preform, followed by iteratively milling and remelting (Romano et al. 2014). This method is old and it was known also earlier in history for the production of colored glass. It is a common procedure for the homogenization of the dopant and co-dopant distribution in active glasses for laser rods.

However, the required high temperature in the (firing) furnace can lead to contamination of the glass with impurities present in the furnace. This problem can be mitigated or even eliminated by melting the glasses on the basis of a laser treatment. The CO_2 laser is very well suited for this task since its wavelength (10.6 μm) is strongly absorbed and transformed to heat (Romano et al. 2014).

Another successful method is based on the stack-and-draw technique (Velmiskin et al. 2012). In this method thin glassy rods of about 1 mm diameter from a PIT preform on the basis of original doped powder respectively granulate are produced, the outer pure silica layer of the rod stemming from the undoped silica preform tube undoped silica tube is eventually removed before the stacking the rods. The stacked rods are then redrawn again to a rod, which can be used for the core area of a new fiber. This stack-and-draw method can be applied iteratively (Velmiskin et al. 2012). The losses in dependence of the wavelength of a fiber drawn in such a way are shown in Fig. 5.

The effect of the stack-and-draw technique manifests itself in the homogenization of the refractive index profile and is shown in Fig. 6. It is also noteworthy that the central dip present in MCVD fibers because of the evaporation of dopant from the central region when collapsing the fibers is completely absent when using the PIT method.

Refractive Index Control by Simultaneous Addition of Al_2O_3 and P_2O_5

An important issue when drawing fibers is the index step between core and cladding. This determines the numerical aperture of the fiber and allows to determine the modal properties of the guided light. When producing alumino-phosphosilicate fibers, the simultaneous addition of aluminum (Al) and phosphorus (P) can be beneficial in several respects: both aluminum and phosphorus increase the solubility of the RE dopants in silica, and both, individually, raise the index of the resulting material (DiGiovanni et al. 1989) (see Fig. 7).

In such a way, the index raise can be precisely tailored which gives a good control of the modal properties of the fiber.

Fig. 5 Optical loss spectra of fibers (on the basis of stack-and-draw technique for improving the homogeneity): 6, B2 (after one drawing-stacking-consolidation cycle); 20, C2 (two cycles); and 29, D2 (three cycles) (Velmiskin et al. 2012)

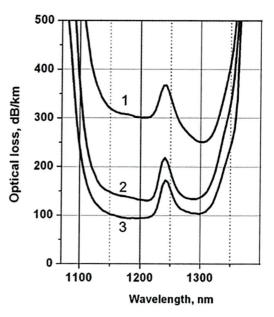

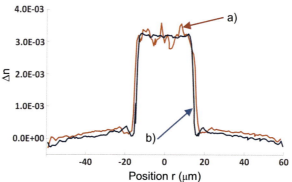

Fig. 6 Refractive index difference profiles across fibers (**a**) B1 and (**b**) C1. Δn is the index difference between the core material and undoped silica glass (Velmiskin et al. 2012)

Producing the Core Material Outside of the Preform

The possibility to produce the core separately from the preform opens up new opportunities in the powder-based production of fiber materials. The core can be produced and vitrified first and then put into a rod-in-tube (Méndez and Morse 2006) assembly to be drawn to a fiber. It is also possible to use the OVD method (Méndez and Morse 2006) to put a pure silica cladding over the core rod and then densify and simultaneously collapse the OVD overcladding. Since the porous OVD material can be collapsed at a much lower temperature than dense tubes (Méndez and Morse 2006), this allows to build the cladding at lower temperatures than are necessary for MCVD tube collapse.

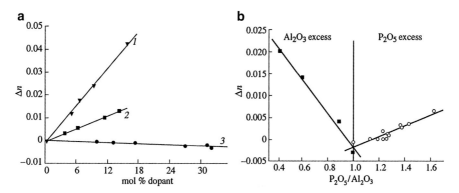

Fig. 7 Effect of doping level on the refractive index of silica glass: (**a**) 6, Al_2O_3; 20: P_2O_5; 29, equimolar amounts of P_2O_5. (**b**) Effect of the P_2O_5/Al_2O_3 molar ratio on the refractive index of silica glass and Al_2O_3 (DiGiovanni et al. 1989)

These two production methods justify the search for powder-based methods to produce high-quality fiber cores.

Core Material Production by Suspension Doping of Fine Silica Powder: REPUSIL

Recently, the Leibniz Institute of Photonic Technology (IPHT) in collaboration with the company Heraeus Quarzglas has presented a new technology for the preparation of doped glasses based on a powder sintering method (Schuster et al. 2012, 2014):

> This preparation method, the so-called REPUSIL process, can be considered as a modification of the solution doping process for production of the Al/RE-doped silica layer for fiber preforms. However, the doping and purification process is achieved outside the silica tube by using a suspension doping step. Because of this outside step, larger amounts of uniformly doped silica can be produced. Starting materials for the REPUSIL process are high-purity gas phase formed silica nanoparticles and water-soluble compounds of the doping components, e.g., $AlCl_3 \times 6H_2O$, RE chlorides and ammonium tetra borate. Defined amounts of the doping solution are mixed into a silica suspension under controlled adjustment of the pH value. As a result, the doping ions are precipitated as pure ore mixed hydroxides on the surface and in the pores that are intrinsically tied to the suspended silica particles. After drying the doped solution, a moldable granulate is produced with the help of isostatic pressure mostly in cylindrical shapes. Next, this porous green body is purified by chlorine at elevated temperatures to remove impurities from raw materials like iron, other 3d elements and, most importantly, bonded water. These cleaned glass precursors are fed into a matched fused silica tube. With the help of typical MCVD equipment, the doped, but not yet glassy silica body, is then transformed into transparent glass samples of up to 50g in weight.
>
> The sintering and vitrification process is carried out at temperatures of up to 2200 °C and is controlled by different adapted runs of an oxygen hydrogen burner. An electrical furnace with a very small heating zone can be used for this step of the process as well. .... Many modifications are available for this process. The prepared glass can be used directly for fiber drawing, or the undoped silica tube can be removed by grinding. Also, pure doped glass is available as rods for stack and draw technology for microstructured fiber preforms. The concept of powder sintering technology is also of special interest in providing fluorine doped glasses (reduced refractive index) since it is difficult to incorporate larger fluorine

amounts homogeneously via the classic MCVD solution doping process. Here, large losses in fluorine by diffusion and evaporation during the high temperature step were observed. The control of fluorine incorporation in alumina and RE-co-doped glasses enables new fiber optic applications, especially in the development of largecore fiber lasers with the highest brightness.

At present, the following parameters for high-purity doped silica glasses can been achieved following the REPUSIL process:

- content of alumina: 0–8mol%
- content of RE oxide in combination with aluminum oxide: 0–0.6mol%
- content of boron oxide in combination with aluminum oxide and RE oxide: 0–10mol%
- best attenuation (fiber loss) in Al/Yb-doped at 1200nm: 15dB/km.
- fluorine content (as mol% SiF_4): 20.0mol% SiF_4 in pure silica, 6.8mol% SiF_4 in Al/RE-doped silica

Doped preform core rods can also be produced by the granulated silica method. The details of this method will be described in the following part (section "General Granulate Considerations"), where we will go into the details of the production of doped alumino-phosphosilicate powders for the production of active materials.

General Granulate Considerations

In this chapter general considerations about the granulated silica, the powder-in-tube (PIT) technique, and doping concentration are presented before two different approaches – oxides and sol-gel – for the granulated silica method (Romano et al. 2014; Pilz et al. 2017) are described in detail in sections "Granulated Silica Material Production: The Oxides Approach" and "Granulated Silica Material Production: The Sol-Gel Approach."

Granulated Silica

Here, a powder is defined to feature an arbitrary grain size and grain size distribution, while a granulate possesses a specific and selected grain size and granulometric distribution. Two different approaches for the fabrication of (doped and undoped) granulated silica in order to use the granulated silica method are presented in detail in sections "Granulated Silica Material Production: The Oxides Approach" and "Granulated Silica Material Production: The Sol-Gel Approach." For glassy material, the main precursor for these two granulate production methods is silicon dioxide SiO_2 (oxides approach) and tetraethyl orthosilicate (TEOS) (sol-gel approach).

Powder-in-Tube (PIT) Technique

Key parameters for the fiber production based on the powder-in-tube (PIT) preform assembly technique (see Fig. 8a) are on the one hand the granulate characteristics

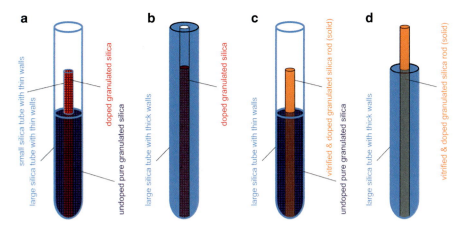

Fig. 8 Schematic of (**a**) the powder-in-tube technique; (**b**) and (**c**) the adapted powder-in-tube, where parts of the preform are solidified and no longer in powdery form; (**d**) the rod-in-tube technique

(material properties such as, e.g., melting temperature T_f and heat conductivity k, the grain size and granulometric distribution, filling height of the granulate in the preform tube) and on the other hand the preform tube characteristics (material properties such as, e.g., onset melting temperature T_{xf} and heat conductivity k, tube diameter, wall thickness of the tube) as well as the vacuum applied during drawing. Fibers made from the powder-in-tube technique respectively made by the granulated silica method feature intrinsically higher background losses than fiber made with MCVD or CVD (see section "Introduction") (Oh and Paek 2012; Unger et al. 2014). The main sources for higher background losses of fibers based on the power-in-tube technique are primarily microbubbles and other scattering centers (arising, e.g., from unmolten granulate or crystalline structure) in the fiber.

During the fiber drawing of a powder-in-tube preform, a vacuum must be applied. A good evacuation of the gaseous part from the preform is based on channels, where the gas in the granulate can escape. A different grain size respectively a broad size distribution can close most of these channels and thus makes it more complicated in general for the gaseous part to be evacuated from the preform. A bad evacuation (e.g., collapsed channels) will lead to the formation of microbubbles, which in turn results in higher scattering and thus higher background losses. However, the channels in the granulate are depending on the interaction of the grain size, the granulometric distribution, the filling height of the granulate in the preform, the diameter of the tube, and the level of vacuum. For example, a 17×21 mm (inner-diameter x outer-diameter) preform tube and a grain size of 100 um till 1 mm with a distribution of a few 100 um are suitable for a filling height of about 10 cm and a vacuum in the range of 10^{-3} mbar. For preforms with a smaller diameter (e.g., for the additional vitrification, see subsection "Additional Vitrification of

Thin Powder-in-Tube Preform" in section "Granulated Silica Material Production: The Sol-Gel Approach"), smaller grain sizes and a narrower size distribution are normally better suited for the same filling height of the granulate and the same vacuum.

For the best mechanical robustness of the fiber, it should be drawn at the lowest possible drawing temperature. In addition, for a powder-in-tube preform, the granulate inside the preform tube should be fully molten before the tube itself becomes soft (depending on the onset melting temperature T_{xf}, heat conductivity k, and wall thickness of the tube) and a droplet is formed, which will draw the fiber by gravity. If the granulate is not fully molten before the droplet is formed, unmolten granulate will be incorporated into the fiber. These unmolten granulate areas are then scattering centers in the fiber and lead to higher background losses. However, the melting during drawing of powder-in-tube preform is depending on the heat conductivity of the granulate k_{gr} and the tube k_{tu} as well on the melting temperature of the granulate $T_{f, gr}$ respectively on the onset melting temperature of the tube $T_{xf, tu}$. While $T_{f, gr}$ and $T_{xf, tu}$ are in turn depending on the composition and homogeneity of the granulate and the tube, the heat conductivity of the granulate k_{gr} is also depending on the grain size (smaller particle features a lower head conductivity). The heat conductivity influences the exposure time of the material to the drawing temperature in the drawing furnace till it is molten. So, after all it is an interaction of applied drawing temperature and exposure time. Just increasing the drawing temperature is not favorable for the production of mechanical strong fibers and does normally also not solve the different melting behaviors of the granulate and the preform tube, since an increase in drawing temperature will decrease the exposure time, with the result that the granulate does not have enough time to be totally molten before the droplet is formed. However, the issue of the unmolten granulate can be addressed by influencing respectively lowering the melting temperature $T_{f, gr}$ of the granulate in virtue of incorporating co-dopant(s). By adding, e.g., the co-dopant phosphorus, the melting temperature can be lowered to the region or beneath the onset melting temperature of the tube $T_{xf, tu}$ so that the granulate totally melts inside the tube before the droplet is formed.

Here, two different approaches for the granulated silica production are presented: the oxides approach section "Granulated Silica Material Production: The Oxides Approach" and the sol-gel approach section "Granulated Silica Material Production: The Sol-Gel Approach." The post-processing of the powder made by these two approaches into a granulate in order that it can be used for the granulated silica method differs (Romano et al. 2014). In subsections "Post-Processing: Oxides Derived Granulate" and "Post-Processing: Sol-Gel Derived Granulate," the particular post-processing steps of powder derived from these two approaches are presented in detail. In these sections, it is also shown that by the post-processing steps for the particular approach, one can get the control over the grain size and the granulometric distribution, which is very important for the powder-in-tube technique, as mentioned before.

Doping Concentration

The doping concentration is one of the most important parameters of optical passively or actively doped fibers. There can be found many different ways in the literature for indicating the doping concentration based on atomic or molar considerations (see Table 1). The definitions based on chemical elements respectively on atoms or ions such as the atomic percentage (Y_i in at.%), the atomic weight percentage (W_i in at.wt.%), and the number density (D_i in ions/cm^3) are explicit and leave no room for ambiguities. This is in contrast to the definitions based on molecules (e.g., molar percentage y_i in mol% and molar weight percentage w_i in mol wt.%), which are depending on the molecular configuration which in turn is depending on the number of atoms of the dopant or host element V_i in the specific molecule. The molecular configuration might change during the (fiber) production process (the dopant precursors usually feature different molecular configurations than the dopants in the final glass (oxides configurations)), so that there is no straightforward and direct comparison between the doping concentration based on molecules of, e.g., the precursors and the final glass. For doping concentration based on elements rather than molecules, this change does not exist, so that the doping concentration can be directly compared in, e.g., atomic percentage at.% over the whole fiber production process. Furthermore, most of the time, only the ratio of the host element (for glassy material (Si)) to the dopant elements is of interest and not the relation of the specific molecules. In this context only the host element and dopant elements are normally taken into account, and the remaining elements such as the oxygen (O) are neglected for the doping considerations. Another reason why it is better to consider the doping concentration based on elements rather than on molecules is that the atomic consideration is not depending on the ionic configuration of the element in the glass, since in the glass the valence respectively the oxidation state of the elements can be different (e.g., for ytterbium (Yb), Yb^{3+} and/or Yb^{2+}, etc.; for aluminum (Al), Al^{3+} and/or Al^{2+}, etc.) (Xia et al. 2012; Zhang et al. 2017; Wang et al. 2014; Marchi et al. 2005). Thus, the common assumption that glass consists only of the same ionic configuration as their most

Table 1 Overview of several common doping concentrations based on atomic or molar considerations

Physical quantity	Symbol	Unit	Basis
Number/amount of atoms	N_i	–	Element/atom
Atomic percentage	Y_i	at.%	Element/atom
Atomic weight percentage	W_i	at.wt.%	Element/atom
Number/particle density	D_i	ions/cm^3	Element/atom
Number/amount of moles	n_i	–	Molecule
Molar percentage	y_i	mol%	Molecule
Molar weight percentage	w_i	mol wt.%	Molecule

commonly encountered oxide form holds not always true (e.g., Yb^{3+} for Yb_2O_3, Al^{3+} for Al_2O_3, etc.).

The indication based on elements/atom is the most explicit way to indicate the doping concentration and is often written by the chemical symbols S_i and the atomic percentage Y_i in the form of $S_1 : S_2 : S_2 : \ldots = Y_1 : Y_2 : Y_3 \ldots$ (see Example 1). The atomic percentage Y_i itself is calculated from the numbers of atoms N_i by:

$$Y_i = 100 \frac{N_i}{\sum_i N_i} \qquad (1)$$

The corresponding atomic weight percentage W_i is determined by the atomic mass $A_{m,i}$ with the following equation:

$$W_i = 100 \frac{Y_i \cdot A_{m,i}}{\sum_i Y_i \cdot A_{m,i}} \qquad (2)$$

On the basis of atomic respectively ionic consideration, the number/particle density D_i is also a very usable specification, since many other important laser parameters such as gain and absorption can be directly calculated together with the transition cross sections. Here an approximation for the number density is presented which is valid for low doping concentrations, where it is assumed that the density respectively the volume of the glass is not changed by the addition of dopants. Furthermore, the before mentioned assumption that the final glass consist only of the same ionic configuration as their most commonly encountered oxide form is also used for this approximation. In this context the number density of pure silica D_{Si} can be calculated by $D_{Si} = N_A \frac{\rho_{SiO_2}}{M_{m,SiO_2}}$, where ρ_{SiO_2} and M_{m,SiO_2} are the density and molar mass of SiO_2 and N_A is the Avogadro constant. By the assumption that the density is not changed for low doping concentrations, the following approximation can be used:

$$D_i \approx D_{Si} \frac{Y_i}{Y_{Si}} \qquad (3)$$

However, the doping definitions based on molecules have its practical relevance in context of determining the initial weight (in molar weight percentage: mol wt.%) for mixing the precursors for, e.g., the two approaches of the granulate silica method presented in sections "Granulated Silica Material Production: The Oxides Approach" and "Granulated Silica Material Production: The Sol-Gel Approach." The indication for the molar consideration is often written by the molecular formula s_i and the molar percentage y_i in the form of $s_1 : s_2 : s_2 : \ldots = y_1 : y_2 : y_3 \ldots$ (see Example 1). The molar percentage is calculated by:

$$y_i = 100 \frac{N_i / V_i}{\sum_i N_i / V_i} \qquad (4)$$

where V_i is the number of atoms of host or dopant elements in the molecule. The corresponding molar weight percentage w_i is calculated by the molar/molecular mass $M_{m,i}$:

$$w_i = 100 \frac{y_i \cdot M_{m,i}}{\sum_i y_i \cdot M_{m,i}} \qquad (5)$$

Based on this doping discussion, the doping labeling must always be clear if it is an atomic or molar consideration and must correspond to the correct chemical symbols or molecular formula (general terms like silica powder, where it is not clear if the element Si is meant or the molecule SiO_2 should not be used). Furthermore, for a production process, the doping concentration must always be linked to the specific production processes step (e.g., precursor, sintered powder, final glass, etc.), since the doping composition must not be conserved over the whole production process.

Example 1: Yb-/Al-/P-doped glass

Atomic mass $A_{m,i}$:

Element	Atomic mass / u
Si	28.0855
Yb	173.04
Al	26.982
P	30.974

Molar/molecular mass $M_{m,i}$:

Oxide molecule	Molar mass / g/mol
SiO_2	60.1
Yb_2O_3	394.08
Al_2O_3	101.96
P_2O_5	141.96

→ Atomic percentage Y_i:

$Si : Yb : Al : P$
$=$
$94.73 at.\% : 0.3 at.\% : 3.11 at.\% : 1.86 at.\%$

→ Atomic weight percentage W_i:

$Si : Yb : Al : P$
$=$
$93.22 at.wt.\% : 1.82 at.wt.\% : 2.94 at.wt.\% : 2.02 at.wt.\%$

→ Approximated number/particle density D_i:

Oxide molecule	Number density / ions/cm^3
SiO_2	$2.206 \cdot 10^{22}$
Yb_2O_3	$6.978 \cdot 10^{19}$
Al_2O_3	$7.243 \cdot 10^{20}$
P_2O_5	$4.332 \cdot 10^{20}$

→ Molar percentage y_i:

$SiO_2 : Yb_2O_3 : Al_2O_3 : P_2O_5$
$=$
$97.29 mol\% : 0.15 mol\% : 1.60 mol\% : 0.96 mol\%$

→ Molar weight percentage w_i:

$SiO_2 : Yb_2O_3 : Al_2O_3 : P_2O_5$
$=$
$94.21 mol\,wt.\% : 0.98 mol\,wt.\% : 2.62 mol\,wt.\% : 2.18 mol\,wt.\%$

Granulated Silica Material Production: The Oxides Approach

Powder Synthesis by Mixing Oxides

The most straightforward and simplest approach for the production of (doped and undoped) granulate for the granulated silica method is the oxides approach (Ballato and Snitzer 1995; Renner-Erny et al. 2007; Di Labio et al. 2008a, b; Neff et al. 2008, 2010; Neff 2010; Braccini et al. 2012; Pilz et al. 2012). This approach is based on mixing pure silicon dioxide SiO_2 with the oxide form of the desired dopant(s) (e.g., rare earth (RE) element oxides, Yb_2O_3, Er_2O_3, Nd_2O_3, etc. (Di Labio et al. 2008a, b; Pilz et al. 2012); metal oxides, Al_2O_3, Bi_2O_3, etc.; transmission metal oxides, V_2O_5, Cu_2O, Mn_2O_3, etc. (Neff et al. 2008, 2010; Neff 2010; Braccini et al. 2012); nonmetal oxides, P_2O_5, etc. (Neff 2010; Pilz et al. 2012)) in dry powder form.

First of all, the powder mixture of oxide precursors consists of silicon dioxide grains and dopant grains, so that the dopant is not homogeneously distributed. For a two-component mixture, one specific grain from this mixture is either doped or undoped, so not every grain is doped. Furthermore, the local accumulation of the dopant will in turn lead to clustering of the dopant (see section "Solubility and Homogeneity of Rare Earth Elements") (Romano et al. 2014; Pilz et al. 2017, 2016). The solubility of the dopant (see section "Solubility and Homogeneity of Rare Earth Elements") in the melt during fiber drawing can be enhanced for the oxides approach by addition of a co-dopant in oxide form (e.g., Al_2O_3 as co-dopant and solubility enhancer for RE element dopant (Digonnet 2001)) leading to a multicomponent powder. Furthermore by using, e.g., Al_2O_3 and P_2O_5 simultaneously as co-dopants, it is possible to tailor the refractive index of the final glass by the ratio of Al to P, since P can compensate the refractive index change arising from the Al (both dopants applied for itself would increase the refractive index) (Vienne et al. 1996; Unger et al. 2009). Moreover, the phosphorus is beneficial in order to reduce the photo darkening (Engholm and Norin 2008). However, the grain sizes of SiO_2 and the dopant oxides are normally different, which in turn facilitates the formation of microbubble (as described in subsection "Powder-in-Tube (PIT) Technique"). Moreover, not only the preform tube and the oxide-derived powder mixture feature different material properties (e.g., thermal conductivity k and melting temperature T_f, also within the inhomogeneous powder mixture of SiO_2 and the dopant oxide(s) itself (some of the grains are doped and some undoped)), but also the material properties differ. This is even more pronounced for a multicomponent mixture. The inhomogeneity makes it even more challenging to totally melt the powder mixture inside the preform tube before the droplet is formed. However, as mentioned before the melting temperature T_f of the granulate can be influenced by co-doping. Phosphorus, for example, not only reduces the photo darkening, but it also lowers the melting temperature, so that the granulate can totally melt before the droplet is formed.

Using this oxide-derived powder mixture direly for the powder-in-tube technique will lead to fibers with inhomogeneous doping distribution and high scattering due

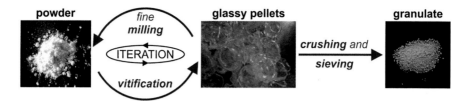

Fig. 9 Schematic of the post-processing for granulated based on the oxides approach

to microbubbles (Romano et al. 2014; Pilz et al. 2017) and incorporated residual unmolten powder. The inhomogeneity, which in turn results in clustering of the dopant(s) in the fibers, is the main reason why these fibers cannot be used for fiber laser and fiber amplifier setups.

By using different post-processing steps (presented in detail in section "Post-Processing: Oxides Derived Granulate"), the oxide-derived powder mixture is transferred into a granulate with increased homogeneity. Furthermore, the possibility of selecting the grain size and the granulometric distribution is gained by these post-processing steps. It should be already mentioned that for the sol-gel approach, the post-processing steps (see section "Post-Processing: Sol-Gel Derived Granulate") are different.

Post-Processing: Oxides Derived Granulate

The entire post-processing for the powder derived from the oxides approach is schematically shown in Fig. 9 and is based on an iteration of a vitrification and fine milling process, which is ultimately followed by a crushing and sieving process in order to get control of the grain size and the granulometric distribution of the final granulate (Romano et al. 2014; Pilz et al. 2016, 2017).

Iterative Vitrification and Fine Milling

In order to increase the homogeneity of the oxide-derived powder mixture and to bypass the issues that are arising from the different grain sizes and material properties within the powder mixture itself, the powder is vitrified (melted and quickly cooled, so that the crystalline phase (e.g., formation of cristobalite) is not formed during cooling) by a CO_2-laser treatment. During this treatment, the powder mixture is exposed to an CO_2-laser beam so that all grains in the powder mixture undergoes the transition from powder to glass, forming a so-called glassy pellet with none residual unmolten powder within (see Fig. 9, Romano et al. 2014). This vitrification process is depending along with the powder (grain size and its distribution) and material properties (melting temperature and heat conductivity) on the laser characteristics (laser power, exposure time, and beam diameter). The intensity of the CO_2-laser beam must be high enough to totally melt the powder mixture but not too high to evaporate the doping from the mixture.

In a next post-processing step, this glassy pellet is milled into fine powder. Due to the laser treatment and the fine milling, the dopant(s) are better distributed in the fine-milled powder (the grains are no longer just doped or undoped; the doping is better distributed over partially doped grains). Thus, the grains feature more similar material properties such as the melting temperature and heat conductivity.

The process of vitrification and fine milling is repeated multiple times till the glassy pellets are transparent and homogeneous by the naked eye (Romano et al. 2014). The iteration of vitrification and fine milling furthermore increases the homogeneity and makes also the material properties of the grains more similar from iteration to iteration.

The improvement in homogeneity due to the iterative vitrification and fine milling is manifested among others in a reduction of clustering of the dopants, which was demonstrated on the basis of an erbium-/aluminum-doped fiber derived from the oxides approach (Romano et al. 2014; Neff 2010). If erbium (Er) ions are too closely located to each other (like in clusters), they feature a green fluorescence under an excitation in the range of 980 nm (see Fig. 10), which will not appear for a homogeneous distribution. The green fluorescence is due to ion-ion interactions and is based on energy transfer up-conversion (ETU) and/or excited state absorption (ESA) (Neff 2010; Quimby et al. 1994). As Fig. 10 demonstrates, the green fluorescence is strongly reduced for a fiber, where the vitrification and fine milling are applied twice for the oxide-derived powder. This reduction of the green fluorescence is synonymous with the reduction of clusters and thus also a proof that the homogeneity has been improved by the iterative vitrification and fine milling post-processing step. However, not all clusters can be removed by the iterative post-processing, so that the oxides approach is still lacking in homogeneity (see Table 2).

Coarse Crushing and Sieving

Subsequently and as a final post-processing step, the glassy pellets are not finely milled; they are coarsely crushed and sieved in order to control and to select the

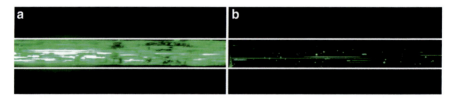

Fig. 10 Er-/Al-doped single-clad fiber (365/450 μm) based on the oxides approach for the granulated silica method. If the Er ion groups in clusters, they feature a green fluorescence ($4S_{3/2} \rightarrow 4I_{15/3}$ transition of Er) under excitation at 980 nm in contrast to a homogeneous distribution of the Er ions. (**a**) For a fiber produced without the iterative vitrification and fine milling post-processing step, the Er shows a green fluorescence under excitation at 980 nm due to ETU and/or ESA of the Er clusters. (**b**) When the iterative vitrification and fine milling post-processing step is applied twice, the green fluorescence and thus the Er clusters are strongly reduced (white lines indicate the cladding boundaries) (Romano et al. 2014)

Table 2 Overview of the advantages and drawback of the oxides approach and the sol-gel approach

	Oxides approach	Sol-gel approach
Flexibility with respect to the geometry(structural freedom)	+	+
Compositional flexibility with respect to the dopant	+	+
High doping concentrations	+	+
Homogeneity	−	+
Fiber background losses	−	(+)

grain size and the granulometric distribution (see Fig. 9). As mentioned earlier in section "Powder-in-Tube (PIT) Technique," the grain size and the granulometric distribution are important for the evacuation of the preform in order to prevent the formation of microbubbles.

Fibers Based on the Oxides Approach

Based on the oxides approach for the granulated silica method, various fibers have been realized and produced with many different granulate compositions (including different doping concentrations) and different fiber geometries (Scheuner et al. 2016, 2017; Di Labio et al. 2008a; b; Neff et al. 2008, 2010; Neff 2010; Braccini et al. 2012; Pilz et al. 2012, 2016). While the feasibility of such fibers has been demonstrated and proven, they still suffer from high background losses in the order of a few dB/m (0.75–5 dB/m at 632 nm) (Romano et al. 2014) due to microbubbles and inhomogeneity.

The advantages and drawbacks of the oxides approach for the granulate silica method are summarized in Table 2.

Granulated Silica Material Production: The Sol-Gel Approach

Powder Synthesis Using the Sol-Gel Process

The sol-gel process is a wet-chemical technique which is mainly applied to produce thin (dense) films (Brinker and Scherer 1990) and coatings (Armelao et al. 2005) (e.g., anti-reflection and protective coatings of optical elements), as well as ceramics (Brinker and Scherer 1990), aerogels, and monolithic glass (Scholze 1988; Kirkbir et al. 1996), but can also be used for the preparation of (multicomponent) homogenous powder.

For the fiber production, the sol-gel process can be used for coating the preform tube (Pedrazza 2006, 2007; Li et al. 2008; Matejec et al. 1998; Wu et al. 1993), for the rod-in-tube method on basis of glass rods derived from the gel (Wang et al. 2013, 2016; Susa et al. 1990) or on homogeneous silica powder (Lou et al. 2014), e.g., for the granulated silica method (Romano et al. 2014; Pilz et al. 2016, 2017). However, here only the production of (multicomponent) homogenous powder for the granulated silica method is presented and treated in detail.

The normal glass production is based on fusion respectively on glass melting at the melting temperature T_f (elevated temperature in the range of 1900 °C for silica glass) and followed by quick cooling to achieve an amorphous condition and in order to avoid the formation of a crystalline structure. By using the sol-gel synthesis for the glass production, the amorphous network of the glass can already be achieved below the glass transition temperature T_g at low temperatures (e.g., room temperature) (Scholze 1988) and is therefore ideally suitable for the production of pure and doped silica material of high purity (Shelby 1997). The low temperature to achieve the amorphous condition is not the only advantage of the sol-gel method. The sol-gel process features a high degree of compositional freedom with respect to the dopants (see Table 2), which in turn allows to precisely tune the chemical composition and furthermore allows to incorporate dopants that cannot or only hardly be incorporated by other doping methods (e.g., phosphorus to reduce the photo darkening). Also, high doping concentration (Ballato and Dragic 2013), even for multicomponent material, can be achieved. Another benefit of the sol-gel process and the final glass derived from the sol-gel process is its high homogeneity (in nanoscale); see section "Solubility and Homogeneity of Rare Earth Elements."

The sol-gel process starts with a colloidal dispersion of solid particles in a liquid, the so-called sol, and undergoes a sol-gel transition, evolving into a gel (see Fig. 11). During this process, the viscous sol is merged into a viscoelastic solid. The homogeneous and isotropic characteristics of the sol are preserved during the sol-gel transition, so that the resulting gel is also homogeneous and amorphous.

As precursor for the sol-gel process, numerous organic or inorganic, metallic (including rare earth (RE) metals and transition metals) or nonmetallic chemical bonds are suitable. Most often salts and metal alkoxides are used. For the glass production, Si-precursor is needed. The most commonly applied and most intensively investigated Si-precursors are the alkoxides TEOS (tetraethyl orthosilicate) and TMOS (tetramethyl orthosilicate), since for these two precursors, the sol-gel synthesis already sets in with the addition of water. For optical active and passive

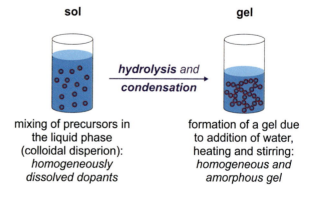

Fig. 11 Schematic of the sol-gel process: the sol transforms into a gel under hydrolysis and condensation

doping various RE elements (standard dopant for optical active fibers), metals and transition metals can be used. The sol is prepared by solving and mixing all precursors together at room temperature. As solvent for the sol, mostly ethanol (C_2H_5OH), methanol (CH_3OH), or water (H_2O) are used (Kirkbir et al. 1996), depending on the solubility of the precursors. Co-dopants are used to increase the solubility of the main doping element (e.g., Al for RE-doping, see section "Solubility and Homogeneity of Rare Earth Elements"). Furthermore, as mentioned before, co-doping affects the melting temperature, and by, e.g., a phosphorus co-doping, the melting temperature can be lowered beneath the onset melting temperature of the preform tube.

The sol-gel synthesis is based on two elementary chemical reactions: hydrolysis and (multiple) condensations. With the addition of water and moderate heating (<80 °C), the homogeneous sol transforms into a homogeneous and amorphous gel, by undergoing hydrolysis and polycondensation (multiple water and alcohol condensations). Here the hydrolysis and polycondensation of TEOS is presented as an example. The chemical formula of TEOS is $Si(OC_2H_5)_4$ which can also be written in a more compact form as $Si(OR)_4$ with R being the alkyl group of $R = C_2H_5$. TEOS gradually undergoes hydrolysis with the addition of water, by a stepwise ($n \leq 4$) separation of the organic part R in the form of an alcohol ROH, here as ethanol:

$$\underbrace{Si(OR)_4}_{TEOS} + 4 \underbrace{H_2O}_{water} \rightarrow Si(OR)_{4-n}(OH)_n + n \underbrace{ROH}_{ethanol} \qquad (6)$$

The degree (partially or complete) of hydrolysis can be influenced by the amount of the additional water and on the presence of a catalyst (Brinker and Scherer 1990). For a complete hydrolysis ($n = 4$), TEOS and the additional water are transferred into silicic acid and ethanol:

$$\underbrace{Si(OR)_4}_{TEOS} + 4 \underbrace{H_2O}_{water} \rightarrow \underbrace{Si(OH)_4}_{silicic\ acid} + 4 \underbrace{ROH}_{ethanol} \qquad (7)$$

The condensation reactions of un-hydrolyzed and/or partially hydrolyzed molecules can link these molecules together by Si-O-Si bounds (formation of Si-containing molecules) and under the multiple separation of water (water polycondensation) or alcohol (alcohol polycondensation) (Brinker and Scherer 1990). For example, the individual water condensation of two partially hydrolyzed ($n = 6$) TEOS molecules separates water (Brinker and Scherer 1990):

$$\underbrace{(OR)_3Si\text{-}OH}_{part.hydrolyzed\ (n=1)} + \underbrace{HO\text{-}Si(OR)_3}_{part.hydrolyzed\ (n=1)} \rightarrow (OR)_3Si\text{-}O\text{-}Si(OR)_3 + \underbrace{H_2O}_{water} \qquad (8)$$

while, e.g., the individual alcohol condensation of un-hydrolyzed TEOS and partially hydrolyzed ($n = 6$) TEOS molecules separate alcohol, herein forming ethanol (Brinker and Scherer 1990):

$$\underbrace{(OR)_3Si\text{-}OR}_{TEOS} + \underbrace{HO\text{-}Si(OR)_3}_{part.hydrolyzed\ (n=1)} \rightarrow (OR)_3Si\text{-}O\text{-}Si(OR)_3 + \underbrace{ROH}_{ethanol} \qquad (9)$$

During these polycondensation reactions, Si-containing molecules are formed and are growing based on polymerization (Brinker and Scherer 1990). With progressing of the condensation process, these molecules are growing so big in size that the sol stiffens into a gel. Normally the speed of the hydrolysis is faster than the speed of condensation ($v_{hydrolysis} > v_{condensation}$), so that both processes are overlapping (Brinker and Scherer 1990). However, these speeds can be influenced by the choice of type and concentration of the catalyst (Brinker and Scherer 1990). The type of catalyst has also a strong influence on the aggregation of the monomers into the polymer. While an acidic catalyst (slower hydrolysis speed) tends to preferentially form linear polymer chains, a basic catalyst (faster hydrolysis speed) preferentially forms branched polymer chains (Scholze 1988).

The sol-gel synthesis of a multicomponent material is more complicated, and the additional precursor must be carefully selected, since the individual components respectively precursors can have different hydrolysis and polycondensation speeds (Scholze 1988). However, this can be affected by a controlled and slow addition of the water for the hydrolysis or by pre-condensing the slower component(s) (Scholze 1988). In conclusion, the hydrolysis and polycondensation processes are strongly depending on (Brinker and Scherer 1990):

- The choice of precursors
- The choice of solvent
- The amount of additional water
- The temperature
- The type and concentration of the catalyst

Since the resulting gel still consists – besides the gel network – of a major part of liquid (residual solvents), it must be post-processed by, e.g., drying and/or sintering to transform it into a powder respectively granular material (see Fig. 12), film, coating, or monolith. In the following subsection, the post-processing into powder and ultimately into a granulate is treated (the low-temperature drying of the gel into a monolithic glass (Brinker and Scherer 1990; Scholze 1988; Kirkbir et al. 1996) is not dealt with here). As mentioned earlier, the post-processing for the sol-gel approach differs from the post-processing for the oxides approach. For the sol-gel post-processing, other techniques (such as sintering) are used to get control over the grain size and the granulometric distribution in order to use it for the granulated silica method.

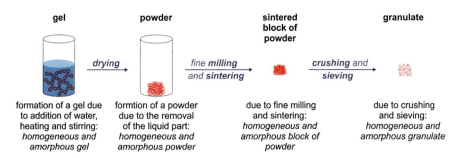

Fig. 12 Schematic of the post-processing for granulated based on the sol-gel approach

Post-Processing: Sol-Gel-Derived Granulate

The entire post-processing for the gel derived from the sol-gel approach is schematically shown in Fig. 12. For the gel, the first post-processing step is based on drying the gel into a powder, followed by a fine milling and sintering process of the powder, and a final crushing and sieving process in order to get control of the grain size and the granulometric distribution of the final granulate (Pilz et al. 2016, 2017). By an additional vitrification process, the fiber background losses based on this approach can be decreased (Pilz et al. 2016, 2017).

Drying into Powder (Low Temp Range)

During the first drying process (at low temperature, slightly above the evaporation temperature of the solvents), most of the liquid part (solvents, including OH-groups) of the gel is evaporated, resulting in a porous powder, where the pores are still filled with residual liquid (xerogel). During this drying process, the gel structure shrinks and is densified into a loose amorphous and homogeneous powder. For doped sol-gel material, every grain of the resulting powder is doped. This is in contrast to the oxide-based approach, where at the beginning of this approach, not every grain is doped and the homogeneity can be increased by an iterative vitrification and fine milling process (see subsection "Iterative Vitrification and Fine Milling" in section "Post-Processing: Oxides Derived Granulate"). Furthermore, the mechanical properties (e.g., thermal conductivity) of the grains are the same, since all the grains are identical in composition. By the removal of the liquid part, also the intrinsic OH-groups content of sol-gel-derived material is reduced. However, the higher OH-groups content of sol-gel material compared to other preform production techniques such as MCVD is not of concern for the targeted visible and NIR (near infrared) spectral range, since in this range the absorption of the OH-groups is low enough.

Fine Milling and Sintering

In order to reduce or eliminate the trapped residual liquid part (solvents, including OH-groups) in the pores of the powder, the powder is finely milled, so that the pores are broken up and the included liquid can be removed by an additional heat treatment, the sintering process (Romano et al. 2014; Pilz et al. 2016, 2017; Etissa et al. 2012). However, the fine milling process leads to a point where the powder is so fine that it cannot be used with large power-in-tube preforms (e.g., 17 × 21 mm) for the granulated silica method (see subsection "Powder-in-Tube (PIT) Technique").

During the sintering process, the loose powder is further densified, and furthermore by the right heat treatment, it can be transformed into a solid block of powder (Pilz et al. 2016, 2017). This block of powder is needed to get the control over the grain size and the granulometric distribution, since this block of powder can be coarsely crushed and sieved. So, the vitrification process for the oxides approach (which is also used to improve the homogeneity) is replaced by the sintering process to get control of the grain sizes. Transforming the loose powder into a block of powder by sintering can also change its structure from amorphous to crystalline (formation of cristobalite) (Romano et al. 2014), depending on the applied temperature and the thermodynamic properties of the specific powder (see section "Thermodynamic Properties of Rare Earth Ion-Doped Silica Powder (T_g, T_x, T_c)"). While it is important for many other glassy applications that during sintering the powder is not changed from an amorphous into a crystalline structure, it is believed that it is not that important for the fiber production since the possible developed crystalline structure can be removed during the fiber drawing (see Fig. 13, Romano et al. 2014). In order to remove the crystalline structure during the fiber drawing, the granulate must completely melt before the droplet is formed. This can be influenced by lowering the melting temperature of the granulate by co-doping or by an increased drawing temperature (with the condition that the droplet is not formed before the granulate is totally molten). During the melting of the granulate, the structure becomes again amorphous. The melting is followed by a fast cooling of the fiber in the drawing tower, where the amorphous structure is "frozen," which is referred to as vitrification. However, it is always better to bypass the crystalline structure if possible, when the densification into a block of powder and the control over the grain size and the granulometric distribution are not needed.

Coarse Crushing and Sieving

By crushing and sieving the sintered block of powder, the size of the granulate and its desired granulometric distribution can be chosen. However, using this sol-gel-derived granulate directly for the powder-in-tube preform assembly technique (rapid prototyping, see Fig. 14) will lead to fibers with scattering centers arising from unmolten granulate (see subsection "Powder-in-Tube (PIT) Technique."), which in turn are the major source for the background losses of these fibers. These background losses can be decreased by an additional vitrification step before drawing (Pilz et al. 2016, 2017). Thus, the sol-gel-derived granulate can either be directly used for the preform assembly based on the powder-in-tube technique ("fiber rapid prototyping") or with an additional vitrification step based on an adapted powder-in-tube preform assembly technique (see Figs. 8 and 14). By the means of these both preform assembly techniques and since this sol-gel-derived granulate can be used for the granulate silica method, one gains the structural respectively the geometrical freedom, explained in section "Rediscovering Powder Techniques for Fiber Production."

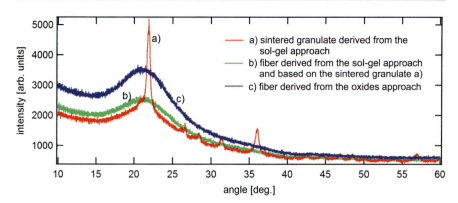

Fig. 13 X-ray diffraction (XRD) patterns of (**a**) sol-gel-derived sintered Yb-/Al-/P-doped granulate; (**b**) Yb-/Al-/P-doped fiber drawn from the sol-gel-derived sintered Yb-/Al-/P-doped granulate(a)); (**c**) Yb-/Al-/P-doped fiber based on the oxides approach (interactively vitrified and milled). The crystalline structure (cristobalite) developed during sintering is removed during fiber drawing (Romano et al. 2014)

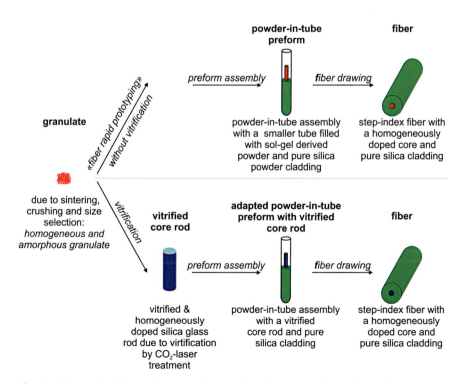

Fig. 14 Schematic of the preform assembly technique for the granulated silica method based on (**a**) the powder-in-tube technique, which is referred to as "fiber rapid prototyping"; (**b**) adapted powder-in-tube technique with a vitrified core rod (Pilz et al. 2017)

Fig. 15 Vitrified core rod based on an additional vitrification on the basis of a CO2-laser treatment

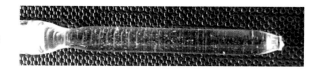

Additional Vitrification of Thin Powder-in-Tube Preform

The fiber rapid prototyping (directly drawing of a powder-in-tube preform filled with granulate) suffers from high background losses, which can be reduced if (at least) the core area is additionally vitrified before preform assembly and drawing (see Fig. 14). This additional vitrification process is based on a CO2-laser treatment of silica tubes (up to a diameter of 5 mm) filled with the granulate and results in vitrified rods (see Fig. 15, Pilz et al. 2016, 2017). These rods can be used for an adapted powder-in-tube technique (see Fig. 14) or the rod-in-tube technique (see Fig. 8).

Fibers Based on the Sol-Gel Approach

By means of this additional vitrification technique, the background losses of Yb-/Al-/P-doped fibers based on the sol-gel approach for the granulated silica method have been reduced to 0.20 dB/m at 633 nm (measurement based on cutback method) or 0.024 dB/m at 1550 nm (measurement based on HR-OTDR) (Pilz et al. 2016). The background losses are in principle low enough to build up fiber laser and fiber amplifier setups.

The advantages and drawbacks of the sol-gel approach for the granulate silica method are also summarized in Table 2.

Solubility and Homogeneity of Rare Earth Elements

Pure silica network can accept extremely small amounts of rare earth (RE) elements before microscopic ion clustering and ion-ion interactions occur, while at higher RE concentrations, crystalline phases can also be formed (Ainslie et al. 1987; Digonnet 2001). RE ions need large coordination numbers. This means insufficient non-bridging oxygens in pure silica network (Fig. 16a) to coordinate isolated RE ions can result in ion clustering and sharing of non-bridging oxygens by RE ions. In this context, host glasses well suited with relatively high concentration of REs without clustering require an open and chain-like structure (Fig. 16b). Thus, an open silica network is essential for improving the solubility of RE ions. Based on this fact, the role of modifier ions or co-dopants (e.g., Al, P, Ge) is defined in order to open the silicate structure and enhance solubility and homogeneity of RE elements. In general, there are three major duties of co-dopants in glass composition as following (Digonnet 2001):

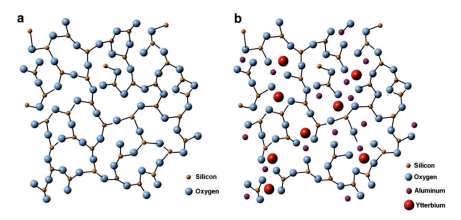

Fig. 16 Schematic silica network structures of (**a**) pure silica and (**b**) doped silica with Al (network modifier) and Yb. Doped silica network shows more non-bridging oxygen to prevent RE clustering

- Reduction of RE clustering as a result of enhancement in RE solubility due to the formation of large number of non-bridging oxygen (NBO) compared to pure silica glass matrix. In other words, opening the silica network is a key role of the co-dopants.
- Enhancement of the refractive index (RI) due to larger sizes of co-dopants as well as due to their strong ionic polarizability.
- Tuning of the melting temperature of glass by tailoring co-dopant concentration.

Arai et al. (1986) have interpreted that RE dissolves in Al_2O_3 however not in SiO_2, where Al_2O_3 dissolves in SiO_2. Therefore, Al_2O_3 forms a solvation shell around the RE ion, and this complex can be merged into the silica network. However, this hypothesis arises an open question regarding the presence of stoichiometric oxides in the silica network. The correlation between the RE positions and Al ions implied by this hypothesis is consistent with the minimum Al/RE ratio of 10 required to prevent clustering reported by Massicott et al. (1991).

Therefore, the properties of the RE dopants and the waveguide can be adjusted by Al-doping and other co-dopants such as phosphorus (P), germanium (Ge), and cerium (Ce), respectively. It results in avoiding the clustering and high background loss problems often encountered in RE-doped medias. As an example, when P and Ge are employed as index-raising dopants, the threshold of RE incorporation before the onset of quenching effect is around 0.1 at.% for Nd^{3+} (Namikawa et al. 1982). The addition of Al as a network modifier has improved the solubility of RE, whereas RE concentrations of 0.3 at.% have been achieved without clustering in host glasses with 9 at.% of Al (Arai et al. 1986). In the case of Er^{3+} as a RE where silica was co-doped with Ge and Al, an Al/Er ratio of 50 was reported to eliminate clustering and quenching effects (Massicott et al. 1991).

It is worthwhile to mention that these critical values are different depending on which RE is used in the glass network. For instance, Nd^{3+} concentration for the onset of clustering in silica network is reported to be 0.1 at.%, whereas Er^{3+}-doped fiberglasses showed significant Er^{3+} ion-ion interactions at concentrations about 0.01 at.% (Namikawa et al. 1982; Shimizu et al. 1990). Since Nd^{3+} and Er^{3+} have an equal solubility in silica network, the order-of-magnitude difference in the onset of quenching could be related to either a stronger interaction between Er^{3+} ions or a higher sensitivity to dissipative processes in amplifier experiments (Digonnet 2001). As mentioned earlier, the addition of Al significantly reduces the clustering problem, and Nd^{3+}-doped and Er^{3+}-doped alumina-silica networks have been reported to show a great performance at the RE levels of 0.3 at.% and 0.08 at.%, respectively (Laming et al. 1990; Arai et al. 1986).

From abovementioned, a compromise has to be met between the concentration of RE and glass composition. On one hand, high concentration of RE can result in reduction of the length of the fiber and consequently the cumulated attenuation. However, it leads to luminescence quenching mechanisms that reduce the fiber efficiency. On the other hand, to avoid such quenching effects, glass modifiers like Al^{3+}, Ge^{3+}, or P^{5+} are employed in the fabrication of fiberglasses; however, these glass modifiers could also be sources of radiation-induced structural defects and thus of optical losses. Thus, this compromise is extremely important in the optimum design of RE-doped fiberglasses.

Consequently, the composition of the applied powder, preform, and fiberglasses or, in other words, the dopant concentrations are key and have to be analyzed and quantified by extremely precise techniques. The main criteria for selecting a trustable technique are precision and also no interference between the involved elements in quantification, i.e., the resolution of the employed technique. X-ray fluorescence (XRF) is one of the most advanced techniques for analysis of powders and glasses. There are two general types of XRF used for elemental analysis applications, i.e., energy dispersive x-ray fluorescence (EDXRF) and wavelength dispersive x-ray fluorescence (WDXRF). In EDXRF, all elements are excited simultaneously, and an energy-dispersive detector is utilized to collect the fluorescence radiation and separate the characteristic energies of each element. The resolution is dependent on the detector and ranges between 150 and 300 eV. WDXRF employs crystals to disperse the fluorescence spectrum into individual wavelengths of each element, which results in higher resolution and low background spectra for more accurate determination of elemental concentrations. It can provide the working resolution between 5 and 20 eV. The higher resolution of WDXRF results in reduced spectral overlaps, where the complex samples such as doped multicomponent glasses can be more accurately analyzed. Additionally, with higher resolution and reduced background, the detection limits and sensitivity are improved. For instance, WDXRF can perfectly detect Yb atoms down to 30 ppm in the powder and also Al quantity without any peak overlap with Yb. An example of WDXRF spectra for the Yb-/Al-/P-doped silica powder has been shown in Fig. 17. Using WDXRF technique, a perfect separation can be achieved between Al and Yb energy peaks at 1.486 and 1.521 KeV, respectively.

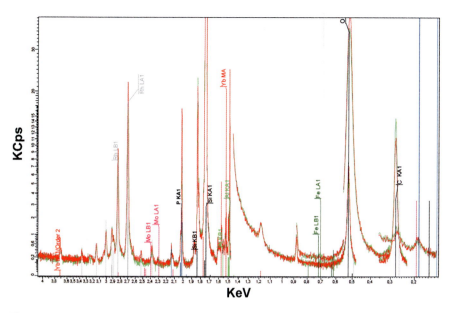

Fig. 17 An example of WDXRF spectra for the Yb-/Al-/P-doped silica powder derived from the sol-gel approach showing a perfect separation of characteristic energies of each element, i.e., Si, O, Al, Yb, and P. As an example, Al KA1 (1.486 KeV) and Yb MA (1.521 KeV) have an energy differential of 35 eV so that the WDXRF with resolution of 5 eV can perfectly identify the peaks without overlapping

The second key aspect of the dopants is their homogeneity in the glass structure. The structure and homogeneity of the fiberglasses can be observed by means of scanning/transmission electron microscope (S/TEM. 80–300 kV) (Najafi et al. 2016). The different feature of scanning transmission electron microscopy (STEM) compared to conventional transmission electron microscopy (TEM) is that it can focus the electron beam into a narrow spot, which is scanned over the sample, and diffracted electrons can be detected by different detectors and in particular by high-angle annular dark-field (HAADF). According to the "Z contrast" imaging theory, the brightness of each dot in the STEM-HAADF image mode could be roughly proportional to the Z^2, while Z is atomic number (Najafi et al. 2016; Liu 2011). Indeed, scanning transmission electron microscopy (STEM-EDX) can provide information in nanoscale chemistry and homogeneity.

Using STEM technique, an ultraprecise chemical mapping can be performed to individualize the different chemical phases, even if they are restricted to ultrasmall quantities where we have an electron beam with a HAADF-STEM resolution of 0.18 nm. As an example of Yb-doped fiberglasses with Al and P as co-dopants, a great progress in homogeneity could be identified by STEM-HAADF technique in the granulated silica derived from the sol-gel process compared to the oxides approach (Pilz et al. 2016), as shown in Fig. 18. Furthermore, Fig. 19 demonstrates the elemental distribution images of Yb-/Al-/P-doped fibers produced by granulated silica derived from the sol-gel process.

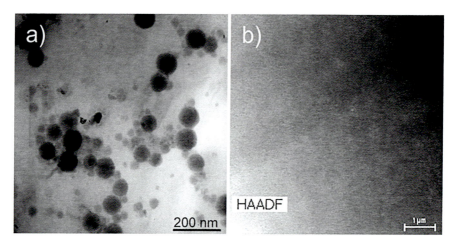

Fig. 18 Progress in homogeneity from (**a**) oxide-based powder mixing to (**b**) sol-gel-based granulated silica method (Pilz et al. 2016)

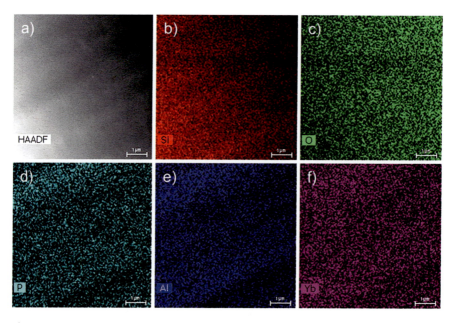

Fig. 19 STEM micrograph showing the nanostructure and elemental distribution of a doped fiber core produced by granulated silica derived from the sol-gel process, (**a**) HAADF image (**b–f**) elemental mapping of the fiberglass (Najafi et al. 2016)

The images are associated with Si, O, P, Al, and Yb, respectively, which can provide useful information on the dopant and co-dopant localization in the nanostructure. In the corresponded study, it is noteworthy that an extremely homogeneous distribution of dopants is observed in the structure in nanoscale (Fig. 19d–f, Najafi

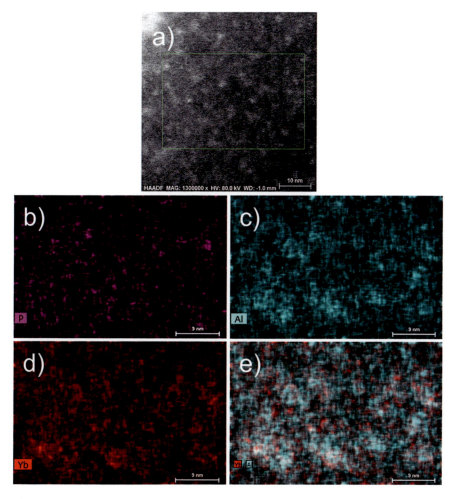

Fig. 20 (a) HAADF-STEM micrograph and (b–e) elemental mapping acquired at ultrahigh magnification from a fiber core (Najafi et al. 2016)

et al. 2016). In order to obtain further details of dopant homogeneity in atomic scale, Fig. 20a shows acquired HAADF image in an extremely high magnification. As mentioned earlier, the brightness of each dot in the STEM-HAADF image mode is proportional to the Z^2 where Z is atomic number.

Considering the rational hypothesis of a flat investigated area in an extremely thin TEM sample within the field of view at such high magnification, it can be concluded that the bright areas on the micrographs in Fig. 20a represent Yb atoms and clusters since the Yb atom species are the only heavy dopant in the doped fiber core (Najafi et al. 2016).

The atomic numbers of other elements and dopants (i.e., Al, P, Si, and O) are much lower than Yb. Furthermore, Fig. 20a shows that at such atomic scale

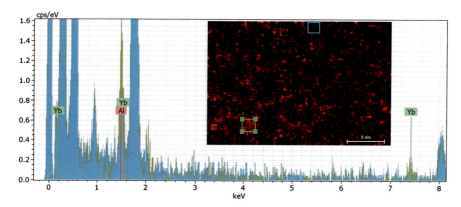

Fig. 21 EDX spectra from a clustered area (green) and a free cluster zone (blue): the visible signal from Yb atoms within the clustered area (Najafi et al. 2016)

magnification, some Yb atoms tend to cluster together, while some are randomly distributed. Given the fact that Yb ionic diameter is around 0.2 nm and the average cluster size in the observation area is around 4 nm. It can be concluded that Yb clusters are formed by the average of twenty atoms, or in other words, Yb clusters are formed by less than twenty atoms in 2D scale (Najafi et al. 2016). In order to validate this hypothesis, the EDX spectrum was performed in both suspected clustered area and free cluster zone (Fig. 21). The green color shows the EDX spectrum of the clustered area, while the blue spectrum indicates that of the free cluster area. The strong Si and O signals come from the fiber host, SiO_2. The only distinction in the two spectra is the visible signal from Yb elements within the clustered area.

It results in the dominating contrast (Fig. 20a) in STEM-HAADF image given the fact of its larger atomic number, Z.

Thermodynamic Properties of Rare Earth Ion-Doped Silica Powder (T_g, T_x, T_c)

In general, glass formation will be favored by its resistance to overall crystallization (Animesh 2016). In this context, the cooling rate during glass formation is a key factor. The following conditions have to be taken into account for optical glass material engineering and fiber drawing according to the T.T.T (Transformation-Time-Temperature) curve of each particular glass (Fig. 22, Animesh 2016):

- A cooling rate that should be at least equal to a critical cooling rate (C.C.R) whereas lower than C.C.R crystallization could start to occur at different levels. C.C.R is defined as the slope of the line drawn between T_{liquid} and T_{nose} in Fig. 22:

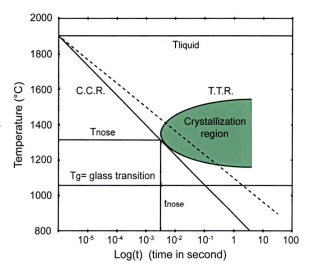

Fig. 22 Glass forming and crystallization of a particular RE-doped SiO2 material: A critical cooling rate (C.C.R) should be met according to the time-temperature-transformation (T.T.T) curve for a glass-forming liquid in order to have fully amorphous glass. Any slower cooling rates (dash line) could result in partial crystallization

$$C.C.R = \frac{T_{\text{liquid}} - T_{\text{nose}}}{t_{\text{nose}}} \qquad (10)$$

- Minimal nucleation and growth in the critical temperature range, which may be possible by engineering compositions that show a high viscosity.
- The absence of heterogeneous nucleation sites (e.g., microbubbles, impurities).
- In multicomponent systems such as SiO_2-Al_2O_3 or SiO_2-P_2O_5, the existence of a eutectic liquid can help in postponing both the nucleation and growth and lead to glass formation by a proper cooling path (lower C.C.R).

Since the formation of crystals in glass media (cristobalite) results in the light scattering, the determination of characteristic temperatures such as glass transition (T_g), onset crystallization temperature (T_x), crystallization temperature (T_c), onset melting temperature (T_{xf}), and melting temperature (T_f) is important in estimating the thermal stability of a glass and its susceptibility to temperature-induced changes during the fiber drawing (Animesh 2016; Han et al. 2013).

Thus, the characterization of these temperatures using differential thermal analysis (DTA) technique would be a great help in thermal treatment design of the fiber manufacturing process.

A schematic representation of a DTA measurement is shown in Fig. 23. The first event is the glass transition (T_g), which is an endothermic transformation and is accompanied by a significant change in the physical state of a glass. The diffusion and viscous flow increase with the temperature, allowing the material to undergo the crystallization (an exothermic transformation) where the onset crystallization (T_x) and crystallization (T_c) peaks can be identified. The last event is when melting starts (T_{xf}) and completes at T_f, which is known as the liquidus temperature.

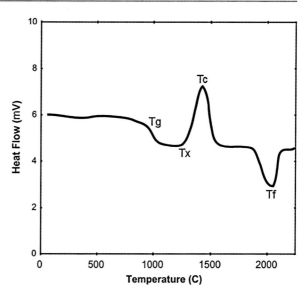

Fig. 23 Schematic representation of a DTA scan for a glass sample, showing the glass transition (T_g), onset crystallization (T_x), crystallization (T_c), onset melting (T_{xf}), and melting (T_f) events when heating at a constant heating rate

These characteristic temperatures are then used for evaluating the resistance of the material composition against vitrification or in some cases devitrification during the fiber manufacturing by following factors:

$$\Delta T_x = T_x - T_g \tag{11}$$

$$\Delta T_c = T_c - T_x \tag{12}$$

$$H = \frac{\Delta T_x}{T_g} \tag{13}$$

$$S = \frac{\Delta T_x \Delta T_c}{T_g} \tag{14}$$

In (Eq. 11), ΔT_x, which defines the temperature differential between the onset of crystallization and glass temperature, is related to the first stability parameter of the material and the barrier to crystal nucleation according to nucleation model. ΔT_c (Eq. 12) is linked to the thermal stability in the glass in the regime above T_x, when both the nucleation and growth are dominant. The larger the temperature width of ΔT_c, the slower the glass to crystal transformation. This width is high in silica glasses (Animesh 2016). In addition, it is necessary to introduce two new critical thermal parameters to evaluate the thermodynamic properties of glasses.

H in (Eq. 13) defines the thermal stability parameter of the glass. The distance between glass transition (T_g) and the onset of crystallization is relevant to the stability of glassy phase. The Saad-Poulain factor S in (Eq. 14) combines the aspects of both nucleation and growth rates by including ΔT_x and ΔT_c. Due to the introduction of crystallization peak width (ΔT_c), which has a higher affinity with the rate of crystallization, the value of S is more reasonable to be adopted to predict thermal stability of the glass. The higher S indicates the higher level of anti-crystallization factor in the investigated glass (Han et al. 2013). In other words, with higher S, the glass has good thermal stability against crystallization in preform for drawing fiberglasses.

From the abovementioned facts, thermodynamic properties of the rare earth (RE) ion-doped silica powders including characteristic temperatures such as glass transition temperature (T_g), onset crystallization temperature (T_x), and crystallization temperature (T_c) have to be taken into account during the fiberglass manufacturing. Evaluations on thermodynamic properties can indicate which specific composition in doped silica glasses and which powder processing conditions are more reasonable for drawing optical fibers.

As an example, Fig. 24 shows thermodynamic properties (RT-1400 °C) of Yb-/Al-/P-doped silica glass powder synthesized by granulated silica derived from the sol-gel process. The first event is related to the evaporation of OH-groups from the powder at 180 °C. The transition temperature (T_g) of Yb-/Al-/P-doped silica glasses is derived to be 1065 °C. The onset crystallization temperature (T_x) and crystallization temperature (T_c) of this particular powder are obtained to be 1360 °C and ≈1450 °C, respectively.

In this particular glass composition, T_g=1065 °C and T_x=1360 °C indicate that an annealed temperature in the range of 1065–1360 °C, such as 1200 °C, can be adopted to achieve the internal stress relaxation and avoid crystallization, thermal shock stress, and other stress conditions.

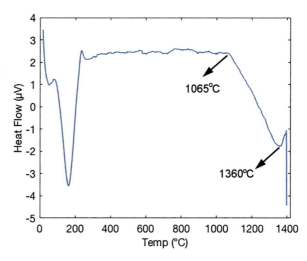

Fig. 24 An example of DTA analysis in Yb-/Al-/P-doped silica glass powder synthesized by granulated silica derived from the sol-gel process

References

B.J. Ainslie et al., J. Mater. Sci. Lett. **6**(11), 1361–1363 (1987)

A.J. Animesh, P. Capper, S. Kasap, A. Willoughby, Inorganic Glasses for Photonics Fundamentals: Engineering and Applications. Wiley Series in Materials for Electronic and Optoelectronic Applications (Wiley, 2016). ISBN 9780470741702. https://www.wiley.com/en-us/Inorganic+Glasses+for+Photonics:+Fundamentals,+Engineering,+and+Applications-p-9780470741702

K. Arai et al., J. Appl. Phys. **59**(10), 3430–3436 (1986)

L. Armelao et al., Surf. Coat. Technol. **190**, 218–222 (2005)

J. Ballato, P. Dragic, J. Am. Ceram. Soc. **96**(9), 2675–2692 (2013)

J. Ballato, E. Snitzer, Appl. Opt. **34**(30), 6848–6854 (1995)

S. Braccini, et al., J. Instrum. **7** (2012). http://iopscience.iop.org/article/10.1088/1748-0221/7/02/T02001/meta#citations

C.J. Brinker, G.W. Scherer, *Sol-Gel Science* (Academic, Waltham, 1990)

E. Desurvire et al., Opt. Lett. **12**(11), 888 (1987)

L. Di Labio et al., Opt. Lett. **33**(10), 1050–1052 (2008a)

L. Di Labio et al., Appl. Opt. **47**(10), 1581–1584 (2008b)

D.J. DiGiovanni et al., J. Non-Cryst. Solids **113**(1), 58–64 (1989)

M. Engholm, L. Norin, Opt. Expr. **16**(2), 1260–1268 (2008)

D. Etissa, et al., Proc. SPIE Micro-Struct. Spec. Opt. Fibres **8426**(84261I) (2012)

I. Fanderlik, *Optical Properties of Glass*. 1983. As cited in Nagel (1987)

X. Han et al., Thermodynamic Properties of Rare-earth Ions Doped Lithium-yttrium-Aluminium-silicate Glasses. *Advanced Materials Research* (Trans Tech Publications, 2013)

J. Hecht, *City of Light: The Story of Fiber Optics* (Oxford University Press, 1999). https://www.goodreads.com/book/show/419395.City_of_Light

J. Hegarty et al., Phys. Rev. Lett. **51**(22), 2033 (1983)

K.C. Kao, G.A. Hockham, Proc. Inst. Electr. Eng. **113**(7), 1151–1158 (1966)

F.P. Kapron et al., Appl. Phys. Lett. **17**, 423–425 (1970)

F. Kirkbir et al., J. Sol-Gel Sci. Technol. **6**, 203–217 (1996)

R.I. Laming et al., Proc. Opt. Amplif. Appl. Conf. **13**, 16–19 (1990)

Y. Li et al., J. Lightwave Technol. **26**(18), 3256–3260 (2008)

H. Liu, Ytterbium-doped fiber amplifiers: Computer modeling of amplifier systems and a preliminary electron microscopy study of single ytterbium atoms in doped optical fibers, Master thesis, Department of Engineering Physics, McMaster University, 2011

F. Lou et al., Opt. Mater. Expr. **4**(6), 1267–1275 (2014)

J. Marchi et al., J. Non-Cryst. Solids **351**, 863–868 (2005)

J.F. Massicott et al., Proc. SPIE Fiber Laser Sources Amplif. **II**(1373), 93–102 (1991)

V. Matejec et al., J. Sol-Gel Sci. Technol. **13**, 617–621 (1998)

R.J. Mears et al., Electron. Lett. **26**, 1026 (1987)

A. Méndez, T. Morse (eds.), *Specialty Optical Fibers Handbook*, 1st edn (Academic, London/Oxford/Boston/New York/San Diego, 2006). ISBN 9780123694065. https://www.sciencedirect.com/science/book/9780123694065

J.F.M. Digonnet, *Rare-Earth-Doped Fiber Lasers and Amplifiers*, 2nd edn. (Marcel Dekker, NewYork, 2001)

T. Miya et al., Electron. Lett. **15**(4), 106–108 (1979)

S. Nagel, IEEE Commun. Mag. **25**(4), 33–43 (1987)

S.R. Nagel et al., IEEE Trans. Microwave Theo. Tech. **30**(4), 305–322 (1982)

H. Najafi, et al., Proc. SPIE Micro-Struct. Speci. Opt. Fibres IV**9886**(98860Z) (2016)

H. Namikawa et al., J. Appl. Phys. **21**, L360–L362 (1982)

M. Neff, Metal and transition metal doped fibers, Doctoral thesis, Institute of Applied Physics, University of Bern, Switzerland, 2010

M. Neff et al., Opt. Mater. **31**(2), 247–251 (2008)

M. Neff et al., Opt. Mater. **33**(1), 1–3 (2010)

K. Oh, U. C. Paek, *Silica Optical Fiber Technology for Devices and Components: Design, Fabrication, and International Standards*, vol. 240 (Wiley, 2012). ISBN 9780471455585

U.C. Paek, C.R. Kurkjian, J. Am. Ceram. Soc. **58**(7–8), 330–335 (1975)

U. Pedrazza, Doped Sol-Gel Materials for the Production of optical Fibers, Doctoral thesis, Institute of Imaging and Applied Optics, Ecole polytechnique federal de Lausanne (EPFL), Switzerland, 2006

U. Pedrazza et al., Opt. Mater. **29**(7), 905–907 (2007)

B.A. Philippe, O.J. Simpson, *Erbium-Doped Fiber Amplifiers – Fundamentals and Technology*, 1st edn (Academic, London/Oxford/Boston/New York/San Diego, 1999). ISBN 9780080505848 https://www.elsevier.com/books/erbium-doped-fiber-amplifiers/becker/978-0-12-084590-3

S. Pilz et al., ALT Proc. **2012** (2012)

S. Pilz et al., Proc. SPIE Micro-Struct. Spec. Opt. Fibres IV **9886**(988614) (2016)

S. Pilz et al., MDPI J. Fibers **5**(24) (2017)

S.B. Poole et al., Electron. Lett. **21**(17), 737–738 (1985)

R.S. Quimby et al., J. Appl. Phys. **76**(8), 4472–4478 (1994)

S.C. Rasmussen, *How Glass Changed the World* (Springer, Heidelberg/New York/Dordrecht/London, 2012)

R. Renner-Erny et al., Opt. Mater. **29**(8), 919–922 (2007)

V. Romano et al., Int. J. Modern Phys. B **28**, 1442010 (2014)

G.F. Sagredo To Galileo **XII**, 417–418 (1618)

J. Scheuner et al., Proc. SPIE Micro-Struct. Spec. Opt. Fibres IV **9886**(988613) (2016)

J. Scheuner, et al., Advances in optical fibers fabricated with granulated silica. In *Optical Fiber Communications Conference and Exhibition (OFC)*, 1–3 (IEEE 2017). https://www.osapublishing.org/abstract.cfm?URI=OFC-2017-M2F.3

H. Scholze, *Glas: Natur, Struktur und Eigenschaften* (Springer, Berlin/Heidelberg, 1988)

K. Schuster, et al., Proc. SPIE Opt. Components Mater. XII **9359**(935914) (2012)

K. Schuster et al., Adv. Opt. Technol. **3**(4), 447–468 (2014)

J.E. Shelby, *Introduction to Glass Science and Technology*, 2nd edn. (Royal Society of Chemistry, Cambridge, 1997)

M. Shimizu et al., IEEE Photon. Technol. Lett. **2**, 43–45 (1990)

E. Snitzer, J. Opt. Soc. Am. A **51**(5), 491–498 (1961)

E. Snitzer, H. Po, F. Hakimi, R. Tumminelli, B. C. McCollum, Double clad, offset core Nd fiber laser, in *Optical Fiber Sensors*. OSA Technical Digest Series, vol 2 (Optical Society of America, 1988), paper PD5

K. Susa et al., J. Non-Cryst. Solids **119**, 21–28 (1990)

S. Unger et al., Proc. SPIE Opt. Components Mater. VI **7212**(72121B) (2009)

S. Unger, et al., Laser Phys. **24**(3) (2014)

V. Velmiskin, et al., Proc. SPIE Micro-Struct. Spec. Opt. Fibres **8426**(84260I) (2012)

G.G. Vienne et al., Opt. Fiber Technol. **2**(4), 387–393 (1996)

S. Wang et al., Opt. Mater. **35**(9), 1752–1755 (2013)

S. Wang et al., J. Mater. Chem. C **2**(22), 4406–4414 (2014)

S. Wang et al., Opt. Mater. Expr. **6**(1), 69–68 (2016)

F. Wu et al., Mater. Res. Bull. **28**(7), 637–644 (1993)

C. Xia et al., Opt. Mater. **34**(5), 769–771 (2012)

W. Zhang et al., Opt. Quant. Electron. **49**, 27 (2017)

Progress in Mid-infrared Fiber Source Development

19

Darren D. Hudson, Alexander Fuerbach, and Stuart D. Jackson

Contents

Introduction	724
Background on Lasers in the Mid-IR	725
Carbon Dioxide and Monoxide Lasers	725
Solid-State Lasers Based on Cr:ZnSe/S	726
Optical Parametric Amplifiers and Oscillators	726
Optical Parametric Chirped-Pulse Amplifiers	727
Mid-IR Fiber Lasers: Overviews and Challenges	727
Fibers and Glasses for the Mid-IR	732
Silicates	733
Fluorides	733
Chalcogenides	735
Spectroscopy of the Significant Rare-Earth Transitions Used for Mid-IR Fiber Lasers	736
Spectroscopy and Lasing of Er^{3+} Ion	737
Spectroscopy and Lasing of Ho^{3+} Ion	740
Mid-IR Fiber Laser Architectures	741
Single-Longitudinal-Mode Systems	741
High-Power cw Systems	743
Tunable cw Systems	744
Ultrafast Systems	745
Supercontinuum Generated in Mid-IR Transparent Fibers	746
Supercontinuum Generated via Optical Parametric Amplification Systems	747
Supercontinuum Generation via Near-IR Fiber Laser Pumping	748

D. D. Hudson · A. Fuerbach
MQ Photonics Research Centre, Department of Physics and Astronomy, Macquarie University, North Ryde, NSW, Australia

S. D. Jackson (✉)
Department of Engineering, MQ Photonics Research Centre, School of Engineering, Macquarie University, North Ryde, NSW, Australia
e-mail: stuart.jackson@mq.edu.au

© Springer Nature Singapore Pte Ltd. 2019
G.-D. Peng (ed.), *Handbook of Optical Fibers*,
https://doi.org/10.1007/978-981-10-7087-7_53

Supercontinuum Generation via Diode Lasers................................... 750
Supercontinuum Generation via Mid-IR Fiber Laser Pumping.................... 750
Conclusion.. 752
References.. 752

Abstract

Mid-infrared fiber lasers have made significant strides in the last few decades and are now beginning to compete with and, in some cases, supersede traditional laser systems in terms of performance. In this chapter, we briefly review the current state of the field of lasers in the mid-infrared including traditional workhorse systems based on parametric amplification, transition metals, and quantum cascading. This is followed by a look at the materials science of mid-infrared compatible optical fiber and rare-earth doping in soft glasses. The advances in fiber laser sources operating near 3 μm are then explored including progress in narrow-linewidth and ultrafast systems. As an example of applications of these laser sources, we also present recent results in supercontinuum generation based on the 3 μm class fiber laser.

Introduction

The mid-infrared (mid-IR) region (2–20 μm) of the electromagnetic spectrum has long held promise for applications ranging from molecular detection and sensing to atmospheric laser propagation. Over the course of the last 40 years, great strides have been made in the field of mid-infrared photonics, with demonstrations of airborne, laser-based countermeasure systems, medical lasers, and thermal imaging systems. The last decade, in particular, has seen the field gain traction in industrial applications with commercial mid-IR detectors enjoying a compound annual growth rate of 20%. Simultaneously, research in this field has led to a range of high-brightness laser sources including the ubiquitous quantum cascade laser, optical parametric amplifiers emitting from near-IR to mid-IR, and more recently fiber lasers that use "soft glass" in order to emit in the mid-IR.

As the field has grown, a wide range of academic and industrial entities have participated in creating the current knowledge base of mid-IR photonics, and various definitions of the exact wavelength range have emerged. The current Federal Communications Commission standard defines near-infrared as 0.780–1.4 μm, the shortwave-infrared as 1.4–3 μm, and the mid-infrared as 3–50 μm. In the literature, however, the research community has to a large extent done away with the shortwave-infrared distinction, instead opting to simply set the near-infrared long wavelength terminating somewhere near 2–3 μm, before the onset of the mid-IR. This definition has some vagueness to it, however, with many researchers in the telecommunications space considering spectral regions longer than 1600 nm as mid-IR. Indeed, the rise of thulium lasers has resulted in a proliferation in the literature of the mid-IR beginning at 2 μm.

In this chapter, we propose a definition of the mid-IR based on the optical properties of materials used to make optical fiber. Our argument for this approach, which defines the mid-IR as 2.5–20 µm, is that it is essentially the material properties that define the emission wavelength and that dictate the technological requirements. The multiphonon edge of amorphous silicate glass occurs around 2.5 µm. This edge is perhaps the most important factor separating the performance between fiber laser systems operating at shorter wavelengths and those operating at longer wavelengths beyond this edge. At wavelengths longer than 2.5 µm, the soft-glass families such as the fluorides, chalcogenides, and tellurites must be employed to allow sufficient optical transparency. These materials are less robust than the well-established silicate family of glasses that has underpinned optical sources at the shorter wavelengths. In-fiber components (i.e. wave-plates, isolators, polarizers, modulators, etc) are commonplace at shorter wavelengths, while at wavelengths where soft-glasses are necessary, these components are only now beginning to emerge. As the field of mid-IR fiber lasers gains traction, this situation will change. However, given the fundamental physical differences between the silicates and soft glasses, there will always be a difference in handling and processing procedures.

This chapter opens with a review of currently available light sources in the mid-IR and then provides an overview of the challenges that exist for the development of rare-earth-doped fiber lasers operating in the mid-IR. In particular, the importance of choosing the right glass host for the mid-IR is emphasized, and the spectroscopic properties of the important dopant materials used for generating mid-IR light are discussed. The following sections review various fiber laser architectures for continuous-wave (cw) and pulsed operation of mid-IR fiber lasers, and the chapter concludes with an overview of existing schemes for the generation of mid-IR supercontinuum radiation.

Background on Lasers in the Mid-IR

Infrared laser sources are finding application in a growing range of fields including remote sensing, medical diagnostics, laser surgery, aircraft countermeasures, and LIDAR. The invention and subsequent commercialization of quantum cascade lasers (QCLs) have made a huge impact in this space. The wide availability of a source of semiconductor laser radiation in the range 4–300 µm combined with the maturation of silicon and chalcogenide glass as platform materials for integrated optics has made mid-IR photonics a fast-growing research field. However, QCLs are typically limited to relatively low power ($\sim$5 W), and ultrafast operation in these systems has proven difficult.

Carbon Dioxide and Monoxide Lasers

One of the best known mid-IR lasers is the CO_2 laser, invented by Kumar Patel in1964 at Bell Labs (Patel 1964). This gas laser, which emits at 10.6 µm, has reached maturity with commercial offerings yielding >10 kW of average power. A close relative of this system is the CO laser, which emits at 5 μm and also exhibits >1 kW average power. These systems have had significant impact in industrial laser cutting and welding and cosmetic tissue surgery. While high average powers have been achieved, the narrow bandwidth of the lasing transition has prevented ultrashort operation and wavelength tuning in these systems.

Solid-State Lasers Based on Cr:ZnSe/S

Solid-state vibronic lasers based on chromium- (λ_c ~2.5 μm) or iron (λ_c ~4 to 4.5 μm)-doped zinc sulfide or selenide lasers have attracted great interest due to their broad gain bandwidth. Similar to titanium-doped sapphire that emits in the near-IR, chromium and iron are transition metals with large phonon broadened laser levels, which leads to gain bandwidths of several hundred nanometers. Chromium lasers have been mode-locked using the Kerr-lens-type saturable absorption first observed in Ti:Al$_2$O$_3$ lasers, with pulse widths reaching sub-100 fs at peak powers of several kW (Cizmeciyan et al. 2009). Recently, ultrafast Cr:ZnSe lasers have entered the commercial domain (http://www.ipgphotonics.com/en/products/lasers/mid-ir-hybrid-lasers/1-8-3-4-micron/clpf-and-clpft-20-150-fs-up-to-6-w). Figure 1 shows the absorption and emission spectra for the chromium- and iron-doped chalcogenides. Vibronic lasers have many available pump wavelengths and offer broad tunability and bandwidth for ultrashort pulse generation, but average power levels have remained less than 5 W, a power level too low for many applications.

Optical Parametric Amplifiers and Oscillators

Optical parametric conversion of light from near-IR laser sources into the mid-IR utilizes second-order optical nonlinearity to convert photons at a particular pump wavelength to both a longer wavelength (idler) and a shorter wavelength (signal).

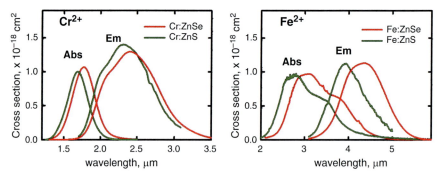

Fig. 1 (Left) Absorption and emission cross sections for Cr:ZnSe/S and (right) Fe:ZnSe/S. The broad bandwidth of the transition metal lasing medium allows for good ultrafast performance

The three photons involved in this interaction obey both energy and momentum conservation:

$$\omega_p = \omega_s + \omega_i$$
$$\mathbf{k}_p = \mathbf{k}_s + \mathbf{k}_i, \tag{1}$$

For a given pump wavelength, the above equations are satisfied at particular signal and idler wavelengths based on matching the index of refraction. Thus, a method of tuning the index of refraction allows for tuning the output wavelength. The most common ways to achieve this index tuning are by rotating a birefringent or temperature tuning a nonlinear crystal. In this manner, optical parametric devices offer smooth tuning of the output laser wavelengths over a wide range. By using this process in an optical cavity, optical parametric oscillation (OPO) can be achieved with narrow-linewidth, continuous-wave, or pulsed emission. Alternatively, single-pass optical parametric amplification (OPA) can be employed by using a high-peak power pump laser, such as an ultrashort laser.

OPA systems have recently seen great progress in the mid-IR wavelength range. Various commercial options are now available, with some systems offering multi-watt average power, ~100 fs operation with wavelength tuning spanning from the UV to >10 μm. OPAs lead the way in performance and tunability but still suffer from high costs and complexity due to the required free-space pump lasers, nonlinear crystals, and, in some cases, temperature-tuned elements.

Optical Parametric Chirped-Pulse Amplifiers

A further improvement to the performance of OPA performance is the addition of a chirped-pulse amplifier (CPA). In this arrangement, the femtosecond idler pulse is first temporally stretched using a diffraction grating stretcher, then amplified in a second OPA stage, and finally recompressed in a free-space grating compressor. While these systems add complexity to the OPA system, they offer unmatched performance in the mid-IR in terms of pulse width and peak power. With these improvements, OPCPA has experimentally demonstrated 3.9 GW peak power, sub-100 fs pulses at 3.25 μm (Elu et al. 2017). These systems offer superior output parameters at the cost of complexity, size, and reliability.

Mid-IR Fiber Lasers: Overviews and Challenges

Rare-earth-doped fiber lasers are playing an increasingly important role in this field; they offer a wide variety of output characteristics that complement and in some cases supersede the specifications provided by QCLs and other mid-IR laser systems. In this section, we will cover the various considerations and relevant physics behind soft-glass fiber lasers operating in the mid-IR (i.e., beyond 2.5 μm).

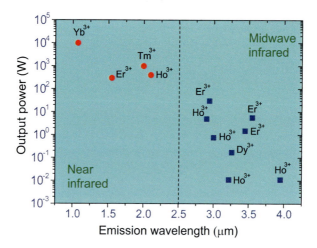

Fig. 2 The reported measured output power from rare-earth-doped fiber lasers as a function of the emission wavelength for fiber lasers based on fiber made from silicate glass (red dots) or fluoride glass (blue squares)

Figure 2 shows the current maximum output power from single (transverse)-mode fiber lasers as a function of the emission wavelength from the laser. The gradual decline in the output power is attributable to both the fiber itself and the rare-earth transition being used to create the emission. The fibers used for mid-infrared sources are less pure than silicate glass fiber and have comparatively weaker thermomechanical properties. These factors create thermal management problems for mid-IR fiber lasers at much lower power levels compared to silicate glass-based systems. The second major problem is the growing difference between the energy of the pump photon and the energy of the mid-IR photon itself. For efficiency, compactness, and cost, diode lasers are the preferred pump source for mid-infrared fiber lasers. Since the cheapest and most powerful diodes emit in the 1 um region, the energy difference can be substantial, and the difference is almost entirely converted to heat. Thus, given the characteristics of the soft glass combined with the large amounts of heat deposited, thermal management of the heat load in optical fibers creating the mid-IR emission is critical. Over the years, there have been a number of important developments that have combatted this problem leading to the record-breaking performance of Er^{3+}-doped ZBLAN fiber lasers operating on one of two rare-earth ion transitions.

The various components comprising the field of mid-IR fiber laser development can be summarized in a block diagram (see Fig. 3). Firstly, the most important aspect is the host glass because it must have a number of important attributes. Of course it must be able to transmit light with low loss in the mid-IR. Of the glasses that have been studied, the two most important to date are the fluoride and the chalcogenide glass families. There have been plenty of books and review papers on these materials, and we will only be examining them briefly in this chapter.

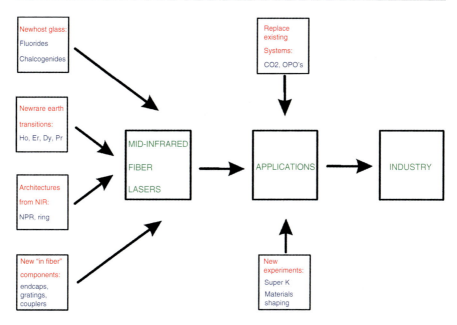

Fig. 3 Schematic diagram showing the inputs and outputs to the field of mid-IR fiber source development and application

The glass material comprising the fiber must support phonons of low to moderate energy; this allows the rare-earth transitions that comprise of relatively closely spaced energy levels to fluoresce and hence provide the necessary light for mid-IR fiber lasers to operate. Importantly, the glasses making the fiber must be able to provide optical fiber of low enough loss (which usually arises from crystallization or impurity incorporation during drawing). This final characteristic has been the most challenging. The solid-state precursor materials used to make the glasses are never 100% pure, and they are difficult to purify to a sufficient level for the fabrication of ultralow-loss fiber. It has taken decades of state-of-the-art engineering and chemistry to remove the most of the impurities in the starting materials. In fact, suppliers of the best fibers themselves purify even further the precursor materials required to make the fiber. As the number of demonstration of mid-IR fiber lasers and applications of mid-IR fiber sources grow, the demand for more mid-IR transparent fiber will grow, and the suppliers of the starting materials will offer higher-purity materials ready to be made to glass with further purification.

The next vital aspect inputting into the field is the rare-earth transitions themselves that create the fluorescence that is optically stimulated to produce laser emission. The rare-earth transitions allow the conversion from the low-beam-quality light from near-IR diode lasers to the typically lower-energy mid-IR (but higher beam quality) light from fiber lasers. When the phonon energy of the glass host is sufficiently low (typically it needs to be <650 cm^{-1}), a number of rare-earth transitions can fluoresce. Figure 4 shows the main transitions that have to date

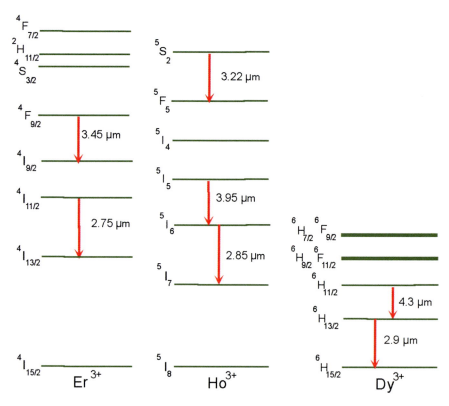

Fig. 4 The primary rare-earth transitions that have underpinned the development of mid-IR fiber lasers. The Er^{3+}, Ho^{3+}, and Dy^{3+} ions have at least two transitions available and have all been shown to fluoresce in fluoride glass optical fiber

underpinned the field of mid-IR fiber lasers. The rare-earth ions Er^{3+}, Ho^{3+}, and Dy^{3+} are the most important, although Pr^{3+} offers transitions that could be exploited in future systems. An important aspect to accessing these transitions is the initial energy level – it cannot be too energetic, or very short pump wavelengths will be required (to excite the upper laser level) that are in themselves less efficient and can cause color centers which are absorptive at the pump and sometimes the laser wavelength itself. In recent years, the use of the dual wavelength pump technique (Henderson-Sapir et al. 2014), in which two infrared pump lasers are combined to excite the initial energy level of the mid-IR transition, has shown to be a very effective approach. To date, 5.56 W from a dual wavelength pumped Er fiber laser at 3.55 μm has been demonstrated (Maes et al. 2017).

The next vital aspect is the architecture of the mid-IR fiber lasers that are used to create the output. To date, most arrangements have focused on the generation of high-power, Q-switched, mode-locked, tunable, or single-longitudinal-mode output. All the fiber laser designs that have to date been tested for mid-IR emission are based on architectures developed in the near-IR. The near-IR has boomed because

Fig. 5 Photomicrograph of the end of a fluoride glass optical fiber showing complete failure of the fiber tip as a result of OH ingress, the absorption of 3 μm, and the subsequent heating leading to melting of the fiber (Martin Gorjan, private communication)

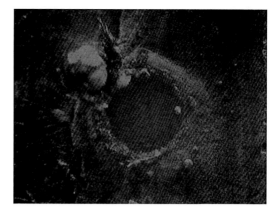

silicate glass fibers, which are very robust and can be made using very high-purity gasses, have their lowest loss (<1 dB/km) in the near-IR (usually around 1.5 μm). Of course, owing to the relatively large maximum phonon energy of silicate glasses, they cannot be used for the mid-IR. Many arrangements including standing-wave oscillators, ring oscillators, nonlinear polarization rotation systems, and nonlinear amplifying loop mirrors are being developed using soft-glass mid-IR transparent optical fiber for the generation of unique forms of mid-IR output.

The final input to the field of mid-IR fiber lasers are the in-fiber-based components that are being developed because of the urgent need to make mid-IR fibers robust and reliable. It is well-known that silicate-based fiber lasers developed for the near-IR are used in an ever-growing number of applications because they can be constructed "all fiber." The main developments required for all-fiber mid-IR fiber sources are fiber Bragg gratings and couplers (WDMs), as well as techniques for splicing and end capping of soft glasses. Figure 5 (Martin Gorjan, private communication) shows the damage caused during high-power testing of an Er^{3+}-doped zirconium fluoride (ZBLAN) fiber laser when the end of the ZBLAN fiber has been cleaved only. The facet of the fiber laser has been severely damaged because, as was clearly shown in a paper studying the output power from 2.8 μm Er-doped ZBLAN fiber lasers as a function of the time (Caron et al. 2012), water ingress from the atmosphere is the primary cause for the breakdown of the ends of the fiber. The basic mechanism behind this process is shown in Fig. 6. The solution to this problem is splicing the ends of the fiber using multimode aluminum fluoride (AlF_3) fiber which is known to be more resistant to attack by water and water vapor. The end capping of fluoride fiber in laser applications could be said to be one of the most important engineering developments in the field that has led to a large number of record-breaking achievements, in particular in terms of the average power emitted.

Perhaps the second most important development is the introduction of fiber Bragg gratings (FBGs) in mid-IR fiber laser research that are written directly into the soft-glass fiber. The inclusion of FBGs removes the need for mirrors based on

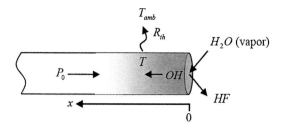

Fig. 6 Schematic diagram showing the end of a fluoride optical fiber and the process of the OH ingress from atmospheric water vapor (Caron et al. 2012)

bulk optics for defining a higher Q resonator. The inclusion has a twofold effect on the performance of the fiber laser. First, the overall robustness is significantly improved because the reflection optic is directly integrated into the gain medium itself. Second, when bulk mirrors are used, it is likely that an air gap will persist between the end of the fiber and the mirror which causes changes in the reflection coefficient and occasionally etalon effects (typically the dielectric mirror coating also fails before significant power levels are achieved). The phase-mask (Bernier et al. 2007) and direct-write (Hudson et al. 2013) techniques have both been used for the inscription of FBGs into fluoride glass fibers.

The successful splicing of soft-glass fibers over recent years has added another important dimension to the landscape. The splicing of end caps to ZBLAN fiber, in combination with FBGs, has underpinned the output power records that have been set (see Fig. 2). Isolating the ZBLAN glass from the environment in order to avoid glass breakdown in combination with the Bragg gratings that define the mirrors of the laser resonator has been shown to be essential for high efficiency and scaling the output power while improving the reliability. From this standpoint, it is clear that all future soft-glass fiber laser systems designed for a wide range of applications will require end caps, in-fiber Bragg gratings, and all-fiber arrangements.

Fibers and Glasses for the Mid-IR

The choice of the fiber material in the mid-IR involves a number of considerations: the maximum phonon energy, the environmental durability, the draw ability, the rare-earth solubility, and ultimately the purity of the starting materials. The maximum phonon energy of the glass sets the overall infrared transparency range of the fiber, and the multiphonon-relaxation rates influence the quantum efficiency of radiative electronic transitions. The multiphonon-relaxation rates for common glasses used for optical fibers as a function of the energy gap between energy levels are shown in Table. 1. The optical transparency window relates to the position of the bandgap at short wavelengths and the infrared absorption cutoff wavelength at longer wavelengths. In a first approximation, the long-wavelength edge relates to the vibrational frequency v of the anion-cation bonds of the glass. For an ordered structure

$$v = (1/2\pi)\sqrt{k/M}, \qquad (2)$$

Table 1 Properties of popular fiber materials

Fiber material	Max. phonon energy (cm^{-1})	Infrared transparency (μm)	Propagation losses (λ at minimum) (dB/km)	Thermal conductivity (W/K m)
Silica	1100 (Harrington 2004)	<2.5	0.2 (1.55 μm)	1.38 (Harrington 2004)
ZBLAN	550 (Tran et al. 1984)	<6.0	0.05 (2.55 μm)	0.7–0.8
GLS	425 (Sanghera et al. 2009)	<8.0	0.5 (3.50 μm)	0.43–0.5

where $M = m_1 m_2 / (m_1 + m_2)$ is the reduced mass for two bodies m_1, m_2 vibrating with an elastic restoring force k. While for disordered structures like glass, this is not an accurate expression; it does nevertheless highlight the important contributions (and hence limits) to the glass transparency. The relative cation-anion bond strength is intimated by the field strength Z/r^2, where Z is the valence state of the cation or anion and r is the ionic radius. Generally, glasses composed of large anions and large cations with low field strengths display the highest transparency in the mid-IR spectral region. The important physical properties of the popular glasses used for optical fibers are shown in Table 1.

Silicates

Silica glass (SiO_2) has a long history as the standard material for optical fiber (http://www.ipgphotonics.com/en/products/lasers/mid-ir-hybrid-lasers/1-8-3-4-micron/clpf-and-clpft-20-150-fs-up-to-6-w; Hudson et al. 2013). The maximum phonon energy of 1100 cm^{-1}, however, sets the multiphonon edge at relatively short wavelengths (of approximately 2.2 μm for optical fiber) (Fortin et al. 2015). Using modified chemical vapor deposition (MCVD), silica can be purified to an extremely high level and drawn into robust optical fibers. Reducing the OH^- content in the glass, which has two main absorption peaks in the range 1.3–2.0 μm (Schneider 1995), improves the near-to-mid-IR utility of the glass. Rare-earth ions such as Nd^{3+} and Er^{3+}, which have high field strengths, have low solubility in silicate glass, which can lead to clustering and microscale phase separation.

Fluorides

The fluoride glass family has enabled great advances in infrared optical fibers over the last several decades. With excellent host properties for rare-earth ions, fluoride-based optical fiber has been fundamental to the development of mid-IR fiber laser sources. The most widespread fluoride fiber is ZBLAN, with a mixture of 53 mol.%

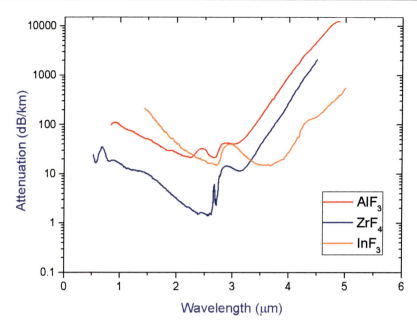

Fig. 7 Measured optical loss in state-of-the-art single-mode fluoride fiber as a function of the wavelength (Figure courtesy of Le Verre Flouré (http://leverrefluore.com))

ZrF_4, 20 mol.% BaF_2, 4 mol.% LaF_3, 3 mol.% AlF_3, and 20 mol.% NaF. The large atomic weight of zirconium combined with the relatively weak bond strength yields a maximum phonon energy of ~ 550 cm^{-1}, which is roughly half that of the glass of the silicate family. The lower phonon energy allows the multiphonon transparency edge of the glass to extend out to around 6 μm. However, multiphonon relaxation of excited ions is significantly beyond 3 μm and can present a problem to achieving efficient lasing. Mechanically, ZBLAN is weaker than the silicates and has a lower optical damage threshold. While ZBLAN has a lower fundamental scattering loss limit compared to the silicates, in practice impurities and fabrication tolerances have prevented ZBLAN from reaching the 0.2 dB/km propagation loss of silicate fibers.

Figure 7 shows the loss of single-mode fluoride fiber manufactured by Le Verre Fluore (France) (http://leverrefluore.com) as a function of the wavelength. There are a number of features that can be observed in this graph. Firstly, fluoride glasses based on the ZrF_4 are the most developed and thus have the lowest loss, of approximately 1 dB/km. The minimum loss for AlF_3-and InF_3-based glasses is close to each other at about 10–20 dB/km, which is about an order of magnitude higher than the background loss for ZrF_4-based glass. This difference primarily relates to the degree of maturity and refinement of the glass and preform and subsequent fiber-drawing process. The ZrF_4-based glass was the first fluoride glass to be fabricated into fiber form and is responsible for the majority of the laser demonstrations in the

Table 2 The magnitude and width of the fundamental OH absorption peak in a number of Er^{3+}-doped glasses

Glass	Peak absorption [cm^{-1}]	FMHW [cm^{-1}]
GeGaS	3420 (2.92 μm)	530 ± 50
Tellurite	2980 (3.356 μm)	900
Fluoride	3480 (2.87 μm)	410
Silica	3680 (2.72 μm)	130

field of mid-IR fiber lasers. The InF$_3$ glass, on the other hand, has a longer infrared cutoff wavelength because of the heavier atomic weight of indium compared with zirconium. The atomic weight of the primary element of these glasses is Al = 26.96, Zr = 91.22, and In = 114.82. Common to all these fibers is the appearance of the OH absorption peak at approximately 3 um. (Note that the OH absorption peak in InF$_3$ glass is red-shifted slightly compared to the other glasses and suggests that the heavier In atom is lowering the energy of the vibrational stretching mode of the OH radical.) In modern applications, the AlF$_3$ glass has been used as the end cap for ZBLAN-based fiber systems owing to its relatively better resistance to water. The InF$_3$ glass, on the other hand, has been primarily developed for applications requiring longer wavelengths than those produced from ZrF$_4$-based glass. The lower phonon energy of this glass compared to the other fluoride glasses creates longer luminescence lifetimes which allows transitions that are quenched in the other glasses to possibly lase. As the fabrication methods get more refined, leading to background loss values that are equivalent or even lower than ZBLAN, InF$_3$-based optical fiber could replace a large proportion of the ZBLAN systems because it also displays much better OH resistance.

Table 2 lists the primary properties of the fundamental OH absorption in mid-IR transparent optical fiber. To date, the ZBLAN glass optical fibers are still the most widely used as the primary host for demonstrating new transitions, new modes of operation, and the majority of the fiber laser system performance records.

Chalcogenides

Chalcogenides are composed of the chalcogen elements S, Se, and Te (Jackson 2009a; Aydin et al. 2017; Librantz et al. 2008). They are environmentally durable, have a low toxicity, and have reasonably large glass-forming regions. When the rare-earth ions are doped into these glasses (Jackson 2004), the radiative transition probabilities and, therefore, the absorption and emission cross sections are high as a result of the high (~2.6) refractive index of the glass and the high degree of covalency of the rare-earth ion with the surrounding medium. Maximum phonon energies of 300–450 cm^{-1} produce low rates of multiphonon relaxation (see Table 1) and therefore reasonably high quantum efficiencies for mid-IR transitions. The low thermal conductivity (see Table 1) is however an important factor to be considered in the design of chalcogenide-based lasers. Of the large number of rare-earth-doped chalcogenides studied for luminescent emission, the most important glasses are the sulfide glasses GaLaS (GLS) (Jackson 2009b) and GeGaS (Sumiyoshi and Sekita 1998) because of the reasonably high rare-earth solubility.

Spectroscopy of the Significant Rare-Earth Transitions Used for Mid-IR Fiber Lasers

Fiber lasers that operate in the mid-IR typically operate on transitions that are well above the ground state (see Fig. 8). This carries with it a number of issues. Firstly, the pump wavelength is usually required to be quite energetic relative to the laser wavelength. Secondly, since the fluorescent lifetime of energy levels decreases as the energy of the energy levels increases, then some engineering of the lasing process is required to prevent bottlenecking on the lower laser level (or an excited energy level below this level). Bottlenecking of the population causes bleaching of the pump absorption that saturates output laser power and the amount of pump light absorbed. Thirdly, since more metastable energy levels are now involved in the entire lasing process (since there are typically other energy levels below the laser transition itself), a number of energy transfer processes might need to be taken into account especially when the rare-earth concentration is high as is required for double-clad fiber configurations. Thus, a great deal of research has involved the spectroscopy of soft glasses doped with one or perhaps two rare-earth ions.

In this section, we will explore the primary systems that have been studied spectroscopically for the development of mid-IR fiber laser transitions. While there are many other rare-earth transitions that have produced cw output, only the primary transitions of the Er^{3+} ion and the Ho^{3+} ion have underpinned Q-switched, mode-locked, and narrow-linewidth emission. It is for this reason alone that we concentrate on the spectroscopy of these ions in this chapter. The interested reader can find many reviews on mid-IR fiber lasers that discuss the other transitions.

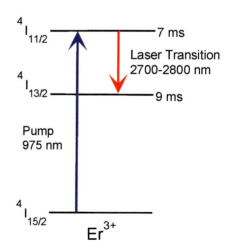

Fig. 8 Simplified energy level diagram for the Er^{3+} ion showing the most well-known mid-IR laser transition and the corresponding pump wavelength used to excite the upper laser level directly

Spectroscopy and Lasing of Er^{3+} Ion

The Er^{3+} ion has been studied the most for the development of mid-IR systems. Before we start the in-depth investigation of the transitions of the Er^{3+} ion that are relevant to the mid-IR, Table 3 lists the vital transition parameters for the $^4I_{13/2} \rightarrow {}^4I_{15/2}$ transition at 1550 nm for a range of glasses used for optical fibers. We can observe that as the refractive index of the glass increases, the stimulated emission cross section increases but the luminescent lifetime correspondingly decreases. This relates to the spontaneous emission probability that increases with increasing refractive index. What is interesting is the fact that the product of the stimulated emission cross section with the lifetime is approximately the same for all glasses, with the fluoride glass having the highest value and the gallium lanthanum sulfide (GLS) glass the lowest.

The most developed mid-IR laser transition is the $^4I_{11/2} \rightarrow {}^4I_{13/2}$ transition at 2750 nm (in fluoride glass). Figure 8 shows the simplified energy level diagram that displays the standard pump transition, the laser transition, and the lifetimes, in fluoride glass of the upper and lower laser levels. As mentioned above, the upper laser level lifetime is longer than the lower laser level, and hence much work over the years has been dedicated to optimizing a number of energy transfer process and laser processes so that self-saturation of the output is avoided.

Figure 9 displays the ways in which self-saturation of the output has been avoided. Figure 9a shows how the introduction of the Pr^{3+} co-dopant into the glass can allow energy transfer to provide the necessary desensitization of the lower laser level (Golding et al. 2000). The $^3F_4$ and $^3F_3$ energy levels of the Pr^{3+} ion are resonant with the $^4I_{13/2}$ level of Er^{3+} allowing very efficient energy transfer to the Pr^{3+} ion, and, by way of nonradiative decay (at least for fluoride glass), the original excitation from the lower laser level of the laser transition is returned to the ground state. Of course, the $^1G_4$ level of the Pr^{3+} ion is slightly resonant with the upper laser level and will reduce the lifetime of the upper laser level of our desired laser transition somewhat. As a result of this issue and the fact that for any detailed numerical modelling analysis, the energy transfer parameters are required, it was necessary to carry out a detailed spectroscopic investigation to measure the effect on the laser transition of adding Pr^{3+} desensitizer ion to the Er^{3+}-ZBLAN

Table 3 Typical emission cross sections and luminescent lifetimes (for the 1550 nm transition) in popular Er^{3+}-doped glasses used for optical fiber

Glass composition	Peak emission cross section [$\times 10^{-21}$ cm^2] $^4I_{13/2} \rightarrow {}^4I_{15/2}$	Luminescent lifetime [ms] $^4I_{13/2} \rightarrow {}^4I_{15/2}$	$\sigma\tau$ [$\times 10^{-21}$ ms cm^2]	Typical refractive index
GaLaS	15.7	2.3	36.1	2.4
PBG	10.3	4.16	42.8	2.3
ZBLAN	5.87	9.0	52.8	1.5
Silicate	4.4	10.2	44.9	1.46

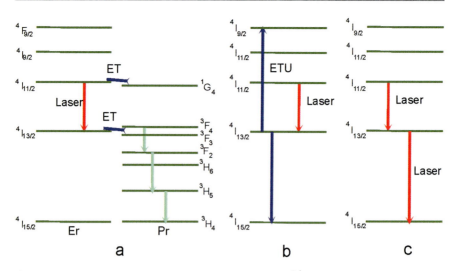

Fig. 9 Mechanisms by which the lower laser level of the Er^{3+} can be depopulated using (**a**) energy transfer to a nearby Pr^{3+} desensitizer ion, (**b**) energy transfer upconversion which can recycle excitation back to the upper laser level, or (**c**) cascade lasing in which two adjacent laser transitions can be made to lase simultaneously

system (Bernier et al. 2007). Overall, for Pr^{3+} ion concentrations in the range 2000–3000 ppm, sufficient quenching of the lower laser level occurred with only a factor of two reductions of the upper laser level lifetime. With this ion, maximum output powers approaching 10 W have been obtained (Zhu and Jain 2007).

Figure 9b shows an alternative but very successful method of using energy transfer upconversion (ETU) to "recycle" energy from the lower laser level to the upper laser level. The Er^{3+} ion has energy levels that are spaced so that a number of energy transfer (ET) processes are quite resonant presenting fast rates of ET. For desensitizing the $^4I_{13/2}$ level, the ETU process $^4I_{13/2}, ^4I_{13/2} \rightarrow {}^4I_{9/2}, ^4I_{15/2}$ populates the $^4I_{9/2}$ level after the energy from one excited Er^{3+} ion transfers to an adjacent excited Er^{3+} ion. This mechanism is the basis for the current output power record from a mid-IR fiber laser (Fortin et al. 2015) in which 30 W was generated at a wavelength of 2.94 μm.

Figure 10 shows measurements of the macroscopic ET parameter for the ET to Pr^{3+} co-dopant and ETU methods of depopulating the lower laser level (Golding et al. 2000). There are a number of important features to understand. First, as expected, the macroscopic ET parameter for all processes increases with increasing Er^{3+} concentration. This relates to the decreasing separation between the rare-earth ions that increase the rate of energy transfer according to the dipole-dipole ET process of the Foster-Dexter model. Importantly, the competing gain-lowering ETU process (ETU1): $^4I_{11/2}, ^4I_{11/2} \rightarrow {}^4F_{7/2}, ^4I_{15/2}$ is much weaker than the ETU process (ETU2): $^4I_{13/2}, ^4I_{13/2} \rightarrow {}^4I_{9/2}, ^4I_{15/2}$. This allows quite high (up to 7 mol.%) concentrations of Er^{3+} ions in double-clad fluoride fibers to be used. This Er^{3+}

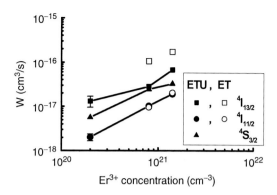

Fig. 10 Measured macroscopic rate parameter as a function of Er^{3+} concentration from (Golding et al. 2000). Reprinted with permission

ion concentration improves energy recycling and increases pump absorption which means that less fiber can be quite expensive to be used. Lastly, the macroscopic parameter relating to ET from the $^4I_{13/2}$ level to the Pr^{3+} ion is observed to be very large compared to all the other parameters. This suggests that the process is quite resonant and that some flexibility in the Er^{3+} ion to Pr^{3+} ion concentration ratio exists while still supporting sufficient lower laser level depopulation.

The final depopulation method for the lower laser level employs cascade lasing. This method was first demonstrated quite some time ago (Schneider 1995), but it was not until 2009 (Jackson 2009a) that a diode-pumped version (hence involving double-clad fluoride fiber) was demonstrated. This more recent work highlighted a number of key findings that allowed cascade lasing to be further studied. One key finding was the lessening of the requirement for high Q cavities, which offers great flexibility in the design space for Er^{3+}-cascade fiber lasers. The other major factor that resulted from this work was the fact that a comparatively low-brightness pump source, such as a diode laser, could be used in cascade fiber laser systems. These pioneering studies laid the groundwork for the many subsequent studies in high-power cascade fiber laser research.

In a recent publication (Aydin et al. 2017), it was shown that cascading the adjacent transitions of the erbium ion at wavelengths of 2.8 μm and 1.6 μm could produce >10 W output power and the slope efficiency reaches a high level of 50% (thus exceeding the Stokes limit by 15%). This remarkable result was due to the combination of low-loss fluoride fiber with an Er^{3+} concentration of just 1 mol.% and the discovery that ESA from the shorter (1.6 μm) wavelength transition recycles the energy accumulating in the lower laser level back to the upper laser level. Of critical importance was the need to highly resonate this 1.6 μm light in order to achieve the output power and slope efficiency that was measured. First, the ESA process at 1.6 μm is actually highly nonresonant, which means that to access the ESA process itself with some degree of efficiency, the 1.6 μm light within the fiber needed to be intense. In addition, it was observed that for lower Q cavities, e.g., when Fresnel reflection was used as a feedback, a self-pulsing process caused the fiber to catastrophically fail. While the output power from this system was not as high as the record levels that can be achieved with double-clad ZBLAN fiber, this demonstration represents a significant advance and could be the approach used for future 100 W power-level demonstrations.

Spectroscopy and Lasing of Ho^{3+} Ion

The $^5I_6 \rightarrow {}^5I_7$ transition of the Ho^{3+} ion (see Fig. 11) is the second most important mid-IR transition because it can be diode pumped to moderate power, and it has been used for many demonstrations of pulsing and tuning. All mid-IR fiber laser demonstrations with this ion have involved fluoride glass fiber. Compared to the transition of the Er^{3+} ion discussed above, the fluorescence spectrum for Ho^{3+}-doped ZBLAN is centered at a wavelength that is approximately 100 nm longer – this has the advantage of avoiding atmospheric water vapor absorption and leads to potentially more stable output. The most studied diode pump wavelength for the 2.9 μm transition of Ho^{3+} is 1150 nm, and, importantly, the quantum efficiency limit increases to 40% compared to 35% for the Er^{3+} system. Like the Er^{3+} ion, however, the lower laser level has a longer lifetime (12 ms) compared to the upper laser level (3.5 ms), and the transition can also be self-saturating. Thus, like the Er^{3+} ion, similar techniques to create true four-level emission on this transition have been tested.

Co-doping Ho^{3+} ions with Pr^{3+} ions has to date been the most successful approach to achieving unsaturated output. The ET process $Ho^{3+}(^5I_7) \rightarrow Pr^{3+}(^3F_2, {}^3H_6)$

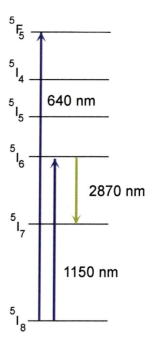

Fig. 11 Simplified energy-level diagram for the Ho^{3+} ion showing the pump transitions that have been reported for fiber lasers operating on the $^5I_6 \rightarrow {}^5I_7$ transition

is resonant, and the $^3F_2$ and $^3H_6$ energy levels of the Pr^{3+} ion are quickly depopulated via multiphonon emission to the ground state. Fortunately, there are no energy levels in the Pr^{3+} ion system that degenerate with the upper laser level of Ho^{3+} making co-doping less of a concern. Nevertheless, detailed studies of energy transfer and excited-state absorption in the Ho^{3+}-doped ZBLAN system have been carried out (Librantz et al. 2008; Jackson 2004). One important result from these studies was the discovery that ETU will not assist in recycling the energy as is the case for the Er^{3+} ion. The macroscopic rate parameter for ETU from the $^5I_6$ level of the Ho^{3+} ion is larger than the corresponding ETU rate parameter for the $^5I_7$ level; thus high-concentration Ho^{3+} fibers will not create slope efficiencies that exceed the Stokes limit. Using Pr^{3+} as a desensitizer, a maximum output power of 2.5 W was produced when the pump wavelengths of 1100 nm from an Yb^{3+}-doped silicate fiber laser (Jackson 2009b) or multiplexed diode lasers operating at 1150 nm (Sumiyoshi and Sekita 1998) were used.

Cascade lasing of both the adjacent $^5I_6 \rightarrow {}^5I_7$ and $^5I_7 \rightarrow {}^5I_8$ laser transitions of a Ho^{3+}-doped fluoride fiber laser has been demonstrated (Sumiyoshi and Sekita 1998; Sumiyoshi et al. 1999) with a total slope efficiency of 65% being created in these studies. Using double-clad Ho^{3+}-doped ZBLAN optical fiber, the first diode-pumped fiber laser with an emission wavelength longer than 3 μm was demonstrated (Li et al. 2011).

Mid-IR Fiber Laser Architectures

As with virtually all other types of lasers, mid-IR fiber lasers can broadly be classified into continuous-wave (cw) lasers and pulsed lasers. The latter group can be further subdivided into gain-switched, q-switched, and mode-locked (ultrafast) lasers. Of those, mode-locked mid-IR fiber lasers are of fundamental significance as they hold promise to become an enabling technology for important spectroscopic applications like trace-gas sensing or early cancer detection. Different approaches to realize ultrafast mid-IR fiber lasers as well as the generation of broadband supercontinuum radiation that can be seeded by those sources will be discussed later in this chapter.

Within the group of cw lasers, three particularly important kinds of laser systems stand out, single-longitudinal-mode lasers, high-power cw lasers, as well as tunable cw lasers, and each of those types will be discussed in the following sections.

Single-Longitudinal-Mode Systems

Lasers that oscillate on a single longitudinal resonator mode can feature linewidths of as narrow as a few kHz and correspondingly long coherence lengths that can exceed 100 km. Applications that exploit those unique characteristics include spectroscopy, Light Detection and Ranging (LIDAR), and coherent beam combining.

In order to suppress lasing from additional longitudinal modes, extremely narrow band-pass filters are required. In the context of fiber lasers, fiber Bragg gratings (FBGs) (Kashyap 2010) are the preferred choice as they represent narrowband reflectors that can directly be integrated into the active laser fiber. In a distributed Bragg reflector (DBR) laser, one or two (typically short) FBGs are inscribed into either one or in both ends of a relatively long actively doped fiber. The resulting resonator geometry resembles the classical textbook-laser resonator that is confined by two cavity end mirrors. In contrast, a distributed feedback (DFB) laser consists of a single π-phase-shifted FBG that is written into a short piece of extremely highly doped piece of active fiber and that constitutes the entire laser oscillator. Although it is typically easier to achieve single-longitudinal-mode operation in a DFB laser, the requirement for high doping concentrations and the typically relatively high pump thresholds mean that DBR lasers are still often the preferred geometry when it comes to single-longitudinal-mode mid-IR fiber lasers.

While techniques for the fabrication of FBGs in silica fibers are well established (Jovanovic et al. 2012), it is less straightforward to realize narrowband fluoride-fiber laser cavities that are based on FBGs in those fibers. One possible method for the fabrication of FBGs in a fluoride fiber is to tightly focus femtosecond laser pulses into its core. The high laser intensity can initiate efficient nonlinear absorption processes that facilitate a highly localized deposition of energy within the core of the fluoride fiber. This energy deposition can, in turn, lead to a highly localized modification of the refractive index which forms the basis of the so-called femtosecond laser direct-write (FLDW) technique (Gattass and Mazur 2008). By periodically repeating the process of inducing a localized refractive index modulation within the core of a fluoride fiber, a periodic refractive index modulation that represents a FBG can directly be inscribed.

In 2013, Hudson et al. used a particular variant of the FLDW method, the femtosecond laser point-by-point technique, to inscribe a 20 mm long FBG into one end of a Ho^{3+}, Pr^{3+}-co-doped ZBLAN fiber. A dichroic mirror that was butt-coupled to the other end of the active fiber completed the resonator (Hudson et al. 2013). With this setup, a maximum output power of 11 mW could be obtained at a wavelength of 2914 nm. Single-longitudinal-mode operation was confirmed using a Mach-Zehnder interferometer. In 2015, Bernier et al. (2015) used the femtosecond dithering phase-mask technique to inscribe a 30 mm long π-phase-shifted FBG into a heavily erbium-doped fluoride fiber to realize a single-longitudinal-mode DFB laser operating at the slightly shorter wavelength of 2794 nm with almost identical output power (12 mW). Most recently, Behzadi et al. have analyzed the design and potential performance characteristics of mid-IR DFB π-phase-shifted Raman fiber lasers in appropriately chosen low-phonon-energy glasses, such as chalcogenides and tellurites (Behzadi et al. 2017). The group has shown that the threshold pump power for optimized devices operating at 3.6 μm could be as low as 50 mW, and the authors have further proposed that their design should pave the way for the realization of a new class of narrow-linewidth lasers that can target virtually any wavelength between 2.5 and 9.5 μm based on a cascaded approach. However, more work is required to experimentally verify these claims.

High-Power cw Systems

Virtually all applications for lasers operating in the mid-IR exploit the fact that this particular spectral region overlaps with very unique and typically very strong absorption lines of a vast majority of molecules. While for sensing and spectroscopic applications, the spectral characteristic of the mid-IR laser is the most critical parameter that defines its suitability for a given task (e.g., extremely narrow-linewidth or ultrabroadband spectral output and/or tunability), for applications that fall into the broad category of material processing, high average output power levels are crucial. Laser radiation around 3 μm is strongly absorbed in liquid water as this wavelength is resonant with symmetric stretch vibrations of the OH radical. Lasers operating in that spectral region are thus highly suited for the cutting of soft and hard tissue in biomedical applications like dentistry, orthopedics, dermatology, or ophthalmology. Laser radiation in the spectral region between 3 and 4 μm on the other hand coincides with the fundamental stretching frequency of the C-H covalent bonds found in many chemical species like polymers, and thus high-precision cutting or welding of materials that fall into this group becomes feasible.

As mentioned previously (section: **Mid-IR Fiber Lasers: Overviews and Challenges**), a major problem that occurs when operating a fluoride fiber laser at very high average power levels is the diffusion of OH impurities inside the fiber according to Fick's laws which will degrade the tip of a ZrF_4 fiber within seconds of exposure to ∼30 W power at 2.94 μm (Caron et al. 2012). One solution to this problem is to splice short sections of coreless fibers to one or both ends of an active ZBLAN fiber, thus effectively sealing the end facets with those so-called end caps. A fully integrated all-fiber design can be achieved by directly inscribing FBGs into both ends of an active laser fiber, splicing a fiber-coupled pump diode to the input end and by finally end capping the other end with a short section of coreless AlF_3 fiber (AlF_3-based glasses are two orders of magnitude less permeable to OH contamination than ZrF_4). Average output power levels in excess of 30 W at 2.94 μm from an erbium-doped laser have been demonstrated with this approach (Fortin et al. 2015). At longer wavelengths, using a similar setup but employing a dual pumping scheme by combining two pump sources at 974 nm and 1976 nm, a maximum output power of 1.5 W at a wavelength of 3.44 μm has been demonstrated from an erbium-doped zirconium fluoride fiber laser (Fortin et al. 2016). Based on the rapid progress that has recently been made in erbium-doped ZBLAN fibers, Raman laser technology is thought to be a promising route to target specific wavelengths that are not readily accessible by rare-earth ion transitions. Numerical simulations of tellurite fiber Raman lasers pumped by erbium-doped ZBLAN fiber lasers have shown that fiber laser sources operating at any wavelength within the 3–5 μm spectral region can be achieved by employing first- and second-order Raman scattering of tellurite fibers (Zhu et al. 2015).

Tunable cw Systems

For spectroscopic and sensing applications, the ability to tune the operating wavelength of a fiber laser over a broad spectral region is a crucial feature. In addition, for the further development of mid-IR fiber lasers themselves, a broadly tunable laser enables the characterization of optical components, for example, the measurement of optical fiber loss which is often unknown in fluoride fibers in contrast to well-characterized silica-based fibers. However, in order to achieve tunable operation, free-space sections typically have to be inserted into the fiber laser cavity. For example, by collimating the output of an angle-cleaved holmium-praseodymium co-doped ZBLAN fiber and by inserting a plane ruled diffraction grating into the resulting free-space section, a laser that produced up to 7.2 W over a tuning range from 2825 to 2975 nm has been demonstrated (Crawford et al. 2015). Using a similar cavity arrangement with a bulk-optic diffraction grating, a tunable dysprosium laser was later demonstrated that operated at longer wavelengths. The laser could be tuned from 2950 to 3350 nm yet at lower output power levels in the tens of mW range (Majewski and Jackson 2016). To access even longer wavelengths, the broadband $^4F_{9/2} \rightarrow {}^4I_{9/2}$ transition in erbium can be targeted. Utilizing a dual wavelength pumping approach and again a bulk diffraction grating, a tuning range extending from 3330 to 3780 nm has been demonstrated (Henderson-Sapir et al. 2016). The long-wavelength edge of this tuning range also represents the longest emission from a fiber laser operating at room temperature. As an alternative to the diffraction grating, a volume Bragg grating (VBG) can also be used to provide wavelength-selective feedback in a fiber laser cavity (Liu et al. 2018). While VBGs offer some advantages over diffraction gratings in terms of high diffraction efficiency, narrow spectral width, low insertion losses, high damage threshold, and good thermal stability, the requirement to incorporate a free-space section into the fiber laser cavity remains an issue as it reduces the overall robustness and versatility of the laser system. An alternative approach is to inscribe fiber Bragg gratings into the active laser fiber and apply stress and compression to the grating to realize wavelength tuning. As described earlier, the femtosecond laser direct-write technique is a highly flexible and reproducible method for the fabrication of FBGs in virtually all kinds of fibers. However, in order to facilitate the inscription process, the protective coating of the fiber typically has to be removed first. This, in turn, reduces the mechanical strength of the grating which is a nuisance, in particular with respect to stress/compression tuning in soft-glass fibers like ZBLAN. To circumvent this problem, Bharathan et al. (Bharathan et al. 2017) have demonstrated that inscription of FBGs directly through the polymer coating of a 480 μm diameter holmium-praseodymium fiber is possible. As this allows one to maintain the mechanical strength of the fiber, a tuning range from 2850 to 2887 nm could be achieved.

Ultrafast Systems

Ultrafast (femtosecond) lasers that are operating in the near-IR have enabled the development of a large number of applications that have resulted in important breakthroughs in both fundamental science and industry. In this regard, ultrafast fiber lasers have an ever-increasing role. Due to material constraints, the unavailability of off-the-shelf in-fiber components, as outlined in previous sections, is a problem for the mid-IR. As a result, it is not straightforward to extend the wavelength coverage of ultrafast systems from the near-infrared to the mid-IR. As late as 2015, two groups almost simultaneously reported the successful demonstration of a femtosecond fiber laser operating near 3 μm. Utilizing an erbium-doped ZBLAN fiber in a ring laser cavity that was mode-locked via nonlinear polarization rotation (NPR), Hu et al. (Hu et al. 2015) demonstrated the generation of 497 fs pulses with a peak power of 6.4 kW, while Duval et al. (Duval et al. 2015) generated pulses with a duration as short as 207 fs with a peak power of 3.5 kW in an almost identical setup. A major problem that was identified in these demonstrations was the presence of strong water vapor absorption that ultimately limited the achievable pulse duration. In order to realize an effective saturable absorption mechanism based on NPR, it is necessary to include a free-space optical section in the fiber ring cavity. The erbium emission spectrum overlaps with several strong water vapor absorption lines which restrict the width of the mode-locked spectrum. A solution for this problem is to move from erbium to holmium as the active laser ion as the resulting wavelength shift of about 100 nm shifts the optical spectrum away from those absorption lines. This was experimentally confirmed by Antipov et al. (2016) who demonstrated the generation of pulses as short as 180 fs with a record-high peak power of 37 kW in a holmium-praseodymium NPR mode-locked fiber laser. Despite the unrivalled performance achievable with NPR mode-locked fiber lasers, a drawback is that lasers based on NPR usually require careful alignment and frequent adjustment. In recent years, a series of two-dimensional nanomaterials have emerged as saturable absorbers for mode-locking lasers because of their ultra-broad absorption band, low saturation intensity, and ultrafast recovery time. By depositing a multilayer graphene saturable absorber on a gold mirror that served as one of the cavity end mirrors, Zhu et al. (2016) have demonstrated the generation of pulses with a duration of 42 ps and an average power of 18 mW at 2800 nm in an erbium-doped ZBLAN fiber laser. While this level of performance is nowhere near the performance of the NPR systems introduced before, it demonstrated the suitability of graphene as a saturable absorber material for the mid-IR. Another candidate material for this task is black phosphorus, very similar in its physical structure to graphene but composed of phosphor atoms instead of carbon atoms. By depositing mechanically exfoliated black phosphorus onto a gold mirror and by focusing onto this saturable absorber mirror that is located in a free-space section of the erbium ZBLAN fiber laser, pulses with an identical duration of 42 fs yet at higher average power levels of 613 mW have been obtained (Qin et al. 2016). Using a liquid exfoliation process for the black phosphorus saturable absorber and switching from erbium to holmium as the active laser ion, pulses as short as 8.6 ps have been demonstrated by Li et al. (2016).

The combination of tunable and ultrafast operation is possible by inserting both a diffraction grating and a saturable absorber into a fiber laser cavity. Shen et al. (2017) have implemented this approach by extending both ends of an erbium fiber laser cavity by short sections of free-space propagation. This allowed the authors to place a diffraction grating into one of those sections and a semiconductor saturable absorber mirror (SESAM) into the other section. This enabled the generation of pulses as short as 6.4 ps with a tunable center wavelength from 2710 to 2820 nm.

A similar approach but based on a holmium-praseodymium-doped ZBLAN fiber resulted in the generation of pulses with a duration of 22 ps and a tuning range from 2842 to 2876 nm (Wei et al. 2017).

Supercontinuum Generated in Mid-IR Transparent Fibers

The creation of broadband laser light sources has been a topic of both research and application since the discovery of "supercontinuum generation" in bulk glasses in 1970 (Alfano and Shapiro n.d). The development of low-loss optical fiber by Keck and co-workers in the 1970s made possible material-light interaction lengths that were orders of magnitude longer than that allowed by bulk optics. Thus optical fibers made from silica, with a relatively low $\chi^{(3)}$ nonlinearity, could still exhibit large cumulative nonlinear effects. Experiments demonstrating this feature of optical fibers for nonlinear optics soon followed (Stolen and Ashkin 1973; Hill et al. 1978; Stolen and Lin 1978).

In the 40 years since this original fiber-based supercontinuum work, a tremendous body of literature has been generated on this topic. We now have a deep understanding of the rich physics that occurs in supercontinuum generation; this knowledge has in turn driven our understanding of seemingly unrelated phenomena such as rogue waves in the ocean (Solli et al. 2007). Meanwhile, the bandwidth of the supercontinua has progressively expanded to cover larger regions of the electromagnetic spectrum, from the UV to the mid-IR. These demonstrations were made possible by advances in materials science, laser technology, and optical fiber development. Indeed, there are now several commercial supercontinuum sources offering high-brightness, broadband light in a compact fiber package. Uses of these sources include optical coherent tomography, characterization of optical components, broadband spectroscopy, and hyper-spectral imaging.

Much of the initial work was focused on supercontinuum generation in the visible (400–700 nm) and near-IR range (700–2500 nm). In this wavelength span, silicate-based fiber performs brilliantly, simultaneously offering low-loss propagation and tunable dispersion with reasonable nonlinearity (through the use of photonic crystal structures). In fact, the optical frequency comb was directly enabled by a silicate photonic crystal fiber (Jones et al. 2000), which allowed the output of a Ti:sapphire laser to be broadened to an octave of bandwidth. Supercontinuum sources in the near-IR have also benefited from the high level of laser development in the near-IR such as Ti:sapphire, thin-disk, and the family of rare-earth-doped fiber lasers such as mode-locked ytterbium, erbium, and thulium.

Supercontinuum sources in the mid-IR have historically lagged behind those in the near-IR due to a scarcity of both high-performance pump laser sources and nonlinear media. However, in recent years, this story has changed dramatically on both fronts. In terms of optical fiber, the commercial emergence of fluoride and chalcogenide fibers has played a big part in driving down both loss and cost. As mid-IR light is more susceptible to molecular absorption, the fiber precursor materials must be carefully purified to high levels to avoid loss from unwanted optical absorption. Progress on this front has resulted in low losses for ZBLAN, as discussed above, and losses <1 dB/m for chalcogenide fiber. To create a broadband supercontinuum, however, not only must the loss be low, but the dispersion and nonlinearity need to be optimal for the given wavelength of the pump pulses.

In the following section, we review the performance of supercontinuum demonstrations for various pump sources and mid-IR compatible fibers. We will not cover supercontinuum generation from a theoretical standpoint as that has been well-encapsulated by several reviews (Dudley et al. 2006; Dudley and Taylor 2009; Price et al. 2012; Moselund et al. 2012; Hudson et al. 2012; Swiderski 2014; Yin et al. 2014) and books (Dudley and Taylor 2016; Alfano 2016). Rather, we will focus on the state-of-the-art results from recent experiments including low-threshold supercontinuum (Gaeta 2004), wide-bandwidth supercontinuum spanning (Cheng et al. 2016; Yu et al. 2015; Hudson et al. 2017), all-fiber supercontinuum (Gattass et al. 2012), and on-chip supercontinuum (Yu et al. 2013; Singh et al. 2015).

Supercontinuum Generated via Optical Parametric Amplification Systems

The most common approach to achieving wide-bandwidth supercontinuum in the mid-IR is to employ an optical parametric oscillator/amplifier (OPO/OPA) as the producer of the pump pulse. Typically, this is achieved using a difference frequency generation between two near-IR lasers to create mid-IR pulses with durations on the order of a few hundred femtoseconds with peak powers ranging from 10 kW to 1 MW. Alternatively, optical parametric chirped-pulse amplification (OPCPA) systems (Hemmer et al. 2013) have been developed that can deliver peak powers up to 4 GW (Elu et al. 2017). Owing to these high peak powers, this class of pump source has seen the highest level of uptake within the research community. However, each of these systems involves several laser systems and a nonlinear crystal and typically requires an expert user to operate. This high threshold of use has prevented the commercial development of mid-IR supercontinuum systems based on OPAs.

The wavelength tunability of this class of pump source combined with the typically high peak power has allowed these systems to be successfully combined with a wide range of nonlinear media including step-index mid-IR fibers (Cheng et al. 2016; Petersen et al. 2014; Yu et al. 2015; Zhao et al. 2016), suspended-core fibers (Møller et al. 2015), photonic crystal fiber (Domachuk et al. 2008), bulk

Table 4 Demonstrations of fiber-based supercontinuum generation in the mid-IR

Year	Pump laser, pulse width	Pump wavelength [μm]	Coupled peak power [kW]	NL media	Span [μm]	Avg. power [mW]	Reference
2016	OPA, 150 fs	4.5	20,000	Ge-As-Se-Te step-index	1.5–14	<1	Zhao et al. (2016)
2016	DFG, 170 fs	9.8	2900	$As_2Se_3/AsSe_2$ step-index	2–15	<1	Cheng et al. (2016)
2015	OPA, 330 fs	4.0	1.7	$Ge_{12}As_{24}Se_{64}/Ge_{10}As_{24}S_{66}$ step-index	1.8–10	1.3	Yu et al. (2015)
2015	OPA, 320 fs	4.2	2.5	Silicon-on-sapphire, planar waveguide	2–6	<1	Singh et al. (2015)
2015	OPA, 320 fs	4.4	5.2	Suspended-core $As_{38}Se_{62}$	1.7–7.5	15.6	Møller et al. (2015)
2014	OPA	6.3	2290	$As_{40}Se_{60}/Ge_{10}As_{23.4}Se_{66.6}$ step-index	1.4–13.3	0.2	Petersen et al. (2014)
2013	OPO, 180 fs	1.6	1,100,000	Bulk ZBLAN glass	0.2–8	200	Liao et al. (2013)
2008	OPO, 110 fs	1.55	8	Tellurite PCF	0.8–4.8	90	Domachuk et al. (2008)

crystals (Liao et al. 2013), and planar waveguides [ref]. In Table 4, we list some of the best results over the last few years in experiments using OPO/OPA/OPCPA sources to drive supercontinua in the mid-IR.

Supercontinuum Generation via Near-IR Fiber Laser Pumping

To alleviate the need for a complex pump laser to create the mid-IR supercontinuum, significant research effort has been aimed at employing robust, high-performance near-IR fiber laser systems. In these arrangements, a Yb^{3+}-based ultrafast fiber laser (emitting in the 1 μm region) or an Er^{3+}-based ultrafast fiber laser (emitting in the 1.5 μm region) is amplified, and typically soliton self-frequency shifted to longer wavelengths, where the pulse is either amplified/shifted again or directly enters a nonlinear fiber stage to generate the supercontinuum (Gattass et al. 2012; Shaw et al. 2011; Salem et al. 2015) (see Table 5). While the chain of amplification stages

Table 5 Demonstrations of supercontinuum generation using near-IR fiber laser pumping

Year	Pump laser, pulse width	Pump wavelength [μm]	Coupled peak power [kW]	Nonlinear media	Span [μm]	Avg. power [mW]	Reference
2015	Shifted-Er FL	2.1	114	InF_3 step-index fiber	1.3–4.6	250	Salem et al. (2015)
	100 fs			(7 μm core)			
2014	Tm MOPA	1.96	/	ZBLAN step-index	0.9–3.8	21,800	Liu et al. (2014)
	24 ps			(9 μm core)			
2013	Tm MOPA	1.96	/	ZBLAN step-index	1.9–3.9	7110	Yang et al. (2013)
	26 ps			(8 μm core)			
2012	Shifted-Er FL,	2.45	~1500	As_2S_3 step-index fiber	1.9–4.8	565	Gattass et al. (2012)
	40 ps			(10 μm core)			

and soliton shifting stages adds complexity and cost, each stage can be completely fiber connected allowing an overall level of robustness to be maintained.

One of the first all-fiber versions to employ Er^{3+}-based fiber lasers was demonstrated by the Naval Research Lab (Gattass et al. 2012; Shaw et al. 2011). In these experiments a picosecond 1550 nm (Er^{3+}-doped silicate) seed laser was first amplified and then self-frequency shifted to 2 μm. This step was followed by a Tm^{3+} amplifier and a highly nonlinear fiber that leads to self-frequency shifting of the pulse to 2.45 μm. At the end of this chain of components, a 1.4 W average power, 10 MHz pulse train with 100 nm bandwidth at 2.45 μm was delivered to a step-index chalcogenide fiber, which generated a 1.9–4.8 μm supercontinuum spectrum. Currently there is at least one company offering a supercontinuum source (Thorlabs) based on pumping a dispersion-engineered InF fiber with an ultrafast Er^{3+}-based fiber seed laser, which generates a spectrum spanning 1.3–4.5 μm.

Alternatively, a Tm^{3+}-doped silicate ultrafast fiber laser emitting at 2 μm can be used as a pump source directly. The Tm^{3+} ion in silica has one of the largest gain bandwidths (~200 nm) of any rare-earth ion, allowing for sub-50 fs pulses without nonlinear spectral broadening. Although fiber component loss is typically higher at 2 μm than at 1 or 1.5 μm, the field of ultrafast thulium lasers has advanced to commercialization, with at least five companies (Novae, AdValue, NP Photonics,

Menlo Systems, PolarOnyx) offering sub-ps, watt-level average power performance. In fact, there is now a commercial mid-IR supercontinuum source that uses a Tm^{3+} ultrafast seed laser (Novae).

Experiments have shown that by directly pumping a mid-IR fiber with a Tm^{3+}-doped ultrafast laser, supercontinua can be generated covering 1–3.9 μm with average power of >1 W. With a Tm^{3+} fiber master oscillator power amplifier (MOPA) pumping a ZBLAN fiber, a recent experiment (Liu et al. 2014) achieved 21.8 W of average power contained in a supercontinuum spanning from 1.9 to 3.8 μm.

Supercontinuum Generation via Diode Lasers

Yet another approach that has been extensively explored relies on modulated diode seed lasers operating near 1550 nm (Xia et al. 2009; Petersen et al. 2016; Yin et al. 2017; Kulkarni et al. 2011). In these schemes, a few-nanosecond pulse from a DFB diode laser is typically amplified to the kW level using an EDFA in either one or several stages before being coupled into an anomalous dispersion fiber. Modulation instability (MI) in the anomalous dispersion fiber (e.g., SMF-28) splits the quasi-cw pulses into much shorter solitons that then undergo soliton self-frequency shifting due to stimulated Raman scattering (Dekker et al. 2011). With appropriate conditions, the long-wavelength soliton can be shifted to 2 μm or beyond. Early experiments (Xia et al. 2009) simply shifted the soliton to long wavelengths for direct pumping of a nonlinear fiber for supercontinuum generation. Subsequent experiments have also demonstrated high performance by wavelength shifting to 2 μm and then coupling the pulse into a thulium-doped fiber amplifier (TDFA) before performing a second soliton self-frequency shift.

This approach again adds complexity and cost in the addition of various discrete stages. However, given that each stage consists of relatively cheap and robust technology, the overall end-user complexity can be low while maintaining moderate costs. In fact, the majority of currently available commercial mid-IR supercontinuum sources are based on this approach, offering turnkey operation with watt-level average power for 50,000–100,000 USD (Table 6).

Supercontinuum Generation via Mid-IR Fiber Laser Pumping

The demonstrations of ultrafast 3 μm class fiber lasers in 2014 (Hu et al. 2014) and the subsequent progress to sub-ps performance (Duval et al. 2015; Hu et al. 2015) have created an exciting pump source for mid-IR supercontinuum (see Table 7). Rare-earth-doped ZBLAN fiber lasers utilizing nonlinear polarization rotation are capable of delivering pulses with durations a few hundred fs, with multiple nJ energies. Since the initial demonstrations in 2015, the performance has increased drastically with demonstrations of 37 kW peak power (Antipov et al. 2016) directly from the laser cavity and over 200 kW in an amplified system (Duval et al. 2016).

Table 6 Supercontinuum generation using diode laser pumping

Year	Pump laser, pulse width	Pump wavelength [μm]	Coupled peak power [kW]	Nonlinear media	Span [μm]	Avg. power [mW]	Reference
2017	DFB laser 1 ns	1.55	2.0	ZBLAN step-index (9 μm core)	1.9–4.2	15,200	Yin et al. (2017)
2011	DFB laser diode 1 ns	1.55	2.5	ZBLAN step-index (8 μm core)	1.9–4.5	2600	Kulkarni et al. (2011)
2009	DFB laser diode 400 ps	1.54	10	ZBLAN step-index (7 μm core)	0.8–4	10,500	Xia et al. (2009)

Table 7 Supercontinuum generation using mid-IR fiber laser pumping

Year	Pump laser, pulse width	Pump wavelength [μm]	Coupled peak power [kW]	Nonlinear media	Span [μm]	Avg. power [mW]	Reference
2017	Ho:ZBLAN 230 fs	2.86	4.2	As_2Se_3 microwire (3 μm core)	2.0–12.0	40	Hudson et al. (2017)
2016	Er:ZBLAN	2.80	/	ZBLAN	2.8–3.6 (soliton shift)	2600	Duval et al. (2016)

While comparatively few supercontinuum experiments have been performed using these lasers, the performance achieved already surpasses most other approaches in terms of bandwidth, but not yet in power spectral density. In Hudson et al. (2017), we combined an ultrafast holmium ZBLAN laser with a tapered chalcogenide microwire device. Owing to the large input core on the microwire, efficient pulse coupling was achieved (with >4 kW coupled to the nonlinear microwire), and in the taper waist region, the pulses propagated in a waveguide with a nonlinear parameter of >4 $W^{-1}m^{-1}$. Nonlinear broadening through self-phase modulation, optical wave breaking, Raman scattering, and four-wave mixing processes resulted in a supercontinuum spanning from 2 to 12 μm. This wavelength coverage competes with the best results from OPO-/OPA-driven systems and can in principle be made into an all-fiber package, thus rivalling the robustness offered by the cascading approaches previously discussed.

Conclusion

Creating sources in the mid-IR continues to be a hot topic in photonics research with experiments demonstrating a range of narrow-line, high-power, tunable, ultrafast, and supercontinuum sources. In particular, the fluoride-based fiber ZBLAN has been used the most extensively as a mid-IR laser medium (as well as a supercontinuum generation fiber). This is due to the reasonably high damage threshold of ZBLAN and its increasing availability. However, the phonon energy of these glasses (~ 550 cm^{-1}) has limited the lasing sources to around 4 μm. The chalcogenide glass fibers, with phonon energies ~ 350 cm^{-1}, allow for low non-radiative decay rates further into the IR than the fluorides; however rare-earth solubility has been a major roadblock in creating lasers in this fiber.

Supercontinuum sources in the mid-IR have become a very important market as they offer the potential to increase brightness in FTIR measurements by many orders of magnitude. Commercial systems have mostly been based on robust diodes emitting pulses on the ns timescale, which are subsequently transferred to the pico- or femtosecond timescale through modulation instability and then used to initiate the standard supercontinuum dynamics. While these sources provide an incoherent supercontinuum, their brightness is many orders of magnitude higher than traditional incandescent light sources (i.e., globars).

Although the majority of commercial supercontinuum sources at the moment are based on fluoride fiber, this situation is likely to change in the near future. A major roadblock for these chalcogenide-based sources has been the incompatibility of chalcogenide glass with near-IR ultrafast pumps. Through nonlinear absorption processes in intense near-IR electric fields, chalcogenide glass can degrade in various ways, ultimately leading to high loss or even catastrophic failure of the fiber. Thus, pumping the chalcogenide fiber in the mid-IR is key to achieving long-term robustness and high performance. With continued advances in mid-IR OPA/OPO/DFG technology, ultrafast tunable pump sources are now becoming more common, and in a laboratory setting, they offer a great solution to driving a wide-bandwidth supercontinuum. Beyond the laboratory domain, the recent invention of the 3 μm class fiber laser could enable chalcogenide-based supercontinuum in a robust package.

References

R. Alfano, *The Supercontinuum Laser Source*. Springer-Verlag New York Inc., US (2016)
R. Alfano, S.L. Shapiro, Emission in the region 4000 to 7000 A via four-photon coupling in glass. PRL **31**(2), 584 (1970)
S. Antipov, D.D. Hudson, A. Fuerbach, S.D. Jackson, High-power mid-infrared femtosecond fiber laser in the water vapor transmission window. Optica **3**(12), 1373 (2016)
Y.O. Aydin, V. Fortin, F. Maes, F. Jobin, S.D. Jackson, R. Vallée, M. Bernier, Diode-pumped mid-infrared fiber laser with 50% slope efficiency. Optica **4**(2), 6–9 (2017)
B. Behzadi, m. Aliannezhadi, M. Hossein-zadeh, R.K. Jain, Design of a new family of narrow-linewidth mid-infrared lasers. J. Opt. Soc. Am. B **34**(12), 2501–2513 (2017)

M. Bernier, D. Faucher, R. Vallée, A. Saliminia, G. Androz, Y. Sheng, S.L. Chin, Bragg gratings photoinduced in ZBLAN fibers by femtosecond pulses at 800 nm. Opt. Lett. **32**(5), 454–456 (2007)

M. Bernier, V. Michaud-belleau, S. Levasseur, V. Fortin, J. Genest, R. Vallée, All-fiber DFB laser operating at 2.8 μm. Opt. Lett. **40**(1), 81–84 (2015)

G. Bharathan, R.I. Woodward, M. Ams, D.D. Hudson, S.D. Jackson, A. Fuerbach, Direct inscription of Bragg gratings into coated fluoride fibers for widely tunable and robust mid-infrared lasers. Opt. Express **25**(24), 30013 (2017)

N. Caron, M. Bernier, D. Faucher, R. Vallée, Understanding the fiber tip thermal runaway present in 3μm fluoride glass fiber lasers. Opt. Express **20**(20), 22188 (2012)

T. Cheng, K. Nagasaka, T.H. Tuan, X. Xue, M. Matsumoto, H. Tezuka, T. Suzuki, Y. Ohishi, Mid-infrared supercontinuum generation spanning 2.0 to 15.1 μm in a chalcogenide step-index fiber. Opt. Lett. **41**(9), 2117 (2016)

M.N. Cizmeciyan, H. Cankaya, A. Kurt, A. Sennaroglu, Cr 2 + : ZnSe laser at 2420 nm. Opt. Lett. **34**(20), 3056–3058 (2009)

S. Crawford, D.D. Hudson, S.D. Jackson, High-power broadly tunable 3 μm fiber laser for the measurement of optical fiber loss. IEEE Photonics J. **7**(3), 1–9 (2015)

S.A. Dekker, A.C. Judge, R. Pant, I. Gris-s, J.C. Knight, C.M. De Sterke, B.J. Eggleton, Highly-efficient, octave spanning soliton self-frequency shift using a specialized photonic crystal fiber with low OH loss. Opt. Express **19**(18), 17766–17773 (2011)

P. Domachuk, N.A. Wolchover, M. Cronin-Golomb, A. Wang, A.K. George, C.M.B. Cordeiro, J.C. Knight, F.G. Omenetto, Over 4000 nm bandwidth of mid-IR supercontinuum generation in sub-centimeter segments of highly nonlinear tellurite PCFs. Opt. Express **16**(10), 7161–7168 (2008)

J.M. Dudley, J.R. Taylor, Ten years of nonlinear optics in photonic crystal fibre. Nat. Photonics **3**, 85 (2009)

J. M. Dudley, J. R. Taylor, *Supercontinuum Generation in Optical Fibers*. Cambridge University Press. UK (2016)

J.M. Dudley, G. Genty, S. Coen, Supercontinuum generation in photonic crystal fiber. Rev. Mod. Phys. **78**(4), 1135–1184 (2006)

S. Duval, M. Bernier, V. Fortin, J. Genest, M. Piché, R. Vallée, Femtosecond fiber lasers reach the mid-infrared. Optica **2**(7), 623 (2015)

S. Duval, J.-C. Gauthier, L.-R. Robichaud, P. Paradis, M. Olivier, V. Fortin, M. Bernier, M. Piché, R. Vallée, Watt-level fiber-based femtosecond laser source tunable from 2.8 to 3.6 μm. Opt. Lett. **41**(22), 5294 (2016)

U. Elu, M. Baudisch, H. Pires, F. Tani, M.H. Frosz, F. Köttig, A. Ermolov, P.S.J. Russell, J. Biegert, High average power and single-cycle pulses from a mid-IR optical parametric chirped pulse amplifier. Optica **4**(9), 1024 (2017)

V. Fortin, M. Bernier, S.T. Bah, R. Vallée, 30 W fluoride glass all-fiber laser at 294 μm. Opt. Lett. **40**(12), 2882 (2015)

V. Fortin, F. Maes, M. Bernier, S.T. Bah, M. D'Auteuil, R. Vallée, Watt-level erbium-doped all-fiber laser at 344 μm. Opt. Lett. **41**(3), 559 (2016)

M. F. A. Gaeta, Ultra-low threshold supercontinuum generation in sub-wavelength waveguides. Opt. Express, **12**, 3137–3143 (2004)

R.R. Gattass, E. Mazur, Femtosecond laser micromachining in transparent materials. Nat. Photonics **2**(4), 219–225 (2008)

R.R. Gattass, L. Brandon Shaw, V.Q. Nguyen, P.C. Pureza, I.D. Aggarwal, J.S. Sanghera, All-fiber chalcogenide-based mid-infrared supercontinuum source. Opt. Fiber Technol. **18**(5), 345–348 (2012)

P. Golding, S. Jackson, T. King, M. Pollnau, Energy transfer processes in Er3+−doped and Er3+,Pr3+−codoped ZBLAN glasses. Phys. Rev. B **62**(2), 856–864 (2000)

J. A. Harrington, *Infrared Fibers and Their Applications* (SPIE Press Book) (2004)

M. Hemmer, M. Baudisch, A. Thai, A. Couairon, J. Biegert, Self-compression to sub-3-cycle duration of mid-infrared optical pulses in dielectrics. Opt. Express **21**, 28095–28102 (2013)

O. Henderson-Sapir, J. Munch, D.J. Ottaway, Mid-infrared fiber lasers at and beyond 3.5um using dual-wavelength pumping. Opt. Lett. **39**(3), 493–496 (2014)

O. Henderson-Sapir, S.D. Jackson, D.J. Ottaway, Versatile and widely tunable mid-infrared erbium doped ZBLAN fiber laser. Opt. Lett. **41**(7), 1676 (2016)

K. O. Hill, D. C. Johnson, B. S. Kawasaki, R. I. Macdonald, CW three-wave mixing in single-mode optical fibers. J. Appl. Phys. **49**, 5098 (1978)

T. Hu, D.D. Hudson, S.D. Jackson, Stable, self-starting, passively mode-locked fiber ring laser of the 3 μm class. Opt. Lett. **39**(7), 2133–2136 (2014)

T. Hu, S.D. Jackson, D.D. Hudson, Ultrafast pulses from a mid-infrared fiber laser. Opt. Lett. **40**(18), 4226 (2015)

D. Hudson, E. Mägi, A. Judge, S. Dekker, B. Eggleton, Highly nonlinear chalcogenide glass micro/nanofiber devices: design, theory, and octave-spanning spectral generation. Opt. Commun. **285**(23), 4660–4669 (2012)

D.D. Hudson, R.J. Williams, M.J. Withford, S.D. Jackson, Single-frequency fiber laser operating at 2.9 μm. Opt. Lett. **38**(14), 2388 (2013)

D. Hudson, S.A. Antipov, L. Lizhu, I. Alamgir, T. Hu, M. Amraoui, Y. Messaddeq, M.R. Rochette, S. Jackson, A. Fuerbach, Toward all-fiber supercontinuum spanning the mid-infrared. Optica **4**(10), 1163–1166 (2017)

S.D. Jackson, Single-transverse-mode 2.5-W holmium-doped fluoride fiber laser operating at 2.86 μm. Opt. Lett. **29**, 334 (2004)

S.D. Jackson, High-power erbium cascade fibre laser. Electron. Lett. **45**(16), 830 (2009a)

S.D. Jackson, High-power and highly efficient diode-cladding- pumped holmium-doped fluoride fiber laser operating at 2 . 94 m. Opt. Lett. **34**(15), 2327–2329 (2009b)

D. Jones, S. Diddams, J. Ranka, A. Stentz, R. Windeler, J. Hall, S. Cundiff, Carrier-envelope phase control of femtosecond mode-locked lasers and direct optical frequency synthesis. Science **288**(5466), 635 (2000)

N. Jovanovic, A. Fuerbach, G. D. Marshall, M. Ams, M. J. Withford, *Fibre Grating Inscription and Applications*. Topics in applied physics. Heidelberg, vol. 123. 2012

R. Kashyap, *Fiber Bragg Gratings*. Academic Press, US (2010)

O.P. Kulkarni, V.V. Alexander, M. Kumar, M.J. Freeman, M.N. Islam, F.L. Terry, M. Neelakandan, A. Chan, Supercontinuum generation from ∼1.9 to 4.5 μ m in ZBLAN fiber with high average power generation beyond 3.8 μm using a thulium-doped fiber amplifier. J. Opt. Soc. Am. B **28**(10), 2486–2498 (2011)

J. Li, D. Hudson, S. Jackson, High-power diode-pumped fiber laser operating at 3 μm. Opt. Lett. **36**(18), 3642–3644 (2011)

J. Li, H. Luo, B. Zhai, R. Lu, Z. Guo, H. Zhang, Y. Liu, Black phosphorus: A two-dimension saturable absorption material for mid-infrared Q-switched and mode-locked fiber lasers. Sci. Rep. **6**(1), 30361 (2016)

M. Liao, W. Gao, T. Cheng, X. Xue, Z. Duan, D. Deng, H. Kawashima, T. Suzuki, Y. Ohishi, Five-octave-spanning supercontinuum generation in fluoride glass. Appl. Phys. Express **6**(3), 4–7 (2013)

A.F.H. Librantz, S.D. Jackson, L. Gomes, S.J.L. Ribeiro, Y. Messaddeq, Pump excited state absorption in holmium-doped fluoride glass. J. Appl. Phys. **103**(2), 1–9 (2008)

K. Liu, J. Liu, H. Shi, F. Tan, P. Wang, High power mid-infrared supercontinuum generation in a single-mode ZBLAN fiber with up to 21.8 W average output power. Opt. Express **22**(20), 24384–24391 (2014)

J. Liu, M. Wu, B. Huang, P. Tang, C. Zhao, D. Shen, D. Fan, S.K. Turitsyn, Widely wavelength-tunable mid-infrared fluoride fiber lasers. IEEE J. Sel. Top. Quantum Electron. **24**(3), 0900507 (2018)

F. Maes, V. Fortin, M. Bernier, R. Vallée, 5.6 W monolithic fiber laser at 3.55 μm. Opt. Lett. **42**(11), 2054 (2017)

M.R. Majewski, S.D. Jackson, Tunable dysprosium laser. Opt. Lett. **41**(19), 4496 (2016)

U. Møller, Y. Yu, I. Kubat, C.R. Petersen, X. Gai, L. Brilland, D. Méchin, C. Caillaud, J. Troles, B. Luther-Davies, O. Bang, Multi-milliwatt mid-infrared supercontinuum generation in a suspended core chalcogenide fiber. Opt. Express **23**(3), 3282 (2015)

P.M. Moselund, C. Petersen, S. Dupont, C. Agger, O. Bang, S.R. Keiding, Supercontinuum: broad as a lamp, bright as a laser, now in the mid-infrared. Proc. SPIE **8381**, 83811A–83811A–6 (2012)

C.K.N. Patel, Interpretation of CO2 optical maser experiments. Phys. Rev. Lett. **12**(21), 588–590 (1964)

C.R. Petersen, U. Møller, I. Kubat, B. Zhou, S. Dupont, J. Ramsay, T. Benson, S. Sujecki, N. Abdel-Moneim, Z. Tang, D. Furniss, A. Seddon, O. Bang, Mid-infrared supercontinuum covering the 1.4–13.3 μm molecular fingerprint region using ultra-high NA chalcogenide step-index fibre. Nat. Photonics **8**(11), 830–834 (2014)

C.R. Petersen, P.M. Moselund, C. Petersen, U. Møller, O. Bang, Spectral-temporal composition matters when cascading supercontinua into the mid-infrared. Opt. Express **24**(2), 749–758 (2016)

J.H.V. Price, X. Feng, A.M. Heidt, G. Brambilla, P. Horak, F. Poletti, G. Ponzo, P. Petropoulos, M. Petrovich, J. Shi, M. Ibsen, W.H. Loh, H.N. Rutt, D.J. Richardson, Supercontinuum generation in non-silica fibers. Opt. Fiber Technol. **18**(5), 327–344 (2012)

Z. Qin, G. Xie, C. Zhao, S. Wen, P. Yuan, L. Qian, Mid-infrared mode-locked pulse generation with multilayer black phosphorus as saturable absorber. Opt. Lett. **41**(1), 56 (2016)

R. Salem, Z. Jiang, D. Liu, R. Pafchek, D. Gardner, P. Foy, M. Saad, D. Jenkins, A. Cable, P. Fendel, Mid-infrared supercontinuum generation spanning 1.8 octaves using step-index indium fluoride fiber pumped by a femtosecond fiber laser near 2 μm. Opt. Express **23**(24), 30592 (2015)

J.S. Sanghera, L.B. Shaw, I.D. Aggarwal, Chalcogenide glass-fiber-based mid-IR sources and applications. IEEE J. Sel. Top. Quantum Electron. **15**(1), 114–119 (2009)

J. Schneider, Mid-infrared fluoride fiber lasers in multiple cascade operation. IEEE Photon. Technol. Lett. **7**(4), 354–356 (1995)

L.B. Shaw, R.R. Gattass, J. Sanghera, I. Aggarwal, All-fiber mid-IR supercontinuum source from 1.5 to 5 μm. Fiber Lasers VIII: Technol. Syst. Appl. **7914**, 79140P–79140P–5 (2011)

Y. Shen, Y. Wang, H. Chen, K. Luan, M. Tao, J. Si, Wavelength-tunable passively mode-locked mid-infrared Er3+−doped ZBLAN fiber laser. Sci. Rep. **7**(1), 14913 (2017)

N. Singh, D.D. Hudson, Y. Yu, C. Grillet, S.D. Jackson, A. Casas-Bedoya, A. Read, P. Atanackovic, S.G. Duvall, S. Palomba, B. Luther-Davies, S. Madden, D.J. Moss, B.J. Eggleton, Midinfrared supercontinuum generation from 2 to 6 μm in a silicon nanowire. Optica **2**(9), 797 (2015)

D.R. Solli, C. Ropers, P. Koonath, B. Jalali, Optical rogue waves. Nature **450**(7172), 1054–1057 (2007)

R.H. Stolen, A. Ashkin, Optical Kerr effect in glass waveguide. Appl. Phys. Lett. **294**(1973), 20–23 (2003)

R. Stolen, C. Lin, Self-phase modulation in silica optical fibers. Phys. Rev. A **17**(4), 1448 (1978)

T. Sumiyoshi, H. Sekita, Dual-wavelength continuous-wave cascade oscillation at 3 and 2 μm with a holmium-doped fluoride-glass fiber laser. Opt. Lett. **23**(23), 1837–1839 (1998)

T. Sumiyoshi, H. Sekita, T. Arai, Cascade Ho : ZBLAN fiber laser and its medical applications. Quantum **5**(4), 936–943 (1999)

J. Swiderski, High-power mid-infrared supercontinuum sources: current status and future perspectives. Prog. Quantum Electron. **38**(5), 189–235 (2014)

D.C. Tran, G.H. Sigel, B. Bendow, Heavy metal fluoride glasses and fibers: a review. J. Lightwave Technol. **2**(5), 566–586 (1984)

C. Wei, H. Shi, H. Luo, H. Zhang, Y. Lyu, Y. Liu, 34 nm-wavelength-tunable picosecond Hoˆ3+/Prˆ3+−codoped ZBLAN fiber laser. Opt. Express **25**(16), 19170 (2017)

C. Xia, Z. Xu, M.N. Islam, F.L. Terry, M.J. Freeman, A. Zakel, J. Mauricio, 10.5 W time-averaged power mid-IR supercontinuum generation extending beyond 4 μm with direct pulse pattern modulation. IEEE J. Sel. Top. Quantum Electron. **15**(2), 422–434 (2009)

W. Yang, B. Zhang, K. Yin, X. Zhou, J. Hou, High power all fiber mid-IR supercontinuum generation in a ZBLAN fiber pumped by a 2 μm MOPA system. Opt. Express **21**(17), 19732–19742 (2013)

S. Yin, P. Ruffin, C. Brantley, E. Edwards, C. Luo, Mid-IR supercontinuum generation and applications: a review. Proc. SPIE 9200, 92000U (2014)

K. Yin, B. Zhang, L. Yang, J. Hou, 15.2 W spectrally flat all-fiber supercontinuum laser source with > 1 W power beyond 3. 8 μm. Opt. Lett. **42**(12), 2334 (2017)

Y. Yu, X. Gai, T. Wang, P. Ma, R. Wang, Z. Yang, D.-Y. Choi, S. Madden, B. Luther-Davies, Mid-infrared supercontinuum generation in chalcogenides. Opt. Mater. Express **3**(8), 1075–1086 (2013)

Y. Yu, B. Zhang, X. Gai, C. Zhai, S. Qi, W. Guo, Z. Yang, R. Wang, D. Choi, S. Madden, B. Luther-davies, 1.8-10 μm mid-infrared supercontinuum generated in a step-index chalcogenide Fiber using low peak pump power. Opt. Lett. **40**(6), 1081 (2015)

Z. Zhao, X. Wang, S. Dai, Z. Pan, S. Liu, L. Sun, P. Zhang, Z. Liu, Q. Nie, X. Shen, R. Wang, 1.5-14 μm midinfrared supercontinuum generation in a low-loss Te-based chalcogenide step-index fiber. Opt. Lett. **41**(22), 5222–5225 (2016)

X. Zhu, R. Jain, 10 W level diode pumepd compact 2.78 μm ZBLAN fiber laser. Opt. Lett. **32**(1), 26–28 (2007)

G. Zhu, L. Geng, X. Zhu, L. Li, Q. Chen, R.A. Norwood, T. Manzur, N. Peyghambarian, Towards ten-watt-level 3-5 μm Raman lasers using tellurite fiber. Opt. Express **23**(6), 7559 (2015)

G. Zhu, X. Zhu, F. Wang, S. Xu, Y. Li, X. Guo, K. Balakrishnan, R.A. Norwood, N. Peyghambarian, Graphene mode-locked fiber laser at 2.8 μm. IEEE Photon. Technol. Lett. **28**(1), 7–10 (2016)

Crystalline Fibers for Fiber Lasers and Amplifiers

20

Sheng-Lung Huang

Contents

Introduction	758
Crystalline Fiber Core	760
The LHPG Method	761
The Growth Mechanism	762
Crystal Fiber Host and Dopant Characterization	766
Glass Cladding	768
The Co-drawing LHPG Method	769
The Residual Strain in Glass-Clad Crystalline Fiber	772
Crystalline Core and Glass Clad Interface	774
Light Transmission Characteristics	779
Crystalline Fiber-Based Broadband Spontaneous Emission	781
Ce:YAG as Crystalline Core	782
Ti:sapphire as Crystalline Core	791
Cr:YAG as Crystalline Core	795
Crystalline Fiber Laser and Amplifier	802
Tunable Cr^{4+}:YAG Crystalline Fiber Laser	803
Crystalline Fiber Amplifier	808
Conclusion	815
References	816

Abstract

Fiber lasers and amplifiers have revolutionary advancements in the past 20 years. Using crystalline core for fiber based devices is an extension to explore what glass fibers may have limitations. The mechanical strength and heat dissipation capability of crystalline materials makes them eminently suitable for high power

---

S.-L. Huang (✉)
Graduate Institute of Photonics and Optoelectronics, and Department of Electrical Engineering, National Taiwan University, Taipei, Taiwan
e-mail: shuang@ntu.edu.tw

or high brightness applications. To utilize these crystalline advantages, it is crucial to have high quality cladding for low transmission loss and low core/clad interface defects. Various glass-cladded crystalline fiber configurations and formation mechanisms are described. After cladding formation, the heterogeneous crystal/glass interface could result in residual strain in the crystalline core, which may deteriorate the active ion emission cross section. Proper design of the crystal waveguide structure with thermal treatment could effectively mitigate the strain-induced degradation. In this chapter, the growth thermodynamics, ion segregation, and optical transmission and amplification modeling are addressed. As an introduction, the yttrium aluminum garnet (YAG) and sapphire crystalline hosts with broadband active ion dopants will be emphasized even though quite a variety of crystals have been grown into fibers since the inception of the crystalline fiber technology 40 years ago. At present, broad and bright continuous-wave light sources with a center wavelength from visible to near infrared range have been well developed. Application wise, they could be adapted as active devices for biomedical imaging, optical metrology, as well as optical communications.

Introduction

It has been over 40 years since crystalline fiber (CF) was first used as the laser gain media (Burrus and Stone 1975; Burrus et al. 1976). Through the advancement of the growth techniques on the crystalline core (Fejer et al. 1984; Feigelson 1986), high quality CFs enable its gradual penetration into various dopants, hosts, and applications. In terms of geometric shape, it is well known that fiber structure has the largest surface to volume ratio, which is advantageous for light source to dissipate heat from quantum defect and other nonradiative losses. Along with the superior mode confinement, flexibility, immune to environmental variation, active fiber devices using silica, silicate, or other glass material as the core are now widely used in various areas such as optical communications, industrial material processing, as well as sensing. CF could make possible optical devices that cannot be realized in glass fibers or bulk crystalline media.

In 1967, growth of continuous sapphire filaments was successfully demonstrated (LaBelle and Mlavsky 1967, 1971). It is due to the crystalline perfection and small dimensions, which minimize the occurrence of the defects that are responsible for the low strength of materials in bulk form. Such property favors the fibers as reinforcing agents in structural components. The development of the optical waveguiding in the 1970s activated the further growth of single CFs for diversified applications. The broad transmission window and high melting point of many crystalline materials make them attractive for lasers and high brightness light sources. Though the optical applications for CF fall into passive, active, and nonlinear categories, this chapter is mainly focused on the active light emitting devices, especially the broadband devices. Incoherent broadband light sources from spontaneous emission are in general low brightness because of the 4π radiation.

The light collection efficiency, η, by a lens with a numerical aperture of NA can be written as $\eta = \left(1 - \sqrt{1 - NA^2}\right)/2$. Even with a relatively large numerical aperture of 0.5 to collect the broadband emission, η is only 6.7% for a 4π radiator.

Transition-metal ion doped laser gain media have broadband nature because of the nonscreened electronic configurations. Ruby ($Cr^{3+}:\alpha-Al_2O_3$) is remarkable due to the $^2E \rightarrow {}^4A_2$ emission of Cr^{3+} in octahedral symmetry (i.e., R lines), which is one of the representative transitions in development of laser history (Abella and Townes 1961; Hughes and Young 1962). Broadly tunable laser gain media have been a particular interest since the advent of the alexandrite ($Cr^{3+}:BeAl_2O_4$) and Ti:sapphire ($Ti^{3+}:\alpha-Al_2O_3$) lasers in the 700–1000 nm visible range (Walling et al. 1980; Moulton 1986). However, in the 1200–1600 nm optical communication band, Cr^{4+} doped lasers have limited progress because of the low concentration of tetrahedrally positioned Cr^{4+} ions and the thermal problem aroused from the excited state absorption of pump light. Among all the Cr^{4+} doped gain media, Cr^{4+}:YAG ($Y_3Al_5O_{12}$) has been shown high concentration of tetrahedrally coordinated Cr^{4+} ions and high emission cross section in fiber communication bands (Borodin et al. 1990; Eilers et al. 1993; Sorokina et al. 1999; Sennaroglu et al. 1994; Lo et al. 2005), although most reports of bulk Cr^{4+}:YAG lasers showed lasing actions at low temperature or with rather high thresholds from 0.4 to 4 W. The threshold pump power increases as bulk temperature rises due to the decrease in fluorescence lifetime. The unprecedented growth of the telecommunications in the past 20 years has resulted in the development of mature glass fiber technology. For future broadband fiber communication system, a broadly tunable laser is essential to offer wavelength-on-demand, dynamic wavelength ports, and simplified inventory managements. In additional to the transition metal ions, such as Cr and Ti ions, for most phosphors doped with Ce^{3+} also exhibit broadband nature with a parity-allowed 5d – 4f emission (Blasse and Bril 1967; Autrata et al. 1978; Jacobs et al. 1978; Lin et al. 2010). The emissions ranging from ultraviolet to red color have been demonstrated depending upon the host lattice and the site size, site symmetry, and coordination number. The emission color of Ce^{3+} can also be controlled in the desired visible region of the spectrum by changing the crystal field strength (Dorenbos 2000).

The laser-heated pedestal growth (LHPG) method is now a well-established technique for the growth of single-crystal fibers (Fejer et al. 1984; Feigelson 1986). It is crucible free and can therefore produce high-purity, low-defect density single crystals with small diameter and long length. Cladding the crystal fiber reduces both scattering loss and the number of propagating modes. Various cladding techniques have been developed over the past 20 years (Lo et al. 2004; Lin et al. 2006; Huang et al. 2008a; Lai et al. 2010; Hsu et al. 2012; Wang et al. 2015). Following the historical path of silica fibers, few mode, single mode, as well as double-clad CFs are advancing rapidly in recent years using the codrawing LHPG method (Hsu et al. 2013). With a well-developed crystal core and clad in terms of crystal quality, it is expected that novel applications could be evolved to meet the broadband need in the future. Incoherent broadband and bright light generation is one of the key advantages that CF can deliver for applications. Optical coherence tomography

(OCT) is a novel technology capable of acquiring in vivo and three-dimensional tomographic images of biological tissues and various kinds of microdevices with a micrometer-scale resolution (Huang et al. 1991). The axial scan is realized by applying a low coherence light source to Michelson interferometer. Interference can only take place when the optical path difference is comparable to the coherence length of the light source. The axial resolution l_a of an OCT system using a broadband light source with an ideal Gaussian spectral shape is:

$$l_a = \frac{2 \ln 2}{\pi} \cdot \frac{\lambda_0^2}{\Delta \lambda} \approx 0.44 \cdot \frac{\lambda_0^2}{\Delta \lambda} \qquad (1)$$

where λ_0 and $\Delta \lambda$ are the center wavelength and 3-dB bandwidth of the broadband light source, respectively. Since the axial resolution is inversely proportional to the bandwidth, a broadband light source with its emission bandwidth as wide as possible is desired. In addition to the bandwidth, the spectral shape of the light source is an important factor for the OCT image since it is related to the pixel cross talk in the axial direction. A Gaussian shape is desired for the ideal light source with the lowest axial image cross talks. The most commonly used broadband light sources of commercial OCT systems are the superluminescent diodes (SLDs) with bandwidths of several tens of nanometers. The axial resolution was around 10–20 μm and was not enough to achieve cellular image resolution. By using multiple SLDs as the light source, the axial resolution was improved by the extended bandwidth but with inferior cross talks due to the humpy spectra. Tungsten halogen lamp had a large bandwidth but the brightness is low, which limits the imaging speed. In recent years, the supercontinuum light sources excited by mode-locked lasers are able to generate broadband emissions at the expense of high peak powers and expensive costs. The continuous-wave light sources using CFs as the gain media have the advantages of high brightness, large bandwidth, near-Gaussian spectral shape, and simple configuration, which make CFs preferred for clinical applications.

Crystalline Fiber Core

Several growth techniques for fabricating single CFs, such as solidification in capillary tubes (Stevenson and Dyott 1974), edge-defined film-fed growth (Chalmers et al. 1972), μ-CZ (Ohnish and Yao 1989), and μ-PD (Yoon et al. 1994), have been reported. Common disadvantages of these techniques are crucible contamination, scanty flexible growth parameters. The LHPG method involves melting one end of a source material, followed by dipping a seed crystal into the free end of the molten zone. With a speed ratio between pulling seed and feeding source rod, various diameters of crystal fibers can be obtained. The contamination problems from crucibles and furnace components are virtually eliminated with this technique, offering the potential to produce pure single CFs.

The LHPG Method

The growth method of the CF by using a CO_2 laser as a heating source was first reported by Stone and Burrus in 1975 (Burrus and Stone 1975). A CO_2 laser beam was split into two and focused onto the CF from its two sides. The LHPG method was modified by Fejer in 1984 by using a reflaxicon consisting of a pair of cones to convert the CO_2 beam into a donut ring and then focused onto the perimeter of the crystal fiber with a better axial symmetry (Fejer et al. 1984). A typical setup of the LHPG system is shown in Fig. 1 (Hsu 2011). A He-Ne laser was used to align the CO_2 laser beam. The CO_2 laser beam was expanded by a pair of ZnSe lenses to 3 cm in diameter and then launched into the growth chamber. A motorized polarizer was used to fine-tune the laser power. A thermopile detector was used to detect the laser power and provide a feedback signal to the motorized polarizer for laser power stabilization. The source and the seed rods were mounted on the upper and lower motor stages. A LabVIEW program was used for controlling the movement of the two stages, monitoring of the laser power from the thermopile detector, laser power stabilization by controlling the polarizer through a feedback loop, and in situ monitor of the grown fiber diameter.

The expanded CO_2 laser beam then entered the growth chamber. The expanded beam was converted into a donut ring by the reflaxicon, reflected upward by the

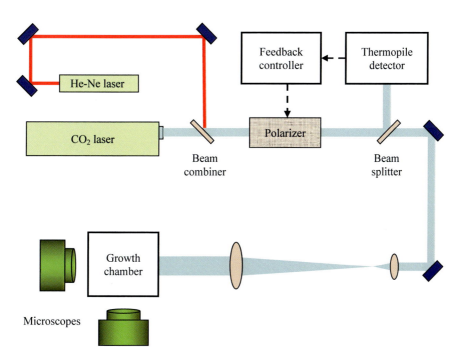

Fig. 1 Schematic of the LHPG system

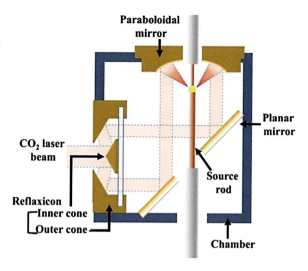

Fig. 2 Architecture of the LHPG growth chamber. A CO_2 laser beam is shaped to a hollow cylinder by reflaxicon and is focused on the top of the source rod by planar and paraboloidal mirrors

planar mirror, and focused by the paraboloidal mirror onto the perimeter of the CF, as shown in Fig. 2. The inner cone was held by several metal rods in Fejer's design (Fejer et al. 1984). For better axial symmetry of the CO_2 heat ring, the inner cone of the reflaxicon can be held by a ZnSe plate which is transparent to CO_2 laser beam (Lo 1994). The 360° axial symmetry prevented cold spots in the growth zone. The source crystal rod and the seed crystal rod are fixed on the lower and upper motorized stages. The longest length of the grown CF is typically only limited by the traveling length of the motorized stages.

The Growth Mechanism

For microfloating zone created using the LHPG method, convections can be induced by the buoyancy force (natural convection), mass transfer, as well as by the surface-tension gradients on the melt/air interface as thermocapillary convection. The natural convection is determined by the density distribution. The mass-transfer convection is changed at various reduction ratios according to the continuity equation. The thermocapillary convection is significantly and locally enhanced by high power laser input at the free surface. There could have high convection rate inside the floating zone due to mass-transfer and thermocapillary convections. The variation in pressure induced by the stable convections can be expressed by the shape, volume, and stability of the melt at various laser powers.

Figure 3 illustrates the microfloating zone formed using the LHPG method. The laser power is absorbed on the constant azimuthal area of the free surface at normal incidence without fluctuations in power and beam pointing. The azimuthal area, determined by the source-rod diameter and the reduction ratio, is fixed to be $2\pi R\ L_i \sec(\theta_L)$ where L_i is the axial length of the azimuthal area whose value is

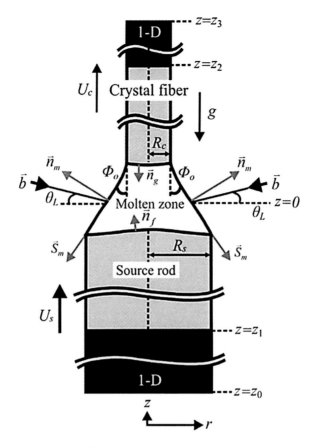

Fig. 3 The schematic sketch of the microfloating zone formed using the LHPG method. z is the axial coordinate along the growth direction, r is the radial coordinate, z_0 and z_3 refer the axial positions at infinite distance, z_1 and z_2 are the lower and upper boundaries of the simulation region far away from the melt, $z = 0$ is the axial position where the laser beam projects on, Φ_0 is the contact angle at the air/melt/solid tri-junction, θ_L is the incident angle of the laser beam related to radial axis, g is the gravitational acceleration, U_s and U_c are the feed speed and growth speed at the melt/solid interfaces, respectively, R_s and R_c are the radius of the source rod and the crystal fiber, respectively, $\vec{n}_f$ and $\vec{n}_g$ are the unit normal vectors, respectively, at the feed and growth front toward the melt. $\vec{n}_m$ is the unit normal vector described by 1×2 matrix at the free surface. $\vec{S}_m$ is the unit tangential vector described by 1×2 matrix at the free surface and $\vec{b}$ is the unit vector along the laser propagation direction

15–30 μm and R is the radius of the molten zone. Owing to the small azimuthal area, the heat re-absorption of the multiple-reflection lights inside the chamber can be neglected. Therefore, the ambient temperature can be considered constant without Gaussian distribution (Duranceau and Brown 1986). As in the floating-zone method for growing bulk crystal, thermal convection is assumed to be symmetric axially, laminar, and in a pseudo steady state. The oscillatory thermocapillary convection is not considered. The gravity is considered and the cylindrical coordinates are used in 2D simulation (Lan and Kou 1990). Simplified two-dimensional cylindrical coordinate system of the stream function $\psi(r, z)$ can be used to track the direction and at the same time that the axis of the radial velocity u and the axial velocity v. Under simplified 2D form in cylindrical coordinates, the governing equations can be expressed as:

Stream equation
$$\frac{\partial}{\partial z}\left(\frac{1}{\rho_L r}\frac{\partial \psi}{\partial z}\right) + \frac{\partial}{\partial r}\left(\frac{1}{\rho_L r}\frac{\partial \psi}{\partial r}\right) + \omega = 0 \quad (2)$$

Motion equation
$$\frac{\partial}{\partial r}\left(\frac{\omega}{r}\frac{\partial \psi}{\partial z}\right) - \frac{\partial}{\partial z}\left(\frac{\omega}{r}\frac{\partial \psi}{\partial r}\right) + \frac{\partial}{\partial r}\left[\frac{1}{r}\frac{\partial}{\partial r}(\mu_m r \omega)\right]$$
$$+ \frac{\partial}{\partial z}\left[\frac{1}{r}\frac{\partial}{\partial r}(\mu_m r \omega)\right] - \rho_m \beta_m g \frac{\partial T}{\partial r} = 0. \quad (3)$$

Energy equation
$$\frac{\partial}{\partial r}\left(C_p T \frac{\partial \psi}{\partial z}\right) - \frac{\partial}{\partial z}\left(C_p T \frac{\partial \psi}{\partial r}\right) + \frac{\partial}{\partial z}\left(rk\frac{\partial T}{\partial z}\right)$$
$$+ \frac{\partial}{\partial r}\left(rk\frac{\partial T}{\partial r}\right) = 0 \quad (4)$$

Mass-transfer equation
$$\frac{\partial}{\partial r}\left(\frac{C_A}{\rho_L}\frac{\partial \psi}{\partial z}\right) - \frac{\partial}{\partial z}\left(\frac{C_A}{\rho_L}\frac{\partial \psi}{\partial r}\right) + \frac{\partial}{\partial z}\left(rD\frac{\partial C_A}{\partial z}\right)$$
$$+ \frac{\partial}{\partial r}\left(rD\frac{\partial C_A}{\partial r}\right) = 0 \quad (5)$$

where the vorticity, $\omega = \frac{\partial u}{\partial z} - \frac{\partial v}{\partial r}$ (Chen et al. 2009).

Table 1 summarizes the parameters used in 4 simulation cases: A, B, C, and D. Two kinds of source rods are used; they are square with a cross-section of 400 × 400 μm^2 in case A and rod with a diameter of 300 μm in other cases. The ambient temperature is 575 K in case A and 650 K in other cases. Case A is employed to verify the source rod whose dimension approximates that of the bulk crystal.

Figure 4 shows the streamlines and isotherms of the microfloating zone at H_i and L_i for all cases (Chang et al. 2001). The patterns of the double eddy flow are almost symmetric when the reduction ratio is 100% as shown in Fig. 4c and d. When the upper eddy flow is confined by the melt volume, it becomes asymmetric as shown in Fig. 4a, b, and e–h. The temperature distribution is not influenced by fluid flow

20 Crystalline Fibers for Fiber Lasers and Amplifiers

Table 1 Sample specifications for cases A, B, C, and D in the experiment

Sample index	A	B	C	D
Source rod size (μm)	400 × 400	ϕ300	ϕ300	ϕ300
Crystal fiber diameter (μm)	290	300	105	75
Feed speed (mm/min)	4	1.2	1.2	1.2
Growth speed (mm/min)	9.6	1.2	9.6	19.2
Reduction ratio (%)	64	100	35	25
Higher allowed power (W)	6.7	2.20	1.68	1.5
Lower allowed power (W)	5.5	2.05	1.55	1.3

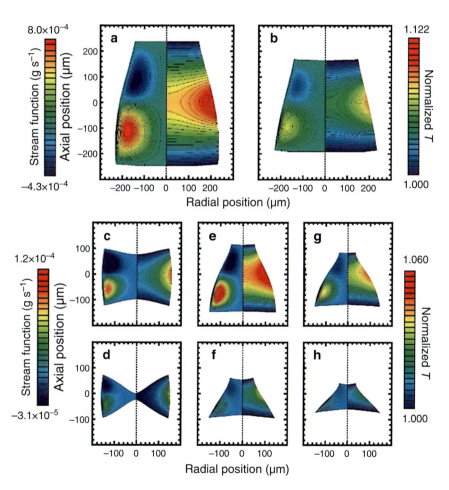

Fig. 4 The streamlines (left) and isotherms (right) of the microfloating zone at (**a**) H_A, (**b**) L_A, (**c**) H_B, (**d**) L_B, (**e**) H_C, (**f**) L_C, (**g**) H_D, and (**h**) L_D. H_i and L_i are the higher and lower allowed laser powers and its subscript i represents case A, B, C, or D

significantly because heat conduction is dominant rather than heat convection and radiation due to the small melt volume. In case B, the melt/solid interface is in deep convex toward the melt at L_B because the heat transfer efficiency from laser incident position to the melt/solid interface is poor due to the weak thermal capillary convection.

Crystal Fiber Host and Dopant Characterization

For structural characterizations, an X-ray diffractometer was employed to determine the degree of crystallinity of the as-grown CF by examining FWHM of the 2θ angle. With the use of $K_{\alpha 1}$ and $K_{\alpha 2}$ of copper, Fig. 5 shows the normalized X-ray diffraction patterns of a raw material grown by the CZ technique and a 400-μm diameter CF grown by the LHPG technique under 0.5 mm/min growth speed. The solid points are the measured data with 0.02° scanning step, and the solid lines were fitted to find out the peak positions and their FWHM values. The respective FWHM values of $K_{\alpha 1}$ are 0.049° and 0.065° for the crystal grown by LHPG and CZ methods. The peak differences between $K_{\alpha 1}$ and $K_{\alpha 2}$ of LHPG and CZ methods are 0.071° and 0.111°, respectively. The narrower peak implies a higher degree of crystallinity. It should be noted that the growth speed by CZ method is quite slow, which is typically about a few μm/hour. Therefore, the CF grown by the LHPG method can maintain high crystal quality under high growth speed.

Since the dopant concentration is critical to the device performance in terms of optical gain and/or guided mode, it should be monitored and controlled at each diameter-reduction step. Figure 6 shows the dopant profiles of Cr_2O_3 and CaO by EPMA measurement with CFs of various diameters, which are grown from a 2-mm-diameter raw material ($\alpha = 4.5$ cm^{-1}). The concentrations of Cr_2O_3 and CaO are uniform in the raw material. In Fig. 6a, the concentration of Cr_2O_3 shows the gradient distributions in CFs with diameters of 920, 530, and 300 μm. The concentration of Cr_2O_3 at the fiber center is lower than that at the edge.

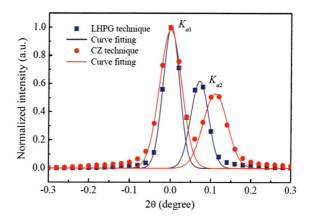

Fig. 5 XRD measurements for a 500-μm-diameter CZ-grown Cr:YAG (red) and a 400-μm-diameter LHPG-grown Cr:YAG CF under 0.5 mm/min growth speed (blue)

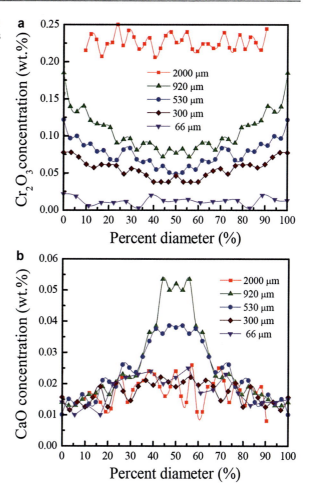

Fig. 6 Dopant profiles of (**a**) Cr_2O_3 and (**b**) CaO at various diameter-reduction steps

After multiple regrowth steps to down size the CF to 66 μm in diameter, the concentration of Cr_2O_3 becomes uniform again. In Fig. 6b, the distribution of CaO is contrary to that of Cr_2O_3. It tends to aggregate at the center during growth. The distribution difference between Cr_2O_3 and CaO is attributed to the signs of the segregation coefficient. It can be found that there is only slight change in the average concentration of CaO throughout the steps, but that of Cr_2O_3 decreases quickly when reducing the diameter of the CF. It is estimated that the concentrations of Cr_2O_3 are 20-fold lower from 2000 μm to 66 μm. It is due to that Cr ions tend to diffuse outward and evaporate during the growth. Figure 7 shows the concentration profiles of Cr^{3+} and Cr^{4+} in CF with various diameters. To explain the phenomenon of the distribution difference between Cr^{3+} and Cr^{4+} ions, the distribution of Ca^{2+} needs to be taken into account. In the 920 μm sample, the distribution of Cr^{4+} ion is flatter than that of Cr_2O_3. The reason is that the concentration of Cr^{4+} ion depends

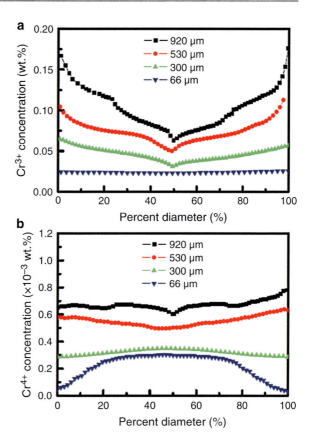

Fig. 7 The measured distribution of (**a**) Cr^{3+} and (**b**) Cr^{4+} at various diameter-reduction steps

on both the concentrations of Cr_2O_3 and CaO. Although the concentration of Cr_2O_3 at fiber center is lower than that at the edge, the concentration of CaO tends to accumulate near the fiber center, which results in the flatter distribution of Cr^{4+} ion. In the 66-μm diameter sample, the distribution of Cr_2O_3 is quite flat, while that of CaO gathers near the fiber center. Therefore, the concentration of Cr^{4+} ion peaks at the fiber center.

The effect of segregation may lead to degradation of the crystal quality or even break in the crystal facets that result from the interior stress of the crystal when the distribution is uneven. Nevertheless, an axially well-controlled dopant gradient may benefit the optical mode guiding and thus improve the active device performance.

Glass Cladding

To form cladding on the crystalline core, a number of techniques have been attempted, such as outdiffusion of dopants (Burrus and Coldren 1977), extrusion/dip coating (Burrus et al. 1976; Digonnet et al. 1987), indiffusion (Sudo et al. 1987), high-energy ion implantation (Saini et al. 1991), and crystal/glass co-drawing (Lo et al. 2004; Huang et al. 2008a, b; Hsu et al. 2012, 2013). Due to the long CO_2 laser wavelength, it is difficult to reduce the fiber diameter down to 10 μm by direct heating. The co-drawing approach has shown core diameter in the order of 10 μm. For cladding pump laser applications, the co-drawing approach has also shown double glass cladding capability. In this section, the co-drawing technique will be introduced, and the issues related to such a heterogeneous core/clad structure will be discussed.

The Co-drawing LHPG Method

In the co-drawing method, the crystalline core was prepared by the LHPG technique, while the claddings were made from various glass capillaries, such as borosilicate, aluminosilicate, flint glass, or even high temperature fused silica. Using YAG double-clad crystal fiber (DCF) growth as an example, a schematic of the co-drawing LHPG technique is depicted as shown in Fig. 8. A 68-μm-diameter YAG single CF was initially prepared from a 0.5 mol.% doped Cr:YAG source rod in <111> crystal orientation with a cross section of 500 μm × 500 μm. With two diameter reduction steps by the LHPG technique, the 68-μm YAG core was grown and then inserted into a fused silica capillary with 76- and 320-μm inner and outer diameters for the co-drawing process by the same LHPG system to form the double-clad structure. The 1970 °C melting temperature of the YAG is comparable to the 1600 °C soften temperature of the fused silica. The heating of the co-drawing LHPG method caused a strong interdiffusion between the YAG core and the fused-silica

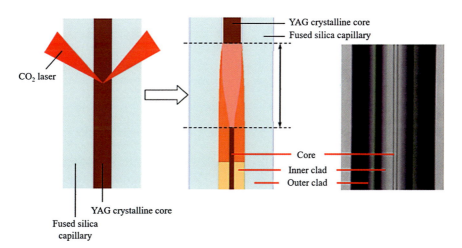

Fig. 8 Left, schematic of the co-drawing LHPG technique for YAG DCF growth; center, illustration of the molten zone during growth; right, side-view photograph of the as-grown YAG DCF. The Core, inner, and outer cladding can be clearly observed

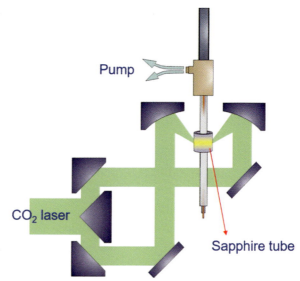

Fig. 9 Schematic of sapphire tube assisted co-drawing LHPG system

capillary, resulting in an inner cladding layer made of the mixtures. In particular, the DCF core diameter can be well controlled by just altering the CO_2 laser power and the relative growth speed.

The co-drawing LHPG technique is flexible to incorporate glass network modifiers during fiber growth (Lai et al. 2008), which is not attainable by conventional modified chemical-vapor deposition or sol-gel processes, which have relatively low doping concentrations (Ostby et al. 2007). To improve the core uniformity, a sapphire tube served as a heat capacitor has shown to be effective in growing the DCF with a core diameter down to 10 μm or below. The capillary with a CF inside was inserted into the sapphire tube for DCF growth. The donut-shape CO_2 laser beam was focused around the outer wall of the sapphire tube to heat up and generate a strong thermal radiation for melting the filled silica capillary. A negative pressure of 200 torrs prevented the generation of bubbles inside the waveguide during the growth process. The schematic of sapphire tube assisted co-drawing configuration is shown in Fig. 9.

The molten zone of the YAG DCF was below the sapphire tube upper edge, the result showed that the sapphire tube assisted growth had the longer melting range than that without sapphire tube growth. For the case of the 10-μm core, the core variation could be reduced from 58% to 17% by using the sapphire tube in the growth system. From the reduced core variation, the thermal capacitance effect of the sapphire tube to alleviate the fluctuation of CO_2 laser power was confirmed.

In addition to the heating laser power stabilization with the thermal capacitance effect, the sapphire tube also provides a longer heating zone with a reduced vertical temperature gradient. The scanning electron microscope (SEM) images of the core end faces using growth methods without and with sapphire tube are shown in Figs. 10 and 11. As shown in Fig. 10, the core shape changed as the core

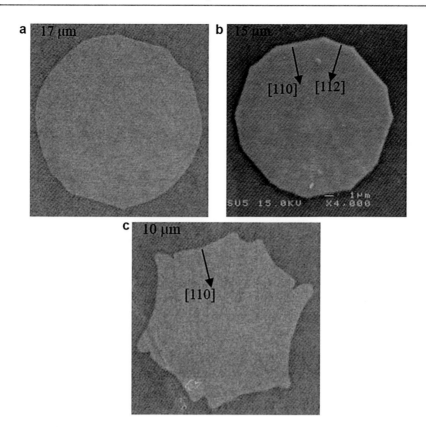

Fig. 10 Cr^{4+}:YAG DCF cores of (**a**) 17-μm, (**b**) 15-μm, and (**c**) 10-μm diameters without the use of sapphire tube

diameter reduced. The core shapes of 17-, 15-, and 10-μm-core fibers are circular, dodecagonal, and hexagonal in sequence. It was because of the constitutional melt supercooling at the growth interface (Lan and Tu 2001). Due to the high temperature gradient, the interfacial kinetic effects near the solid-melt interface become prominent. Without supercooling, the flow and thermal fields as well as dopant fields at the interface are axisymmetric. With a few degrees of supercooling, the interface is no longer at the melting point, and the lowest temperature occurs inside the facets. When the core diameter grew smaller, the higher temperature led the lower viscosity of surrounding melt and higher extension rate of crystal core to enhance the {110} and {112} facets. With large temperature gradient along YAG [111] axis, the dendrite forms due to the strong competition between {110} and {112} faces. Fibers grown by the sapphire tube assisted method showed smoother interfaces between the core and the inner clad. Comparing Figs. 10 and 11, the cores show similar circular, dodecagonal, and hexagonal shapes, but the cores grown with sapphire tube are 6 μm smaller than those grown without using sapphire tube when

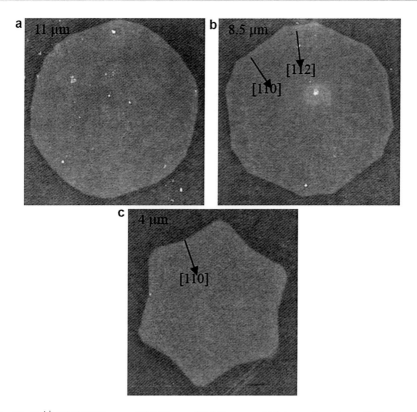

Fig. 11 Cr^{4+}:YAG DCF cores of (**a**) 11-μm, (**b**) 8.5-μm, and (**c**) 4-μm diameters with the use of sapphire tube

having similar shapes. The hot zone lengths that with and without sapphire tube assisted growth were around 1000 μm and 400 μm, respectively. The temperature gradient of the sapphire tube assisted growth was 2.5 times smaller so the growths in {110} and {112} facets were delayed. The circular shape is apparently more favorable than the dodecagonal and hexagonal shapes for coupling with SMFs.

The Residual Strain in Glass-Clad Crystalline Fiber

Strain field is known to have a crucial role in determining the optoelectronic properties of semiconductors. Prominent examples are the quantum well semiconductor lasers and Si/SiGe semiconductor heterostructures. As a result, control of the strain is essential to optimize both the fluorescence lifetime and device performance. Extensive piezospectroscopic investigations of transition-metal ions in crystal lattices have also been studied. More specifically, the influence of the strain dependence on octahedral Cr^{3+} and tetrahedrally coordinated Cr^{4+} in YAG has been widely studied (Shen et al. 1997). The current understanding of the strain distributions in CF-based devices is very limited. It is owing to the lack of high-resolution strain measurement instrument and the complexity of specimen preparation for such a heterogeneous structure.

The R_1 line of Cr^{3+} ions is widely used because of its sharp line width. One can readily obtain reasonable estimation on the strain variation K within the Cr:YAG core related to the R_1-line shifts, $\Delta\lambda$. It can be expressed as (Chi et al. 1990).

$$\Delta\lambda = \lambda_0 \left\{ K^{-S} exp\left[(1/2)\, D(1-K)^2 + (1/3)\,(D+E)(1-K)^3 \right] - 1 \right\} \quad (6)$$

where λ_0 is the R_1 center wavelength at ambient pressure, in our case $\lambda_0 = 687.86$ nm, determined by an undoped YAG single crystal. S, D, and E are constants of 0.3527, −4.266, and 668.7, respectively. The strain distribution in two different orientations as a function of positions for various core diameters can, therefore, be obtained, as presented in Fig. 12. From this calculation, it can be seen that at the early stage of interdiffusion process (i.e., larger core size), the DCF core encountered a compressive strain toward a tensile strain as the fused silica diffused into the YAG core deeper and deeper (i.e., smaller core size). For 25-μm-core edge, it is compressively strained along the radial direction, and this compression is relieved in the core center. For others, tensile strain fields were identified in the entire core as well as a compressively strained core edge to the core center. Another

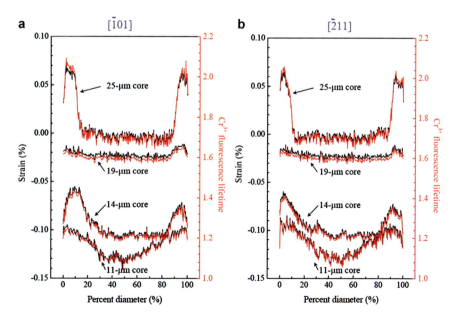

Fig. 12 Strain distribution in (**a**) [$\bar{1}01$] and (**b**) [$\bar{2}11$] orientations as a function of position for various core diameters

important feature of this figure worthy of emphasizing is that there exists a nearly zero strain cross the entire core with a diameter of ~20 μm.

The related positions of the energy levels of transition-metal ions depend on the crystal field strength. The metastable level can be either ^1E or $^3T_2$, depending on the strength of the crystal field in the tetrahedral site. The strong or weak crystal field is the field above or below the cross-over point of the two states, respectively. In a weak crystal field, the transition is expected to be between $^3T_2 \rightarrow {}^3A_2$ states, with a short fluorescence lifetime (spin-allowed transition) and broadband emission due to strong electron-phonon coupling. In a strong crystal field, where the electron-phonon coupling is weak and the transition is spin forbidden, a long lifetime and narrowband emission due to the $^1E \rightarrow {}^3A_2$ transition is expected. However, the strong orbital splitting of $^3T_2$ into ^3E and $^3B_2$ brings the $^3B_2$ level below the ^1E level, and this makes the Cr^{4+} behave as a weak crystal field system. In addition, the crystal field analysis for the tetrahedrally coordinated Cr^{4+} ion is difficult, because the transitions to the crystal field components of the $^3T_2$ energy level are either weak and overlapping each other – like in Cr^{4+}:YSO, or they are hidden under the strong absorption of the $^3T_1$ level – like in Cr^{4+}:YAG or other garnets. In the tetrahedrally coordinated Cr^{4+} ion, due to the strong electron-phonon coupling and the distortion of the tetrahedral symmetry, it results in the broadband emission, which is $^3T_2 \rightarrow {}^3A_2$ transition. Various types of Cr^{4+}-doped crystals have shown wavelength span in the range from 1150 nm to 1850 nm.

Crystalline Core and Glass Clad Interface

Figure 13a shows the cross sectional HRTEM image along YAG [111]-zone axis that was taken in a very thin and large electron-transparent area of the specimen. The HRTEM images were obtained with a FEI Tecnai G2 F20 TEM equipped with a field emission gun with a point resolution of 0.23 nm operated at 200 kV. This image reveals two distinct regions exhibiting an atomically sharp core/inner-clad interface. In the left side, there is a polycrystalline layer that contains several γ-Al_2O_3 nanocrystals with different lattice spacings in the inner-cladding; in the right side, there is a high-quality single crystal viewed from [111]-zone axis of the YAG core. The interface is very distinct, thus indicating that the growth-induced strain gave rise to the lattice distortion at the interface had already been released to the polycrystalline layer (i.e., inner cladding). No structural defects like dislocations or lattice distortions were observed at the vicinity of the interface and the interior of the core region, as it would be expected for crystal growth with interdiffusion by the LHPG technique. Figure 13b shows a HRTEM image of a representative γ-Al_2O_3 nanocrystal in a diameter of ~2.8 nm with a nanotwin structure. A Fourier transform of the corresponding HRTEM image is shown in Fig. 13c, revealing an obvious twin boundary (TB) within the nanocrystal as indicated by red arrows.

Twinning is a typical characteristic feature of the microstructure of the coalescent nanocrystals. Atomic structure of the core region was also examined by HRTEM combined with computer image simulations. Lattice image simulations based on

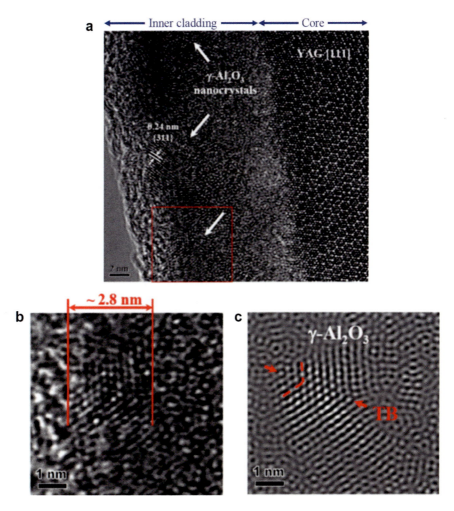

Fig. 13 (**a**) [111] zone axis HRTEM image of core/inner-clad interface with several γ-Al_2O_3 nanocrystals located in the inner cladding, as marked by the arrows. (**b**) Enlargement of the red rectangle area in (**a**). (**c**) Fourier transform filtering of (**b**). Within the 2.8-nm-diameter nanocrystal, a TB marked by red arrows is depicted. The red dashed lines represent two different lattice planes

the multislice method were performed with the MAC TEMPAS software package for comparison with the experimentally obtained images. The computer simulated image is optimally fitted to the experimental image with a lattice parameter of 12.008 Å (Joint committee on powder diffraction standards (JCPDS) file 33–0040), as shown in Fig. 14. Together with atomic column positions of Y, Al, and O, it demonstrates that the core region has a nearly perfect YAG single crystal structure. The inner cladding structure is more complicated due to the interdiffusion between core (YAG) and outer cladding (SiO_2); as a result, the change of interface

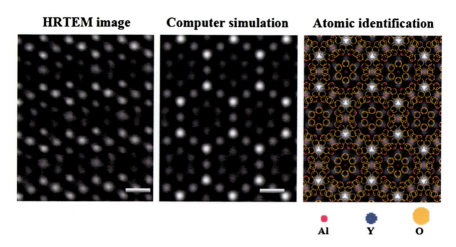

Fig. 14 Left, HRTEM image of the core with the electron beam along the [111] direction; center, a computer simulated image (thickness = 60 Å and defocus = 23 Å) that optimally fits the experiment; right, atomic column positions of Y, Al, and O. Scale bars of all images represent 5 Å

composition affects the microstructures and also the optical transmission quality of the fiber. In Fig. 13a, another interesting observation is that the shapes of most of the γ-Al_2O_3 nanocrystals are irregular, strongly indicating a coalescence of several smaller nanocrystals occurred during interdiffusion.

The observations mentioned above suggest that the surface energy is significantly reduced as the nanocrystal/nanocrystal interface is eliminated due to a strong thermodynamic driving force for the coalescence. Two different locations at the interface were investigated using the HRTEM images and the fast Fourier transformation (FFT) processed images along the [111] and [011] YAG zone axes followed by dislocation analysis. In Fig. 15, clearly distinguished orientation relationship between the γ-Al_2O_3 nanocrystal and YAG core is shown, where the nanometer-sized γ-Al_2O_3 crystals with [013]-zone axis exist at the core/inner-clad interface. This indicates that there is a preferred crystalline growth between the γ-Al_2O_3 nanocrystals and the YAG core region during the SiO_2 and YAG interdiffusion. The inset of Fig. 15a is the corresponding FFT pattern of the interface structure. It can be discriminated that the ($\bar{1}$ 3 1) plane of the γ-Al_2O_3 nanocrystal is slightly deviated from the YAG (1 $\bar{4}$ 3) plane by an angle of 2°. The diffraction spots connected by solid and dashed lines are from the YAG core and the γ-Al_2O_3 nanocrystals. The lattice between the growth direction of the interfacial γ-Al_2O_3 ($\bar{1}$ 3 1) nanocrystal and the (1 $\bar{4}$ 3) YAG plane is well matched with only a 1.48% difference, suggesting that the nanocrystals play an important role in the interdiffusion mechanism. This mechanism could be explained by the full alignment through the coherence in the 2D plane between the nanocrystal/matrix interface. Note that the γ-Al_2O_3 nanocrystal and YAG core have significantly different atomic structures and lattice constants. The lattice constant of γ-Al_2O_3

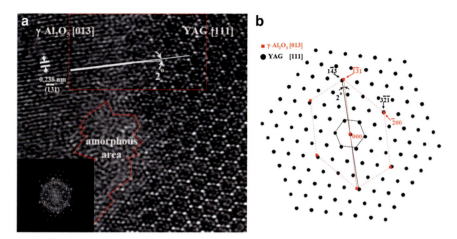

Fig. 15 (**a**) HRTEM image of core/inner-clad interface. The inset is the corresponding FFT pattern. (**b**) Schematic representation of the FFT pattern with the YAG reciprocal lattice. The orientation relationship between the γ-Al_2O_3 nanocrystal and the YAG core is depicted

is about 0.79 nm (JCPDS file 10-0425) and that of YAG is 1.2008 nm (JCPDS file 33-0040). Matching the two lattices over a unit-cell dimension could result in a > 30% lattice difference. However, when the lattice difference is considered in this case, 2 γ-Al_2O_3 lattices match 1 YAG lattice with the lattice difference of only ~2.86%, resulting in edge dislocations can be seen over the whole γ-Al_2O_3/YAG interface, giving rise to the preservation of the high single crystallinity of the YAG core. In order to identify the local orientation relationship more accurately, another region at the interface was also examined along the [011]-zone axis of the YAG core. Figure 16 shows a cross sectional HRTEM images recorded along [011] YAG at the core/inner-clad interface comprising the crystal core and a polycrystalline layer. The similar behavior that a γ-Al_2O_3 nanocrystal is located at the core/inner-clad interface in crystallographic orientation with respect to the YAG core is observed. Inhomogeneous contrast is found to appear in the YAG side near the interface and is typical for diffraction contrast due to the local lattice strain. Figure 16b shows the reciprocal space analysis of Fig. 16a, illustrating the components associated with the reciprocal view of the YAG [011] and γ-Al_2O_3 [011] zone axes. The inset is the corresponding FFT pattern of the interface structure in Fig. 16a, and the diffractions from both YAG core and γ-Al_2O_3 nanocrystals can be clearly distinguished. It is also apparent that the ($\bar{3}$ 1 $\bar{1}$) plane of the γ-Al_2O_3 nanocrystal rotates by 154° counterclockwise to the YAG (4 2 $\bar{2}$) plane. The theoretical lattice plane spacing between the interfacial γ-Al_2O_3 ($\bar{3}$ 1 $\bar{1}$) nanocrystal and the YAG (4 2 $\bar{2}$) matrix is just ~2.82%. However, the measured lattice difference is significantly reduced to ~1.13%, strongly revealing that the lattice planes of γ-Al_2O_3 nanocrystal are still in the transformation phase. Meanwhile, the distance between lattice planes of these γ-Al_2O_3 nanocrystals located in the inner cladding,

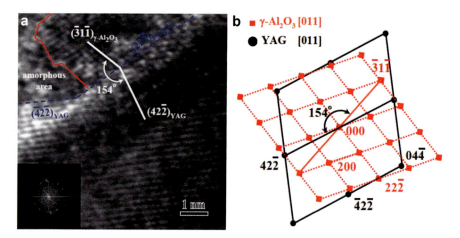

Fig. 16 (a) Highly magnified HRTEM image of the core/inner-clad interface showing a γ-Al_2O_3 [011] nano-crystal with a local orientation relationship to the YAG [011] core. The inset is the corresponding FFT pattern. (b) Schematic representation of the FFT pattern and the YAG reciprocal lattice

as marked in Fig. 13, is 0.24 nm. These values are nearly the same as the interplanar spacings of the {311} planes (i.e., 0.239 nm). Further, the small nanocrystals have a larger surface area-to-volume ratio and a higher total energy. The higher-energy surface in these small γ-Al_2O_3 nanocrystals is not compatible with the high-energy defects. Therefore, it is speculated that the lattice difference between the γ-Al_2O_3 nanocrystal and YAG matrix in Fig. 16a appears to be released by modifying the γ-Al_2O_3 nanocrystal structure through a gradual structural relaxation instead of the conventional dislocation mechanism, as proven by the HRTEM observations. Under this condition, it results in the coherent planes with a preferred orientation relationship during the interdiffusion process.

The formation of coherent planes at higher surface energy of ($\bar{3}$ 1 $\bar{1}$) and (4 2 $\bar{2}$) may be ascribed to the nonequilibrium state, i.e., initially the nanocrystal has just finished re-crystallizing in the inner cladding near the interface, and then contacted with the YAG matrix. The γ-Al_2O_3 nanocrystal finally has its corresponding planes well aligned with the YAG matrix nearly identical, while the interdiffusion process terminated. In other words, repeat of this structural relaxation process (nanocrystals undergo deformation) brought gradually the nanocrystal well aligned the matrix with finally less lattice difference, while the nanocrystals contact with the matrix. At this point, the experimental results have clearly showed that the γ-Al_2O_3 nanocrystals were initially re-crystallized from the amorphous phase, i.e., inner cladding, and randomly oriented because of the structure similarity between the amorphous phase and the γ-Al_2O_3. The same phenomenon arose near the core/inner-clad interface, but subsequently the γ-Al_2O_3 nanocrystals contacted and aligned with the YAG matrix during the interdiffusion process. In addition to the alignment with the YAG matrix, coherence in 2D planes at the nanocrystal/matrix interface occurs instead of generating dislocations in the nonequilibrium state. This facilitates the accommodated strain by the lattice difference and finally results in the preserved high single crystallinity of the YAG core through the formation of dislocations at core/inner-clad interface.

Light Transmission Characteristics

As indicated earlier, the core variation could be significantly improved by using the sapphire tube in the growth system. The variation of core diameter may still lead to a loss of propagating power because of the coupling of a low-order mode to a high-order mode. The tapered angle, Ω, derived from the core variation should meet the adiabatic criterion as follows (Love et al. 1991):

$$\Omega \leq \frac{\rho(\beta_1 - \beta_2)}{2\pi} \qquad (7)$$

where ρ is the core radius and β_1 and β_2 are the respective propagation constants of the modes before and after the tapering. For a 10-μm-core fiber, the adiabatic criterion requests that the tapering angle of the core must be within +/−1.35-degree to avoid the mode leakage. Figure 17a shows the core tapering angle along the propagation axis measured by the LabVIEW vision program. All the measured core tapering angles of the fiber grown with sapphire tube meet the +/−1.35-degree requirement. Without mode leakage, the scattering loss from the random fluctuation of core diameter may become the dominant factor for light attenuation source (Marcuse 1991). The autocorrelation length of the core diameter variation as a function of position was found to be ~1.7 mm, as shown in Fig. 17b. It is 2 orders of magnitude longer than the core diameter and thus alleviates the scattering loss. Such a long length is expectable since the core variation due to the heating fluctuation power becomes smaller, with the aid of the sapphire tube as a heat capacitor.

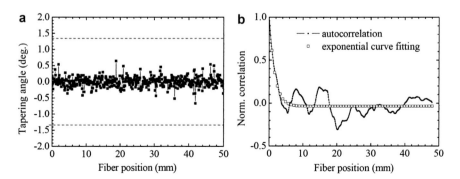

Fig. 17 (a) Tapering angle of the 10-μm-core fiber fabricated with sapphire tube during LHPG growth. (b) Autocorrelation curve of the core diameter for a 10-μm-core DCF

The measured fiber propagation losses are typically from 0.02 to 0.08 dB/cm. Compared with those fibers grown without sapphire tube, the propagation loss was around 0.6 dB/cm.

It is well known that the number of guided modes in an optical fiber is related to the normalized frequency V:

$$V = \frac{2\pi}{\lambda_0} a \sqrt{n_{co}^2 - n_{cl}^2} \tag{8}$$

where λ_0 is the optical wavelength in free space. $NA = \sqrt{n_{co}^2 - n_{cl}^2}$ is the numerical aperture of the fiber. When V is smaller than 2.405, the fiber is single-mode, and only the fundamental HE_{11} (LP_{01}) mode is guided. For a fiber to be single-mode, both the core radius a and the numerical aperture NA are required to be small. For example, the core radius and numerical aperture of Corning SMF-28e fiber are 8.2 μm and 0.14, respectively. The advantage of the YAG DCF fabrication by the co-drawing LHPG process is the capability of fabricating a CF with a core diameter of 10 μm or smaller. But the numerical aperture of a DCF is as large as 0.75 due to the large index difference between the core and the inner cladding. For a YAG DCF with a diameter of 10 μm, the normalized frequency V is about 16.7. And an impractical reduction of the core diameter is required to make it single mode. To reduce the number of glass-clad CF modes, the index of the inner cladding should be increased. Titanium dioxide (TiO_2) with a high refractive index of 2.61 (rutile form) at 589 nm has been widely used as the coating material. It also has a high melting temperature of 1843 °C. If TiO_2 is mixed in the inner cladding composition, the index of the inner cladding can be increased and the index difference between the core and inner cladding of YAG DCF could be reduced. This idea can be realized by depositing a TiO_2 layer on the circumference of the YAG CF. By using the co-drawing LHPG process, TiO_2 is intermixed in the inner cladding layer of the grown DCF. The required TiO_2 layer thickness for achieving equal index of the inner cladding and the core of YAG double-clad crystal fiber is estimated using the lever rule, as shown in Fig. 18a. The required TiO_2 thickness ranges between 5.8 and 5.4 μm for a core diameter of 10 to 30 μm. Such thickness is pretty large for the thin film deposition process. Two deposition processes including sputtering and E-gun were used. The deposition rate of sputtering was pretty slow. A 1-μm thick TiO_2 film was deposited on the 68-μm YAG single CF with a process time of 40 hours by sputtering. The TiO_2-deposited YAG CF was annealed at 800 °C under 99.9% O_2 for one hour and then grown into a DCF using the co-drawing LHPG process. To shorten the process time, E-gun deposition was used to deposit a 3-μm TiO_2 layer with a process time of 6 hours. The index profiles of the YAG DCFs with and without TiO_2 side-deposition are shown in Fig. 18b. The core diameters were 37 and 35 μm for the sputtering and E-gun samples. Both samples show increased indices of the inner cladding layers. The indices of the cores also increased for both samples with TiO_2 side-deposition, possibly due to the in-diffusion of TiO_2. The sputtering process shows an index difference of 0.085 with only 1-μm TiO_2

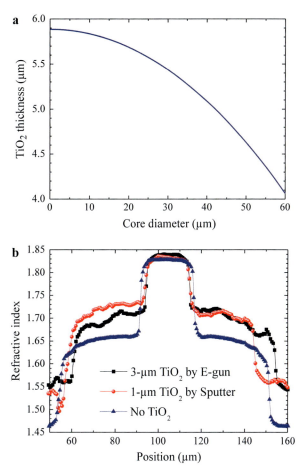

Fig. 18 (a) Simulation of the required TiO$_2$ thickness for achieving equal index of the inner cladding and the core of the Cr^{4+}:YAG double-clad crystal fiber. (b) Reduction of the index difference between the core and inner cladding by TiO$_2$ side-deposition using sputtering and E-gun processes

thickness. But further increasing the film thickness is difficult considering the long process time. Although the TiO$_2$ thickness was 3 μm, the index difference of the E-gun sample was 0.11. The increase of the inner cladding index was smaller than that by sputtering (Hsu 2011).

Crystalline Fiber-Based Broadband Spontaneous Emission

Spontaneous emission is the major source for most of the light we see all around us. It is so ubiquitous that there are many associated names with essentially the same quantum process. Spontaneous emission usually exhibits low brightness because of its 4π radiation nature. Directive spontaneous emission with broad spectra is desired for applications such as OCT and optical metrology. Take OCT as an example, a continuous wave (CW) source can exhibit better biological safety as compared to pulsed light sources because the brightness of the OCT image is mainly contributed from the average power of the broadband emission. Using laser diode as the pump source, various Ce^{3+}-, Ti^{3+}-, and Cr^{4+}-based broadband and bright emissions have been demonstrated using CFs as the host. The purpose of this section is to review the existing broad and bright CF-based light sources.

Ce:YAG as Crystalline Core

Cerium is a rare-earth element with the electronic configuration of $[Xe]4f^1 5d^1 6s^2$. At trivalent state, there is a single 4f electron outside the xenon core. The energy diagram of Ce^{3+} ions doped in YAG is shown in Fig. 19. The broadband yellow emission comes from the transition between the 4f and 5d orbitals. The 4f orbital is not sensitive to the crystal field due to the shielding of the outer $5s^2 5p^6$ orbitals. The spin-orbit coupling splits the 4f states into $^2F_{5/2}$ and $^2F_{7/2}$ with the latter unoccupied at room temperature. However, the crystal field effect has strong impact on the excited 5d state. The 5d orbitals are split by the crystal field and appear as broad bands at room temperature. Many of the 5d bands lie above the conduction band of the YAG crystal except the lowest two. The second lowest 5d bands partially overlaps with the conduction band.

The broadband fluorescence of Ce^{3+}:YAG consists of two unresolved peaks which correspond to the transition from the lowest 5d band to $^2F_{5/2}$ and $^2F_{7/2}$ levels. The fluorescence lifetime, τ_f, is 68 ns at room temperature (Hamilton et al. 1989).

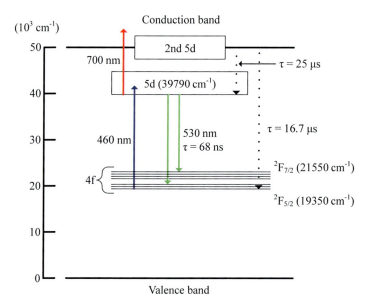

Fig. 19 Energy diagram of Ce^{3+}:YAG crystal (Hamilton et al. 1989)

Because the Ce^{3+}:YAG has a quantum yield very close to unity at room temperature (Bachmann et al. 2009), the radiative lifetime τ_r is quite close to the fluorescence lifetime. The 3-dB bandwidth of the fluorescence is typically around 100 nm. The emission cross section spectrum, σ_e, as a function of wavelength λ can be derived from the bulk fluorescence spectrum $I(\lambda)$ with the Füchtbauer–Ladenburg equation (sometimes referred to as the McCumber method) (Kück 2001):

$$\sigma_e(\lambda) = \frac{\lambda^5 I(\lambda)}{8\pi n^2 c \tau_r \int I(\lambda)\,\lambda\,d\lambda} \qquad (9)$$

where n is the refractive index and c is the speed of light in vacuum. Note that the λ in the integration in the denominator was often mistakenly omitted in many literatures. Assume the shape of the line shape function, $g(\nu)$, is Gaussian in the frequency domain, then the shape of $\sigma_e(\lambda)$ in the wavelength domain can be expressed as (Yoon et al. 1994):

$$\sigma_e(\lambda) = \frac{c^2}{8\pi n^2 \nu^2 \tau_r} g(\nu) \propto \lambda^2 \exp\left[-\left(\frac{\lambda-\lambda_0}{K\lambda}\right)^2\right] \qquad (10)$$

where λ_0 is the peak wavelength of g and K is a bandwidth-related factor. This spectrum has a long tail in the long-wavelength side and does not appear as Gaussian due to the nonlinear wavelength and frequency domain conversion. This equation indicates a blue shift of the fluorescence spectrum with respect to the $\sigma_e(\lambda)$. The physical origin behind this blue shift is the higher emission probability for short wavelength photons due to higher optical mode density. The Ce^{3+}:YAG fluorescence can be fitted by a combination of two fluorescence line shape functions with different weightings on λ_0 and K values. The fitted result shows λ_0 are 522.6 nm and 570.7 nm for each line shape function. The radiative lifetime and the refractive index used in the calculation are 68 ns and 1.835, respectively. The fitted fluorescence spectrum and the calculated spectrum of σ_e are also shown in Fig. 20. The calculated σ_e has a peak value of 48×10^{-19} cm^2 at the wavelength of 573 nm.

To formulate the rate equations of electron and photon interaction in Ce^{3+}:YAG, a simplified energy diagram is shown in Fig. 21. The ground state is the $^2F_{5/2}$ level and the excited metastable state is the lowest 5d band. Due to the existence of the usually unoccupied $^2F_{7/2}$ level, the system can be represented as a 4-level system. The small overlap between the ground-state absorption spectrum and the emission spectrum also supports this assumption. The pump and signal ESA transitions to the conduction band and the nonradiative relaxations from the conduction band are also included in this model. In a 4-level system, the terminating levels for both absorption and stimulated emission are rapidly depopulated to fast nonradiative relaxations:

$$N_1 \approx N_3 \approx 0 \qquad (11)$$

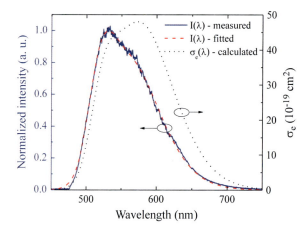

Fig. 20 Ce^{3+}:YAG fluorescence spectrum at room temperature. Solid line: measured fluorescence spectrum. Dashed line: fitted fluorescence spectrum. Dotted line: calculated emission cross section

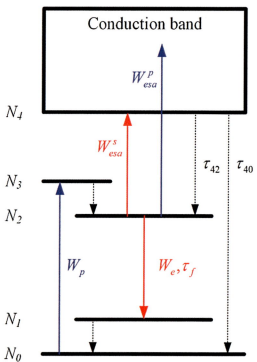

Fig. 21 Simplified energy diagram of Ce^{3+}:YAG. Solid arrows: absorptive or emissive transitions. Dotted arrows: nonradiative relaxations

So we have

$$N_T \approx N_0 + N_2 + N_4 \qquad (12)$$

where N_T is the doping concentration of Ce^{3+} ions and N_0 to N_4 are the populations at the levels shown in Fig. 21.

The rate equations can be written as:

$$\frac{dN_0}{dt} = -N_0 W_p + N_2 W_e + \frac{N_2}{\tau_f} + \frac{N_4}{\tau_{40}} \quad (13)$$

$$\frac{dN_2}{dt} = N_0 W_p - N_2 W_e - \frac{N_2}{\tau_f} - N_2 \left(W_{esa}^p + W_{esa}^s\right) + \frac{N_4}{\tau_{42}} \quad (14)$$

$$\frac{dN_4}{dt} = N_2 \left(W_{esa}^p + W_{esa}^s\right) - \frac{N_4}{\tau_{40}} - \frac{N_4}{\tau_{42}} \quad (15)$$

where τ_f if the fluorescence lifetime, τ_{40} is the lifetime of the transition between the conduction band and ground state, and τ_{42} is the lifetime of the transition between the conduction band and the excited state. W_p, W_e, W_{esa}^p, and W_{esa}^s are the transition probabilities of the ground state absorption, the stimulated emission, pump ESA, and signal ESA, respectively. These probabilities are expressed as following:

$$W_p = \frac{\sigma_a I_p \lambda_p}{hc} \quad (16)$$

$$W_e = \frac{\sigma_e I_s \lambda_s}{hc} \quad (17)$$

$$W_{esa}^p = \frac{\sigma_{esa}^p I_p \lambda_p}{hc} \quad (18)$$

$$W_{esa}^s = \frac{\sigma_{esa}^s I_s \lambda_s}{hc} \quad (19)$$

where σ_a is the GSA cross section at the pump wavelength λ_p, σ_e is the stimulated emission cross section at the signal wavelength λ_s, σ_{esa}^p is the ESA cross section at the pump wavelength, σ_{esa}^p is the ESA cross section at signal wavelength, I_p is the pump intensity, I_s is the signal (or fluorescence) intensity, c is the speed of light in vacuum, and h is Planck's constant.

By letting the Eqs. 13, 14, and 15 equal to zero, the steady-state solutions can be obtained as follows:

$$N_2 = N_T \frac{W_p \tau_f}{1 + W_p \tau_f + W_e \tau_f + W_{ESA} \tau_f \tau_4 \left(W_p + \tau_{40}^{-1}\right)} \quad (20)$$

$$N_0 = N_T - N_2 \left(1 + W_{ESA} \tau_4\right) \quad (21)$$

$$N_4 = N_2 W_{ESA} \tau_4 \quad (22)$$

The parameters of Ce^{3+}:YAG are summarized in Table 2. The pump wavelength is chosen to be 446 nm used in the experiments. The signal wavelength is chosen to be the peak wavelength of the emission cross section spectrum, which is 573 nm.

Comparing to the single crystalline core, the glass-clad CF provides a more robust waveguide. The guiding properties of the crystalline core are immune from the environmental variation, such as the application of the thermal grease or metal packaging. The glass cladding also reduces the propagation loss of the CF. Corning 7740, which is widely known as the Pyrex glass, was used as the glass capillary. The refractive index of Pyrex is 1.474 at the wavelength of 589.3 nm. Since the YAG crystal has a refractive index of 1.833 at the same wavelength, the fluorescence emission within the cone angle of 36.5° can be captured by the waveguide.

After the cladding process, the CF was placed in an aluminum holder, and molten tin was poured into the holder to fix the CF. This tin packaging method improves the thermal dissipation of the CF. The packaged fibers were then cut into pieces with desired lengths. Both ends of each section of CF were ground and polished for later optical measurement. In order to verify the effectiveness of the annealing, two CFs were prepared for measuring the optical conversion efficiency. One of the samples was as-grown and the other was annealed with the soaking time of 1 hour. The lengths were 8.3 and 8.5 mm for the as-grown sample and the annealed sample, respectively. The core diameters of both samples were 18 μm.

To measure the spontaneous emission as shown in Fig. 22, a 1.4-W 450-nm laser diode (PL-TB450, Osram) was used as the pump light source. The pump light was

Table 2 Summary of optical parameters of Ce^{3+}:YAG

Parameters	Symbol	Value
Refractive index (at 446 nm)	n	1.850
Refractive index (at 573 nm)	n	1.834
Absorption cross section (at 446 nm)	σ_a	16.8×10^{-19} cm^2
Emission cross section (at 573 nm)	σ_e	48×10^{-19} cm^2
Pump excited state absorption (at 446 nm)	σ_{esa}^p	20×10^{-19} cm^2
Signal excited state absorption (at 573 nm)	σ_{esa}^s	71.6×10^{-19} cm^2
Fluorescence lifetime (at 300 K)	τ_f	68 ns
Radiative lifetime (at 300 K)	τ_r	68 ns
Conduction band lifetime	τ_4	10 μs
Conduction band lifetime (to ground state)	τ_{40}	16.7 μs

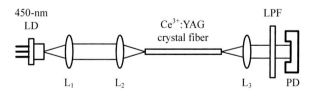

Fig. 22 Schematic of measuring the forward spontaneous emission. LD: laser diode; L_1–L_3: aspheric lenses; LPF: long-wavelength-pass filter; PD: photo diode

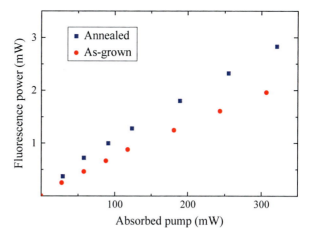

Fig. 23 Fluorescence power for an as-grown (red) CF and an annealed CF with 1-h soaking time (blue)

collimated by an aspheric lens (L_1) with f = 4.5 mm and NA = 0.5 (5722-H-A, New Focus) and then coupled into the CF by another aspheric lens (L_2) with f = 2.8 mm and NA = 0.65 (5721-H-A, New Focus). The forward fluorescence output from the other end of the CF was collimated with another aspheric lens, L_3, which was identical to L_2. A long-wavelength-pass filter (LPF) (BLP01-488R-25, Semrock) was used to block out the residual pump light. For measuring the residual pump power, the LPF was changed to a short-wavelength-pass filter to block out the spontaneous emission. The conversion efficiency was determined by the ratio of the spontaneous emission power and the absorbed pump power, which was defined as difference between the incident pump power and the residual pump power.

The measured performance is shown in Fig. 23. The fluorescence of the annealed CF was apparently higher than the as-grown sample. The conversion efficiency under 100-mW incident pump power was enhanced from 0.79% of the as-grown sample to 1.24% of the annealed sample. The theoretical conversion efficiency is 2.05% with 98% transmittances for both L_2 and L_3. There are two possible reasons for the large difference between the measurement and the theory: one is the annealing condition which had not been optimized and the other is the low pump coupling efficiency.

To optimize the soaking time, a batch of 12 CFs were fabricated and divided into 3 groups with different soaking times of 1 h, 4 h, and 16 h. The lengths of these fibers were all about 2 cm long, and the core diameters were increased to 25 µm for better coupling efficiency. The conversion efficiencies of these samples under 100-mW pump were measured with the same setup. The results are shown in Fig. 24. The group with the 4-h soaking time shows the best efficiency. The highest efficiency among all samples was 1.95%, which was very close to the theoretical limit. The estimated quantum yield was 95%. This indicates that both the pump coupling and the crystal quality were nearly optimized. The residual pump of the sample was about 30% of the incident pump power.

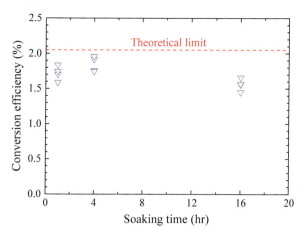

Fig. 24 Conversion efficiency versus the soaking time at a temperature of 1000 °C

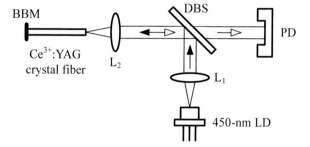

Fig. 25 The backward fluorescence measurement setup. BBM: broadband mirror; L_1 and L_2: aspheric lenses; DBS: dichroic beam splitter; LD: laser diode; PD: photo diode. Solid arrow: pump propagation direction. Hollow arrow: fluorescence propagation direction

To increase the brightness of the broadband emission, a double-pass scheme, as shown in Fig. 25, was attempted. The pump light was reflected by a dichroic beam splitter (LM01-466, Semrock), DBS, and focused into the CF. The collimation lens and the focusing lens were identical to the ones used in the Fig. 22. A broadband dielectric mirror (CVI BBD1-PM-1037-C), BBM, was attached on the opposite end of the CF to reflect both the pump light and the forward fluorescence light. The reflectance of the broadband mirror was >99% from 488 to 694 nm and about 95% at 450 nm. The reflected forward fluorescence, together with the backward fluorescence, was collimated by L_2 and transmitted through the dichroic beam splitter. Ideally, assuming perfect alignment and no propagation loss, attaching the BBM can increase 20% pump absorption and increase 100% fluorescence by reflecting the forward fluorescence. The expected overall increase would be $(1 + 20\%) \times (1 + 100\%) - 100\% = 140\%$. For comparison between the forward and the backward fluorescence power, the forward fluorescence was also measured by removing the BBM and used the same measurement setup as Fig. 22. The CF used in this experiment was the one measured to have the highest conversion efficiency in the previous experiments. The fiber was 2-cm long and had a core diameter of 25 μm.

The measured fluorescence powers are shown in Fig. 26a. The powers under 1.4-W incident pump power were 14.2 mW and 13 mW for the backward fluorescence without BBM and the forward fluorescence, respectively. The difference can be explained by smaller propagation loss experienced by the backward fluorescence than the forward one. Because most of the fluorescence was emitted near the pumping end where the pump intensity is higher than the opposite end, the averaged propagation distance before exiting the CF would be shorter for the backward fluorescence. The conversion efficiency dropped with the increasing pump power, from about 2% under 100-mW pump to 1.45% under 1.4-W pump. The reason of this efficiency decay was likely to be thermal quenching. After attaching the BBM, the highest output power under 1.4-W incident pump power came to 19.9 mW. The enhancement with respect to the backward case without the BBM is 40%, much lower than the ideal value of 140%. The results indicate that the contact between the BBM and the CF end face was not good. Inspection of the CF end face under the optical microscope revealed that although the packaged CF had been polished to optical quality, the surrounding tin surface is still not flat within the 1 cm × 1 cm

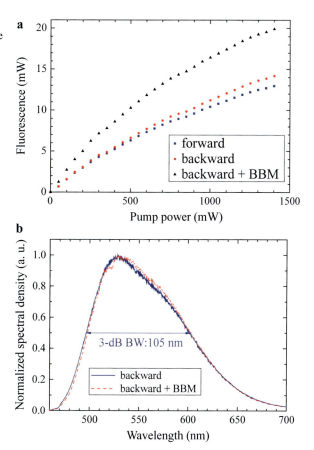

Fig. 26 (a) The output fluorescence power versus the incident pump power. (b) Normalized fluorescence spectra of the Ce^{3+}:YAG backward fluorescence with and without BBM

area. This problem can be overcome by using a bare CF, but the thermal dissipation may not be as good as the tin package even if thermal grease were used. The backward fluorescence spectra with and without the BBM are measured with a miniature spectrometer (USB4000-UV-VIS, Ocean Optics), as shown in Fig. 26b. The normalized spectra were very similar because of the flat reflectance spectrum of the BBM in the fluorescence wavelength range. The 3-dB bandwidth of both spectra is 105 nm.

By assuming 98% transmittances for the collimation lens L_2, the radiance at the CF output was calculated to be 21.57 mW·sr^{-1}. The calculated brightness (radiance) distribution is shown in Fig. 27. The maximum radiance is 32.6 W·mm^{-2}·sr^{-1} at 53.3 degree away from the axial direction and 30.3 W·mm^{-2}·sr^{-1} in the axial direction.

The output powers and radiances of the Ce^{3+}:YAG CF light source and several commercially available high-power LED light sources are listed in Table 3. The viewing angle was defined as the full-width-half-maximum of the radiance intensity distribution. Although the power of the LEDs are high, the CF light source is still 53 times brighter than the brightest LED due to the much smaller emitting size of the CF.

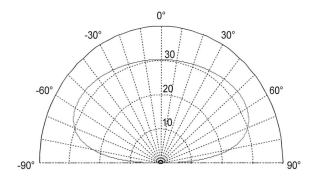

Fig. 27 Radiance distribution of the high-brightness crystal fiber light source. The unit of the radial scale is W·mm^{-2}·sr^{-1}

Table 3 Characteristics of Ce:YAG crystal fiber light source and commercially available high-power LED light sources

Light source	λ_p (nm)	Power (mW)	Emitting area (mm^2)	Viewing angle (degree)	Radiance (W mm^{-2} sr$^{-1)}$)
Ce^{3+}:YAG	530	19.9[a]	4.9 × 10^{-4}	124	30.3
MWWHL3	635	550[b]	1	120	0.175
MCWHL5	450	840[b]	1	80	0.571
MBB1L3	560	80[b]	1	120	0.025
M565 L3	565	979[b]	1	125	0.289

[a]Collimated output power measured with a 0.65-NA collimation lens
[b]Total output power measured with an integrating sphere

Ti:sapphire as Crystalline Core

Ti:sapphire is one of the most commonly used broadband fluorescence material. It is known for being used in a variety of applications that employ tunable lasers and mode-locked lasers (Moulton 1986). The central wavelength of Ti:sapphire, 760 nm, is in a region with a low tissue-scattering loss and low water absorption. Thus, it has been recently applied to a multitude of biological measurement systems. However, Ti:sapphire possesses two drawbacks: a low-absorption cross section and a short lifetime. Thus, Ti:sapphire does not exhibit strong broadband emission below the laser threshold and can hardly be directly used as an amplified spontaneous emission (ASE) light source. Many research teams have proposed crystal waveguide structures to enhance Ti:sapphire fluorescent characteristics. During the LHPG process, Ti^{3+} ions are oxidized to Ti^{4+} ions and Ti^{3+}-Ti^{4+} pairs form, which reduce fluorescence efficiency. After the LHPG process to produce a 18-μm-core-diameter CF, a follow-up reducing-annealing process is necessary. Table 4 lists the annealing conditions reported in the literature and employed in this work. A previous study reported that H_2 can be used as an effective reducing atmosphere (Wang et al. 2015; Kokta 1986, 1989; Aggarwal et al. 1988; Wu et al. 1995).

The reducing-annealing process was achieved using a high-temperature furnace (Lenton furnace LTF-18). Figure 28 shows the reducing-annealing setup. A CF grown through LHPG was placed on an Al_2O_3 boat and inserted into the furnace. To avoid an explosive reaction of O_2 and H_2, the gas tightness of the furnace-tube system needs to be carefully inspected. During the annealing process, a gas mixture

Table 4 Annealing conditions of $Ti:Al_2O_3$ crystal in literature

Growth method	Annealing temperature	Atmosphere	Diameter	Time	Reference
Czochralski	1850 ~ 2000 °C	Vacuum of 1×10^{-6} torr	1.5 in	> 48 h	(Kokta 1986)
VGF	1600 °C	Ar-H_2 mixture	2.5 cm	Several days	(Aggarwal et al. 1988)
Czochralski	1750 ~ 2025 °C	Inert atmosphere containing 20% H_2	1–1.5 in	> 48 h	(Kokta 1989)
LHPG	1800 °C	Reducing atmosphere	350 μm	3 days	(Wu et al. 1995)
LHPG	1600 °C	Ar-H_2 (5%) mixture	18 μm	18 h	(Wang et al. 2015)

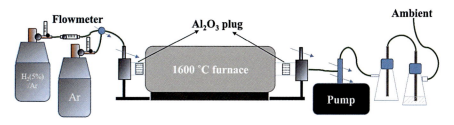

Fig. 28 Reducing annealing setup

(5% H_2/95% Ar) was continually pumped into the chamber to supply H_2 for the chemical reaction.

To improve the ASE efficiency with reducing annealing, the flow of the gas mixture (5% H_2/95% Ar) was well controlled at an annealing temperature of 1600 °C. Dwelling times of target temperatures were tested sequentially, and the annealed CFs were characterized by a confocal fluorescence microscope. The measurement results for different dwelling times are shown in Fig. 29. Successful reducing annealing was achieved with an 18-h dwelling time. The annealing treatment enhanced the fluorescence by 2 orders of magnitude.

The energy level of Ti^{3+} ions in α-Al_2O_3 is a four-level system. Since there are no excited-state absorptions, the absorption mechanism of the pump is mainly by the ground-state absorption. Because of the small absorption cross-section and short lifetime, the saturation pump power of Ti:Al_2O_3 is high. It is, therefore, inefficient to use low intensity pumps for Ti:Al_2O_3 (e.g., flash-lamps and laser diodes).

Using the cut-back method, the waveguide's attenuation coefficients with crossed polarization directions could be obtained. The waveguide's attenuation

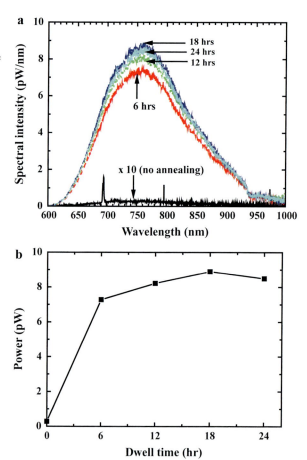

Fig. 29 (a) Fluorescence measurements with different dwelling times. (b) Impact of dwelling time on fluorescence power

Table 5 Attenuation coefficients of the glass-clad crystal fibers (Wang et al. 2015)

Sample		Pure sapphire	Ti:sapphire	
Attenuation coefficient (cm^{-1})			Polarization	
			π	ρ
Wavelength	520 nm	0.060	0.946	0.587
	532 nm	0.042	0.896	0.496
	780 nm	0.014	0.017	0.017

Table 6 Attenuation coefficients of Ti:sapphire crystalline waveguides at 780 nm

Author	Structure	Fabrication process	[a]Attenuation coefficient (cm^{-1})	Reference
L. Wu et al.	Single CF	LHPG	0.806	(Wu et al. 1995)
A. A. Anderson et al.	Planar	Pulsed laser deposition	0.414	(Anderson et al. 1997)
A. Crunteanu et al.	Rib	Reactive ion etching	0.691	(Crunteanu et al. 2002)
C. Grivas et al.	Rib	Ar$^+$-Milled	0.391	(Grivas et al. 2004)
L. Laversenne et al.	Buried	Proton implantation	0.161	(Laversenne et al. 2004)
V. Apostolopoulos et al.	Channel	Femtosecond-irradiation	0.530	(Apostolopoulos et al. 2004)
L. M. B. Hickey et al.	Channel	Thermal-diffusion	0.161	(Hickey et al. 2004)
C. Grivas et al.	Rib	PLD-grown	0.276	(Grivas et al. 2005)
C. Grivas et al.	Buried-channel	Proton implantation	0.230	(Grivas et al. 2006)
C. Grivas et al.	Channel	Written with pulse laser	0.150	(Grivas et al. 2012)
S. C. Wang et al.	Glass-clad CF	LHPG	0.017	(Wang et al. 2015)

[a]Attenuation coefficients at 780 nm include the propagation loss and re-absorption of Ti^{3+}-Ti^{4+} ion pairs. Propagation loss includes the loss of reflection, diffraction, and scattering

includes the propagation loss and dopant absorption (from Ti^{3+} ions as well as Ti^{3+}-Ti^{4+} ion pairs). Table 5 summarizes the absorption and loss measurements at various exciting wavelengths. At 780 nm, the signal loss of glass-clad Ti:Al$_2$O$_3$ CF was 0.017 cm^{-1}, which includes the propagation loss and re-absorption of the Ti^{3+}-Ti^{4+} pair. The loss was markedly lower than that of the Ti:sapphire waveguides in the literature, which were typically in the range of 0.15–0.806 cm^{-1}, as shown in Table 6. Distinguishing the polarization dependency is difficult because of the low re-absorption loss of the Ti^{3+}-Ti^{4+} pair. Based on the propagation loss of the glass-clad un-doped (pure) sapphire CF, the calculated absorption at 520 nm of glass-clad Ti:sapphire CF was 0.886 cm^{-1} (π) and 0.527 cm^{-1}(ρ), and the re-absorption of the Ti^{3+}-Ti^{4+} pair was 0.003 cm^{-1}. With the results, the figure of merit $\left(\alpha_{520}/\alpha_{780}\right)$ was 295. With the measured absorption coefficient at 520 nm (α_{520}), the 0.049 wt.% Ti^{3+} concentration of the glass-clad Ti:Al$_2$O$_3$ CF was obtained. At such a low Ti^{3+} concentration, the absorption cross section and emission cross section were comparable to those of the commercial bulk material.

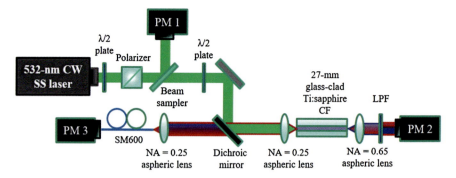

Fig. 30 A glass-clad Ti:sapphire CF BLS setup. LPF: long-wavelength pass filter; PM: power meter

To generate broadband emission, a frequency-doubled 532-nm solid-state laser was used as the pump source. A single-pump and single-pass setup is shown in Fig. 30. With a 16× aspheric lens (5720-a, New Focus, f = 16 mm, NA = 0.25), the pump laser was focused into a tin-packaged a-cut glass-clad Ti:sapphire CF with a length of 27 mm to absorb 90% of the pump power. A 60× aspheric lens (5720-B, New Focus, f = 2.8 mm, NA = 0.65) was used for fluorescence light collection. A long-wavelength-pass filter (NG610, CVI, HR: 532 nm, T: 95% approximately 650–1100 nm) was utilized for measuring fluorescence excluding pumps; in addition, a short-wavelength-pass filter (Semrock FF01-529, HR: 546–1000 nm, T: 95% approximately from 513–546 nm) was applied to reduce fluorescence in order to facilitate residual pump power measurement. After filtering the reflected pumped light with a dichroic mirror (Semrock LPD01-633RS-25), the ASE was coupled to a single-mode fiber (SM600) by using a 16x aspheric lens (5720-b, New Focus, f = 16 mm, NA = 0.25).

At room temperatures and without the need for active cooling, the measurement results are shown in Fig. 31. At an input power of 1.625 W, 18.3 mW of ASE power was generated. The optical conversion efficiency is 1.1%. With an 18-μm core diameter, the measured radiance was 47.68 W·mm^{-2} sr^{-1}. As shown in Fig. 31b, 31.2 μW of the ASE was coupled into an SM600 single mode fiber. It should be noted that this coupled fluorescence power was three times as high as that ever obtained from a bulk Ti:sapphire crystal. The optical convention efficiency from pump power to single-mode fluorescence power was 1.8×10^{-5}.

The emission spectra were monitored using a spectrometer (Anritsu MS9740A). Figure 32a shows the spectrum at 1.625-W pump power. By using the Gaussian fit, the peaks and bandwidths were obtained. Slight red shift of the center wavelength was observed when increasing the pump power. The 3-dB bandwidth also decreased gradually when the pump power was increased. The measured 3-dB bandwidth of 148 nm was narrower than the typical 180-nm bandwidth. It was due to chromatic aberration (by aspherical collecting and coupling lens) and the gain effect at high pump power. The point spread response of the ASE for OCT application was 1.71 μm, as shown in Fig. 32b.

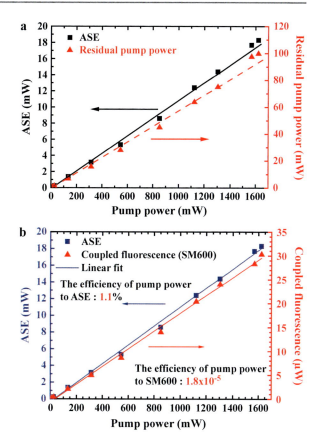

Fig. 31 (a) The ASE and residual pump powers. (b) The ASE powers just out of the CF and coupled in a SM600 SMF

A double-pass configuration was attempted to enhance the ASE power. Nichia NDG7475 520-nm LDs were used as the pump sources. The backward ASE was collected through a dichroic filter to block the pump light, as shown in Fig. 33. A silver thin film was deposited at one end of the Ti:sapphire CF to reflect the residual pump power and the ASE. The measurement results are shown in Fig. 34. In π-polarization exciting, the max ASE power and slope efficiency were 24.6 mW and 2.34%, respectively. The total maximum ASE power was increased from 24.6 to 42 mW with the LD_2 excitation in the σ polarization. The different ASE efficiencies by the two 520-nm LDs were mainly caused by the different excitation polarizations. A radiance of 77.59 W·mm^{-2}·sr^{-1} was achieved at a total pump power of 2.1 W.

Cr:YAG as Crystalline Core

Cr^{4+}:YAG with a broadband emission spectrum that just covers the silica fiber transmission window has a potential to develop optical communication components, such as ASE light source, optical amplifier, and tunable laser. It can also be used

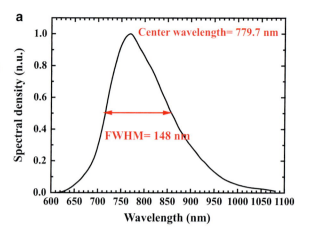
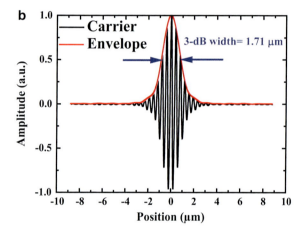

Fig. 32 (**a**) Fluorescence spectrum of a glass-clad Ti:sapphire CF BLS at 1.625-W pump power. The measured FWHM was limited by the chromatic aberration of using collection and coupled lens. (**b**) Point spread response of the glass-clad Ti:sapphire CF BLS with aspherical coupling lens

as light source in OCT technology for deep tissue penetration. In this section, we demonstrate and discuss the key factors to its optical performance. Cross sections of absorption, emission, and ESAs were also determined by experimental curve fitting. Further improvement is also discussed.

Cr^{4+}:YAG rectangular rods with absorption constant of 4.5 cm^{-1} were used as the source materials for crystalline core growth. The orientation of grown CF followed the [111] orientation of the seed crystal. As mentioned earlier, core-reduction processes were developed and evolved from the LHPG method to sapphire tube assisted co-drawing LHPG technique. When using sapphire tube assisted co-drawing LHPG to fabricate Cr^{4+}:YAG DCF, core diameter around 10 μm can be obtained. The grown CF has a double-clad structure with a core made of pure crystal and an outer cladding made of pure fused silica. In between there is an inner cladding layer made of mixtures of crystal and fused silica. The refractive indices of the core, inner, and outer claddings as shown in Fig. 35 are 1.82, 1.66,

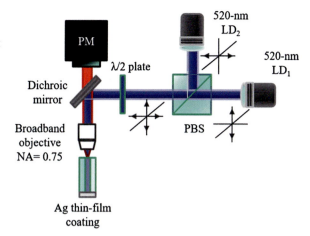

Fig. 33 A dual pump and double-pass setup. PM: power meter; PBS: polarizing beam splitter

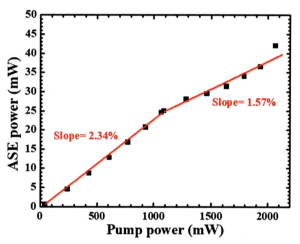

Fig. 34 ASE power generated by a Ti:sapphire CF with dual pump and double pass scheme. Different slopes were caused by the π and σ excitation polarizations

and 1.46, respectively. The large index difference leads to a 0.75 numerical aperture of the DCF. The fluorescence is concentrated in the core region. The fluorescence intensity in the inner cladding region was weak because of the weak fluorescence generation ability from the Cr^{4+}:YAG and fused silica mixture.

With a single pass scheme, the ASE output powers from fibers of different core diameters were measured in an end-pump scheme. The grown CFs were packaged with Pb-Sn alloy to improve the heat dissipation. The fiber endfaces were grinded and polished to obtain optical-quality input and output faces. A 1064-nm Yb-fiber laser with its wavelength at the peak of the Cr^{4+}:YAG absorption spectrum was used to pump the Cr^{4+}:YAG CFs with 920, 100, 25, and 10-μm core diameters. The pump light was launched into the core of the CFs. The corresponding fiber lengths were 1.5, 4.7, and 5.5 cm, respectively. Figure 36a shows the measured and simulated ASE powers as a function of the absorbed pump power. In the experiments, the ASE efficiency of the 25-μm-core fiber with the strongest optical

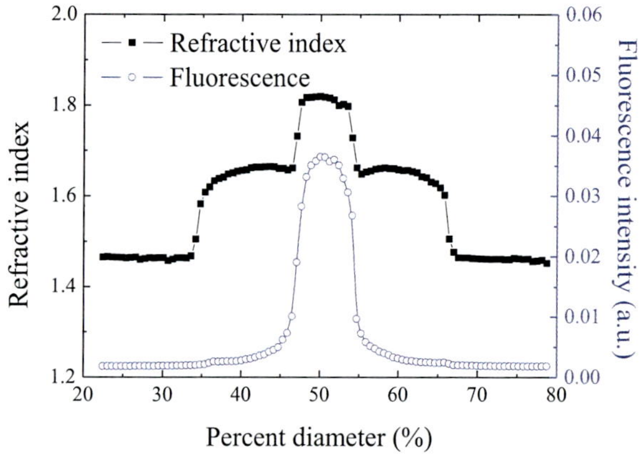

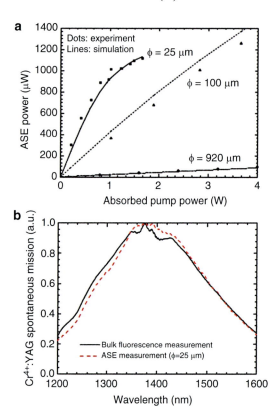

Fig. 35 Refractive index profile and fluorescence mapping of a Cr^{4+}:YAG DCF

Fig. 36 (a) ASE powers versus pump powers for the samples with core diameters of 920, 100, and 25 μm. The dots and the lines represent the measured and simulated data. (b) The comparison of fluorescence spectra between Cr^{4+}:YAG bulk and a CF output

confinement obtained the highest efficiency. The rising ASE efficiency as the core diameter decreasing was in agreement with our simulation. The roll-off in the $\varphi = 25$ μm curve was due to the limited CDF length. The ASE power can be further improved by both reducing the core diameter and increasing the fiber length. The normalized spectra of Cr^{4+}:YAG fluorescence and ASE output are shown in

Fig. 36b. The bandwidth is only slightly decreased from 277 nm to 265 nm. It should be noted that the output light has a large portion of spontaneous emission and a small portion of amplified spontaneous emission. At present, the ASE efficiency is still low due to the small cross section ratio of ESA to GSA and pump ESA loss.

The net gain coefficient, $g_{net}(z, v_i)$, is defined as the subtraction of the fluorescence propagation loss, α, of the core:

$$g_{net}(z, v_i) = g(z, v_i) - \alpha = [\sigma_e(v_i) - \sigma_{esa}(v_i)] N_2(z) - \alpha \qquad (23)$$

where σ_e and σ_{esa} are the emission and excited-state absorption cross sections, respectively.

The simulated net gain coefficients over the spectra of interest for the complete population inversion ($N_2 = N_T$) and the half inversion ($N_2 = N_T/2$) cases are shown in Fig. 37. The material gain barely covers the propagation loss of the CF over the spectra under the complete inversion case. The peak value for the net gain coefficient was 0.03 cm^{-1}. But for the half inversion case it became net loss outside the 1.3 to 1.5 µm band. With such a low gain, the output from the Cr^{4+}:YAG DCF is mostly from the spontaneous emission with a slightly amplification by the stimulated emission. For the fluorescence propagating in the inner cladding, there is no gain but only the propagation loss.

For dual-pump scheme, two pump laser beams were launched into the gain fiber from both ends, so the long gain fiber could be sufficiently pumped. The experimental setup of the dual-pump Cr^{4+}:YAG DCF broadband light source is shown in Fig. 38. The pump laser used is a 1064-nm ytterbium fiber laser (IPG Photonics PYL-10 M). The collimated free-space pump beam was split by a polarization beam splitter. The two pump beams were focused and coupled into the gain fiber via the L_1 and L_2 lenses. L_2 also collected and collimated the output beam. The forward and backward directions referred to the direction of the light source output. A long-wavelength pass filter (Thorlabs FEL1200) was used as the

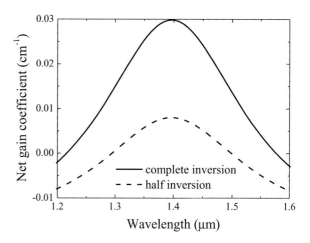

Fig. 37 Net gain coefficient for the complete inversion and the half inversion cases

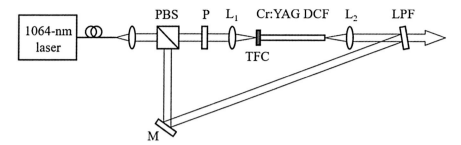

Fig. 38 Experimental setup of the dual-pump double-pass Cr^{4+}:YAG DCF scheme. PBS: polarization beam splitter; P: polarizer; L1, L2: aspheric lenses; M: mirror; LPF: long-wavelength pass filter; TFC: thin film coating

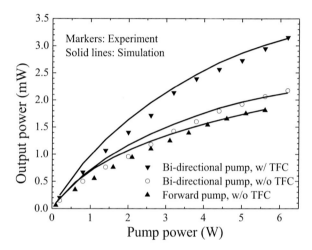

Fig. 39 Output powers of various pump and coating schemes. The solid lines are the simulation results

pump and fluorescence WDM filter so the short-wavelength pump was reflected and the long-wavelength fluorescence passed through. Transmission of this filter is >80% for fluorescence band between 1250 and 1600 nm at normal incidence. The backward pump was incident on this filter with a small angle. The reflectance only slightly degraded.

The broadband output powers from the Cr^{4+}:YAG DCF using the forward pump scheme and the bi-directional pump scheme with and without the coating on the sample endface are shown in Fig. 39. For the bi-directional pump scheme, the pump power was defined as the sum of forward and backward pump powers. The ratio of forward to backward pumps was 1:1. The dual pump scheme is more efficient than the forward pump scheme. The output power enhancement of the bi-directional pump scheme is about 22% greater than the forward pump scheme because of the better pump efficiency and also the reduction of the thermal problem, despite the reduced light collection efficiency of the L_2 lens.

To make use of the backward fluorescence, a multilayer thin film coating which is highly transmissive for the pump and highly reflective for the fluorescence was coated directly by E-gun deposition at 280 °C on one fiber endface. The backward fluorescence was reflected by the coating so the Cr^{4+}:YAG DCF output contained both the forward and backward fluorescence. With the coating on the fiber endface for collection of both directions, almost 20% of the generated spontaneous emission is captured by the core and inner cladding of the DCF structure.

For the coated sample, the ratio of forward and backward pumps was also set at 1:1. The forward pump power was clamped at 1.9 W to avoid coating damage. For higher pump power, only the backward pump power was tuned up. The forward pump power was kept at 1.9 W by using a polarizer. The maximum output power increased 45% from 2.17 mW to 3.15 mW with the coating. Compared with the 1.8 mW maximal power using the forward pump scheme, the enhancement was 74%. The reason why the increase was not as expected was attributed to the imperfection of the endface coating and the propagation losses of core and inner cladding of the long DCF, especially the high inner cladding loss. The fitted reflectivity of the coating on the fiber endface by simulation was 0.7. The measured broadband output spectrum of the coated sample is shown in Fig. 40. The center wavelength was around 1380 nm and the 3-dB bandwidth was 222 nm. The ripples around the peak were due to the water absorptions. The bandwidth was not narrowed much by the coating and the gain effect.

Using the broadband emission for OCT application, the interference signal was calculated by taking the Fourier transform of the measured spectrum. Figure 41a shows the normalized interference fringe. The axial resolution is the full width half maximum of the interference signal. The calculated axial resolution was 3.62 μm. This axial resolution is much better than the 10-μm order resolution of the most often used SLD light sources and is comparable to the 2.76-μm resolution of the Ni-doped fiber.

The advantages of using Cr^{4+}:YAG DCF as a light source for OCT not only come from its broad spectral bandwidth, but also from its Gaussian-like spectral shape. To elucidate this, the envelope of the interference signal H(z) was obtained by taking

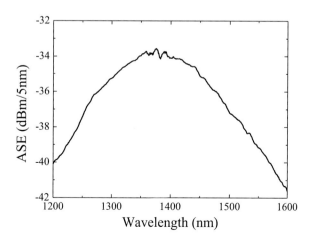

Fig. 40 ASE spectrum of a Cr^{4+}:YAG DCF. The center wavelength is around 1380 nm and the 3-dB bandwidth is 222 nm

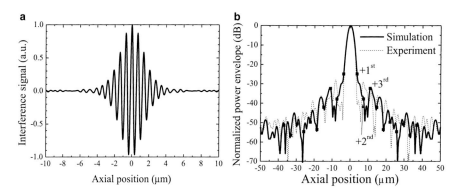

Fig. 41 (a) Interference fringe. The axial resolution is 3.62 μm. (b) Envelope of the normalized interference signal. Squares: image cross talks for the longitudinal pixels

Table 7 Image axial cross talk. Adj.: adjacent

Pixel No.	Relative position (μm)	Amplitude	Cross talk (dB)	Remark
+1	3.62	0.058	−24.7	+1 adj. Pixel
+2	7.25	0.014	−37.2	+2 adj. Pixel
+3	10.87	0.023	−32.9	+3 adj. Pixel

the Hilbert transform. The calculated and experimentally measured normalized power envelopes of the interference signal 20·log[H(z)] are shown in Fig. 41b. The dots represent the image cross talks of the neighboring pixels by calculation in the axial direction. The calculated power envelope of the interference signal is an even function since the spectrum is real. The sidelobes were significantly low, as a result of the Gaussian-like shape of the spectrum of the broadband light source. The adjacent image pixel cross talk could be defined as the relative magnitude of the power envelope of the interference signal of the adjacent pixels to the main peak. As shown in Table 7, the adjacent pixel cross talk was −24.7 dB. Nonadjacent image pixel cross talk can be defined similarly. In this case, they were − 37.2 and − 32.9 dB for the second and third pixels, respectively. The experimentally measured power envelope of the interference signal was very close to the calculated curve and confirmed the calculation. The advantages of Cr^{4+}:YAG DCF broadband light source not only include a high axial resolution but also the excellent axial image pixel cross talks.

Crystalline Fiber Laser and Amplifier

Tunable solid-state lasers have been widely used in applications for scientific research, medicine, ultrafast lasers, and optical communication systems. Most tunable laser gain media use 3d-3d transitions of transition-metal ions, such as Ti^{3+}, Cr^{2+}, Cr^{3+}, Cr^{4+}, Co^{2+}, and Ni^{2+}, to cover the spectral range from 700 to 2800 nm.

In this section, the structural and optical characteristic of Cr ions are described. Cr^{4+}:YAG crystal can generate broadband emission from 1200 to 1600 nm. It can be utilized as a broadly tunable fiber laser for ultrahigh capacity fiber communication systems. The detailed properties of the Cr^{4+}:YAG crystal as an optical amplifier are also given.

Tunable Cr^{4+}:YAG Crystalline Fiber Laser

Broadly tunable Cr^{4+}:YAG CF lasers have been demonstrated with tuning elements, such as pellicle beam splitter, diffraction grating, or birefringent filter (BRF). The diffraction grating was mounted in Littrow configuration, thus itself serves as an output coupler. The output coupler for the lasers tunable with a pellicle beam splitter or the BRF was replaced by HR mirrors in order to maximize the tuning range in the tunable laser experiments.

Wavelength Tuning by Pellicle Etalon

The uncoated pellicle beam splitter (BP108, Thorlabs) is a nitrocellulose film with a refractive index about 1.50 and a thickness about 2 μm. When used as an etalon at wavelengths around 1.5 μm, the pellicle exhibits a free spectral range larger than 300 nm at normal incidence. However, the modulation depth on the loss spectrum is small due to the low reflectance of the air/nitrocellulose interface. It is obvious that the s polarization is better than the p polarization for the tunable filter due to higher peak transmittance and larger modulation depth. The laser wavelength can be tuned by adjusting the tilt angle of the pellicle etalon in the laser cavity.

Both the hemispherical and the collimated cavities were used in this experiment. The output coupler used in the hemispherical external-cavity crystalline fiber laser (ECCFL) experiment was a broadband HR mirror with a radius of curvature of 50 mm. The experimental setup was shown in Fig. 42. The CF used in this experiment was 4.4-cm long and with a core diameter of 19 μm. The CF was

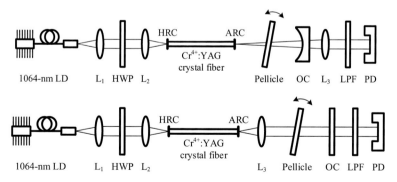

Fig. 42 Setups of tunable ECCFLs with uncoated pellicle beamsplitter. Upper: hemispherical cavity. Lower: collimated cavity

rotated so that the laser polarization was s-polarized with respect to the pellicle beam splitter. At some pellicle tilt angles, there were multiple lasing peaks in the laser spectra, due to the small loss modulation of the pellicle. Discrete tuning from 1354 to 1476 nm was obtained, corresponding to a tuning range of 122 nm. The tuning was stepwise with respect to the pellicle tilt angle. One possible reason for the discrete tuning is the mismatch between the multimode CF modes and the free space resonator modes. The multimode interference could lead to periodic modulation in the cavity loss spectrum. Similar discrete tuning curve is obtained by inserting the pellicle into the collimated ECCFL with a tuning range from 1368 to 1490 nm. The tuning curves for both cavities are shown in Fig. 43.

Wavelength Tuning by Diffraction Grating

Contrary to the pellicle beam splitter, the diffraction grating provides excellent ability to reject the unwanted wavelength. The setup of the grating tunable CF laser is depicted in Fig. 44. The diffraction grating (53004BK02-246H, Richardson Gratings) used in this experiment was a planar holographic grating designed for use in the Littrow configuration, which means the -1st order diffraction light goes back in the direction reversed to the incident light. The grating has 1050 grooves per mm

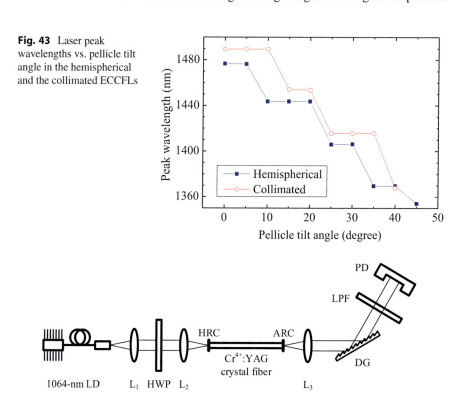

Fig. 43 Laser peak wavelengths vs. pellicle tilt angle in the hemispherical and the collimated ECCFLs

Fig. 44 Setup of the tunable ECCFL with a Littrow-mounted diffraction grating. DG: diffraction grating

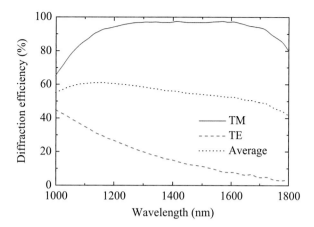

Fig. 45 Diffraction efficiency of the grating. Measured near the Littrow configuration

and the area is 30 × 30 mm. The diffraction efficiency provided by the manufacturer is shown in Fig. 45. The diffraction efficiency is 96.9% at the wavelength of 1452 nm for TM polarization.

The CF used in this experiment is the same 6.2-cm sample used in section 4.3. The CF was rotated so that the laser polarization was parallel to the optical table. The diffraction grating was mounted with its groove in the vertical direction, so that the intracavity laser beam was TM-polarized. The wavelength tuning was accomplished by rotating the grating in the table plane. The zero order diffraction of the grating was taken as the laser output. The direction of the output beam changed when the grating was rotated. The measured output power at the peak wavelength of 1452 nm is shown in Fig. 46a. The slope efficiency is 9.9% and the threshold pump power is 187 mW. The wavelength tuning spectrum recorded with the "maximum hold" function of the optical spectrum analyzer is shown in Fig. 46b. The laser was discretely tunable from 1399 nm to 1514 nm. The possible cause of the discrete tuning will be discussed in the next section.

Wavelength Tuning by Birefringent Filter

BRFs have been extensively used in broadband tunable solid-state and dye lasers due to their broad tuning range and low insertion loss. A BRF made of single-plate quartz was designed for tuning the laser as shown in Fig. 47. The BRF has a thickness of 888 μm and the optic axis lies in the plate plane. The calculated single-pass p-polarized transmittance of the BRF is shown in the background of Fig. 48. The CF used in this experiment was 4.4-cm long and with a core diameter of 19 μm. The CF was rotated so the laser polarization is in the TM direction of the BRF. No laser operation was observed after inserting the BRF into the hemispherical ECCFL, because the Brewster-oriented BRF introduces large aberration loss for the intracavity divergent beam.

Tunable laser was successfully demonstrated with the insertion of the BRF into the collimated ECCFL. The laser wavelengths vs. the in-plane rotation angle of the

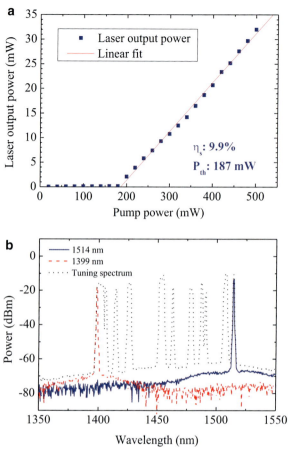

Fig. 46 (a) Output power of the tunable ECCFL with a Littrow-mounted grating at the laser peak wavelength of 1452 nm. (b) The laser tuning spectrum and the laser output spectra at the two extreme wavelengths. The laser tuning spectrum was recorded using the "maximum hold" function of the optical spectrum analyzer while rotating the diffraction grating

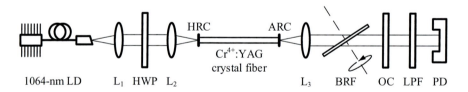

Fig. 47 Setup of the tunable ECCFL with a birefringent filter

BRF are plotted as the dots in Fig. 48 and the tuning spectrum is shown in Fig. 49a. The length of the collimated beam was about 3.8 cm. The wavelength tuning was accomplished only by rotating the BRF, without re-aligning the laser cavity. The tuning was from 1353 to 1509 nm, corresponding to a tuning range of 156 nm. The lowest threshold pump power was 70 mW at the peak wavelength of 1452 nm. The tuning was also discrete but with denser lasing peaks in the tuning

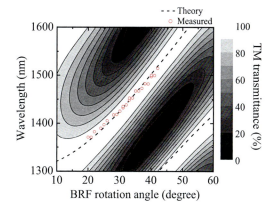

Fig. 48 Laser wavelength vs. the rotation angle of the BRF. Dashed line: Theoretical tuning curves of the BRF. Dots: Measured laser wavelengths. Background contours: Single-pass p-polarized transmittance of the BRF

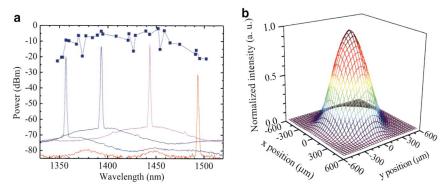

Fig. 49 (**a**) The laser output power vs. laser output wavelengths under 400-mW pump power (dots). Lines: four example spectra lasing at 1357, 1393, 1443, and 1494 nm. The OSA resolution bandwidth is 1 nm. (**b**) Output beam profile at the laser wavelength of 1477 nm

spectrum, as a consequence of better wavelength selectivity of the BRF than that of the pellicle etalon due to the larger modulation depth in the loss spectrum.

To investigate the relations between discrete tuning and the multimode nature of the CF, the output beam profile of the tunable collimated ECCFL was recorded with an infrared charge-coupled device (CCD) (CamIR[1550], Applied Scintillation Technologies). This infrared image sensor was a silicon CCD coated with up-conversion phosphors with the responsive wavelength range from 1460 to 1600 nm. However, the distribution of the phosphors was quite nonuniform. To remove this nonuniformity, multiple images of the laser beam were recorded by moving the CCD in the transverse direction using a one-axis linear stage. All images were then re-aligned and averaged. Figure 49b shows the recorded beam profile when the wavelength was at 1477 nm, which was an averaged result from 71 images with a 30-μm interval in the x direction.

During the wavelength tuning operation, the intensity pattern of the output beam changed abruptly whenever the laser wavelengths jumped. Adding an iris at the collimated part in the cavity as a spatial filter to select the fundamental mode may not work because the modal size difference between the fundamental and the higher-order transverse modes was not large enough. As a result, a combination of multiple transverse modes with the lowest round trip loss was selected by the laser oscillation. This combination could be very sensitive to the wavelength due to different phase velocity of each mode. Thus, when the tuning element is adjusted, the laser seeks the next combination of modes with the lowest loss, which corresponds to a specific wavelength. One solution for the discrete tuning problem is to develop a single-mode CF by using high-index cladding materials, such as high-index glasses or YAG ceramics.

Crystalline Fiber Amplifier

In addition to the application in the broadband tunable laser, the Cr^{4+}:YAG CF is also an attractive candidate as an optical amplifier providing broadband gain over the entire communication wavelengths. The important optical parameters of the chromium-doped fiber (CDF) and a typical erbium-doped fiber (EDF) are summarized in Table 8. The active ion concentration of a typical EDF is much higher than the CDF's. This is because the Erbium ions in the silica host are naturally in the desired trivalent state. The highest Erbium doping concentration is limited by the interactions between the Erbium ions like up-conversion and concentration quenching. Unlike the EDF case, the chromium ions in the YAG host are initially in the trivalent state. A charge compensation mechanism is used to turn the oxidation state of some of the chromium ions into the quadrivalent state. The charge compensation efficiency for generating Cr^{4+} ions out of the Cr^{3+} ions is typically less than 6%. Most of the chromium ions still remain in the trivalent state. This charge compensation efficiency is the limiting factor of the Cr^{4+} concentration. As described in the previous section, the chromium ions evaporate out of the melt during the fiber growth process so the Cr^{4+} concentration of the fiber further decreases.

The pump absorption cross section and the signal stimulated emission cross section of the CDF are several orders of magnitude larger than those of the EDF. The absorption and the gain are related to the products of the Cr^{4+} ion concentration and the corresponding cross sections. The products $\sigma_a N_T$ and $\sigma_e N_T$ of the CDF are 35 and 5 times larger than those of the EDF. This implies that the required CDF length would be shorter, which is favorable with consideration of the significant propagation losses of the CDF. The CDF length could be even shorter by increasing the Cr^{4+} doping concentration. The $\sigma_e N_T$ product related to the gain constant of the CDF is large as compared to that of the EDF. However, for the application of fiber broadband fluorescence light source, the relative magnitude of $\sigma_e N_T$ compared to its own $\sigma_a N_T$ product is more suitable for evaluation of its performance. The $\sigma_e N_T$ product of the CDF is about one third of its $\sigma_a N_T$ product. If including the effect of the ESAs of the pump and fluorescence, the ratio of the gain constant to the pump absorption constant becomes even smaller. As a result, before the fluorescence

Table 8 The parameters of the CDF and a typical Erbium doped fiber (Fejer et al. 1984; Dorenbos 2000)

Parameter	CDF	EDF	Unit
N_T	3.2×10^{17}	80×10^{17}	cm^{-3}
σ_a	22×10^{-19}	0.025×10^{-19} @ 980 nm 0.018×10^{-19} @ 1480 nm	cm^2
σ_e	2×10^{-19}	0.050×10^{-19} @ 1552 nm	cm^2
σ_e/σ_a	0.091	2 @ 980n nm pump 2.78 @ 1480 nm pump	
σ_{esa}^p	4.2×10^{-19}	–	cm^2
σ_{esa}^f	1.2×10^{-19}	–	cm^2
σ_a^f	–	0.023×10^{-19} @ 1552 nm	cm^2
τ_f	4.5	~10,000	μs
$\Delta\lambda$	265	30	nm
I_p^{sat}	18.8×10^3	8.1×10^3 @ 980 nm 7.5×10^3 @ 1480 nm	W/cm^2
I_f^{sat}	157.5×10^3	2.56×10^3	W/cm^2

acquires sufficient gain, the pump light in CDF already decays below its pump saturation intensity and has no ability to achieve population inversion within a short fiber length. Using the EDF as a reference, its $\sigma_e N_T$ product is two times of the $\sigma_a N_T$ product for 980-nm pump. Therefore, the fluorescence can acquire enough gain while it propagates along the gain fiber. The σ_e/σ_a ratio is a useful parameter in evaluating an optical material for a fiber broadband fluorescence source or amplifier. Although the wide emission bandwidth is very attractive, the light amplification ability of the Cr^{4+}:YAG is not so strong as its σ_e/σ_a ratio is only 0.318. For EDF, the σ_e/σ_a ratios are 2 and 2.78 for 980-nm and 1480-nm pumps, respectively. Measures to further improve the fluorescence efficiency of CDF are necessary. In order to extent the CDF length that the pump light can deliver its power along and thus improve the fluorescence powers, the bi-directional pumping scheme is suitable. In addition, since CDF could have a double-cladding structure, cladding pump scheme is useful in relieving the pump absorption while keeping the same gain constant since both the forward and backward fluorescence are only propagating in the core. Incorporating a high reflectivity mirror at the input end of the CDF could improve the fluorescence efficiency because the backward fluorescence is reflected and can experience more gain as it travels again through the CDF.

The lifetime of the CDF is about three orders of magnitude shorter than that of the EDF. The pump power for EDF can be as low as a few tens of milliwatts but still maintaining a good optical efficiency due to the long lifetime of the erbium ion. For CDF the much shorter lifetime of the excited Cr^{4+} ions leads to a faster spontaneous emission rate. Consequently, a faster pumping rate is required for the CDF for maintaining the excited state population.

As shown in Table 8, the pump saturation intensity of the CDF is about two times larger than that of the EDF. The much larger cross-sections of CDF compensate the drawback of its short lifetime. So the much shorter lifetime of the CDF does not cause a serious issue in achieving population inversion. But when taking into consideration that the core diameter of the CDF is much larger than the EDF's, which can be as small as 3~4 μm, the required pump power of CDF is much larger. The pump power for the CDF should be at least in the order of watt level to have a better fluorescence output power. The low fluorescence efficiency of the CDF can be attributed to the relatively small emission cross-section, compared with its own pump absorption cross-section.

As a preliminary trial, the Cr^{4+}:YAG DCF was used as the optical gain fiber for amplifier with a small core diameter (10 μm) and low propagation loss (0.02~0.07 dB/cm). The small-core DCF was also favored to combine with a single-mode fiber (SMF), since it has a good match in mode field diameter with that of SMF 28. Reducing the insertion loss between Cr^{4+}:YAG DCF and SMF is an important issue toward its practical utilizations in almost all applications. The refractive index profile of the Cr^{4+}:YAG DCF was obtained by measuring the Fresnel reflection of the end face by LSCM with a 635-nm distributed feedback laser. The 1.46 refractive index of the fused silica outer clad was used as the reference. The measured refractive indices of the core, inner clad, and outer clad were 1.82, 1.66, and 1.46, respectively. The fiber is multimode since the refractive index difference between core and inner clad is quite large. The mode coupling efficiency of the DCF fundamental mode to SMF 28 was simulated with superposition integral at 1064-nm pump wavelength, 1400-nm emission center wavelength, and 1550-nm communication wavelength. The simulation results are shown in Fig. 50, and the core diameters for optimum mode coupling efficiency at 1064 nm, 1400 nm, and 1550 nm are 11.5 μm, 13.5 μm, and 14.5 μm, respectively. All the optimum mode coupling efficiencies exceed 96%. Based on the simulation result, a 13-cm-long fiber in 13.5-μm core was chosen to measure the insertion loss from a SMF to DCF and to another SMF by butt coupling scheme. As shown in Fig. 50a, the insertion loss was measured from 1260 nm to 1640 nm by using 4 sets of lasers as the light source. The measured insertion loss varies from 1.97 dB to 2.88 dB. The insertion loss includes the Fresnel losses, mode coupling loss, and propagation loss of the DCF in the scheme. The insertion loss was estimated at various wavelengths, as shown in Fig. 50b. It is clear that the insertion loss is mainly from the Fresnel losses at the uncoated end faces.

To have better understanding of the Cr^{4+}:YAG crystal fiber amplifier, its performance is numerically simulated. The propagation losses of the CF were determined by the measurements. All interfaces are assumed to be lossless. Three different designs of the crystal fiber amplifier were considered: the single-pump and single-pass (SPSP) scheme, the single-pump and double-pass (SPSP) scheme, and the dual-pump and double-pass (DPDP) scheme. Illustrations of these schemes are shown in Fig. 51. In the SPSP scheme, the signal light and the pump light are coupled into the CF from the same end face, and the output signal is detected at the opposite end. A wavelength-division multiplexer or a dichroic mirror is required for combining the pump light and the signal light. In the SPDP scheme, the signal light is reflected at the opposite end so the signal pass through the gain medium twice. This approach can effectively double the gain at the expense of more

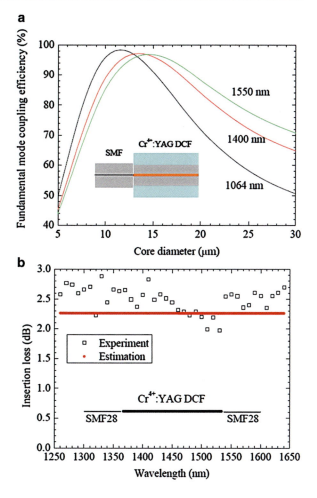

Fig. 50 (a) The simulation results of mode coupling efficiencies between the SMF28 and the Cr^{4+}:YAG DCF using different signal wavelengths. (b) The measured and simulated insertion losses. The inset shows the measurement scheme

complicated structures. An optical circulator is required to separate the input signal and the output signal. The pump light can also be reflected to enhance the pump power utilization. The DPDP scheme is similar to the SPDP scheme, but with the incident pump coupled into the CF from both ends. The signal reflector in the DPDP scheme was designed to have high transmittance for pump. The ratio between the forward and backward pump in the DPDP scheme was fixed at 1:1 in the following discussion.

Simulations were carried out for comparing these three schemes. The wavelengths of pump and signal were 1064 nm and 1431 nm, respectively. The input signal power was −30 dBm. The simulation result is shown in Fig. 52a. The DPDP scheme provides the highest optical gain. The SPDP scheme can achieve similar performance when the CF is well saturated by the pump power. The

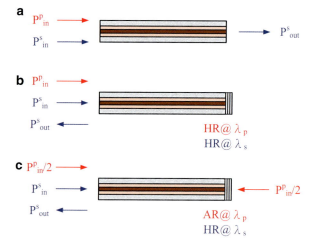

Fig. 51 Schematic plots of different Cr^{4+}:YAG crystal fiber amplifier schemes. (**a**) SPSP, (**b**) SPDP, and (**c**) DPDP

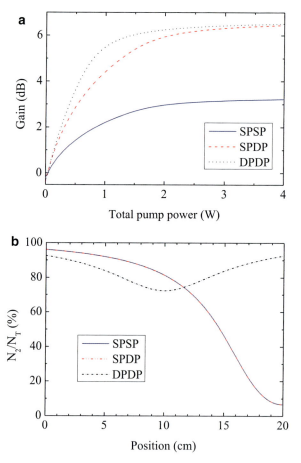

Fig. 52 (**a**) Performance of different schemes of the Cr^{4+}:YAG crystal fiber amplifier. The input signal power is −30 dBm and the fiber length is 10 cm. (**b**) Excited state population of different schemes of Cr^{4+}:YAG crystal fiber amplifiers. The pump power is 1 watt and the CF length is 20 cm. For both (**a**) and (**b**), the CF core diameter was 17 μm. The SPSP and the SPDP results are almost indistinguishable

DPDP scheme can offer higher gain than the SPDP scheme due to a more uniform population inversion over the entire CF. The excited state population under 1-watt pump is shown in Fig. 52b. For the single-pump schemes, the excited state population decays to <10% of the total concentration at the tail of the CF, because the pump is almost fully absorbed. The double-pump scheme has a more uniform population distribution and the lowest excited-state population is 72% of the total concentration. Assuming complete inversion, the small-signal gain constant would be $N_T \left(\sigma_e - \sigma_{esa}^s \right)$, which is about 0.4 dB/cm. σ_{esa}^s is the excited-state absorption cross section of signal. Since the 0.01-dB/cm propagation loss is much lower than the small-signal gain, the optical gain can be enhanced by increasing the fiber length and the pump power without introducing too much loss. Increasing the dopant concentration can effectively reduce the required length for achieving the same optical gain. The Cr^{4+} concentration in the Cr^{4+}:YAG CF drops after each LHPG growth. The small-signal pump absorption coefficient at 1064 nm, α_{p0}, was 4.5 cm^{-1} for the source crystal and 0.75 cm^{-1} for the DCF, which means the Cr^{4+} concentration drops to 1/6 of its original value. The Cr^{4+} concentration can be increased by depositing a layer of Cr_2O_3 together with a layer CaO or MgO onto the source rod before the LHPG process. The optical gain of CFs with absorption coefficients of 0.75 cm^{-1} and 4.5 cm^{-1} under a total pump power of 10 W and 1 W is plotted in Fig. 53a. The maximum gain is 17.1 dB at 60 cm for CFs with $\alpha_{p0} = 0.75$ cm^{-1} and 17.7 dB at 10 cm for CFs with $\alpha_{p0} = 4.5$ cm^{-1}. Enhancing the concentration by 6 times results in a 6-fold reduction of the optimized length. The constant decay after the optimized length is due to the 0.01-dB/cm signal propagation loss.

Since enhancing the Cr^{4+} concentration can greatly reduce the required CF length, the use of high-concentration CF with an α_{p0} of 4.5 cm^{-1} with the DPDP scheme is preferred. The simulated optical gain of a 10-cm long CF versus the pump power is shown in Fig. 53b. The small-signal gains are 22.7 dB and 19.4 dB under the pump powers of 20 W and 10 W, respectively. For the input signal power ranging from -30 dBm to -10 dBm, the gain curves are almost identical. The optical gain versus the input signal power is plotted in Fig. 53c. The input signal saturation powers at which the gains drop by 3 dB are 12.6 dBm, 9.2 dBm, and 8.6 dBm at the pump powers of 5 W, 10 W, and 20 W, respectively. These signal saturation powers are much higher than the rare-earth doped fiber amplifiers, because the fluorescence lifetime of Cr^{4+}:YAG is very short compared to those materials. However, the required pump power for saturating the Cr^{4+}:YAG CF is also much higher as a trade-off. The small-signal gain spectra of a 10-cm-long CF with different pump powers are shown in Fig. 53d. It was assumed that the emission/absorption cross sections and the propagation losses are independent of the wavelength. The 3-dB bandwidths are 175, 152, and 139 nm under the total pump powers of 5 W, 10 W, and 20 W, respectively. This amplifier can provide >10-dB gain from 1281 nm to 1622 nm under a 20-W pumping power. This spectral range matches very well to the O-E-S-C-L band (1260 to 1625 nm). The simulation shows that the Cr^{4+}:YAG CF-based fiber amplifier has potential for the future ultrabroadband optical communication systems.

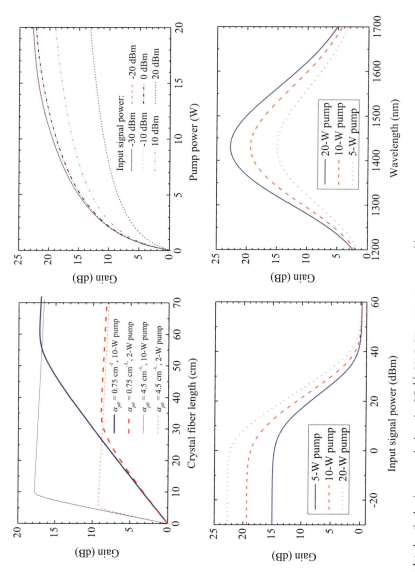

Fig. 53 (**a**) Simulated optical gain versus the length of Cr^{4+}:YAG CFs with different Cr^{4+} concentrations under 10-W and 2-W pumping in the DPDP scheme. (**b**) Simulated optical gain versus total pump power under different input signal power. The length of the CF is 20 cm. (**c**) Gain versus the input signal level for a 10-cm-long CF. (**d**) Small-signal gain spectra of a 10-cm long Cr^{4+}:YAG CF under different pump powers

Conclusion

Forth years have been passed since the first active CF was grown. Due to the superior optical, mechanical and thermal properties, CF-based light sources have shown advantages in areas where high power and high brightness are needed. Compared to silica fibers, the development of CFs is still very primitive. There are a long way to go before they can be widely deployed. In this chapter, the growth of the crystalline core, the glass cladding methods, as well as some active CFs light sources were introduced.

In terms of crystal waveguide, adiabatic propagation was achieved. The waveguide propagation loss in sapphire and YAG can be as low as 0.02 dB/cm and 0.01 dB/cm, respectively. The residual strain can be reduced with proper selection of the cladding glass and postgrowth thermal processes. To reduce the number of transverse modes, it is more viable to reduce the core/clad index difference than reducing the core diameter. A 2D simulation on the fabrication of the micro-floating zone by the LHPG method was successfully modified from that for growing bulk-crystal using the floating-zone method. According to the nonorthogonal body-fitting grid system and the control-volume finite difference method, accurate results of the molten-zone interfaces can be achieved with reasonable computation power. The stable growth condition can be determined numerically, and it matched well with the experiments. Due to the smaller melt volume formed using the LHPG method, natural convection drops several orders of magnitude and conduction rather than convection determines the temperature distribution. The effects caused by thermocapillary and mass convection in the melt and the trade-off between mass transfer and cross-section at the growth front for heat dissipation are discussed. The free surface is estimated and optimized for efficient laser absorption according to curvature radius and the inflection point. There is no significant change if the gravity is varied along the growing direction, when especially the laser power is low. Thermocapillary convection rather than mass convection becomes dominant for mass flow in the melt. The symmetry and mass flow of double eddy pattern are influenced by the molten-zone shape due to diameter reduction. Based on this work, the dopant concentration profile near the growth front can be investigated further.

Several diode-laser pumped CFs generated broad and bright emissions from visible to near infrared wavelength range. High-brightness Ce^{3+}:YAG CFs light source was successfully demonstrated. The Pyrex-cladded CF has a 20-mm length and a 25-μm core diameter. The single-crystalline core was annealed at 1000 °C for 4 hours before the cladding process. The single-pass optical conversion efficiency under low pump power was 1.95%, and the quantum yield was estimated to be 95%. Adding up the backward spontaneous emission, the fluorescence power was enhanced by 40% with attaching a broadband mirror on the opposite end of the CF to reflect both residual pump and the broadband fluorescence. The highest output power was 19.9 mW under 1.4-W pumping with only passive cooling. The output radiance was 30.3 $W \cdot mm^{-2} \cdot sr^{-1}$ in the axial direction. Using $Ti:Al_2O_3$ CFs as the gain medium, up to 42 mW of amplified spontaneous emission was generated with a core diameter of 18 μm. A radiance of 77.59 $W \cdot mm^{-2} \cdot sr^{-1}$ was achieved at a total pump power of 2.1 W.

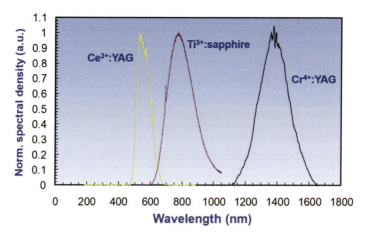

Fig. 54 High brightness light sources generated from Ce^{3+}:YAG, Ti^{3+}:sapphire, and Cr^{4+}:YAG CFs

Efficient operation of the external-cavity Cr^{4+}:YAG DCF laser was demonstrated with a 17.3% slope efficiency and a 56.6-mW threshold pump power. A numerical simulation with the distributed model was carried out to fit the external-cavity laser results with different output couplers. Tunable Cr^{4+}:YAG DCF lasers were demonstrated by using the diffraction grating, the pellicle beam splitter, or the birefringent filter as the wavelength tuning element. The broadest tuning range was 156-nm with the birefringent filter, and the threshold pump power was 70 mW. Wavelength hopping due to multimode interference was observed during the wavelength tuning. The tuning characteristics can be improved by employing a few-mode or single-mode CF. The performance of the Cr^{4+}:YAG crystal fiber amplifier was simulated. A 10-cm long crystal fiber amplifier with enhanced doping concentration can provide >10-dB small-signal gain from 1281 to 1622 nm under a 20-W pumping power. The peak small-signal gain is 22.7 dB at the wavelength of 1431 nm.

In summary, Fig. 54 depicts the wide coverage of the high brightness emissions from Ce^{3+}:YAG, Ti^{3+}:sapphire, and Cr^{4+}:YAG CFs. Even though the development of CFs is far from mature, there does have areas they could start to show the superior performance. Real-time OCTs with cellular-resolution is an example for in vivo biomedical applications in clinics.

References

I.D. Abella, C.H. Townes, Mode characteristics and coherence in optical ruby masers. Nature **192**, 957 (1961)

R.L. Aggarwal, A. Sanchez, M.M. Stuppi, R.E. Fahey, A.J. Strauss, W.R. Rapoport, C.P. Khattak, Residual infrared absorption in as-grown and annealed crystals of Ti:Al$_2$O$_3$. IEEE J. Quantum Electron. **24**, 1003 (1988)

A.A. Anderson, R.W. Eason, L.M.B. Hickey, M. Jelinek, C. Grivas, D.S. Gill, N.A. Vainos, Ti:sapphire planar waveguide laser grown by pulsed laser deposition. Opt. Lett. **22**, 1556 (1997)

V. Apostolopoulos, L. Laversenne, T. Colomb, C. Depeursinge, R.P. Salathé, M. Pollnau, Femtosecond-irradiation-induced refractive-index changes and channel waveguiding in bulk Ti^{3+}:sapphire. Appl. Phys. Lett. **85**, 1122 (2004)

R. Autrata, P. Schauer, J. Kvapil, J. Kvapil, A single crystal of YAG-new fast scintillator in SEM. J. Phys. **E11**, 707 (1978)

V. Bachmann, C. Ronda, A. Meijerink, Temperature quenching of yellow Ce^{3+} luminescence in YAG:Ce. Chem. Mater. **21**, 2077 (2009)

G. Blasse, A. Bril, A new phosphor for flying-spot cathode-ray tubes for color television: yellow-emitting $Y_3Al_5O_{12}$–Ce^{3+}. Appl. Phys. Lett. **11**, 53 (1967)

N.I. Borodin, V.A. Zhitnyuk, A.G. Okhrimchuk, A.V. Shestakov, Izv. Akad. Nauk SSSR, Ser. Fiz. **54**, 1500 (1990.) (Eng)

C.A. Burrus, L.A. Coldren, Growth of single-crystal sapphire-clad ruby fibers. Appl. Phys. Lett. **31**, 383 (1977)

C.A. Burrus, J. Stone, Single-crystal fiber optical devices: a Nd:YAG fiber laser. Appl. Phys. Lett. **26**, 318 (1975)

C.A. Burrus, J. Stone, A.G. Dentai, Room-temperature 1.3 μm CW operation of a glass-clad Nd:YAG single-crystal fiber laser end pumped with a single LED. Electron. Lett. **12**, 600 (1976)

B. Chalmers, H.E. Labelle Jr., A.I. Mlavsky, Edge-defined, film-fed crystal growth. J. Cryst. Growth **13–14**, 84 (1972)

C.L. Chang, S.L. Huang, C.Y. Lo, K.Y. Huang, C.W. Lan, W.H. Cheng, P.Y. Chen, Simulation and experiment on laser-heated pedestal growth of chromium-doped yttrium-aluminum-garnet single-crystal fiber. J. Cryst. Growth **318**, 674 (2001)

P.Y. Chen, C.L. Chang, K.Y. Huang, C.W. Lan, W.H. Cheng, S.L. Huang, Experiment and simulation on interface shapes of an yttrium aluminium garnet miniature molten zone formed using the laser-heated pedestal growth method for single-crystal fibers. J. Appl. Crystallogr. **42**, 553 (2009)

Y. Chi, H. Yang, S. Liu, M. Li, L. Wang, G. Zou, Compression ratio and red shift of the R_1 line for YAG:Cr. High Pressure Res. **3**, 153 (1990)

A. Crunteanu, M. Pollnau, G. Jänchen, C. Hibert, P. Hoffmann, R.P. Salathé, R.W. Eason, C. Grivas, D.P. Shepherd, Ti:sapphire rib channel waveguide fabricated by reactive ion etching of a planar waveguide. Appl. Phys. B Lasers Opt. **75**, 15 (2002)

M.J.F. Digonnet, C.J. Gaeta, D. O'Meara, H.J. Shaw, Clad Nd:YAG fibers for laser applications. IEEE J. Lightwave Technol. **LT-5**, 642 (1987)

P. Dorenbos, The 5d level positions of the trivalent lanthanides in inorganic compounds. J. Lumin. **91**, 155 (2000)

J.L. Duranceau, R.A. Brown, Thermal-capillary analysis of small-scale floating zone: steady-state calculations. J. Cryst. Growth **75**, 367 (1986)

H. Eilers, W.M. Dennis, W.M. Yen, S. Kuck, K. Peterman, G. Huber, W. Jia, Performance of a Cr:YAG laser. IEEE J. Quantum Electron. **29**, 2508 (1993)

R.S. Feigelson, Pulling optical fibers. J. Cryst. Growth **79**, 669 (1986)

M.M. Fejer, J.L. Nightingale, G.A. Magel, R.L. Byer, Laser-heated miniature pedestal growth apparatus for single-crystal optical fibers. Rev. Sci. Instrum. **55**, 1791 (1984)

C. Grivas, T.C. May-Smith, D.P. Shepherd, R.W. Eason, M. Pollnau, M. Jelinek, Broadband single-transverse-mode fluorescence sources based on ribs fabricated in pulsed laser deposited Ti:sapphire waveguides. Appl. Phys. A Mater. Sci. Process. **79**, 1195 (2004)

C. Grivas, D.P. Shepherd, T.C. May-Smith, R.W. Eason, Single-transverse-mode Ti:sapphire rib waveguide laser. Opt. Express **13**, 210 (2005)

C. Grivas, D.P. Shepherd, R.W. Eason, L. Laversenne, P. Moretti, C.N. Borca, M. Pollnau, Room-temperature continuous-wave operation of Ti:sapphire buried channel-waveguide lasers fabricated via proton implantation. Opt. Lett. **31**, 3450 (2006)

C. Grivas, C. Corbari, G. Brambilla, P.G. Lagoudakis, Tunable, continuous-wave Ti:sapphire channel waveguide lasers written by femtosecond and picosecond laser pulses. Opt. Lett. **37**, 46302 (2012)

D. Hamilton, S. Gayen, G. Pogatshnik, R. Ghen, Optical-absorption and photoionization measurements from the excited states of $Ce^{3+}:Y_3Al_5O_{12}$. Phys. Rev. B **39**, 8807 (1989)

L.M.B. Hickey, V. Apostolopoulos, R.W. Eason, J.S. Wilkinson, Diffused Ti:sapphire channel-waveguide lasers. J. Opt. Soc. Am. B **21**, 1452 (2004)

K. Y. Hsu, Glass-clad crystal fibers based broadband light sources, Ph.D. dissertation, 2011

K.Y. Hsu, D.Y. Jheng, Y.H. Liao, T.S. Ho, C.C. Lai, S.L. Huang, Diode- laser-pumped glass-clad Ti:sapphire crystal fiber based broadband light source. IEEE Photon. Technol. Lett. **24**(10), 854 (2012)

K.Y. Hsu, M.H. Yang, D.Y. Jheng, C.C. Lai, S.L. Huang, K. Mennemann, V. Dietrich, Cladding YAG crystal fibers with high-index glasses for reducing the number of guided modes. Opt. Mater. Express **3**, 813 (2013)

D. Huang, E.A. Swanson, C.P. Lin, J.S. Schuman, W.G. Stinson, W. Chang, M.R. Hee, T. Flotte, K. Gregory, C.A. Puliafito, J.G. Fujimoto, Optical coherence tomography. Science **254**, 1178 (1991)

K.Y. Huang, K.Y. Hsu, D.Y. Jheng, W.J. Zhuo, P.Y. Chen, P.S. Yeh, S.L. Huang, Low-loss propagation in Cr^{4+}:YAG double-clad crystal fiber fabricated by sapphire tube assisted CDLHPG technique. Opt. Express **16**, 12264 (2008a)

K.Y. Huang, K.Y. Hsu, S.L. Huang, Analysis of ultra-broadband amplified spontaneous emissions generated by Cr^{4+}:YAG single and glass-clad crystal fibers. IEEE/OSA J. Lightwave Technol. **26**, 1632 (2008b)

T.P. Hughes, K.M. Young, Mode sequences in ruby laser emission. Nature **196**, 332 (1962)

R.R. Jacobs, W.F. Krupke, M.J. Weber, Measurement of excited-state-absorption loss for Ce^{3+} in $Y_3Al_5O_{12}$ and implications for tunable 5d→4f rare-earth lasers. Appl. Phys. Lett. **33**, 410 (1978)

M. R. Kokta, Process for enhancing $Ti:Al_2O_3$ tunable laser crystal fluorescence by annealing, US Patent No. 4,587,035, 1986

M. R. Kokta, Process for enhancing fluorescence of $Ti:Al_2O_3$ tunable laser crystals, US Patent No. 4,836,953, 1989

S. Kück, Laser-related spectroscopy of ion-doped crystals for tunable solid-state lasers. Appl. Phys. B Lasers Opt. **72**, 515 (2001)

H.E. LaBelle Jr., A.I. Mlavsky, Growth of Sapphire filaments from the melt. Nature **216**, 574 (1967)

H.E. LaBelle Jr., A.I. Mlavsky, Growth of controlled profile crystals from the melt: part I – Sapphire filaments. Mater. Res. Bull. **6**, 571 (1971)

C.C. Lai, H.J. Tsai, K.Y. Huang, K.Y. Hsu, Z.W. Lin, K.D. Ji, W.J. Zhuo, S.L. Huang, Cr^{4+}:YAG double-clad crystal fiber laser. Opt. Lett. **33**, 2919 (2008)

C.C. Lai, Y.S. Lin, K.Y. Huang, S.L. Huang, Study on the core/cladding interface in Cr:YAG double-clad crystal fibers grown by the co-drawing laser heated pedestal growth method. J. Appl. Phys. **108**, 054308 (2010)

C.W. Lan, S. Kou, Thermocapllary flow and melt/solid interfaces in floatinf-zone crystal growth under microgravity. J. Cryst. Growth **102**, 1043 (1990)

C.W. Lan, C.Y. Tu, Three-dimensional simulation of facet formation and the coupled heat flow and segregation in Bridgman growth of oxide crystals. J. Cryst. Growth **233**, 523 (2001)

L. Laversenne, P. Hoffmann, M. Pollnau, P. Moretti, J. Mugnier, Designable buried waveguides in sapphire by proton implantation. Appl. Phys. Lett. **85**, 5167 (2004)

Y.S. Lin, C.C. Lai, K.Y. Huang, J.C. Chen, C.Y. Lo, S.L. Huang, T.Y. Chang, J.Y. Ji, P. Shen, Nanostructure formation of double-clad Cr^{4+}:YAG crystal fiber grown by co-drawing laser-heated pedestal. J. Cryst. Growth **289**, 515 (2006)

Y.S. Lin, T.C. Cheng, C.C. Tsai, K.Y. Hsu, D.Y. Jheng, C.Y. Lo, P.S. Yeh, S.L. Huang, High-luminance white-light point source using Ce,Sm:YAG double-clad crystal fiber. IEEE Photon. Technol. Lett. **22**, 1494 (2010)

C. Y. Lo, Growth, characterization, and applications of doped-YAG single-crystal fibers, Ph.D. dissertation, 1994

C.Y. Lo, K.Y. Huang, J.C. Chen, S.Y. Tu, S.L. Huang, Glass-clad Cr^{4+}:YAG crystal fiber for the generation of superwideband amplified spontaneous emission. Opt. Lett. **29**, 439 (2004)

C.Y. Lo, K.Y. Huang, J.C. Chen, C.Y. Chuang, C.C. Lai, S.L. Huang, Y.S. Lin, P.S. Yeh, Double-clad Cr^{4+}:YAG crystal fiber amplifier. Opt. Lett. **30**, 129 (2005)

J.D. Love, W.M. Henry, W.J. Stewart, R.J. Black, S. Lacroix, F. Gonthier, Tapered single-mode fibres and devices. IEE Proc. J. Optoelecton. **138**, 343 (1991)

D. Marcuse, *Theory of dielectric optical waveguides* (Academic Press, New York, 1991)

P.F. Moulton, Spectroscopic and laser characteristics of $Ti:Al_2O_3$. J. Opt. Soc. Am. B: Opt. Phys. **3**, 125 (1986)

N. Ohnish, T. Yao, A novel growth technique for single-crystal fibers: the micro-Czochralski (μ-CZ) method. Jap. J. Appl. Phys. **28**, L278 (1989)

E.P. Ostby, L. Yang, K.J. Vahala, Ultralow-threshold Yb^{3+}:SiO_2 glass laser fabricated by the solgel process. Opt. Lett. **32**, 2650 (2007)

D.P.S. Saini, Y. Shimoji, R.S.F. Chang, N. Djeu, Cladding of a crystal fiber by high-energy ion implantation. Opt. Lett. **16**, 1074 (1991)

A. Sennaroglu, C.R. Pollock, H. Nathel, Continuous-wave self-mode-locked operation of a femtosecond Cr^{4+}:YAG laser. Opt. Lett. **19**, 390 (1994)

Y.R. Shen, U. Hömmerich, K.L. Bray, Observation of the ^1E state of Cr^{4+} in yttrium aluminum garnet. Phys. Rev. B Condens. Matter **56**, R473 (1997)

I.T. Sorokina, S. Naumov, E. Sorokin, E. Wintner, A.V. Shestakov, Directly diode-pumped tunable continuous-wave room-temperature Cr^{4+}:YAG laser. Opt. Lett. **24**, 1578 (1999)

J.L. Stevenson, R.B. Dyott, Optical fiber waveguide with a single-crystal core. Electron. Lett. **10**, 449 (1974)

S. Sudo, A. Cordova-Plaza, R.L. Byer, H.J. Shaw, $MgO:LiNbO_3$ single-crystal fiber with magnesium-ion in-diffused cladding. Opt. Lett. **12**, 938 (1987)

J.C. Walling, O.G. Jenssen, H.P. Jenssen, R.C. Mirris, E.W. O'Dell, Tunable alexandrite lasers. IEEE J. Quantum Electron. **QE-16**, 1702 (1980)

S.C. Wang, T.I. Yang, D.Y. Jheng, C.Y. Hsu, T.T. Yang, T.S. Ho, S.L. Huang, Broadband and high-brightness light source: glass-clad Ti:sapphire crystal fiber. Opt. Lett. **40**, 5594 (2015)

L. Wu, A. Wang, J. Wu, L. Wei, G. Zhu, S. Ying, Growth and laser properties of Ti:sapphire single crystal fibres. Electron. Lett. **31**, 1151 (1995)

D.H. Yoon, I. Yonenaga, T. Fukuda, N. Ohnishi, Crystal growth of dislocation-free $LiNbO_3$ single crystals by micro pulling down method. J. Cryst. Growth **142**, 339 (1994)

Cladding-Pumped Multicore Fiber Amplifier for Space Division Multiplexing

21

Kazi S. Abedin

Contents

Introduction	822
Multicore Fiber Amplifier	823
Multicore Erbium-Doped Fiber	824
Signal/Pump Coupler for MCFA	826
Pump Dump	829
Cross Talk Among the Spatial Channels	829
Numerical Simulation	830
Gain and NF of Cladding-Pumped MC-EDFA	833
Effect of Enlarging the Core Size	834
Cross-Gain Modulation Due to Gain Depletion	835
Power Conversion Efficiency	837
Experimental Demonstration of Cladding-Pumped Multicore Fiber Amplifiers	838
Cladding-Pumped MC-EDFA with End Pumping	839
Cladding-Pumped MC-EDFA Employing Side-Coupled Pumping	839
Experimental Results	840
End-Coupled Pumping	840
Side-Coupled Pumping	840
Comparison Between Core- and Cladding-Pumped Amplifiers	842
Electrical Power Consumption	843
Recent Advancements	844
Summary	846
References	846

Abstract

Space division multiplexing (SDM), a way of transmitting data over multiple cores embedded in single strand of fiber or multiple modes of a few-mode fiber, has generated much interest as a potential means for enhancing the

---

K. S. Abedin (✉)
OFS Laboratories, Somerset, NJ, USA
e-mail: kabedin@ofsoptics.com

© Springer Nature Singapore Pte Ltd. 2019
G.-D. Peng (ed.), *Handbook of Optical Fibers*,
https://doi.org/10.1007/978-981-10-7087-7_50

capacity of optical transmission systems. During the past several years, in addition to developing high-performance specialty optical fibers for SDM, worldwide research efforts have made possible realization of SDM signal amplifiers with remarkable optical performances. Multicore fiber amplifiers with different number of cores, both single and few moded, and pumping schemes have been developed to pump the cores individually or through a common cladding. This chapter describes state-of-the-art cladding-pumped multicore fiber amplifiers developed for amplifying space division multiplexed signals. Different structures and underling key components, design procedures, experimental demonstration, and evaluation of optical properties of these amplifiers will be described.

Introduction

The ever-increasing need for bandwidth has driven us to explore novel means for data multiplexing. Space division multiplexing (SDM) is a method of multiplexing data whereby optical signals are transmitted using multiple cores or multiple modes in a fiber and has been regarded as a potential means of increasing transmission capacity beyond that is achievable using conventional multiplexing technologies such as wavelength division multiplexing, polarization division multiplexing, and multilevel modulation formats (Morioka 2009; Chraplyvy 2009; Essiambre and Mecozzi 2012; Richardson et al. 2013; Morioka et al. 2012).

Various multicore fibers exhibiting low loss, low cross talk, and with different core arrangements have been developed for use in SDM systems. Attenuation and cross talks among cores in multicore fibers have been reduced to as low as 0.18 dB/km (for seven-core fiber), −55 dB (over a length of 17.6 km span), respectively (Hayashi et al. 2011). Besides enhancing the transmission capacity, the stimulus behind pursuing SDM systems involves reducing size through integration, reducing power consumption, and minimizing the cost of deployment and operation. In order to further increase the span of transmission link, suitable means to amplify all the channels simultaneously becomes essential (Richardson 2013; Krummrich 2011). To this end, different forms of rare-earth-doped SDM amplifiers, bundled (Yamada et al. 2012), multielement (Jain et al. 2013), and multicore (Abedin et al. 2011, 2012, 2014a, b; Tsuchida et al. 2012a, b; Sakaguchi et al. 2014; Mimura et al. 2012; Tsuchida et al. 2014; Ono et al. 2013; Jin et al. 2015a), have already been proposed and experimentally demonstrated. Figure 1 shows a schematic diagram of a point-to-point SDM transmission system based on multicore fiber. At the transmission side, the various channels are multiplexed in the spatial domain using a suitable SDM multiplexer (S-MUX), and the SDM signal is transmitted using multicore fiber to the receiving end. The transmission link may incorporate a number of SDM amplifiers to overcome the transmission loss following each span. At the receiving end, the different channels are extracted using SDM-demultiplexer (S-DEMUX).

Fig. 1 Schematic of a multicore SDM transmission link

In addition to high gain, broad bandwidth, and low noise figure (NF), properties that are common in conventional single-mode erbium-doped fiber amplifiers (EDFAs) used in telecommunications systems, multicore amplifiers call for several other critical attributes, which include low inter-core cross talk, gain flatness among spatial channels, low cross-gain modulation between cores sharing common pump, minimum design complexity, low cost, and low electrical power consumption. Multicore erbium-doped fiber amplifiers, where individual cores are pumped separately using single-mode pump diodes (Abedin et al. 2011; Tsuchida et al. 2012a, b; Sakaguchi et al. 2014) or using a shared single-mode pump (Krummrich 2011) and cores simultaneously pumped through the cladding using a single multimode pump laser diode (Abedin et al. 2012, 2014a; Mimura et al. 2012; Tsuchida et al. 2014; Ono et al. 2013; Jin et al. 2015a), have been demonstrated.

Cladding-pumped SDM amplifier where all the cores are pumped simultaneously is advantageous as it requires fewer optical components and has the potential to use a low-cost, energy-efficient multimode laser diode. This chapter details the recent advancements in the development of cladding-pumped multicore erbium-doped fiber amplifiers (MC-EDFA). Various gain fibers and passive key components required for constructing these amplifiers are discussed. Procedures to optimize performance of cladding-pumped MC-EDFAs using numerical simulation will be explained. Finally, construction of cladding-pumped MC-EDFA employing end and side pumping and their evaluation of optical performances are shown.

Multicore Fiber Amplifier

Multicore fiber amplifiers developed for SDM systems can be configured in two different ways. In one form, the SDM amplifier is provided with passive multicore fibers at the input and output, which can be directly connected to multicore transmission fibers. Alternatively, multicore fiber amplifier can be built with multiple single-mode fibers at the input and output, which allow amplification of multiple signal channels carried by single-mode fibers. It is desirable that spatially multiplexed signal from passive multicore transmission fibers be directly launched into the multicore amplifier as shown in Fig. 2a. Currently, however, in the absence of suitable passive optical components, e.g., isolator, gain flattening filter, and tap coupler with proper multicore fiber pigtails, required for building fully integrated SDM system, it is necessary to provide access to individual cores using suitable

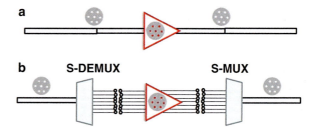

Fig. 2 Structures of multicore amplifiers with input and output configured with (**a**) multicore fiber, (**b**) single-mode fibers

fan-in/fan-out devices. To utilize MC amplifier with such multiple single-mode fiber input in SDM systems, optical signal arriving from multicore transmission fiber is first demultiplexed into individual channels using a fan-out device, and each channel is then connected to individual single-core fiber input of multicore fiber amplifier. After amplification, the different signal channels are then multiplexed using fan-in device for launching into multicore transmission fiber. A schematic of a multicore fiber amplifier connected to MC transmission fiber using fan-out and fan-in devices as S-DEMUX and (S-MUX), respectively, is shown in Fig. 2b.

Multicore Erbium-Doped Fiber

The key element of a MC amplifier is the gain fiber that has multiple cores embedded within a common cladding. The cores are essentially doped with rare-earth element, erbium, in order to achieve gain in the telecom band around 1.55 μm. The cores are sufficiently spaced so that electric field of signal within one core interferes minimally with the field of the neighboring cores. Such cores are called "uncoupled" cores. In order to initiate population inversion and thus gain, the cores are pumped at suitable wavelengths, typically at 980 nm or 1480 nm, where erbium exhibits high absorption. Multicore amplifiers can be pumped mainly in two ways; one is "core pumping," where the cores are pumped separately by launching single-mode pump radiation into each cores, and the second approach is called "cladding pumping," where all cores are being pumped simultaneously using multimode pump light launched into the cladding. The cores are often co-doped with ytterbium in order to increase pump absorption, which is often desirable in cladding-pumped amplifiers for efficient utilization of pump.

To ensure desired optical performance, it is important that the cores have appropriate rare-earth doping concentration, size and numerical aperture (NA), core-to-core spacing, and suitable arrangement within a cladding. Erbium-doped fibers developed for use in conventional single-mode amplifiers have a core diameter and a numerical aperture such that single-mode propagation is ensured for both the signal and the pump waves. For example, the commercial telecom EDF optimized for 980 nm core pumping, MP980 (OFS), has a core diameter of 3.2 μm and a NA of 0.23, which gives V-number of 1.49 and 2.36, for 1550 nm signal and 980 nm pump, respectively. However, in a cladding pump amplifier, the core needs to be single moded only for pump, since the multimoded pump is guided through the cladding. Thus, for the same NA, the core diameter can be increased to ~5.2 μm while still ensuring single-mode operation at 1550 nm. Further enhancement of the core diameter would be possible by lowering the NA of the core. By lowering the NA of the core to 0.13, diameter of the core can be enlarged to 9.1 μm while ensuring single-mode propagation at the signal wavelength of 1550 nm. Such enlargement of the core size is beneficial as it could enhance multimode pump absorption (proportional to the ratio of cross-sectional areas of core to cladding) and increase the amplifier saturation power.

In order to increase absorption of pump guided through the cladding, it is also desirable to minimize the cladding size, possible through choosing a close-packed lattice for the cores and reducing the core-to-core separation. The erbium-doped MC-EDF used in the first demonstration of a cladding-pumped amplifier had seven cores, hexagonally arranged inside a cladding 100 μm in diameter with a core-to-core spacing of 41 μm (Abedin et al. 2012). A cross-sectional view of such seven-core EDF fiber is shown in Fig. 3a. The mode field diameter (MFD) at 1550 nm is ~6 μm. The erbium-doped core has an absorption coefficient of ~6.5 dB/m at 1530 nm. The cladding is coated with low-index polymer coating (~NA: 0.45) for guiding multimode pump light.

Figure 3b shows cross section of an Er/Yb co-doped 12-core fiber, where cores are arranged in the periphery of a hexagon with an average pitch of 37.2 μm (Ono et al. 2013). The cladding diameter is 216 μm, which is surrounded by low-index polymer coating for cladding pumping. Core count has been increased further by incorporating additional layer/ring of cores. Sakaguchi et al. (2014) reports on

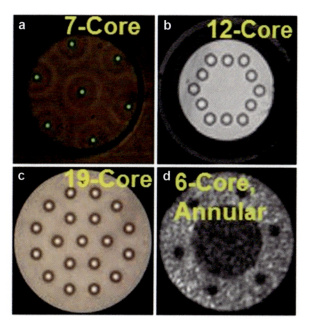

Fig. 3 Cross section of different MC-EDFs used for constructing SDM amplifiers. (**a**) A 7-core EDF for cladding pumping (Abedin et al. 2012), (**b**) 12-core EDF for cladding pumping (Ono et al. 2013), (**c**) 19-core EDF for core pumping (Sakaguchi et al. 2014), (**d**) 6-core EDF with annular cladding for cladding pumping (Jin et al. 2015a)

a multicore core-pumped EDFA that has 19 cores (MFD: 6.6 μm) arranged in three layers within a cladding size of 220 μm. Cross section of the 19-core EDF is shown in Fig. 3c. Multicore erbium-doped fiber with enlarged core supporting few modes has also been reported recently. As shown in Fig. 3d, the Er-doped fiber has six cores, each supporting three modes (LP01, LP11a, and LP11b) (Jin et al. 2015a). The cores are embedded within an annular cladding region, and the distance between two neighboring cores is 62 μm. The outer and inner diameters of the annular cladding are 170 and 85 μm, respectively. The NA between the core and annular cladding is 0.104, and the NA between the annular cladding and inner circular cladding is 0.11. The inner circular region having a refractive index lower than the cladding causes the multimode pump radiation to be confined within the annular region, which results in an increased pump intensity.

All the erbium-doped multicore fibers used so far for cladding pumping applications are made with cylindrical cladding surfaces, which is in contrast to that observed in single-core double-clad fiber. When a multimode pump is launched into the cladding region of fiber with cylindrical surface, there generate certain cladding modes that propagate like helical rays, which have no overlap with the central region of the fiber. In single-core gain fiber, those modes remain unabsorbed, and it could result in nonuniform pump distribution across the cladding area. Such modes with poor overlap are avoided by breaking the structural symmetry, such as by choosing off-centered core (Snitzer et al. 1988) or by choosing non-cylindrical cladding surface like D shape (Jeong et al. 2007) or star shape (Grubb and Welch 2000). Fortunately, in the case of multicore fiber, the off-centered cores close to the cladding surface help brake the cylindrical symmetry and prevent the formation of helical modes and promote better mixing of modes resulting in uniform pump distribution across the cladding (Abedin et al. 2012; Jin et al. 2015b).

Signal/Pump Coupler for MCFA

In constructing MC-EDFA, one critical task is devising a suitable means for launching SDM signal and pump into the gain fiber. Various fan-in, fan-out devices have been proposed to couple optical signal, which includes all-fiber-based tapered fiber bundled (TFB) coupler (Abedin et al. 2011, 2012; Zhu et al. 2010), bulk-optic couplers (Sakaguchi et al. 2011), reduced-cladding bundled coupler (Tsuchida et al. 2012b), as well as 3-D waveguide-based coupler (http://www.optoscribe.com/products/3d-optofan-series/).

For cladding pumping, in addition to signal being coupled into the different cores, the multimode pump needs to be launched into the cladding. Different types of couplers/combiners have been developed to couple signal and pump simultaneously, such as bulk-optic coupler based on multicore collimators and miniature dichroic filter (Tsuchida et al. 2014), modified TFB pump/signal combiner (Abedin et al. 2012), and also side-coupled pump coupler (Abedin et al. 2014a). Schematics of these pump signal combiners for multicore erbium-doped fibers are shown in Fig. 4.

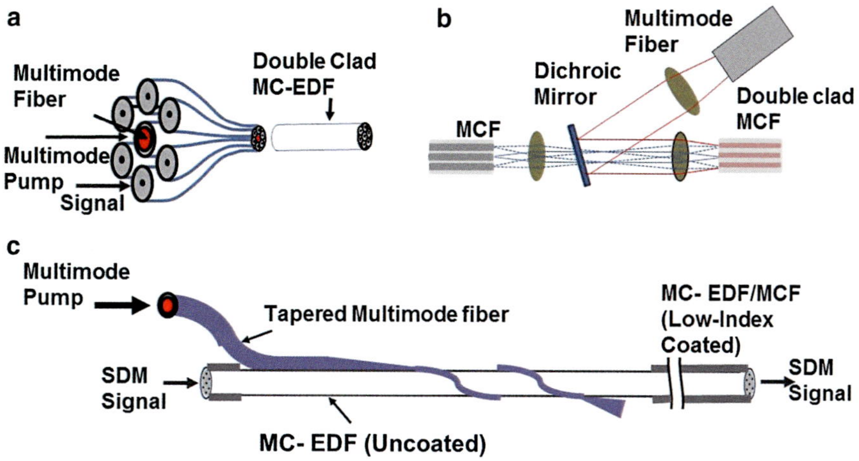

Fig. 4 Pump/signal coupler for use in cladding-pumped MC-EDFAs. (**a**) Modified TFB pump/signal combiner (Abedin et al. 2012), (**b**) bulk-optic coupler (Ono et al. 2013), (**c**) also side-coupled pump coupler (Abedin et al. 2014a)

Figure 4a shows a schematic of modified TFB pump/signal coupler for cladding pumping MC-EDF, which are fabricated by tapering a bundle of specially designed single-mode fibers by a predetermined ratio so that the core-to-core pitch at the tapered end matches with that of the MC-EDF (Zhu et al. 2010). The TFB consists of a central multimode fiber surrounded by six single-mode fibers. The refractive index profiles of the single-mode fibers are specially designed so that the mode field diameter changes adiabatically in the tapered region and at the output matches well with that of the multicore gain fiber when tapered to the required taper ratio (DiGiovanni and Stentz 1999). Figure 5 shows a plot of core MFD at the tapered end of the bundle plotted as a function of tapered ratio (Abedin et al. 2011, 2014b). The input fibers (at the un-tapered end) of TFBs could be spliced to SMFs with minimal loss. The tapered end, with an appropriate taper ratio, could allow efficient splicing to MC-EDF. This particular TFB allowed launching signals into the outer six cores and multimode pump into the cladding through the central fiber.

Multimode light propagation through the central multimode fiber of the TFB is substantially different from what explained above for single-mode signal propagation. When a bundle of multimode fiber is subjected to gradual taper, the multimode light propagating through the fiber undergoes a change in its numerical aperture. If NA_i and NA_o are the numerical apertures of light in the input multimode fiber and at the output of the tapered fiber bundle, and $\sum A_i$ is the summation of the cross-sectional areas of input fibers, and A_o is the area at the tapered end of the fiber, the brightness conservation theorem ensures that (DiGiovanni and Stentz 1999; Kosterin et al. 2004):

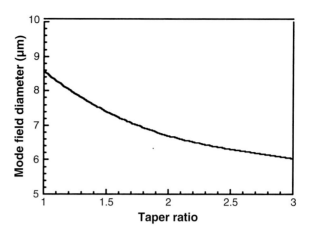

Fig. 5 MFD at the tapered end of the TFB plotted as a function of the taper ratio

$$\sum A_i \cdot (NA_i)^2 = A_o \cdot (NA_o)^2 \qquad (1a)$$

$$\left(\frac{NA_o}{NA_i}\right)^2 = \frac{\sum A_i}{A_o} \qquad (1b)$$

Inside the TFB coupler, as the multimode fiber gets tapered, the numerical aperture of the pump field increases. It is to be noted that during the tapering process, the fibers melt and deform in shape to fill in the interstice. Thus, when a bundle of seven fibers with the same diameter is tapered down to a diameter the same as the original fiber, the center fiber needs to be tapered by a factor of $\sqrt{7}$, i.e., 2.646. This will result in an increase in NA by a factor of 2.646.

Bulk-optic-based free-space couplers have also been reportedly used for cladding pumping application. Figure 4b shows a bulk-optic coupler that consists of multicore (12 cores) passive fiber, two spherical lenses and a dichroic mirror (transmitting signal and reflecting pump wave), and a double-clad multicore output fiber. The core-to-core pitch of MCF at the input and outputs is the same (~36.6 μm), and the end faces are AR coated to suppress back reflection. This device is capable of coupling signal from multicore passive fiber to multicore double-clad fiber with an insertion loss in the range of 0.4–1.8 dB, while the inter-core cross talk ranged between −39 and −59 dB (Ono et al. 2013).

One problem associated with using fan-in and fan-out devices at the amplifier ends is that the throughput loss results in attenuation of the signal and an increase in NF of the amplifier. This problem can be overcome by coupling the multimode pump through the side of the gain fiber, without interruption of the core waveguide. Side coupling with high efficiency has been demonstrated in single-core double-clad fibers in various forms, such as V-groove side-coupling combiners (Ripin and Goldberg 1995), embedded-mirror combiners (Koplow et al. 2003), and direct fusion of tapered multimode fiber (Gapontsev and Samartsev 1996; Theeg et al. 2012), and also in the form of distributed coupling using GT-wave fiber assembly

(Grudinin et al. 2004; Yla-Jarkko et al. 2003; Jain et al. 2014). Indeed recently, all-fiber combiner for coupling multimode pump light from the side of multicore fiber has been developed and utilized for constructing compact cladding-pumped multicore fiber amplifier.

Figure 4c shows such a coupler that launches multimode pump from the side of a multicore fiber. In this side-coupled combiner, a short (few cm long) section of the gain fiber near the input end is stripped of its low-index coating, and the exposed section is brought into optical-contact with a tapered multimode pump fiber (Abedin et al. 2014a). This pumping geometry make the two ends of the gain fiber available for splicing directly to passive multicore fibers. Note that although in this demonstration pump is coupled directly into the gain fiber, it is also possible to use passive double-clad multicore fiber for making a coupler, which can be spliced to the MC gain fiber. The pump coupling efficiency has been around 65%, suggesting scope for further improvement through optimization of tapering process. For example, side coupling with an efficiency as high as 95% has been demonstrated by the use of coreless fiber in the tapered region to guide pump (Theeg et al. 2012).

Pump Dump

In a cladding-pumped amplifier designed for broadband amplification, the unabsorbed fraction of the pump needs to be removed from the gain fiber to prevent it from entering the signal path. This is achieved by incorporating at the output end of the MC-EDF a "pump dump," where short segment, few cm long, is stripped of its low-index coating and is covered with thermally conductive silicone elastomer paste to remove the residual pump (Abedin et al. 2012). To avoid absorption of the amplified signal while it propagates through the short section of unpumped gain fiber that follows the pump dump, one could splice a short piece of MC passive fiber at the output of the gain fiber and create a pump dump in the passive fiber section. The fiber section comprising the pump dump is further affixed to a suitable metallic cooling plate or surface for efficient dissipation of heat.

Cross Talk Among the Spatial Channels

Due to close proximity of the cores in multicore amplifier, different spatial channels can interfere with each other, which results in cross talk among spatial channels. The cross talk could result from coherent coupling between the cores along the multicore gain fiber and also from fan-in/fan-out devices used at the fiber ends. To measure cross talk, light is launched into each of the cores at the input end, one at a time, and light outputting from each cores at the output is measured. Thus, attenuation between the corresponding cores represents the insertion loss. The difference (in dB) in attenuations between corresponding and different cores is considered as a measure of cross talk (Abedin et al. 2011). Figure 6 shows the loss and cross-talk

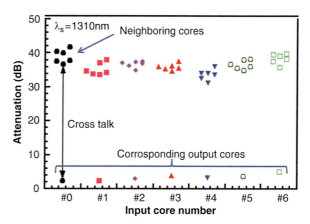

Fig. 6 Cross talk between the various cores at the input and output of the TFB-MCEDF-TFB module. #0, central core, #1–#6, outer corresponding cores. (Abedin et al. 2011)

properties measured for an amplifier module consisting of a seven-core EDF and two TFBs spliced at the ends (Abedin et al. 2011). The numbers in the horizontal axis represent the core in which light is launched, and the vertical axis shows the attenuation in the signal, measured at seven outputs of the second TFB. The loss between an input and corresponding output core of the gain assembly remains within 2.5–4.9 dB. The cross talk averaged over six cores varies between 30.2 and 36.6 dB for the seven channels.

Numerical Simulation

Numerical simulations can be performed to study the amplification and noise properties of a cladding-pumped amplifier and optimize the amplifier design. Numerical studies of multicore erbium-doped fiber amplifiers have been conducted for both single- (Abedin et al. 2012) and few-moded (Abedin et al. 2014b; Jin et al. 2015b) multicore amplifiers, and amplifications and noise properties are predicted fairly accurately by using such models. In the following, a generalized numerical model applicable to few-moded cladding-pumped MC-EDFA, followed by a simplified one for single-mode MC-EDFA, is presented.

Assuming uniform pump distribution across the cladding of a cladding-pump MC-EDFA, the evolution of signal propagating as $LP_{m,n}$ mode in the i-th core and corresponding ASE and the pump wave can be expressed as (Abedin et al. 2012; Bai et al. 2011; Kang et al. n.d.; Ip 2012):

$$\frac{dP_{s,i}}{dz}(z) = \int_0^a \int_0^{2\pi} \left(\sigma_{vs}^e N_{2,i} - \sigma_{vs}^a N_{1,i} \right) P_{s,i} I_{mn(vs)}^N r\, d\phi\, dr \qquad (2a)$$

$$\frac{dP^{\pm}_{ASE(vj),i}}{dz}(z) = \pm \int_0^a \int_0^{2\pi} \left(\sigma^e_{vj} N_{2,i} - \sigma^a_{vj} N_{1,i}\right) P^{\pm}_{ASE(vj),i} I^N_{mn(vj)} r d\phi dr \quad (2b)$$

$$\pm \int_0^a \int_0^{2\pi} \sigma^e_{vj} N_{2,i} (2hv\Delta v) I^N_{mn(vj)} r d\phi dr$$

$$\frac{dP_P}{dz}(z) = \sum_i \int_0^a \int_0^{2\pi} \left(\sigma^e_{vp} N_{2,i} - \sigma^a_{vp} N_{1,i}\right) \left(P_p/A_{cladding}\right) r d\phi dr \quad (2c)$$

where σ^e and σ^a are the emission and absorption cross sections. $A_{cladding}$ is the area of the cladding, and a is the radius of the doped core. $I_{mn(vs)}(r, \theta)$ and $I_{mn(vj)}(r, \theta)$ are the normalized intensity distribution of $LP_{m,n}$ mode of the signal with frequency v_s and ASE with frequency v_j and are identical for all cores. Thus:

$$\int_0^\infty \int_0^{2\pi} I^N_{mn} r d\phi dr = 1 \quad (3)$$

$N_{2,i}(r,\phi,z)$ and $N_{1,i}(r,\phi,z)$ are the upper and lower state populations in the i-th core (assuming a two-level system where pump state population is close to 0), respectively. The integration is performed over the doped region for each of the cores. The radial dependence of N_2 and N_1 can be expressed as:

$$N_{2,i}(r,\phi,z) = N_0 \cdot \frac{\frac{\sigma^a_{vs}}{hv_s} \cdot P_{s,i} I^N_{mn(vs)} + \sum_j \frac{\sigma^a_{vj}}{hv_j} \cdot P_{ASE(vj),i} I^N_{mn(vj)} + \frac{\sigma^a_{vp}}{hv_p} \cdot (P_p/A_{cladding})}{\frac{(\sigma^a_{vs} + \sigma^e_{vs})}{hv_s} \cdot P_{s,i} I^N_{mn(vs)} + \sum_j \frac{(\sigma^a_{vj} + \sigma^e_{vj})}{hv_j} \cdot P_{ASE,(vj),i} I^N_{mn(vj)} + \frac{(\sigma^a_{vp} + \sigma^e_{vp})}{hv_p} \cdot \frac{P_p}{A_{clad}} + \frac{1}{\tau}} \quad (4a)$$

$$N_{1,i}(r,\phi) = N_0(r) - N_{2,i}(r,\phi) \quad (4b)$$

Here, N_0 is the concentration of erbium ions. P_p is the pump power at a distance z from the input of doped fiber. $P_{ASE(vj),i}$ is the sum of the forward ($P^+_{ASE(vj),i}$) and backward ($P^-_{ASE(vj),i}$) propagating ASE component at frequency v_j integrated over a bandwidth of Δv_j. τ is the lifetime of the upper energy level.

The gain G_i and NF_i for the i-th core can be calculated from the following equations:

$$G_i = P_{s,i}(z=L)/P_{s,i}(z=0) \tag{5a}$$

$$NF_i(dB) = 10 \cdot \log\left(\frac{P^+_{ASE,i}}{h\nu\Delta\nu G_i} + \frac{1}{G_i}\right) \tag{5b}$$

where $P^+_{ASE,i}$ represents the forward propagating ASE in the i-th core at the amplifier output, for a bandwidth of $\Delta\nu$ calculated at a wavelength the same as the signal wavelength.

For multicore fiber with single-mode cores, normalized signal (and also ASE) distribution can be expressed by Gaussian field distribution, with mode field radius ω_0. In such case, the normalized intensity distributions $I^N_{(\nu s)}$ and $I^N_{(\nu j)}$ can be approximated by $\Gamma_{\nu s}/(\pi a^2)$ and $\Gamma_{\nu j}/(\pi a^2)$, respectively, where $\Gamma_{\nu s}$ and $\Gamma_{\nu j}$ are the effective overlap factors (EOF) between signal and ASE, respectively, with the doped core. EOFs are expressed as $\Gamma_{\nu s} = 1 - \exp\left(-2a^2/\omega^2_{o(\nu s)}\right)$ and $\Gamma_{\nu j} = 1 - \exp\left(-2a^2/\omega^2_{o(\nu j)}\right)$. Under these assumptions, the Eqs. 2a, 2b, and 2c can be transformed into:

$$\frac{dP_{s,i}}{dz}(z) = \left(\sigma^e_{\nu s} N_{2,i} - \sigma^a_{\nu s} N_{1,i}\right) P_{s,i} \Gamma_{\nu s} \tag{6a}$$

$$\frac{dP^\pm_{ASE(\nu j),i}}{dz}(z) = \pm\left(\sigma^e_{\nu j} N_{2,i} - \sigma^a_{\nu j} N_{1,i}\right) P^\pm_{ASE(\nu j),i}\Gamma_{\nu j} \pm \sigma^e_{\nu j} N_{2,i} (2h\nu\Delta\nu) \Gamma_{\nu j} \tag{6b}$$

$$\frac{dP_P}{dz}(z) = \sum_i \left(\sigma^e_{\nu p} N_{2,i} - \sigma^a_{\nu p} N_{1,i}\right)(A_{core}/A_{cladding}) P_p \tag{6c}$$

where

$$N_{2,i} = N_0 \cdot \frac{\frac{\sigma^a_{\nu s}}{h\nu_s} P_{s,i} \frac{\Gamma_{\nu s}}{\pi a^2} + \sum_j \frac{\sigma^a_{\nu j}}{h\nu_j} P_{ASE(\nu j),i} \frac{\Gamma_{\nu j}}{\pi a^2} + \frac{\sigma^a_{\nu p}}{h\nu_p}(P_p/A_{cladding})}{\frac{(\sigma^a_{\nu s}+\sigma^e_{\nu s})}{h\nu_s} \cdot P_{s,i} \frac{\Gamma_{\nu s}}{\pi a^2} + \sum_j \frac{(\sigma^a_{\nu j}+\sigma^e_{\nu j})}{h\nu_j} P_{ASE,(\nu j),i} \frac{\Gamma_{\nu j}}{\pi a^2} + \frac{(\sigma^a_{\nu p}+\sigma^e_{\nu p})}{h\nu_p} \cdot \frac{P_p}{A_{cladding}} + \frac{1}{\tau}} \tag{7}$$

The summation in Eq. 6c accounts for pump absorption in the multiple cores. When EDF is pumped near 980 nm, the emission cross section, $\sigma^e_{\nu p} = 0$. In regions close to the input end of the amplifier, where the intensity of signal and ASE are typically much lower than pump intensity, the upper state population becomes:

$$N_2 = N_0 \cdot \frac{\frac{\tau\sigma_{vp}^a}{hv_p}(P_p/A_{cladding})}{\frac{\tau(\sigma_{vp}^a+\sigma_{vp}^e)}{hv_p} \cdot \frac{P_p}{A_{cladding}} + 1} = N_0 \cdot \frac{I_p/I_{po}}{I_p/I_{p0}+1} \qquad (8)$$

Here, $I_p = P_p/A_{cladding}$ and $I_{p0} = hv_p/(\tau\sigma_{vp}^a)$ represent the pump power density which results in 50% population of the upper level. Using typical values for erbium-doped fiber amplifier cladding pumped at 980 nm, $\tau = 10$ ms, $\sigma_{vp}^a = 1.8 \times 10^{-25}$ m², and $\sigma_{vp}^e = 0$, I_{po} becomes 112 MW/m², which will correspond to a pump power of 0.88 W for a cladding of 100 μm in diameter. Note that at this power level, upper state population will be 50%, level at which amplification can be achieved only for signal wavelengths with $\sigma_{vs}^e > \sigma_{vs}^a$.

In cladding-pumped amplifiers, as the signal grows, its intensity can become comparable with or even larger than the pump intensity I_p. This in effect decreases the population inversion, making it harder to achieve amplification at shorter wavelengths, where $\sigma_s^a > \sigma_s^e$. In practice, much higher population inversion and thus pump intensities much larger than I_{po} are required to achieve gain at shorter wavelengths.

Gain and NF of Cladding-Pumped MC-EDFA

Gain and NF of a seven-core erbium-doped fiber amplifier cladding pumped using a 980 nm multimode pump can be calculated using the numerical model presented in section "Numerical Simulation." In the following, results of numerical simulations based on the following parameters of the gain fiber are presented: core diameter, 3.2 μm; cladding diameter, 100 μm; lifetime of the upper energy level τ, 10 ms; and erbium concentration, 7.1×10^{24}/m³. The ASE is considered by introducing 181 wavelength components with 1 nm separation over a wavelength range of 1450–1630 nm. The overlap factor Γ_{vs} (Γ_{vj}) between the signal (ASE) field and the doped core is calculated from their respective mode field diameters (MFDs) using the equation $\Gamma = 1 - \exp(-2a^2/\omega_o^2)$, where a is the radius of doped core and ω_o represents spot size ($= MFD/2$). It was found to vary linearly from 0.539 to 0.436 over the range 1450–1630 nm. A constant overlap factor of 0.428 (value calculated at 1550 nm) is chosen for signal and all the ASE components considered in the simulation.

Figure 7 shows gain and NF calculated at different signal wavelengths for a constant input power level of 7.6 W and for different length of the multicore doped fiber. Gain over 40 dB is obtained at wavelength of 1560 nm for input signal of −20 dBm. It can be seen that as the length is increased, the bandwidth of amplification becomes narrower and the gain peak shifts toward the longer wavelength. For a length of 50 m, the NF remains below 6.4 dB when the signal wavelength is longer than 1560 nm.

In Fig. 8 the gain obtained in a 50-m-long fiber is plotted as a function of input signal power, while the pump is held constant at 7.6 W. The peak gain decreases

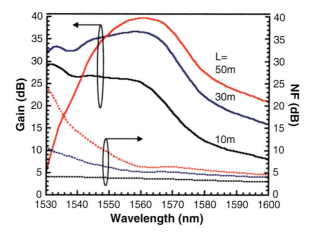

Fig. 7 Gain and NF plotted as a function of wavelength for different lengths of doped multicore fiber. The pump power is 7.6 W, and the input signal power is −20 dBm. (Abedin et al. 2012)

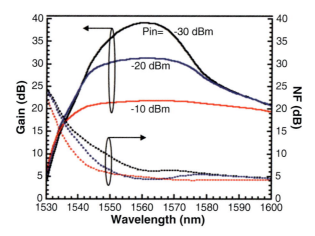

Fig. 8 Gain plotted as a function of wavelength for different signal power level. The pump power is 7.6 W, and the length of the fiber is 50 m. (Abedin et al. 2012)

with an increase in the input signal power, which indicates saturation of gain of the amplifier. On both sides of the peak-gain region, the gain remains almost independent of the input signal power. For an amplifier length of 50 m and 0 dBm input signals, the NF remains below 5 dB for wavelengths longer than 1560 nm.

Effect of Enlarging the Core Size

In a cladding-pumped erbium-doped multicore fiber amplifier, even when several watts of power is used, the pump intensity is relatively small compared with core pumping. In these amplifiers, the signal, as it is amplified along the length, can deplete the upper state population, making it difficult to achieve gain in the C-band with low NF, especially wavelengths below 1530 nm, where the absorption cross section is larger than the emission cross section (Becker et al. 1999). Results of numerical simulation (Abedin et al. 2012) and experiments (Abedin et al. 2014a) show that the gain spectrum can be extended to C-band by choosing a shorter length of the erbium-doped fiber, although at the expense of reduced amplifier saturation power. A simple solution to increase the saturation power would be making the core size bigger.

Gain and NF are calculated numerically for cladding pump amplifier with a larger core size of 5.2 μm with different lengths and signal input power, and the results are shown in Fig. 9a, b. The other parameters of the fiber are chosen to be the same. It can be seen that small-signal gain over 30 dB can be obtained throughout the C-band using 20 m fiber, and for 0 dBm signal input, a saturated output power over 20 dBm from each cores can be obtained. Corresponding curves for MC-EDFA with core size of 3.2 μm are shown in Fig. 9c, d for comparison. These results clearly show the potential advantages of increasing the core size in enhancing the amplifier saturation power. A further increase in gain should be possible by increasing the core size and proportionately reducing the core NA. It is to be noted however that, while larger core size could yield higher gain, at the same time, it could cause an increase in NF. Therefore, the fiber parameters need to be carefully chosen so that NA remains within the acceptable limit over the wavelength range of interest.

Cross-Gain Modulation Due to Gain Depletion

In a cladding-pumped MC-EDFA, since multiple cores share the same pump, it is important to investigate whether there is any cross-gain modulation due to pump depletion. When one or more signal channel(s) are added, multimode pump in the cladding could deplete further, depending on the intensity of the newly added channel(s), which could lower the gain of the existing channels. Alternatively, when one or more of the spatial channels are dropped, it can cause the gain experienced by the remaining channels to increase by a certain extent. Figure 10 shows the gain spectrum when the number of channels is $N = 1$ and $N = 7$, calculated for a range of signal power levels in a seven-core EDFA. The core and cladding diameter is assumed to be 5.2 and 100 μm, respectively. The length of gain fiber used in simulation is 20 m, and the launched multimode pump at 980 nm is 7.6 W.

The plot shows that the change in gain spectrum as the number of operating channels is switched between lowest 1 to highest 7 is rather insignificant. This is due to the fact that, in a cladding-pumped multicore fiber amplifier, a large fraction of the multimode pump remains unabsorbed, particularly when a shorter length is chosen to maintain a broad gain spectrum. However, if a long length of gain fiber is used for operation in the L-band, pump depletion can be significant enough to induce cross-gain modulation among the spatial channels.

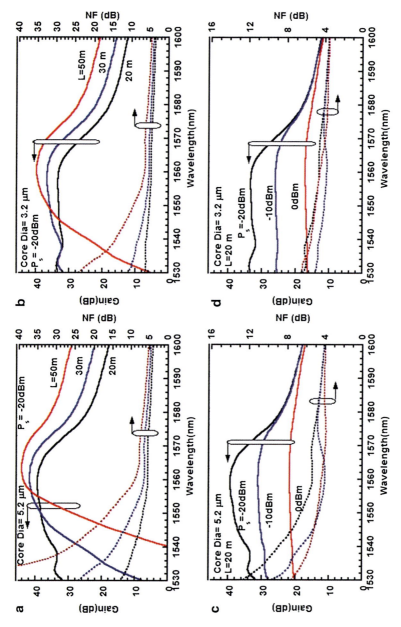

Fig. 9 Gain and NF of MC-EDFA compared between core size of 5.2 and 3.2 μm (Abedin et al. 2014b). The pump power was 7.6 W. (**a**) Core diameter 5.2 μm, a fixed signal power (−20 dBm), and (**b**) core diameter 3.2 μm, for a fixed fiber length (20 m). (**a**) Core diameter 3.2 μm, a fixed signal power (−20 dBm), and (**b**) core diameter 3.2 μm, for a fixed fiber length (20 m)

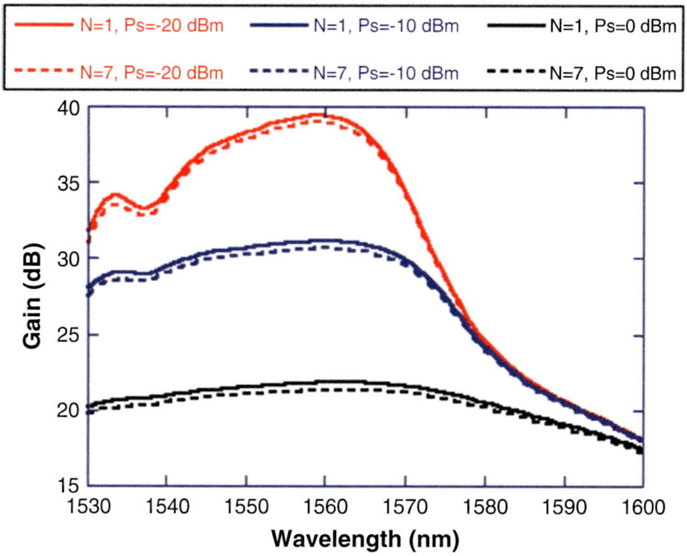

Fig. 10 Gain spectrum when the number of channels operating is N = 1 and N = 7, calculated for a seven-core EDFA

Power Conversion Efficiency

Due to the small (effective) absorption coefficient of the pump radiation in a cladding pump EDFA, a gain fiber with relatively long length is required to absorb the pump. However, it is not possible to increase the length arbitrarily, because, as the signal strengthens along the fiber, the signal intensity in the core (signal power/effective area of core) becomes larger compared with the pump intensity. Consequently, after a certain length of gain fiber, population inversion becomes too low, and the signal output power begins to decrease, even though a large amount of pump still remains unabsorbed. This results in inefficient utilization of pump and poor power conversion efficiency (PCE).

The presence of multiple cores in MC-EDFA can help utilize pump power more efficiently. Since the signal is amplified through individual cores, a greater fraction of pump is converted to signal radiation. For a MC-EDFA, the PCE is defined as (Abedin et al. 2014b):

$$PCE = \sum_{i}^{N} (P_{s,i}(z=L) - P_{s,i}(z=0))/P_p(z=0) \qquad (9)$$

Figure 11 shows a comparison between the PCE calculated for cladding-pumped single-core (N = 1) and multicore (N = 7) fiber amplifiers with different lengths of gain fiber. The enhancement in PCE as a result of multiple cores, defined as

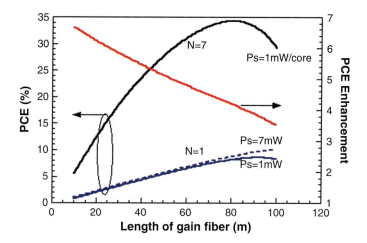

Fig. 11 Calculated PCE of erbium-doped fiber with single core (N = 1) and multiple cores (N = 7) and PCE enhancement

PCE(N = 7)/PCE(N = 1), is also shown in the figure. It is seen that a larger PCE can be achieved for a seven-core amplifier compared with a single-core EDFA. It is to be noted also that even if the signal power is increased by seven times in the case of single-core EDFA, the PCE is increased by only a small amount. In the case of single-core EDFA, the maximum achievable power conversion efficiency was only about 8.6% corresponding to a length of 90 m, following which the efficiency starts to decline. For a seven-core EDFA, maxiom PCE is 34% (occurring at a length of 85 m), resulting in PCE enhancement of ~4. A gradual decrease in the PCE enhancement can be seen as the length of gain fiber is increased, which is due to pump depletion.

Experimental Demonstration of Cladding-Pumped Multicore Fiber Amplifiers

A cladding-pumped multicore amplifier essentially consists of a length of Er- (or Er/Yb-) doped multicore double-clad fiber, a pump/signal combiner, and a multimode pump source. Depending on whether the input signals are being launched from a multicore fiber or from multiple SMFs, pump/signal combiner of appropriate configuration needs to be employed. In addition to signal, the multimode pump light needs to be added which can be achieved by launching from the end or side of the gain fiber. Different types of pump/signal combiners, as shown in Fig. 4, have already been realized to serve these purposes. In the following, MC-EDFAs employing end pumping and side-coupled pumping are described.

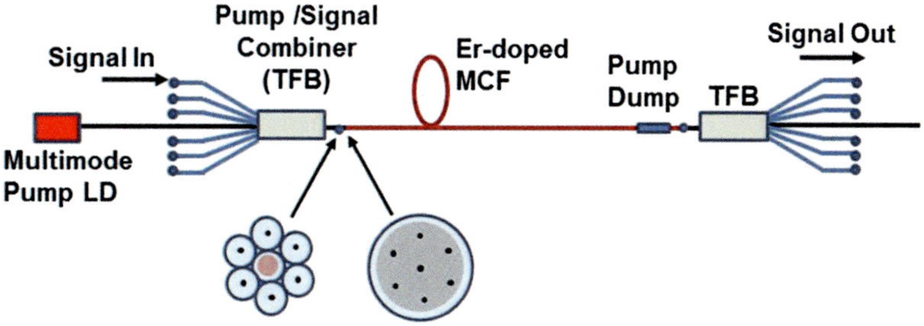

Fig. 12 Schematic diagram of a cladding pump multicore fiber amplifier with end-pumping architecture

Cladding-Pumped MC-EDFA with End Pumping

The schematic diagram of a MC-EDFA cladding-pumped through the end is shown in Fig. 12 (Abedin et al. 2012). The amplifier, as described in Abedin et al. (2012), consists of a seven-core low-index coated MC-EDF and two TFB couplers spliced at both ends of the gain fiber. The TFB at the input is used to simultaneously couple optical signals from six single-mode fibers and to launch the multimode pump radiation. The TFB couplers at the output acted as a demultiplexer to output the amplified signals through six single-mode fibers.

The central multimode fiber at the input of the TFB coupler was spliced to a 980 nm multimode pump laser diode. The input signals are launched through the TFB coupler into the outer six cores of the MC-EDF; the central core remains unused. The length of the MC-EDF is 50 m. At the output end of the MC-EDF, near the output TFB coupler, a pump dump is installed to remove the residual pump and prevent it from entering the output single-mode fibers.

Cladding-Pumped MC-EDFA Employing Side-Coupled Pumping

Fan-in and fan-out devices at the amplifier ends result in attenuation of the signal and prevent direct coupling of signals from a passive SDM fiber into the amplifier. This problem is overcome by using the side coupling technique. Side pumping allows coupling of multimode pump radiation into the cladding through the side of the gain fiber without interruption of the core waveguide. The schematic of a seven-core fiber amplifier with side-coupled pumping (Abedin et al. 2014a) is shown in Fig. 13. The pump is coupled from the side of the gain fiber, closed to the input end, using side coupling technique as depicted in Fig. 4c. In the pump coupling scheme shown in Fig. 13, the short length of MC-EDF existing before the coupler remains unpumped and could lead to absorption of input signal. This problem could be overcome by the use of a side-coupled pump/signal combiner built from passive multicore fiber that can be conveniently spliced to the MC-EDF.

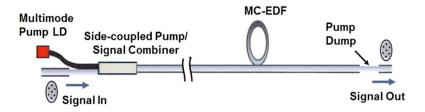

Fig. 13 Schematic diagram of a cladding pump multicore fiber amplifier with side-coupled pumping architecture

Experimental Results

Gain and NF properties of cladding pumping MC-EDFAs are reported for with both end- (Abedin et al. 2012) and side-coupled pumping architectures (Abedin et al. 2014a). The results are summarized in the following.

End-Coupled Pumping

Figure 14a shows the net (external) gain versus wavelength plotted for the six cores, when the pump power is 7.6 W and input signal power levels of -20 and 0 dBm (Abedin et al. 2012). For a small signal of -20 dBm, a maximum gain of 32 dB is obtained from the amplifier. Some variations in the gain in different cores are observed, which are due to differences in the passive losses (splice and input/output TFBs). Small-signal gain larger than 20 dB is available for signal amplification over a wavelength range of 32 nm. Considering the total passive loss (two TFBs and splice loss of $\sim$7.0 dB), the net gross gain in the amplifier is about 39 dB, sufficiently close to the calculated gain. The internal NF for different cores, as a function of wavelength, is shown in Fig. 14b. Average NF is about 6 dB for wavelength longer than 1560 nm.

Note that although the peak gain is over 30 dB at 1560 nm, the gain at 1530 nm and at shorter wavelength is quite small, in consistence with the simulation results ($L = 50$ m). In order to extend the gain spectrum to C-band, a shorter length of erbium-doped fiber is required, which is validated in the following section on amplifier with side-coupled pumping using a shorter length of gain fiber.

Side-Coupled Pumping

Figure 15a shows the gross gain versus wavelength measured in the seven cores for input signal powers of -20 and 0 dBm and a coupled pump power of 4.7 W (Abedin et al. 2014a). The length of the gain fiber is 34 m. Due to the absence of TFB couplers, signal insertion loss is reduced significantly, resulting in higher external gain in comparison to the end-pumped amplifier, even at a lower pump power.

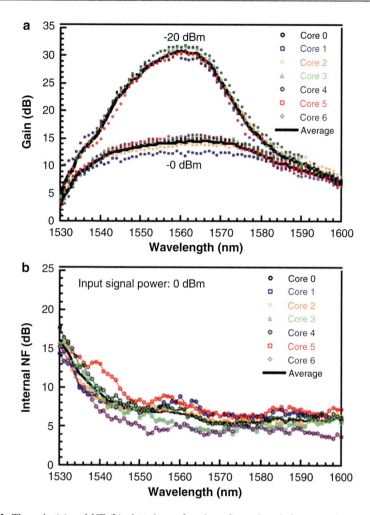

Fig. 14 The gain (**a**) and NF (**b**) plotted as a function of wavelength for outer six cores of the multicore EDF amplifier, cladding pumped through the end. The input pump and coupled pump power are 10.6 and 7.6 W, respectively

Here, gross gain represents the gain that signals from a multicore transmission fiber will experience when spliced directly to the multicore gain fiber. The maximum gross gain is about 36 dB near 1560 nm, and gain over 25 dB is obtained over a bandwidth of ∼40 nm. Shortening the length of the gain fiber to 34 m has resulted in a significant expansion of the bandwidth into the C-band. At 1530 nm, the small-signal gain is ∼20 dB. The internal NF for different cores as a function of wavelength is shown in Fig. 15b for input signal power of 0 dBm. As shown in the figure, the NF becomes high for wavelength shorter than 1540 nm, while it tends to decrease for longer wavelengths. The NF is ∼5 dB for the signal wavelength of 1560 nm, and it increased to about 8 dB at 1530 nm.

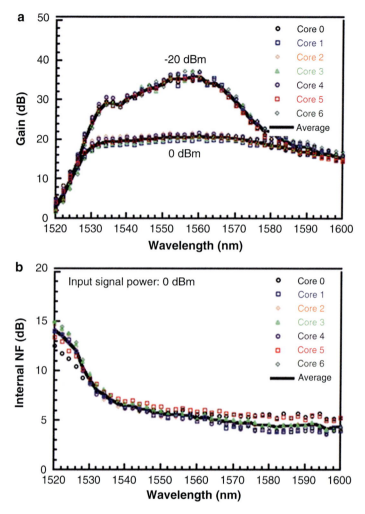

Fig. 15 Gain (**a**) and NF (**b**) measured for the seven different cores of the MC-EDFA, with side-coupled cladding pumping. The input pump and coupled pump power are 7.6 and 4.7 W, respectively. The central core is represented by core 0

Comparison Between Core- and Cladding-Pumped Amplifiers

Table 1 shows a comparison of optical performances between core and cladding-pumped (end- and side-coupled) all-fiber-based MC-EDFAs. In core-pumped amplifiers, the internal amplification and noise properties are quite similar to conventional amplifiers (Abedin et al. 2011). However, the available gain can be small due to the passive losses incurred in fan-in/fan-out devices used at the input and output end. Moreover, NF can be increased by the signal attenuation due to the input fan-out.

Table 1 Comparison between core- and cladding-pumped amplifiers (Abedin et al. 2014b)

	Cladding pumped End-coupled/Side-coupled	Core pumped
Number of cores	6/7	7
Fan-out	TFB/not required	TFB
Amplifier length	50 m/34 m	15 m
Pump power	7.6 W/4.7 W	1.0 W
Net internal gain	39 dB (Abedin et al. 2012)/35 dB	30 dB
Net available small-signal gain	32 dB/35 dB	25 dB
Bandwidth of gain >20 dB	32 nm/64 nm	45 nm
NF (internal)	<6 dB/<5 dBFor λ > 1560 nm	<4 dB (C-band)
Cross talk	/ < −45 dB (Abedin et al. 2014a)	< −30.2 dB (Abedin et al. 2011)

The gain and NF properties can be improved by using fan-in/fan-outs with lower insertion loss.

In cladding-pumped amplifiers, due to lower pump intensity (consequently low pump absorption), it is difficult to achieve gain in the short wavelengths, e.g., 1530 nm, unless the length is shortened, which can significantly reduce the PCE of the amplifier. It is worth noting that for L-band operation, PCE can be as high as 34% in a seven-core amplifier made with a long length of gain fiber (Fig. 11). Thus, side-coupled cladding pumping can be potentially useful in realizing low-cost and energy-efficient booster MC-EDFA for L-band operation.

Electrical Power Consumption

Cladding pumping has the potential to use low-cost, energy-efficient multimode diodes. The electrical to optical power conversion efficiency of a multimode pump laser can be as high as 46%, which is more than twice the efficiency of single-mode pump laser diodes ($\sim$20%). This indicates that in multimode pumps, a smaller fraction of electrical power is wasted as heat, which is beneficial from the context of cooling. Table 2 shows a comparison of electrical power consumption of cladding-pumped and core-pumped multicore (seven-core) EDFA. Output power per core is assumed to be the same, i.e., 20 dBm. Numerical simulation shows that a cladding-pumped seven-core amplifier, with a core and a cladding diameter of 5.2 and 100 μm, 100 mW of output can be obtained by launching 8.5 W of 976 nm pump into a 20-m-long gain fiber, which corresponds to a pump-to-signal conversion efficiency of about 8%. Consequently, an uncooled multimode pump with a rated output of 9 W can be used in this purpose; the total electrical power consumption is $\sim$17.6 W (http://www.bwt-bj.com/en/product/list_38_50.html). On the other hand, assuming a pump-to-signal conversion efficiency of about 50% for core pumping, two 700 mW fiber Bragg grating stabilized 980 nm pump modules can be used

Table 2 Comparison of electrical power consumption in core and cladding-pumped multicore fiber amplifiers

	1 × 9 W, Multimode LDλ-stabilized, uncooled	2 × 700 mW, single-mode LDλ-stabilized, semi-cooled (45 °C)
Total pump power (W)	9 W	1.4 W
Pump-to-signal conversion efficiency	~8%	~50%
Output power/core	100 mW	100 mW
Total electrical power consumption	17.6 W	15.4 W Including TEC

by sharing 1.4 W of pump among seven cores. To minimize the consumption of power for thermoelectric cooling, semi-cooled LDs operated at 45 °C can be used. The total power consumption (including cooling) will be ~15.4 W (7.73 W, each) (https://www.lumentum.com/en/products/700-mw-fiber-bragg-grating-stabilized-980-nm-pump-modules-low-power-consumption), similar to what is required for cladding pumping. It can be seen that although the pump-to-signal conversion is rather low for cladding pumping, significant electrical energy saving achieved by using uncooled multimode pump diodes can render it as an attractive alternative for application in SDM systems.

Recent Advancements

In the last few years, several research groups have reported on improving the performance of cladding-pumped multicore fiber amplifiers. A list of multicore amplifier developed in recent years along with their optical performances are shown in Table 3. A seven-core, cladding-pumped, fiber amplifier for L-band operation, with saturation power as much as 20.6 dBm per core has been demonstrated (Tsuchida et al. 2012b). A MCF amplifier has also been developed using Er/Yb co-doped, 12-core fiber, which allowed reducing the length to 5 m (Ono et al. 2013). Very recently, cladding pumped, multicore, multimode amplifiers with six cores and three modes have been reported (Jin et al. 2015a). Six, few-moded cores, 16–17 μm in diameter, relatively heavily doped (Er concentration: 2.8×10^{25} ions/m^3) are arranged in an annular cladding region, occupying about 8% of cladding area, reportedly offering 15 dBm output per mode. Due to the use of confined cores uniformly doped with erbium, there remains the problem of differential modal gain, which could be eliminated by using ring doping (Lim et al. 2014) or extended rare-earth-doped cores (Abedin et al. 2015). A multicore core-pumped amplifier with as many as 19 cores with bulk-optic-based WDM and isolator has been reported in Sakaguchi et al. (2014). Such approaches may result in the realization of multicore amplifiers with higher core, mode counts, and pump-to-signal conversion efficiencies as well as with improved amplification

Table 3 Optical performance of different multicore fiber amplifiers (Abedin et al. 2017)

Type of amplifier	Number of cores	Mode number	Pump power	Peak gain	Bandwidth	Psat/core	NF	Length	References
Core pumped	7	1	146 mW/core	23–27 dB	1530–1565 nm	10 dBm	4 dB	14 m	Abedin et al. (2011)
Core pumped	7	1	40 mW/core	20 dB	1530–1565 nm		<7 dB	16 m	Tsuchida et al. (2012b)
Core pumped	19	1	400 mW/core	23 dB	1520–1560 nm (@20 dB)	12–17 dBm	6–7 dB	–	Sakaguchi et al. (2014)
Cladding pumped	6	1	10.6 W	32 dB	1542–1576 nm (@20 dB)	15 dBm	6 dB	50 m	Abedin et al. (2012)
Cladding pumped	7	1	7.6 W	25 dB	1532–1582 nm @20 dB	15 dBm	5–7 dB	34 m	Abedin et al. (2014a)
Cladding pumped	7	1	17 W	20 dB	1578.4–1608.1 nm	20.6 dBm	10–5 dB	107 m	Maeda et al. (2015)
Cladding pumped	12	1	1.9 W (launched)	18.3 dB	1534–1561.4 nm @11 dB			5 m(Er/Yb)	Ono et al. (2013)
Cladding pumped	6	3	25 W	20 dB	1530–1560 nm	15 dBm/mode/core	6–9 dB	2.1 m	Jin et al. (2015a)

and noise properties that would be more suitable for deployment in practical SDM systems.

Summary

This chapter describes state-of-the-art cladding-pumped multicore EDFAs useful for amplifying SDM signals. Cladding pumping requires fewer optical components and has the potential to use low-cost, energy-efficient multimode diodes. Cladding-pumped amplifiers with different core and mode counts, fiber structures, and pumping configurations have been developed, which promises for potential applications in SDM transmission systems.

References

K.S. Abedin, T.F. Taunay, M. Fishteyn, M.F. Yan, B. Zhu, J.M. Fini, E.M. Monberg, F.V. Dimarcello, P.W. Wisk, Amplification and noise properties of an erbium-doped multicore fiber amplifier. Opt. Exp. **19**, 16715–16721 (2011)

K.S. Abedin, T.F. Taunay, M. Fishteyn, D.J. DiGiovanni, V.R. Supradeepa, J.M. Fini, M.F. Yan, B. Zhu, E.M. Monberg, F.V. Dimarcello, Cladding-pumped erbium-doped multicore fiber amplifier. Opt. Express **20**, 20191–20200 (2012)

K.S. Abedin, J.M. Fini, T.F. Thierry, B. Zhu, M.Y. Fan, L. Bansal, F.V. Dimarcello, E.M. Monberg, D.J. DiGiovanni, Seven-core erbium-doped double-clad fiber amplifier pumped simultaneously by side-coupled multimode fiber. Opt. Lett. **39**, 993–996 (2014a)

K.S. Abedin, J.M. Fini, T.F. Thierry, V.R. Supradeepa, B. Zhu, M.F. Yan, L. Bansal, E.M. Monberg, D.J. DiGiovanni, Multicore erbium doped fiber amplifiers for space division multiplexing systems. J. Lightwave Technol. **32**, 2800–2808 (2014b)

K.S. Abedin, M.F. Yan, J.M. Fini, T.F. Thierry, L.K. Bansal, B. Zhu, E.M. Monberg, D.J. DiGiovanni, Space division multiplexed multicore erbium-doped fiber amplifiers. J. Opt. **45**(3), 231–239 (2015)

K.S. Abedin, M.F. Yan, T.F. Taunay, B. Zhu, E.M. Monberg, D.J. DiGiovanni, State-of-the-art multicore fiber amplifiers for space division multiplexing. Opt. Fiber Technol. **35**, 64–71 (2017)

N. Bai, E. Ip, T. Wang, G. Li, Multimode fiber amplifier with tunable modal gain using a reconfigurable multimode pump. Opt. Express **19**(17), 16601–16611 (2011)

P.C. Becker, N.A. Olsson, J.R. Simpson, *Erbium Doped Fiber Amplifiers: Fundamentals and Technology* (Academic Press, San Diego, 1999)

A.R. Chraplyvy, The coming capacity crunch, ECOC plenary talk (2009)

D.J. DiGiovanni, A.J. Stentz, Tapered fiber bundles for coupling light into and out of cladding-pumped fiber devices. U.S. Patent 5,864,644, 1999

R. Essiambre, A. Mecozzi, Capacity limits in single mode fiber and scaling for spatial multiplexing, in *Optical Fiber Communication Conference*. OSA Technical Digest (Optical Society of America, 2012), Paper OW3D.1

V.P. Gapontsev, I. Samartsev, Coupling arrangement between a multimode light source and an optical fiber through an intermediate optical fiber length. U.S. Patent 5,999,673, 1996

S.G. Grubb, D.F. Welch, Double-clad optical fiber with improved inner cladding geometry. U.S. Patent 6,157,763 A, 2000

A.B. Grudinin, D.N. Payne, P.W. Turner, L.J.A. Nilsson, M.N. Zervas, M. Ibsen, M.K. Durkin, U.S. Patent 6,826,335, 2004

T. Hayashi, T. Taru, O. Shimakawa, T. Sasaki, E. Sasaoka, Low-crosstalk and low-loss multi-core fiber utilizing fiber bend, in *Optical Fiber Communication Conference/National Fiber Optic*

Engineers Conference 2011. OSA Technical Digest (CD) (Optical Society of America, 2011), Paper OWJ3
http://www.bwt-bj.com/en/product/list_38_50.html
https://www.lumentum.com/en/products/700-mw-fiber-bragg-grating-stabilized-980-nm-pump-modules-low-power-consumption
http://www.optoscribe.com/products/3d-optofan-series/
E. Ip, Gain equalization for few-mode fiber amplifiers beyond two propagating mode groups. IEEE Photon. Technol. Lett. **24**(21), 1933–1936 (2012)
S. Jain, T.C. May-Smith, A. Dhar, A.S. Webb, M. Belal, D.J. Richardson, J.K. Sahu, D.N. Payne, Erbium-doped multi-element fiber amplifiers for space-division multiplexing operations. Opt. Lett. **38**, 582–584 (2013)
S. Jain, Y. Jung, T.C. May-Smith, S.U. Alam, J.K. Sahu, D.J. Richardson, Few-mode multi-element fiber amplifier for mode division multiplexing. Opt. Exp. **22**, 29031–29036 (2014)
Y. Jeong, J. Nilsson, J.K. Sahu, D.N. Payne, R. Horley, L.M.B. Hickey, P.W. Turner, Power scaling of single-frequency ytterbium-doped fiber master-oscillator power-amplifier sources up to 500 W. IEEE J. Sel. Top. Quantum Electron **13**(3), 546–551 (2007)
C. Jin, B. Huang, K. Shang, H. Chen, R. Ryf, R.J. Essiambre, N.K. Fontaine, G. Li, L. Wang, Y. Messaddeq, S. LaRochelle, Efficient annular cladding amplifier with six, three-mode cores. Presented at the European Conference Exhibition Optical Communication, Valencia, 2015a, Paper PDP.2.1
C. Jin, B. Ung, Y. Messaddeq, S. LaRochelle, Annular-cladding erbium doped multicore fiber for SDM amplification. Opt. Express **23**, 29647–29659 (2015b)
Q. Kang, E.-L. Lim, Y. Jung, J. K. Sahu, F. Poletti, C. Baskiotis, S. Alam, D.J. Richardson, Accurate modal gain control in a multimode erbium doped fiber amplifier incorporating ring doping and a simple LP01 pump configuration (n.d.), Opt. Express **20**(19), 20835–20843 (2012)
J.P. Koplow, S.W. Moore, D.A.V. Kliner, A new method for side pumping of double-clad fiber sources. IEEE J. Quantum Electron. **39**, 529–540 (2003)
A. Kosterin, V. Temyanko, M. Fallahi, M. Mansuripur, Tapered fiber bundles for combining high-power diode lasers. Appl. Opt. **43**, 3893–3900 (2004)
P.M. Krummrich, Optical amplification and optical filter based signal for cost and energy efficient spatial multiplexing. Opt. Exp. **19**, 16636–16652 (2011)
E.L. Lim, Y. Jung, Q. Kang, T.C. May-Smith, N.H.L. Wong, R. Standish, F. Poletti, J.K. Sahu, S. Alam, D.J. Richardson, First demonstration of cladding pumped few-moded EDFA for mode division multiplexed transmission, in *Optical Fiber Communication Conference.* OSA Technical Digest (online) (Optical Society of America, 2014), Paper M2J.2
K. Maeda, Y. Tsuchida, S. Takasaka, T. Saito, K. Watanabe, T. Sasa, R. Sugizaki, K. Takeshima, T. Tsuritani, *Cladding Pumped Multicore EDFA with Output Power Over 20dBm Using a Fiber Based Pump Combiner.* OECC2015, Paper PWe.30, 2015
Y. Mimura, Y. Tsuchida, K. Maeda, R. Miyabe, K. Aiso, H. Matsuura, R. Sugizaki, Batch multicore amplification with cladding-pumped multicore EDF, in *European Conference and Exhibition on Optical Communication.* OSA Technical Digest (online) (Optical Society of America, 2012), Paper Tu.4.F.1
T. Morioka, *New Generation Optical Infrastructure Technologies: "EXAT Initiative" Towards 2020 and Beyond.* OECC2009, Paper FT4, 2009
T. Morioka, Y. Awaji, R. Ryf, P.J. Winzer, D. Richardson, F. Poletti, Enhancing optical communications with brand new fibers. IEEE Commun. Mag. **50**(2), s31–s42 (2012)
H. Ono, K. Takenaga, K. Ichii, S. Matsuo, T. Takahashi, H. Masuda, M. Yamada, 12-core double-clad Er/Yb-doped fiber amplifier employing free-space coupling pump/signal combiner module. Presented at the 39th European Conference Exhibition Optical Communication, London, Sept 2013, Paper We.4.A.4
D.J. Richardson, Optical amplifiers for space division multiplexed systems, in *Optical Fiber Communication Conference/National Fiber Optic Engineers Conference 2013.* OSA Technical Digest (online) (Optical Society of America, 2013), Paper OTu3G.1

D.J. Richardson, J.M. Fini, L.E. Nelson, Space-division multiplexing in optical fibres. Nat. Photonics **7**, 344–362 (2013)

D.J. Ripin, L. Goldberg, High efficiency side-coupling of light into optical fiber using imbedded V-groove. Electron. Lett. **31**, 2204–2205 (1995)

J. Sakaguchi, Y. Awaji, N. Wada, T. Hayashi, T. Nagashima, T. Kobayashi, M. Watanabe, Propagation characteristics of seven-core fiber for spatial and wavelength division multiplexed 10-Gbit/s channels, in *Optical Fiber Communication Conference/National Fiber Optic Engineers Conference 2011*. OSA Technical Digest (CD) (Optical Society of America, 2011), Paper OWJ2

J. Sakaguchi, W. Klaus, B.J. Puttnam, J.M.D. Mendinueta, Y. Awaji, N. Wada, Y. Tsuchida, K. Maeda, M. Tadakuma, K. Imamura, R. Sugizaki, T. Kobayashi, Y. Tottori, M. Watanabe, R.V. Jensen, 19-core MCF transmission system using EDFA with shared core pumping coupled via free-space optics. Opt. Express **22**, 90–95 (2014)

E. Snitzer, H. Po, F. Hakimi, R. Tumminelli, B.C. McCollum, Double clad, offset core Nd fiber laser, in *Optical Fiber Sensors*, (Optical Society of America, Washington, DC, 1988), p. PD5

T. Theeg, H. Sayinc, J. Neumann, L. Overmeyer, D. Kracht, Pump and signal combiner for bi-directional pumping of all-fiber lasers and amplifiers. Opt. Express **20**, 28125–28141 (2012)

Y. Tsuchida, K. Maeda, Y. Mimura, H. Matsuura, R. Miyabe, K. Aiso, R. Sugizaki, Amplification characteristics of a multi-core erbium-doped fiber amplifier, in *Optical Fiber Communication Conference*. OSA Technical Digest (Optical Society of America, 2012a), Paper OM3C.3

Y. Tsuchida, K. Maeda, K. Watanabe, T. Saito, S. Matsumoto, K. Aiso, Y. Mimura, R. Sugizaki, Simultaneous 7-Core pumped amplification in multicore EDF through fibre based fan-in/out, in *European Conference and Exhibition on Optical Communication*. OSA Technical Digest (online) (Optical Society of America, 2012b), Paper Tu.4.F.2

Y. Tsuchida, M. Tadakuma, R. Sugizaki, Multicore EDFA for space division multiplexing by utilizing cladding-pumped technology, in *Optical Fiber Communication Conference*. OSA Technical Digest (online) (Optical Society of America, 2014), Paper Tu2D.1

M. Yamada, K. Tsujikawa, L. Ma, K. Ichii, S. Matsuo, N. Hanzawa, H. Ono, Optical fiber amplifier employing a bundle of reduced cladding erbium-doped Fibers. IEEE Photon. Tech. Lett. **24**, 1910–1913 (2012)

K. Yla-Jarkko, S.U. Alam, P.W. Turner, J. Moore, J. Nilsson, R. Selvas, D.B. Soh, C. Codemard, J.K. Sahu, *Optical Amplifiers and Their Applications*. OSA Technical Digest Series (Optical Society of America, 2003), Paper TuC1

B. Zhu, T.F. Taunay, M.F. Yan, J.M. Fini, M. Fishteyn, E.M. Monberg, F.V. Dimarcello, Seven-core multicore fiber transmissions for passive optical network. Opt. Express **18**, 11117–11122 (2010)

Optical Amplifiers for Mode Division Multiplexing

22

Yongmin Jung, Shaif-Ul Alam, and David J. Richardson

Contents

Introduction	850
Current State of the Art in SDM Amplifiers	851
SDM Optical Components	853
Fiber-Optic Collimator Assembly	853
Pump Coupler	855
Mode-Field Diameter Adaptor	856
Mode-Dependent Loss Equalizer	858
Design of Few-Mode Fiber Amplifiers	859
The Importance of Differential Modal Gain (DMG) Control	859
Design Strategies for Reducing DMG of FM-EDFAs	860
Core-Pumped 6-Mode EDFA	868
Cladding-Pumped 6-Mode EDFA	870
Challenges and Future Development	871
Conclusion	871
References	872

Abstract

Space division multiplexing (SDM) has attracted considerable attention from the optical fiber communication community as a promising means to increase the transmission capacity per fiber and, more importantly, to reduce the associated cost per transmitted information bit by utilizing multiple spatial channels within a single strand of glass. Various SDM transmission fibers, for example, few-mode fibers, multicore fibers, and few-mode multicore fibers, have been proposed, and the possibility to support capacities beyond that of conventional single-mode fiber technology has now been proven. However, in order to realize the potential

---

Y. Jung (✉) · S.-U. Alam · D. J. Richardson
Optoelectronics Research Centre (ORC), University of Southampton, Southampton, UK
e-mail: ymj@orc.soton.ac.uk; sua@orc.soton.ac.uk; djr@orc.soton.ac.uk

© Springer Nature Singapore Pte Ltd. 2019
G.-D. Peng (ed.), *Handbook of Optical Fibers*,
https://doi.org/10.1007/978-981-10-7087-7_49

energy and cost savings offered by SDM systems, the individual spatial channels should be simultaneously multiplexed, transmitted, amplified, and switched with associated SDM components and subsystems. In particular, SDM amplifiers can simultaneously amplify multiple spatial channels in a single device and can provide significant space, cost, and energy savings from component sharing and device integration. In this chapter, we will review the current state of the art in optical amplifiers for the various SDM approaches under investigation – with particular focus on few-mode fiber amplifiers for mode division multiplexing. In these amplifiers, differential modal gain (or mode-dependent gain) is the most important property, and various mitigation strategies are introduced/highlighted in this chapter. Afterwards, we will focus in particular on both core-pumped and cladding-pumped 6-mode erbium-doped fiber amplifiers as practical implementation examples and will further discuss differential modal gain control for both pumping configurations. Finally, the remaining challenges to realizing practical few-mode fiber amplifiers are discussed, and future prospects for the few mode fiber amplifier are envisioned.

Introduction

About 48% of the world's population now use the Internet, and the resulting data traffic on the world's optical communication networks is growing at a rate of 25–40% year-on-year. Over the last 30 years, optical fiber networks based on single-mode fiber (SMF) have played a vital role in the development of high-quality, high-speed telecommunication systems. However, recent ultrahigh capacity experiments and theoretical predictions strongly indicate that we are getting close to the fundamental information carrying capacity limit imposed by fiber nonlinearity (generally known as the "nonlinear Shannon limit") (Ellis et al. 2010; Essiambre et al. 2010) and the maximum bandwidth of current optical amplifiers. Network providers are already looking ahead to find ways to deliver future applications for the next 20 years, and it will almost certainly mean radical changes in the sorts of optical fibers and amplifiers used in future optical networks. This issue is currently a very hot topic of research within the international optical communications community. There are several ways to increase the optical transmission capacity of a fiber over a fixed bandwidth, but most of them (i.e., polarization, time, frequency and quadrature multiplexing) are already being used commercially. In fact, the only unused physical dimension that remains unexploited is "space," and consequently space division multiplexing (SDM) (Richardson et al. 2013; Winzer 2014), i.e., the use of multiple spatial data pathways through the same fiber, has emerged as the most promising long-term solution to the problems described above. There are two basic strategies for achieving multiple spatial channels within the same fiber, namely, multicore fiber whereby multiple cores (rather than just one as now) are incorporated in the fiber cross section and few-mode fiber that utilizes multiple spatial modes in a single large core (2–3 times the size of an existing single-mode fiber core). Both of these fundamental approaches are being investigated intensively around the globe, and a tenfold increase in overall fiber capacity (1 Petabit/s) (Takara et al. 2012) has already been achieved using the multicore fiber approach. More recently, few-mode multicore fibers combining both of these technological approaches to SDM have been introduced to further increase the data carrying capacity of a single optical fiber. This approach has the potential for a 100-fold capacity increase and 10 Petabit/s transmission already achieved (Soma et al. 2018). Using SDM technology, all individual spatial channels (each of which it is to be appreciated carries the same number of wavelength channels as in current single-mode systems) can be simultaneously amplified within a single SDM optical amplifier, and the signals routed through the network using individual SDM reconfigurable optical add/drop multiplexers, providing great potential for cost reduction (relative to the use of multiple parallel single-mode systems carrying the same overall capacity).

In this chapter, we first review the current state of the art in SDM amplifier technology, tabulating progress in terms of SDM approaches adopted and the number of spatial channels so far achieved. Following this, several key SDM components (e.g., isolators, pump combiners) are introduced and which are key in order to provide practical, field-deployable SDM amplifiers. We will then focus our discussions on the development of few-mode fiber amplifiers offering low differential modal gain (DMG) between all the supported spatial modes. In few-mode fiber amplifiers, DMG directly affects the system outage probability and can also limit system reach, and so we will discuss in detail various techniques to mitigate it. We will explain in more detail both core-pumped and cladding-pumped 6-mode erbium-doped fiber amplifiers (EDFAs) to provide a better understanding of the practical implementation of low-DMG few-mode fiber amplifiers in both pumping configurations. In this instance, recently developed fully integrated, few-mode fiber amplifiers will be highlighted. Finally, the challenges and a future outlook to SDM amplifiers are presented.

Current State of the Art in SDM Amplifiers

Over the last 5–7 years, many different types of SDM amplifier have been proposed and investigated in order to address the need for optical amplification in long-haul transmission systems over the new generation of SDM transmission fibers. Figure 1 shows the classification of the primary SDM amplifiers reported to date. There are two main streams of SDM amplifier for achieving multiple spatial channel amplification within a single device: multicore fiber (MCF) amplifiers and few-mode fiber (FMF) amplifiers. In the case of multicore erbium-doped fiber amplifiers (MC-EDFAs), both core-pumped (Abedin et al. 2011) and cladding-pumped 7-core EDFAs (Abedin et al. 2012) have been demonstrated in 2011 and 2012, respectively. The core-pumped MC-EDFA was then scaled to 19 cores in 2013 (Sakaguchi et al. 2014) and the cladding-pumped MC-EDFA to 32 cores in 2016 (Jain et al. 2016), which represents the highest core-density MC-EDFA to date. Improved sharing of the optical components and significant device integration were also demonstrated in this 32-core MC-EDFA, as required to obtain the anticipated cost reduction

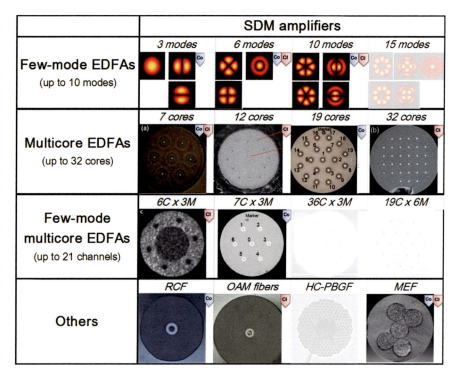

Fig. 1 The current state of the art in SDM amplifiers reported to date. Faded images are used to indicate instances where suitable amplifier solutions have yet to be demonstrated (*cl*, cladding-pumped, and *co*, core-pumped amplifier)

benefits of SDM. For example, two integrated SDM isolators (Jung et al. 2016b) were used to simultaneously prevent unwanted reflections from all 32 cores, and two side couplers were used to couple pump beams into the cladding of the active MCF. In the case of few-mode erbium-doped fiber amplifiers (FM-EDFAs), both core-pumped (Jung et al. 2014b) and cladding-pumped 6-mode EDFAs (Jung et al. 2014a) were demonstrated in 2014. More recently both core-pumped (Wada et al. 2016) and cladding-pumped 10-mode EDFA (Fontaine et al. 2016) were reported representing the highest-mode-count FM-EDFAs reported so far. The cladding-pumped FM-EDFA, which was originally end-pumped using free-space optics, has recently been upgraded to incorporate a fully fiberized side-pumping configuration, and this functionality makes it possible to construct a fully integrated SDM amplifier capable of providing stable modal amplification without the need for free-space optics. Moreover, few-mode multicore fiber amplifiers combining the two SDM fiber technologies have been introduced to further increase the spatial channel density and a 3-mode, 7-core SDM amplifier (Ono et al. 2016) was presented in a core-pumped configuration, and a 3-mode, 6-core SDM amplifier (Chen et al.

2016) was also demonstrated in a cladding-pumped configuration. Other interesting forms of SDM amplifiers, e.g., ring-core fiber (RCF) amplifiers (Kang et al. 2014a; Jung et al. 2017a), orbital angular momentum (OAM) fiber amplifiers (Jung et al. 2017c), and multielement fiber (MEF) amplifiers (Jain et al. 2014), have also now been demonstrated.

SDM Optical Components

In order to fulfill the vision of fully integrated SDM amplifiers, the development of a wide range of passive SDM components offering benefits of relatively low material costs, simple fabrication procedures, and highly integrated solutions is very important. Unlike widely used single-mode fiber systems, the essential components needed to build SDM transmission systems are still not commercially available, and basic prototype systems used in many experiment to date have been implemented only with the aid of numerous free-space optical components on relatively large optical benches. These are not only bulky but also expensive and introduce high optical losses. From this perspective a fiber-optic approach will inevitably be the preferred route forward, and the fabrication of fully fiberized components is a prerequisite for the realization of practical SDM systems. In this section, we will focus on several key SDM components that are essential to realize fully integrated SDM amplifiers, for example, fiber-optic collimator assemblies, pump couplers, mode-field diameter adaptors, and mode-dependent loss equalizers.

Fiber-Optic Collimator Assembly

An optical fiber collimator (i.e., fiber-optic collimation and focusing assembly) represents an important platform for the realization of many commonly used fiber-optic components and devices including optical isolators, circulators, gain flattening filters, WDM couplers, and switches. Here, SDM fibers (e.g., FMFs or MCFs) can be incorporated in a micro-optic collimator assembly, and compact SDM components can be realized (Jung et al. 2016a). Due to the simple fabrication process, good beam transfer quality, and low cost can be achieved and these devices represent an extremely attractive means to reducing the cost of building and operating SDM systems. As shown in Fig. 2a, compact fiber-optic collimators (typically GRIN lens or C-lens) are used to transform the emergent light from an input FMF into a collimated free-space beam that can then be refocused into another FMF using a second identical assembly but in reverse. Optical elements (e.g., isolator core, bandpass filter, gain flattening filter, and so on) can then be inserted into the free space region to provide in-line functionality. The mode-dependent coupling loss from a pair of FMF collimators is relatively low for 10–30 mm working distances (see Fig. 2b), and this allows the incorporation of most functional optical elements in the free space region between the two collimators.

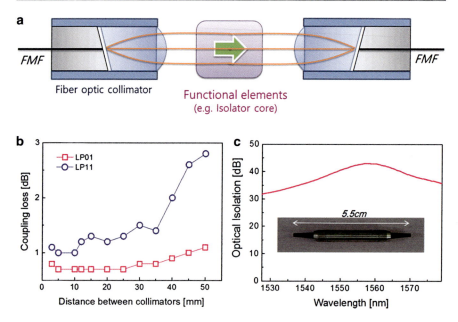

Fig. 2 (**a**) Schematic of a few-mode fiber isolator using compact fiber-optic collimators, (**b**) mode-dependent coupling loss as a function of distance between collimators, and (**c**) optical isolation of the packaged FMF isolator

To illustrate the capability we note that excellent optical performance in terms of average insertion loss (<1 dB) and mode-dependent loss (<1 dB) has readily been demonstrated for both 3-mode and 6-mode FMF isolators. The maximum isolation peak was more than 40 dB and optical isolation over the full C-band was more than 30 dB as shown in Fig. 2c. These results show that a free-space optical isolator can be integrated in FMF collimator-based subcomponents such that optical isolation can be achieved on all spatial modes simultaneously and with near identical performance for all spatial modes. This approach can be applicable to other optical components such as circulators, beam splitters, WDM filters, routing switches, and so on. Using slightly different lens choices and by adopting bulk components with a slightly larger clear aperture, this approach can be applied to other types of SDM fiber, for example, a high-core-count 32-core multicore fiber isolator has recently been demonstrated (Jung et al. 2016b). Unlike FMF collimators, MCF collimators intrinsically require rotational alignment, and a multi-axis precision micro-stage (offering translation, tilt, and rotation adjustment) was required to align the collimators prior to gluing. The optical performance is most encouraging with an average insertion loss of ∼1.5 dB, core-to-core variation of ∼2 dB and an intercore cross talk of less than −40 dB. This indicates that the fiber-optic collimator assembly concept can be extended to SDM fibers, providing the opportunity for an array of new and practical packaged components with performance, in terms of functionality and insertion loss, comparable to the equivalent existing single-mode fiber devices while at the same time ensuring low levels of interchannel cross talk.

Pump Coupler

Pump couplers (or WDM couplers) have been widely used to efficiently couple pump light ($\lambda = 980$ or 1480 nm) into the active fiber along with the signal ($\lambda = 1525-1565$ nm). Fused fiber coupler and dichroic mirror coupler technologies are most commonly used to fabricate SMF pump couplers. However, the fused fiber coupler approach is not considered as suitable for SDM applications because each spatial mode (core) experiences different coupling efficiencies, resulting in mode (core)-dependent loss. Therefore, the dichroic mirror-based fiber-optic collimator assembly is the preferred approach for SDM applications, and low mode (core)-dependent loss can be achieved with this technique. The schematic of a typical dichroic mirror coupler is shown in Fig. 3a. A dichroic mirror is placed in the optical path between the two collimators (i.e., signal input and output collimators) without disturbing the signals and combines the pump and signal light onto the same optical path. Note that this approach is very efficient for core-pumped amplifiers but less so for cladding-pumped amplifiers as it requires a high power handling capability (up to a few tens of Watts). Recently, a fully fiberized side-pump coupler (Abedin et al. 2014) has been employed to couple the multimode pump radiation into the active SDM fiber as shown in Fig. 3b. A multimode pump delivery fiber (core/clad diameter = 105/125 μm, numerical aperture = 0.22 which is compatible with that used on most multimode 976 nm pump diodes) was tapered down to 15 μm with a uniform taper length of 20 mm and wound with a slight tension around a short-stripped section of the active SDM fiber. Although a pump coupling efficiency of ~70% can easily be achieved, it is possible to increase the coupling efficiency to values in excess of 90% using slightly more sophisticated approaches. A UV-curable low-index acrylate polymer (or silicone rubber) can be applied to the tapered section and then cured to ensure robust stable optical contact within the

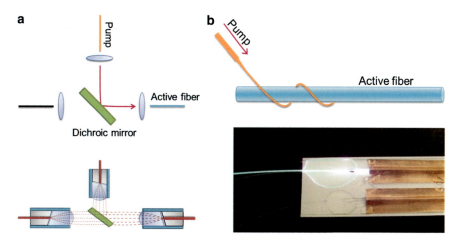

Fig. 3 Schematic of pump couplers for (**a**) core-pumped and (**b**) cladding-pumped SDM amplifiers

pump combiner. There is no observable change in pump coupling efficiency as a result of this packaging. This approach provides a very practical means of pump coupling and is capable of delivering many Watts of pump power into the device if required. Indeed such approaches have allowed coupling of many 100s of Watts of pump power into high-power fiber lasers and are already industrially deployed, validating the commercial readiness and suitability. Also, note that multiple side-pump couplers can be employed without access to the core of the signal fiber, and this can be used to substantially improve the uniformity of population inversion along the fiber length.

Mode-Field Diameter Adaptor

Another important issue for successful SDM transmission is the optical interconnection between dissimilar SDM fibers. For example, even in the same category of few-mode fibers, multiple fiber designs exist with different mode-field diameters (MFDs). Generally, the MFD mismatch results in a mode-dependent splice loss for each spatial mode, and this splice loss variation directly results in an optical signal to noise ratio variation between spatial modes, ultimately limiting the overall SDM transmission performance. To resolve this issue, a very simple and compact all-fiber structure for low-loss optical interconnection of dissimilar FMFs is highly desirable. Recently, a graded index fiber (GIF)-based MFD adaptor has been introduced (Jung et al. 2017b), and the mode-dependent splice loss was significantly reduced. Use of a GIF-based MFD adaptor was previously introduced to accommodate the MFD mismatch between two dissimilar single-mode fibers and to reduce the splice loss, and the same technology can be employed for low-loss optical interconnection between dissimilar FMFs. Figure 4a shows the basic geometry of the all-fiber mode adaptor using a graded index fiber and a coreless fiber (CSF), where the GIF is used as an effective lens element and a segment of CSF used as a spacer. By choosing an appropriate length of both fibers, a compact all-fiber MFD adaptor can easily be realized with a simple cleaving and fusion splicing procedure. To prove the technique two FMFs having significantly different MFDs were taken, and the mode-dependent splice loss was investigated. FMF1 has a core diameter of 20 μm and an NA of 0.12, and FMF2 has a core diameter of 10 μm and an NA of 0.22. As shown in Fig. 4b, the direct splice loss is expected to be ~2.2 dB for the LP_{01} and 4.0 dB for LP_{11} modes, respectively; however using the proposed all-fiber mode adaptor, these values can be reduced to less than 0.15 dB for both spatial modes. Moreover, this technique can also be used for dissimilar multicore fiber interconnection, where a substantial core-pitch difference between dissimilar MCFs can be effectively compensated by employing this simple all-fiber structure (Jung et al. 2017b). These examples clearly show that such all-fiber mode adaptors can provide great control and efficiency in optical coupling for a plurality of different SDM fibers and applications.

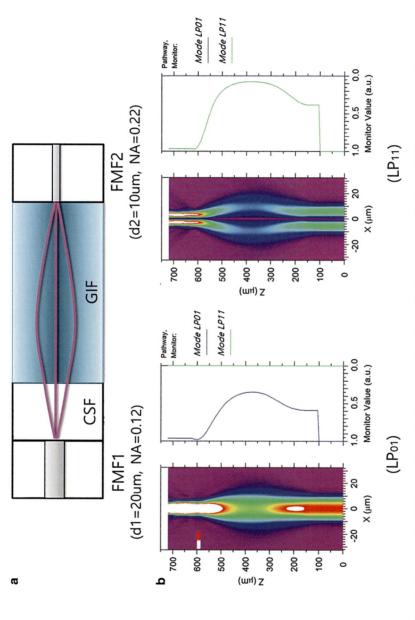

Fig. 4 (**a**) All-fiber mode-field diameter adaptor for low mode-dependent splice loss between dissimilar FMFs and (**b**) beam propagation method (BPM) simulation of the proposed mode adaptor

Mode-Dependent Loss Equalizer

One of the critical issues in FMF-based SDM systems is mitigation of the mode-dependent loss (MDL) which can lead to significant reduction in overall performance. Generally, higher-order modes (HOMs) in a FMF have a larger MFD and typically suffer higher attenuation and coupling losses in FMFs and FMF components. Also, FM-EDFAs typically exhibit a preferential modal gain for the lower-order modes (LOMs). Consequently, in most FMF transmission system, HOMs experience greater optical losses than LOMs, and a MDL equalizer is highly desirable. However, realization of a suitable LOM filter is not simple because LOMs are well confined within the core compared to HOMs. To address this issue, the first dynamic MDL equalizer has been demonstrated (Weiss et al. 2014) which worked by blocking the central portion of the beam with a spatial light modulator as shown in Fig. 5a. The fundamental mode was selectively attenuated over a 0–10 dB MDL equalization range. However, the use of a spatial light modulator is an expensive way to realize MDL compensation, and an all-fiber approach would be preferable – offering potential benefits in terms of low insertion loss, low cost,

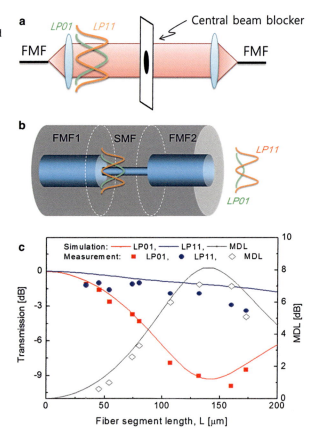

Fig. 5 (**a**) Schematic of a MDL equalizer implemented by blocking the central portion of the beam, (**b**) an all-fiber MDL equalizer for lower-order mode filtering, and (**c**) the spatial filter response as a function of small-core SMF length

and ready integration within fiber systems. Recently, a simple all-fiber solution for spatial mode filtering has been proposed (Jung et al. 2015) that relies on splicing a short segment of small-core SMF between two FMFs, as shown in Fig. 5b. The small-core SMF in the middle creates two spatial optical pathways (i.e., a guided mode along the core and a radiated diffracting beam within the fiber cladding). These beams serve to define a Mach-Zehnder interferometer which can be used to ensure that the LOMs experience higher loss relative to HOMs through destructive interference as the light is coupled back into the second section of FMF. As shown in Fig. 5c, the spatial filter provides excellent lower-order mode filtering performance with an 8 dB MDL equalization range and a minimum insertion loss of the HOM (<2 dB) in a two-mode group fiber. This all-fiber spatial mode filter can provide a cost-effective solution to help mitigate the effect of MDL, and it should be possible to extend this concept to fibers supporting a much larger number of modes. However, it is a fixed MDL equalization device, and a low-cost but dynamically tuneable MDL equalizer should still be further developed for a fully flexible SDM system.

Design of Few-Mode Fiber Amplifiers

The Importance of Differential Modal Gain (DMG) Control

For conventional single-mode EDFAs, there are multiple design parameters that define the optical performance including the gain, noise figure, output power, gain ripple, and the optical signal to noise ratio (OSNR). In current wavelength division multiplexing (WDM) or dense WDM transmission system, multiple wavelength channels of information can be transmitted/amplified simultaneously, and the spectral gain flatness of the amplifier becomes a critical parameter. When WDM signals are amplified multiple times over a series of EDFAs, the OSNR for certain wavelength signals can be seriously degraded due to tilt of the gain spectrum, and the total system performance is mostly limited by these worst performing channels. Therefore, for WDM applications, minimizing the spectral gain tilt (or spectral gain ripple) is one of the most important parameters of the single-mode EDFA design. Similarly, if multiple spatial modes are employed to carry information in an MDM transmission system, each spatial mode should be co-amplified uniformly. For such an MDM system, the most important design factor of the amplifier is the differential modal gain (DMG) control required to ensure that all spatial channels are equally amplified. Figure 6 shows a schematic illustration of the impact of DMG on MDM transmission performance. In most MDM transmission systems, lower-order spatial modes (LOMs) experience lower propagation loss in transmission fibers and lower insertion loss in optical components but higher gain in few-mode EDFAs compared to higher-order spatial modes (HOMs). Therefore, unless precautions are taken, the optical power in higher-order modes decreases gradually over the transmission link, causing the OSNR to be degraded. In the case of a mode-selective MDM transmission system (i.e., weakly coupled regime), the worst quality channel limits the total transmission performance, and consequently DMG of the amplifier turns

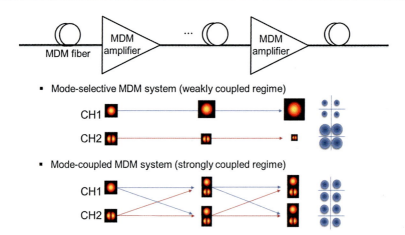

Fig. 6 Schematic illustration of the impact of mode-dependent gain on the MDM transmission performance in two scenarios: (i) mode-selective and (ii) mode-coupled MDM system

out to be a truly critical parameter. In the case of mode-coupled MDM transmission system (i.e., strongly coupled regime), power is exchanged between each spatial mode through mode coupling, and the effect of mode-dependent loss or gain can be averaged out over the transmission link mitigating the effect of DMG. However, note that although the system impact of the amplifier's DMG can be mitigated to an extent in a mode-coupled as compared to a mode-selective MDM system, it is still a significant performance limiting concern even there. Therefore, DMG control of MDM amplifiers is a most important design consideration for successful MDM transmission systems.

Design Strategies for Reducing DMG of FM-EDFAs

As shown in Fig. 7, the DMG is a function of the overlap integrals of the dopant distribution, the pump mode profile, and signal mode profile. Consequently, there are three generic strategies to minimize the DMG, i.e., by:

(i) Controlling the transverse pump field distribution
(ii) Tailoring the dopant distribution of the active fiber
(iii) Engineering the signal mode profiles

A brief description of each approach is provided in the following subsections.

Controlling the Transverse Pump Field Distribution

The first approach to address DMG is based on pump mode control. Under a particular transverse pump mode condition, each signal spatial mode experiences a different overlap with the pump profile (and hence overlap with the population

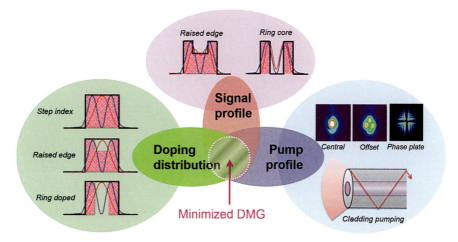

Fig. 7 Key design strategies to minimize the DMG of FM-EDFAs. The DMG can be engineered by (i) tailoring the dopant distribution within the active fiber, (ii) controlling the pump field intensity distribution, and (iii) engineering the signal mode profiles

inversion), and it induces DMG in a FM-EDFA. In order to gain intuitive insight into the effect of pump mode control, a simple uniformly doped, step-index EDF supporting two spatial mode groups (LP_{01} and LP_{11} modes) was simulated under two different pump field distributions (LP_{01-p} and LP_{11-p} modes). For the present discussion assume the fiber has a step-index refractive index profile, a core diameter of 16 μm, and a numerical aperture of 0.11. The core section is fully doped with erbium ions with a concentration of 1.5×10^{25} m^{-3}. In this exemplary EDF, two spatial mode groups (LP_{01} and LP_{11}) can be guided at the signal wavelengths (around 1550 nm), but four spatial mode groups (LP_{01-p}, LP_{11-p}, LP_{21-p}, and LP_{02-p}) are guided at the pump wavelength of 980 nm. Figure 8 shows the modal gain evolution along the fiber for two different pump distributions. Under an LP_{01-p} pump mode condition (Fig. 8a), the LP_{01} signal experiences much higher gain than the LP_{11} signal due to its greater overlap with the erbium ions excited by the LP_{01-p} pump. The DMG is more than 8 dB at L = 4 m. Under the LP_{11-p} pump condition, however, the LP_{11} signal mode has a better overlap with the pump mode and thus has larger gain. Therefore, it should be possible to control the DMG of the FM-EDFAs by adjusting the pump field distribution, and the minimum DMG can be obtained by choosing an appropriate splitting ratio between these two pump spatial modes.

In the first MDM amplifier experiment in 2011 (Jung et al. 2011), a uniformly doped step-index EDF supporting two mode groups was examined, and it was observed that the amplifier preferentially amplifies the LP_{01} modes compared to higher-order modes as expected. Indeed >10 dB DMG was observed with an on-axis pump launch condition. On axis, launch of the pump light into the active fiber generally creates an LP_{01}-like pump intensity profile which tends to result in preferential amplification of the LP_{01} signal mode. In contrast for an offset

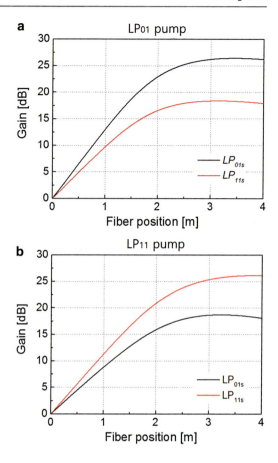

Fig. 8 Modal gain evolution against fiber position for different pump spatial modes: (**a**) LP_{01} pump and (**b**) LP_{11} pump

pump launch (the pump offset relative to the fiber axis or launching the pump light into well-defined HOMs) tends to result in preferential gain for the HOMs. A systematic investigation of DMG mitigation by pump mode profiling/control has been recently experimentally demonstrated for a four-mode group FM-EDFA in a core-pumped configuration. These results will be discussed in more detail in the next section. Subsequently, as the number of spatial modes is scaled up to 10 or more, it becomes increasingly challenging to meet the associated pump power requirements with single-mode pump diodes. Even though this is in principle technically possible, multiplexing a number of single-mode pump diodes to generate sufficient pump power is an expensive way of pumping such an amplifier. In this sense, cladding pumping represents a more promising way to address these issues which allows the use of just one or two high-power, but low brightness, multimode pump laser diodes whose output can be fiber-coupled directly into the amplifier in a very straightforward and low-cost manner. Moreover, the electrical power required to power and control a single multimode pump laser is generally less than that required to drive multiple single mode pumps – particularly as the number of spatial channels is increased. In a cladding-pumped configuration, more than a 100 spatial pump modes are guided by the inner cladding of the active fiber, and the pump power will be close to evenly distributed over all guided modes. In these circumstances, we can neglect the impact of the pump mode profiling, and two other strategies (i.e., ion doping distribution and signal mode profiling) become more important in cladding-pumped FM-EDFAs. Importantly, the cladding-pumped configuration offers a convenient route toward a fully integrated FM-EDFA system by adopting a side coupling approach. Here, the pump radiation can easily be coupled into the active fiber through a fully fiberized pump coupler, and the FM-EDFA can be directly spliced to the passive few-mode transmission fiber. However, the performance of a cladding-pumped amplifier is inevitably compromised at some level (in terms of noise figure) by the reduced pump brightness and hence maximum achievable level of population inversion. To achieve adequate population inversion under cladding-pumped operation at a pump wavelength of 976 nm, it is necessary to reduce the core-to-clad area ratio and/or to adopt an Er/Yb co-doped gain medium. The area ratio between the core and inner cladding is an important parameter to increase the pump intensity in the core, and a reduced cladding fiber (e.g., 80 μm outer diameter) is preferred. The use of ytterbium as a co-dopant provides far stronger pump absorption (∼6 times stronger) than that can be achieved using pure Er doping alone. The excited Yb ions transfer their energy to the erbium ions, resulting in a shorter device length and promote short wavelength operation in a cladding-pumped amplifier.

Tailoring the Radial Dopant Distribution of the Active Fiber

The second approach is to control the dopant distribution within the FM-EDFA. In cladding-pumped FM-EDFAs, the pump field distribution is essentially uniform across the whole inner cladding of the fiber, and dopant distribution control is a more efficient way to address the DMG of the FM-EDFA. In core-pumped FM-EDFAs, it is hard to address DMG control solely by either pump mode shaping or by dopant distribution tailoring, and a combination of both strategies is frequently required in practice. While use of a HOM pump mode is frequently used to minimize the DMG, it is to be appreciated that the generation of a HOM involves relatively high conversion/coupling loss in itself. In that sense, a simple LP_{01} pump mode is the preferred choice of pump mode profile for reducing the pump power loss and avoiding complexity. Under $LP_{01\text{-}p}$ pump launch conditions, a uniformly doped step-index FM-EDFA is expected to have a large differential modal gain for the LP_{01} mode relative to HOMs as discussed in Fig. 8a. To address this, both raised-edge (or center depressed) and ring-doped erbium dopant profiles have been introduced as a means to reduce the overlap of the dopant with the field of the fundamental mode in the central core region. Confining the erbium ions within a ring around the fiber axis can be particularly effective and can drastically enhance HOM gain. To illustrate the benefits of the ring-doped fiber design, we have simulated and investigated the DMG for different relative ring thickness to core radius values (a ratio of t/a in Fig. 9a). As shown in Fig. 9b, it is possible to obtain a relatively higher gain for the LP_{11} mode with a thinner ring design (t/a = 0.52), while higher gain for the LP_{01}

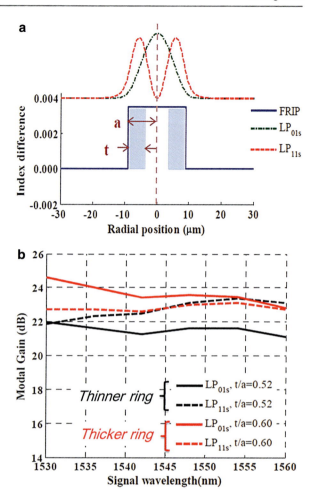

Fig. 9 (a) Ring-doped FM-EDFA design and (b) modal gain dependence for different ring thickness

mode is achieved with a thicker ring design (t/a = 0.60). Therefore, ring-doped fiber designs provide an efficient means to address DMG control in a FM-EDFA. For large-mode-count FM-EDFAs, a multi-ring dopant distribution can be considered for gain equalization (Kang et al. 2014b). However, it is certainly more complex and very challenging to realize multiple ring-doped profiles with accurate radial control of the dopant distribution using conventional modified chemical vapor deposition (MCVD) and solution doping techniques. Also, dopant diffusion is another limiting factor in the fabrication of such a complex multilayer structure because some degree of dopant diffusion is inevitable during the fiber fabrication process due to the heat treatment during tube collapse and subsequent fiber drawing.

More recently, several other interesting fiber designs have been introduced to address the DMG of FM-EDFAs by tailoring the doping distributions. The first example is to employ an extra doping layer in the cladding region as depicted in Fig. 10a. In this fiber structure, the erbium-doped area is larger than core area

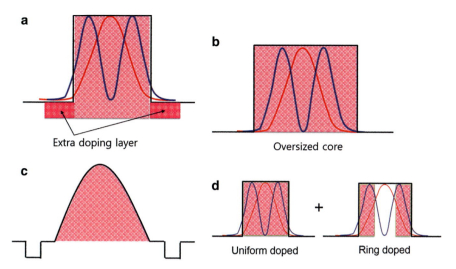

Fig. 10 Other interesting fiber designs to address the DMG: (**a**) FM-EDF with an extra doping layer in the cladding, (**b**) oversized core FM-EDF, (**c**) trench-assisted graded-index FM-EDF, and (**d**) serial concatenation of several different FM-EDFs

which can improve the overlaps between the Er-doped area and spatial modes, resulting in a reduction of the DMG. This fiber design can be particularly beneficial in a cladding-pumped configuration, and DMG values of less than 3.3 dB have been demonstrated in the C-band (Wakayama et al. 2016). In a core-pumped configuration, however, the amplifier may suffer from a low population inversion in the extra doping area, which can result in a decrease in the gain efficiency and an increase in the NF. Therefore, this scheme is only really suitable for cladding-pumped FM-EDFAs. A second approach is the use of an oversized core FM-EDFA, as shown in Fig. 10b. By intentionally oversizing the core diameter to support a greater number of spatial modes than required, the desired subset of required amplified modes can be engineered to have better confinement inside the core, such that the overlap integral between the signal modes and the dopant can be improved. A 10-spatial-mode EDFA with $\sim$2 dB DMG has recently been demonstrated with this approach (Fontaine et al. 2016). In this exemplary amplifier the core diameter was 24 μm and can support more than 28 spatial modes. However, amplification of signals on only ten spatial modes is used. The mode coupling into the unused spatial modes was small (less than 25 dB over the short length of amplifier fiber) and neglected in the report. In fact, use of a smaller-core but higher-NA EDF will provide for similarly increased optical confinement in the core and can also be considered as a means to reduce DMG in a FM-EDFA; albeit in this case there will be a large MFD mismatch between the passive and active fiber which can lead to large splice losses, and efficient MFD adaptors should be included in this approach to make it practical. The third approach is to use a trench-assisted graded-index FM-EDF as shown in Fig. 10c. A graded-index erbium-doped fiber can improve the mode matching and coupling efficiency with passive transmission fiber, which can be useful to reduce the mode-dependent splice loss and to suppress the modal cross talk at the splice joint. However, the doping concentration in the central region of the core is relatively high compared to that of the outer region, and LOMs will exhibit higher gain compared to HOMs. Therefore, a GI profile is not a very good idea from the perspective of DMG. Moreover, the active fiber fabrication process can become complex and expensive for graded-index profiles. Typically EDFs are fabricated using the conventional MCVD process coupled with solution doping. However, the solution doping method dictates use of a simple step-index-like profile, and only a few silica layers can be doped with rare-earth ions at a time. A multiple layer solution doping profile would be essential to realize a graded-index fiber, which is both impractical and uneconomic. The gas phase deposition technique has the potential for realizing complex graded-index profiles, but it is not well established yet. However, note that the trench-assisted fiber structure may be beneficial to enhance mode confinement, and this in itself can help to reduce DMG. A fourth approach is based on the serial concatenation of several FM-EDFs with different erbium dopant distributions. Generally, uniformly doped FM-EDFs show relatively higher gain for the LOMs, whereas ring-doped FM-EDFs offer relatively higher gain for the HOMs. Consequently, a combination of these two EDFs can reduce the total DMG of a FM-EDFA. In a recent report (Salsi et al. 2012), a concatenation of 1 m of uniformly doped step-index EDF and 3.5 m of ring-doped EDF were used to mitigate the DMG of a 6-mode EDFA, and a ~4 dB DMG was experimentally obtained. Also, 1.5 m of uniformly doped EDF and 7.5 m of ring-doped EDF were used in a recent core-pumped 10-mode EDFA with a MDG <3.5 dB achieved (Wada et al. 2016).

Engineering the Signal Mode Profiles

The third technique to reduce DMG is based on signal mode profiling which can be realized by modifying the fiber refractive index profile. Figure 11a shows a center-depressed core FM-EDF design (Wada et al. 2017; Jung et al. 2011) where the fiber has an intentionally designed central refractive index dip in its core. As depicted in the mode profile, the mode shape of the LP_{11} mode is almost the same as that of the step-index fiber, but the intensity distribution of the LP_{01} mode is considerably distorted relative to the well-known Gaussian distribution. The LP_{01} mode profile becomes more similar to the LP_{11} mode profile when increasing the depth of the central dip, resulting in reduced DMG between the two spatial modes. However, this modal distortion can result in a high splice loss between passive and active fibers, and a suitable mode-field adaptor should be employed in this instance. When the refractive index of the central dip is reduced and becomes the same as that of the cladding, the fiber becomes a ring-core fiber as shown in Fig. 11b. This ring-core fiber amplifier can provide nearly identical gain for all guided signal modes owing to the fact that similar overlap factors can be achieved between the erbium-doped core and all the signal spatial modes. Ring-core fiber amplifiers are thus very attractive for MDM amplifiers in terms of having low DMG, and a more detailed description can be found in the following reference (Kang et al. 2014a).

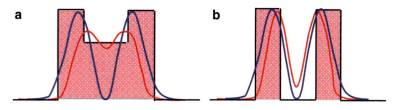

Fig. 11 (a) A center-depressed core design and (b) a ring-core fiber design to minimize the DMG

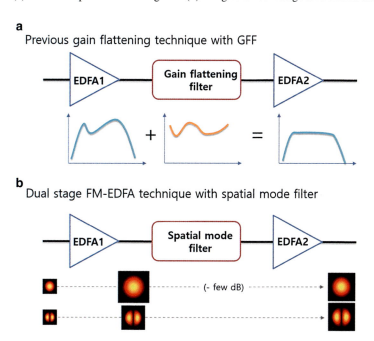

Fig. 12 (a) Previous gain flattening technique with a gain flattening filter and (b) dual-stage FM-EDFA technique with spatial mode filter

Other Approaches

Although multiple different mitigation strategies have been demonstrated to address the DMG of FM-EDFAs, many involve complex dopant distribution tailoring, pump mode profiling, or peculiar fiber refractive index profiles. In this aspect, rather than targeting complex fiber structures with low DMG (<1 dB), adopting a less complicated fiber design offering moderate DMG (2–5 dB) and using this in conjunction with a spatial modal filtering device offering controllable modal loss is perhaps a more realistic approach. The spatial filter (fixed or preferably dynamic) may be inserted in the middle of a dual-stage FM-EDFA as shown in Fig. 12, and the total DMG can be effectively improved without an appreciable NF degradation in an analogous manner to the gain flattening filters used in many current single-mode amplifiers.

Core-Pumped 6-Mode EDFA

Based on these various DMG mitigation techniques, we will now discuss both core-pumped and cladding-pumped 6-mode EDFAs in the following two sections to understand how these can be implemented in practice. Figure 13a shows a schematic of a core-pumped 6-mode EDFA (Jung et al. 2014b) for the simultaneous amplification of six spatial modes in four mode groups, namely, LP_{01}, LP_{11a}, LP_{11b}, LP_{21a}, LP_{21b}, and LP_{02}. Two dichroic mirrors are used for combining 980 nm pump light with the 1550 nm input signals, and two polarization-insensitive free-space isolators are included to prevent unwanted parasitic lasing within the amplifier. In order to mitigate the DMG, a ring-doped EDF was fabricated to provide dopant distribution control, and bidirectional HOM pumping was employed to provide pump mode control. Note that this fiber was originally designed to be a ring-doped EDF, but undesired dopant diffusion took place from the highly doped ring regions into the undoped central region (see the inset figure of Fig. 13a). This dopant diffusion is to be expected during high-temperature glass treatment (such as the final tube collapse or fiber drawing process itself). For the reconfigurable pump mode shaping, LP_{21} pump phase plates (designed to operate at 980 nm) were inserted in the optical path of the pump beam, where a uniform phase sector can be positioned for LP_{01} excitation, a half-sector for LP_{11} excitation, and a quadrant sector for LP_{21} excitation. Figure 13b shows the measured mode-dependent gain as a function of launched pump power for different spatial pump modes (e.g., LP_{01-p}, LP_{11-p}, and LP_{21-p}). With an LP_{01-p} pump launch condition, the LP_{01} signal experiences much higher gain than other HOMs as expected due to its greater overlap with the erbium ions excited by the LP_{01-p} pump. The measured DMG was more than 4.5 dB for the LP_{01-p}, but it decreased to 1.5 dB for LP_{11-p} and less than 1 dB for a LP_{21-p} pump. It can be clearly seen that the use of HOM pump profiles enables us to significantly reduce the DMG of the FM-EDFA. It is to be noted that the pump coupling loss is dependent on the particular pump mode under consideration and that HOMs experience higher coupling loss. In this particular 6-mode EDFA experiment, an LP_{21} pump mode was used and a 3.5 dB ($\sim$55%) coupling loss was observed. Therefore a single-pump laser diode (LD) with a maximum output power of 750 mW (i.e., the most powerful, commercially available single-mode pump LD at this moment) is not sufficient to provide enough gain for all the spatial modes, and two pump LDs are consequently used in a bidirectional pump configuration to get acceptable signal gain. In this particular amplifier, both fiber dopant distribution control and pump mode control were used to address the DMG. Small signal gain of >20 dB was obtained for all the spatial modes with a DMG < 2 dB. Nevertheless, it clearly shows that a low-loss, highly efficient pump mode converter is essential to further scale the amplifier to a larger number of modes for core-pumped FM-EDFAs. As the number of spatial modes is further scaled to 10 and beyond, it will become increasingly challenging to meet the associated pump power requirements with single-mode pump diodes. For this reason, a cladding-pumped scheme represents an attractive way to provide enough pumplight by utilizing low-cost, powerful multimode pump diodes.

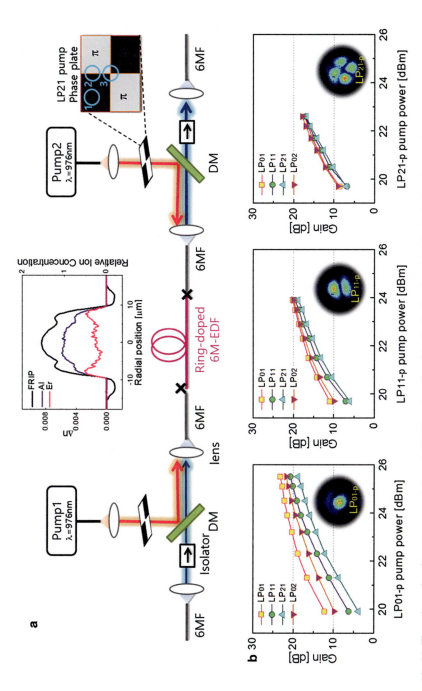

Fig. 13 (**a**) The schematic of a core-pumped 6-mode EDFA gain tailored by both ring-doped EDF and bidirectional adjustable pump mode. (**b**) Mode-dependent gain as a function of different pump spatial modes

Cladding-Pumped 6-Mode EDFA

Figure 14a shows a schematic of a fully integrated cladding-pumped FM-EDFA (Jung et al. 2014a). The basic configuration of the amplifier is very similar to the conventional single-mode fiber amplifier, containing input and output optical isolators, a side-pump coupler as a WDM coupler, and a pump diode. The only difference is that all of the in-line components are few-mode fiber compatible devices, and the pump diode and WDM coupler are multimode devices. More importantly, a side coupling approach is used so that the active fiber can be directly spliced to the transmission FMFs – opening a route toward fully integrated MDM transmission lines. For efficient side coupling, a passive multimode pump delivery fiber was tapered and wound around a short-stripped section of the active FM-EDF as described before. More than 70% pump coupling efficiency can be readily achieved with this side coupling method. Any residual pump power can be removed at the output of the amplifier by applying a high-index polymer to a further section of stripped fiber and removing the associated heat generated. This approach provides a very practical means of pump coupling and is capable of delivering high pump power into the device. A 2 m-long 6-mode ring-doped EDF was used as a gain medium, and 6-mode FMF isolators were included at the input/output of the amplifier (no gain flattening filter was used in this experiment). As shown in Fig. 14b, a small signal gain of >20 dB was achieved across the C-band with a DMG of ~3 dB among the mode groups. The amplified signal mode profiles and fully

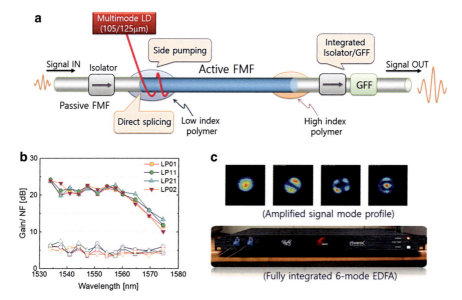

Fig. 14 (**a**) Schematic of a fully integrated cladding-pumped FM-EDFA, (**b**) measured modal gain/noise figure of the cladding-pumped 6-mode EDFA, and (**c**) amplified signal mode profile and developed rack mountable amplifier module

integrated 6 M-EDFA package (i.e., 1 inch rack mountable amplifier module) are presented in Fig. 9c. Note that there are no free-space optical elements in the FMF amplifier module and the physical dimension of the 6 M-EDFA is almost the same as that of a conventional single-mode EDFA. To date many FM-EDFAs have been implemented in a relatively bulky system configuration incorporating a substantial number of free-space components, and a fully integrated cladding-pumped amplifier design is an important step in the mode scalability of few-mode EDFAs, offering cost-effective and efficient amplification of a large number of spatial data channels in a single device.

Challenges and Future Development

- Development of few-mode fiber amplifiers with very low DMG (less than 1 dB): Even though various DMG mitigation techniques have been demonstrated, low-DMG FM-EDFAs have been realized only in a 3-mode FM-EDFA. Therefore further experimental demonstrations of low-DMG FM-EDFAs supporting more spatial modes are needed.
- Increasing the number of spatial modes in few-mode or few-mode multicore fiber amplifiers: Currently, the largest number of spatial mode channels amplified is only 10 in FM-EDFAs and 21 in few-mode multicore fiber amplifiers. The scalability with respect to the number of modes is an essential feature for the penetration of MDM in next-generation optical transmission systems. In these amplifiers, both intercore cross talk and the DMG are the important design considerations, and careful fiber design and optimization of the amplifier is indispensable.
- Independent spatial channel control of the amplifier: In recent cladding-pumped configurations, pump light is delivered into the cladding of the active fiber, and population inversion can be readily achieved in a cost-effective way. However, although fully integrated FMF amplifiers in a cladding-pumped configuration are very attractive, it inevitably implies a reduction in independent spatial information channel control – an important requirement in today's optical networks. Therefore, this needs to be a major focus of research in the coming years.
- Gain dynamics or transient control of FM-EDFAs: The intermodal cross gain and associated transient effects in a FM-EDFA should be considered under a range of different add/drop conditions, and dynamic gain control systems should be investigated for practical SDM transmission application.

Conclusion

Significant research advances have been made in the development of few-mode fiber amplifiers, with comparable gains and noise figures per spatial channel to conventional single-mode EDFAs. Amplification of up to 10 spatial modes has been demonstrated in few-mode fiber amplifiers and up to 21 channels in few-mode multicore fiber amplifiers. Moreover, fiberized side-pumping schemes have been demonstrated, along with integrated passive components such as isolators and filters, which promise significant cost savings relative to the use of an equivalent numbers of conventional single-mode amplifiers. However, although very good performance has been shown both from a device and system perspective, the impact of the associated loss of independent control per spatial channel has yet to be properly assessed, and the envisaged cost and power savings have yet to be properly quantified. A concerted research effort will be required in the years ahead to address these important issues.

References

K.S. Abedin, T.F. Taunay, M. Fishteyn, M.F. Yan, B. Zhu, J.M. Fini, E.M. Monberg, F.V. Dimarcello, P.W. Wisk, Opt. Express **19**, 16715 (2011)

K.S. Abedin, T.F. Taunay, M. Fishteyn, D.J. DiGiovanni, V.R. Supradeepa, J.M. Fini, M.F. Yan, B. Zhu, E.M. Monberg, F.V. Dimarcello, Opt. Express **20**, 20191 (2012)

K.S. Abedin, J.M. Fini, T.F. Thierry, B. Zhu, M.F. Yan, L. Bansal, F.V. Dimarcello, E.M. Monberg, D.J. DiGiovanni, Opt. Lett. **39**, 993 (2014)

H. Chen, C. Jin, B. Huang, N.K. Fontaine, R. Ryf, K. Shang, N. Grégoire, S. Morency, R.-J. Essiambre, G. Li, Y. Messaddeq, S. Larochelle, Nat. Photonics **10**, 529 (2016)

A.D. Ellis, J. Zhao, D. Cotter, J. Lightwave Technol. **28**, 423 (2010)

R.-J. Essiambre, G. Kramer, P.J. Winzer, G.J. Foschini, B. Goebel, J. Lightwave Technol. **28**, 662 (2010)

N.K. Fontaine, B. Huang, Z. Sanjabieznaveh, H. Chen, C. Jin, B. Ercan, A. Velázquez-Benetez, S.H. Chang, R. Ryf, A. Schülzgen, J. Carlos Alvarado, P. Sillard, C. Gonnet, E. Antonio-Lopez, R. Amezcua Correa, in *Optical Fiber Communication Conference* (OSA, Washington, DC, 2016), p. Th5A.4

K. Igarashi, D. Souma, Y. Wakayama, K. Takeshima, Y. Kawaguchi, T. Tsuritani, I. Morita, M. Suzuki, in *Optical Fiber Communication Conference* (OSA, Washington, DC, 2015), p. Th5C.4

S. Jain, V.J.F. Rancaño, T.C. May-Smith, P. Petropoulos, J.K. Sahu, D.J. Richardson, Opt. Express **22**, 3787 (2014)

S. Jain, T. Mizuno, Y. Jung, Q. Kang, J.R. Hayes, M.N. Petrovich, G. Bai, H. Ono, K. Shibahara, A. Sano, A. Isoda, Y. Miyamoto, Y. Sasaki, Y. Amma, K. Takenaga, K. Aikawa, C. Castro, K. Pulverer, M. Nooruzzaman, T. Morioka, S. Alam, D.J. Richardson, in *European Conference Optical Communication* (2016), p. Th3.A1

Y. Jung, S. Alam, Z. Li, A. Dhar, D. Giles, I.P. Giles, J.K. Sahu, F. Poletti, L. Grüner-Nielsen, D.J. Richardson, Opt. Express **19**, B952 (2011)

Y. Jung, E.L. Lim, Q. Kang, T.C. May-Smith, N.H.L. Wong, R. Standish, F. Poletti, J.K. Sahu, S.U. Alam, D.J. Richardson, Opt. Express **22**, 29008 (2014a)

Y. Jung, Q. Kang, J.K. Sahu, B. Corbett, J. O'Callaghan, F. Poletti, S.-U. Alam, D.J. Richardson, IEEE Photon. Technol. Lett. **26**, 1100 (2014b)

Y. Jung, S.-U. Alam, D.J. Richardson, in *Optical Fiber Communication Conference* (OSA, Washington, DC, 2015), p. W2A.13

Y. Jung, S. Alam, D.J. Richardson, in *Optical Fiber Communication Conference* (OSA, Anaheim, 2016a), p. W2A.40

Y. Jung, S. Alam, Y. Sasaki, D.J. Richardson, in *European Conference Optical Commununication* (IEEE, Dusseldorf, 2016b), p. W2.B4

Y. Jung, Q. Kang, L. Shen, S. Chen, H. Wang, Y. Yang, K. Shi, B.C. Thomsen, R. Amezcua Correa, Z. Sanjabi Eznaveh, J. Carlos Alvarado Zacarias, J. Antonio-Lopez, P. Barua, J.K. Sahu,

S.U. Alam, D.J. Richardson, in *European Conference Optical Communication* (IEEE, Denmark, 2017a), p. P1.SC1.18

Y. Jung, J. Hayes, Y. Sasaki, K. Aikawa, S. Alam, D.J. Richardson, in *Optical Fiber Communication Conference* (OSA, Washington, DC, 2017b), p. W3H.2

Y. Jung, Q. Kang, R. Sidharthan, D. Ho, S. Yoo, P. Gregg, S. Ramachandran, S. Alam, D.J. Richardson, J. Lightwave Technol. **35**, 430 (2017c)

Q. Kang, E. Lim, Y. Jun, X. Jin, F.P. Payne, S. Alam, D.J. Richardson, in *European Conference Optical Communication* (IEEE, Cannes, 2014a), p. P.1.14

Q. Kang, E.-L. Lim, Y. Jung, F. Poletti, C. Baskiotis, S. Alam, D.J. Richardson, Opt. Express **22**, 21499 (2014b)

H. Ono, Y. Amma, T. Hosokawa, M. Yamada, in *IEEE Photonics Society Summer Topicals Meeting Series* (IEEE, Newport Beach, 2016), pp. 74–75

D.J. Richardson, J.M. Fini, L.E. Nelson, Nat. Photonics **7**, 354 (2013)

J. Sakaguchi, W. Klaus, B.J. Puttnam, J.M.D. Mendinueta, Y. Awaji, N. Wada, Y. Tsuchida, K. Maeda, M. Tadakuma, K. Imamura, R. Sugizaki, T. Kobayashi, Y. Tottori, M. Watanabe, R.V. Jensen, Opt. Express **22**, 90 (2014)

M. Salsi, D. Peyrot, G. Charlet, S. Bigo, R. Ryf, N.K. Fontaine, M.A. Mestre, S. Randel, X. Palou, C. Bolle, B. Guan, G. Le Cocq, L. Bigot, Y. Quiquempois, in *European Conference Optical Communication* (OSA, Washington, DC, 2012), p. Th.3.A.6

D. Soma, Y. Wakayama, S. Beppu, S. Sumita, T. Tsuritani, T. Hayashi, T. Nagashima, M. Suzuki, M. Yoshida, K. Kasai, M. Nakazawa, H. Takahashi, K. Igarashi, I. Morita, and M. Suzuki, J. Lightwave Technol. **36**, 1362 (2018)

H. Takara, A. Sano, T. Kobayashi, H. Kubota, H. Kawakami, A. Matsuura, Y. Miyamoto, Y. Abe, H. Ono, K. Shikama, Y. Goto, K. Tsujikawa, Y. Sasaki, I. Ishida, K. Takenaga, S. Matsuo, K. Saitoh, M. Koshiba, T. Morioka, in *European Conference Optical Communication* (OSA, Washington, DC, 2012), p. Th.3.C.1

M. Wada, T. Sakamoto, S. Aozasa, T. Mori, T. Yamamoto, K. Nakajima, in *European Conference Optical Communication* (VDE, Dusseldorf, 2016), p. M2.A4

M. Wada, S. Aozasa, T. Sakamoto, T. Yamamoto, T. Mori, K. Nakajima, J. Lightwave Technol. **35**, 762 (2017)

Y. Wakayama, K. Igarashi, D. Soma, H. Taga, T. Tsuritani, in *European Conference Optical Communication* (IEEE, Dusseldorf, 2016), p. M.2.A.3

I. Weiss, J. Gerufi, D. Sinefeld, M. Blau, M. Bin Nun, R.B. Lingle, L.E. Gruner-Nielsen, D. Marom, in *Optical Fiber Communication Conference* (OSA, Washington, DC, 2014), p. Th4A.8

P.J. Winzer, Nat. Photonics **8**, 345 (2014)